APPLICATIONS OF INFRARED SPECTROSCOPY IN BIOCHEMISTRY, BIOLOGY, AND MEDICINE

Applications of

INFRARED SPECTROSCOPY

IN BIOCHEMISTRY, BIOLOGY, AND MEDICINE

FRANK S. PARKER
Department of Biochemistry
New York Medical College
New York, New York

ℚ PLENUM PRESS • NEW YORK • 1971

Library of Congress Catalog Card Number 70-131882
SBN 306-30502-X

© 1971 Plenum Press, New York
A Division of Plenum Publishing Corporation
227 West 17th Street, New York, N.Y. 10011

Printed in the United States of America

To my wife, Gladys,
whose love and devotion helped to write this book,
and to my children, Judith and George

PREFACE

This book is not intended to be a *basic* text in infrared spectroscopy. Many such books exist and I have referred to them in the text. Rather, I have tried to find applications that would be interesting to a variety of people: advanced undergraduate chemistry students, graduate students and research workers in several disciplines, spectroscopists, and physicians active in research or in the practice of medicine. With this aim in mind there was no intent to have exhaustive coverage of the literature.

I should like to acknowledge my use of several books and reviews, which were invaluable in my search for material:

G. H. Beaven, E. A. Johnson, H. A. Willis and R. G. J. Miller, *Molecular Spectroscopy*, Heywood and Company, Ltd., London, 1961.

J. A. Schellman and Charlotte Schellman, "The Conformation of Polypeptide Chains in Proteins," in *The Proteins, Vol. II*, 2nd Ed. (H. Neurath, ed.), Academic Press, New York, 1964.

R. T. O'Connor, "Application of Infrared Spectrophotometry to Fatty Acid Derivatives," *J. Am. Oil Chemists' Soc.* **33**, 1 (1956).

F. L. Kauffman, "Infrared Spectroscopy of Fats and Oils," *J. Am. Oil Chemists' Soc.* **41**, 4 (1964).

W. J. Potts, Jr., *Chemical Infrared Spectroscopy, Vol. I, Techniques*, Wiley, New York, 1963.

R. S. Tipson, *Infrared Spectroscopy of Carbohydrates*, National Bureau of Standards Monograph 110, Washington, D.C., 1968.

C. N. R. Rao, *Chemical Applications of Infrared Spectroscopy*, Academic Press, New York, 1963.

H. R. Mahler and E. H. Cordes, *Biological Chemistry*, Harper and Row, New York, 1966.

T. R. Kasturi, "Steroids," in C. N. R. Rao, *Chemical Applications of Infrared Spectroscopy*, Academic Press, New York, 1963, p. 403.

R. I. Dorfman and F. Ungar, *Metabolism of Steroid Hormones*, Academic Press, New York, 1965.

R. N. Jones and C. Sandorfy, "The Application of Infrared and Raman Spectrometry to the Elucidation of Molecular Structure," in *Chemical Applications of Spectroscopy* (W. West, ed.), Interscience Publishers, New York, 1956, p. 247.

S. Schwartz, M. H. Berg, I. Bossenmaier and H. Dinsmore, "Determination of Porphyrins in Biological Materials," *Methods Biochem. Anal.* **8**, 221 (1960).

M. Tsuboi, "Application of Infrared Spectroscopy to Structure Studies of Nucleic Acids," *Appl. Spectrosc. Rev.* **3**, 45 (1969).

M. Tichý, "The Determination of Intramolecular Hydrogen Bonding by Infrared Spectroscopy and Its Applications in Stereochemistry," in *Advances in Organic Chemistry, Methods and Results, Vol.* 5 (R. A. Raphael, E. C. Taylor and H. Wynberg, eds.), Interscience Publishers, New York, 1965, p. 115.

R. D. Stewart and D. S. Erley, "Detection of Volatile Organic Compounds and Toxic Gases in Humans by Rapid Infrared Techniques," in *Progress in Chemical Toxicology, Vol.* 2 (A. Stolman, ed.), Academic Press, New York, 1965, p. 183.

J. J. Katz, R. C. Dougherty, and L. J. Boucher, "Infrared and Nuclear Magnetic Resonance Spectroscopy of Chlorophyll," in *The Chlorophylls* (L. P. Vernon and G. Seely, eds.), Academic Press, New York, 1966, p. 185.

W. P. Jencks, C. Moore, F. Perini, and J. Roberts, "Infrared Spectra of Activated Acyl Groups in Deuterium Oxide Solution," *Arch. Biochem. Biophys.* **88**, 193 (1960).

M. Jacobson, *Insect Sex Attractants*, Interscience Publishers, New York, 1965.

R. A. Kekwick, ed., "The Separation of Biological Materials," *British Medical Bulletin* **22**, 103 (1966).

Besides the titles listed above are all the papers, brochures, books, etc., which I have referred to in the various chapters. It is a pleasure to thank all those authors who kindly sent me reprints of their papers.

F.S.P.

ACKNOWLEDGMENTS

I appreciate very much the help given to me by many friends and associates during the preparation of this book. Dr. K. R. Bhaskar, Dr. Irving Rappaport, and Martin Stryker have read various chapters critically and have made valuable suggestions. My colleagues on the staff of the Department of Biochemistry have been very co-operative. Marji Gold, Randy Mound, George Parker, and Nancy Rodriguez have assisted in various ways. I thank Juanita O'Brien and Ruth Prunty for their readiness to help at all times, and want especially to express my appreciation to Anne Zver for her willing cooperation and exceptional typing throughout all phases of the writing of the manuscript. Also, I want to thank the staff of the Lillian Morgan Hetrick Library of New York Medical College for their help.

To two other persons whose help in the past years has been very meaningful to me, I give my sincere thanks: Dr. J. Logan Irvin, for introducing me in his kind and thorough way to spectroscopy; and Dr. Murray Lieberman, for his assistance in many ways, including the writing of this book.

It is a pleasure to acknowledge here the various research grants I have received from the United States Public Health Service, National Institutes of Health. I appreciate also the support of my research and a Career Scientist Award from the Health Research Council of the City of New York.

F.S.P.

CONTENTS

xi

APPENDIXES

Chapter 1

INTRODUCTION AND BRIEF THEORY

There are many texts available which deal with the theory of molecular vibrations and infrared spectra. Complete treatments are given in the books by Herzberg (1945) and by Wilson *et al.* (1955). Other texts that can be highly recommended are referred to in Chapter 20, which deals with information sources.

Infrared spectra are directly involved with the vibrations of the atoms or groups of atoms in a molecule (as distinguished, e.g., from rotations of the whole molecule, or from electronic motions characteristic of ultraviolet and visible absorption spectra). When we examine infrared absorption characteristics we are studying group vibrations. As a first approach to the study of these vibrations let us consider the vibrations of small molecules and the common ways of looking at such systems.

Theory

Solutions and liquids (and usually gases and solids) have spectra whose origins arise from transitions between the vibrational energy levels of the molecules under examination. If a molecule has N atoms, there are $3N - 6$ (or $3N - 5$, for a linear molecule) vibrational frequencies. However, they may not all be different and may not all be present in the infrared spectrum. Let us look briefly at a simple system as a model for the $3N - 6$ vibrational motions of a general molecule. We shall examine the motions which the diatomic molecule A—B (represented by two balls connected by a spring) can undergo along the bond between atoms A and B. The mass of each ball is proportional to its atomic mass, and the spring has strength proportional to the chemical forces bonding atoms A and B together as a molecule. A stretching vibration occurs if the spring is compressed or stretched and the whole molecule then allowed to relax on its own. If the periodic oscillation is moderate the system follows Hooke's law to a first approximation, and the frequency (cycles/sec) of the stretching vibration ($\tilde{v}$) can be expressed by the following equation, which holds for simple harmonic vibrations:

$$\tilde{v} = \frac{1}{2\pi}\sqrt{\frac{k}{M}}$$

where k is the force constant of the bond in dynes per centimeter, and M is the reduced

1

mass defined as $m_A m_B / (m_A + m_B)$; m_A and m_B are the masses in grams of atoms A and B, respectively. Dividing $\tilde{v}$ by c, the velocity of light (3×10^{10} cm/sec), one obtains waves per centimeter, usually referred to as wave numbers (cm^{-1}).

Of course, the equation relates only to one pair of atoms joined by a bond of a particular strength. Stretching vibrations of individual bonds of more complex molecules may be examined similarly, but other vibrations are possible and absorption band frequencies are influenced by other factors. A polyatomic molecule has many ways in which groups of atoms may vibrate, bend, rock, or twist relative to each other and relative to other groups of atoms.

Substitution in the equation for simple harmonic vibrations of the proper values of c, k, and M for the C—H bond gives a frequency of 3040 cm^{-1}, in fair agreement with methyl group C—H stretching vibration frequencies of 2975–2950 cm^{-1} and 2885–2860 cm^{-1}.

A C≡N group absorbs near 2250 cm^{-1}, a C=N group absorbs near 1650 cm^{-1}, and a C—N group absorbs near 1050 cm^{-1}. The differences in the positions of these group frequencies are due to the differences in the force constants, k, of the triple, double, and single bonds attaching the same elements.

A C—H group absorbs near 3000 cm^{-1}, a C—C group near 1000 cm^{-1}, and a C—I group near 500 cm^{-1}. Here the bond strengths remain approximately the same, but the mass of one atom changes in each group.

Wave Number and Wavelength Units

Absorption band positions are given in wave number units (v), which are expressed as cm^{-1}, or alternatively in wavelength units (λ), in microns (μ) (also called micrometers, μm). The true unit of frequency ($\tilde{v}$) is measured in reciprocal seconds (sec^{-1}). The relations of these units to each other are: v [cm^{-1}] $= 1/\lambda$ [cm] $= 10,000/\lambda$ [μ]; $\tilde{v} = c/\lambda$; thus, 2000 cm$^{-1} = 5.00 \mu$, and 1250 cm$^{-1} = 8.00 \mu$. The reader is referred to Appendix 2 for a Table of Reciprocals, which is useful in converting wavelengths to wave numbers, and vice versa.

The near-infrared region extends from about 12,500 to 4000 cm^{-1}. Most infrared spectra in the literature have been recorded in the region from 4000 to 650 cm^{-1}. The range from 4000 to 200 cm^{-1} is called "mid-infrared," and that from 200 cm^{-1} down to about 15 cm^{-1} "far-infrared," although values from 667 cm^{-1} downward are often included in the latter category. The region from about 667 to 400 cm^{-1} is often referred to as the "potassium bromide" region, but it is included in the "cesium bromide" region, which goes down to 286 cm^{-1}.

The Double-Beam Recording Spectrophotometer

The sample absorption is determined directly from the difference in energy between two beams, one of which passes through the solution while the other passes through a solvent blank. This energy difference is usually measured by either the optical null or the ratio-recording method. In the *optical null* method, a wedge is pushed automatically (servo system) into the reference beam until the two beams give signals of equal strength. The energy for the wedge movement comes from the difference between the beams and is therefore zero when the beams are equal. The movement of the wedge is accompanied by a simultaneous and proportional movement of the recorder pen and therefore gives a percent transmittance or percent absorbance recording. In this method a light-chopping device is introduced (e.g., a semicircular mirror cutting the beam at approximately ten cycles per second). The amplifier is so designed that only oscillating (e.g., 10 cps) signals are amplified. The sample and reference beams can be distinguished since they impinge on the detector alternately. In a prism instrument the spectrum is scanned by rotating a Littrow mirror (see Fig. 1.1) (a mechanical cam is necessary to provide a linear frequency or wavelength scale). The dispersed radiation moves across the exit slit, only a narrow band of wavelengths passing through to the detector. The widths of the slits are controlled by a slit program (a cam on the wavelength drive, or an electrical device which monitors the energy in the reference beam), and an approximately constant

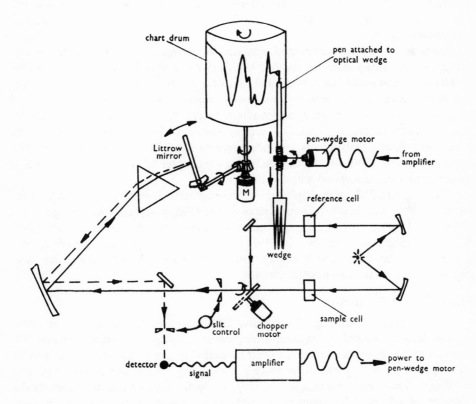

Fig. 1.1. Typical arrangement of a double-beam recording prism spectrophotometer. M is the motor driving the Littrow mirror and the chart drum. (Bladon and Eglinton, 1964.)

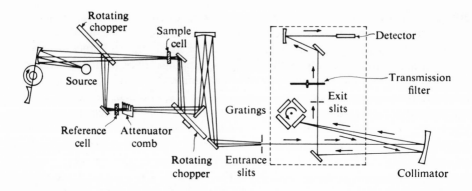

Fig. 1.2. Typical optical path of a grating spectrophotometer. [R. T. Conley, 1966 (courtesy Beckman Instruments, Inc.).]

level of radiation is maintained in the reference beam in spite of big variations in the source intensity with wavelength. Figure 1.2 shows a typical optical path of a grating spectrophotometer.

Since this is a book concerned primarily with applications, no further details are given concerning instrumentation. The reader is referred to Alpert *et al.* (1970), in which are discussed: an optical diagram of a double-beam spectrophotometer; operating variables (resolution, photometric accuracy); components of infrared spectrophotometers (sources, types of photometers, dispersing elements, detectors, amplifiers, and recorders); special operating features, such as optimization of scan time; and available instruments and their specifications. The books by Martin (1966), Conn and Avery (1960), and Potts (1963), and the chapter by Herscher (1966) are also recommended for details on some of these topics.

The potentialities of multiple-scan interferometers have been explored by Low and coworkers, who in several papers have covered the subjects of emission spectra of pesticides in the microgram range (Coleman and Low, 1966), human skin (Low, 1966), reflectance spectra of minerals (Low, 1967), and absorption spectra of gas-chromatographic fractions (Low and Freeman, 1967). More recently Low (1969) has also discussed Fourier transform spectrometers.

The Infrared Spectrum

When infrared radiation of successive frequencies (or wavelengths) is incident on a molecule, some of the radiation frequencies correspond to characteristic frequencies of the molecule, and under such natural resonant conditions energy can be exchanged from one system to another if a coupling mechanism is available. The coupling mechanism in this case is the change in electric dipole moment caused by the vibration of a bond. The scanning of a molecular sample with successive infrared frequencies shows frequency values for which the radiation is absorbed, and these values correspond qualitatively to mechanical vibration frequencies of the molecule. The infrared spectrum is an analysis of the molecular vibrations.

In an infrared spectrum the amount of radiation absorbed (or transmitted) is plotted as a function of the wavelength. The intensity of radiation of one wavelength transmitted by the sample (I) is related to the intensity incident on this sample (I_0), to the path length in the sample (b), and to the concentration of the sample (c) by the equation

$$\log_{10}(I_0/I) = abc$$

where a is a constant for a particular substance at a given wavelength. This equation is a form of the Beer–Lambert or Beer–Bouguer law. The relationship and its use are presented in greater detail in Chapter 4. Two commonly used terms derive from this law, the transmittance, T, expressed by

$$T = I/I_0$$

and the absorbance, A, expressed by

$$A = \log(1/T) = \log(I_0/I)$$

An infrared spectrum can be a plot of either absorbance A or transmittance T *versus* wavelength or wave number. The convention in this country is that the ordinate scale is usually set up in such a way that absorption peaks appear as valleys in the curve, regardless of whether absorbance or transmittance is plotted on the ordinate. Foreign laboratories often plot spectra with the scale arranged differently, and these appear upside down compared to ours; but both types give equivalent information, and one should be familiar with them both.

Calibration of Wave Number and Wavelength

Table 1.1 lists many of the commonly used calibration standards. A film of polystyrene is frequently employed because it is easily handled and can be stored permanently. Water vapor and carbon dioxide are widely used for single-beam calibration, but are usually not used for double-beam measurements (Martin, 1966). A better standard than polystyrene is a mixture of indene with 0.8 wt % of camphor and cyclohexanone to provide carbonyl stretching frequencies (Jones *et al.*, 1961). Eleven bands of this mixture providing calibration points over the range 650–300 cm^{-1} have also been recorded (Jones *et al.*, 1959).

Table 1.1 Calibration Standards (Alpert *et al.*, 1970)

(All lines refer to vapor absorption bands unless otherwise noted)

λ, μ (in air)	v, cm^{-1} (*in vacuo*)	Material
1.0140	9859.4	Hg (emission)
1.1287	8857.0	Hg (emission)
1.3673	7311.5	Hg (emission)
1.5296	6535.9	Hg (emission)
1.7073	5855.6	Hg (emission)
2.3253	4299.3	Hg (emission)
2.605	3837.9	H_2O

Table 1.1 (*Continued*)

λ, μ (in air)	v, cm^{-1} (*in vacuo*)	Material
2.913	3432.0	NH_3
3.302	3027.1	Polystyrene film
3.420	2924.0	Polystyrene film
3.507	2850.7	Polystyrene film
4.254	2349.9	CO_2
4.258	2347.6	CO_2
5.142	1944.0	Polystyrene film
5.146	1942.6	H_2O
5.348	1869.4	H_2O
5.421	1844.2	H_2O
5.549	1801.6	Polystyrene film
5.577	1792.6	H_2O
5.640	1772.6	H_2O
5.708	1751.4	H_2O
5.763	1734.6	H_2O
5.988	1669.4	H_2O
6.074	1646.0	H_2O
6.184	1616.7	H_2O
6.211	1609.6	Indene liquid
6.243	1601.4	Polystyrene film
6.315	1583.1	Polystyrene film
6.342	1576.2	H_2O
6.414	1558.5	H_2O
6.436	1553.3	Indene liquid
6.824	1464.9	H_2O
6.958	1436.7	H_2O
7.044	1419.3	H_2O
7.176	1393.2	Indene liquid
7.344	1361.3	Indene liquid
8.244	1212.7	NH_3
8.366	1195.0	NH_3
8.493	1177.1	NH_3
8.626	1158.9	NH_3
8.661	1154.3	Polystyrene film
8.765	1140.6	NH_3
9.060	1103.4	NH_3
9.217	1084.6	NH_3
9.292	1075.9	NH_3
9.725	1028.0	Polystyrene film
9.814	1018.6	Indene liquid
10.072	992.6	NH_3
10.503	951.8	NH_3
11.007	908.2	NH_3
11.026	906.7	Polystyrene film
11.607	861.3	Indene liquid
12.380	807.5	NH_3
13.693	730.1	Indene liquid

Table 1.1 (*Continued*)

λ, μ (in air)	v, cm^{-1} (*in vacuo*)	Material

The following bands are not established standards but are commonly used for calibrating infrared spectrophotometers in the longer-wavelength regions

13.680	730.8	Polyethylene film
13.890	719.7	Polyethylene film
14.986	667.1	CO_2
16.178	617.9	H_2O
17.400	574.5	1,2,4-Trichlorobenzene liquid
18.160	550.5	1,2,4-Trichlorobenzene liquid
19.008	525.9	H_2O
19.907	502.2	H_2O
21.161	472.4	H_2O
21.790	458.8	1,2,4-Trichlorobenzene liquid
21.860	457.3	H_2O
21.872	457.1	H_2O
22.617	442.0	H_2O
22.760	439.2	1,2,4-Trichlorobenzene liquid
23.860	419.0	H_2O
25.140	397.7	H_2O
26.620	375.5	H_2O
29.830	335.1	H_2O
30.510	327.7	H_2O
33.010	302.8	H_2O

Examples of the Interpretation of Spectra

Figure 1.3 shows a spectrum of 3,4-dimethyl-5-hexen-3-ol for which the group frequencies have been designated. Symbols used are: v, stretch; δ_{as}, asymmetric bend; δ_s, symmetric bend; β, in-plane bend; γ, out-of-plane bend (Szymanski, 1966).

Figure 1.4 shows the spectrum of a compound with molecular formula $C_7H_7N_3$. The interpretation, according to Bentley *et al.* (1968), is as follows:

3060 cm^{-1}: Aromatic C—H stretching vibration.

2955 cm^{-1}: CH_3 asymmetric stretching vibration.

2000–1600 cm^{-1}: Characteristic pattern of overtone and combination bands for 1,2-disubstituted benzenes.

1568 cm^{-1}: Possibly a C=N stretching vibration.

1450 cm^{-1}: Aromatic skeletal C—C frequency and/or CH_3 deformation.

748 cm^{-1}: CH out-of-plane deformation characteristic of 1,2-disubstituted benzenes.

502 cm^{-1}: Suggests either an in-plane ring deformation of a 1,2-disubstituted benzene or a tertiary amine.

442 cm^{-1}: An out-of-plane ring deformation of a 1,2-disubstituted benzene.

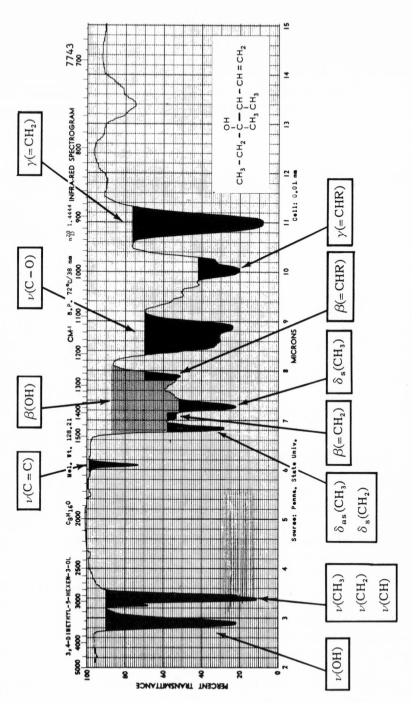

Fig. 1.3. 3,4-Dimethyl-5-hexen-3-ol spectrum. (Szymanski, 1966.)

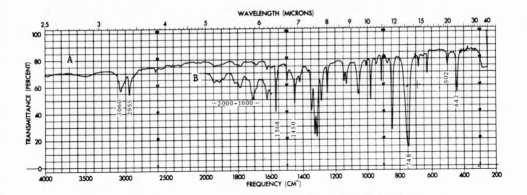

Fig. 1.4. Spectrum of $C_7H_7N_3$. Curve *A*, capillary film; curve *B*, 0.025 mm. (Bentley *et al.*, 1968.)

A methyl group is evident from the sodium chloride region, and lack of an absorption band between 388 and 260 cm^{-1} indicates that the methyl is not attached to the benzene ring. The lack of absorption bands between 3500 and 3300 cm^{-1} eliminates primary and secondary amines from consideration. According to the known molecular formula, the only possible structural formulas are

Since the absorption band at 1568 cm^{-1} is probably a $C=N$ stretching vibration, the first structure is more likely. An $N=N$ stretching vibration also would occur near here (1630–1575 cm^{-1}); however, one would expect the intensity of the band to be much weaker. The first structure is 2-methylbenzotriazole.

Other examples of the interpretation of spectra are given in various parts of this book, under specific topics.

Several very useful infrared correlation charts appear in the *Handbook of Biochemistry* (Sober, 1968, pp. J-4–J-12), and a typical correlation chart for the region from ~700 to 300 cm^{-1}, taken from the same source, is presented in Table 1.2. Correlations for the 4505–450 cm^{-1} region are also given in Chapter 6.

Isotopes are frequently used in attempts to correlate spectra with structure. For instance, ^{18}O substitution has been employed by a number of investigators (Pinchas *et al.*, 1961; Becker *et al.*, 1963; and Karabatsos, 1960). An example of its application is the synthesis by Howard and Miles (1964) of inosine specifically labeled with ^{18}O as an aid in interpreting the band at 1673 cm^{-1} in inosine-6-^{16}O. Spencer (1959) has used water containing 8.7 atom % excess of ^{18}O to investigate the mode of action of an arylsulfatase. (For details of these investigations, see Chapter 15.)

The isotopes ^{15}N and D are also used frequently in establishing assignments of absorption bands. (See work by Suzuki *et al.* on Polyglycines I and II, cited in Chapter 10.)

Table 1.2. Chart of Characteristic Frequencies between ~700 and 300 cm^{-1} (Sober, 1968)

(Prepared by Freeman F. Bentley, Lee D. Smithson, and Adele L. Rozek)

This chart summarizes the characteristic frequencies known to occur between approximately 700–300 cm^{-1}. Those who anticipate using this region of the spectrum should consult *Infrared Spectra and Characteristic Frequencies ~700–300 cm^{-1}* (Wiley-Interscience, New York), which not only contains a complete discussion of the characteristic frequencies summarized in this chart, but also a large collection of infrared spectra (700–300 cm^{-1}) of most of the common organic and inorganic compounds, and an extensive bibliography of references to infrared data below ~700 cm^{-1}.

In this chart the black horizontal bars indicate the range of the spectrum in which the characteristic frequencies have been observed to occur in the compounds investigated. The number of compounds investigated is given immediately to the right of the names or structures of the compounds. Obviously those characteristic frequency ranges based upon a limited number of compounds should be used with caution.

The letters above the bars indicate the relative intensities of the absorption bands. These intensities are based upon the strongest band in the spectra (700–300 cm^{-1}) of specific classes of compounds investigated, and they cannot be compared accurately with the intensities given for other classes.

When known, the specific vibration giving rise to the characteristic frequency is printed in abbreviated form immediately to the right of the bar indicating the frequency range except when lack of space prevents this. When there can be no ambiguity, this information may be printed other than to the right. In doubtful cases, arrows are used for clarification.

Naturally, the characteristic frequencies vary in their specificity and analytical value. The user is, therefore, cautioned to use this chart with some reserve. After reviewing this chart, the reader should be aware that there are many characteristic frequencies in the 700–300 cm^{-1} region. Used cautiously, this chart can be of considerable value in the elucidation of structures of unknown compounds.

It is important to emphasize that the region of the infrared between ~700–300 cm^{-1} should be used in conjunction with the more conventional 5000–700 cm^{-1} region. Much of the value of the 700–300 cm^{-1} region can only be realized after interpreting the spectrum between 5000–700 cm^{-1}.

The following symbols and abbreviations are used:

Symbol or Abbreviation	Definition	Symbol or Abbreviation	Definition
αCCC	In-plane bending of benzene ring	$\perp$	Perpendicular
Antisym.	Antisymmetrical (Asymmetrical)	ϕ	Phenyl
~	Approximately	ϕCC	Out-of-plane bending of
β	In-plane bending of ring		aromatic ring
	substituent bond	r	Rocking
δ	In-plane bending	s	Strong
γ	Out-of-plane bending	sh	Shoulder
i.p.	In-plane	Sym.	Symmetrical
m	Medium	v	Variable
ν	Stretching	w	Weak
ν_s	Symmetrical stretching	"X" Sensitive	An aromatic vibrational mode
ν_{as}	Antisymmetrical stretching		whose frequency position is
o.p.	Out-of-plane		greatly dependent on the nature
$\parallel$	parallel		of the substituent.

Table 1.2 (*Continued*)

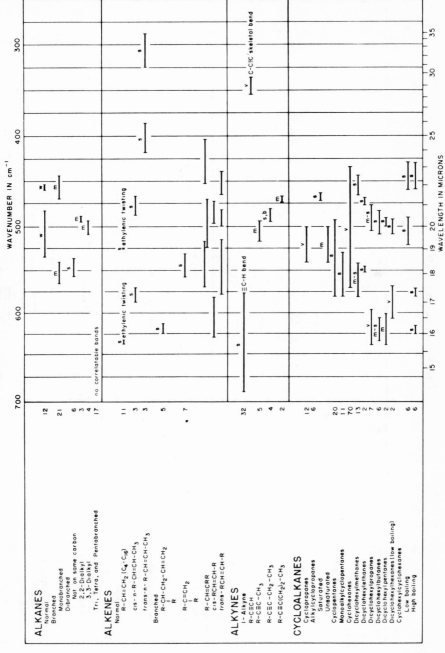

Table 1.2 (*Continued*)

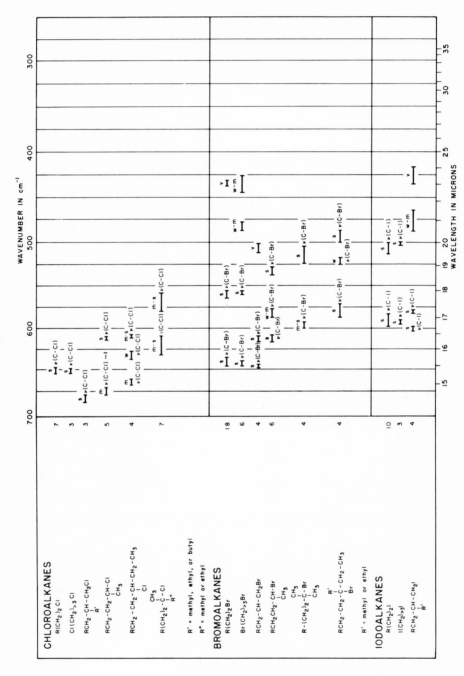

Table 1.2 (*Continued*)

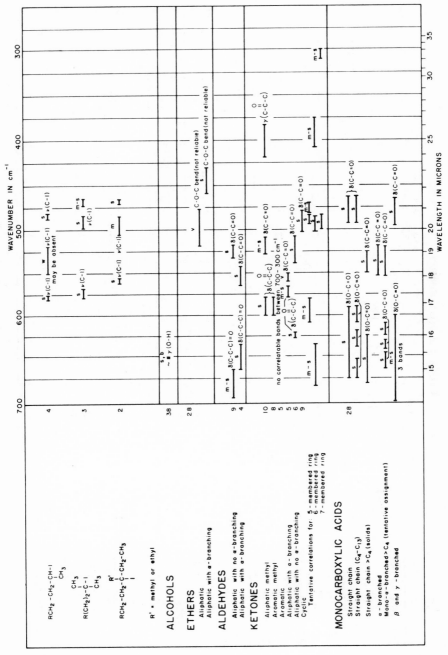

Table 1.2 *(Continued)*

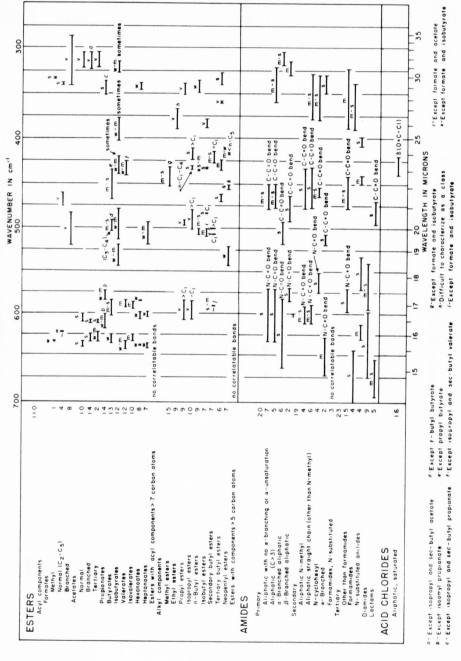

a - Except isopropyl and sec - butyl acetate   d - Except t - butyl butyrate   g - Except formate and isobutyrate

b - Except isoamyl propionate   e - Except propyl butyrate   h - Difficult to characterize as a class

c - Except isopropyl and sec - butyl propionate   f - Except isopropyl and sec - butyl valerate   i - Except formate and isobutyrate

j - Except formate and acetate

k - Except formate and isobutyrate

Table 1.2 *(Continued)*

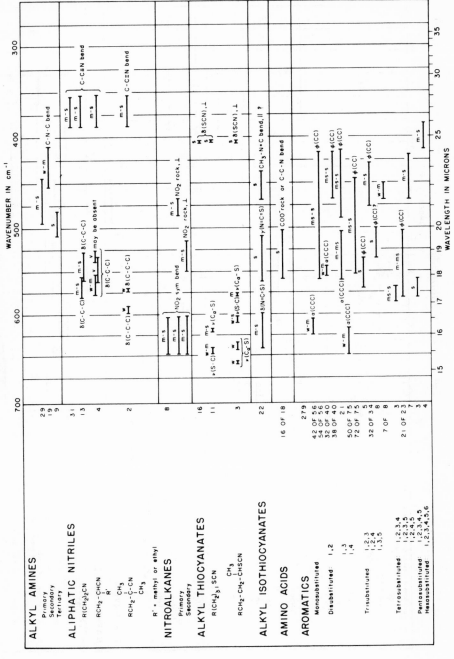

Table 1.2 *(Continued)*

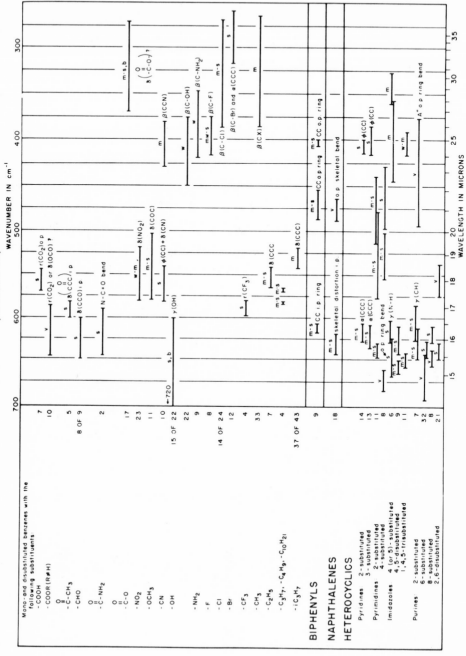

Table 1.2 *(Continued)*

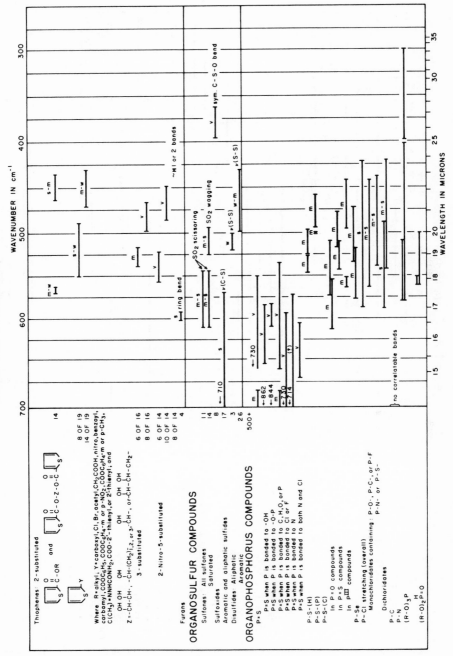

Table 1.2 (Continued)

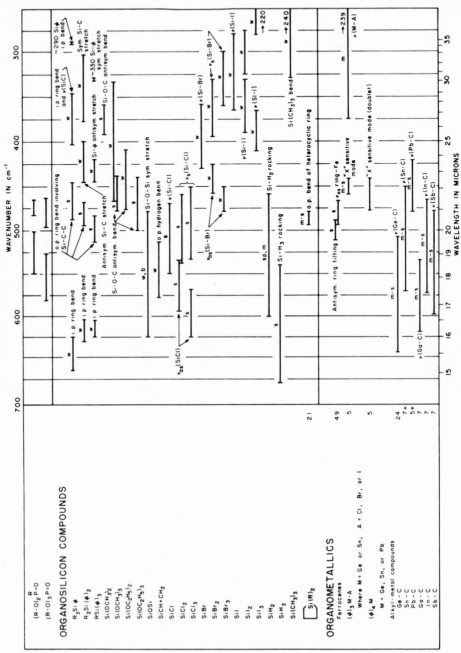

Table 1.2 *(Continued)*

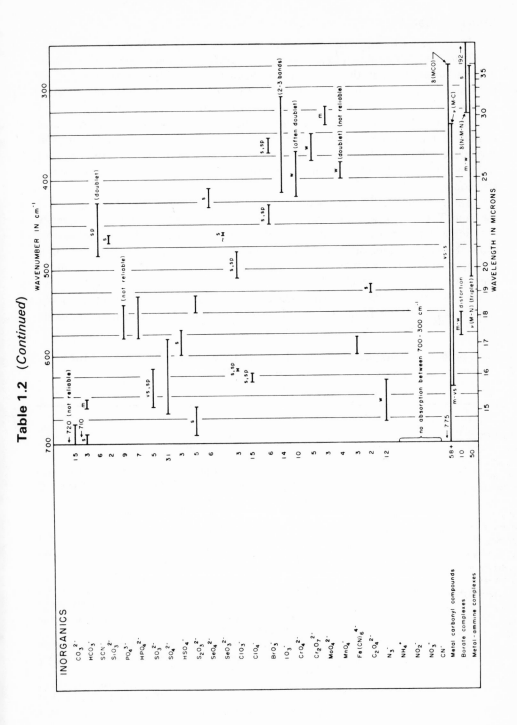

Spurious Bands

In analyzing the spectrum of a substance, one occasionally finds a band or a shoulder that is very difficult to interpret. This band may, upon further consideration, be found to be spurious, that is, it does not belong to the sample under analysis, but is caused by an instrumental effect, the method of handling the sample, or some unexpected phenomenon. For example, it has been pointed out (Inchiosa, 1965) that certain disposable syringes used in clinical and laboratory medicine yielded water-soluble extractives. One of these substances was identified as 2-(methylthio)benzothiazole, which has fungistatic and insecticidal activity. Such extractives or their reaction products could show up in the infrared spectrum during a laboratory procedure.

Table 1.3 gives a list of spurious bands, arranged according to the frequency of the band. Table 1.4 presents a list of spurious bands that might occur at any frequency.

Sometimes, optically polished materials that one obtains for the laboratory show absorption bands of an unknown origin. These bands could have been caused by impurities within the material itself or by impurities left in the surface from the grinding and polishing process. To help workers identify the source of such absorption bands, McCarthy (1968) has published the spectra of aluminum oxide, barnesite, cerium oxide, glassite, rouge, sodium thiosulfate, stannic oxide, and titanium oxide.

Table 1.3. Spurious Bands That Occur at Specific Frequencies (Launer, 1962)

Band position, cm^{-1}	Compound or structural group	Comments
3704	H_2O	In the near-infrared, where thick cells are used, a trace of moisture in carbon tetrachloride or hydrocarbon solvents gives an O—H band at 3704 cm^{-1}.
3650	H_2O	Some fused silica windows show a sharp band at 3650 cm^{-1} produced by occluded water.
3571	?	The earliest lithium fluoride prisms contained an impurity that gave a strong band at 3571 cm^{-1} in single-beam recordings. Later lithium fluoride prisms did not show this absorption, so now the 3571 cm^{-1} ghost probably has only historical interest.
3448–3333	H_2O	Coblentz, in his pioneering infrared studies, pointed out that entrained water in samples gives rise to a band near 3333 cm^{-1}. A band is present at 3448 cm^{-1} in nearly all KBr-disk spectra. (See 1639 cm^{-1} band.)
2347	CO_2	If the beams of a double-beam spectrometer are not properly balanced, a band due to atmospheric carbon dioxide will appear. (See 667 cm^{-1} band.)
2326	CO_2	Liquid samples that have been stored over dry ice sometimes show an extraneous band near 2326 cm^{-1} due to dissolved carbon dioxide. In methyl silicone fluids the exact band position is 2331 cm^{-1}.

Table 1.3 (*Continued*)

Band position, cm^{-1}	Compound or structural group	Comments
1996	BO_2^-	Some artificial NaCl crystals obtained from Harshaw in the past few years have shown a sharp absorption line at 1996 cm^{-1} due to a trace of metaborate ion.
2000–1429	H_2O	Atmospheric water vapor has many sharp, strong bands in this range. Even on a double-beam spectrometer, where the water absorption is balanced out, the pen may go dead momentarily in passing through one of these water bands. This explains some of the "anomalous" shoulders seen on the side of C=O bands. The effect is more pronounced on grating instruments than on prism instruments. It can be corrected by slowing the scanning speed or by flushing the spectrometer with dried air.
1812	$COCl_2$	Chloroform, which has been prepared for use as an infrared solvent by removing the alcohol inhibitor, oxidizes to phosgene on exposure to air and sunlight. A band due to the C=O group of phosgene appears at 1812 cm^{-1} and increases slowly in intensity with time.
1754	Phthalic anhydride	Phthalic anhydride, unexpectedly obtained during a chromatographic separation, was found to be a decomposition product of the dialkyl phthalate plasticizer in tubing attached to the chromatographic column. (See 1724 cm^{-1} band.)
1754–1695	C=O	The coating or glaze on some kinds of bottle cap liners is dissolved by solvents and gives rise to false C=O bands.
1724	Phthalates	The dialkyl phthalate plasticizer in flexible tubing made from polyvinyl chloride is leached readily from the tubing by organic solvents and appears occasionally as a contaminant in samples. (See 1754 cm^{-1} band.)
1639	H_2O	Entrained liquid water is present in many materials. It is particularly troublesome in some inorganic salts, minerals, cellulosic materials, carbohydrates, and alkylaryl sulfonate surfactants. (See 3448–3333 cm^{-1} region.)
1613–1515		Alkali halide windows (particularly KBr) can react with carboxylic acid or metal carboxylates to produce an alkali salt of a carboxylic acid and to give a spurious band due to the carboxylate anion. This sort of thing also happens in KBr disks. Dicarboxylic acids like maleic acid seem to be the worst offenders.
1429	CO_3^{2-} (?)	Some alkali halide crystals and many KBr disks show a weak band near 1429 cm^{-1}. Perhaps this band is due to an inorganic carbonate impurity.
1355	$NaNO_3$	A spurious band at 1355 cm^{-1}, along with a weaker but sharper band at 836 cm^{-1}, was traced to a deposit of sodium nitrate on a sodium chloride window near a Nernst glower. The deposit was attributed to the reaction of nitric acid with the window, the nitric acid originating from nitric oxide produced by direct union of nitrogen and oxygen at ca. 1900°C on the Nernst glower.
1325	$Na_2C_2O_4$	A sharp 1325 cm^{-1} band appearing in some spectra of minerals was traced to the sodium oxalate dispersing agent used in sedimentation of the infrared sample.

Table 1.3 (*Continued*)

Band position, cm^{-1}	Compound or structural group	Comments
1266	Si—CH$_3$	Silicone stopcock grease is attacked by aromatic and chlorinated solvents. The presence of a sharp 1266 cm^{-1} band along with broad absorption in the 1111–1000 cm^{-1} range is characteristic of the silicones in stopcock greases. Silicone greases should be rigorously excluded in all careful separations work.
1111	?	In KBr disks, a shallow band at 1111 cm^{-1} is evidence of an impurity that is present even in some samples of infrared-grade potassium bromide.
1111–1053	Glass	Some workers prepare mulls by grinding the sample with mineral oil between the surfaces of a ground-glass joint. In a few instances we have observed in the resulting spectrum a broad band due to powdered glass.
1111–1000	Si—O—Si	Unless special precautions are taken, chlorosilanes hydrolyze readily in air to form siloxane polymers and hydrogen chloride. Siloxane deposits on absorption cell windows give strong broad Si—O—Si bands in the region 1111–1000 cm^{-1}.
1099	Si—O—Si	In semiconductor research, a trace of oxygen in a silicon crystal gives a strong Si—O—Si band in the vicinity of 1099 cm^{-1}.
1087–1075	SiO$_2$	After a gas cell has been used for samples containing silicon tetrachloride, a broad background absorption appears due to silica on the cell windows.
980	K$_2$SO$_4$	An extraneous band sometimes is observed in KBr-disk spectra of inorganic sulfates. It is caused by the double decomposition reaction of potassium bromide with the inorganic sulfate to give potassium sulfate.
935	Paraformaldehyde	A white deposit of paraformaldehyde, $(CH_2O)_x$, whose strongest band is at 935 cm^{-1}, forms on the windows of a gas cell containing formaldehyde.
907	CCl$_2$F$_2$	Used refrigerator oils frequently show a surprisingly strong band at 907 cm^{-1} due to dissolved Freon 12 (CCl$_2$F$_2$). Freon 12, which is a gas at room temperature (b.p. $-29.2°C$), can remain dissolved in the oil even after prolonged heating.
836	NaNO$_3$	See 1355 cm^{-1} band.
823	KNO$_3$	This band sometimes appears in KBr-disk spectra of inorganic nitrates. It is due to a double decomposition reaction of potassium bromide with inorganic nitrate to give potassium nitrate.
794	CCl$_4$ (gas)	Carbon tetrachloride vapor—coming from leaky cells containing carbon tetrachloride solutions has been known to lie in "pools" trapped in a spectrometer housing.
787	CCl$_4$	After a carbon tetrachloride solution has been examined in a cell, incomplete flushing of the cell may leave behind some carbon tetrachloride, and there will be false bands at 787 and 763 cm^{-1}. At the 787 cm^{-1} band (the stronger of the two) 0.1 vol % carbon tetrachloride gives an absorbance of ca. 0.1 in a 0.1-mm cell.

Table 1.3 (*Continued*)

Band position, cm^{-1}	Compound or structural group	Comments
730	Polyethylene	Polyethylene now is so widely used for making all kinds of laboratory ware that it appears occasionally as a contaminant in samples. Polyethylene has a strong doublet with components at 730 and 719 cm^{-1}.
727	Na_2SiF_6	Silicon tetrafluoride attacks a sodium chloride window forming sodium silicofluoride.
719	Polyethylene	See 730 cm^{-1} band.
697	Polystyrene	Polystyrene vials, used for mixing solid samples with KBr on mechanical vibrators, are easily abraded. The resulting KBr disk often shows a polystyrene band at 697 cm^{-1}. We should expect similar troubles with mixing vials made of other plastics.
667	CO_2	If the beams of a double-beam spectrometer are not properly balanced, a band due to atmospheric carbon dioxide will appear. (See 2347 cm^{-1} bands.)

Table 1.4. Spurious Bands That Might Occur at Any Frequency (Launer, 1962)

Cause	Comments
Christiansen effect	A spurious absorption band on the short-wavelength side of a true absorption band sometimes is observed in slurries or mulls of crystals when the particle size is the same order of magnitude as the wavelength of the radiation.
Crystal orientation	Crystal orientation can have some striking effects on a spectrum. In an oriented crystalline film, for example, a vibrating group may be fixed in such a position that it will not interact with the radiation from the source. Another difficulty is that the incident radiation in a spectrometer is partially polarized, and the relative intensities of absorption bands may change as a crystalline sample is rotated.
Differential analysis	In differential analysis, with a solvent-filled cell in the reference beam, the pen is dead in regions of strong solvent absorption. Because someone looking over the spectrogram later may think he sees a "band" here, it is a good plan to label clearly such unusable regions before filing away the spectrogram.
Foreign gases	Because of the double-beam feature of modern spectrometers, traces of foreign gases in the air of the laboratory rarely cause any trouble. In one instance, however, during a standard test run (single beam, test signal) some unexpected bands were encountered. It turned out that they were due to Freon from a leaking refrigerator unit.
Interference fringes	Thin unsupported solid sheets often show interference fringes that could be mistaken for absorption bands. So do organic coatings on metals when examined by the reflection technique. Interference fringes can be a nuisance, but there are ways of eliminating them.

Table 1.4 (*Continued*)

Cause	Comments
Molten solids	The examination of a melt has its pitfalls. For example, sudden crystallization (or even a phase change after crystallization) can result in a rapid drop of transmission that might be mistaken for an absorption band.
Optical attenuator	A nick or irregularity in the teeth of the optical wedge will give rise to a false step or shoulder on the side of an absorption band. This effect is most pronounced in the 0–10% transmission range.
Polymorphism	Different crystalline forms of the same substance show differences in their infrared spectra. This should be kept in mind constantly when interpreting infrared spectra of crystalline solids. Sometimes a different form of the crystal can be obtained merely by dissolving the crystal in a solvent and then depositing it on a plate by evaporation of the solvent. Making a mull or a KBr disk can produce a change in the crystal form of the starting material.

REFERENCES

Alpert, N. L., Keiser, W. E., and Szymanski, H. A. *IR—Theory and Practice of Infrared Spectroscopy*, 2nd Ed., Plenum Press, New York, 1970.

Becker, E. D., Ziffer, H., and Charney, E. *Spectrochim. Acta* 19, 1871 (1963).

Bentley, F. F., Smithson, L. D., and Rozek, A. L. *Infrared Spectra and Characteristic Frequencies* ~700–300 cm^{-1}, Wiley-Interscience, New York, 1968.

Bladon, P. and Eglinton, G. in *Physical Methods in Organic Chemistry* (J. C. P. Schwarz, ed.), Holden-Day, Inc., San Francisco, 1964.

Coleman, I. and Low, M. J. D. *Spectrochim. Acta* 22, 1293 (1966).

Conley, R. T. *Infrared Spectroscopy*, Allyn and Bacon, Boston, 1966.

Conn, G. K. T. and Avery, D. G. *Infrared Methods*, Academic Press, New York, 1960.

Herscher, L. W. in *Applied Infrared Spectroscopy* (D. N. Kendall, ed.), Reinhold, New York, 1966.

Herzberg, G. *Molecular Spectra and Molecular Structure: Infrared and Raman Spectra of Polyatomic Molecules*, D. Van Nostrand, New York, 1945.

Howard, F. B. and Miles, H. T. *Biochem. Biophys. Res. Commun.* 15, 18 (1964).

Inchiosa, M. A., Jr. *J. Pharm. Sci.* 54, 1379 (1965).

Jones, R. N., Faure, P. K., and Zaharias, W. *Revue Universelle des Mines* 15, 417 (1959).

Jones, R. N., Jonathan, N. B. W., MacKenzie, M. A., and Nadeau, A. *Spectrochim. Acta* 17, 77 (1961).

Karabatsos, G. J. *J. Org. Chem.* 25, 315 (1960).

Launer, P. J. *Perkin-Elmer Instrument News* 13 (No. 3), 10 (1962).

Low, M. J. D. *Experientia* 22, 262 (1966).

Low, M. J. D. *Appl. Opt.* 6, 1503 (1967).

Low, M. J. D. *Anal. Chem.* 41 (No. 6), 97A (1969).

Low, M. J. D. and Freeman, S. K. *Anal. Chem.* 39, 194 (1967).

McCarthy, D. E. *Appl. Spectrosc.* 22, 66 (1968).

Martin, A. E. *Infrared Instrumentation and Techniques*, Elsevier, Amsterdam, 1966.

Pinchas, S., Samuel, D., and Weiss-Broday, M. *J. Chem. Soc.* 1961, 2382, 3063, and references cited therein.

Potts, W. J., Jr. *Chemical Infrared Spectroscopy, Vol. I: Techniques*, Wiley, New York, 1963.

Sober, H. A., ed. *Handbook of Biochemistry, Selected Data for Molecular Biology*, The Chemical Rubber Company, Cleveland, Ohio, 1968.

Spencer, B. *Biochem. J.* 73, 442 (1959).

Szymanski, H. A. *Interpreted Infrared Spectra, Vol. 2*, Plenum Press Data Division, New York, 1966.

Wilson, E. B., Decius, J. C., and Cross, P. C. *Molecular Vibrations*, McGraw-Hill, New York, 1955.

Chapter 2

NEAR-INFRARED SPECTROSCOPY

Introduction

In the near-infrared region the absorption process is associated with the overtone and combination bands of vibrational transitions. This region can be studied with the same equipment that is used for electronic spectra, apart from changing from photocell to photoconductive detectors above 1 μ. A review by Kaye (1954; see also, 1955) defines the near-infrared region to be from 0.7 μ to 3.5 μ, and quartz prisms are usable over much of this range. Goddu (1960), Wheeler (1959), McCallum (1964), and Whetsel (1968) have more recently presented reviews of near-infrared spectrophotometry. Wheeler's is a comprehensive review with references to near-infrared spectra of somewhere near 500 organic compounds of many classes. Whetsel's review also contains extensive references to organic applications.

The spectra observed in this region involve mainly hydrogen stretching vibrations in, for example, C—H, N—H, and O—H bonds. A fundamental absorption band at a given frequency may be accompanied by bands at all multiples of this frequency. These additional bands are called overtones. The first overtone is much weaker than the fundamental, and successive overtones are progressively weaker still. Absorption bands may also occur at a frequency which is the sum or difference of two fundamental frequencies, or the sum or difference of an overtone and a fundamental frequency. These are called combination frequencies. Overtone and combination bands are most readily observed on comparatively thick samples in the region between the visible and 2.5 μ, where there are no fundamental absorption bands, and all the bands arise from this cause.

The vibration of the X—H group is large in amplitude because of the low atomic weight of hydrogen, and consequently, deviates appreciably from true harmonic motion. The overtone and combination bands are therefore relatively intense. The phenomena most studied with near-infrared spectroscopy have been intermolecular associations, the type most familiar to biochemists being hydrogen bonding.

For two fundamental frequencies, a and b, first overtones will occur near $2a$ and $2b$, second overtones near $3a$ and $3b$, etc., and combination bands can appear at $a + b$ and $a - b$ cm^{-1}. Summation bands are commonly observed in the near-infrared spectra of many molecules, but combination bands arising from difference tones are improbable in the near-infrared region at room temperature (Kaye, 1954).

25

We can calculate the expected absorption bands for n-octane as an example. For a CH_2 group, if we use 1460 cm^{-1} as the value of the symmetrical deformation frequency (δ) and 2900 cm^{-1} as the value of the stretching frequency (v), we should expect some of the bands listed in Table 2.1. The approximate values of observed frequencies are also recorded in the table, and are seen to support the calculated results.

Many near-infrared absorption bands occur with sufficient regularity to allow the characterization of certain molecular groups, just as is done with fundamental bands in the $3–15\ \mu$ range.

Table 2.1. Overtone and Combination Bands of n-Octane (Willis and Miller, 1961)

Overtone or combination		Calculated, cm^{-1}	Observed, cm^{-1}
$\delta + v$	(strong)	4360	~ 4350
3δ		4380	
$2v$	(strong)	5800	
$2\delta + v$		5820	~ 5800
4δ		5840	
$\delta + 2v$	(strong)	7260	
$3\delta + v$		7280	~ 7200
5δ		7300	
$3v$	(strong)	8700	
$2(\delta + v)$		8720	~ 8300
$4\delta + v$		8740	
6δ		8760	
$\delta + 3v$	(strong)	10,160	
$5\delta + v$		10,200	$\sim 10,800$
7δ		10,220	
$4v$	(strong)	11,600	

Applications

Figure 2.1 (Kaye, 1954) is a chart of characteristic frequencies of purified organic compounds. It shows positions, ranges, and types of characteristic bands [see also, Goddu (1960)]. The length of the lines indicates the normal range within which the bands are usually found. Most of the absorptions indicated in Fig. 2.1, except for the fundamental bands between 2.6 and $3.6\ \mu$, are due to overtone bands.

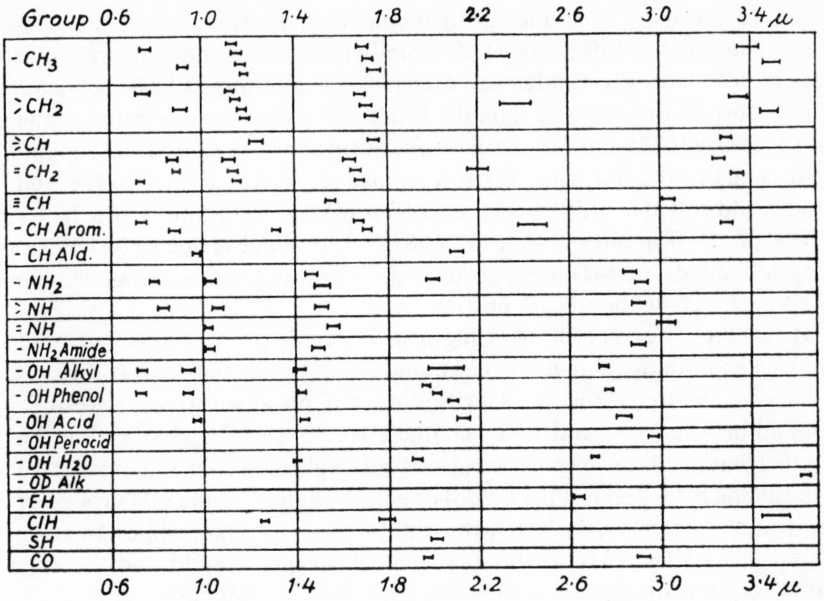

Group 0·6   1·0   1·4   1·8   2·2   2·6   3·0   3·4 μ

Group	0·6	1·0	1·4	1·8	2·2	2·6	3·0	3·4 μ
- CH₃								
⋗CH₂								
⋛CH								
= CH₂								
≡ CH								
- CH Arom.								
- CH Ald.								
- NH₂								
⋗ NH								
≡ NH								
- NH₂ Amide								
- OH Alkyl								
- OH Phenol								
- OH Acid								
- OH Peracid								
- OH H₂O								
- OD Alk								
- F H								
C/H								
SH								
CO								

0·6   1·0   1·4   1·8   2·2   2·6   3·0   3·4 μ

Fig. 2.1. Assignments for the near-infrared region. (Kaye, 1954.)

Goddu (1960) has presented another chart with additional spectra–structure correlations and molar absorptivity data between 1 and 3.1 μ. Among the groups for which additional data have been given are vinyloxy ($-OCH=CH_2$); terminal

$-CH-CH_2$; terminal $-CH-CH_2$; terminal $\equiv CH$; *cis* $-CH=CH-$;

$\overset{\diagdown}{}O\overset{\diagup}{}$ $\overset{\diagdown}{}CH_2\overset{\diagup}{}$

CH_2

C   O (oxetane); $-CH$ formate; $-NH_2$ amine, aromatic and aliphatic;

CH_2

$-NH_2$ hydrazine; $-OH$ hydroperoxide, aromatic and aliphatic; $-OH$ phenol, free and intramolecularly bonded; $-OH$ glycol, 1,2, 1,3, 1,4; $=NOH$ oxime; HCHO (possibly hydrate); $-SH$; $>PH$; $>C=O$; $-C\equiv N$; N$-$H, primary amide; and a few others.

Among the analytical applications using near-infrared spectroscopy have been the following determinations: water in hydrocarbons; water in alcohols; water in carboxylic acids; alcohols in hydrocarbons; alcohols in acids; acids in hydrocarbons; acids in anhydrides; amines in hydrocarbons; benzene in hydrocarbons; and olefins in hydrocarbons. By differential techniques it is possible, for example, to lower the sensitivity limits (detectability) by another factor of 10 over the usual limits. The usual limit for water in alcohols ranges from ~ 0.05 to 0.2%. Examples of special interest to biochemists are the applications of Klotz and Frank (1965) of near-infrared spectroscopy to studies of hydrogen–deuterium exchange (see Chapter 11 for other examples of this process) of N-methylacetamide and of Scarpa *et al.* (1967) to studies of hydrogen–deuterium exchange in a non-α-helical polyamide. The polymer, poly-N-isopropylacrylamide, in chloroform shows essentially one N—H peak, at 1.487 μ (6725 cm^{-1}). The position of this peak is shifted only slightly if the solvent is changed from chloroform (to dioxane) to water. In dioxane and water, however, additional bands appear at 1.53 μ (6536 cm^{-1}) and 1.57 μ (6369 cm^{-1}). Scarpa *et al.* (1967) point out that the 1.487 μ absorbance is due to N—H groups *not* hydrogen bonded to C=O, whereas the 1.53–1.57 μ peaks are seen under conditions where N—H···O=C bonds are present.

Mizushima *et al.* (1950), by using near-infrared spectroscopy, had observed in dilute CCl$_4$ solutions of N-methylacetamide absorption maxima at 2.86, 2.88, and 1.48 μ characteristic of the N—H vibration but no absorption between 2.70 and 2.82 μ and between 1.39 and 1.46 μ characteristic of the O—H vibration. They had concluded that in dilute nonpolar solutions the single molecule of N-methylacetamide is in the amide form and not the imidol or enol form having the OH group. At higher concentrations, however, the 2.88 μ absorption became weaker while new absorptions appeared at 2.97 to ~3.03 and 3.22 μ, and in the pure liquid and the crystal the 2.88 μ absorption disappeared. This was explained by assigning the 2.88 μ band to the N—H vibration of the free molecule and the 2.97–3.03 and 3.22 μ bands to that of the hydrogen-bonded, associated molecule. Similar explanations were given for the observation in the 1.5 μ region. Pimentel and McClellan (1960) have discussed and referred to many situations in which use of the overtone region is made to study hydrogen bonding. Murthy and Rao (1964; 1968) have also cited many cases of the use of infrared spectroscopy (including the near-infrared region) to study hydrogen bonding.

Near-infrared measurements of hemoglobin derivatives have revealed (Hayer, 1941) differences between the spectra of oxyhemoglobin and reduced hemoglobin and methemoglobin, metcyanohemoglobin, cyanohemoglobin, and carbonyl hemoglobin. The changes in the latter were sufficient to analyze for carbon monoxide in the blood to less than 1%. Some of the other substances of biochemical interest mentioned by Wheeler (1959) are wool, silk, gelatin, other peptides, synthetic polymers, various oils of natural origin, cellulose, and formyl and deuteroformyl free radicals. That review also discusses work by others on the use of near-infrared spectroscopy for determination of the number of groups (e.g., CH$_3$) in a molecule, possible analysis of alkaloids, estimation of the number of double bonds with *cis* α,β-hydrogen atoms, and the estimation of the *cis* form content of fatty acids.

Proteins and Polypeptides

Band assignments in the 1.4–1.6 μ region for the position of the first overtone of the NH stretching vibration were based on an extensive series of investigations (Klotz and Franzen, 1962; Hanlon *et al.*, 1963; Klotz *et al.*, 1964; Hanlon and Klotz, 1965) conducted with N-methylacetamide and with two polyamino acids, poly-L-alanine and poly-γ-benzyl-L-glutamate (Table 2.2).

Table 2.2. Band Assignments of the First Overtone of the NH Stretching Vibration (Klotz and Franzen, 1962; Hanlon et al., 1963; Klotz et al., 1964; Hanlon and Klotz, 1965)

Band assignments	Chemical state of −NH in polypeptides	Position of −NH overtones, λ
Peptide H-bond	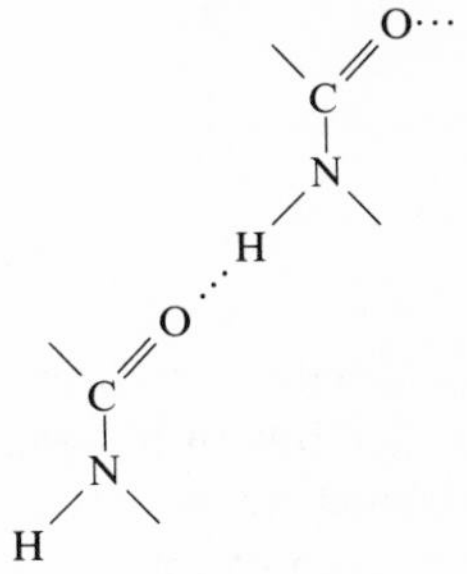	1.54 μ, 1.58 μ
Protonated peptide	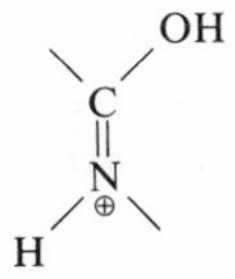	1.51 μ
Peptide–acid H-bond	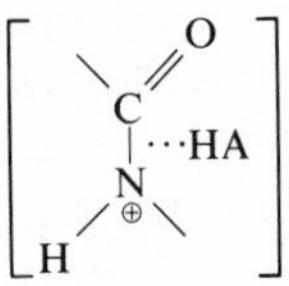	1.49 μ

In an unassociated state, solutions of N-methylacetamide in an inert solvent exhibit a band at 1.47 μ. This band is due to the first overtone of the NH stretching vibration, as is evident from its complete absence in the spectrum of N,N-dimethylacetamide. As the concentration of the amide is increased, linear aggregates form due to NH···O=C hydrogen bonds. The spectrum of N-methylacetamide in this aggregated state displays a double band with maxima at 1.525 and 1.57 μ. Spectra of poly-γ-benzyl-L-glutamate in the solvents which cause helix formation, chloroform and pyridine, reveal bands of similar shape and absorptivities, although the maxima are shifted to higher wavelengths, 1.54 and 1.58 μ. These latter wavelengths are presumably characteristic of the peptide NH involved in an intramolecular peptide hydrogen bond.

The spectrum of N-methylacetamide in a relatively weak acid such as acetic acid shows a band at 1.485 μ. Undoubtedly this reflects hydrogen bonding of the NH in the amide to the undissociated acid, there being no evidence for extensive protonation of the amide in such solutions. In a stronger acid, such as trifluoroacetic acid, however, conductivity and density studies indicate that the amide is protonated by the solvent. The spectra then exhibit two bands, one at 1.51 μ and the other at 1.525 μ.

Only the 1.51 μ band is predominant in the much stronger sulfuric and hydrochloric acids, as well as in a mixed solvent containing perchloric acid and dioxane. Under the conditions of the latter experiment, NMR data showed that the amide is protonated (Berger *et al.*, 1959):

$$
\begin{array}{c}
\diagdown \quad OH \\
\diagdown\diagup \\
C \\
\parallel \\
N \\
\diagup \; \oplus \; \diagdown \\
H
\end{array}
$$

Spectra of poly-L-alanine in this same perchloric acid solvent also show a single major band at 1.51 μ, thereby demonstrating that the incorporation of the amide residue into the polypeptide structure results in no pronounced changes in the position of the maximum due to the protonated species. Therefore, a band at 1.51 μ was assigned to the protonated peptide species.

Hanlon (1966) has presented spectra for poly-γ-benzyl-L-glutamate in systems of varying solvent composition of dichloroacetic acid and ethylene dichloride. The 1.51 μ band indicates protonation of the peptide residues. Poly-γ-benzyl-L-glutamate exhibits a unique protonation pattern characterized by two transition regions: one at low concentrations of acid, and the other between concentrations of 75% and 80% acid; only the latter transition can be reversed by an increase in temperature within the range 11 to 55°C. This unique protonation pattern parallels that found with measurements of other physical properties of poly-γ-benzyl-L-glutamate in this solvent system.

The infrared data show that about 60% of peptide residues are already protonated at the point where the major transition in the other physical properties begins to appear (Hanlon, 1966).

To understand the forces responsible for establishing a particular protein conformation we must know about the stability of the various types of bonds that might exist in a polypeptide. The common feature of all polypeptides is the amide group, with its ability to form the hydrogen bond. Klotz and Russo (1966) have studied the state of the peptide group in simple model amides and synthetic polypeptides in a variety of solvent environments. Klotz and Franzen (1962) have used N-methylacetamide as the simplest model of a peptide CONH group and near-infrared spectroscopy as a versatile probe of its state of hydrogen bonding. The overtone region of the infrared has been very useful in connection with studies in aqueous solvents. In the overtone region (1.4–1.6 μ) bands tend to be sharper than in the fundamental region. Consequently there is minimal overlap between the absorption due to O—H of the solvent and that due to N—H of the solute. Also, in the overtone range absorptivities are smaller, making it possible to use cells of 0.1–10 cm light path, which can be accurately manufactured and reproduced, whereas cell paths of $\sim$0.01 mm needed in the fundamental range are difficult to reproduce precisely. The glass cells used in the overtone range are transparent and one can use conventional absorption cells.

From studies in a variety of solvents Klotz and Franzen (1962) have found that an amide N—H group that is not hydrogen-bonded to C=O shows a peak at 1.47–1.48 μ (Fig. 2.2) whereas the NH$\cdots$O=C state is revealed by an alternative double

Fig. 2.2. Schematic summary of positions of bands in overtone infrared region for amide N—H group. (Klotz and Franzen, 1962.)

peak at 1.53–1.57 μ. A specific example of the spectrum of a solution containing N-methylacetamide partly in the monomeric N—H form and partly in the aggregated NH$\cdots$O=C array is shown (Hanlon and Klotz, 1965) in Fig. 2.3 to illustrate the position of the peaks characteristic of each state.

It is possible to make quantitative measurements of the absorbances of these bands as a function of the concentration of N-methylacetamide and thereby to evaluate the equilibrium constant and other thermodynamic parameters.

Figure 2.4 (Klotz and Russo, 1966) illustrates the variation of degree of hydrogen bonding with solute concentration for a number of solvents; Table 2.3 summarizes

Table 2.3. Thermodynamics of N—H $\cdots$ O=C Hydrogen Bond Formation of N-Methylacetamide at 25°C (Klotz and Russo, 1966)

Solvent	ΔG°, kcal/mole	ΔH°, kcal/mole	ΔS°, gibbs/mole
Carbon tetrachloride	−0.92	−4.2	−11
N-Methylmorpholine	−0.38	—	—
Acetonitrile	0.17	−0.7	−3
Dioxane	0.39	−0.8	−4
Water	3.1	0.0	−10

the corresponding energy quantities. Figure 2.4 shows that carbon tetrachloride is the least effective competitor for NH and CO hydrogen bonding sites while water is the best competitor. Acetonitrile, dioxane, and N-methylmorpholine fall into a group occupying a middle ground and quite distinct from either carbon tetrachloride or water.

The association reaction for N-methylacetamide can be expressed as follows:

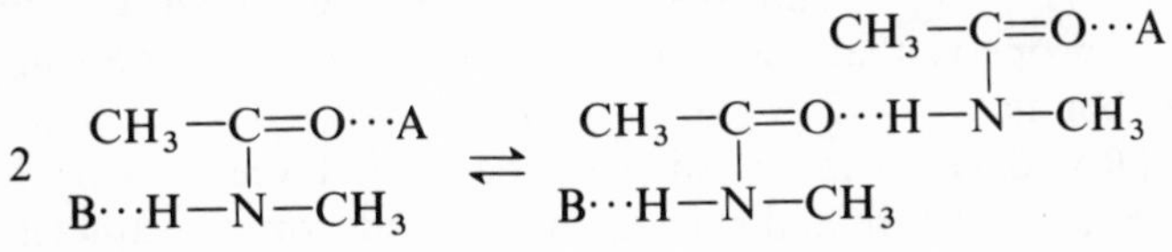

$$2 \quad \begin{matrix} CH_3-C=O\cdots A \\ | \\ B\cdots H-N-CH_3 \end{matrix} \rightleftharpoons \begin{matrix} CH_3-C=O\cdots A \\ | \\ CH_3-C=O\cdots H-N-CH_3 \\ | \\ B\cdots H-N-CH_3 \end{matrix}$$

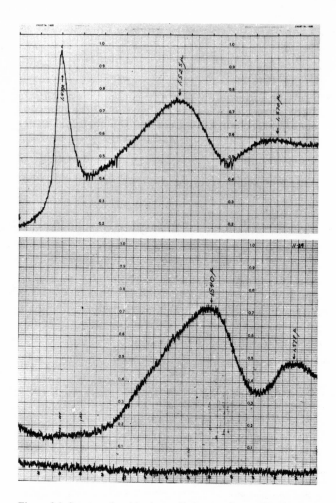

Figure 2.3. Spectra of amides in a self-associated state (Hanlon and Klotz, 1965). Upper curve, N-methylacetamide (3M) in CCl$_4$ *versus* CCl$_4$; lower curve, poly-γ-benzyl-L-glutamate (0.053 residue molar) in pyridine *versus* CCl$_4$ in pyridine. Ordinates, actual absorbance × 10; abscissa, wavelengths in microns.

where A is a solvent donor group and B a solvent acceptor for a hydrogen bond. Carbon tetrachloride cannot act as a donor or acceptor molecule. Methylmorpholine, dioxane, and acetonitrile are all bases and each can accept a hydrogen bond, but cannot donate one. Therefore each is more effective than carbon tetrachloride in breaking the N—H···O=C bond. Water can act *both* as a hydrogen bond acceptor and a donor, a property that accounts for its special effectiveness in dissociating N—H···O=C bonds.

Klotz and Russo have also carried out association studies of N-methylacetamide as a function of temperature and calculated thermodynamic characteristics for the

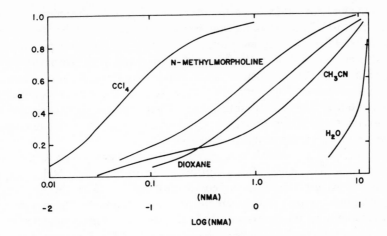

Fig. 2.4. Fraction (α) of total amide groups in hydrogen-bonded state, N—H$\cdots$O=C, as a function of molar concentration of N-methylacetamide (NMA) dissolved in carbon tetrachloride, N-methylmorpholine, dioxane, acetonitrile, and water, respectively. (Klotz and Russo, 1966.)

formation of the N—H$\cdots$O=C bond (Table 2.3). A comparison of these quantities among the solvents used shows that the value of ΔG° is determined largely by ΔH°. This is quite evident when we compare carbon tetrachloride and water, where the ΔS° values are -11 and -10, respectively. Thus, the free energy of formation of the hydrogen bond in different environments appears to be dominated by the change in internal energy, and the latter is fixed by the ability of the solvent to compete for hydrogen bonds with the N—H and C=O groups.

Near-infrared measurements (6600–7200 cm^{-1} region) have been used (Franzen and Stephens, 1963) to study hydrogen bonding between N-methylacetamide and ϵ-caprolactam. The data support the viewpoint that interamide hydrogen bonds of proteins are stable only in the well-shielded, nonpolar, hydrocarbon-rich regions of protein molecules.

Using near-infrared spectroscopy, Susi et al. (1964) have investigated the interaction of amide groups through hydrogen bonding in aqueous solution. δ-Valerolactam was used as a model compound. Absorption measurements were made as a function of temperature and concentration in the region from 1.3 to 1.7 μ. Data obtained between 0° and about 40°C could be consistently interpreted in terms of the dimerization of cis amide groups. An overall enthalpy change of -5.5 ± 1 kcal/mole was obtained for the reaction

$$2 \quad \begin{matrix} \diagdown \\ C=O \\ | \\ N-H \\ \diagup \end{matrix} \text{(aq.)} \rightleftarrows \begin{matrix} \diagdown \qquad \diagup \\ C=O\cdots H-N \\ | \qquad\quad | \\ N-H\cdots O=C \\ \diagup \qquad\quad \diagdown \end{matrix} \text{(aq.)}$$

Hanlon and Klotz (1968) have discussed the use of near-infrared spectroscopy for studying structural problems of biochemistry. In particular, they have considered the equilibrium state of the peptide unit in a number of synthetic polyamino acids (poly-L-alanine, poly-L-leucine, poly-L-methionine, and poly-γ-benzyl-L-glutamate) as a function of solvent composition under conditions where the transitions in other physical properties of these polymers have been interpreted as simple peptide, hydrogen-bonded, helix-to-coil transitions. Their spectral data demonstrate that these conversions involve protonated peptide species and are far more complicated than investigators of these processes had assumed.

Klotz and Mueller (1969) have used near-infrared spectroscopy to show that the rate–pH profile of hydrogen–deuterium exchange of an amide NH in a polymer (poly-N-isopropylacrylamide) is shifted toward higher pH in the presence of an anionic detergent, sodium lauryl sulfate, which is bound by the polymer. These workers accounted for the shift in terms of a change in local environment of the amide group which influences the acid–base character of the CONH moiety. This extrinsic effect is analogous to that produced intrinsically by electron-withdrawing substituents attached to an amide group.

Nonheme Iron Proteins

Wilson (1967) has reported on the use of low temperatures ($-196°C$) to sharpen the absorption bands of some nonheme iron proteins, e.g., ferredoxins of spinach and of *Clostridium acidi-urici*, and an adrenal protein. All three of these nonheme proteins have absorption maxima near 700 nm which disappear on reduction by dithionite. The oxidized spinach ferredoxin has two additional absorptions, a shoulder near 830 nm and a weak maximum at 930 nm. Wilson (1967) investigated the possibility that the absorbance of these bands, particularly the 714 nm band of spinach ferredoxin, is sensitive to protein conformational changes in a manner analogous to the 695 nm band of ferricytochrome *c* (Schejter and George, 1964).

Protein Molecule Excluded Volume

McCabe and Fisher (1965) have measured the excluded volume of protein molecules by means of differential spectroscopy in the near-infrared region. The excluded specific volumes of bovine serum albumin and lysozyme were found to be 0.74 ml/g and 0.68 ml/g, respectively, values which are close to the values of apparent specific volume reported in the literature (Dayhoff *et al.*, 1952; Sophianopoulos *et al.*, 1962). The authors suggested that excluded specific volume as measured by their method is closely associated with the partial specific volume of the protein molecule.

Lipids

Holman and Edmondson (1956) have recorded the infrared spectra of fatty acids and other lipids (about 60 compounds) between 0.9 and 3.0 μ. They were able to distinguish *cis* double bonds, terminal double bonds, hydroxyl groups, amine groups, the acyloin group, the hydroperoxide —OH group, methyl and ethyl esters, acids, and CH_2 and CH_3 groups.

The overtone band of the O—H stretching vibrations at $7143\,\text{cm}^{-1}$ has been used to obtain the hydroxyl value of alcohols (Kauffman, 1964). Studies of compounds containing a terminal epoxide group indicate that bands at $6061\,\text{cm}^{-1}$ and $4545\,\text{cm}^{-1}$ can be used to quantitate these groups. The $6061\,\text{cm}^{-1}$ band is the first overtone of the fundamental C—H stretching vibration ($3058\,\text{cm}^{-1}$) of terminal epoxides. The band at $4545\,\text{cm}^{-1}$ is a combination band characteristic of terminal epoxides, and can be used for differentiation in the presence of other oxygen-containing rings, such as oxetanes, furans, and dioxanes.

Isolated double bonds of the —CH=CH$_2$ type and the *cis* type can be determined in the near-infrared region, the terminal methylene type at 6250 and $4762\,\text{cm}^{-1}$, and the *cis* type at $4762\,\text{cm}^{-1}$. *Trans*-unsaturated and saturated C—H groups give no interference. Therefore, mixtures of *cis*, *trans*, and terminal double bonds can be analyzed readily for *cis* and terminal bonds (Kauffman, 1964).

The stretching vibration of the OH group of monoglycerides has been examined in the region 7692–$6667\,\text{cm}^{-1}$ (Kauffman, 1964). 1-Mono- and 2-monoglycerides can be quantitated by plots of absorbances of the $6993\,\text{cm}^{-1}$ band *versus* concentration.

The C—H stretching bands of 1,2- and 1,3-diglycerides have been investigated and the spectra of these isomers showed possibilities of three main configurations (Kauffman, 1964), one internally hydrogen-bonded, and two with no hydrogen bond involvement. The fraction of hydrogen-bonded molecules seemed to be relatively low. The hydrogen-bonded form of 1,3-diglycerides was believed to have a *trans-gauche* configuration around the glyceride C—C bond, and the non-hydrogen-bonded forms to have probably a *trans-trans* configuration around the glyceride backbone, OH being either *trans* or *gauche* with respect to the nearest C—H group. In the 1,2-diglycerides, hydrogen bonds appeared to be mainly formed between groups attached to the α–γ carbon atoms of the glyceride residue, with a *trans-trans* configuration around the backbone. Two bands due to stretching of non-hydrogen-bonded hydroxyl groups were associated with rotational isomerism around the C—O bond of the alcoholic group.

Porphyrin Compounds

The reader is also referred to Chapter 14.

Recently considerable interest has been centered on the near-infrared absorption spectra of biological pigments such as cytochrome *c* and cytochrome oxidase. Ferricytochrome *c* has an absorption band at 695 nm which is strongly dependent on the protein structure (Schejter and George, 1964), while oxidized cytochrome oxidase has an 830 nm band which is associated with the copper moiety of the enzyme (Griffiths and Wharton, 1961).

Purified preparations of cytochrome oxidase in the oxidized state display an absorption band (maximum ∼830 nm) in the near-infrared region which disappears on reduction and reappears on reoxidation at rates commensurate with those of the band at 605 nm (Wharton and Tzagoloff, 1964). This chromophore can be gradually and irreversibly destroyed by treating the oxidase with the copper-specific chelator bathocuproinesulfonate in the presence of acetate buffer below pH 5 or by dialyzing the enzyme against cyanide or alkaline buffers, such as borate, glycine, and phosphate at specific concentrations and pH values. Treatment with urea and sodium dodecyl sulfate also results in a loss of the chromophore. The decrease in the intensity of the near-infrared band is accompanied by a parallel decrease in the specific activity of the enzyme and also of copper. The available evidence indicated that the chromophore absorbing near 830 nm is an integral component of cytochrome oxidase, and suggested that it may be copper bound to the protein in some specific manner.

Day *et al.* (1967*a*) have obtained the crystal spectra (range $\sim 16,000$ to ~ 5000 cm^{-1}) of mesoporphyrin IX iron (III) methoxide, ten myoglobin derivatives, and cytochrome c peroxidase. The comparison between these spectra and the corresponding solution spectra show that in all cases there is a reduction in intensity of absorption bands in the solid state, but that in most cases the energy of the spectrum is virtually unchanged. Myoglobin fluoride and cytochrome c peroxidase are exceptions.

These workers have proposed that when heme proteins are crystallized, low-spin forms are favored in the equilibrium high-spin form $\rightleftharpoons$ low-spin form, and changes in conformation can also be involved. [A discussion of spin forms has been given by Kotani (1964).] The use of polarized spectra of single crystals permitted an assignment of all the absorption bands in the spectra.

The polarization of all the bands in the heme spectra from 25,000 to 5000 cm^{-1} is in-plane. The proposed character of the transitions at 20,000, 16,000, and 10,000 cm^{-1} as being mixed charge-transfer and ligand bands (Day *et al.*, 1964) was supported by the polarization data of Day *et al.* (1967*a*), for the symmetry of all the transitions was of necessity the same and, as the transitions of the ligand were in-plane, was required to be in-plane. Besides these bands there is a series of very weak bands at 7200, 8300, and 9400 cm^{-1} in the spectra of the mesoporphyrin IX iron (III) methoxide crystals. These are possibly spin-forbidden components of the lowest electronic excited state, all polarized in the same direction, i.e., in the heme plane. Sharp, weak bands at lower wave numbers were assigned to vibrational overtones.

Although the spectral changes are small when myoglobin crystals are compared with myoglobin solutions, nevertheless these changes are always discernible, and whatever causes these changes does bring about marked changes in reactivity. The changes in the spectra of cytochrome c peroxidase are much larger, but the effect of phase change on reactivity is not yet known. Day *et al.* (1967*a*) feel that unless a determined effort is made to prove that other proteins have the same geometry in solution and in the crystalline state, it does not seem to be an essentially safe conclusion to assume that no such differences exist.

Day *et al.* (1967*b*) have presented polarized crystal spectra of the low-spin compounds ferricytochrome c and ferrimyoglobin cyanide. In both spectra a weak band at about 8000 cm^{-1} was found to be polarized in the heme plane. Ferricytochrome c also displays another band at 10,700 cm^{-1}, which probably indicates that the heme complex contains a few percent of the high-spin form since such a band is found in this region in high-spin ferric hemoproteins.

When the spectrum of reduced purified cytochrome c oxidase was recorded, it showed the disappearance of a broad absorption band at 12,048 cm^{-1} (830 nm), which had been present for the oxidized form (Smith and Stotz, 1954; Griffiths and

Wharton, 1961). Chance (1966) has discussed various aspects of this absorption band and has measured the kinetics of the 12,048 cm^{-1} band in mitochondrial suspensions. He observed an absorbance change during the transition from the oxidized to the reduced state in suspensions of intact pigeon heart mitochondria that agreed qualitatively and quantitatively with that observed in purified cytochrome c oxidase.

Gibson and Greenwood (1965) have reported experiments on the kinetic behavior of the 820 nm band of cytochrome c oxidase and have shown that the species producing it is kinetically distinct from both cytochrome a and cytochrome a_3. In the reaction of reduced cytochrome oxidase with oxygen the rate of appearance of the near-infrared band of the oxidized form at 820 nm increased with oxygen concentration up to a limiting rate of 6×10^3 sec^{-1}. This rate is greater than the rate of oxidation of cytochrome a (1×10^3) but below the maximal rate of oxidation of cytochrome a_3 $(30 \times 10^3$ sec$^{-1})$ which could be reached with their preparations. These workers attributed the appearance of the band at 820 nm to a change other than the oxidation of cytochromes a and a_3.

Brill and Sandberg (1968) have studied cyanide, fluoride, and azide complexes of bacterial micrococcus catalase (BMC) and correlated the near-infrared absorption bands with magnetic susceptibility values of similar complexes of other heme proteins. The fluoride complex of horse heart ferrimyoglobin has a band at ~ 850 nm with a shoulder at ~ 750 nm, while the cyanide complex of ferrimyoglobin has no bands in this region. Similarly, free BMC has a band at ~ 870 nm with a shoulder at 750 nm, while the cyanide complex of BMC has no bands in this region (Fig. 2.5). Furthermore, the azide complex of BMC also has a band at 880 nm with a shoulder at ~ 760 nm. Extension of the magnetic data from horse blood catalase (HBC) complexes to BMC complexes has support in the behavior of the near-infrared spectra. For free HBC at pH 7.0 the magnetic moment was 5.7 Bohr magnetons. For the fluoride complex it was also 5.7, and for the azide and cyanide complexes it was 5.5 and 2.4, respectively.

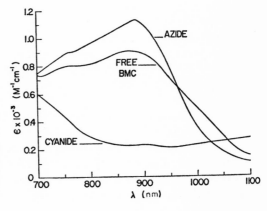

Fig. 2.5. Near-infrared spectra of free BMC and azide and cyanide complexes (700–1100 nm). For the azide complex, the azide/heme ratio was 28; for the cyanide complex, the cyanide/heme ratio was 17. In all these measurements, the heme concentration was between 340 and 360 μM and the pH was 7.0. (Brill and Sandberg, 1968.)

Flavo Compounds

Ehrenberg and Hemmerich (1964) have investigated the near-infrared absorption of metal complexes with flavin free radicals. These complexes are important in the catalytic action of metalloflavoproteins. The metal chelates of the flavin radicals possess an intense absorption with a maximum in the region of 810 to 840 nm (Fig. 2.6) in the case of valence-stable transition metal ions Zn^{2+}, Cd^{2+}, Co^{2+}, Ni^{2+}, and Mn^{2+}. This absorption band was established as due to the complex between the metal ion and the flavin free radicals by showing that for a diamagnetic metal ion (Zn^{2+}) a linear relationship exists between the electron spin resonance and light absorption (Fig. 2.7). The observed sharp absorption band of the chelated radical was easily distinguished from the broader bands of the dimolecular flavin complexes. The sharp absorption band was not observed with such ions as Fe^{2+}, Cu^{+}, or MoO^{3+}.

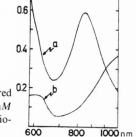

Fig. 2.6. Light absorption of tetraacetyl riboflavin (7 mM) in the near-infrared region dissolved in CHCl$_3$, 14 mM (C$_2$H$_5$)$_3$N, 5.5 M CH$_3$CN. (a) With 1.8 mM Cd(NO$_3$)$_2$ added; (b) with no metal ion added. Samples half reduced with dithionite. (Ehrenberg and Hemmerich, 1964.)

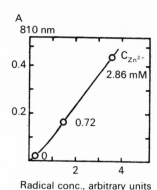

Fig. 2.7. Comparison of light absorption at 810 nm and the integrated ESR absorption for different concentrations of added Zn^{2+} ions. Conditions as in Fig. 2.6. (Ehrenberg and Hemmerich, 1964.)

Tissues

Bakerman and Mitchell (1960) have examined bovine corneas in the near-infrared region shortly after excision, after drying for 5 hr and for 2 days, and after attempted regeneration of the tissue by soaking in water for 30 min and for 2 days. These workers concluded that the withdrawal of water led to structural changes in the cornea, but they only considered the presence of protein and not of other constituents known to be present in the cornea.

Besic *et al.* (1962) have examined undecalcified tooth sections in an infrared microscope in the range 0.76–1.20 μ. Hypocalcification spots can be distinguished by infrared microscopy from sections of early caries when they are imbibed in media (water, glycerol, or eugenol) that show them to be similar in appearance when viewed only with visible-light and polarizing microscopes.

Moisture Determination

The moisture content of seeds has been determined (Hart *et al.*, 1962) by near-infrared spectrophotometry of their methanol extracts. The absorbance of water at 1.93 μ was used. The standard deviation of the results from those obtained by titration with Karl Fischer reagent was $\pm 0.24\%$. The infrared method was more rapid and less exacting than the Karl Fischer method.

Near-infrared spectrophotometry (1.9 μ band) has also been used to determine the moisture in particulate matter of cigarette smoke (Crowell *et al.*, 1961).

REFERENCES

Bakerman, S. and Mitchell, C. *Nature* **187**, 1033 (1960).

Berger, A., Loewenstein, A., and Meiboom, S. *J. Am. Chem. Soc.* **81**, 62 (1959).

Besic, F. C., Zimmerman, S. O., and Wiemann, M. R., Jr. *J. Dental. Res.* **41**, 718 (1962).

Brill, A. S. and Sandberg, H. E. *Biophys. J.* **8**, 669 (1968).

Chance, B. in *Biochemistry of Copper* (J. Peisach, P. Aisen, and W. E. Blumberg, eds.), Academic Press, New York, 1966, p. 293.

Crowell, E. P., Kuhn, W. F., Resnik, F. E., and Varsel, C. J. *Tobacco, New York* **152**, 20 (1961).

Day, P., Scregg, G., and Williams, R. J. P. *Biopolymers Symp.* **1**, 271 (1964).

Day, P., Smith, D. W., and Williams, R. J. P. *Biochemistry* **6**, 1563 (1967a).

Day, P., Smith, D. W., and Williams, R. J. P. *Biochemistry* **6**, 3747 (1967b).

Dayhoff, M. O., Perlmann, G. E., and MacInnes, D. A. *J. Am. Chem. Soc.* **74**, 2515 (1952).

Ehrenberg, A. and Hemmerich, P. *Acta Chem. Scand.* **18**, 1320 (1964).

Franzen, J. S. and Stephens, R. E. *Biochemistry* **2**, 1321 (1963).

Gibson, Q. H. and Greenwood, C. *J. Biol. Chem.* **240**, 2694 (1965).

Goddu, R. F. in *Advances in Analytical Chemistry and Instrumentation, Vol. 1* (C. N. Reilley, ed.) Interscience, New York, 1960, p. 347.

Griffiths, D. E. and Wharton, D. C. *J. Biol. Chem.* **236**, 1850 (1961).

Hanlon, S. *Biochemistry* **5**, 2049 (1966).

Hanlon, S. and Klotz, I. M. *Biochemistry* **4**, 37 (1965).

Hanlon, S. and Klotz, I. M. "Near Infrared Spectroscopy in Structural Problems of Biochemistry," in *Developments in Applied Spectroscopy, Vol. 6* (W. K. Baer, A. J. Perkins, and E. L. Grove, eds.), Plenum Press, New York, 1968, p. 219.

Hanlon, S., Russo, S. F., and Klotz, I. M. *J. Am. Chem. Soc.* **85**, 2024 (1963).

Hart, J. R., Norris, K. H., and Golumbic, C. *Cereal Chemistry* **39**, 94 (1962).

Hayer, H. *Z. Elektrochem.* **47**, 451 (1941).

Holman, R. T. and Edmondson, P. R. *Anal. Chem.* **28**, 1533 (1956).

Kauffman, F. L. *J. Am. Oil Chemists' Soc.* **41**, 4 (1964).

Kaye, W. *Spectrochim. Acta* **6**, 257 (1954).

Kaye, W. *Spectrochim. Acta* **7**, 181 (1955).

Klotz, I. M. and Frank, B. H. *J. Am. Chem. Soc.* **87**, 2721 (1965).

Klotz, I. M. and Franzen, J. S. *J. Am. Chem. Soc.* **84**, 3461 (1962).

Klotz, I. M. and Mueller, D. D. *Biochemistry* **8**, 12 (1969).

Klotz, I. M. and Russo, S. F. in *Protides of the Biological Fluids, Vol. 14* (H. Peeters, ed.), Elsevier Publ. Co., Amsterdam, 1966, p. 427.

Klotz, I. M., Russo, S. F., Hanlon, S., and Stake, M. A. *J. Am. Chem. Soc.* **86**, 4774 (1964).

Kotani, M. in *Advances in Chemical Physics, Vol. 7: The Structure and Properties of Biomolecules and Biological Systems* (J. Duchesne, ed.), Interscience, New York, 1964, p. 159.

McCabe, W. C. and Fisher, H. F. *Nature* **207**, 1274 (1965).

McCallum, J. D. in *Progress in Infrared Spectroscopy, Vol. 2* (H. A. Szymanski, ed.) Plenum Press, New York, 1964, p. 227.

Mizushima, S., Simanouti, T., Nagakura, S., Kuratani, K., Tsuboi, M., Baba, H., and Fujioka, O. *J. Am. Chem. Soc.* **72**, 3490 (1950).

Murthy, A. S. N. and Rao, C. N. R. "Hydrogen Bonding," *Tech. Report No. 1 of the Department of Chemistry*, Indian Institute of Technology, Kanpur, India, 1964.

Murthy, A. S. N. and Rao, C. N. R. *Appl. Spectrosc. Rev.* **2**, 69 (1968).

Pimentel, G. C. and McClellan, A. L. *The Hydrogen Bond*, W. H. Freeman, San Francisco, 1960.

Scarpa, J. S., Mueller, D. D., and Klotz, I. M. *J. Am. Chem. Soc.* **89**, 6024 (1967).

Schejter, A. and George, P. *Biochemistry* **3**, 1045 (1964).

Smith, L. and Stotz, E. *J. Biol. Chem.* **209**, 819 (1954).

Sophianopoulos, A. J., Rhodes, C. K., Holcomb, D. N., and Van Holde, K. E. *J. Biol. Chem.* **237**, 1107 (1962).

Susi, H., Timasheff, S. N., and Ard, J. S. *J. Biol. Chem.* **239**, 3051 (1964).

Wharton, D. C. and Tzagoloff, A. *J. Biol. Chem.* **239**, 2036 (1964).

Wheeler, O. H. *Chem. Revs.* **59**, 629 (1959).

Whetsel, K. B. *Appl. Spectrosc. Rev.* **2**, 1 (1968).

Willis, H. A. and Miller, R. G. J. in *Molecular Spectroscopy* (G. H. Beaven, E. A. Johnson, H. A. Willis, and R. G. J. Miller), Heywood and Company, Ltd., London, 1961, p. 171.

Wilson, D. F. *Arch. Biochem. Biophys.* **122**, 254 (1967).

Chapter 3

SAMPLING METHODS

The methods for handling samples to record infrared spectra in a biochemical laboratory are essentially those used in the everyday testing by the organic chemist, with certain exceptions.

Solids

One can readily record the spectra of many solid compounds as mulls and as alkali halide pellets (or even without the use of a suspending medium, as in films, or by means of attenuated total reflection). The solid particles should be extremely fine (5 μ or less) or excessive loss of energy due to scattering of light will result. One can minimize scattering losses by dispersing the solid in a medium having a similar refractive index, but scattering losses can be considerable even when the solid is dispersed in mineral oil (as a mull) or in alkali halide (as a disk).

Crystalline samples sometimes produce spectra with distorted band shapes, an effect known as the Christiansen effect [see Potts (1963) and Table 1.4]. Also, polymorphic forms of the same substance frequently show differences in infrared spectra. An example is *N*-benzoyl-2,3,4,6-tetra-*O*-benzoyl-β-D-glucosylamine, a compound that exists in a form with melting point 113–115°C which, when heated to 117–120°C and allowed to crystallize from the melt, gives a form with melting point 184°C having a somewhat different spectrum in Nujol (Tipson, 1968). Also, different crystal habits (same melting point) of a compound may display partially differing spectra, especially if examined as mulls, in which little pressure is applied. Shifts of up to 20 cm^{-1} for certain bands have been observed (Barker *et al.*, 1956) for crystalline and amorphous forms of some carbohydrates. In all such instances, however, spectra of samples of each of the forms, recorded after dissolution in the same solvent, or as a molten substance, are identical.

The Mull Technique

The proper preparation of a mull or paste is an excellent way to get a sample ready for qualitative infrared spectroscopy. One grinds vigorously about 3 to 10 mg of substance with a hard pestle in a hard and smooth mortar (e.g., agate) for 1 to 5 minutes until the powder is so fine that its caked surface takes on a "glossy" appearance. Then *a small drop* of mulling fluid is added, and the vigorous grinding

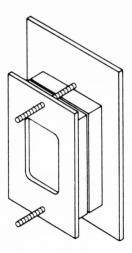

Fig. 3.1. A simple device for holding a capillary film between salt plates. (Potts, 1963.)

continued until the slurry has the consistency of cold cream. Another small drop of mulling fluid may be added if the slurry seems too dry, and grinding is continued just beyond the point at which the last substance caked on the mortar becomes dispersed in the slurry. A rod with a rubber tip, known to the chemist as a "policeman," is then used to scrape the material and transfer it to a *flat* sodium chloride or potassium bromide plate. Another *flat* salt plate is placed on the slurry, and the slurry is squeezed to form a thin uniform film by a gentle rotary motion. The "sandwich" formed is placed in a holder, as shown in Fig. 3.1, for mounting in the spectrophotometer. After recording, plates are easily cleaned by rubbing on a flat polishing cloth wetted with acetone, and then dried by rubbing on a dry part of the cloth.

Mulling agents commonly used are Nujol (mineral oil), perfluorokerosene (Fluorolube), and hexachlorobutadiene. Figure 3.2 shows spectra of the first two substances. Fluorolube is a polymer of $-(CF_2-CFCl)-$ units. One can obtain a spectrum essentially without interfering bands from mulling agents by recording the spectrum of a Fluorolube mull from 4000 to $\sim 1330\ cm^{-1}$ and then recording from 1330 to $400\ cm^{-1}$ with a Nujol mull. Hexachlorobutadiene absorbs strongly in the regions $1640–1510\ cm^{-1}$, $1200–1140\ cm^{-1}$, and $1010–760\ cm^{-1}$. Nujol is not usable in the near-infrared region because of interference from C—H overtone and combination bands, and has no appreciable absorption bands in the far-infrared region.

Frequently, it is quite difficult to grind up plastic or rubbery substances, and such materials lend themselves more readily to casting as a film on a sodium chloride or potassium bromide plate. This is done by dissolving the sample in a volatile solvent, pouring the solution onto a salt plate, evaporating off the solvent, and mounting the plate in a holder, as shown in Fig. 3.1.

Biological polymers such as fibrous proteins, cellulose, resins, and other natural materials are not soluble in volatile solvents, but certain types of proteins and

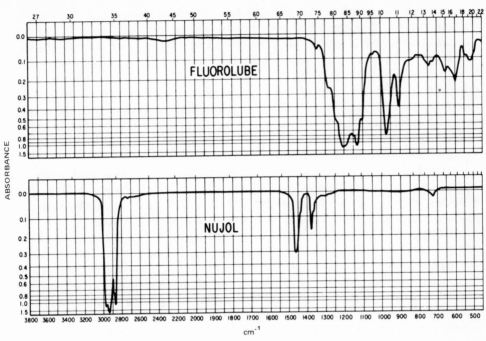

Fig. 3.2. Spectra of Fluorolube and Nujol (capillary films). (Potts, 1963.)

polysaccharides have frequently been cast as films from aqueous solutions onto silver chloride plates. To obtain mull spectra, however, one must use other methods. For example, a little finely powdered sodium chloride (particles $<2\,\mu$ in size) admixed with the sample in the mortar can act as an abrasive. If the material is quite hard, a fingernail file may be used to produce a powder which can be collected in the mortar. The powdered form often makes mulls readily. One should use a "dry box" or some other enclosed space purged of moisture for making mulls of hygroscopic materials, since they will otherwise produce poor spectra with prominent absorption by water.

Quantitative analysis by the mull technique is difficult, but internal standards can reduce such problems. Internal standards such as calcium carbonate and lead thiocyanate have been used in quantitative analysis with mulls (Barnes *et al.*, 1947; Bradley and Potts, 1958).

The Pellet Technique

The use of alkali halide pellets in which the organic sample is uniformly suspended has become very popular for recording spectra. The method of forming the pellets consists of grinding a few milligrams of the sample with about a gram of an alkali halide, placing the mixture in a die, evacuating air to remove moisture, and compressing the mixture to form a transparent disk about 1 mm thick under a pressure of about 80,000 lb/in². (Special dies and presses can be obtained from manufacturers.) The disk is then removed from the die, placed in a holder of suitable dimensions, and scanned directly in the spectrophotometer. Micropellets ($\sim 5\,\text{mm}^2$)

have also been prepared by various methods. A common way is to place a piece of blotting paper with a small central hole into the die, put the desired amount of mixture into the hole, and apply pressure. The microdisk can then be examined while supported by the paper in a holder. A beam condenser may be used for working with micropellets to insure that adequate energy impinges on the sample. The very pure nonhygroscopic matrix material must be stable and have high transmittance in the spectral range examined. It must also have a low sintering pressure and the proper refractive index. The best matrix material down to 400 cm^{-1} is potassium bromide. Sodium chloride, potassium chloride, potassium iodide, and cesium bromide are also used, the last being suitable down to 290 cm^{-1}.

Owing to moisture in the atmosphere, pellet spectra often show weak hydroxyl bands when there is no hydroxyl group present in the sample. Care should be used when one interprets spectra of pellets in the O—H stretching region. A check on this problem is to record the spectrum of a disk of the alkali halide alone.

A very convenient and good method for grinding and mixing the sample with the matrix is to use a mechanical device, such as the "Wig-L-Bug" amalgamator, which is sold by the Crescent Dental Manufacturing Co., Chicago, Illinois. The powder and several steel balls are put into a stainless steel capsule and then placed in the machine, where they are shaken vigorously for a timed period (timing helps to achieve reproducibility in the spectra). This process yields satisfactory mixing and grinding, and the technique can be used for quantitative work.

There are other ways of mixing sample with matrix material. Usually they involve hand grinding with a mortar and pestle or a ball mill. The sample is sometimes ground separately in the presence of a highly volatile solvent to obtain improved distribution of particle size. A solution of the sample in a volatile solvent may be added to the matrix powder. During the grinding of the matrix powder, the solvent evaporates and leaves a fine dispersion of the sample on the matrix particles. A difficulty encountered with methods involving evaporation of solvent is that moisture is absorbed during the process.

A very useful technique for preparing a mixture of a biochemical substance and alkali halide is to freeze rapidly a solution of the mixture and remove the water under vacuum (lyophilization). Although this method is not used in many organic and spectroscopic laboratories because of the need for special apparatus and the length of time required for sample preparation, it is frequently used to advantage in biochemistry laboratories, most of which own lyophilization equipment.

When used properly, the mull technique and the pellet method both yield good results for solid-state spectra. The pellet technique does, however, have certain advantages: interfering bands are absent; concentration and homogeneity of sample are easier to control; small samples are easier to handle; and samples can be stored for future use. On the other hand, the pellet technique has the disadvantage of chemical interaction between certain samples and the alkali halide, thereby producing spectra not characteristic of the sample. Such anomalous spectra may come from chemical and physical alterations produced by grinding the sample and halide together. One often observes such changes for polar and ionic compounds. Frequently, the pellet spectrum of a compound is not identical with the mull spectrum, and these anomalies in pellet spectra are caused by exchange of ions, formation of solid solutions, orientation effects, relaxation of stress (effects of pressure), and crystallization (polymorphism).

When certain crystalline sugars are suspended in alkali halide pellets they are changed to amorphous forms (Farmer, 1957 and 1959). The progressive changes in the spectra of some sugars, for example α-D-glucose, prepared as a pellet, are due to complex formation with trace amounts of sodium bromide or sodium iodide present as contaminants of the corresponding potassium halide (Farmer, 1959). Moist alkali halide may allow the sugar to mutarotate or form a hydrate while the pellet is stored (Barker *et al.*, 1954).

Tipson and Isbell found that, even though specimens were examined immediately after preparation of pellets, 8 of 24 aldopyranosides (1960) had interacted with the halide, and 6 of 27 free sugars (1962) displayed spectra with different characteristics in potassium iodide and in Nujol.

Other effects which may cause differences between pellet and mull spectra are particle size effects (Christiansen effect), differences in the refractive index of the medium, lattice energies of the sample and the matrix material, adsorption of the sample on the matrix, and atmospheric interactions with the matrix (Rao, 1963). It should be kept in mind that the crystal lattice is distorted during grinding and pelleting. When pressure is removed the distorted structure undergoes degrees of reversion to the original structure with time, and as a result the pellet spectrum at various intervals of time after the removal of pressure shows gradual changes during the recrystallization of the sample from a disordered to an ordered state. Attention has been called to the effect of the matrix material (KBr, KI, CsBr, CsI) on band contours in the pellet technique (Schiele, 1966), and to the effect of this type of sample preparation on the distribution of rotational isomers (Park and Wyn-Jones, 1966).

Use of Solid Films

Solid films have been examined frequently for infrared analysis in biochemical work, for example, in structural studies of proteins, polypeptides, and polysaccharides. Such films have been of particular value for studying polarization spectra of macromolecules in intact films and in oriented ones (stretched, rolled, or stroked), thereby permitting knowledge to be gained concerning spatial arrangements within the molecule and conformational effects among molecules. (See *The Use of Polarized Infrared Radiation and the Measurement of Dichroism*, p. 73, for a detailed discussion.) A few workers have discussed the film technique (Lecomte, 1948; Randall *et al.*, 1949; Hacskaylo, 1954).

To prepare a film one allows a solution of the sample to evaporate slowly on the face of a suitable window material. The plate may be gradually heated if necessary. For many compounds this means dissolution in a volatile organic solvent, and an alkali halide plate is used. Films of such substances as proteins and mucopolysaccharides have frequently been cast from aqueous solution on a silver chloride plate. IRTRAN-2 (ZnS) and KRS-5 (thallium bromoiodide) plates can also be used.

To obtain uniform films the plate may be placed on a platform that can be leveled properly. Thickness is controlled by adjusting the concentration of the solution.

Poor spectra are obtained from layers of grainy crystals. The orientation of molecules in some preferred alignment leads to difficulties in film spectra. Interference fringes found frequently in the spectra of films cause difficulties with analysis. The spectrum of a polystyrene film, for example, shows interference fringes in the 4000–2000 cm^{-1} region. These fringes are absent in the spectrum determined by the attenuated total reflection technique. Interference fringes can be prevented by placement of a film of Nujol between the sample film and the window material (Lutinski, 1958). Melts of compounds can be deposited on windows, but gradual warming of the windows is also necessary to prevent cracking. The examination of powders rubbed into polyethylene film has been described (Schwing and May, 1966).

Solution Techniques

The biochemist is quite familiar with ultraviolet and visible spectroscopy, in which a compound is frequently dissolved in an aqueous solution, a good spectrum is obtained, and quantitative analysis can be readily applied. In the case of infrared spectroscopy a common method of obtaining a spectrum is to dissolve the sample in an appropriate solvent, place the solution in a suitable cell, and record the spectrum. Certainly the solvent must have reasonable transparency to infrared radiation in the region to be used. This method is used widely in qualitative analysis, and is the most commonly used method in quantitative analysis.

If it is possible to use a solution for recording the spectrum of a substance, there are definite advantages to be had by doing so. It is easiest to interpret a spectrum when the molecule is in the simplest, least complicated, and most reproducible environment. With such conditions established, one can more facilely make useful correlations between vibrational frequencies and molecular structure. The best way to maintain materials in similar and simple surroundings is to dissolve them in dilute solutions in an inert nonpolar solvent.

Because water dissolves alkali halides and has intense interfering absorption bands, it has not been used much by spectroscopists for infrared work; but the easy availability of heavy water (D_2O) has made it possible to use both forms of water in conjunction to obtain adequate spectra for many biochemical substances in cells which are insoluble in water. A discussion of the use of water as a solvent is given later under *Aqueous Solutions*.

When studying spectra of solutions one should be aware that interactions between solute and solvent have effects on band position and intensity, but it seems preferable to put up with the relatively mild interactions between solute and nonpolar solvent than to try to identify frequencies from the much stronger interactions between adjacent polar molecules in a pure liquid or the even stronger interactions between juxtaposed molecules in a crystal lattice. Moreover, in a pure nonpolar liquid the interactions of adjacent molecules are just as important as interactions between these molecules and nonpolar molecules of solvent in a dilute solution.

Without dilution, many of the absorption bands of pure substances are so intense that spectra must be recorded from very thin layers to obtain bands of appropriate intensity. It is difficult to reproduce very thin optical-path lengths in cells for thin layers of sample.

Hydrogen bonding between compound and solvent may cause band shifts or alterations. In solution, molecules of a compound may form intermolecular hydrogen bonds. With the concentration less than $0.005M$, very little intermolecular hydrogen bonding occurs for solutions in such solvents as carbon tetrachloride (Kuhn, 1952, 1954).

Using standard solution techniques allows one to develop a sense of the relationship of band intensity to molecular structure and to apply this information in determining the structure. The problem that arises with the solution technique is that a nonpolar solvent does not exist which is transparent to infrared radiation throughout the ranges of interest and which dissolves every organic substance. However, several good solvents exist which have adequate transparency in useful ranges of the infrared region. A very satisfactory method commonly used for recording spectra of solutions is to use one solvent in certain regions and another solvent in other regions. The two solvents most often used because of their transparent properties and because many organic compounds dissolve in them are carbon tetrachloride and carbon disulfide (Fig. 3.3). The useful range for carbon tetrachloride is 4000 to $\sim 1330 \text{ cm}^{-1}$, except for a narrow range between 1475 and 1600 cm^{-1}, where tetrachloroethylene may be used. The useful range for carbon disulfide is ~ 1330 to 450 cm^{-1}. Various concentrations of solute may be used in a variety of cells with different thicknesses (optical paths), the best compromise for having adequate concentration of solute,

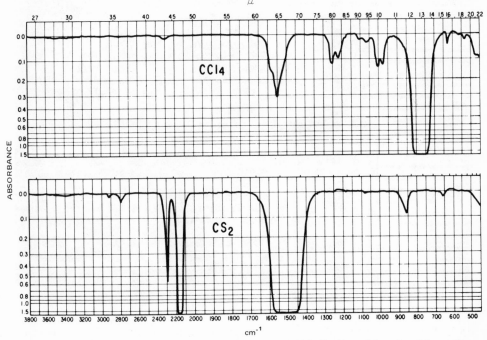

Fig. 3.3. The spectra of carbon tetrachloride and carbon disulfide liquids, cell length ~ 0.1 mm. (Potts, 1963.)

satisfactory transparency of solvent, and ease in cleaning cells being the use of 10%
solutions (weight per volume) in cells with 0.1 mm optical path.

Many compounds are not soluble enough in either of these two solvents, for
example, organic and inorganic substances of ionic nature, highly polar substances,
and most polymers. Other solvents can be used for dissolving these materials, but
most of these solvents have strong infrared absorption or low transmission, and they
are usually not satisfactory for use in the identification of compounds or for structural

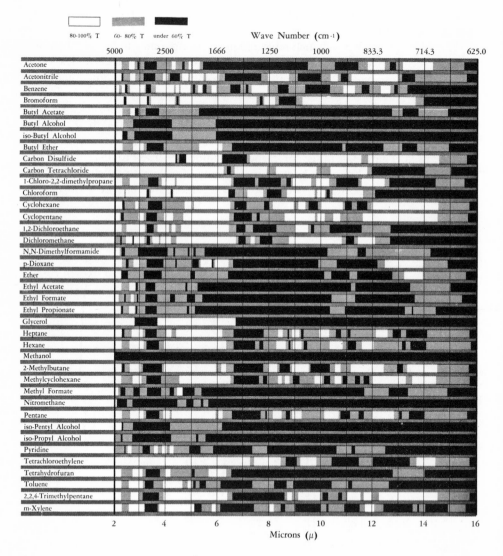

Fig. 3.4. Solvents for the 5000–625 cm⁻¹ region. Maximum transmittance in the infrared is shown in
three ranges, below 60% transmittance, 60–80%, and 80–100% transmittance, obtained with a 0.10-mm
cell path. (Courtesy of Matheson Coleman and Bell Division of the Matheson Company, Inc.)

determinations. Therefore, spectra of polar substances for such uses are recorded by techniques mentioned earlier. Many solvents are available, however, which dissolve polar substances and do have adequate transmission in some regions, so that it is still possible to do quantitative analysis by the preferred solution methods.

A chart showing the useful regions of solvents in the 5000–625 cm^{-1} region is presented in Fig. 3.4. Figure 3.5 shows the useful regions of solvents from 667 to 286 cm^{-1} (Szymanski, 1964).

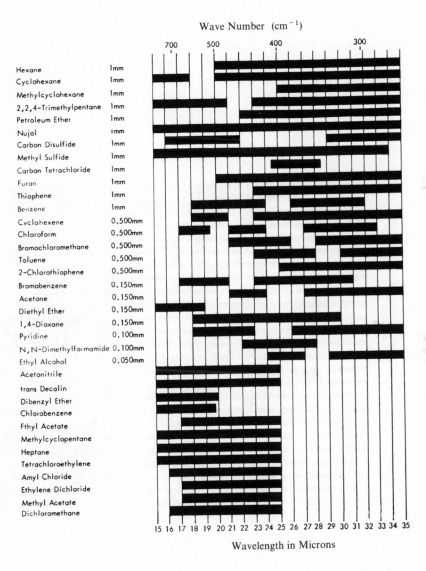

Fig. 3.5. Solvents for the 667–286 cm^{-1} region. The black lines represent useful regions. (Szymanski, 1964.)

Chloroform has transmission properties such that it is not quite as useful as carbon tetrachloride or carbon disulfide, but it dissolves polar substances better. Methylene chloride is also a much better solvent for polar compounds and its transmission properties are almost as good as those of chloroform.

A wide range of polar materials is soluble in acetone, acetonitrile, dioxane, dimethyl formamide, and nitromethane, all of which have some spectral regions free of absorption where measurements are feasible.

McNiven and Court (1970) have published the spectra of twelve solvents in their deuterated and undeuterated forms, along with a chart showing regions of transmittance over 10% (Fig. 3.6) for all of them. These deuterated solvents will provide additional "windows" for studying the infrared-frequency shifts of various functional groups in solvents [see reviews by Rao (1963) and Hallam (1961)]. The transparency in certain regions will also allow the deuterated solvents to be used for quantitative analyses of dissolved solute mixtures where absorption bands are obscured by non-deuterated solvents. Data for several solvents in the near-infrared region are presented in Fig. 3.7 (Goddu and Delker, 1960).

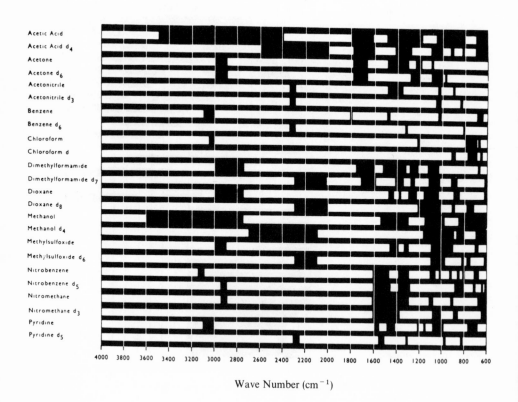

Wave Number (cm⁻¹)

Fig. 3.6. Solvents showing regions (white) of infrared transmittance over 10%. (McNiven and Court, 1970.)

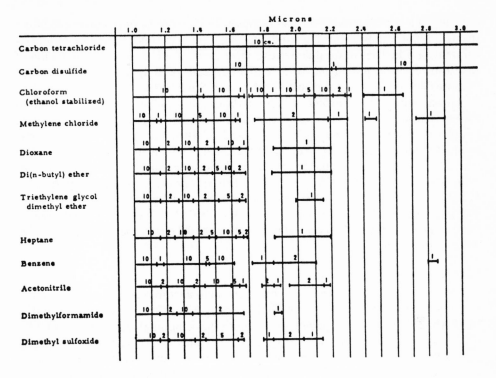

Fig. 3.7. Data on solvents in the near-infrared region. Maximum path length (cm) which can be tolerated in qualitative work in each case is also indicated. (Goddu and Delker, 1960.)

Compensation of Solvent Absorption Bands

It is very advantageous, of course, to be able to "remove" solvent absorption bands in a spectrum. In double-beam infrared spectrophotometers, atmospheric absorption bands are readily "blanked out." These instruments can also be used to blank out solvent absorption bands, thus producing an unperturbed baseline. This is done by placing in the reference beam of the spectrophotometer another cell of matched optical path containing pure solvent. The low-intensity bands of carbon tetrachloride (Fig. 3.3) in a 0.1-mm cell are blanked out easily in this way. This is true for other solvents having some bands of low intensity (e.g., acetone at ~ 2600 and 2000 cm^{-1}, and dimethyl formamide at $\sim 1950 \text{ cm}^{-1}$, in 0.1-mm cells).

One must avoid attempting to cancel out any solvent band having *total* absorption, for example, the broad band in the region $1600-1400 \text{ cm}^{-1}$ in the carbon disulfide spectrum (Fig. 3.3). If one attempts to blank out this band and to obtain the spectrum of a solute that absorbs in this region, no band(s) would be recorded, since in this region there is no infrared radiation at all impinging on the detector (except perhaps scattered light), and the servo system is completely inactive. Thus, the region $1600-1400 \text{ cm}^{-1}$ would display a meaningless straight line in the spectrum, with perhaps some upward or downward movement of the pen due to drift of the servo system. A difficulty of a similar nature will also be observed if compensation is attempted on a strong but not totally absorbing band. One will obtain misinformation or very inaccurate data if he tries to do quantitative measurements on a solute in the region of such a solvent absorption band, or even to interpret it correctly. The reader is referred to more detailed discussions of such problems in Potts (1963).

Many aliphatic and aromatic hydrocarbons are useful as solvents in the 700 to 290 cm^{-1} range of the far-infrared region. For example, there are 1,4-dioxane, furan, thiophene, benzene, hexane, cyclohexane, and 2,4,4-trimethylpentane. Benzene has adequate transparency ($>40\%$) in the 555–285 cm^{-1} region and dioxane has greater than 60% transmittance from 555 to 322 cm^{-1}. Acetone transmits well in the regions from $\sim$475 to 425 and $\sim$375 to 300 cm^{-1}. Ether transmits well in the regions from $\sim$700 to $\sim$525, $\sim$415 to $\sim$390, and 350 to 325 cm^{-1}. By using a combination of acetone, ether, and dioxane one can examine the whole cesium bromide region (Bentley *et al.*, 1958). Cells of optical path as large as 0.2 mm are satisfactory for polar solvents such as acetone.

Cells for the Use of Liquids

Many types of cells are available from manufacturers for use with liquids. Various window materials are available (NaCl, KBr, CsBr, AgCl, CaF$_2$, BaF$_2$, IRTRAN-2, thallium bromoiodide, etc.) in both the demountable and the sealed cells. For these cells there are available spacers of various thicknesses of lead, Teflon, polyethylene, etc. Since the fixed-thickness sealed cells often contain lead spacers cemented to the windows by a mercury amalgam (and these metals are biological poisons) it is wise to consider whether the use of such cells is advisable in a particular instance (e.g., an aqueous enzymic system in one such cell). Cells are available for semimicro, micro, and ultramicro work. Variable path cells of the micrometer or wedge type are available, and these are particularly useful for solvent compensation.

Figure 3.8 shows one type of completely demountable cell for liquids. The parts are laid out in order of assembly as follows: cell frame, supporting gasket, window spacer(s), window with holes, gasket, metal plate with holes for syringes and Teflon stoppers, O rings, and the screw-on cover for tightening the cell.

Fig. 3.8. Complete demountable cell in order of assembly. (Courtesy of Barnes Engineering Co.)

Obviously, any cell material should be insoluble in the solvent(s) being used. Most nonpolar solvents like carbon disulfide, carbon tetrachloride, chloroform, tetrachloroethylene, benzene, etc., can be employed in cells of any material, while polar solvents like methyl alcohol cannot be used in NaCl cells.

Aqueous Solutions

Only in recent years has the use of water as an infrared solvent become fairly routine in the biochemical laboratory. However, Coblentz (1905) had used water as an infrared solvent as early as 1905, and Gore *et al.* (1949) had studied aqueous solutions of several amino acids in 1949. Blout has published many infrared spectra of biochemical polymers in water and D_2O solution, examples of which can be found in Blout and Lenormant (1953) and Blout (1957). Figure 3.9 (Blout, 1957) shows absorption spectra of water and D_2O (with and without compensation) of 0.025 mm thickness in the region 4000–600 cm^{-1}. It can be seen that D_2O transmits where water absorbs and vice versa, thus making the combination of these solvents useful for examining aqueous solutions. The O—H deformation modes of water are present between 1700 and 1600 cm^{-1} and the O—D deformation of D_2O lies at ~1200 cm^{-1}. Except for these regions the spectra show better than 40% transmittance and satisfactory compensation is readily obtained in a double-beam spectrometer. The optimum concentration of a solute is from 5 to 20%. Two percent solutions have been used (Blout, 1957) and even lower concentrations are possible with a suitable solute, for example, 0.45% phenol in water (Parker and Kirschenbaum, 1959).

More recent reviews have appeared on the subject of aqueous solution infrared spectroscopy, each giving emphasis to different applications (Nachod and Martini, 1959; Goulden, 1959; Jencks, 1963; Parker, 1962, 1967). Table 3.1 shows some infrared frequency assignments in D_2O solution (see Jencks, 1963, for details).

Table 3.1. Some Infrared Frequency Assignments in Aqueous Solution (Jencks, 1963)

Bond	Frequency,[a] cm^{-1}
A. X—H	
O—H of H_2O (solvent)	2800–3800, 2100, 1600–1800
O—H of HDO	3380, 1455(b)
O—D of D_2O (solvent)	2200–2850, 1550(w), 1150–1250
CH_3	2950–2980(m), 1420–1470(m), 1365–1390(m)
CH_2	2915–2950(m), 1430–1480
CH_2, α to C=O	2915–2950(m), 1405–1435
C—H, aromatic and olefinic	3010–3130(m)
B. C=O (all strong or very strong)	
Esters	
Normal	1710
Ethyl carbamate	1681
Acetylcholine chloride	1735
Alanine ethyl ester hydrochloride	1743
Ethyl trifluoroacetate	1786

Table 3.1 (*Continued*)

Bond	Frequency,[a] cm^{-1}
Acids	
Normal	1710
α-Keto and amino acids	1720–1730
Acyl phosphates	
Acetyl phosphate dianion	1713
Acetyl phosphate monoanions	1735–1750
Carbamyl phosphate	1670
Thiol esters	1670–1680
Amides	
Normal	1625–1680
Urea	1604
Acetylimidazole	1740
Aldehydes and ketones	1665–1730
Carboxylate ions	
Normal	1560–1590, 1405–1420
α-Keto and amino acids	1610–1630, 1400–1410
Bicarbonate	1630, 1363
Carbamate	1540, 1441
Carbonate	1416

C. C—O

Esters and acids	1100–1350(s)
Alcohols and ethers	1050–1200(s)

D. C=C

General	1600–1680(w–m)
Aromatic	∼1600, ∼1500, ∼1450 (variable intensity)
Nucleotides (principal ring absorptions at neutrality)	
AMP	1626(s)
UMP	1645–1658(s), 1678–1692(s)
IMP	1674(s)
CMP	1493–1505(s), 1610(s), 1649(s)
TMP	1620(s), 1650(s)

E. Phosphates

Dianions	975–985(s), 1070–1090(s). 1105–1120 (variable)
Monoanions	1050–1090(s), 1205–1240(s), 915–940(?)

[a] b = broad, w = weak, m = medium, s = strong.

The reader will find many references in this book to spectroscopy done on biochemical substances in aqueous solution. Chapters 6, 10, 11, 12, and 15 on carbohydrates, proteins and polypeptides, hydrogen–deuterium exchange, nucleic acids, and enzymology contain representative discussions of such studies.

Examples of only a few of the types of compounds conveniently studied in water (H_2O) in the region between ∼1550 and 950 cm^{-1} are: organic acids of the tricarboxylic acid cycle (Parker, 1958), amino acids (Parker and Kirschenbaum, 1960; Goulden, 1959), and carbohydrates (Goulden, 1959; Parker, 1960).

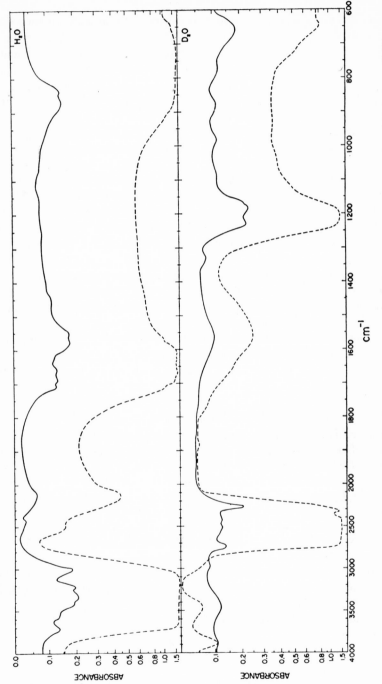

Fig. 3.9. Absorption curves of H_2O and D_2O in 0.025-mm cells: Broken curves, spectra with no compensation in the reference beam. Solid curves, spectra showing balance obtainable in a double-beam spectrometer. (Blout, 1957.)

Cells for Aqueous Solutions

Jencks (1963) has discussed the types of cell materials available for work with aqueous solutions in the mid-infrared region. Calcium fluoride and barium fluoride cells are frequently used. The former is transparent down to ~ 1111 cm^{-1} and the latter to ~ 769 cm^{-1}. These materials are unstable to acid and concentrated ammonium salts, and are quite expensive, with BaF_2 nearly twice the cost of CaF_2. Zinc sulfide (IRTRAN-2) has reasonably good transmission from 10,000 to ~ 769 cm^{-1}, is resistant to dilute acids, moderately concentrated bases, and organic solvents, and cells made of it cost less. An example has been given of the use of a zinc sulfide cell to study an enzymic reaction (Parker, 1964). Inexpensive silver chloride cells can be constructed (Nachod and Martini, 1959). Silver chloride is transparent down to 400 cm^{-1} and has been used in many studies, but it deteriorates with exposure to light, is attacked by amines, and is too pliable to give cells of reproducible path length. Cells made of polyethylene supported on sodium chloride plates have been described (Antikainen, 1958). Other cells made of polyethylene (which is both inert and cheap) have also been mentioned (Robinson, 1959; Gordon, 1963). Teflon-on-NaCl cells were used to study acidic and basic solutions (Fogelberg and Kaila, 1957) and adequate spectra were obtained except in the 1200 cm^{-1} region where there is strong absorption from the Teflon. Various other types of polyethylene and Teflon cells have been used, but their main difficulties are that they are flexible, it is hard to distribute fluid evenly, and it is difficult to obtain matched cells.

For cases in which it is not necessary to use a matching cell in the reference beam, an attenuator (supplied by manufacturers) may be used in the reference beam, or a screen may be employed, in order to obtain adequate spectra from the available energy.

Calibration of Infrared Absorption Cells

For quantitative work it is essential to obtain accurately the length (optical path) of each cell so that comparisons can be made between substances not examined in identical cells. The interference-fringe method is used for this purpose. It is based on the interference of light which is directly transmitted through a cell with light that has been twice reflected internally in the cell. The fringe method is carried out in the following manner:

1. Place the dry, empty cell in the spectrophotometer in the usual sample beam. Do not place a cell in the reference beam.

2. Operate the spectrophotometer as near as possible to the 100% transmittance line on the paper to produce fringes of the greatest amplitude, and obtain a spectrum containing from 20 to 50 fringes as in Fig. 3.10.

3. Calculate the cell thickness by the following equation:

$$b \text{ (in millimeters)} = \frac{n\lambda_1\lambda_2}{(\lambda_2 - \lambda_1)(2000)}$$

where b is the cell thickness in millimeters, λ_1 is the starting wavelength and λ_2 the final wavelength (in microns), and n is the number of fringe maxima between λ_1 and

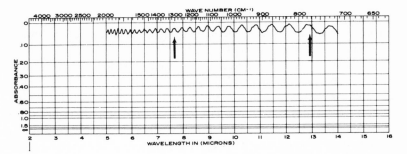

Fig. 3.10. Fringes obtained with a 0.1-mm sealed cell. (Courtesy of Barnes Engineering Co.)

λ_2. When measurements are recorded in wave numbers, the following equation is used:

$$b \text{ (in millimeters)} = \frac{5n}{(f_1 - f_2)}$$

where f_1 and f_2 are the starting and finishing wave number values. In Fig. 3.10 there are 11 maxima between 1305 and 775 cm^{-1}. A cell thickness of 0.104 mm is then calculated.

Cells for Gases

There are fewer molecules in a given volume of gas than in the same volume of liquid, and a greater sample thickness is needed. A commonly used cell has a 10-cm thickness, which yields a reasonable absorbance level for most gases and vapors at convenient values of partial pressure (for example, 5–40 mm Hg). Figure 3.11 shows

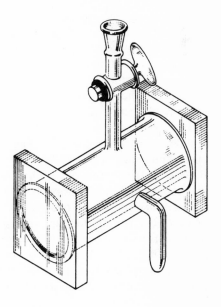

Fig. 3.11. A simple gas absorption cell. (Potts, 1963.)

a simple gas cell with a 5-cm path length. A side-arm is put on the cell so that a small, convenient amount of sample may be frozen out for later expansion into the cell chamber. Windows are made of NaCl or KBr.

Cells of different path lengths and windows can be purchased. Cells with multiple reflections have effectively long path lengths and these are used to permit the study of dilute gases and for measurement of both strong and weak absorption bands in the same sample. Several types of multiple-pass gas cells have been described (White and Alpert, 1958). A good discussion of the use of a 40-meter gas cell for atmospheric studies has included lower detection limits for many gases (Hollingdale-Smith, 1966). The most commonly used design of a multiple-reflection gas cell for the analysis of trace organic vapors is described in another section of this book (see Chapter 18).

Infrared spectroscopy can be used to perform qualitative analysis of gases. Pierson *et al.* (1956) have given a chart to be used for this purpose. Also, infrared spectroscopy and gas chromatography are a useful combination for characterizing volatile substances (see Appendix 1). Sometimes fractions from a vapor fractometer are collected as vapors in a gas cell. White *et al.* (1959) have described a microgas cell of 1 m path for use with a beam condenser to observe the spectra of gas chromatographic fractions less than 0.05 % of the total charge.

There are many devices used for trapping gaseous fractions from conventional gas chromatographs, to be transferred to the spectrometer (Anderson, 1959; Senn and Drushel, 1961; Flett and Hughes, 1963; Chang *et al.*, 1961). Many descriptions of liquid traps which can be used in conjunction with gas chromatographic instruments are available in the literature [see Grasselli and Snavely (1967) and references therein]. Stewart (1958) has discussed the use of a low-temperature microcell for the study of condensed gases.

Micromethods

It is sometimes necessary to work with very small samples of solids, liquids, or gases, and for such occasions various types of cells and equipment have been devised, many of which are available commercially.

When the sample size is reduced, the area or thickness of the sample is also decreased. A thinner sample decreases the absorbance, and ordinate-scale expansion is required. The reduced sample area caused by the use of a small cell decreases the amount of energy going to the detector. A beam-condensing device (mentioned earlier) overcomes this problem by reducing the size of the beam going through the sample. When the beam size must be reduced greatly, a reflecting microscope or micro-illuminator is used. Workers have described several devices of this kind; Barer *et al.* (1949) used one of the first of such attachments. A reflecting microscope that can be used in the sample beam of a double-beam spectrophotometer (Fig. 3.12) has been described by Blout and Abbate (1955). A gas cell is placed in the reference beam to compensate for the increased path length in the sample beam. Reflecting microscopes have been used for samples of 0.1 μg or less. Anderson and coworkers (1953a; 1953b) have described another beam condenser with silver chloride lenses that can illuminate 10–100 μg, and White *et al.* (1958) have presented a better model with potassium bromide lenses. Infrared spectra of very small samples can be obtained

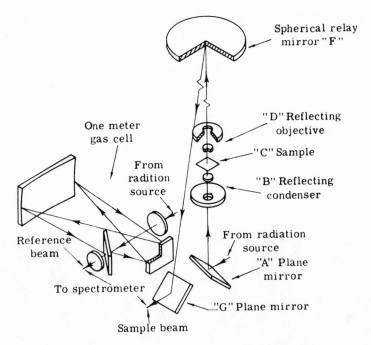

Fig. 3.12. Optical path of an infrared microspectrometer with a reflecting microscope in the sample beam and a multireflection gas cell in the reference beam. (Blout and Abbate, 1955.)

with reflecting microscopes, but a larger amount of sample is needed since there are losses in preparation and handling. Special cells and methods have been used to lessen these difficulties. Micromethods have their greatest application with solid samples. The use of micromulls (Lohr and Kaier, 1960) and micropellets (Anderson and Miller, 1953a; White *et al.*, 1958; Clark and Boer, 1958; Bisset *et al.*, 1959; Dinsmore and Edmondson, 1959; Resnik *et al.*, 1957) has been discussed. One to 20 μg of sample can be used to make pellets of cross section 0.5–5 mm, weighing from a fraction of a milligram to a few milligrams of the pelleting material.

Many microsamples come from thin-layer chromatographic (TLC) separations. The sample is separated by chromatography, the adsorbent removed from the area of the chromatogram containing the separated material, the sample eluted with a suitable solvent, filtered to remove the suspended adsorbent, and finally mixed with KBr. Garner and Packer (1968) describe a timesaving technique. By using a porous triangle of pressed KBr ("Wick-Stick," Harshaw Chemical Co., Cleveland) in a small glass vial capped so that evaporation is restricted to the center of the vial, the filtration of adsorbent and deposition of the sample on KBr can be accomplished in a single step. The adsorbent containing the sample is scraped from the TLC plate and transferred to a glass vial containing a Wick-Stick by means of a thin-stemmed funnel to prevent the adsorbent from dusting on the top half of the Wick-Stick. A suitable eluting solvent is added and a vented cap is placed on the vial.

The pressed KBr triangle is 2.5 cm high, 0.8 cm wide at its base, and 0.2 cm thick. The vial is 3.5 cm high with a 1.0-cm inside diameter. The stainless-steel cap has a vent hole 0.3 cm in diameter. A stainless steel spring clip holds the KBr triangle upright and centered in the vial. The solvent climbs the KBr by capillary action and evaporation takes place preferentially at the apex of the triangle, depositing the sample (Fig. 3.13).

Next, 1 to 2 mm of the tip is cut off with a sharp scalpel, mashed on a clean metal surface, and pressed into a transparent disk with a microdie. Satisfactory spectra for qualitative analysis are obtained from 10 to 50 μg of sample when a 1.5-mm-diameter micro-KBr-die and a beam-condensing unit are used.

(a)   (b)

Fig. 3.13(a). Wick-Stick is shown here ready for use. The triangular-shaped KBr Wick-Stick is placed in a holder clip and set inside the glass vial. Solvent and sample can now be carefully put into the vial and solvent allowed to evaporate from the tip of the Wick-Stick. (b) Sample has migrated to the tip of the Wick-Stick. (Note darkened tip area.) When the solvent has evaporated, the tip of the Wick-Stick containing the sample is sliced off, placed in a KBr die and pressed into a micropellet. (Courtesy of Harshaw Chemical Company.)

Fig. 3.14. Infrared microcell. The item at the left is an adapter to fit the instrument. Filling takes place through the needle by capillarity or suction. (Courtesy of Perkin–Elmer Corp.)

Figure 3.14 shows a type of fixed-thickness microcell used for liquids. One can use as little as ~ 0.02 ml of liquid in cells of regular design. Cells holding $\sim 2\,\mu$l and little dead space have been made with 0.1-mm spacers. Demountable cells holding 0.1–0.5 μl can be used with nonvolatile liquids. As little as 0.02 μl of material can be used for recording spectra if slits are masked and spacers reduced to 0.02 mm. The spectrum of the 727 cm^{-1} band of toluene has been recorded by means of 90–100% ordinate scale expansion (Stewart, n.d.) on only 86×10^{-12} liter in a cell of 0.02 mm thickness and 1 mm^2 area.

Figure 3.15 (Potts, 1963) shows a "cavity cell" which is convenient for micro-samples, and although the machining process used to produce the cavity in a small solid block of sodium chloride does not yield a cell of precise optical path length, it does not matter for qualitative work with microsamples. A mounting device is used for placing the cell in a beam-condensing system.

A microcell has been made by drilling holes in a salt crystal (Price et al., 1967). The spectrum of 5.04 μg of benzyl acetate run in the microcell compared favorably with the spectrum of 2283 μg run in a macrocell.

Low-Temperature Work

There have been a few thorough studies of the temperature dependence of the infrared spectra of simple compounds, but not many such studies with solids have been done. The effect of temperature on vibrational spectra has been discussed by several workers (Walsh and Willis, 1950; Richards and Thompson, 1945; Hainer and King, 1950). Generally, if no change in phase occurs, the effects of the lowering of temperature are: a slight narrowing of the absorption bands, a slight increase in the

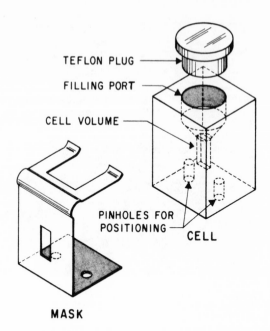

TEFLON PLUG

FILLING PORT

CELL VOLUME

PINHOLES FOR
POSITIONING

CELL

Fig. 3.15. A cavity cell for obtaining spectra of
microsamples. (Potts, 1963.)

MASK

intensity of bands, and, sometimes, some band splitting (Dows, 1963). Hainer and
King (1950) have reported the effect of temperature on a portion of the spectrum of
cholesterol. They suggested that investigation of the spectra of complex materials at
low temperature should be pursued. Zhbankov *et al.* (1966) studied the spectra of
some sugars at liquid-nitrogen temperature, but did not study the whole spectral
region. These two studies covered only the region $1500–1400 \text{ cm}^{-1}$, in which the
effects of temperature are not so pronounced as in other regions. Katon *et al.* (1969)
have recently discussed critically what they consider to be erroneous conclusions of
Zhbankov *et al.* on low-temperature effects.

Caspary (1968a) indicated the usefulness of low temperature for the identification
of materials having room-temperature spectra which are similar to each other (e.g.,
n-hexyl bromide and *n*-heptyl bromide; *n*-butylamine and *n*-hexylamine; and sebacic
acid di-*n*-butyl ester and azelaic acid di-*n*-butyl ester [Fig. 3.16]). He compared other
compounds at room temperature and low temperature, such as various ketones
(1968b) and aldehydes (1965). These substances, all of which are liquids at room
temperature, show many differences at the low temperature.

Omori and Kanda (1967) published similar spectra of several carbonyl compounds at about 100°K, e.g., aldehydes, acid halides, and ketones. They found that the
carbonyl vibration decreased markedly in frequency—about 11 to 25 cm^{-1} in
aldehydes and acid halides, and about 17 to 50 cm^{-1} in ketones—upon changing
from the vapor to the solid state.

These effects and those reported by Caspary, which involve differences in the
spectra due to phase change from liquid to solid (or gas to solid) are not the same as
those referred to by Katon *et al.* (1967, 1969), in which all materials are solids at room
temperature, and in which no phase change occurs as the temperature is lowered.

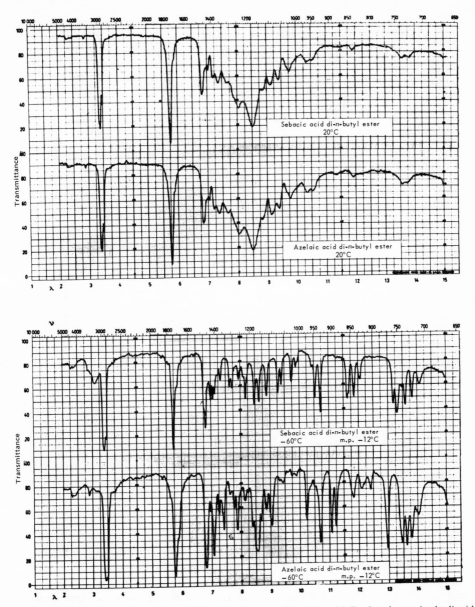

Fig. 3.16. Infrared spectra of sebacic acid di-*n*-butyl ester and azelaic acid di-*n*-butyl ester in the liquid state (top) and solid state, $-60°C$ (bottom). (Caspary, 1968*a*.)

Katon *et al.* (1967; 1969) used liquid-nitrogen temperature to investigate the detailed structure of crystalline sugars and the usefulness of infrared spectroscopy in the differentiation of sugars. Some of their spectra cover the range from 4000 to 33 cm^{-1}. Figure 3.17 shows the infrared spectrum of α,α-trehalose dihydrate at both room temperature and 113°K recorded from a Nujol mull. (Similar results are

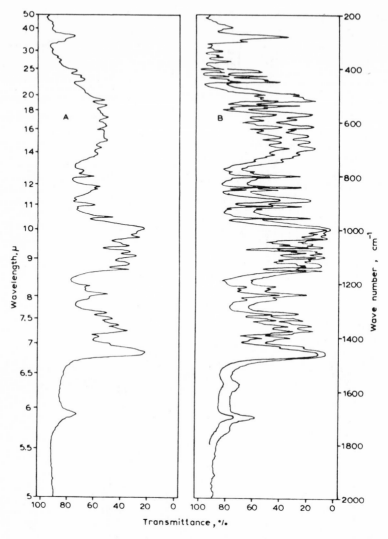

Fig. 3.17. Part of the infrared spectrum of a Nujol mull of α,α-trehalose dihydrate.
(A, at room temperature, B, at ∼110°K.) (Katon *et al.*, 1969.)

obtained when the KBr-pellet technique is used, but this method can give unrepro-
ducible results.) There is marked improvement in band definition as well as the appearance of new bands. X-ray diffraction patterns of this compound at these two
temperatures show that there is no gross phase change, but the low-temperature
diffraction pattern shows a striking line intensification. The absence of a phase change
is confirmed by the fact that the infrared spectrum changes continuously and smoothly
as the temperature is lowered.

These facts can be explained by an order–disorder transition involving the hydrogen atoms bonded to the oxygen atoms. A similar situation occurs in ammonium

chloride and is caused by the hindered rotation of the ammonium ion in the crystal lattice (Garland and Schumaker, 1967). In sugars the disorder can be due to internal rotation about the C—O bond, a motion which is hindered by a potential barrier of the order of magnitude of that in ammonium chloride (5–6 kcal/mole). The continuous nature of the temperature dependence is then due to the fact that the number of molecules possessing sufficient energy to surmount this barrier and lead to disorder follows the usual Boltzmann distribution and is a smooth function of temperature. It is therefore expected that further cooling of the sample will lead to further reduction of background of the spectrum due to further ordering of the hydrogen atoms.

Figure 3.18 shows the spectra of lactose monohydrate at three temperatures—298, 113, and $\sim 20°K$ (temperature of boiling hydrogen). The spectrum at $20°K$ has the best resolution and the greatest band intensities. Figure 3.19 shows spectra for the same compound at higher frequencies, and it is seen that, on cooling, the bonded-OH band breaks up and shows much fine structure. The monomer-OH stretching-frequency appears at $\sim 3525\ cm^{-1}$. Apparently, one hydroxyl group of lactose is not hydrogen-bonded in the crystal, as is the case with sucrose, which has one hydroxyl group unbonded.

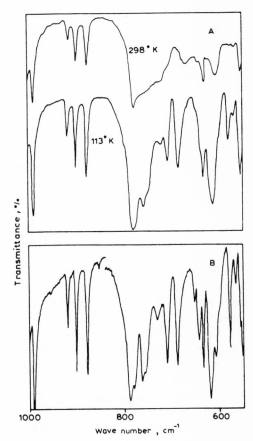

Fig. 3.18. The infrared spectrum of a Nujol mull of lactose monohydrate in the 1000–550 cm^{-1} region. (A, at 298°K and 113°K: B, at $\sim 20°K$.) (Katon *et al.*, 1969.)

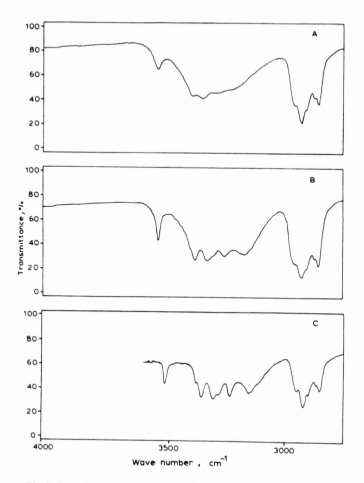

Fig. 3.19. The infrared spectrum of a Nujol mull of lactose monohydrate in the 3600–2800 cm^{-1} region. (A, at room temperature; B, at $\sim 110°$K; C, at $\sim 20°$K.) (Katon *et al.*, 1969.)

The effects seen above with sugars are to be expected with other complex molecules which possess internal motions that can be described by a potential function possessing more than one minimum and which are hindered by a relatively small potential barrier in the solid state.

Katon *et al.* (1968) have extended their low-temperature work to other types of compounds and have given a description of the low-temperature cell they used. Other carbohydrates were examined (e.g., raffinose, sucrose, fructose, arabinose, xylose, lactose, mannose, maltose, galactose, rhamnose, cellobiose, and melibiose) as well as a noncrystalline trypsin (little or no change at the lower temperature), urea, diphenyl, stearoyl chloride, cholesterol (Fig. 3.20), cholesteryl acetate, serotonin creatinine sulfate, sodium creatinine phosphate hexahydrate, daunomycinone, carnosine, and

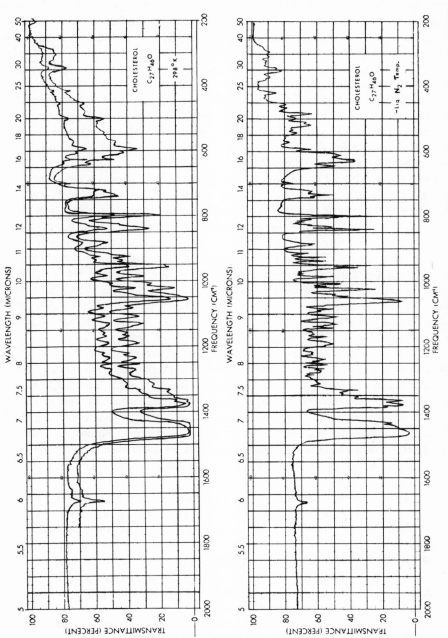

Fig. 3.20. The partial infrared spectrum of a Nujol mull of cholesterol at 298° K and at about liquid-nitrogen temperature. (Katon *et al.*, 1968.)

others. Some of the spectra (lactose and sucrose) were run at liquid-*hydrogen* temperatures at 20°K by E. R. Lippincott. Most of the low-temperature spectra have very much improved resolution and often extra absorption bands. These striking results obtained on lowering the temperature have only been found with compounds that can exhibit hydrogen bonding. Unfortunately, hydrogen bonding is not completely understood, and, as a result, it is difficult to determine what characteristics of the hydrogen bond are important to this effect (Katon *et al.*, 1968, 1969).

Michell (1970) has reported the low-temperature spectra of some polysaccharides, including cellulose. More recently, McCall *et al.* (1971) have studied the low-temperature spectra of algal cellulose, hydrocellulose I (from cotton), and hydrocellulose II (from regenerated cellulose). The changes which were observed on cooling were associated with hydrogen bonding effects.

Bradbury and Elliott (1963), in an interesting conformational study of peptide group interactions, examined the infrared spectra of orthorhombic *N*-methylacetamide with the electric vector of the normally incident radiation along each of the three crystal axes in turn. Their measurements were made at room temperatures (when the crystal structure was not completely ordered) and at temperatures down to that of liquid nitrogen. The use of the low temperatures to study peptide group interactions in the unit cell allowed them to reach different conclusions regarding some of the assignments of Miyazawa *et al.* (Miyazawa *et al.*, 1956, 1958; Miyazawa, 1960*a*, 1960*b*). (See discussion concerning Miyazawa's theory under *Conformation Studies*, in Chapter 10.) Also, the liquid-nitrogen temperatures enhanced the fine structures of the NH and ND stretching bands.

Various types of low-temperature cells are available commercially. Rochkind (1968) has presented a new low-temperature (20°K) technique, which provides a practical and sensitive method of infrared quantitative analysis of all infrared-absorbing gases and volatile liquids. The method, called pseudomatrix isolation spectroscopy (PMI), also provides a tool for the analysis of complex gas mixtures. The PMI method distinguishes between molecular isotopes, for example, Figure 3.21 shows a PMI spectrum of a mixture of isotopic d_2-ethylenes (condensed on a 20°K CsI window). Rochkind claims that equivalent distinguishability has not been demonstrated with gas chromatography.

Hermann *et al.* (1969) have reviewed the subject of infrared spectroscopy at subambient temperatures, including: an extensive literature review with over 600 references; and discussions of work on pure molecules, and on molecules and molecular fragments within matrices. Hermann and coworkers (Hermann and Harvey, 1969; Hermann *et al.*, 1969; Hermann, 1969) have also discussed the design of cells for use at low temperatures. These cells belong in four categories: conventional transmission cells; matrix isolation cells; pseudomatrix isolation cells; and multiple internal reflection cells (ATR).

Ford *et al.* (1969) have described a new technique for accurately calibrating the temperature of low-temperature infrared cells. A liquid of known melting point is introduced into the cell as a capillary film between NaCl or AgCl windows. The cell is assembled with the thermocouple in the usual position, and the temperature lowered until the liquid freezes. The monochromator is set at the frequency at which the

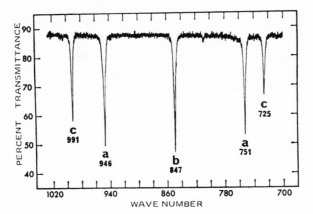

Fig. 3.21. Infrared PMI spectrum, 20°K, of a mixture of d_2-ethylenes at 1% in nitrogen (700–1025 cm^{-1}). (a) (1,1) ethylene-d_2; (b) *cis*-(1,2) ethylene-d_2; (c) *trans*-(1,2) ethylene-d_2. Twelve micromoles of ethylene mixture were deposited. Spectral slit, ~ 0.9 cm^{-1}. (Rochkind, 1968.)

transmittance of the liquid and the solid show the greatest difference. The transmittance at this frequency is recorded as a function of time as the sample warms up at the rate of 1.5 to 3.3 degrees/min. Thermocouple emf readings are marked on the chart at frequent intervals and are converted later to temperatures by the use of thermocouple calibration tables. Figure 3.22 shows that the transmittance of the sample (in this case, ice) is nearly constant until the solid begins to melt and becomes constant again when it has completely melted. The midpoint of the melting curve (which usually extends over a range of 1° or less) is taken as the observed melting point. The melting points of the following sharply melting substances have been useful for calibration: *n*-butyl ethyl ether (-124°C); methanol (-97.8°C); ethyl acetate (-83.6°C); anisole (-37.3°C); H_2O(0.0°C); and D_2O (4.0°C).

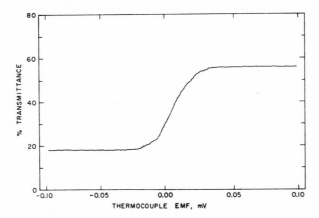

Fig. 3.22. Changes of transmittance at 2650 cm^{-1} during the melting of a thin film of ice. Recorded on a Perkin–Elmer model 521 spectrophotometer. (Ford *et al.*, 1969.)

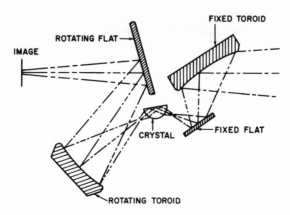

Fig. 3.23. Optical schematic diagram, top view.
(Courtesy of Perkin–Elmer Corp.)

Attenuated Total Reflection

Sometimes the sample to be studied transmits infrared radiation poorly, or the material is difficult to prepare by the usual methods, or one does not want to change the specimen in any way before studying it. The technique known as attenuated total reflection or ATR (also called multiple internal reflection, MIR, and FMIR, frustrated multiple internal reflection) can be applied in such cases. It was developed independently by Fahrenfort (1961) and Harrick (1960). To produce a useful ATR spectrum the basic requirement is that the infrared radiation enter a crystal of high refractive index, strike a sample of lower refractive index one or more times at an angle above the critical angle at the reflecting interface, and emerge through the crystal into the monochromator. The spectrum obtained is very much like that of a conventional spectrum produced by transmission techniques, and is entirely independent of the sample thickness. The depth of penetration of the radiation beam into the sample is 5 μ or less. Figure 3.23 is an optical schematic diagram in which an ATR hemicylindrical crystal (top view) is shown in place where it receives radiation from the spectrometer source by means of a fixed toroid mirror and a reflecting flat mirror. The sample is not shown, but it rests on the flat face of the crystal. Other types of crystals and arrangements are also used, in which more than one internal reflection is obtained. Figure 3.24 is a diagram of a trapezoidal shaped crystal (top view) in which the path of the light is shown entering on the left and emerging on the right. In this case, the sample can be placed on one face of the crystal, or on both faces, to achieve more reflections from the sample and consequently greater absorption band intensities.

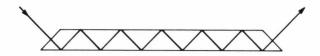

Fig. 3.24. Multiple reflection ATR plate.

The most useful crystals for ATR work are thallium bromoiodide (KRS-5), silver chloride, zinc sulfide (IRTRAN-2), and germanium, in decreasing order of usefulness. A variety of instrumentation and ATR accessories are available from Wilks Scientific Corporation, Perkin–Elmer Corporation, Barnes Engineering Company, and Harrick Scientific Corporation.

Harrick's (1967) book is a comprehensive study of internal reflection spectroscopy. Polchlopek (1966) has discussed some theoretical aspects, instrumentation, sampling techniques, and applications of the method. Wilks and Hirschfeld (1967) have discussed advantages of the ATR method, materials for crystals, sampling techniques, quantitative analysis, microtechniques, and numerous applications. Pawlak *et al.* (1967) have also reviewed the subject of ATR.

Fujiyama *et al.* (1970), in a discussion of some systematic errors in infrared absorption spectrophotometry of liquid samples, have concluded that workers interested in serious study of integrated intensities or of band shapes would do well to avoid reflection–interference troubles by using the ATR technique, or to take careful and detailed account of these effects in any conventional absorption spectrophotometry.

A misleading statement that ATR of water solutions requires concentrations of 20% or higher was made in the review by Pawlak *et al.* (quoting a paper by Katlafsky and Keller, 1963). Parker (1963) has shown that suitable ATR spectra can be obtained on 5% solutions of glycyl-L-alanine, and evidence points to the detection of other substances at concentrations much less than 20% (Hermann, 1965b).

The author of this book has had considerable experience with the ATR technique and finds it particularly useful for work with biological specimens. The optical properties of tissues are such that they must be very thinly sectioned to obtain *transmission* spectra, whereas the ATR method does not depend on sample thickness. Useful spectra have been recorded from normal and diseased human arterial tissue (Fig. 3.25) and heart tissue (chicken, rat, calf) (Parker and Ans, 1967). Examinations

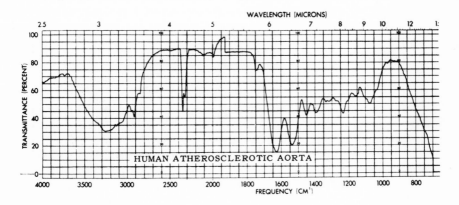

Fig. 3.25. Single reflection spectrum of adult human aorta containing atherosclerotic plaque. The bands near 2320 and 2350 cm^{-1} were caused by CO_2. (Parker and Ans, 1967.)

of the biochemical lesions in diseased corneas and lenses of the human eye (Parker and D'Agostino, 1967) and in human xanthelasma tissue (Parker *et al.*, 1969) have been carried out by use of ATR. A 45° angle of incidence was used for this work. This fact is mentioned since Pawlak *et al.* (1967) say that a 45° incidence angle unit cannot be used with water solutions, and this statement might be incorrectly interpreted to mean that water-containing tissues would not display useful spectra. Also, Robinson and Vinogradov (1964) have successfully used a 45° angle in examining amino acids in aqueous solutions. Hermann (1965*b*) has recorded useful ATR spectra of rat muscle, kidney, and stomach tissues. In one case he claims to have shown the presence of chlorpromazine or a metabolite of it in tissue.

The fact that useful spectra can be obtained from polymers in various forms, from fibers which cannot be studied by transmission techniques, from other intractable materials, from aqueous solutions, etc., should make this technique useful in many disciplines. The use of ATR for the study of the chemistry of surfaces should be further explored in biochemical applications, for example, deposition of monolayers from solution (see, for example, Sharpe, 1961, 1965). The ATR technique has been used for analysis of bacterial cultures (Johnson, 1966) and in forensic science (Denton, 1965). It has also been applied to a great variety of substances: molecular species present at electrode interfaces (Hansen *et al.*, 1966; Mark and Pons, 1966); carbohydrates (Parker and Ans, 1966); a single crystal of pentaerythritol (Tsuji *et al.*, 1970); cosmetics on the skin (Wilks Scientific Corp., 1966); pesticidal traces (Hermann, 1965*a*); water–alcohol mixtures (Malone and Flournoy, 1965); nitrate ion (Wilhite and Ellis, 1963); leather (Pettit and Carter, 1964); and blood spectra from within the human circulatory system (Kapany and Silbertrust, 1964). The last-mentioned application requires special equipment.

Submicrogram quantities of organic phosphorus compounds in air have been monitored by ATR spectroscopy (Prager and LaRosa, 1968). Adsorptive platinum films (50–80 nm thick) on germanium crystals (57 × 14 × 2 mm) were used in order to increase the sensitivity of the method. Adsorption was reversible by air flowing through the germanium cell, thus permitting repeated use of the platinum-coated

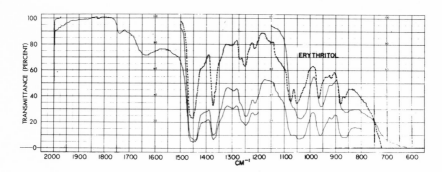

Fig. 3.26. ATR spectra of erythritol (Nujol mull). Solid line, room temperature; dotted line, liquid-nitrogen temperature, on 1-mm KRS-5 plate. 1375 and 1450 cm^{-1} bands are caused by Nujol. (Parker, 1969.)

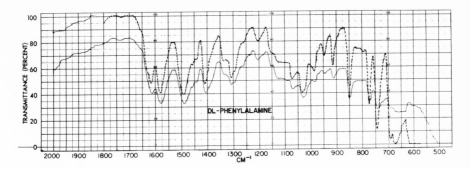

Fig. 3.27. ATR spectra of DL-phenylalanine (powder, run on 1-mm KRS-5 plate directly). Solid line, room temperature; dotted line, liquid-nitrogen temperature. (Parker, 1969.)

crystals. At the optimum thickness and temperature it was possible to detect 0.1 μg of dimethylmethylphosphonate per liter of air in 30 seconds.

Parker (1969) has shown ATR spectra for erythritol (Fig. 3.26) and DL-phenyl-alanine (Fig. 3.27), at room temperature and at liquid-nitrogen temperature. The resolution increases in both cases at the lower temperature. The reader will find discussions of additional work in which ATR was used in other sections of this book.

The Use of Polarized Infrared Radiation and the Measurement of Dichroism

Polarized infrared radiation is used to obtain information about the direction of transition moments of normal modes of vibration in solid oriented compounds. If one knows the molecular orientation in a solid, he can use polarization studies in making band assignments. (See Chapters 6 and 10 for such applications in carbohydrate and polypeptide chemistry.) The measured direction of the transition moment of the vibration producing a band must coincide with the direction deduced from the structure if the assignment is correct. On the other hand, knowing the band assignment but not the molecular orientation in the solid, one can deduce some knowledge of the molecular orientation.

In the crystalline state the normal modes of the unit cell, not the molecule, interact with radiation impinging on it. The normal modes of a unit cell are comprised of the normal modes of the individual molecules which exist in as many different phases as the number of molecules in a unit cell. For example, in a unit cell consisting of two molecules, each having a C=O group, the two C=O stretching modes are comprised of in-phase and out-of-phase stretching of the two C=O groups.

Polarizers

To obtain a beam of polarized infrared radiation one can use silver chloride (Newman and Halford, 1948) or selenium (Elliott *et al.*, 1948*a*) polarizers. Several plates of silver chloride or a thin film of selenium are tilted at the polarizing angle relative to the unpolarized infrared beam of radiation. Figure 3.28 (Colthup *et al.*, 1964) describes the optical effects in schematic fashion.

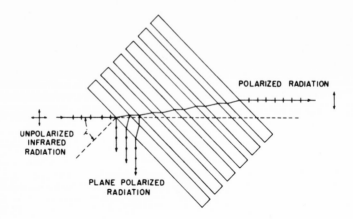

Fig. 3.28. Plane polarization of infrared radiation by several plates. If
I is the incident parallel beam of radiation striking at the angle of the
incidence, i, the reflected beam R is polarized. The incident plane is the
plane of the paper, therefore beam R is vibrating normal to the paper.
This reflection–polarization process occurs at each face of the stack of
plates. After a sufficient number of reflections the transmitted beam T
has been depleted of vibrations normal to the plane of incidence and
consists almost completely of vibrations in this plane. Thus, two beams
of polarized light are produced. (Colthup *et al.*, 1964.)

A single crystal or group of crystals oriented the same way may be used for
crystalline compounds. Various conditions for producing crystallization exist:
between salt plates from the melted compound (Halverson and Francel, 1949); out
of the vapor phase (Zwerdling and Halford, 1955); or from solution (Hallman and
Pope, 1958). Many spectra of crystalline compounds have been recorded but mostly
as mulls or potassium halide pellets, techniques not very useful for polarization
studies because of the finely ground state of the sample and the random distribution
of the individual crystals. For work with crystals, a platelet of 5 to 20 μ thickness is
usually needed and, unless an infrared microscope is used, the platelet should be large
enough to cover the light beam of the spectrophotometer. The use of polarized single
crystals has been described (Chance *et al.*, 1967) for obtaining infrared spectra of
cytochrome *c* and myoglobin compounds. Not many nonpolymeric crystalline
compounds have been analyzed by polarization techniques (see, for example, Couture,
1944; Pimentel *et al.*, 1955; and Zbinden, 1953).

Orientation of thin polymer films may be accomplished by stretching, which
results in uniaxial orientation where a crystallographic axis, usually the long chain
axis, tends to line up parallel to the stretching direction but no preferred orientation
of crystallites about this axis takes place. Another way to orient a polymer is to roll
the film between rollers. This can also be accomplished by placing the polymer film
between silver chloride plates (Elliott *et al.*, 1948*b*) and rolling, removing the silver
chloride with a sodium thiosulfate solution if necessary. Orientation in the direction

of rolling occurs, but sometimes double orientation results (orientation of certain crystallographic planes parallel to the plane of the film).

Measurement of Dichroic Ratio

The stretched polymer is mounted in a spectrophotometer with the stretching (or rolling) direction parallel to the slit. The spectrum is then recorded for polarized light, first with the electric vector parallel and then perpendicular to the stretching direction. For any particular absorption band one determines the absorbances $A_\parallel$ and $A_\perp$ for the two polarization directions, respectively. The ratio R of these two values,

$$R = \frac{A_\parallel}{A_\perp}$$

is called the dichroic ratio for the particular absorption band. A sample which is not uniform in thickness and shape need not be moved. Instead, the polarizer is rotated, thus exposing the same part of the sample to the beam. Special care must be taken to avoid errors caused by this method as a result of partial polarization in the spectrometer itself. This polarization is caused by reflections at various mirrors and especially at the prism faces. To avoid these difficulties when rotating the polarizer, one can

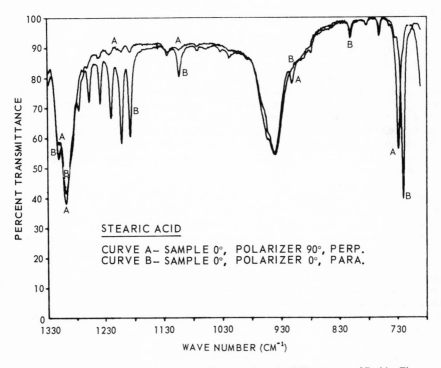

Fig. 3.29. Polarization spectra for stearic acid. (Hannah *et al.*, 1968, courtesy of Perkin–Elmer Corp.)

make measurements with the polarizer oriented 45° with respect to the slit, and then rotated 90° to a position 45° on the other side. If the sample has enough uniformity, one leaves the polarizer stationary in the position producing highest transmission while rotating the sample 90°. This is the most satisfactory way to measure polarization properties (Colthup *et al.*, 1964). Polarization effects due to the spectrometer itself have also been noted in grating instruments (George, 1966). Hannah *et al.* (1968) studied oriented stearic acid and polyvinyl alcohol using a wire-grid polarizer in an instrument equipped with an analytical data system composed of a digital data recorder, a paper tape punch, and a manual keyboard. With the polarizer placed between the sample and the monochromator, the monochromator polarization did not contribute to errors in the dichroic ratio. Figure 3.29 shows polarization spectra for stearic acid oriented by melting between two sodium chloride plates and then cooling one end, thus providing a temperature gradient during the crystallization. Koenig *et al.* (1967) have discussed the use of three-dimensional measurements for doing dichroic studies.

Bird *et al.* (1958) and Bird and Blout (1959) have developed a streaming technique for the observation of dichroism in solution. Their results constituted supportive evidence for the α-helical structure of poly-γ-benzyl-L-glutamate and poly-L-glutamic acid in solution. The method does not, however, involve the use of polarized radiation.

REFERENCES

Anderson, D. H. and Miller, O. E. *J. Opt. Soc. Am.* **43**, 777 (1953*a*).
Anderson, D. H. and Woodall, N. B. *Anal. Chem.* **25**, 1906 (1953*b*).
Anderson, D. W. *Analyst* **84**, 50 (1959).
Antikainen, P. J. *Suomen Kemistilehti* **B31**, 223 (1958).
Barer, R., Cole, A. R. H., and Thompson, H. W. *Nature* **163**, 198 (1949).
Barker, S. A., Bourne, E. J., Neely, W. B., and Whiffen, D. H. *Chem. Ind.* (*London*) **1954**, 1418.
Barker, S. A., Bourne, E. J., and Whiffen, D. H. *Methods Biochem. Anal.* **3**, 213 (1956).
Barnes, R. B., Gore, R. C., Williams, E. F., Linsley, S. G., and Petersen, E. M. *Ind. Eng. Chem., Anal. Ed.*, **19**, 620 (1947).
Bentley, F. F., Wolfarth, E. F., Srp, N. E., and Powell, W. R. *Spectrochim. Acta* **13**, 1 (1958).
Bird, G. R. and Blout, E. R. *J. Am. Chem. Soc.* **81**, 2499 (1959).
Bird, G. R., Parrish, M., Jr., and Blout, E. R. *Rev. Sci. Instr.* **29**, 305 (1958).
Bisset, F., Bluhm, A. L., and Long, L., Jr. *Anal. Chem.* **31**, 1927 (1959).
Blout, E. R. *Ann. N. Y. Acad. Sci.* **69**, Art 1, 84 (1957).
Blout, E. R. and Abbate, M. J. *J. Opt. Soc. Am.* **45**, 1028 (1955).
Blout, E. R. and Lenormant, H. *J. Opt. Soc. Am.* **43**, 1093 (1953).
Bradbury, E. M. and Elliott, A. *Spectrochim. Acta* **19**, 995 (1963).
Bradley, K. B. and Potts, W. J., Jr. *Appl. Spectrosc.* **12**, 77 (1958).
Caspary, R. *Spectrochim. Acta* **21**, 763 (1965).
Caspary, R. *Appl. Spectrosc.* **22**, 694 (1968*a*).
Caspary, R. *Appl. Spectrosc.* **22**, 689 (1968*b*).
Chance, B., Estabrook, R. W., and Yonetani, T., eds. *Hemes and Hemoproteins*, Academic Press, New York, 1967.
Chang, S. S., Ireland, C. E., and Tai, H. *Anal. Chem.* **33**, 479 (1961).
Clark, D. A. and Boer, A. P. *Spectrochim. Acta* **12**, 276 (1958).
Coblentz, W. W. *Investigations of Infrared Spectra*, Carnegie Institution of Washington, Washington, D.C. (1905).

Colthup, N. B., Daly, L. H., and Wiberley, S. E. *Introduction to Infrared and Raman Spectroscopy*, Academic Press, New York, 1964.

Couture, L. *Compt. Rend.* **218**, 669 (1944).

Denton, S. *J. Forensic Sci. Soc.* **5**, 112 (1965).

Dinsmore, H. L. and Edmondson, P. R. *Spectrochim. Acta* **15**, 1032 (1959).

Dows, D. A. "Infrared Spectra of Molecular Crystals," in *Physics and Chemistry of the Organic Solid State, Vol. I* (D. Fox, M. M. Labes, and A. Weissberger, eds.), Interscience, New York, 1963.

Elliott, A., Ambrose, E. J., and Temple, R. B. *J. Opt. Soc. Am.* **38**, 212 (1948a).

Elliott, A., Ambrose, E. J., and Temple, R. B. *J. Chem. Phys.* **16**, 877 (1948b).

Fahrenfort, J. *Spectrochim. Acta* **17**, 698 (1961).

Farmer, V. C. *Spectrochim. Acta* **8**, 374 (1957).

Farmer, V. C. *Chem. Ind. (London)* **1959**, 1306.

Flett, M. St. C. and Hughes, J. *J. Chromatogr.* **11**, 434 (1963).

Fogelberg, B. C. and Kaila, E. *Paperi Ja Puu* **39**, 375 (1957).

Ford, T. A., Seto, P. F., and Falk, M. *Spectrochim. Acta* **25A**, 1650 (1969).

Fujiyama, T., Herrin, J., and Crawford, B. L., Jr. *Appl. Spectrosc.* **24**, 9 (1970).

Garland, C. W. and Schumaker, N. E. *J. Phys. Chem. Solids* **28**, 799 (1967).

Garner, H. R. and Packer, H. *Appl. Spectrosc.* **22**, 122 (1968).

George, R. S. *Appl. Spectrosc.* **20**, 101 (1966).

Goddu, R. F. and Delker, D. A. *Anal. Chem.* **32**, 140 (1960).

Gordon, J., in Jencks (1963).

Gore, R. C., Barnes, R. B., and Petersen, E. *Anal. Chem.* **21**, 382 (1949).

Goulden, J. D. S. *Spectrochim. Acta* **15**, 657 (1959).

Grasselli, J. G. and Snavely, M. K. in *Progress in Infrared Spectroscopy, Vol. 3* (H. A. Szymanski, ed.), Plenum Press, New York, 1967, p. 55.

Hacskaylo, M. *Anal. Chem.* **26**, 1410 (1954).

Hainer, R. M. and King, G. W. *Nature* **166**, 1029 (1950).

Hallam, H. E. *Unicam Spectrovision* No. 11, 1961.

Hallman, H. and Pope, M. "Technical Report on Preparation of Thin Anthracene Single Crystals," ONR Contract No. 285, (25), June 9, 1958, cited in Colthup *et al.*, 1964.

Halverson, F. and Francel, R. J. *J. Chem. Phys.* **17**, 694 (1949).

Hannah, R. W., Savitzky, A., and Kessler, H. B. Paper delivered at The Pittsburgh Conference, Cleveland, Ohio, Mar. 3–8, 1968.

Hansen, W. N., Osteryoung, R. A., and Kuwana, T. *J. Am. Chem. Soc.* **88**, 1062 (1966).

Harrick, N. J. *Phys. Rev. Letters* **4**, 224 (1960).

Harrick, N. J. *Internal Reflection Spectroscopy*, Wiley, New York, 1967.

Hermann, T. S. *Appl. Spectrosc.* **19**, 10 (1965a).

Hermann, T. S. *Anal. Biochem.* **12**, 406 (1965b).

Hermann, T. S. *Appl. Spectrosc.* **23**, 461, 473 (1969).

Hermann, T. S. and Harvey, S. R. *Appl. Spectrosc.* **23**, 435 (1969).

Hermann, T. S., Harvey, S. R., and Honts, C. N. *Appl. Spectrosc.* **23**, 451 (1969).

Hollingdale-Smith, P. A. *Can. Spectry.* **11**, 107 (1966).

Jencks, W. P. *Methods in Enzymology Vol. 6*, Academic Press, New York, 1963, p. 914.

Johnson, R. D. *Anal. Chem.* **38**, 160 (1966).

Kapany, N. S. and Silbertrust, N. *Nature* **204**, 138 (1964).

Katlafsky, B. and Keller, R. E. *Anal. Chem.* **35**, 1665 (1963).

Katon, J. E., Miller, J. T., Jr., and Bentley, F. F. *Arch. Biochem. Biophys.* **121**, 798 (1967).

Katon, J. E., Miller, J. T., Jr., and Ferguson, R. R. Technical Report AFML-TR-68-169, Air Force Materials Laboratory, Wright–Patterson Air Force Base, 1968.

Katon, J. E., Miller, J. T., Jr., and Bentley, F. F. *Carbohyd. Res.* **10**, 505 (1969).

Koenig, J. L., Cornell, S. W., and Witenhafer, D. E. *J. Polymer Sci., Part A-2*, **5**, 301 (1967).

Kuhn, L. P. *J. Am. Chem. Soc.* **74**, 2492 (1952).

Kuhn, L. P. *J. Am. Chem. Soc.* **76**, 4323 (1954).

Lecomte, J. *Anal. Chim. Acta* **2**, 727 (1948).

Lohr, L. J. and Kaier, R. J. *Anal. Chem.* **32**, 301 (1960).

Lutinski, C. *Anal. Chem.* **30**, 2071 (1958).

Malone, C. P. and Flournoy, P. A. *Spectrochim. Acta* **21**, 1361 (1965).

Mark, H. B., Jr. and Pons, B. S. *Anal. Chem.* **38**, 119 (1966).

McCall, E. R., Morris, N. M., Tripp, V. W., and O'Connor, R. T. *Appl. Spectrosc.* **25**, 196 (1971).

McNiven, N. L. and Court, R. *Appl. Spectrosc.* **24**, 296 (1970).

Michell, A. J. *Austral J. Chem.* **23**, 833 (1970).

Miyazawa, T. *J. Mol. Spectrosc.* **4**, 198 (1960*a*).

Miyazawa, T. *J. Chem. Phys.* **32**, 1647 (1960*b*).

Miyazawa, T., Shimanouchi, T., and Mizushima, S. *J. Chem. Phys.* **24**, 408 (1956).

Miyazawa, T., Shimanouchi, T., and Mizushima, S. *J. Chem. Phys.* **29**, 611 (1958).

Nachod, F. C. and Martini, C. M. *Appl. Spectrosc.* **13**, 45 (1959).

Newman, R. and Halford, R. S. *Rev. Sci. Instr.* **19**, 270 (1948).

Omori, T. and Kanda, Y. *Memoirs of the Faculty of Science, Kyushu University, Series C, Chemistry* **6** (No. 1) 29 (1967).

Park, P. J. D. and Wyn-Jones, E. *Chem. Commun.* **1966**, 557.

Parker, F. S. *Appl. Spectrosc.* **12**, 163 (1958).

Parker, F. S. *Biochim. Biophys. Acta* **42**, 513 (1960).

Parker, F. S. *Perkin–Elmer Instrument News* **13**, No. 4, 1 (1962).

Parker, F. S. *Nature* **200**, 1093 (1963).

Parker, F. S. *Nature* **203**, 975 (1964).

Parker, F. S. in *Progress in Infrared Spectroscopy, Vol. 3* (H. A. Szymanski, ed.), Plenum Press, New York, 1967, p. 75.

Parker, F. S. Mid-America Symposium on Spectroscopy, Chicago, Illinois, Paper 119, May, 1969.

Parker, F. S. and Kirschenbaum, D. M. *J. Phys. Chem.* **63**, 1342 (1959).

Parker, F. S. and Kirschenbaum, D. M. *Spectrochim. Acta* **16**, 910 (1960).

Parker, F. S. and Ans, R. *Appl. Spectrosc.* **20**, 384 (1966).

Parker, F. S. and Ans, R. *Anal. Biochem.* **18**, 414 (1967).

Parker, F. S. and D'Agostino, M. *Bull. N.Y. Acad. Med.* **43**, 418 (1967).

Parker, F. S., Kleinman, C. S., and Mittl, R. *Proceedings of the Eighth Meeting of the Career Scientists of the Health Research Council of N.Y. City*, New York Academy of Medicine, Dec. 1969.

Pawlak, J. A., Fricke, G., and Szymanski, H. A. in *Progress in Infrared Spectroscopy, Vol. 3* (H. A. Szymanski, ed.), Plenum Press, New York, 1967, p. 39.

Pettit, D. and Carter, A. R. *J. Soc. Leather Trades' Chemists* **48**, 476 (1964).

Pierson, R. H., Fletcher, A. N., and Gantz, E. St. C. *Anal. Chem.* **28**, 1218 (1956).

Pimentel, G. C., McClellan, A. L., Person, W. B., and Schnepp, O. *J. Chem. Phys.* **23**, 234 (1955).

Polchlopek, S. E. in *Applied Infrared Spectroscopy* (D. N. Kendall, ed.), Reinhold, New York, 1966, p. 462.

Potts, W. J., Jr. *Chemical Infrared Spectroscopy, Vol. I: Techniques*, Wiley, New York, 1963.

Prager, M. J. and LaRosa, C. N. *Appl. Spectrosc.* **22**, 449 (1968).

Price, G. D., Sunas, E. C., and Williams, J. F. *Anal. Chem.* **39**, 138 (1967).

Randall, H. M., Fuson, N., Fowler, R. G., and Dangl, J. R. *Infrared Determination of Organic Structures*, Van Nostrand, Princeton, New Jersey, 1949.

Rao, C. N. R. *Chemical Applications of Infrared Spectroscopy*, Academic Press, New York, 1963.

Resnik, F. E., Harrow, L. S., Holmes, J. C., Bill, M. E., and Greene, F. L. *Anal. Chem.* **29**, 1874 (1957).

Richards, R. E. and Thompson, H. W. *Trans. Faraday Soc.* **41**, 183 (1945).

Robinson, T. *Nature* **184**, 448 (1959).

Robinson, F. P. and Vinogradov, S. N. *Appl. Spectrosc.* **18**, 62 (1964).

Rochkind, M. M. *Science* **160**, 196 (1968).

Schiele, C. *Appl. Spectrosc.* **20**, 253 (1966).

Schwing, K. J. and May, L. *Anal. Chem.* **38**, 523 (1966).

Senn, W. L., Jr. and Drushel, H. V. *Anal. Chim. Acta* **25**, 328 (1961).

Sharpe, L. H. *Proc. Chem. Soc. (London)* **1961**, 461.

Sharpe, L. H. *Perkin–Elmer Instrument News* **15**, No. 4, 1 (1965).

Stewart, J. E. *Application Data Sheet IR-88-MI*, Beckman Instruments Division.

Stewart, J. E. *Anal. Chem.* **30**, 2073 (1958).

Szymanski, H. A. *IR—Theory and Practice of Infrared Spectroscopy*, Plenum Press, New York, 1964.

Tipson, R. S. *Natl. Bur. Std. Monograph 110*, 1968.

Tipson, R. S. and Isbell, H. S. *J. Res. Natl. Bur. Std.* **64A**, 239 (1960).

Tipson, R. S. and Isbell, H. S. *J. Res. Natl. Bur. Std.* **66A**, 31 (1962).

Tsuji, K., Yamada, H., Suzuki, K., and Nitta, I. *Spectrochim. Acta* **26A**, 475 (1970).

Walsh, A. and Willis, J. B. *J. Chem. Phys.* **18**, 552 (1950).

White, J. U. and Alpert, N. L. *J. Opt. Soc. Am.* **48**, 460 (1958).

White, J. U., Weiner, S., and Alpert, N. L. *Anal. Chem.* **30**, 1694 (1958).

White, J. U., Alpert, N. L., Ward, W. M., and Gallaway, W. S. *Anal. Chem.* **31**, 1267 (1959).

Wilhite, R. N. and Ellis, R. F. *Appl. Spectrosc.* **17**, 168 (1963).

Wilks Scientific Corp., *Model 26 MIR Skin Analyzer* (1966).

Wilks, P. A., Jr. and Hirschfeld, T. *Appl. Spectrosc. Rev.* **1**, 99 (1967).

Zbinden, R. *Helv. Phys. Acta* **26**, 129 (1953).

Zhbankov, R. G., Ivanova, N. V., and Komar, V. P. *Vysokomol. Soedin.* **8**, 1778 (1966).

Zwerdling, S. and Halford, R. S. *J. Chem. Phys.* **23**, 2221 (1955).

Chapter 4

QUANTITATIVE ANALYSIS

Quantitative analysis in the infrared region is based on considerations similar to those applied routinely in visible and ultraviolet spectrophotometry, namely, application of the Beer–Lambert law. The law states that at a given wavelength of light

$$A = abc$$

where A is the absorbance (formerly called optical density), which is the logarithm to the base 10 of the reciprocal of the transmittance ($A = \log_{10} I_0/I$, where I_0 and I are the intensities of the incident and transmitted beams, respectively); a is the absorptivity; c, the concentration, is expressed as g/liter; and b is the sample path length in centimeters. If c is expressed in moles/liter, a value ϵ is substituted for a, and ϵ is called the molar absorptivity.

The Beer–Lambert law states that the absorbance A is a linear function of the concentration of the absorbing substance, and there are many cases where this relationship holds true. An example is given in Figs. 4.1 and 4.2. However, linearity is not always observed. For cases in which a plot of absorbance *versus* concentration is not linear as a result of the association or dissociation of the sample, or interaction of the sample with the solvent, correction can be made by use of a calibration curve based on an experimental plot over the nonlinear region. Corrections can also be made by a change of solvent to reduce the association, dissociation, or interaction. Deviations from the law can also be caused by other factors, but use of the spectrometer under the proper operating conditions will frequently resolve such difficulties.

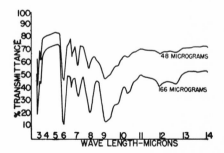

Fig. 4.1. Infrared spectra of deoxyribonucleic acid suspended in solid potassium bromide. The upper curve gives the transmittance of a disk, containing 48 μg of DNA in 50 mg of KBr; the lower curve gives the transmittance of a disk containing 166 μg of DNA in 50 mg of KBr. (Schwarz *et al.*, 1956.)

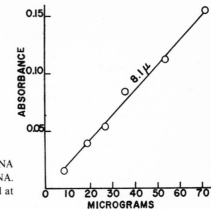

Fig. 4.2. Infrared absorbance *versus* concentration of DNA in disks containing 9, 18, 27, 36, 54, and 72 μg of DNA. The curve was obtained from measurements of the band at 8.1 μ (1235 cm^{-1}). (Schwarz *et al.*, 1956.)

In the case of gases proper adjustment of gas pressures will eliminate deviations due to a phenomenon known as pressure broadening, in which there are changes in the apparent absorbance values.

In the example given in Figs. 4.1 and 4.2, absorbances were measured by a method known as the baseline technique. Several variations of this method are used in quantitative analysis to obtain the accurate measurement of band intensity (Potts, 1963; Alpert *et al.*, 1970; Conley, 1966). This can also be done by measuring the area of the band when certain conditions prevail. Figure 4.3 depicts the baseline method for determining the peak height. The absorbance is measured as the distance from the absorption maximum to the baseline drawn between the two wings of the band. The line *ab* measures the height of the band. Point *c* is at the zero transmittance line. For a series of such bands in separate spectra representing varying concentrations of a sample, the line *ab* will have different values, and these values of absorbance are plotted against concentration to obtain the calibration curve, as in Fig. 4.2. In cases where one is examining mixtures containing substances that produce broad or asymmetric absorption bands, the baseline method is not used since it is difficult to determine the exact baseline. Instead, one can determine the area under the band envelope and take that value as a measure of the concentration of the material. A plot of band area *versus* concentration produces a calibration curve.

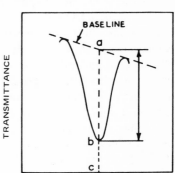

Fig. 4.3. Baseline method for determining the peak height (*ab*). (Rao, 1963.)

The method of quantitative analysis which uses an internal standard has been referred to earlier in Chapter 3. Although this method has been used mainly for solid samples, Beyermann (1967) has determined the protein content of diluted aqueous solutions of a variety of proteins. He used the film from an evaporated solution and measured the intensity of the peptide band at 1538 cm^{-1} and compared it to a thiocyanate band at 2041 cm^{-1}. Protein quantities larger than 5 μg were measured with a standard deviation of 15 %.

The differential method is often used to analyze mixtures containing several substances. It requires that one match the absorbance of the sample against the reference in a double-beam spectrophotometer. If the mixture contains two or more substances that absorb at the same frequency, one can determine what part of the absorbance is due to any single component by preparing a reference containing all the components in the correct proportion except the one being studied. The preparation of several mixtures and references allows one to make a calibration chart of intensity *versus* concentration (Rao, 1963; Potts, 1963; Alpert *et al.*, 1970), which is then applicable to the multicomponent system.

The differential method is often important for the identification and determination of impurities in a natural product, for the assay of a pharmaceutical preparation, or for the detection of adulterants in foods and other substances. One places the impure sample in the sample beam of a double-beam spectrophotometer and the pure major component in the reference beam. One adjusts the thickness of the reference sample until the absorption bands of the major component are blanked out, and obtains a difference spectrum produced by the impurity in the mixture. A calibration curve of absorbance against concentration of impurity can be made from a series of such difference spectra. The advent of ordinate scale expansion in modern instruments has helped considerably in the study of impurity bands and weak bands generally, as well as in microanalysis.

Quantitative differential spectroscopy is used advantageously in biochemical systems. For example, in a typical experiment on the enzyme carbonic anhydrase (Riepe and Wang, 1968), which is discussed in Chapter 15, the infrared sample cell was filled with carbonic anhydrase solution under equilibrium CO_2 pressure. The matched reference cell was similarly filled with the same enzyme solution containing an inhibitor compound, equilibrated under the same CO_2 atmosphere. By accurately measuring the difference spectrum of CO_2-equilibrated carbonic anhydrase against the inhibited enzyme, Riepe and Wang were able to obtain the absorption maximum for the asymmetric stretching vibration of the CO_2 molecule loosely bound to a hydrophobic surface at the active site of the enzyme. The kinetic study of the enzyme β-glucosidase (Parker, 1964) (also discussed in Chapter 15) and the study of the effect of oxidized and reduced nicotinamide adenine dinucleotide on the conformation of yeast alcohol dehydrogenase (Hvidt and Kägi, 1963) (discussed in Chapter 11) are also examples of the application of differential quantitative spectroscopy. The reader will find topics dealing with quantitative analysis throughout this book.

A discussion of recommended practices for general techniques of infrared quantitative analysis has been published by the American Society for Testing and Materials (1966). Perry (1970) has recently reviewed the use of infrared spectrophotometry in quantitative analysis. Rao (1963) has given tables of many organic and inorganic compounds that have been subjected to quantitative analysis.

REFERENCES

Alpert, N. L., Keiser, W. E., and Szymanski, H. A. *IR—Theory and Practice of Infrared Spectroscopy*, *2nd Ed.*, Plenum Press, New York, 1970.

American Society for Testing and Materials, *Manual on Recommended Practices in Spectrophotometry*, Philadelphia, 1966.

Beyermann, K. *Clin. Chim. Acta* **18**, 143 (1967).

Conley, R. T. *Infrared Spectroscopy*, Allyn and Bacon, Boston, 1966.

Hvidt, A. and Kägi, J. H. R. *Compt. Rend. Trav. Lab. Carlsberg* **33**, 497 (1963).

Parker, F. S. *Nature* **203**, 975 (1964).

Perry, J. A. *Appl. Spectrosc. Rev.* **3**, 229 (1970).

Potts, W. J., Jr. *Chemical Infrared Spectroscopy, Vol. I: Techniques*, Wiley, New York, 1963.

Rao, C. N. R. *Chemical Applications of Infrared Spectroscopy*, Academic Press, New York, 1963.

Riepe, M. E. and Wang, J. H. *J. Biol. Chem.* **243**, 2779 (1968).

Schwarz, H. P., Childs, R., Dreisbach, L., and Mastrangelo, S. V. *Science* **123**, 328 (1956).

Chapter 5

HYDROGEN BONDING

"*It is hardly an exaggeration to say that in the chemistry of living systems the H-bond is as important as the carbon–carbon bond.*" (Pimentel and McClellan, 1960.)

Introduction

The hydrogen bonds in biological systems are undoubtedly of great importance, especially since they have a prominent role in determining the gross structure and shape of such molecules as proteins, polypeptides, and nucleic acids. Hydrogen bonding in these types of molecules is discussed in the chapters dealing with them. Besides being of importance in the large biological polymers, hydrogen bonds occur in many other systems, which are also discussed in various chapters. Some of the other topics in which hydrogen bonding is discussed elsewhere in this book are: the structure of water; the chemistry of bones, teeth, and minerals; drugs and pharmaceuticals, including sulfonamides; steroid chemistry; and applications of near-infrared spectroscopy, e.g., hydrogen–deuterium exchange studies of polypeptides.

Infrared spectroscopy is an important tool which is easily used for the investigation of hydrogen bonding. Measurements of spectral shifts offer definite criteria for establishing that hydrogen bonding occurs in various systems. Applications of infrared spectroscopic methods allow the calculation of thermodynamic values, examples of which are referred to in the text.

An excellent book giving comprehensive coverage to the subject of hydrogen bonding is the one by Pimentel and McClellan (1960). For a more recent discussion of this topic the reader is referred to the book by Hamilton and Ibers (1968). Murthy and Rao* (1968) have recently reviewed the subject and have included data from NMR studies of hydrogen bonding. Hydrogen bonding in biological systems has been discussed by Engel (1969) and Löwdin (1969).

A hydrogen bond involves the partial transfer of a proton from a donor, $X-H$, to an acceptor, Y. The strength of the bond may be expected to be proportional to the tendency for complete proton transfer to occur, that is, to the acidity of $X-H$ and the basicity of Y. This proportionality is generally exhibited experimentally, although

---

*This paper reviews the literature from 1958 to 1967. It contains extensive tables of thermodynamic data obtained for various systems by spectroscopic methods.

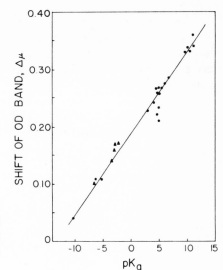

Fig. 5.1. The relationship of basicity to hydrogen-bonding ability, as measured by the shift in the infrared stretching frequency of the O—D band of CH_3OD (● : Gordy and Stanford, 1941; ▲ : Arnett and Wu, 1960). (From Gordon and Jencks, 1963.)

exceptions are known. For example, the shift in the OD-stretching frequency of deuteromethanol (CH_3OD), which is an approximate measure of hydrogen-bond strength, is proportional to the basicity of a variety of proton acceptors of different kinds with basicities ranging from a pK_a of about -10 to about $+10$ (Fig. 5.1) (Gordon and Jencks, 1963).

Different classes of amines show deviations from the line shown in Fig. 5.1, which appear to be due largely to steric and solvation effects (Tamres et al., 1954). Correlations of hydrogen-bonding ability with the acidity of X—H and the basicity of Y have been demonstrated in various systems by measurement of equilibrium constants for hydrogen-bond formation, as well as by measurement of X—H stretching frequencies. (For example, see Baker and Shulgin, 1959; Henry, 1959; Gordon, 1961). A very serious exception to this relationship has been made by Becker (1961), who found that dimethylformamide, which is inactive as a denaturant towards bovine serum albumin in 5.45M urea, is more active than pyridine as a hydrogen acceptor from alcohols, as estimated from equilibrium constants, although less active in terms of O—H frequency shifts.

The limited data available do not suggest that ureas have exceptional hydrogen-bonding ability compared to amides, at least as proton acceptors, since the infrared data of Cook (1958) and of Baker and Harris (1960) show that the hydrogen-bonding ability of tetramethylurea is similar to that of dimethylformamide and dimethylacetamide. Mizushima and Shimanouchi (1957) have reported that acetanilide, dimethylacetamide, and methylacetamide fall in approximately the expected order of both hydrogen-acceptor and hydrogen-donor ability in respect to their basicity and acidity, respectively. Thus, although a precise correlation should not be expected, particularly when hydrogen bonding to different classes of compounds is compared, available data generally support the expected proportionality of hydrogen-bonding ability to the acidity of the hydrogen donor and the basicity of the hydrogen acceptor.

The number of groups which can function as partners in a hydrogen bond, X—H···Y, is quite large. The most common proton donor (X—H) is the OH group, less frequently the NH group, and only occasionally the SH or CH group. Proton acceptors (Y) occur in a wide variety of atoms, or groups, possessing a free electron pair (Table 5.1).

Table 5.1. Atoms and Functional Groups Capable of Acting in a Hydrogen Bond (Tichý, 1965)

Proton donors

    OH, COOH, NH, SH, CH

Proton acceptors

    OH, OR, CO, COOR, NO_2
    NH_2, NHR, NR_2, —N=N—, CN, N_3
    F, Cl, Br, I
    SH, SR
    Multiple bond, aromatic systems, C≡N

Intramolecular Hydrogen Bonding

It is well known that the absorption of the stretching vibration of the O—H bond in a nonassociated (free) alcoholic or phenolic hydroxyl group produces a strong band at 3600 to 3650 cm^{-1} in the fundamental region and near 7100 cm^{-1} in the first overtone (Bellamy, 1954; Nakanishi, 1962). The frequency of the nonassociated hydroxyl ($v_{(OH)}f$) depends partly on the inductive effect of any strongly polar substituents near the hydroxyl, and also on whether the hydroxyl is primary, secondary, or tertiary (Cole and Jefferies, 1956; Cole et al., 1959a). The general steric environment also affects the absorption frequency. For example, in cyclohexanols, the frequency of an axial OH is higher than that of an equatorial OH (Cole et al., 1959a, b; Allsop et al., 1956). The nonassociated (free) OH band of certain primary and secondary alcohols is displayed as a partially resolved doublet on high-resolution instruments. This condition results from the presence of two nonequivalent C—O rotamers (Aaron and Rader, 1963; Oki and Iwamura, 1959a; Piccolini and Winstein, 1959; Dalton et al., 1962; Eddy et al., 1963).

An OH group which acts as proton donor to form a hydrogen bond should have spectral properties different from those of the nonassociated group. Thus, the stretching frequency of such a bonded OH group ($v_{(OH)}b$) shifts towards the lower wave numbers. When the compound contains species with chelated and species with nonchelated OH side by side, the spectrum displays the bands of the associated and of the free hydroxyls. (However, see Tichý (1965) for the special case of two OH groups acting as partners in chelate formation.) The distance, in cm^{-1}, between the bands of the free and the bonded OH is designated as $\Delta v_{(OH)}$, as follows:

$$\Delta v_{(OH)} = v_{(OH)}f - v_{(OH)}b$$

Tichý (1965) has listed hundreds of $\Delta v_{(OH)}$ values and OH stretching frequencies for many types of compounds.

Figure 5.2 shows the spectrum of 3-hydroxypiperidine (Sicher and Tichý, 1958), which contains an intramolecular hydrogen bond.

The experimental technique involves the use of a high-resolution instrument, one with better resolving power than the usual sodium chloride prism instrument. The solvent commonly used is carbon tetrachloride (sometimes carbon disulfide), both of which have practically no absorption in the 3300 cm^{-1} region. Amino alcohols cannot be studied in these solvents, however, since they slowly react with the solvents. Tetrachloroethylene is quite suitable for these compounds, since they are stable in it for extended periods at room temperature.

The infrared spectra are usually recorded for concentrations of about 0.005 mole/liter, a concentration low enough to exclude intermolecular association in compounds other than those known to form exceptionally stable associates (carboxylic acids, amides, etc.) (For primary aliphatic alcohols the limiting concentration for intermolecular association is stated by Flynn *et al.* (1959) to be 0.017 mole/liter; for phenols they give a value of about 0.005 mole/liter.) In any case, care should be taken concerning concentration, since several instances of intermolecular association (mainly dimerization) have recently been reported for certain alcohols at 0.005 mole/liter (examples given in Tichý, 1965).

The best way of determining the character of a bonded hydroxyl band consists of determining the dependence of molar absorptivity of the band on concentration; for intramolecular hydrogen bonds the extrapolation of absorptivity to zero concentration does not go through the origin. If a band due to a carbonyl vibrational overtone may be mistakenly identified as a hydroxyl band, deuteration can be used to establish the identity. Decrease or complete disappearance of the original questionable band and the appearance of a new band of the same character in the O—D stretching region $(2700–2400 \text{ cm}^{-1})$ represents proof that the original band was caused by the O—H bond.

The simple subtraction of the values found in a spectrum usually gives an estimation of $\Delta v_{(OH)}$ as shown above. However, in some cases a simple subtraction does not give a true picture of the conditions as, for example, in certain secondary–tertiary diols where the secondary OH acts as a donor and the tertiary-OH as acceptor. A different procedure has therefore been offered (Cole and Jefferies, 1956), that of

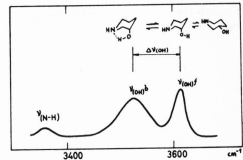

Fig. 5.2. Infrared spectrum of 3-hydroxypiperidine (concentration: 0.005 mole/liter in tetrachloroethylene) (Sicher and Tichý, 1958). (From Tichý, 1965.)

subtracting the experimental $v_{(OH)}b$ value not from the experimental $v_{(OH)}f$ value but from a "standard" value, the latter value being the OH frequency of a compound having the same structure as the one studied but minus the proton-accepting group (e.g., cyclohexanol instead of cyclohexane-diol). The procedure of Cole and Jefferies can be criticized, however, because it does not account for the fact that in vicinal derivatives the presence of the polar proton acceptor can affect the frequency of the free OH group, and it does not consider whether in the cyclohexanols the OH is axial or equatorial. In other geometrically more complex systems, other similar features must also be considered.

Standards must also be used for evaluating $\Delta v_{(OH)}$ in the case of conformationally nonhomogeneous molecules in which the free OH and the associated OH belong to different conformational species.

As a rough approximation we can say that the absorptivity values of the bonded OH and the free OH frequencies reflect the relative concentrations of the bonded and free species; the value of $\Delta v_{(OH)}$ is related to the strength of the hydrogen bridge itself (Pimentel and McClellan, 1960; Bellamy et al., 1961).

For the amino group acting as proton donor (i.e., $N—H\cdots Y$), the $N—H$ stretching band, situated at about 3400 cm^{-1}, is shifted to lower frequencies, as is the case for the hydroxyl group (Hambly, 1961; Badger et al., 1962; Kuhn and Kleinspehn, 1963; Hambly and O'Grady, 1963; Kolinski et al., 1958; Surrey et al., 1959; Castro et al., 1963).

One should be able to study hydrogen bonding by observing the absorption bands of the proton acceptors, e.g., the bands of the ketone or ester carbonyl group or of the ether oxygen of the ester group while these groups function as proton acceptors (Hoyer and Chua, 1963; Flett, 1948; Henbest and Lovell, 1956; Dalton et al., 1963). In fact, these bands are shifted relative to those of the nonchelated groups, but not as much as those of the bands of the donor groups, nor do they tend to be as characteristic.

For a detailed comparison of $O—H\cdots O$ and $O—H\cdots N$ chelates in various substances, see Freedman (1961).

Tichý (1965) has discussed the hydrogen bond and steric conditions for open-chain derivatives, alicyclic compounds (cyclobutane, cyclopentane, and cyclohexane derivatives), phenols, heterocyclic compounds, and compounds with hydroxyl to π-electron interactions. A few examples will be given here.

Valuable information on the steric structures for the cis- and trans-cycloalkane-1,2-diols (Ia) and the cis- and trans-2-aminocycloalkanols (Ib) has been obtained by determining how the $\Delta v_{(OH)}$ values of these compounds vary with ring size.

$$(CH_2)_{n-2} \begin{array}{c} CH—OH \\ | \\ CH—Y \end{array}$$

I

a: Y = OH
b: Y = NH$_2$

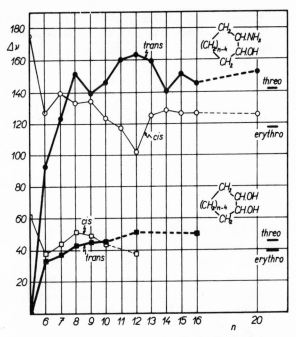

Fig. 5.3. Variation of $\Delta \nu_{(OH)}$ with ring size for the *cis*- and *trans*-cycloalkane 1,2-diols and the *cis*- and *trans*-2-aminocycloalkanols; *threo*- and *erythro*- refer to 3-aminobutan-2-ol and butane-2,3-diol; n is the number of carbon atoms in the ring. (Sicher *et al.*, 1959; Kuhn, 1954.) (From Tichý, 1965.)

In the smaller rings the values of $\Delta \nu_{(OH)}$ of the *cis* isomers are higher than those of the *trans* ones, but the reverse is observed for the medium and large rings (Fig. 5.3). This reversal occurs in the ten-membered ring for the diols (Kuhn, 1954) but in the eight-membered ring for the amino alcohols (Sicher *et al.*, 1959). This fact indicates that in the larger rings *trans*-located vicinal substituents come closer to each other than *cis*-located substituents.

In a study of chair $\rightleftharpoons$ chair conformational equilibria in mobile systems, Piľha *et al.* (1963) have used the conformationally homogeneous *t*-butylated compounds as standards of reference. A "mobile" compound such as *trans*-2-amino-*trans*-5-isopropylcyclohexanol exists as a mixture of two chair conformers (IIa and IIIa)

II

III

a: Y = NH$_2$
b: Y = OH

a: Y = NH$_2$
b: Y = OH

A comparison of the ϵb values found for this "mobile" compound with those found for the "fixed" t-butylated compounds (IVa and Va) made possible the evaluation of the constant of the conformational equilibrium, IIa $\rightleftharpoons$ IIIa. From the value

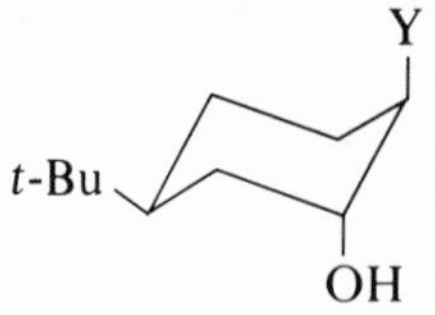

IV

a : Y = NH₂
b : Y = OH

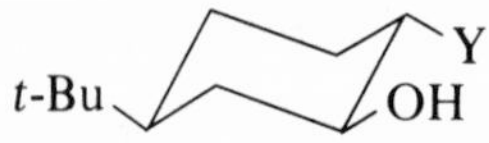

V

a : Y = NH₂
b : Y = OH

of this constant the overall free energy of the interaction of the two vicinal equatorial groups was estimated to be attractive and have a value of about 0.6 kcal/mole. In the same way a value was calculated for the diol system IIb $\rightleftharpoons$ IIIb, with IVb and Vb as reference standards: the free energy of the interaction between the two vicinal equatorial OH groups had a value of 0.8 kcal/mole, again attractive.

A detailed study (Sicher *et al.*, 1963) of diastereoisomeric unsaturated alcohols of the type

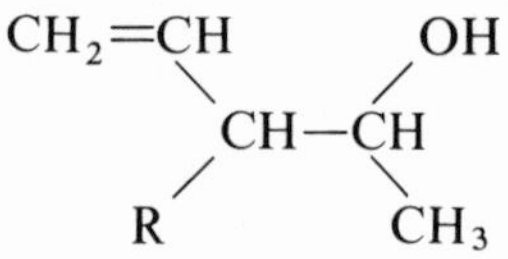

VI

has shown that values of $\Delta v_{(OH)}$ of members of diastereoisomeric pairs do not differ greatly but that the epimers can be distinguished (and configurations assigned) on the basis of $\epsilon b/\epsilon f$ values (Table 5.2) (see also Schleyer *et al.*, 1959).

Table 5.2. Values of $\Delta v_{(OH)}$, in cm⁻¹, and $\epsilon b/\epsilon f$, in parentheses, for Alcohols[a] of Structure (VI)

R	Me	Et	*i*-Pr	*t*-Bu
threo	41 (0.90)	43 (1.0)	53 (1.1)	65 (0.50)
erythro	41 (0.40)	48 (0.45)	53 (0.40)	52 (0.35)

[a]Data from Sicher *et al.* (1963).

As one would expect, the spectrum of epicholesterol (VII) exhibits a band due to intramolecularly bonded hydroxyl ($\Delta v_{(OH)} = 30$ cm⁻¹); the spectrum of cholesterol (VIII) does not (Eddy *et al.*, 1963; Schleyer *et al.*, 1958; Oki and Iwamura, 1959*b*).

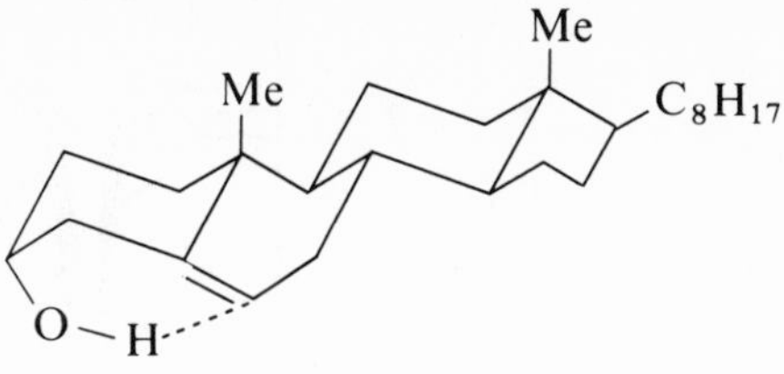

VII

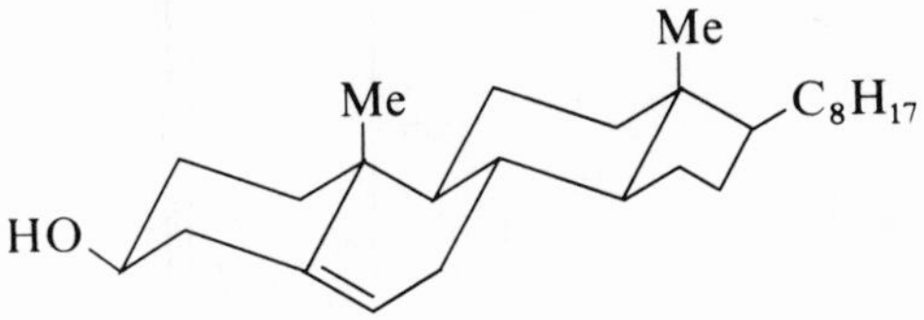

VIII

The spectroscopic method of determining configurational assignment has been used for the alkaloid chelidonine (IX). The band produced by a very strongly bonded hydroxyl ($\Delta v_{(OH)}$ = 414 cm^{-1}) was characteristic of the conformer chelidonine-I (X);

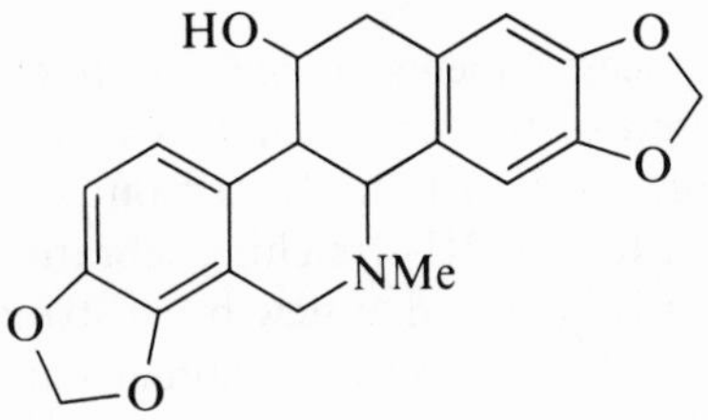

IX

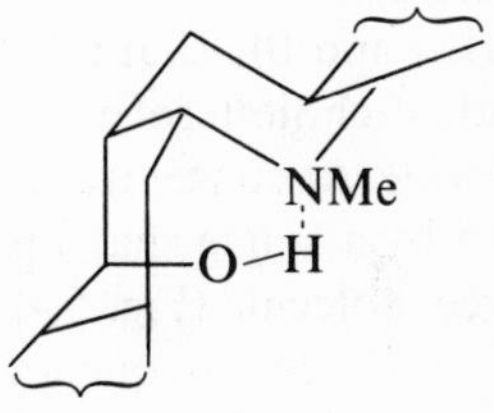

X

the configuration at all three asymmetric centers was thus settled by a single measurement (Bersch, 1958; Šantavý *et al.*, 1960).

Intermolecular Hydrogen Bonding*

Thermodynamics—Calculation of the Energy of an O—H···O Bond

There has been much interest in recent years in the quantitative determination of the energy aspects of hydrogen bonds by spectroscopic methods (Pimentel and McClellan, 1960). Much information is available in the literature on the thermodynamics of association due to hydrogen bonding involving O—H bonds. The

---

*See also various topics mentioned in the *Introduction* to this chapter.

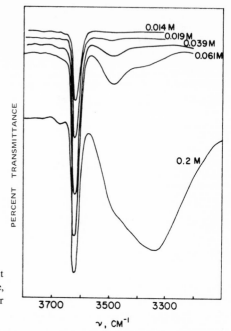

Fig. 5.4. OH-stretching vibration bands of cholesterol at different concentrations. Solvent, CCl_4; temperature, 23°C. The ordinate axis is not drawn to scale. (Parker and Bhaskar, 1968.)

studies include hydrogen bonds formed by self-association of hydroxylic compounds and by their interaction with different bases (Murthy and Rao, 1968, and references cited therein).

Parker and Bhaskar (1968) have used infrared measurements of the OH-stretching band of cholesterol to study the self-association of this compound in carbon tetrachloride. At concentrations below $0.014M$, cholesterol exists only as a monomer, as shown by a single sharp band at $3620 \, \text{cm}^{-1}$ due to the OH-stretching vibration of the free molecule (Fig. 5.4). As the concentration is increased, a new broad band appears on the low-frequency side at $3470 \, \text{cm}^{-1}$ in addition to the monomer band at $3620 \, \text{cm}^{-1}$. The $3470 \, \text{cm}^{-1}$ band is due to the bonded OH group of the dimer. As the concentration is increased, a third band appears at $3330 \, \text{cm}^{-1}$ in addition to the monomer and dimer bands and is probably due to a trimer or higher aggregate. At a concentration of $\sim 0.2M$ the dimer band is observed only as a shoulder on the 3330 cm^{-1} band.

The association for the dimer was calculated from measurements of the intensity of the monomer band at different concentrations by use of the limiting slope method of Liddel and Becker (1957). In this method, the apparent molar absorptivity of the monomer band is plotted against the concentration of the associating solute, and the dimerization constant K_d is obtained from the limiting slope of the resulting curve by use of equation (1), where ϵ_M^0 is the molar absorptivity of the monomer extrapolated to infinite dilution ($c = 0$)

$$\left[\frac{d\epsilon}{dc} \right]_{c \to 0} = - K_d \epsilon_M^0 \tag{1}$$

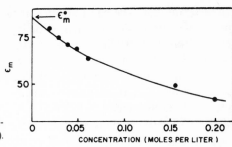

Fig. 5.5. Apparent molar absorptivity vs concentration of cholesterol in CCl_4 (Liddel and Becker plot). (Parker and Bhaskar, 1968.)

This equation is derived for the case of an open-chain dimer, in which the free OH group of the dimer also absorbs in the region of the monomer band. An assumption is made that the absorptivity of the free OH group of the dimer is the same as that of the monomer OH.

A typical curve obtained is shown in Fig. 5.5. The extrapolated value of ϵ_M^0 is 85.5 liters/mole/cm. The slope of the limiting curve is -384.6. Then, $-K_d\epsilon_M^0 = -384.6$, and $K_d = 4.5$ liters/mole at 23°C.

The enthalpy of dimerization was determined from the dimerization constants measured at different temperatures between 5 and 50°C. The spectral and thermodynamic data are summarized in Table 5.3.

Table 5.3. Spectral and Thermodynamic Data of the Self-Association of Cholesterol in CCl_4 (Parker and Bhaskar, 1968)

Associated species	v_{OH}, cm^{-1}	Δv_{OH},[a] cm^{-1}	K,[b] liter/mole	$-\Delta H$, kcal/mole
Dimer	3470	150	4.5	1.8
Trimer	3330	290		

[a]Shift from the v_{OH} of the monomer at 3620 cm^{-1}.
[b]At 23°C.

From the time Badger and Bauer (1937) proposed a relationship between Δv (the shift in frequency of the OH band of the associated molecule from that of the free molecule) and ΔH of association (or hydrogen bonding), this relationship has been tested in several cases and it is generally accepted that Δv is an index of ΔH of hydrogen-bond formation (Pimentel and McClellan, 1960). It can be seen in Table 5.3 that the frequency shift (Δv) of the trimer is almost twice that of the dimer. Based on this an estimate of approximately -3.6 kcal/mole was made for the ΔH of the trimer.

Parker and Bhaskar (1968) also studied the interaction between cholesterol and each of several triglycerides. Infrared spectra of mixed solutions of cholesterol and triacetin, tributyrin, or trilaurin gave evidence of the formation of a 1:1 hydrogen-bonded complex. The equilibrium constants and enthalpies of formation of the complexes of cholesterol with each of the triglycerides were calculated. The $K_{23°C}$ values ranged from 2.4 to 3.7 liters/mole, and the ΔH values from -3.5 to -5.4 kcal/mole. An explanation of the method of calculation follows.

The spectrum of a mixture of cholesterol and tributyrin in CCl_4 in concentrations of $0.01M$ and $0.25M$, respectively, is shown in Fig. 5.6, curve 4. In addition to the OH band at $3620\,cm^{-1}$, a new band is present at $3540\,cm^{-1}$. At this concentration cholesterol exists only as a monomer, as shown in curve 1. Since the spectrum was recorded against a blank solution containing the triglyceride at the same concentration as in the mixed solution, it is obvious that the new band is not due to either component alone. The figure also shows that the intensity of the OH band of cholesterol at $3620\,cm^{-1}$ in the mixture is less than that in the pure solution (curve 1). Thus, the new band was due to a hydrogen-bonded complex between cholesterol and the triglyceride.

The equilibrium constant K for a 1:1 complex is given by equation (2),

$$K = \frac{[DA]}{[D][A]} \tag{2}$$

where $[DA]$ is the equilibrium concentration of the complex, $[D]$ the equilibrium concentration of the free proton donor, and $[A]$ the equilibrium concentration of the free proton acceptor.

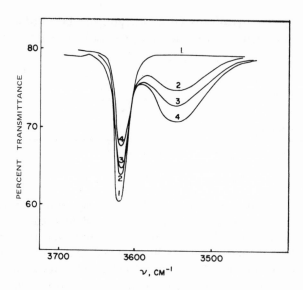

Fig. 5.6. Effect of varying the concentration of tributyrin on the associated OH band of cholesterol–tributyrin complex. Solvent, CCl_4; temperature, 23°C. Curve 1: cholesterol ($0.01M$); curves 2–4: cholesterol ($0.01M$) plus tributyrin (0.10, 0.15, and $0.25M$). (Parker and Bhaskar, 1968.)

Since the acceptor concentration ($\sim 0.25M$) is much larger than that of the donor ($\sim 0.013M$), the equilibrium concentration of the free acceptor is not significantly different from its initial (total) concentration. With this assumption the only unknowns in the above equation are the equilibrium concentrations of the complex and of the free donor. Both are readily determined from measurements of the intensities of OH absorption bands of cholesterol (3620 cm^{-1}) in pure carbon tetrachloride and in the presence of the acceptor. If a_1 is the absorbance at the v_{max} of the free OH band (3620 cm^{-1}) of a dilute cholesterol solution ($\sim 0.013M$) and a_2 the absorbance of an equal concentration of cholesterol in the presence of a large excess of the acceptor, then

$$\frac{a_1 - a_2}{a_2} = \frac{[DA]}{[D]} \tag{3}$$

and

$$K = \frac{a_1 - a_2}{a_2} \frac{1}{\bar{A}} \tag{4}$$

where $\bar{A}$ is the initial concentration of the acceptor.

The enthalpy of complex formation was determined by measuring the equilibrium constants at different temperatures. Since the complex band (3540 cm^{-1}) showed more marked intensity changes with temperature than did the free OH band (3620 cm^{-1}, in Fig. 5.7), the former band was used to determine ΔH. To do this it is necessary to determine the molar absorptivity of the complex band by the following procedure. The concentration of the complex is given (in terms of the absorbance measurements referred to in the preceding paragraph) as

$$[DA] = \frac{a_1 - a_2}{a_1}[\bar{D}]$$

where $[\bar{D}]$ is the initial concentration of cholesterol and the other terms have the same meaning as before. By use of this value of the concentration and the absorbance of the complex band the molar absorptivity (ϵ) of the band can be calculated by the familiar Beer–Lambert law equation for absorbance, A

$$A = \epsilon[DA]l$$

where l is the optical path length. The molar absorptivity thus calculated can in turn be used to determine the concentration of the complex at different temperatures. Some data are given here from measurements of the interaction between cholesterol and tributyrin in CCl$_4$ at 5°C: the absorbance (a_1) of the cholesterol OH band in 0.0124M cholesterol was 0.1494, and the absorbance of the cholesterol OH band (a_2) in the mixture containing 0.012M cholesterol plus 0.2519M tributyrin, was 0.0771. From equation (4)

$$K = \frac{a_1 - a_2}{a_2} \times \frac{1}{\bar{A}} = \frac{0.0723}{0.0771} \times \frac{1}{0.2519} = 3.72$$

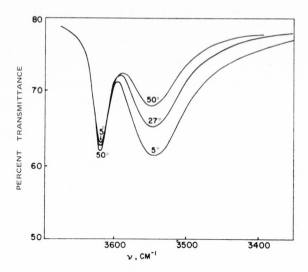

Fig. 5.7. Effect of temperature on the intensities of the free and associated OH bands of cholesterol (0.012M)–tributyrin (0.25M) complex; solvent, CCl_4. (Parker and Bhaskar, 1968.)

The absorbance of the complex band is 0.1012. The concentration of the complex is given by

$$[DA] = ((a_1 - a_2)/a_1)[\bar{D}] - [\bar{D}]$$

where $[\bar{D}]$ is the initial concentration of cholesterol. This calculation leads to a value of 0.006108. The molar absorptivity, ε, of the complex can be calculated from

$$A = \epsilon cl = \epsilon \times 0.006108 \times 0.15 = 0.1012$$

since the measurement was made with a cell of 0.15 cm path length. Therefore, ϵ has a value of 110.5 With this value of ϵ and the measured absorbances of the complex band at different temperatures, the equilibrium constant was calculated as in the following example: The absorbance of the complex band at 27°C is 0.0768. Therefore, the concentration $[DA] = 0.0768/(110.5 \times 0.15) = 0.004633$, the concentration of free cholesterol at equilibrium is $0.01240 - 0.004633 = 0.007767$, and the concentration of free tributyrin is $0.2519 - 0.0046 = 0.2473$. Therefore

$$K = \frac{0.004633}{0.2473 \times 0.007767} = 2.41$$

A typical plot of log K versus $1/T$ obtained for the cholesterol–tributyrin complex is shown in Fig. 5.8. The spectral and thermodynamic data of the cholesterol–triglyceride complexes are summarized in Table 5.4.

Table 5.4. Spectral and Thermodynamic Data of the Cholesterol-Triglyceride Complexes[a] (Parker and Bhaskar, 1968)

Triglyceride	ν_{OH}, assocd. cm^{-1}	$\Delta\nu_{OH}$,[b] cm^{-1}	K,[c] liter/mole	$-\Delta H$, kcal/mole
Triacetin	3540	80	2.9	4.3
Tributyrin	3545	75	2.4	3.5
Trilaurin	3545	75	3.7	5.4

[a]The solvent is CCl_4.
[b]Shift from the free cholesterol peak at $3620\ cm^{-1}$.
[c]At 23°C.

Cole and Macritchie (1959) have used intensity measurements of the hydroxyl stretching absorption band at various temperatures to determine the energy of the hydrogen bond formed between a substituted cyclohexanol, *trans*-dihydrocryptol (XI), and a cyclic ether, dioxane (XII) in tetrachloroethylene solutions. The methods

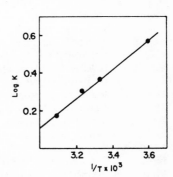

XI   XII

they used were similar to those already described for hydrogen bonding of cholesterol by Parker and Bhaskar (1968). Cole and Macritchie found a value of 3.2 ± 0.3 kcal/mole for the energy of the hydrogen bond in the *trans*-dihydrocryptol-dioxane system.

Fig. 5.8. Log K *versus* $1/T$ plot for the cholesterol–tributyrin complex. (Parker and Bhaskar, 1968.)

Thermodynamics of Hydrogen Bonding of Secondary Amides

Since not much information was available on the quantitative thermodynamics of hydrogen bonding involving the N—H bonds of secondary amides, Bhaskar and Rao (1967) carried out investigations on the self-association of secondary amides as well as their interaction with certain bases by infrared spectroscopy. They studied the thermodynamics of the hydrogen-bonding interaction of *cis* as well as *trans* N—H bonds of secondary amides.

N-methylacetamide is known to exist almost exclusively in the *trans* form whereas N-phenylurethane is found to be 95% *cis* and 5% *trans* (Russell and Thompson, 1956). Bhaskar and Rao (1967) studied the dimerization equilibria of N-methylacetamide and N-phenylurethane by measuring the concentration dependence of the first N—H overtone band. The procedure for determining the dimerization equilibrium constant K_d was similar to that of Liddel and Becker (1957). The enthalpy of dimerization was evaluated from values of K_d at different temperatures. The details of the procedure and uncertainties in the determination of K_d have been described in the paper by Singh and Rao (1967). The values of K_d (liters/mole) at 26°C were 5.4 for N-methylacetamide in CCl_4 and 1.5 for N-phenylurethane. The values of $-\Delta H$ were 4.7 and <3.0 kcal/mole, respectively. The values of K_d and $-\Delta H^0$ for each of these compounds with various bases, e.g., pyridine and benzophenone, were also determined by a method similar to Becker's (1961). The K_d and ΔH^0 values did not show a clear-cut dependence on the configuration of the N—H bond of the secondary amide.

REFERENCES

Aaron, H. S. and Rader, C. P. *J. Am. Chem. Soc.* **85**, 3046 (1963).
Allsop, I. L., Cole, A. R. H., White, D. E., and Willix, R. L. S. *J. Chem. Soc.* **1956**, 4868.
Arnett, E. M. and Wu, C. Y. *J. Am. Chem. Soc.* **82**, 4999 (1960).
Badger, R. M. and Bauer, S. H. *J. Chem. Phys.* **5**, 839 (1937).
Badger, G. M., Harris, R. L. N., Jones, R. A., and Sasse, J. M. *J. Chem. Soc.* **1962**, 4329.
Baker, A. W. and Harris, G. H. *J. Am. Chem. Soc.* **82**, 1923 (1960).
Baker, A. W. and Shulgin, A. T. *J. Am. Chem. Soc.* **81**, 1523 (1959).
Becker, E. D. *Spectrochim. Acta* **17**, 436 (1961).
Bellamy, L. J. *The Infrared Spectra of Complex Molecules*, Methuen, London, 1954, p. 89.
Bellamy, L. J., Eglinton, G., and Morman, J. F. *J. Chem. Soc.* **1961**, 4762.
Bersch, H. W. *Arch. Pharm.* **291**, 491 (1958).
Bhaskar, K. R. and Rao, C. N. R. *Biochim. Biophys. Acta* **136**, 561 (1967).
Castro, A. J., Marsh, J. P., Jr., and Nakata, B. T. *J. Org. Chem.* **28**, 1943 (1963).
Cole, A. R. H. and Jefferies, P. R. *J. Chem. Soc.* **1956**, 4391.
Cole, A. R. H. and Macritchie, F. *Spectrochim. Acta* **15**, 6 (1959).
Cole, A. R. H., Müller, G. T. A., Thornton, D. W., and Willix, R. L. S. *J. Chem. Soc.* **1959***a*, 1218.
Cole, A. R. H., Jefferies, P. R., and Müller, G. T. A. *J. Chem. Soc.* **1959***b*, 1222.
Cook, D. *J. Am. Chem. Soc.* **80**, 49 (1958).
Dalton, F., Meakins, G. D., Robinson, J. H., and Zaharia, W. *J. Chem. Soc.* **1962**, 1566.
Dalton, F., McDougall, J. I., and Meakins, G. D. *J. Chem. Soc.* **1963**, 4068.
Eddy, C. R., Showell, J. S., and Zell, T. E. *J. Am. Oil Chemists' Soc.* **40**, 92 (1963).
Engel, J. "Hydrogen Bonds in Biological Systems," in *Physics of Ice*, Plenum Press. New York, 1969, p. 138.
Flett, M. St. C. *J. Chem. Soc.* **1948**, 1441.
Flynn, T. D., Werner, R. L., and Graham, B. M. *Australian J. Chem.* **12**, 575 (1959).

Freedman, H. H. *J. Am. Chem. Soc.* **83**, 2900 (1961).

Gordon, J. E. *J. Org. Chem.* **26**, 738 (1961).

Gordon, J. A. and Jencks, W. P. *Biochemistry* **2**, 47 (1963).

Gordy, W. and Stanford, S. C. *J. Chem. Phys.* **9**, 204 (1941).

Hambly, A. N. *Rev. Pure Appl. Chem.* **11**, 212 (1961).

Hambly, A. N. and O'Grady, B. V. *Australian J. Chem.* **16**, 459 (1963).

Hamilton, W. C. and Ibers, J. A. *Hydrogen Bonding in Solids*, W. A. Benjamin, Inc., New York, 1968.

Henbest, H. B. and Lovell, B. J. *Chem. Ind. (London)* **1956**, 278.

Henry, L. in *Hydrogen Bonding* (D. Hadži, ed.), Pergamon Press, London, 1959, p. 163.

Hoyer, H. and Chua, M. *Z. Naturforsch.* **18b**, 349 (1963).

Koliński, R., Piotrowska, H., and Urbański, T. *J. Chem. Soc.* **1958**, 2319.

Kuhn, L. P. *J. Am. Chem. Soc.* **76**, 4323 (1954).

Kuhn, L. P. and Kleinspehn, G. G. *J. Org. Chem.* **28**, 721 (1963).

Liddel, U. and Becker, E. D. *Spectrochim. Acta* **10**, 70 (1957).

Löwdin, P. O. "Some Aspects of the Hydrogen Bond in Molecular Biology," in *Ann. N.Y. Acad. Sci.* **158**, Art. 1, 86 (1969).

Mizushima, S. and Shimanouchi, T. in *Festschrift Stoll Arthur*, Birkhäuser, Basel, 1957, p. 305.

Murthy, A. S. N. and Rao, C. N. R. *Appl. Spectrosc. Rev.* **2**, 69 (1968).

Nakanishi, K. *Infrared Absorption Spectroscopy—Practical*, Holden-Day, San Francisco, 1962, p. 30.

Oki, M. and Iwamura, H. *Bull. Chem. Soc. Japan* **32**, 950 (1959a).

Oki, M. and Iwamura, H. *Bull. Chem. Soc. Japan* **32**, 306 (1959b).

Parker, F. S. and Bhaskar, K. R. *Biochemistry* **7**, 1286 (1968).

Piccolini, R. and Winstein, S. *Tetrahedron Letters*, No. 13, 4 (1959).

Pimentel, G. C. and McClellan, A. L. *The Hydrogen Bond*, W. H. Freeman, San Francisco, 1960.

Piťha, J., Sicher, J., Šipoš, F., Tichý, M., and Vašičková, S. *Proc. Chem. Soc.* **1963**, 301.

Russell, R. A. and Thompson, H. W. *Spectrochim. Acta* **8**, 138 (1956).

Šantavý, F., Horák, M., Maturová, M., and Brabenec, J. *Collection Czech. Chem. Commun.* **25**, 1344 (1960).

Schleyer, P. von R., Trifan, D. S., and Bacskai, R. *J. Am. Chem. Soc.* **80**, 6691 (1958).

Schleyer, P. von R., Wintner, C., Trifan, D. S., and Bacskai, R. *Tetrahedron Letters*, No. 14, 1 (1959).

Sicher, J. and Tichý, M. *Collection. Czech. Chem. Commun.* **23**, 2081 (1958).

Sicher, J., Horák, M., and Svoboda, M. *Collection Czech. Chem. Commun.* **24**, 950 (1959).

Sicher, J., Cherest, M., Gault, Y., and Felkin, H. *Collection Czech. Chem. Commun.* **28**, 72 (1963).

Singh, S. and Rao, C. N. R. *J. Phys. Chem.* **71**, 1074 (1967).

Surrey, A. R., Lesher, G. Y., Mayer, J. R., and Webb, W. G. *J. Am. Chem. Soc.* **81**, 2887 (1959).

Tamres, M., Searles, S., Leighly, E. M., and Mohrman, D. W. *J. Am. Chem. Soc.* **76**, 3983 (1954).

Tichý, M. "The Determination of Intramolecular Hydrogen Bonding by Infrared Spectroscopy and Its Applications in Stereochemistry," in *Advances in Organic Chemistry, Methods and Results, Vol. 5* (R. A. Raphael, E. C. Taylor, and H. Wynberg, eds), Interscience, New York, 1965, p. 115.

Chapter 6

CARBOHYDRATES*

Comparison of Samples

After exhaustive studies of infrared spectra of organic compounds, it has been found that every compound with an unique molecular structure displays a characteristic spectrum. Consequently, for a enantiomorphic pair, in which the details of structure are identical (but in mirror image), the spectra of the solids, in the same polymorphic modification, are identical (Barker *et al.*, 1956). Polymorphism may cause slight spectral differences, but for the liquid phases (solution or melt) the spectra are identical.

A slight change in the structure of a large molecule may change the spectrum only slightly. Thus, progressing up a series of oligosaccharides having the same anomeric form of the same monosaccharide residue only, the spectra of the tetraose, pentaose, hexaose, etc., become more and more alike (Kuhn, 1950).

For a carbohydrate, the infrared spectrum is usually a much more specific and characteristic property than the melting point, boiling point, density, refractive index, or the ultraviolet spectrum. Thus, changing the orientation at only one carbon atom in anomers of a sugar or glycoside changes the spectrum markedly (Kuhn, 1950). Anomers have different spectra in the region 667 to 250 cm^{-1} (Parker and Ans, 1966). Also, the spectrum of the furanoid form of a sugar derivative differs from that of the pyranoid form in the same anomeric modification, as with methyl β-D-ribofuranoside and β-D-ribopyranoside (Ellis, 1966).

In 1950, Kuhn recorded the spectra of many carbohydrates, ten of which were crystalline sugars in Nujol mulls. Except for α-D-glucose, the type of anomer used was not mentioned. Nor did Urbański *et al.* (1959) specify anomeric form when they recorded the spectra of six sugars. Tipson and Isbell (1962) have since identified these compounds by comparing their spectra with those of authentic anomeric forms of those sugars. Such studies for the identification of compounds are among the most common applications of infrared spectroscopy.

Interpretation of Spectra

The empirical approach is mainly used in attempting to correlate the structure of a complex molecule with the absorption frequencies it displays. By comparing the spectra of many compounds that contain the same common group, one can usually

*Based in part on a chapter by R. S. Tipson and F. S. Parker in *The Carbohydrates, Vol. IB* (W. Pigman and D. Horton, eds.), Academic Press, New York, 1971; and Tipson (1968).

find an absorption band the frequency of which remains rather constant and fairly independent of the rest of the molecule. For example, the spectra of 24 acetylated glycosides were studied (Tipson and Isbell, 1960*a*), along with those for 21 non-acetylated glycosides (Tipson and Isbell, 1960*b*), and certain conclusions were reached: (a) All the acetates displayed a band at 1773 to 1736 cm^{-1}, the C=O stretching frequency, not shown by any of the nonacetylated compounds. (b) All the nonacetylated glycosides, but not the acetates, exhibited a band at 3413 to 3279 cm^{-1}, —OH stretching. (c) Moreover, all the hydrates (but no anhydrous compounds) had a band at 1664 to 1634 cm^{-1}. Only the positions of the bands were noted, no matter what the relative intensities.

In the absence of disturbing effects, all compounds having the same group absorb infrared radiation at approximately the same frequency. Bands in the region from 5000 to ~ 1430 cm^{-1} are usually characteristic group frequencies; for these, the associated vibration is well localized, as in C=O and X—H stretching vibrations, and the precise positions of the bands are more reliable than for those in the region 1430 to 667 cm^{-1}, which contains the "fingerprint region" (1430–910 cm^{-1}). This region usually furnishes abundant information to characterize a compound; however, the origin of these bands is frequently not easily ascertained, so a detailed study of this region of the spectrum is usually delayed while the rest of the spectrum is analyzed. For carbohydrates and derivatives, the region 1250 to 1000 cm^{-1} contains bands for C—O of esters, ethers, and hydroxyl groups, and may not show highly individual bands for such structures. The bands in the region 1430 to 667 cm^{-1} may be the result of single-bond skeletal stretching vibrations between atoms of similar or identical mass, or from bending deformations. The latter occur in this region because less energy is needed to cause them; some of these bands can be used for group diagnosis, but most of them are greatly affected by changes of structure in the molecule.

The location of a band at a position expected for a certain group does not necessarily mean that the particular group is contained in the compound, since there are many possibilities of interference by other bands. However, absence of a group frequency usually indicates absence of that group from the compound, provided that effects (such as hydrogen bonding) that could shift or even remove the band are not present. Thus, the bands for carbonyl groups are usually very strong. If the spectrum displays only a weak band at 1700 cm^{-1} then: (a) this band is not caused by carbonyl, (b) the molecule is very large and possibly has only one carbonyl group, or (c) a carbonyl compound is present as a contaminant. Certain group absorptions do not have similar intensities in a variety of compounds, although the frequency of the absorption is essentially the same.

Group Frequencies in Carbohydrates and Their Derivatives

Except for particular steric and electrical effects, every organic compound that possesses a certain group displays the corresponding characteristic group frequency in its spectrum. Group-frequency tabulations are available from many sources.*

*The following reference contain group-frequency charts, some quite extensive: Bellamy, 1958; Bellamy, 1968; Jones and Sandorfy, 1956; Rao, 1963; Alpert, Keiser, and Szymanski, 1970; Nakanishi, 1962; Jones, 1959; Cross, 1964; Colthup, Daly, and Wiberley, 1964; Miller, 1953; Brügel, 1962; Bentley, Smithson, and Rozek, 1968; Randall, Fowler, Fuson, and Dangl, 1949; Simon and Clerc, 1967; Conley, 1966; Flett, 1963; Brand and Eglinton, 1965. (Also see listings in Chapter 20.)

Table 6.1 contains a list, and an estimate of the relative intensities, of the group frequencies that investigators of organic compounds (including carbohydrates) are likely to encounter. In the ensuing discussion of some of the more important bands, all values for their positions are approximate unless a range is stated. It should be noted, however, that although frequencies usually fall within the limits indicated, they may, in special cases, lie outside these ranges. Therefore, when the correlations are used, all other available evidence should also be considered.

Table 6.1. Characteristic Infrared Bands of Various Groups (Tipson and Parker, 1971)

Range, cm^{-1}	Intensity^a	Group	Type of vibration^b	Comments
4505–4200	w	C—H	str	aliphatic (combination)
4255–4000	w	C—H	str	aromatic (combination)
3650–3500	var	O—H	str	free OH, oxime
3640–3623	m (sh)	O—H	str	free OH, alcohols
3600–3100	m	O—H	str	water of crystallization
3590–3425	var (sh)	O—H	str	intramolec. bonded OH
3550–3500	m	O—H	str	free OH, carboxylic acid (very dilute solution)
3550–3450	var (sh)	O—H	str	intermolec. bonded OH (dimeric)
3550–3195	w	C=O	str	carbonyl (first overtone)
∼3520	s	N—H	str	primary amide (free)
∼3500	m	N—H	str (asym)	primary amine, free NH (dilute solution)
3500–3300	m	N—H	str	secondary amine, free NH
3500–3060	m	N—H	str	associated NH, amine or amide
∼3400	s	N—H	str	primary amide (free)
∼3400	m	N—H	str (sym)	primary amine, free NH (dilute solution)
3400–3225	s (br)	O—H	str	intermolec. bonded OH (polymeric)
∼3380	m	NH$_3^+$	str	amine salt (soln.)
3355–3145	m	NH$_3^+$	str	amine salt (solid); several bands
∼3350	m	N—H	str	primary amide (bonded)
∼3300	s	C—H	str	≡C—H, acetylenes
3300–2500	w (vbr)	O—H	str	H-bonded carboxylic acid dimers
∼3280	m	NH$_3^+$	str	amine salt (soln.)
∼3175	m	N—H	str	primary amide (bonded)
3155–3050	w	C—H	str	—CH=C—O— and —C=CH—O—
3095–3075	m	C—H	str	RCH=CH$_2$, olefin
3075–3030	w–m	C—H	str	C—H of aromatic ring
3050–2995	w	C—H	str	of epoxide (shifts to 3040–3030 if ring strain increases)
3040–3010	s, m	C—H	str	$\diagdown$C—H; RCH=CH$_2$, RCH=CHR′ (cis or trans), RCR′=CHR″, olefin

Table 6.1 *(Continued)*

Range, cm^{-1}	Intensity^a	Group	Type of vibration^b	Comments
~2960	s	C—H	str (asym)	C-methyl
~2925	s	C—H	str (asym)	$>CH_2$, methylene, Ar—CH$_3$
2900–2880	w	C—H	str	C—H, methine
2900–2705 (two)	w	C—H	str	—C(=O)H, aldehyde
2900–2300 (several)	w	N—H	str	quaternary amine salt, bonded
~2875	s	C—H	str (sym)	C-methyl
~2850	s	C—H	str (sym)	$>CH_2$, methylene
2835–2815		C—H	str	O-methyl
~2825	m	C—H	str	$-\overset{\displaystyle OCH_2-}{\underset{\displaystyle OCH_2-}{CH}}$, alkyl acetal
~2780		C—H	str	—O—CH$_2$—O—
2705–2560	w (br)	P—OH	str	phosphoric ester, H-bonded
2705–2300	s	NH$_2^+$, NH$^+$	str	(may be several bands)
~2580	w	S—H	str	thiol, free
~2400	w	S—H	str	thiol, H-bonded
~2270	vs	N=C=O	asym str	isocyanate
2260–2240	w	—C≡N	str	saturated nitrile
2260–2190	var	C≡C		RC≡CR'; acetylenes
2230–2215	s	—C≡N	str	unsaturated conjugated nitrile
2200–2050	vs	C=S	asym str	—N—C=S, isothiocyanate (2 or more bands)
2200–2000	s			cyanide, thiocyanate, cyanate
2180–2120		C≡N	str	R—N$^+$≡C$^-$
2160–2120	s	N≡N	str	azide
2140–2100	w	C≡C		RC≡C—H; acetylenes
~1810	s	C=O	str	—COCl, aliphatic acid chloride
1780–1740	s	C=O	str	—O—(C=O)—O—, carbonate
~1770	s	C=O	str	γ-lactone
1745–1735	s	C=O	str	saturated esters
~1740	s	C=O	str	δ-lactone
1740–1720	s	C=O	str	—C(=O)H, aldehyde
~1725	s	C=O	str	formic ester
1725–1705	s	C=O	str	ketone
~1720	s	C=O	str	benzoic ester
1720–1700	s	C=O	str	—COOH; aliphatic carboxylic acid (dimer)
1700–1670	s	C=O	str	—CONHR, secondary amide, free (dil. soln.): Amide I
1690–1670	s	C=O	str	—CONH$_2$, primary amide, free (dil. soln.): Amide I
1680–1630	s	C=O	str	secondary amide (solid)
1680–1620	var	C=C	str	nonconjugated C=C
1678–1668		C=C		trans olefin; RHC=CHR'
~1675	s	C=S	str	thioester

Table 6.1 *(Continued)*

Range, cm^{-1}	Intensity^a	Group	Type of vibration^b	Comments
∼1675	s	C=O	str	thioester
∼1670	w	C=N	str	aliphatic oxime
1670–1620	s	C=O	str	primary amide (solid), H-bonded, 2 bands: Amide I
1662–1652		C=C		*cis* olefin; RHC=CHR′
1658–1648	m	C=C		terminal olefin; RR′C=CH$_2$
1650–1620	s	N—H	def	primary amide (solid): Amide II band
1650–1600	s	NO$_2$	asym str	—O—NO$_2$, nitrate
1650–1580	m–s	N—H	def	NH$_2$; primary amine
1650–1550	w	N—H	def	NHR; secondary amine
1648–1638	var	C=C	str	terminal olefin; RHC=CH$_2$
∼1625	s	C=C	str	Ph-conjugated C=C
1625–1585	m	C=C	sk, i-p	aromatic C=C
1620–1590	s	N—H	def	primary amide (dil. soln.)
1620–1560	m–s	NH$_2^+$	def	
1610–1540	vs	C—O	asym str	—COO$^-$, carboxylate
∼1600	s	C=C	str	CO or C=C conjugated with C=C
∼1585	m	NH$_3^+$	asym def	amine salt
1580–1520	m	C=N (plus C=C)	int eff	pyrimidines
1570–1515	s	N—H	def	secondary amide (solid): Amide II band
1550–1510	s	N—H	def	secondary amide (dilute solution)
∼1500	var	C=C	sk, i-p	aromatic C=C
1500–1470	s	C=S	str	—N—C=S
1500–1300	m	NH$_3^+$	sym def	amine salt
∼1468	s	C—H	sc	alkane —CH$_2$—
∼1460	m	C—H	bend (asym)	—CH$_3$
1460–1400	s	C—O	sym str	—COO$^-$, carboxylate
∼1455	s	C—H	sc	alicyclic —CH$_2$—
1450–1400	w	—N=N—	str	azo
1440–1395	w	C—O	str (plus OH def)	carboxylic acid
1440–1350	s	S=O	str	(RO)$_2$SO$_2$, sulfuric ester
1440–1325	m	C—C		aliphatic aldehyde
1420–1406	w	C—H	i-p bend	C=CH$_2$
1420–1330	s	S=O	str	ROSO$_2$R′, sulfonic ester
1418–1400	m	C—N	str	primary amide
∼1410	w	C—N	str	aliphatic amine
1390–1360 doublet	m	C—H	bend (sym)	*gem*-dimethyl
1385–1375	m	C—H	bend (sym)	—CH$_3$
1370–1250		C—O	str	lactone
∼1340	w	C—H	bend	alkane C—H
1340–1280	s	S=O	sym str	R$_2$SO$_2$, sulfone
1340–1180	w	N≡N	str	azide
1320–1210	s	C—O	str	carboxylic acid
1310–1250	s	C—O	str	benzoic ester, phthalic ester

Table 6.1 *(Continued)*

Range, cm^{-1}	Intensity^a	Group	Type of vibration^b	Comments
1305–1200	m	N—H	def	secondary amide, Amide III band
1300–1250	s	NO$_2$	sym str	—O—NO$_2$, nitrate
1300–1200	s	P=O	str	phosphoric ester, free P=O
1270–1150	s	C—O	str	—(O=)C—O—R in esters
1256–1232	s	C—O	str	CH$_3$COOR, acetic ester
~1250		C—O	str	methylene acetal
~1250		C—O	str	epoxide
~1250	vs	Si—CH$_3$	sym CH$_3$ def	Si(CH$_3$)$_3$, trimethylsilyl
1250–1150	vs	P=O	str	phosphoric ester, H-bonded P=O
1235–1212	s	C=S	str	(RO)$_2$C=S, thioketone
1230–1150	s	S=O	str	(RO)$_2$SO$_2$, sulfuric ester
1225–1175 1125–1090 1070–1000 (two)	w	C—H	i-p bend	*p*-substituted phenyl
1220–1020	m	C—N	str	aliphatic amine
1200–1145	s	S=O	str	ROSO$_2$R′, sulfonic ester
1200–1040		C—O	str	C—O—C—O—C, cyclic acetal (4–5 bands)
1200–1170	s	C—O	str	propionic and higher esters
1200–1000	s	C—OH	str	alcohols
1185–1175	s	C—O	str	formic ester
1175–1165 1170–1140	s	C—H	sk	(CH$_3$)$_2$C<, isopropyl
1175–1125 1110–1070 1070–1000	w	C—H	i-p bend	unsubstituted phenyl
1150–1100	s	S=O	asym str	R$_2$SO$_2$, sulfone
1150–1100	s	C—O	str	benzoic ester, phthalic ester
1150–1070	s	C—O—C	asym str	aliphatic ether
~1120	s	C=S	str	—NH—(C=S)—, thioamide
1110–1000	s	C—F	str	monofluoro derivatives
1090–1030	vs	P—O—C		phosphoric ester
1090–1020	vs	Si—O	str	Si—O—C, trimethylsilyl
1058–1053	s	C=S	str	(RS)$_2$C=S, trithiocarbonate
1050–1020	s	S=O	str	>S=O, sulfoxide
~1040		C—O	str	methylene acetal
1005–990 915–910	vs	C—H	bend	C=C—H, vinyl
995–985 910–905		C—H	o-o-p bend	RCH=CH$_2$
980–965		C—H	o-o-p bend	*trans* RHC=CHR′
970–940	br	P—O—P		pyrophosphate
965–960 945–940	s	C—H	bend	vinyl ether
960–930		N—O	str	oxime
950–810		C—O	str	epoxide

Table 6.1 *(Continued)*

Range, cm^{-1}	Intensity[a]	Group	Type of vibration[b]	Comments
~948	s	C—H	bend	vinyl ester
~925		C—O	str	methylene acetal
895–885		C—H	o-o-p bend	RR'C=CH$_2$
~840	vs	Si—CH$_3$	str	Si(CH$_3$)$_3$, trimethylsilyl
840–790		C—H	o-o-p bend	RR'C=CHR''
840–790	m	C—H	sk	(CH$_3$)$_2$C<, isopropyl
840–750		C—O	str	epoxide
833–810	vs	C—H	o-o-p bend	*p*-substituted phenyl
~800	w	NH$_3^+$	rock	amine salt
~800	w	NH$_2^+$	rock	
770–730 710–690	s	C—H	o-o-p bend	unsubstituted phenyl
~755	vs	Si—CH$_3$	str	Si(CH$_3$)$_3$, trimethylsilyl
750–700	s	C—Cl	str	monochloro derivatives
~720	m(br)	N—H	def	secondary amide, bonded: amide V band
705–570	w	C—S	str	thiol, sulfide
~690		C—H	o-o-p bend	*cis* RHC=CHR'
~650	s	C—Br	str	bromo derivatives
600–480	s	C—I	str	iodo derivatives
550–450	vw	S—S	str	disulfide

[a]br, broad; m, medium; s, strong; sh, sharp; v, very; var, variable; w, weak.
[b]asym, asymmetrical; def, deformation; i-p, in-plane; int eff, interaction effects; o-o-p, out-of-plane; sc, scissoring; sk, skeletal; str, stretching; sym, symmetrical.

C—H Bands: Stretching and Deformation Frequencies

All sugars and their derivatives have the methine (methylidyne) grouping —$\overset{|}{\underset{|}{C}}$—H, which exhibits a band at ~2900 cm^{-1}, assigned to C—H stretching. For example, all 28 cyclic acetals of various sugars displayed a band at 3010–2965 cm^{-1} (Tipson *et al.*, 1959).

A strong C—H stretching band of ≡C—H in acetylenic compounds is found at ~3255 cm^{-1}. In 5-hexyne-D-*lyxo*-1,2,3,4-tetrol tetraacetate this band occurs at 3300 cm^{-1} (Hurd and Jenkins, 1966) and in 1,2:3,4-di-*O*-isopropylidene-6-heptyne-D-*gulo*-1,2,3,4,5-pentol it occurs at 3215 cm^{-1} (Horton and Tronchet, 1966). The L-*manno* isomer of the latter compound has the band at 3268 cm^{-1}.

Sugar derivatives show other C—H stretching bands: olefinic =CH$_2$ at 3075 cm^{-1} and 2975 cm^{-1}; Ar —H at 3050 cm^{-1}; olefinic =$\overset{|}{C}$—H at 3020 cm^{-1}; —CH$_2$— at 2925 and 2850 cm^{-1}; and C—CH$_3$ at 2960 and 2865 cm^{-1}. In the case of *O*-methyl, 21 methyl aldopyranosides showed a C—H stretching frequency at 2882–2841 cm^{-1} (Tipson and Isbell, 1960*b*), not shown by C—CH$_3$ or C$_2$H$_5$O— groups

(Henbest *et al.*, 1957) that characterizes the glycosidic methoxyl group, regardless of the type of anomer or of substitution (or lack of it) at C-5.

Among the C—H deformation frequencies are the following: —CH_2— at 1450 cm^{-1} and C—CH_3 at 1375 cm^{-1}. The C—CH_3's of the isopropyl group have bands at 1380 and 1370 cm^{-1} that are quite useful for detecting the isopropylidene acetal structure. The unsubstituted phenyl group shows weak bands for C—H in-plane deformations at 1175–1125, 1110–1070, and 1070–1000 cm^{-1}; and strong bands for C—H out-of-plane deformations at 770–730 and 710–690 cm^{-1}.

The bands due to C—H in-plane and out-of-plane deformation in substituted phenyl groups vary depending on the position and degree of substitution. In *p*-substituted phenyl compounds, found very frequently among sugar derivatives, weak bands (in-plane deformations) occur at 1225–1175, 1125–1090, and 1070–1000 cm^{-1}, and a strong band (out-of-plane deformation) at 860–800 cm^{-1}. For example, the *p*-substituted phenyl group of *p*-toluenesulfonic esters of alditols (Tipson, 1952) and sugars (Guthrie and Spedding, 1960) has a hydrogen out-of-plane deformation at 830–810 cm^{-1}, not displayed by methanesulfonates.

The group CHR=CH_2 in the olefins has a weak band at 1420–1410 cm^{-1}, a band at 1300–1290 cm^{-1}, a medium band at 995–985 cm^{-1}, and a strong band at 915–905 cm^{-1}. A *cis* double bond exhibits a weak band at 1420–1400 cm^{-1} and a strong band at 730–665 cm^{-1}; a *trans* double bond has a weak band at 1325–1290 cm^{-1} and a strong band at 980–960 cm^{-1}. Thus, *trans*-3-hexene-D-*threo*-1,2,5,6-tetrol (Tipson and Cohen, 1966) has bands at 1325 and 976 cm^{-1}; its 1,2:5,6-di-*O*-iso-propylidene derivative has bands at 1307 and 971 cm^{-1}.

The C—H deformation vibrations at the anomeric carbon atom of various sugars have been identified by replacement of the hydrogen atom on C-1 with deuterium (Stacey *et al.*, 1958). Thus, to prepare α-D-glucopyranose-1-C-*d*, Stacey *et al.* (1958) dissolved D-glucono-1,5-lactone in deuterium oxide, reduced the carbonyl group to —CD(OD) with sodium amalgam in deuterium oxide, and then converted the OD
|

groups to OH groups by dissolving in water several times and evaporating; the C—D bond at C-1 did not change. The spectrum of the α-D-glucopyranose-1-C-*d* was then compared with that of α-D-glucopyranose. According to calculations using the simple harmonic-oscillator model, replacement of hydrogen by deuterium should cause the hydrogen deformation frequencies to be approximately $\sqrt{2}$ times the corresponding deuterium frequencies if the deformation is a pure bending or stretching vibration. Stacey *et al.* found corresponding C-1—H and C-1—D bands at 1375 and 1095 cm^{-1}, and at 1284 and 965 cm^{-1}. The ratios of the hydrogen-to-deuterium frequencies were 1.26 and 1.33, compared to the theoretical value of 1.414. They made similar assignments for the *β*-anomer and for the two anomers of other sugars.

N—H Bands: Stretching and Deformation Frequencies

Primary amines at low concentrations in a nonpolar solvent have two bands in the region 3500–3300 cm^{-1} arising from stretching vibrations of the amino group. In the presence of hydrogen bonding, or in the solid, the range is moved to 3500–3100

cm^{-1}. Dilute solutions of secondary amines have only one N—H stretching band, at 3400–3300 cm^{-1}.

Primary amines show an N—H deformation frequency at 1645–1550 cm^{-1}; for example, D-glucosylamine has a band at 1621 cm^{-1}, and 2-amino-2-deoxy-D-glucopyranose has one at 1600 cm^{-1} (Barker et al., 1956). The —NH— group has a band at 1580–1510 cm^{-1}. Primary amides show an NH_2 deformation band at 1650–1620 cm^{-1} for the solid, and at 1620–1590 cm^{-1} for solutions; this is called the amide II band. Secondary amides, having an NH group, display the amide II band at 1570–1515 cm^{-1} for the solid, and at 1550–1510 cm^{-1} for solutions. Hydrogen-bonded secondary amides have an NH deformation mode at $\sim 720\ cm^{-1}$, called the amide V band.

Sixteen 1-acetamido derivatives of sugars (Tipson and Isbell, 1961a) produced spectra with at least one band at 3356–3236 cm^{-1}, owing to N—H stretching, and, at 1575–1541 cm^{-1}, the amide II band. All of 60 1-acylamido derivatives (secondary amides) of aldofuranoid, aldopyranoid, and acyclic sugars (Tipson et al., 1967) had at least one band at 3460–3226 cm^{-1} arising from N—H stretching. In this region, completely esterified compounds could not be distinguished from those having free hydroxyl groups that would show O—H stretching in the same region. All the compounds had a band at 1575–1504 cm^{-1}, i.e., amide II. Two amide II bands were displayed for the acyclic 1,1-bis (acylamido)-1-deoxyalditols, possibly because the two acylamido groups on C-1 of these compounds are not equivalent. They may have a hydrogen-bonded structure, possibly of the following type (XIII), where R is Me, Et, or Ph; and R' is H, Ac, EtCO, or Bz:

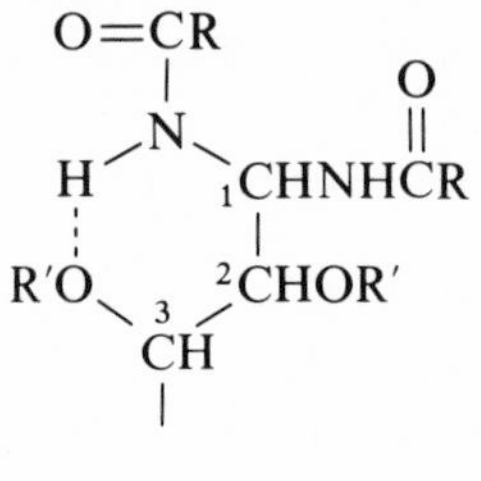

$$
\begin{array}{c}
O{=}CR \\
| \\
N \\
H \quad \diagdown \ _1CHNHCR \\
| \\
R'O \diagdown _3 \ ^2CHOR' \\
CH \\
|
\end{array}
\qquad \begin{array}{c} O \\ \| \end{array}
$$

XIII

O—H Bands

A free hydroxyl group in a compound has a stretching mode at 3730–3520 cm^{-1}. The O—H bond is weakened if the group is hydrogen-bonded, and the band is broadened, with a shift to lower frequency, 3520–3100 cm^{-1}. The band for O—H deformation is located at 1080–1030 cm^{-1}. The O—H frequencies are not useful for distinguishing primary from anomeric and secondary hydroxyl groups, since hydrogen bonding causes frequency shifts.

A recent review (Murthy and Rao, 1968) has discussed in detail the use of infrared and Raman spectroscopy, as well as NMR and electronic spectroscopy, for studying the hydrogen bond in many types of systems. Infrared spectra are frequently used for the detection of hydrogen bonding, which may be either intermolecular or

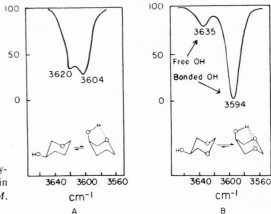

Fig. 6.1. Infrared spectra of (A) tetrahydropy-
ran-3-ol and of (B) 1,3-O-methyleneglycerol in
carbon tetrachloride; both solutions 0.005M.
(Barker *et al.*, 1959.)

intramolecular. For compounds soluble in carbon tetrachloride, intermolecular hydrogen bonding is negligible at concentrations below 0.005 M, and absorptions at
$\sim 3585\ \mathrm{cm}^{-1}$ and $\sim 3625\ \mathrm{cm}^{-1}$ may be associated with bonded and free hydroxyl
groups, respectively (Kuhn, 1952; 1954). By this means, the extent of intramolecular
hydrogen bonding may be measured (Barker *et al.*, 1959; Brimacombe *et al.*, 1958). In
pyranoid (and *m*- or *p*-dioxane) derivatives, a suitably located OH group can form a
hydrogen bond with the oxygen atom of the ring, as shown in Fig. 6.1. The spectra are
consistent with the equilibria shown, assuming that the molecules exist preferentially
in chair conformations. For compounds with low solubility in carbon tetrachloride,
the use of *p*-dioxane has been suggested (Åkermark, 1961). This solvent disrupts intermolecular hydrogen bonds and has no effect on strong intramolecular hydrogen bonds.

In solution, both penta-O-acetyl-*aldehydo*-D-galactose aldehydrol (XIV) and
the corresponding ethyl hemiacetal (XV) display a band (Isbell *et al.*, 1957) at 3597
cm^{-1} for free hydroxyl, and a band at 3483 cm^{-1} for hydrogen-bonded hydroxyl.

$$
\begin{array}{llll}
\mathrm{HO} & \mathrm{HO} & \mathrm{EtO} & \mathrm{EtO} \\
| & |\ \ \mathrm{O-H} & | & |\ \ \mathrm{O-H} \\
\mathrm{HCOH} & \mathrm{HC} & \mathrm{HCOH} \rightleftharpoons & \mathrm{HC} \\
| \rightleftharpoons & | \qquad \mathrm{O} & | & | \qquad \mathrm{O} \\
\mathrm{HCOCOCH_3} & \mathrm{HC}\ \ \mathrm{O-C} & \mathrm{HCOCOCH_3} & \mathrm{HC}\ \ \mathrm{O-C} \\
| & | \qquad\quad \mathrm{CH_3} & | & | \qquad\quad \mathrm{CH_3}
\end{array}
$$

XIV   XV

Tipson *et al.* (1962) examined the O—D bands in the spectra of fully O-deuterated
sugars. The OH band at 3390 cm^{-1} of α,β-D-glucose was shifted, when deuterated,
to 2469 cm^{-1}, a wave number ratio of 1.37, and other new bands appeared.

C≡C and C=C Bands

The stretching absorption of C≡C is weak and has limited diagnostic value; symmetrically disubstituted acetylenes do not show it. A weak stretching band for H—C≡C—R falls at 2150–2100 cm^{-1}; for example, 5-hexyne-D-*lyxo*-1,2,3,4-tetrol tetraacetate exhibits a band (Hurd and Jenkins, 1966) at 2150 cm^{-1}, and two 6-heptynepentol derivatives have a band (Horton and Tronchet, 1966) at 2120–2110 cm^{-1}. In the case of R—C≡C—R', the band lies at 2260–2190 cm^{-1}.

Phenyl-conjugated C=C bonds have a strong stretching band at ~1625 cm^{-1}; in CO or C=C conjugation, the band lies at ~1600 cm^{-1}. Nonconjugated C=C bonds show bands at 1680–1620 cm^{-1}. For olefins, a *cis* double bond in RCH=CHR' lies at 1662–1653 cm^{-1}, and a *trans* double bond at 1678–1653 cm^{-1}. Thus, *trans*-3-hexene-D-*threo*-1,2,5,6-tetrol has a band (Tipson and Cohen, 1966) at 1653 cm^{-1}. A terminal exocyclic double bond, as in H$_2$C=CRR', displays a band at 1658–1648 cm^{-1}, and H$_2$C=CHR displays one at 1648–1638 cm^{-1}.

The unsubstituted phenyl ring exhibits bands of medium variable intensity at 1625–1575, 1590–1575, 1465–1440, and 1525–1475 cm^{-1} for skeletal in-plane stretching vibrations of aryl C=C. Thus, the phenyl ring of the benzoyl group has bands at 1600 and 1584 cm^{-1}. For example, a group of 1-acylamido sugar derivatives containing *N*-benzoyl or *O*-benzoyl groups or both, have bands (Tipson *et al.*, 1967) at 1613–1600, 1587–1567, and 1506–1477 cm^{-1}; and the *p*-substituted phenyl ring of *p*-toluenesulfonic esters of alditols (Tipson, 1952) and sugars (Guthrie and Spedding, 1960) has an aryl C=C band at 1605–1600 cm^{-1} which distinguishes them from methanesulfonates.

C≡N, C=N, and C—N Bands

The C≡N stretching band is found at 2400–2100 cm^{-1}. For R—C≡N, it lies at 2260–2240 cm^{-1}, and for conjugated R—C≡N it is found at 2240–2215 cm^{-1}. For isonitriles, R—N≡C, the absorption is at 2200–2100 cm^{-1}. For thiocyanate, —S—C≡N, the band lies at ~2160 cm^{-1}.

Compounds having C=N— exhibit a band at 1660–1610 cm^{-1} that is useful in determining whether such compounds as *N*-substituted glycosylamines are cyclic or acyclic (Ellis, 1966) (see *Determination of Structure*, in this chapter). Care must be taken to dry the compound completely, since moisture gives a band at 1650–1600 cm^{-1}. Also, if water of crystallization is present, a band appears at 1650–1640 cm^{-1}. For isothiocyanate —N=C=S and carbodiimides —N=C=N— there is a strong band at ~2100 cm^{-1}.

Aliphatic amines have a band of medium intensity for C—N stretching at 1220–1020 cm^{-1} and a weak band at ~1410 cm^{-1}. Nitro compounds have a band of medium intensity for C—N stretching at 920–850 cm^{-1}, and primary amides at 1418–1400 cm^{-1}. Not so useful diagnostically are the C—N band at 1370–1310 cm^{-1} for the *N*-methyl group and the Ph—N stretching band, observed (Kübler *et al.*, 1960; Rosenthal and Weir, 1963) at 1149–1131 cm^{-1} or (Bassignana and Cogrossi, 1964) 1160–1130 cm^{-1} for phenylhydrazones and phenylazo derivatives.

C=O Bands

Aldehydes and Ketones

The carbonyl stretching frequency of aldehydes and ketones is found at 1730–1665 cm^{-1}. For example, the acyclic form of some aldoses and ketoses (in a lyophilizate of the equilibrium mixture of mutarotation) produces a very weak band (Tipson and Isbell, 1962) at 1718 cm^{-1}. Kuhn (1950) attributed the band at 1613 cm^{-1} exhibited by periodate-oxidized methyl α-D-glucopyranoside to aldehydic carbonyl. Periodate-oxidized cellulose has only a very weak band (Rowen et al., 1951), and exists mainly as the hemialdal —CH(OH)—O—CH(OH)— (Spedding, 1960), formed by hydration of two aldehyde groups per oxidized residue.

If the bands displayed by two different kinds of groups are located close together in the spectrum (see Table 6.1), the bands may be resolved, but often one is seen as a shoulder on the other, or one may hide the other. Thus, some anhydrous monomeric *aldehydo* sugar acetates show no strong band (Isbell et al., 1957) for the aldehyde carbonyl group, as it is probably hidden by the acetate carbonyl band. On the other hand, when other information is lacking, the CHO band can be mistaken for OAc or OBz. When there is no such interference, *aldehydo* sugars display the C=O band; for example, 3-*O*-benzyl-1,2-*O*-isopropylidene-α-D-*xylo*-pentodialdo-1,4-furanose shows a band at 1724 cm^{-1} (Wolfrom and Hanessian, 1962).

Un-ionized Carboxylic Acids

The C=O stretching frequency of the COOH of un-ionized carboxylic acids (Orr et al., 1952; Stevenson and Levine, 1952; Levine et al., 1953; and, Orr, 1954) lies at ~1736 cm^{-1}, including polysaccharides such as hyaluronic acid, the chondroitin sulfates, alginic acid, and pneumococcal capsular material.

Lactones

Barker et al. (1958) showed that 22 out of 24 aldono-1,4-lactones displayed a band at 1790–1765 cm^{-1}, and eleven aldono-1,5-lactones all had a band at 1760–1726 cm^{-1}. Thus, a spectrum with a strong 1785 cm^{-1} band suggests that the aldonolactone is 1,4, and a band at 1730 cm^{-1} indicates a possible 1,5 compound. However, a spectrum with a band at 1770–1755 cm^{-1} indicates the need for an additional method to distinguish between the two types. The 6,3-lactones of 1,2-*O*-isopropylidene-α-D-*gluco* (and β-L-*ido*)-furanuronic acid exhibit C=O stretching (Tipson et al., 1959) at 1789–1764 cm^{-1} (see also, Nitta et al., 1962).

Acetates and Other Esters

The C=O stretching vibration of the *O*-acetyl group shows strong absorption at 1748–1724 cm^{-1}. The octaacetates of α-cellobiose, α-gentiobiose, and β-maltose exhibit strong bands (Stephens, 1954) at 1727, 1748, and 1736 cm^{-1}, respectively. Also, six acetates of cyclic acetals of sugars had a band at 1748–1739 cm^{-1} (Tipson et al., 1959); 24 acetylated aldopyranosides had at least one band (Tipson and Isbell, 1960a) in the region 1764–1736 cm^{-1}; eight reducing pyranose acetates had a band (Tipson and Isbell, 1961a) at 1751–1736 cm^{-1}; 20 completely acetylated pyranoses had a band (Tipson and Isbell, 1961b) at 1757–1742 cm^{-1}; and for 14 acetates (and a tetrapropionate) of 1-acylamido derivatives of sugars, all showed a band (Tipson et al., 1967) at 1761–1742 cm^{-1}, except for N-acetyl-2,3,4-tri-O-acetyl-β-D-ribosylamine, showing a band (Tipson, 1961) at 1718 cm^{-1}. Benzoates of the same group of 1-acylamido derivatives had a band (Tipson et al., 1967) at 1745–1727 cm^{-1}, except for 1,1-bis (benzamido)-6-O-benzoyl-1-deoxy-D-glucitol, with a band at 1698 cm^{-1}. Acetate–benzoate mixed esters exhibited two bands in this region.

The five-membered cyclic carbonate group in sugar carbonates possesses an enhanced carbonyl stretching frequency compared with the mixed-ester carbonates of sugars, average values being at 1820 and 1760 cm^{-1}, respectively (Hough et al., 1960). If the cyclic carbonate is fused in a trans conformation in sugar derivatives, a strong carbonyl stretching band is found (Doane et al., 1967) at 1842 cm^{-1}; the band is not found for carbonates fused in the cis conformation.

Primary Amides

The C=O stretching frequency for $-\overset{\overset{\textstyle O}{\|}}{C}-NH_2$ (amide I) is found near 1650 cm^{-1} for solids, and near 1690 cm^{-1} for dilute solutions. The amide I band is also shown by secondary and tertiary amides. (See Chapters 8 and 10.) Three glycopyranuronamide derivatives displayed this band (Tipson and Isbell, 1961a) at 1667–1661 cm^{-1}.

N-Acetyl and S-Acetyl

The amide I band for the monosubstituted amide group, as in $-NH-\overset{\overset{\textstyle O}{\|}}{C}-CH_3$, lies at ~1655 cm^{-1} for solids, and at 1700–1670 cm^{-1} for dilute solutions. For example, the anomers of 2-acetamido-2-deoxy-D-glucopyranose and their tetraacetates have the band (Micheel et al., 1955) at 1675–1616 cm^{-1}. In 16 1-acetamido derivatives of sugars, the band is found (Tipson and Isbell, 1961a) at 1709–1661 cm^{-1}. Hydrates displayed bands at 1664–1642 cm^{-1} that either overlapped amide I bands in the same region, obscured them, or were hidden by them. Sixty 1-acylamido derivatives of sugars had a band (Tipson et al., 1967) at 1681–1626 cm^{-1}. The amide I band lies near 1648 cm^{-1} for such polysaccharides as chitin (Darmon and Rudall, 1950), but at 1618 cm^{-1} for 2-acetamido-2-deoxy-α-D-glucopyranose, and at 1665 cm^{-1} for its tetraacetate, probably because of differences in the hydrogen bonding of the carbonyl group to the OH and NH groups (Barker et al., 1956). Since an ionized carboxyl group absorbs in this region the spectra of polysaccharides like the chondroitin sulfates and hyaluronic acid should be recorded for films cast from acidic solution.

The carbonyl group in the S-acetyl group $-S-\overset{\overset{\textstyle O}{\|}}{C}-CH_3$, has a band (Horton and Wolfrom, 1962) at ~1680 cm^{-1}. The O-acetyl group, lying near 1740 cm^{-1}, and the N-acetyl group, absorbing near 1640 cm^{-1}, are thus readily distinguishable from the S-acetyl group.

C—O Bands

Esters

These compounds show strong bands for the C—O stretching vibrations; thus, acetates show absorption at $1250–1230\ cm^{-1}$, formates at $1200–1180\ cm^{-1}$, and propionates at $1200–1170\ cm^{-1}$. Esters of aromatic acids have two strong bands for C—O stretching at 1300–1250 and $1150–1100\ cm^{-1}$.

Carboxylate Ion

Carboxylic acid salts, such as the lithium and barium salts of 1,2-*O*-isopropyli-dene-α-D-glucofuranuronic acid display a strong C—O stretching band (Tipson *et al.*, 1959) at $1637–1600\ cm^{-1}$ that distinguishes COO⁻ from the C=O stretching band of esters, and a medium band at $1420–1300\ cm^{-1}$.

N≡N, N=N, and NO₂ Bands

Azides have a strong N≡N stretching band at $2160–2120\ cm^{-1}$ and a weak one at $1340–1180\ cm^{-1}$. The N=N group in aromatic diazo compounds shows bands at 1585–1570 and $1420–1381\ cm^{-1}$. Bassignana and Cogrossi (1964) reported the bands to lie at 1582–1558 and $1439–1379\ cm^{-1}$.

Nitric esters of sugars give strong bands for asymmetric NO₂ stretching (Guthrie and Spedding, 1960) at $1667–1613\ cm^{-1}$ and for symmetric NO₂ stretching at $1285–1267\ cm^{-1}$; a broad band (for O—NO₂ stretching) at $871–828\ cm^{-1}$ can be used only for confirmation since it lies in a region that has bands (Types 2a, 2c, and C) for sugars and the C—H out-of-plane deformation of the *p*-substituted phenyl group.

S=O, —SO₂—, and C=S Bands

Esters of sulfuric acid, such as the 6-sulfates of D-glucose, D-galactose, and 2-acetamido-2-deoxy-D-glucose, display a very strong absorption band (Lloyd and Dodgson, 1959) at $1240\ cm^{-1}$, caused by S=O stretching vibrations. A number of sulfated polysaccharides from mammals, cartilaginous fishes, and seaweeds display this band (Lloyd *et al.*, 1961). The position of a band at $850–820\ cm^{-1}$ for the C—O—S frequency of the sulfate group has been correlated with its attachment (primary or secondary) and with the axial or equatorial disposition if secondary (see *Conformational Studies*, this chapter). Note: Turvey *et al.* (1967), Peat *et al.* (1968), and Harris and Turvey (1970) have commented on the dangers inherent in using only infrared data [see Orr (1954), Lloyd and Dodgson (1961), and Lloyd *et al.* (1961)] to assign positions to sulfuric ester groups on sugar rings, and to sulfonic esters of pyranose derivatives [see Onodera *et al.* (1965)].

Sulfones display strong absorption at $1355–1310\ cm^{-1}$ and $1160–1110\ cm^{-1}$ for S=O stretching; for example, the monosulfone obtained by oxidizing penta-*O*-acetyl-*aldehydo*-D-glucose dibenzyl dithioacetal has a band at $1305\ cm^{-1}$ (Whiffen, 1957).

The *p*-toluenesulfonic esters of alditols (Tipson, 1952) have bands at 1370–1350 and $1190–1175\ cm^{-1}$ for the asymmetrical and symmetrical stretching modes of the —SO₂— group (Onodera *et al.*, 1965; Albano *et al.*, 1966). Sulfonic esters of sugars usually display two bands in each region (Guthrie and Spedding, 1960). Bands for C—O—S at 848–840 and 893–877 cm^{-1} were correlated with the axial or equatorial spatial arrangement of the sulfonic ester group (see *Conformational Studies*, this chapter).

The C=S group of thionocarbonates has bands (Carey *et al.*, 1966) at 1330 cm^{-1} and at 1307 cm^{-1}; a band lying at 1190 cm^{-1} also occurs (Albano *et al.*, 1966).

Dimethylthiocarbamates of sugars exhibit, for $-O\overset{\overset{\displaystyle S}{\|}}{C}NMe_2$, a strong band (Horton and Prihar, 1967) at 1540–1515 cm^{-1}. The N,N-dimethyldithiocarbamates have, for $-S\overset{\overset{\displaystyle S}{\|}}{C}NMe_2$, a strong band at 1490–1470 cm^{-1}.

Correlations for the Fingerprint Region and Beyond

The Fingerprint Region

Although the spectra of compounds with similar structures may have characteristics that are alike in the range of 5000 to 1430 cm^{-1} notable spectral differences occur in the fingerprint region from 1430 to 910 cm^{-1}; in this region the bands are caused by certain stretching as well as bending vibrations. Consequently, small structural differences are displayed as large spectral effects. Thus, the region is very useful for identification of a compound, and for distinguishing between isomers, including anomers.

However, the fingerprint region is often less useful for the recognition of certain organic groups than the region above 1430 cm^{-1}, where the presence or absence of bands may supply useful reliable evidence for characteristic groups. For example, the characteristic group frequencies for α-D- or α-L-glucopyranose are essentially the same as those for β-D- or β-L-glucopyranose; thus, their spectra above 1430 cm^{-1} are similar. However, for crystals in which the molecules have the favored chair conformation, the interactions of the hydroxyl group at C-1 with the ring-oxygen atom of a neighboring molecule should be different for the axially attached group of the α-D- or α-L-anomer and the equatorially attached group of the β-D- or β-L-anomer, resulting in the expectation that their spectra will be different in the fingerprint region. In the hydrogen-bonded structure proposed by McDonald and Beevers (1952) for crystalline α-D-glucopyranose, the bonding between the C-1 hydroxyl group of one molecule and the ring-oxygen atom of a neighbor would be different from that for the crystalline β-D-anomer. Also, the hydrogen-bonding arrangements in crystalline α- and β-D-glucose have been shown to be different from each other (Ferrier, 1960; Chu and Jeffrey, 1968; Brown and Levy, 1965), and there is one non-hydrogen-bonded OH group in β-D-glucose which does not occur in α-D-glucose, accounting for the differences in their infrared spectra. Figure 6.2 (Brown, 1970) shows the hydrogen bonds in the crystal structure of α-D-glucose. The combined use of X-ray crystallography, broad-line NMR, and infrared spectroscopy for analyzing such differences should ultimately lead to improved interpretation of the absorption frequencies in the fingerprint region and in the lower wave number range.

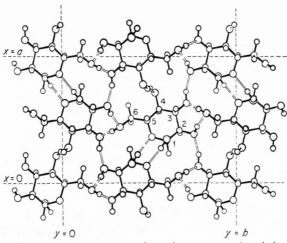

Fig. 6.2. The crystal structure of α-D-glucopyranose viewed along the [00Ī] axis. (There are two hydrogen bonds at each O-4, but one is omitted for clarity.) (Brown, 1970.)

Correlations in the Region 1000 to 667 cm⁻¹

Correlations for Certain Aldopyranose Derivatives. In the range 960 to 730 cm⁻¹, Barker *et al.* (1954*a,b,c*) have found bands characteristic of several aldopyranoses and their derivatives. They identified (1954*a*), for D-glucopyranose derivatives, three principal sets of bands, which are presented in Table 6.2. These were as follows: for α anomers, type 1a at 917 ± 13 cm⁻¹, type 2a at 844 ± 8 cm⁻¹, and type 3a at 766 ± 10 cm⁻¹; and for β anomers, type 1b at 920 ± 5 cm⁻¹, type 2b at 891 ± 7 cm⁻¹, and type 3b at 774 ± 9 cm⁻¹.

Table 6.2. Positions (Mean Values, cm⁻¹) of Various Types of Bands for D-Glucopyranose Derivatives (Barker *et al.*, 1954*a*)

Linkage	Type 1	Type 2a	Type 2b	Type 3
α Anomeric				
Monosaccharides	915	847		767
	900	842		751
Higher saccharides	930	843		761
	917	839		768
β Anomeric				
Monosaccharides	914		896	
	918[a]		891	772[b]
Higher saccharides	921		890	774[c]

[a]Six of ten compounds did not show this band.
[b]Eleven of sixteen compounds did not show this band.
[c]Five of sixteen compounds did not show this band.

If the bands are to be useful for distinguishing (D or L)-glucopyranose derivatives, α anomers should not display type 2b absorption, and β anomers should not display type 2a absorption. However, Barker *et al.* (1954*a*) found that (a) some α anomers show type 1 absorption in the region of type 2b bands, and (b) some β anomers exhibit "weak peaks of type 2a," which they thought were caused by traces of the α anomers. The type 2a band was found to be applicable with confidence to diagnosis of the α-anomeric form, particularly in polymers of glucopyranose. The type 2b band was not found useful to diagnose the β-anomeric form, but the *absence* of the type 2a band was applicable for the recognition of the β-anomeric form. Types 1 and 3 were useful only for determination of linkage points in α-glucopyranose polymers.

When they reported (Barker *et al.*, 1954*b*) the spectra of additional derivatives of glucopyranose, they found slightly different positions for type 2a and 3 bands (see Table 6.3). As before, some of the α anomers displayed type 1 bands in the range of

Table 6.3. Infrared Bands (Mean Values, cm^{-1}) Shown by Five (D or L)-Aldopyranoses and Their Derivatives (Barker *et al.*, 1954*a*,*b*)

Band type	Xylose	Arabinose	Glucose	Mannose	Galactose
		Both anomers			
1	?	?	917	?	?
2c	—	—	—	876	871
3	—	763	770^a 753^a	791^a	752^a
		α anomers only			
2a	—	—	844^a 843^a	833^a	825^a
3	749	—	—	—	—
		β anomers only			
2a	—	843 845	—	—	—
2b	—	—	891^b 890^b	893^c	895^c

^{a}Many derivatives containing a benzene ring absorb here.
^{b}Must be confirmed by absence of absorption at ∼844 cm^{-1}.
^{c}Bands for other types of vibration also occur here.

type 2b bands. Also, some derivatives having a phenyl ring may absorb in the region of the type 2a band, and the acetates may absorb in the region of the type 2b band. The findings (Barker *et al.*, 1954*b*) for bands characteristic of four other aldopyranoses and their derivatives are also summarized in Table 6.3. The type 2a band can be used for diagnosing α anomers in manno- and galactopyranose derivatives; absence of the type 2a band can be used to diagnose the β-anomeric form. A type 2c band at 876 ± 9 cm^{-1} was characteristic of mannopyranose derivatives, and a type 2c band at 871 ± 7 cm^{-1} was characteristic of galactopyranose derivatives. The mean frequency for a particular type of band may vary with the configuration of the group; for example, the mean for type 3 absorption lies at 791 cm^{-1} for the *manno* configuration, but at 752 cm^{-1} for the *gluco* and *galacto* configurations.

Barker *et al.* (1954*c*) also observed that 2- and 3-deoxy derivatives of gluco-, manno-, and galactopyranose absorb at 869–865 cm^{-1}; seven 6-deoxy derivatives of mannopyranose or galactopyranose absorb near 967 cm^{-1}.

Application of these correlations (Barker *et al.*, 1954*a,b,c*) has proved useful (Barker *et al.*, 1956) in the study of many related compounds, including oligo- and polysaccharides. Assignments suggested (Barker *et al.*, 1956; 1954*a,b,c*) for the bands are presented in Table 6.4. In the region 1000–667 cm^{-1}, methyl β-D-xylopyranoside has only three bands (Tipson and Isbell, 1960*b*), namely, at 976, 963, and 898 cm^{-1}; indeed, β-D- or β-L-xylopyranose derivatives cannot be characterized by any of the bands listed in Table 6.3. Thus, none of the bands listed in Tables 6.3 and 6.4 can be considered as characteristic of the pyranoid ring, *per se*, of aldopyranoid derivatives.

Table 6.4. Bands Possibly Characteristic of Various Features of Some Aldopyranose Derivatives (Barker *et al.*, 1956; 1954*a,b,c*)

Band type	Structural feature	Bands,[a] cm^{-1}	References
	Terminal *C*-methyl-group rocking[b]	967	*d*
1	Antisymmetrical ring-vibration[c]	917	*e*
2b	Anomeric C—H axial bond	891	*e*
2c	Equatorial C—H deformation (other than anomeric C—H)	880	*f*
	Ring-methylene rocking vibration (if not adjacent to the ring-oxygen atom)	867	*d*
2a	Anomeric C—H equatorial bond	844	*e*
3	Symmetrical ring-breathing vibration	770	*e*

[a]Mean value.
[b]This band may not have diagnostic value.
[c]For glucopyranose derivatives.
[d]Barker *et al.*, 1954*c*.
[e]Barker *et al.*, 1954*a*.
[f]Barker *et al.*, 1954*b*.

Correlations for Pyranoid and Furanoid Forms of Aldose and Ketose Derivatives. For aldo- and ketofuranose derivatives, Barker and Stephens (1954) found absorption frequencies at the following mean values: type A, 924 cm^{-1}; and type D, 799 cm^{-1}. Type A absorption was not distinguishable from types 1 or 2b of aldopyranoses, and therefore has no diagnostic value for differentiating between furanoid and pyranoid aldoses. Most of the furanoid compounds also displayed type B absorption at 879 cm^{-1} and type C absorption at 858 cm^{-1}. It has been observed (Tipson *et al.*, 1967; Tipson *et al.*, 1959; Tipson and Isbell, 1961*a*) that these correlations are, for the most part, restricted to the compounds they studied, and cannot be extended to have a wider diagnostic applicability to related compounds.

Tipson and Isbell (1962) recorded the spectra of most of the readily available, unsubstituted aldo- and ketopentoses and aldo- and ketohexoses. Shortly thereafter, Verstraeten (9164; see also, 1966) studied these spectra, included some additional

2-ketoses, and observed that most of the common sugars having a cyclic structure, and their derivatives, show type 1 absorption at a mean value of 929 cm^{-1}. Hence, the type 1 (type A) band is not useful for distinguishing between aldoses and ketoses, or between glycofuranoses and glycopyranoses. Besides, since the type 1 band appears (Tipson *et al.*, 1967) for acyclic 1-acylamido derivatives of sugars, it cannot be applied to distinguish between cyclic and acyclic forms of such compounds.

Verstraeten (1964) noted that some ketopyranoses, as well as aldopyranoses, display a type 3 band at 781 cm^{-1}. Thus, this band also has limited diagnostic value. He concluded that type 3 absorption is displayed provided that two conditions are met. First, the sugar must have a pyranoid ring, and second, this pyranoid form must assume a conformation having at least one axial hydroxyl group. If the number of axial hydroxyl groups is increased (resulting in decreased conformational stability), type 3 absorption appears. For example, β-D- or β-L-xylopyranose, which presents no type 3 absorption, has no axial hydroxyl groups, whereas the α anomer in the favored conformation, which has an axial hydroxyl group at C-1, displays a band at 760 cm^{-1}.

It was observed (Verstraeten, 1964; 1966) that 2-ketoses show "type I" bands at 875 cm^{-1} and "type IIA" bands at 817 cm^{-1}, regardless of whether the 2-ketoses are pyranose or furanose. These bands were ascribed to the presence of structural feature XVI and were tentatively assigned to a skeletal vibration. However, six aldoses also

$$\begin{array}{cc} -O & OH \\ \diagdown \mid & \mid \\ & C \\ \mid \diagup & \mid \\ \blacksquare C & CH_2OH \\ \mid & \end{array}$$

XVI

display these bands. The type I band, which appears to be the same as Barker's type B band for aldo- and ketofuranoses at 879 cm^{-1}, has no diagnostic value for 60 aldo-furanoid, aldopyranoid, and acyclic 1-acylamido derivatives (Tipson *et al.*, 1967). The type IIA band is found in about the same region as Barker's type D band for aldo- and ketofuranose derivatives, which is at 799 cm^{-1}. If the hydroxyl groups of a 2-ketofuranose are substituted, or if C-2 of the 2-ketofuranose is attached to a pyranoid or furanoid structure, a type IIB band is presented at 834 cm^{-1}, in addition to the type IIA band or instead of it.

Verstraeten found that only furanoses have "type 2" absorption at 850 cm^{-1}. He stated that his type 2 absorption is the same as the type C absorption of Barker and Stephens (1954), and to avoid confusion, it should be referred to as the latter. The type C band is displayed by both aldo- and ketofuranoses, and therefore cannot be used for distinguishing between them.

Tipson *et al.* (1967) found that, if an *N*-acetyl group (but no ester group) is present, the bands of types C, 3, IIA, and IIB may have diagnostic value; if an *N*-benzoyl group (but no ester group) is present, the bands of types 3, IIA, and IIB may have diagnostic value. For *N*-acetyl-*O*-acetyl derivatives of sugars, the bands of types

IIA and IIB may differentiate between ketoses and nonketoses, but not between cyclic and acyclic compounds.

Conformational Studies

In determining the conformations of sugars and their derivatives, the most direct information is obtained by NMR spectroscopy (Eliel *et al.*, 1965). However, the empirical correlation of infrared spectra has been used to give conformational information (Tipson and Isbell, 1960*b*). Suppose that the spectra of the α and β anomers of the methyl pyranosides of the four aldopentoses and eight aldohexoses were to be determined. There would be recorded 24 spectra of closely related compounds. Each compound has $C-H$, $C-OH$, $C-OCH_3$, and a pyranoid ring, and yet the spectrum of each is unique because the precise positions of the various bands change from one compound to another, depending on interactions arising from configuration and conformation and on whether or not there is a CH_2OH group present.

As an example, the spectra of the α and β anomers of methyl D-xylopyranoside and methyl L-arabinopyranoside were studied (Tipson and Isbell, 1960*b*). All the common bands shared by the four glycosides were ignored. All bands then shown in common by the two xylosides were regarded as characteristic of the *xylo* configuration and were ignored; similarly, all bands appearing in common for the two arabinosides were disregarded. This left a set of bands distinguishing between the anomers of the xylosides, on the one hand, and between the arabinoside anomers, on the other (see Table 6.5). This indicated a similarity between the β-D-xylopyranoside

and the α-L-arabinopyranoside. Inasmuch as the conformation of methyl β-D-xylopyranoside has been shown by X-ray studies (Brown, 1960) to be that shown in

Table 6.5. Anomer-Differentiating Bands, in cm^{-1}, Shown by Four Methyl Pyranosides (Tipson and Isbell, 1960*b*)

D-Xylo		L-Arabino	
β	α	α	β
3448		3460	
1385		1395	
1295		1295	
1218		1227	
1060		1058	
976		973	
645		646	
473		487	
	3333		3322
	2710		2695
	741		744
	437		433

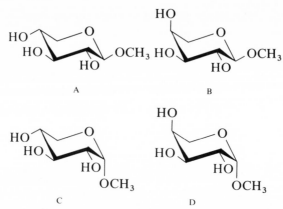

Fig. 6.3. The structures of (A) methyl-β-D-xylopyranoside-CA and (B) methyl α-L-arabinopyranoside-CE compared with those of (C) methyl α-D-xylopyranoside-CA and (D) methyl β-L-arabinopyranoside-CE.

Fig. 6.3, the conformational correlations are as indicated. These formulas are in agreement with the conformation predicted by considerations of interaction energies.

This correlation is purely empirical, but similar comparisons have been made for other pairs of anomers of methyl aldopyranosides (Tipson and Isbell, 1960b), acetylated methyl aldopyranosides (Tipson and Isbell, 1960a), and fully acetylated aldopyranoses (Tipson and Isbell, 1961b). In each case, the empirical correlations made from the spectra agreed with the predicted conformations. Those sugar derivatives for which one chair conformation is not predicted to be strongly favored over the other produced data that did not fit in the correlations. Examples are methyl α- and β-D-lyxopyranoside and their triacetates, methyl α-D-gulopyranoside and its tetraacetate, and penta-O-acetyl-α-D-gulopyranose.

In a group of fully acetylated monosaccharides, those containing an axial OAc at C-1 exhibited a band (Isbell et al., 1957), possibly for a C—O stretching vibration, at 1174–1153 cm^{-1}; if the group was equatorial, a band was displayed at 1127 cm^{-1}. For each region, the other anomer showed the band only weakly or not at all. The data for compounds having the gulo, ido, and talo configurations indicated that they exist in the 1C(D) or C1(L) conformation, or as a mixture of the chair conformations.

Acetylated methyl glycosides (Isbell et al., 1957) possessing an axial OMe at C-1 had bands at 1203–1198 and 1143–1130 cm^{-1}, whereas those with an equatorial OMe did not absorb in either range.

Sherman et al. (1968) have identified myo-inosose-2 (XVII) and scyllo-inositol (XVIII) in mammalian tissues. Scyllo-inositol trimethylsilyl ether has a simpler infrared spectrum than that of the myo-inositol derivative, a fact stated to be consistent with a structure in which all hydroxyls are in an equivalent configuration (equatorial) in the scyllo-compound. Bands present in trimethylsilyl-myo-inositol but not in the scyllo-form occur at 1170, 1130, 1080, 1049, 995, 945, 920, and 680 cm^{-1}.

Orr (1954) made some assignments for the chondroitin sulfates. (See note on p. 113.) Since polysylfated hyaluronic acid (Fig. 6.4), which has equatorial sulfate

OH
HO
HO
HO   OH
    OH

HO
HO   OH
HO   OH
    OH

XVII   XVIII

groups only, absorbs at 820 cm^{-1}, he concluded that the sulfate group of chondroitin sulfate C showing a band at 825 cm^{-1} is equatorially attached, and that the sulfate group of chondroitin sulfate A showing a band at 855 cm^{-1} is axially attached. He ascribed the bands to the C—O—S vibration. Lloyd and Dodgson (1961) later found that the equatorial sulfate group in D-glucose 3-sulfate has a band at 832 cm^{-1}, and that a band at 820 cm^{-1} is displayed by the 6-sulfates of D-galactose, D-glucose, and 2-acetamido-2-deoxy-D-glucose (N-acetyl-D-glucosamine), in which the ester group is on the equatorial primary hydroxyl group. Hence, chondroitin sulfate C (and D) probably has an equatorial sulfate group on C-6, and chondroitin sulfate A (and B) has an axial sulfate on C-4 of the 2-acetamido-2-deoxy-D-galactose residues. The chondroitin sulfates have also been studied by Meyer et al. (1956); Mathews (1958); and Hoffman et al. (1958).

Mathews (1958) has shown that chondroitin sulfates A and C can be distinguished by their infrared spectra (Fig. 6.5) in the 1000–700 cm^{-1} region. The C form has unique bands at 1000, 820, and 775 cm^{-1}; the A form has bands at 928, 852, and 725 cm^{-1}. Chondroitin sulfate B resembles A, with bands at 928, 855, 840, and 712 cm^{-1}.

The mucopolysaccharide from nuclei pulposi, called "chondroitin sulfuric acid B" (Orr, 1954), and the shark cartilage chondroitin sulfate (Nakanishi et al., 1956), may also be regarded as the C sulfate.

Suzuki and Strominger (1960) have indicated that the spectra of acetylgalactosamine 4-sulfate obtained from chondroitin sulfate A and B, and of acetylgalactosamine 6-sulfate from chondroitin sulfate of shark cartilage, corresponded respectively to those of chondroitin 4-sulfate and the 6-sulfate. Therefore, the spectral differences would be attributed to the presence of the C-4 and C-6 sulfate in the galactosamine moieties of chondroitin sulfates. Although chondroitin sulfate from shark cartilage has a high sulfur content of 7.6 percent (1.3 residues of sulfate per acetylgalactosamine residue), the spectra are identical to that of chondroitin C from chordoma (Mathews, 1958; Nakanishi et al., 1956).

In a report by Murata (1962) infrared spectroscopic data showing a high degree of sulfation for chondroitin polysulfate have been related to the anticoagulant activity of the polysulfate. (The chondroitin polysulfate had been prepared by sulfation of shark cartilage chondroitin sulfate with chlorosulfonic acid and pyridine.) Chondroitin polysulfate has sulfate groups at both C-4 and C-6.

Chondroitin sulfates B and C, a small amount of a glucosamine-containing sulfated mucopolysaccharide, and hyaluronic acid were found in human umbilical cord (Danishefsky and Bella, 1966). Characterization of the chondroitin sulfate B was based on characteristic bands at 928, 860, and 840 cm^{-1}; chondroitin sulfate C was identified by bands at 825 and 1000 cm^{-1}.

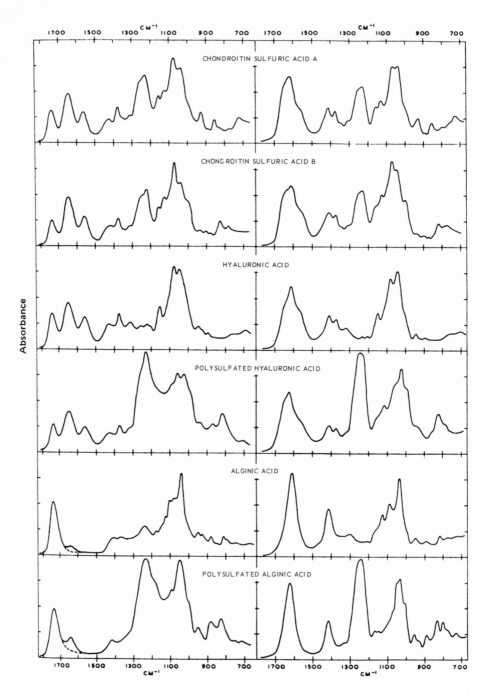

Fig. 6.4. The infrared spectra of the polysaccharides, on the left as free carboxylic acids, and on the right as salts. (Orr, 1954.)

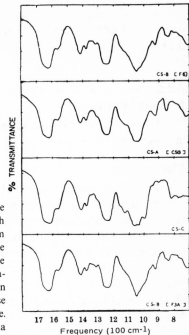

Fig. 6.5. Infrared spectra for isomeric chondroitin sulfates. The sodium salt of each polysaccharide (1.0 mg) was mixed with 150 mg of potassium bromide, formed into a film of 0.45 mm thickness (nominal) and compared with a potassium bromide blank. The A sulfate was prepared from bovine nasal cartilage (Mathews and Dorfman, 1953). Chondroitin sulfate B preparations were obtained by fractionation of a sample of β-heparin from bovine lung (Marbet and Winterstein, 1951). One of these fractions (F-6, top spectrum), contained 20% of the A sulfate. The C sulfate was a preparation from a human chordoma (Meyer *et al.*, 1956). (Mathews, 1958.)

Scheinthal and Schubert (1963) obtained fractionated materials from alkaline or trypsin degradation of the proteinpolysaccharide light fraction (PP-L) from cartilage and from alkaline degradation of proteinpolysaccharide heavy fraction (PP-H), and found absorption bands at 1648 and 1410 cm^{-1} (carboxylate) and 1230 to 1255 cm^{-1} (sulfate) and at 928, 852, and 725 cm^{-1} (axial 4-sulfate). Certain other fractions from alkaline degradation of PP-L and PP-H showed bands at 940, 890, 817, and 775 cm^{-1}, which were assigned to an equatorial 6-sulfate configuration.

Hunt and Jevons (1966) have isolated a polysaccharide (glucan) sulfate from the hypobranchial mucin of the whelk *Buccinum undatum*. The infrared spectra of the glucan sulfate contained intense absorptions at 1240 cm^{-1} and over the range 840–800 cm^{-1}. These two bands disappeared entirely on desulfation, indicating ester sulfate (Hoffman *et al.*, 1958), and other bands (940, 920, and 1000 cm^{-1}) also disappeared. Hunt and Jevons also attribute these to sulfate disappearance.

Terho *et al.* (1966) have investigated a sulfated mucopolysaccharide of human gastric juice. The intensity of the 1260–1230 cm^{-1} band characterized the degree of sulfation of two fractions that had been isolated. Absorption bands at 820 and 775 cm^{-1} found in the spectra of polysaccharide from gastric juice are also present in those of chondroitin sulfate C. Terho *et al.* therefore thought that the sulfate groups might occupy the C-6 position in the gastric mucopolysaccharide.

Rees (1963) found λ-carrageenan to have a broad band at 860–810 cm^{-1}, with a maximum at 827 cm^{-1} which was compatible with the presence of residues of D-galactose 2,6-disulfate (di-equatorial) and D-galactose 4-sulfate (axial).

Sulfonic esters of pyranoid sugars exhibited a strong band at 848–840 cm^{-1} and a weak one at 893–877 cm^{-1}. These bands have been ascribed (Onodera et al., 1965) to the C—O—S vibration of an equatorial and an axial sulfonic ester group, respectively, on a pyranoid ring.

Additional Examples of Applications of Infrared Spectra to Carbohydrates

In addition to those already mentioned, the following are some examples of ways in which infrared spectra may be applied to carbohydrates.

Mutarotation

Tipson and Isbell (1962) used an elementary approach in studying the mutarotation of sugars. For several monosaccharides, an aqueous solution of one crystalline anomer was kept until mutarotation was complete; the solution was then lyophilized, and the spectrum of the lyophilizate was compared with those of the two crystalline anomers. For D-glucose, L-rhamnose, and D-mannose, all the bands in the spectrum of the equilibrium mixture that are not displayed by the α-pyranose anomer are either shown by the β-pyranose anomer or could be produced by overlapping and summation of closely situated bands of the two anomers, indicating that the equilibrium mixture is composed of the anomers of the pyranose form. This conclusion confirms the results of optical rotation studies by Isbell and Pigman (1937). On the other hand, the lyophilizate of D-talose had a spectrum showing bands absent from the spectrum of either anomer of the pyranose form. These observations also agreed with those of the earlier work by Pigman and Isbell (1937) (namely, that the equilibrium mixture contains the α- and β-pyranose and the α- and β-furanose forms) and have since been confirmed and quantified by NMR spectroscopy (Angyal and Pickles, 1967).

Parker (1960) has measured the mutarotation of several sugars quantitatively. In his paper are recorded spectra of 20% aqueous solutions (in H$_2$O, not D$_2$O) of α-D-glucose, β-D-glucose, and β-D-mannose at: (a) 2.5 minutes after dissolution, and (b) at the end of mutarotation. The kinetics of such changes were calculated from changes in absorbance values (at 1143 cm^{-1} for α- and β-D-glucose; and at 1163 cm^{-1} for β-D-mannose) for 10% solutions observed continuously from 2.5 minutes after dissolution until the end of mutarotation. The reason for the use of 10% solutions was that previous workers (Hudson and Dale, 1917; Hudson and Sawyer, 1917) had used solutions of 10% or less in polarimetric studies. The mutarotation constants he determined agreed well with those calculated from the changes in optical rotation by the earlier workers.

Enzymic Studies Involving Carbohydrate

Infrared spectroscopic studies of glucose oxidase (Parker, 1962a), invertase (Parker, 1962b), and β-glucosidase (Parker, 1964) in aqueous solutions have been reported. In the case of the latter two enzymes, kinetic studies were done (see also, Parker, 1967). The substrate used with β-glucosidase was phenyl β-D-glucoside. Rate constants were calculated from absorbance data of the free-glucose band at 1149 cm^{-1}.

Rate constants for the inversion of sucrose were calculated from absorbance values at 990 cm^{-1}. (See Chapter 15.)

Determination of the Extent of Methylation

A widely used method for following the progress of methylation is a measurement of the methoxyl content of the material (Laver and Wolfrom, 1962). This can be accurately determined and is of considerable value. The infrared spectrum is also of great use (for example, Kuhn *et al.*, 1958; Wolfrom *et al.*, 1963; Danishefsky *et al.*, 1963; Bouveng *et al.*, 1963). Methylation is carried out until absorption displayed by the products in the region of 3600 to 3400 cm^{-1} remains constant. The usefulness of this method has been strongly emphasized by Wallenfels *et al.* (1963); as little as 1 % of unmethylated material can be detected and the procedure is applicable to substances in solution in dry carbon tetrachloride or in the solid state with sodium chloride, or it may be used directly with syrups.

Determination of Structure

The determination of the structure of D-talose monobenzoate is an example of the use of infrared spectroscopy for such purposes. This compound was obtained (Pigman and Isbell, 1937) as a byproduct from the action of peroxybenzoic acid on D-galactal. It was thought that it might be a 1,2-(orthobenzoate) (XIX). Therefore its spectrum was compared (Isbell *et al.*, 1956) with that of 1,2-*O*-(1-methoxyethylidene)-L-rhamnose (XX)

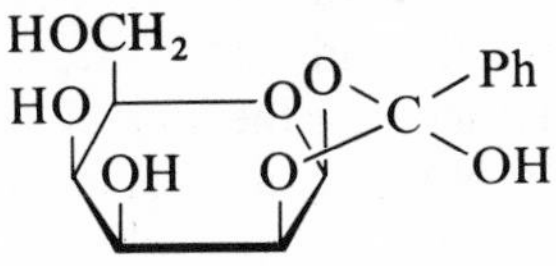

XIX 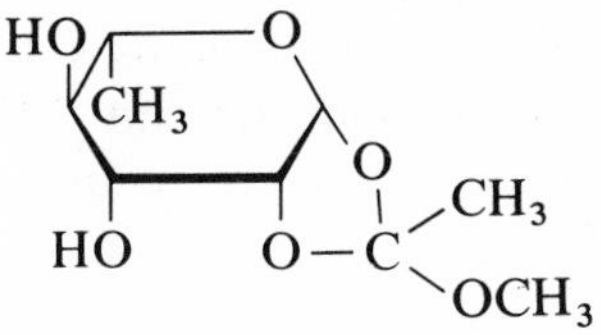   XX

As may be seen from Fig. 6.6, the latter compound has no band at 1733 cm^{-1}, whereas the D-talose monobenzoate has a strong ester carbonyl band there. Hence, the compound of interest here is not an orthobenzoate but a benzoate. It was thought to be 1-*O*-benzoyl-α-D-talose, but was later shown (Wood and Fletcher, 1957) to be the β-D-anomer (XXI).

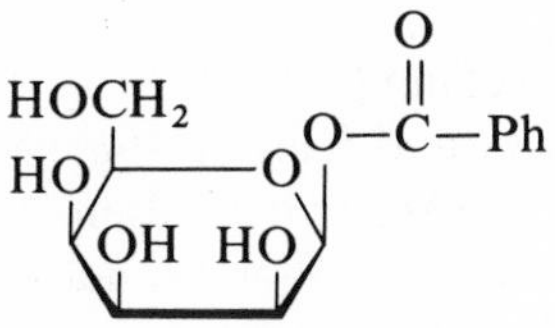

XXI

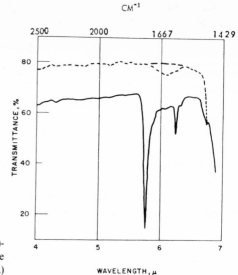

CM⁻¹

Fig. 6.6. Infrared spectra of 1,2-O-(1-methoxyethylidene)-
L-rhamnose (– – – –) and 1-O-benzoyl-β-D-talopyranose
(———) in potassium chloride pellets. (Isbell *et al.*, 1956.)

For sugars in which the hetero-atom of the ring may be nitrogen, the infrared spectra show immediately which form has this structure. For example, for the two ring forms of 5-acetamido-5-deoxy-L-arabinose (Jones and Turner, 1962), one form exhibits bands at 3300, 1630, and 1555 cm⁻¹ for OH and NH, NAc, and NH, respectively, and is therefore the furanose form (XXII), whereas the other form has bands at 3380 cm⁻¹ for OH, and 1616 and 1595 cm⁻¹ for NAc, but no NH absorption near 1555 cm⁻¹ and is therefore the pyranose (XXIII).

XXII   XXIII

In another example, 4-acetamido-4,5-dideoxy-2,3-O-isopropylidene-*aldehydo*-L-xylose has bands (El-Ashmawy and Horton, 1965) at 3226 cm⁻¹ (NH), 1721 cm⁻¹ (CHO), 1639 and 1538 cm⁻¹ (NHAc), and 1370 cm⁻¹ (CMe₂). On treatment of this compound (XXIV) with acetic acid, 4-acetamido-4,5-dideoxy-α,β-L-xylofuranose (XXV) was obtained; this compound, which cannot exist in a pyranoid form, exhibits carbonyl absorption at 1634 cm⁻¹ (amide I).

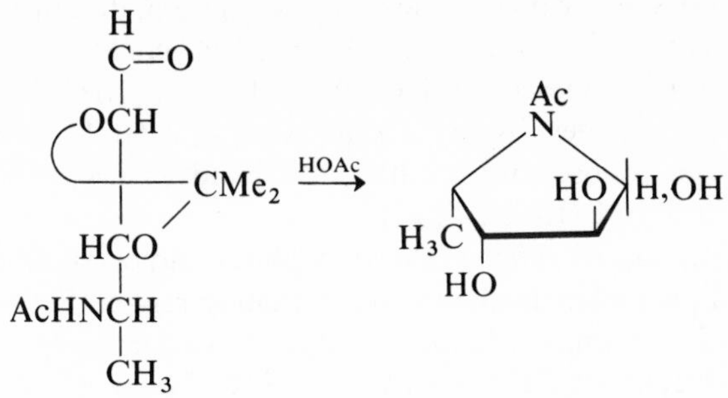

XXIV

XXV

The Schiff-base structure (XXVI) was proposed (Legay, 1952) for N-o-tolyl-D-glucosylamine, because it exhibited a C=N band at 1653 cm⁻¹. However, the pure compound shows no band (Ellis, 1966) at 1653 cm⁻¹, indicating that the structure is cyclic, probably the pyranoid form (XXVII). All the spectra of N-substituted glycosylamines examined by Ellis (1966) indicated the compounds to be cyclic.

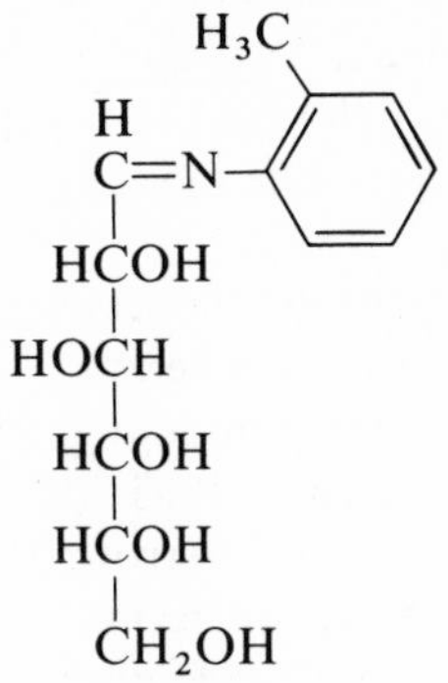

XXVI

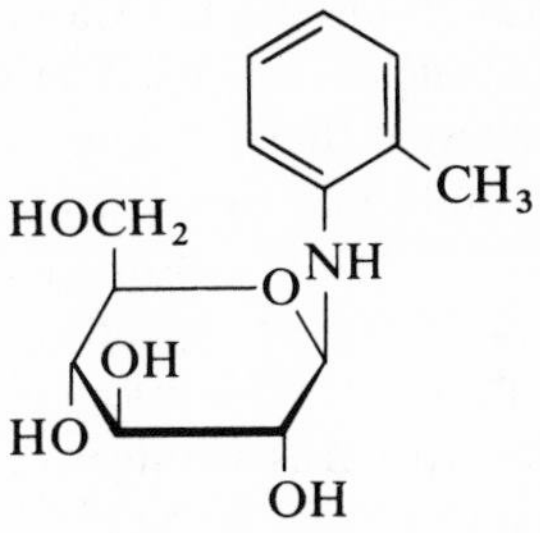

XXVII

The oximes of arabinose, rhamnose (Legay, 1952), and fructose (Bredereck *et al.*, 1956) display a weak band at 1653 cm⁻¹, which may indicate the acyclic form, but it is possible that the N—H of the cyclic form

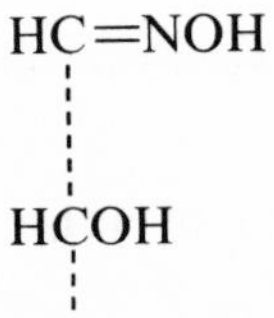

acyclic

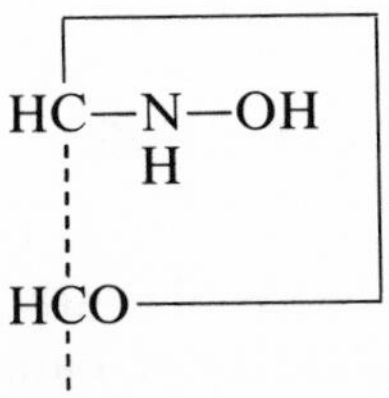

cyclic

might have a weak band of about the same frequency. The D-glucose oxime does not have a band in this region (Legay, 1952; Bredereck *et al.*, 1956) and is, presumably, cyclic. The acetyl derivatives of sugar oximes are undoubtedly cyclic, since they have the characteristic bands for the N-acetyl group (Bredereck, 1956).

Phenylosazones of sugars are known to exist preponderantly in the acyclic form, and they show the C=N band (Otting, 1961) at 1587 cm^{-1}.

Acyclic semicarbazones usually exhibit a band at 1675–1645 cm^{-1} for C=N. Acyclic thiosemicarbazones display a weak band at 1650–1630 cm^{-1}; the thiosemicarbazones of seven aldoses did not have this band, and were therefore considered to have a cyclic structure (Holker, 1964).

Cellulose I, II, and III differ in the amorphous and crystalline regions, and have different spectral properties in the O—H stretching region. For example, cellulose I exhibits five bands (Brown *et al.*, 1951) in the OH region, and when cellulose film is treated with deuterium oxide vapor (Almin, 1952; Marrinan and Mann, 1954), the intensity of one of these, at 3600–3000 cm^{-1}, rapidly decreases and an O—D band at 2700–2400 cm^{-1} appears. The band affected is the OH stretching of the amorphous regions. The four bands that remain belong to hydrogen-bonded hydroxyl groups in the crystalline region. The ratio of the intensities of the OH and OD bands then gives an indication of the proportion of hydroxyl groups that are hydrogen-bonded in a crystalline manner, and this ratio thus provides a measure of the crystallinity (Mann and Marrinan, 1956).

The directions of the hydroxyl groups were determined by means of plane-polarized infrared radiation (Mann and Marrinan, 1958*a,b*; Tsuboi, 1957) (See *Specialized Techniques* below). The OH band at 3309 cm^{-1} was "perpendicular," and it was suggested that some of the hydroxyl bonds lie along the chain direction and form intramolecular hydrogen bonds.

The spectrum of starch shows that the hydroxyl groups are extensively hydrogen-bonded (Samec, 1953). The spectrum of potato starch is different from that of corn starch, particularly in absorption regions for oxygen-containing groups. Thus, in corn starch absorption is stronger (Samec, 1957) at 1681, 1053–952, and 855 cm^{-1}, whereas in potato starch the band at 926 cm^{-1} is stronger. Arrowroot, corn, potato, rice, and wheat starches exhibit bands at 3333, 2105, and 1626 cm^{-1}. When the water content of the starch is changed, the band at 3333 cm^{-1} undergoes a change characteristic of the particular starch, and this property may be used to identify and classify starches (Yovanovitch, 1961).

The crystal of α-D-glucopyranose shows one band at 3405 cm^{-1} and four others in the region of 3347 to 3204 cm^{-1}. These have been correlated (Marrinan and Mann, 1954) with the results of X-ray diffraction studies, which show (McDonald and Beevers,

$$1952)\ \text{that there is one O—H}\cdots\text{O} \overset{\displaystyle \diagup\ C}{\underset{\displaystyle \diagdown\ C}{}}\ \text{bond (between two molecules) and that there}$$

are four O—H$\cdots$OH bonds. Similarly, the directions of the hydroxyl groups in sucrose have been determined (Ellis and Bath, 1938).

Specialized Techniques

Polarized Spectra and Infrared Dichroism of Carbohydrates

When a substance absorbs one of the components of polarized radiation more strongly than the other the substance is said to exhibit dichroism. Absorption of polarized radiation by molecules at a characteristic frequency depends on changes in dipole moments due to particular intramolecular vibrations. The absorption at a characteristic band is proportional to the interaction between the dipole change vector **M** and the electric vector **P** of the vibration. For those bands depending on the direction of **P**, the absorption is maximum when **P** and **M** are parallel and zero when **P** and **M** are perpendicular. Consequently, knowing the groups that a polymer contains and the characteristic wavelengths associated with the vibration of those groups, one can make significant deductions for those substances that display infrared dichroic properties. Thus, when a group displays dichroism with respect to one orientation of the infrared radiation, it may be assumed that there is orientation in the molecular aggregate.

Hyaluronate. Synthetic polymers and polysaccharides have been subjected to dichroism studies for structural information (Liang and Marchessault, 1959*b*; Pearson *et al.*, 1960). Quinn and Bettelheim (1963) used such methods on sodium hyaluronate isolated from umbilical cords. Figures 6.7a and 6.7b show the spectrum of an oriented sodium hyaluronate film which has been elongated by 40%. Table 6.6 contains band assignments made by these workers based on data reported in the literature.

They replotted as absorbance data the spectra of the parallel and perpendicular polarized radiation of the hyaluronate films at various values of elongation

Table 6.6. Assignment and Polarization of Infrared Absorption Bands[a] (Quinn and Bettelheim, 1963)

Frequency, cm^{-1}	Polarization	Interpretation
905	Perpendicular	Ring stretching (Bellamy, 1956)
960	Perpendicular	C—O stretching (Bellamy, 1956)
1055	Perpendicular	C—O—H stretching and bending (Orr, 1954)
1090	Parallel	C—O stretching (Liang and Marchessault, 1959*b*)
1155	Parallel	Asymmetric bridge oxygen stretching (Liang and Marchessault, 1959*b*)
1230	Parallel	Acetyl group (Bellamy, 1956)
1330	Perpendicular	Amide III (Bellamy, 1956)
1395	Perpendicular	CH bending and symmetric CH_3 deformation (Pearson *et al.*, 1960)
1420	Parallel	Ionized carboxyl (Bellamy, 1956)
1560	Perpendicular	Amide II, ionized carboxyl (Bellamy, 1956)
1625	Perpendicular	CO of *N*-acetylamine group (Orr, 1954)
1655	Parallel	Amide I (Pearson *et al.*, 1960)
2900	Perpendicular	CH stretching (Pearson *et al.*, 1960)
2940	Perpendicular	CH_3 stretching (Pearson *et al.*, 1960)
3300	Perpendicular	NH stretching (Pearson *et al.*, 1960)
3480	Perpendicular	OH stretching (Pearson *et al.*, 1960)

[a]The polarizations given in the table represent that orientation of the electric vector which was absorbed more strongly at the highest elongation, 40 %.

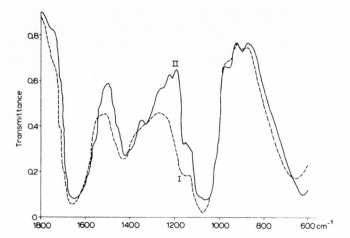

Fig. 6.7a. Infrared absorption spectra of sodium hyaluronate films, 40% elongation, from 1800 to 600 cm^{-1}. Electrical vector of polarized infrared radiation parallel (——) and perpendicular (– – – –) to the direction of elongation. (Quinn and Bettelheim, 1963.)

[absorbance equals log(1/transmittance)]. They resolved the assigned bands by assuming Gaussian distribution curves in each case, and plotted the resolved bands as absorbance index "a" *versus* frequency [$a = 2.303d \times \log(1/\text{transmittance})$, where d is film thickness (in cm) obtained by microscope]. Figure 6.8 is an example of such a plot. The area under the curves was determined by a polar planimeter and the perpendicular dichroic ratios were calculated from the relationship $D_\perp = a_\perp/a_\parallel$. Perpendicular dichroic ratios were calculated for the bands at 1090, 1155, 1420, 1655, and 2940 cm^{-1}.

Quinn and Bettelheim pointed out that although for a random orientation of polymer chains a dichroic ratio of one is expected, not all their absorption bands displayed this ratio at zero elongation. Small differences in the transmittance of parallel and perpendicular polarized radiation were observed in some of the bands, indicating a slight degree of preferential orientation during the preparation of the film. In order to relate the average degree of orientation of a specific group of atoms

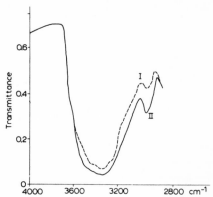

Fig. 6.7b. Infrared absorption spectra of sodium hyaluronate films, 40% elongation, from 4000 to 2800 cm^{-1}. Electrical vector of polarized infrared radiation parallel (——) and perpendicular (– – – –) to the direction of elongation. (Quinn and Bettelheim, 1963.)

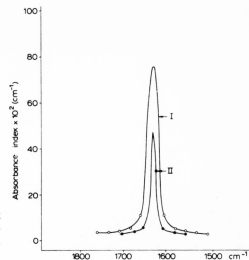

Fig. 6.8. Resolution of the band at $1625 \, cm^{-1}$ at 40% elongation. Parallel and perpendicular polarized absorbance *vs* frequency of radiation. (Quinn and Bettelheim, 1963.)

within the polymer to the degree of elongation, the dichroic ratios obtained at the different elongations were normalized so that each band was assigned a dichroic ratio of one at zero elongation. These normalized perpendicular dichroic ratios are plotted *versus* elongation in Fig. 6.9, which shows the different degree of orientation of the different atomic groups upon stretching. The greatest orientation and the fastest was observed with the $1155 \, cm^{-1}$ band which was assigned to the antisymmetric

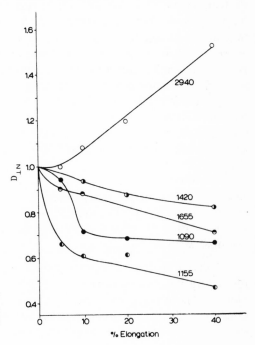

Fig. 6.9. Normalized perpendicular dichroic ratio of sodium hyaluronate as a function of elongation for different absorption bands. (Quinn and Bettelheim, 1963.)

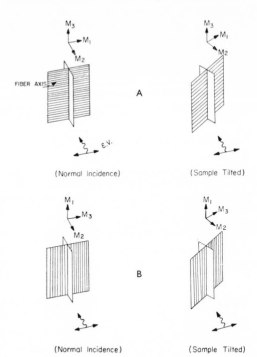

Fig. 6.10. Tilting effects of doubly oriented films showing settings for observing: (A) parallel tilting spectrum, and (B) perpendicular tilting spectrum. The vectors are: E.V., the electrical vector (perpendicular to slit); M_1, the transition moment vector parallel to chain axis; M_2, the vector perpendicular to doubly oriented film surface; and M_3, the vector perpendicular to chain axis and parallel to film surface. (Marchessault *et al.*, 1960.)

stretching vibration of the glycosidic oxygen bridge. Since the perpendicular dichroic ratio decreases with elongation, the vibrations of this group of atoms oriented their dipole-change vector in the direction of stretch. The vector sum of the antisymmetric vibration of the C—O—C bridge is along the long geometrical axis of the polymer molecule. Thus, hyaluronate has stretch properties similar to those of other polymers in the orientation of their long axis in the direction of stretch.

For the 2940 cm^{-1} band the dichroic ratio increases with elongation, indicating that the vibrations of the CH_3 groups of the *N*-acetylamine side chain become progressively more perpendicular to the direction of stretch. The vibrations of C—O, COO^- and amide C=O oriented at a lesser degree parallel to the direction of elongation.

The orientations of the various atomic groups in the hyaluronate molecule with elongation were interpreted as indicating a rather stiff molecular configuration of the chains in the gel form. The elongation was believed to be mainly due to the turning and orientation of the molecular aggregates rather than to uncoiling of a random isotropic chain configuration.

Chitin and Chitan. Marchessault *et al.* (1960) have examined the "tilting" infrared spectra of two chitin samples: crab chitin crystallites and larval cuticle chitin of the blowfly. The general principles for interpreting and the practical procedures for obtaining "tilted" spectra of polymers have been discussed by Liang and Marchessault (1960) and by Pearson *et al.* (1960). Figure 6.10 presents schematically the tilting effects of doubly oriented films with settings for observing a parallel tilting spectrum and a perpendicular tilting spectrum. A tilting spectrum is obtained when the normal to the sample surface is at some angle θ ($0° < \theta < 70°$) relative to the incident beam. A "perpendicular tilting spectrum" is obtained when the fiber axis is parallel to the slit. A "parallel tilting spectrum" is obtained when the fiber axis is perpendicular to the slit. An increase in an absorption band due to tilting is called a "positive tilting effect" and a decrease is a "negative tilting effect."

Figures 6.11 and 6.12 show "tilting" spectra for crab chitin crystallites and blow-fly larval chitin. For the former substance the 100 crystallographic plane is parallel to the film surface while for the latter the 001 plane is parallel to the film surface. Combining the knowledge of these orientations and the tilting effects found, Marchessault *et al.* (1960) confirmed certain amide band assignments made earlier [amide I (1652 cm^{-1}), amide II (1555 cm^{-1}), amide III (1310 cm^{-1}), and amide V (730 cm^{-1})]. They also made suggestions for interpreting other bands, e.g., 3106 cm^{-1} for NH stretching mode; 2962 cm^{-1} for CH$_3$ symmetrical stretching mode; and 1619 cm^{-1} for C=N stretching mode of enol form, or more likely, the amide I band of a rotational isomer of the aminoacetyl group (rotated 180° about the bond linking the nitrogen atom to the sugar ring).

Falk *et al.* (1966) reported the results of chemical, infrared, X-ray diffraction, and NMR studies which established the composition and structure of chitan, and provided information on its macrostructure and the nature of the difference between this polysaccharide and arthropod chitin. Chitan is *pure homogeneously* crystalline β-(1→4)-linked 2-acetamido-2-deoxy-D-glucan, whereas chitin is said to be a mixture

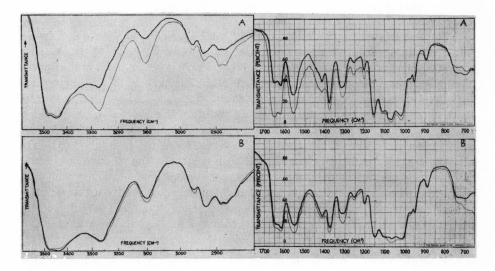

Fig. 6.11. Tilting spectra for crab chitin crystallites (A) and blowfly larval chitin (B) with electric vector and fiber axis both perpendicular to slit. (A) ——, normal incidence; – – – –, sample tilted 60° (by rotating about a vertical axis parallel to the slit). Film thickness $\simeq 6\mu$. (B) ——, normal incidence; – – – –, sample tilted 50°. Film thickness $\simeq 8\mu$. (Marchessault *et al.*, 1960.)

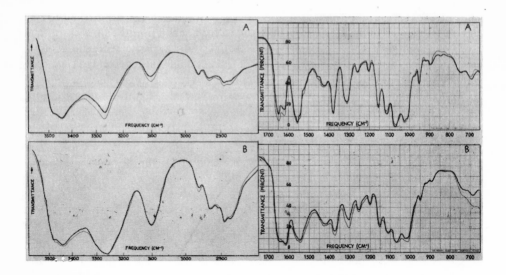

Fig. 6.12. Tilting spectra for crab chitin crystallites (A) and blowfly larval chitin (B) with electric vector parallel to slit and fiber axis perpendicular to slit. (A) ——, normal incidence; – – – –, sample tilted 45°. (B) ——, normal incidence; – – – –, sample tilted 50°. (Marchessault *et al.*, 1960.)

of polysaccharides which contain variable amounts of *N*-acetylglucosamine and glucosamine. (A striking feature of all preparations of chitin, according to Falk *et al.*, is the deviation of the elemental analyses from the theoretical values, and they reserve the name chitin as a group name for those polysaccharides which contain glucosamine in addition to *N*-acetylglucosamine.) The infrared spectra of chitan and chitin are quite different from each other (Fig. 6.13). These workers have given a detailed infrared analysis of their spectra.

It should be noted that the spectrum of chitin reflects mainly the macrostructure of the fully *N*-acetylated regions, because the spectral contributions of a small number of glucosamine residues are lost in the presence of broad and intense bands caused by the bulk of the polysaccharide. Falk *et al.* suspect previous conclusions about the macrostructure of chitin derived from X-ray diffraction and infrared studies, and feel that in any case they are not relevant to the structure of chitan.

The conclusion that chitan, but not chitin, is completely crystalline is supported by hydrogen–deuterium exchange experiments. When mats of chitan were immersed in D_2O at 21°C, only about 25% of the hydroxyl protons and 6% of those in the NH groups were exchanged after 1 month. Lithium thiocyanate treated mats exchanged many times faster (hot 50% LiSCN solutions transform chitan to chitin). Analogous experiments with cellulose (Mann and Marrinan, 1956) have established that the rate of deuterium exchange is much faster for amorphous than for crystalline samples.

Cellulose and Xylan. Detailed assignments have been made of bands in the O—H stretching region of the polarized spectra of the polysaccharides, cellulose I and

II (Mann and Marrinan, 1958; Liang and Marchessault, 1959a; Marchessault and Liang, 1960) and xylan (Marchessault and Liang, 1962) on the basis of their observed dichroisms. However, studies of mixtures of normal and partly deuterated cellulose I suggested that in the spectrum of cellulose I some bands in the O—H stretching region arose from coupled vibrations (Mann and Marrinan, 1958). Such coupling can affect the dichroic behavior of bands.

In the C—H stretching region (3000–2800 cm^{-1}) of the spectrum of cellulose, there are a number of partially overlapping bands some of which cannot be assigned on the basis of group frequency. The C—H bonds of the ring are known to be approximately perpendicular to the chain axis; therefore, the dichroism of the bands associated with them must also be perpendicular. Consequently, the perpendicular bands at ~ 2900 cm^{-1} have been assigned (Marchessault, 1962) to those vibrations. The 2853 cm^{-1} band of cellulose has been assigned to the symmetrical stretch for CH$_2$, and a band at 1430 cm^{-1} to the CH$_2$ symmetrical bending mode.

Marchessault and Liang (1962) have studied the spectra of xylan polysaccharides and applied polarization techniques in an analysis of the structures. For example, Fig. 6.14 shows a polarized infrared spectrum of oriented 4-O-methyl-D-glucuronoxylan. The frequency, relative intensity, and polarization of the absorption bands are given in Table 6.7. From Fig. 6.14 it is seen that the dipole-moment changes of the various CH stretching modes are predominantly perpendicular to the molecular axis. This finding agrees with the hypothetical chain conformation shown in Fig. 6.15 in which the CH$_2$ symmetric and asymmetric and the CH stretching modes are perpendicular to the molecular axis. The CH$_2$ symmetric bending mode at 1465 cm^{-1} also has this polarization.

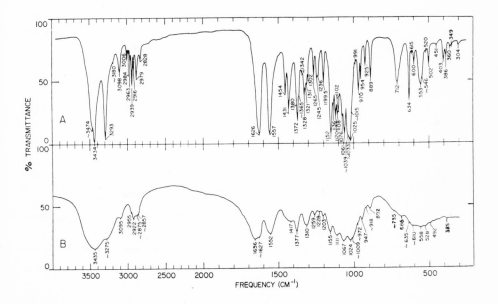

Fig. 6.13. The infrared absorption spectrum of chitan (A) and chitin (B). (Falk *et al.*, 1966.)

Table 6.7. Infrared Spectrum of 4-O-Methyl-D-glucuronoxylan (Marchessault and Liang, 1962)

Frequency, cm^{-1}		Relative		
Xylan	Xylan-OD	intensity[a]	Polarization[b]	Interpretation
650	650	m	⊥ ⎫	
700–800		sh	⊥ ⎭	OH out-of-plane bending
897	897	m	‖	C-1 group frequency or ring frequency
980	990	s	‖	
1045		s	‖ ⎫	
1085	1035–1100	s	‖ ⎬	C—O stretching
1115		s	‖ ⎭	
1130	1130	s	‖	Antisymmetric in-phase ring stretching
1165	1160	s	‖	Antisymmetric bridge C—O—C stretching
1215	1220	vw	⊥	
1245		w	‖	OH in-plane bending
	1260	w	‖ ⎫	
	1270	sh	? ⎬	CH bending
1275		vw	⊥ ⎭	
1317		w	⊥	OH in-plane bending?
	1335	w	‖	CH$_2$ wagging
1350		w	‖ ⎫	
1365		sh	‖ ⎭	OH in-plane bending
1390	1390	m	‖	CH bending
1425	1425	w	‖	$-C\overset{\displaystyle O}{\underset{\displaystyle O}{<}}$ symmetric stretching (salt)
				C—O stretching of free acid
1440		sh	?	
1465	1465	w	⊥	CH$_2$ symmetric bending
∼1600	1600	sh	u	$-C\overset{\displaystyle O}{\underset{\displaystyle O}{<}}$ antisymmetric stretching (salt)
1640		m	u	Water of hydration
1725	1725	m	u	C=O stretching (acid)
	2500	s	υ	OD stretching
2853	2853	w	⊥	CH$_2$ symmetric stretching
2873	2873	m	⊥ ⎫	
2914	2914	m	⊥ ⎭	CH stretching
2935	2935	sh	⊥	CH$_2$ antisymmetric stretching
2957	2957	sh	⊥ ⎫	
2975	2975	w	u ⎭	CH stretching
∼3430		vs	?	OH stretching

[a]m, medium; s, strong; sh, shoulder; v, very; w, weak.
[b]u, unpolarized; ?, not known; ‖, parallel; ⊥, perpendicular.

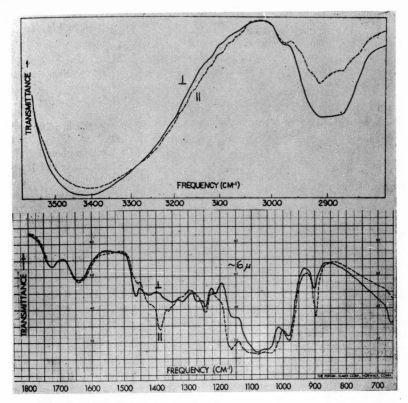

Fig. 6.14. Polarized infrared spectrum of oriented 4-O-methyl-D-glucuronoxylan. (Marchessault and Liang, 1962.)

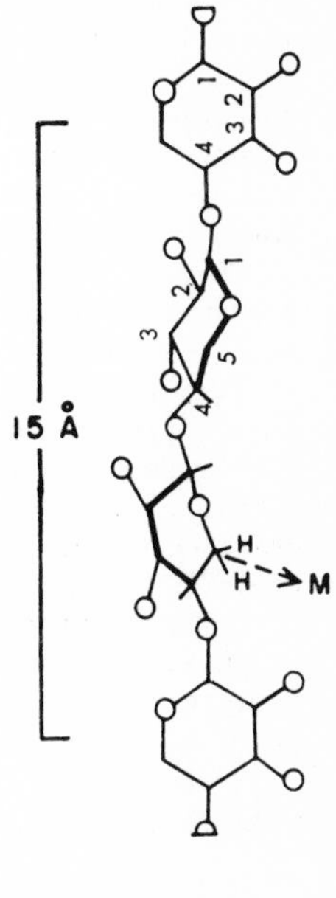

Fig. 6.15. Tentative molecular conformation and repeat unit of the xylan molecule in the hydrate unit cell. (Marchessault and Liang, 1962.)

Low-Temperature Spectroscopy of Carbohydrates

Katon *et al.* (1967; 1969) have reported on the infrared transmission spectroscopy of carbohydrates and other biochemical compounds at liquid-nitrogen temperature ($-196°C$) and liquid-hydrogen temperature ($-253°C$). They found spectra of improved quality at both these temperatures. The technique is useful for identification, characterization, and differentiation of complex compounds of biological interest; and although the effects of temperature could not be explained quantitatively, they appear to be related to hydrogen-bonding effects. Zhbankov *et al.* (1966) have recorded the spectra of several sugars at liquid-nitrogen temperature. Parker (1969) has used the ATR technique to study the infrared spectra of carbohydrates and other substances of biological interest at liquid-nitrogen temperature, and also has found the resolution of the spectra to be better than that at room temperature. The studies by Katon *et al.*, Zhbankov *et al.*, Parker, and others are described in greater detail in *Low Temperature Work* (Chapter 3).

References

Albano, E., Horton, D., and Tsuchiya, T. *Carbohyd. Res.* **2**, 349 (1966).

Almin, K. E. *Svensk. Papperstidn.* **55**, 767 (1952).

Alpert, N. L., Keiser, W. E., and Szymanski, H. A. *IR—Theory and Practice of Infrared Spectroscopy*, 2nd Ed. Plenum Press, New York, 1970.

Åkermark, B. *Acta Chem. Scand.* **15**, 985 (1961).

Angyal, S. J. and Pickles, V. A. *Carbohyd. Res.* **4**, 269 (1967).

Barker, S. A. and Stephens, R. *J. Chem. Soc.* **1954**, 4550.

Barker, S. A., Bourne, E. J., Stacey, M., and Whiffen, D. H. *J. Chem. Soc.* **1954***a*, 171.

Barker, S. A., Bourne, E. J., Stephens, R., and Whiffen, D. H. *J. Chem. Soc.* **1954***b*, 3468.

Barker, S. A., Bourne, E. J., Stephens, R., and Whiffen, D. H. *J. Chem. Soc.* **1954***c*, 4211.

Barker, S. A., Bourne, E. J., and Whiffen, D. H. *Methods Biochem. Anal.* **3**, 213 (1956).

Barker, S. A., Bourne, E. J., Pinkard, R. M., and Whiffen, D. H. *Chem. Ind. (London)* **1958**, 658.

Barker, S. A., Brimacombe, J. S., Foster, A. B., Whiffen, D. H., and Zweifel, G. *Tetrahedron* **7**, 10 (1959).

Bassignana, P., and Cogrossi, C. *Tetrahedron* **20**, 2361 (1964).

Bellamy, L. J. *The Infra-red Spectra of Complex Molecules*, Methuen, London, 1956.

Bellamy, L. J. *The Infrared Spectra of Complex Molecules*, 2nd Ed., Wiley, New York, 1958.

Bellamy, L. J. *Advances in Infrared Group Frequencies*, Methuen, London, 1968.

Bentley, F. F., Smithson, L. D., and Rozek, A. L. *Infrared Spectra and Characteristic Frequencies*, $\sim700-300\ cm^{-1}$, Interscience, New York, 1968.

Bouveng, H. O., Kiessling, H., Lindberg, B., and McKay, J. *Acta Chem. Scand.* **17**, 1351 (1963).

Brand, J. C. D. and Eglinton, G. *Applications of Spectroscopy to Organic Chemistry*, Oldbourne Press, London (in U.S.A., Daniel Davey and Co., Inc.), 1965.

Bredereck, H., Wagner, A., Hummel, D., and Kreiselmeier, H. *Chem. Ber.* **89**, 1532 (1956).

Brimacombe, J. S., Foster, A. B., Stacey, M., and Whiffen, D. H. *Tetrahedron* **4**, 351 (1958).

Brown, C. J. *Acta Cryst.* **13**, 1049 (1960).

Brown, G. M., and Levy, H. A. *Science* **147**, 1038 (1965).

Brown, G. M. Private communication cited in: G. Strahs, "Crystal-Structure Data for Simple Carbohydrates and Their Derivatives," in *Advances in Carbohydrate Chemistry and Biochemistry*, Vol. 25, (R. S. Tipson and D. Horton; eds.). Academic Press, New York, 1970, p. 53.

Brown, L., Holliday, P., and Trotter, I. F. *J. Chem. Soc.* **1951**, 1532.

Brügel, W. *An Introduction to Infrared Spectroscopy* (translated from the German), Methuen, London, 1962.

Carey, F. A., Ball, D. H., and Long, L., Jr. *Carbohyd. Res.* **3**, 205 (1966).

Chu, S. S. C. and Jeffrey, G. A. *Acta Cryst.* **B24**, 830 (1968).

Colthup, N. B., Daly, L. H., and Wiberley, S. E. *Introduction to Infrared and Raman Spectroscopy*, Academic Press, New York, 1964.

Conley, R. T. *Infrared Spectroscopy*, Allyn and Bacon, Boston, 1966.

Cross, A. D. *Introduction to Practical Infrared Spectroscopy*, 2nd Ed., Butterworths, Washington, D.C., 1964.

Danishefsky, I. and Bella, A., Jr. *J. Biol. Chem.* **241**, 143 (1966).

Danishefsky, I., Eiber, H. B., and Williams, A. H. *J. Biol. Chem.* **238**, 2895 (1963).

Darmon, S. E. and Rudall, K. M. *Discussions Faraday Soc.* **9**, 251 (1950).

Doane, W. M., Shasha, B. S., Stout, E. I., Russell, C. R., and Rist, C. E. *Abstracts of the 153rd National Meeting of the American Chemical Society, 1967, Abstract C14.*

El-Ashmawy, A. E. and Horton, D. *Carbohyd. Res.* **1**, 164 (1965).

Eliel, E. L., Allinger, N. L., Angyal, S. J., and Morrison, G. A. *Conformational Analysis*, Interscience, New York, 1965, p. 398.

Ellis, G. P. *J. Chem. Soc.* (B)**1966**, 572.

Ellis, J. W. and Bath, J. *J. Chem. Phys.* **6**, 221 (1938).

Falk, M., Smith, D. G., McLachlan, J., and McInnes, A. G. *Can. J. Chem.* **44**, 2269 (1966).

Ferrier, W. G. *Acta Cryst.* **13**, 678 (1960).

Flett, M. St. C. *Characteristic Frequencies of Chemical Groups in the Infrared*, Elsevier, London, 1963.

Guthrie, R. D. and Spedding, H. *J. Chem. Soc.* **1960**, 953.

Harris, M. J. and Turvey, J. R. *Carbohyd. Res.* **15**, 51 (1970).

Henbest, H. B., Meakins, G. D., Nicholls, B., and Wagland, A. A. *J. Chem. Soc.* **1957**, 1462.

Hoffman, P., Linker, A., and Meyer, K. *Biochim. Biophys. Acta* **30**, 184 (1958).

Holker, J. R. cited in H. Spedding, *Advan. Carbohyd. Chem.* **19**, 23 (1964); see p. 40 and footnote 51 on p. 37.

Horton, D. and Prihar, H. S. *Carbohyd. Res.* **4**, 115 (1967).

Horton, D. and Tronchet, J. M. *J. Carbohyd. Res.* **2**, 315 (1966).

Horton, D. and Wolfrom, M. L. *J. Org. Chem.* **27**, 1794 (1962).

Hough, L., Priddle, J. E., Theobald, R. S., Barker, G. R., Douglas, T., and Spoors, J. W. *Chem. Ind. (London)* **1960**, 148.

Hudson, C. S. and Dale, J. K. *J. Am. Chem. Soc.* **39**, 320 (1917)

Hudson, C. S. and Sawyer, H. L. *J. Am. Chem. Soc.* **39**, 470 (1917).

Hunt, S. and Jevons, F. R. *Biochem. J.* **98**, 522 (1966).

Hurd, C. D. and Jenkins, H. *Carbohyd. Res.* **2**, 240 (1966).

Isbell, H. S. and Pigman, W. W. *J. Res. Natl. Bur. Std.* **18**, 141 (1937).

Isbell, H. S., Stewart, J. E., Frush, H. L., Moyer, J. D., and Smith, F. A. *J. Res. Natl. Bur. Std.* **57**, 179 (1956).

Isbell, H. S., Smith, F. A., Creitz, E. C., Frush, H. L., Moyer, J. D., and Stewart, J. E. *J. Res. Natl. Bur. Std.* **59**, 41 (1957).

Jones, J. K. N. and Turner, J. C. *J. Chem. Soc.* **1962**, 4699.

Jones, R. N. *Infrared Spectra of Organic Compounds: Summary Charts of Principal Group Frequencies*, National Research Council, Ottawa, Canada, 1959.

Jones, R. N. and Sandorfy, C. in *Chemical Applications of Spectroscopy: Technique of Organic Chemistry Vol. 9* (W. West, ed.), Interscience, New York, 1956, Chapter 4.

Katon, J. E., Miller, J. T., Jr., and Bentley, F. F. *Arch. Biochem. Biophys.* **121**, 798 (1967).

Katon, J. E., Miller, J. T., Jr., and Bentley, F. F. *Carbohyd. Res.* **10**, 505 (1969).

Kübler, R., Lüttke, W., and Weckberlin, S. *Z. Elektrochem.* **64**, 650 (1960).

Kuhn, L. P. *Anal. Chem.* **22**, 276 (1950).

Kuhn, L. P. *J. Am. Chem. Soc.* **74**, 2492 (1952).

Kuhn, L. P. *J. Am. Chem. Soc.* **76**, 4323 (1954).

Kuhn, R., Baer, H. H., and Seeliger, A. *Ann. Chem.* **611**, 236 (1958).

Laver, M. and Wolfrom, M. L. *Methods Carbohyd. Chem.* **1**, 454 (1962).

Legay, F. *Compt. Rend.* **234**, 1612 (1952).

Levine, S., Stevenson, H. J. R., and Kabler, P. W. *Arch. Biochem. Biophys.* **45**, 65 (1953).

Liang, C. Y. and Marchessault, R. H. *J. Polymer Sci.* **37**, 385 (1959*a*).
Liang, C. Y. and Marchessault, R. H. *J. Polymer Sci.* **39**, 269 (1959*b*).
Liang, C. Y. and Marchessault, R. H. *J. Polymer Sci.* **43**, 85 (1960).
Lloyd, A. G. and Dodgson, K. S. *Nature* **184**, 548 (1959).
Lloyd, A. G. and Dodgson, K. S. *Biochim. Biophys. Acta* **46**, 116 (1961).
Lloyd, A. G., Dodgson, K. S., Price, R. G., and Rose, F. A. *Biochim. Biophys. Acta* **46**, 108 (1961).
Mann, J. and Marrinan, H. J. *Trans. Faraday Soc.* **52**, 481, 487, 492 (1956).
Mann, J. and Marrinan, H. J. *J. Polymer Sci.* **27**, 595 (1958*a*).
Mann, J. and Marrinan, H. J. *J. Polymer Sci.* **32**, 357 (1958*b*).
Marbet, R., and Winterstein, A. *Helv. Chim. Acta* **34**, 2311 (1951).
Marchessault, R. H. *Pure Appl. Chem.* (*London*) **5**, 107 (1962).
Marchessault, R. H., and Liang, C. Y. *J. Polymer Sci.* **43**, 71 (1960).
Marchessault, R. H., and Liang, C. Y. *J. Polymer Sci.* **59**, 357 (1962).
Marchessault, R. H., Pearson, F. G., and Liang, C. Y. *Biochim. Biophys. Acta* **45**, 499 (1960).
Marrinan, H. J. and Mann, J. *J. Appl. Chem.* (*London*) **4**, 204 (1954).
Mathews, M. B. *Nature* **181**, 421 (1958).
Mathews, M. B. and Dorfman, A. *Arch. Biochem. Biophys.* **42**, 41 (1953).
McDonald, T. R. R., and Beevers, C. A. *Acta Cryst.* **5**, 654 (1952).
Meyer, K., Davidson, E., Linker, A., and Hoffman, P. *Biochim. Biophys. Acta* **21**, 506 (1956).
Micheel, F., van de Kamp, F. P., and Wulff, H. *Chem. Ber.* **88**, 2011 (1955).
Miller, F. A. in *Organic Chemistry, An Advanced Treatise, Vol. 3* (H. Gilman, ed.), Wiley, New York, 1953, Chapter 2.
Murata, K. *Nature* **193**, 578 (1962).
Murthy, A. S. N., and Rao, C. N. R. *Appl. Spectrosc. Rev.* **2**, 69 (1968).
Nakanishi, K. *Infrared Absorption Spectroscopy—Practical*, Holden–Day, San Francisco, 1962.
Nakanishi, K., Takahashi, N., and Egami, F. *Bull. Chem. Soc. Japan* **29**, 434 (1956).
Nitta, Y., Ide, J., Momose, A., and Kuwada, M. *Yakugaku Zasshi* **82**, 790 (1962); *CA* **57**, 14601*i* (1962).
Onodera, K., Hirano, S., and Kashimura, N. *Carbohyd. Res.* **1**, 208 (1965).
Orr, S. F. D. *Biochim. Biophys. Acta* **14**, 173 (1954).
Orr, S. F. D., Harris, R. J. C., and Sylvén, B. *Nature* **169**, 544 (1952).
Otting, W. *Ann.* **640**, 44 (1961).
Parker, F. S. *Biochim. Biophys. Acta* **42**, 513 (1960).
Parker, F. S. *Perkin–Elmer Instrument News* **13**(4), 1 (1962*a*).
Parker, F. S. *Federation Proc.* **21**, 243 (1962*b*).
Parker, F. S. *Nature* **203**, 975 (1964).
Parker, F. S. in *Progress in Infrared Spectroscopy, Vol. 3* (H. A. Szymanski, ed.), Plenum Press, New York, 1967, p. 75.
Parker, F. S. *Mid-America Symposium on Spectroscopy*, Chicago, Illinois, Paper 119, May, 1969.
Parker, F. S. and Ans, R. *Appl. Spectrosc.* **20**, 384 (1966).
Pearson, F. G., Marchessault, R. H., and Liang, C. Y. *J. Polymer Sci.* **43**, 101 (1960).
Peat, S., Bowker, D. M., and Turvey, J. R. *Carbohyd. Res.* **7**, 225 (1968).
Pigman, W. W. and Isbell, H. S. *J. Res. Natl. Bur. Std.* **19**, 189 (1937).
Quinn, F. R. and Bettelheim, F. A. *Biochim. Biophys. Acta* **69**, 544 (1963).
Randall, H. M., Fowler, R. G., Fuson, N., and Dangl, J. R. *Infrared Determination of Organic Structures*, Van Nostrand, Princeton, New Jersey, 1949.
Rao, C. N. R. *Chemical Applications of Infrared Spectroscopy*, Academic Press, New York, 1963.
Rees, D. A. *J. Chem. Soc.* **1963**, 1821.
Rosenthal, A. and Weir, M. R. S. *J. Org. Chem.* **28**, 3025 (1963).
Rowen, J. W., Forziati, F. H., and Reeves, R. E. *J. Am. Chem. Soc.* **73**, 4484 (1951).
Samec, M. *Die Stärke* **5**, 105 (1953).
Samec, M. *J. Polymer Sci.* **23**, 801 (1957).
Scheinthal, B. M. and Schubert, M. *J. Biol. Chem.* **238**, 1935 (1963).
Sherman, W. R., Stewart, M. A., Simpson, P. C., and Goodwin, S. L. *Biochemistry* **7**, 819 (1968).

Simon, W. and Clerc, T. *Strukturaufklärung Organischer Verbindungen Mit Spektroskopischen Methoden*, Akademische Verlagsgesellschaft, Frankfurt am Main, 1967. (Combined tables of IR, UV, NMR, and mass spectra correlations.)

Spedding, H. *J. Chem. Soc.* **1960**, 3147.

Stacey, M., Moore, R. H., Barker, S. A., Weigel, H., Bourne, E. J., and Whiffen, D. H. *Proc. 2nd U.N. Intern. Conf. Peaceful Uses At. Energy, Geneva, 1958* **20**, 251.

Stephens, R., Ph.D. Thesis, University of Birmingham, England (1954).

Stevenson, H. J. R. and Levine, S. *Science* **116**, 705 (1952),

Suzuki, S. and Strominger, J. L. *J. Biol. Chem.* **235**, 2768 (1960).

Terho, T., Hartiala, K., and Häkkinen, I. *Nature* **211**, 198 (1966).

Tipson, R. S. *J. Am. Chem. Soc.* **74**, 1354 (1952).

Tipson, R. S. *J. Org. Chem.* **26**, 2462 (1961).

Tipson, R. S. *Infrared Spectroscopy of Carbohydrates*, National Bureau of Standards Monograph 110, Washington, D.C., 1968.

Tipson, R. S. and Cohen, A. *Carbohyd. Res.* **1**, 338 (1965).

Tipson, R. S. and Isbell, H. S. *J. Res. Natl. Bur. Std.* **64A**, 405 (1960*a*).

Tipson, R. S. and Isbell, H. S. *J. Res. Natl. Bur. Std.* **64A**, 239 (1960*b*).

Tipson, R. S. and Isbell, H. S. *J. Res. Natl. Bur. Std.* **65A**, 31 (1961*a*).

Tipson, R. S. and Isbell, H. S. *J. Res. Natl. Bur. Std.* **65A**, 249 (1961*b*).

Tipson, R. S. and Isbell, H. S. *J. Res. Natl. Bur. Std.* **66A**, 31 (1962).

Tipson, R. S. and Parker, F. S. in *The Carbohydrates*, *Vol. 1B* (W. Pigman and D. Horton, eds.), Academic Press, New York, 1971.

Tipson, R. S., Isbell, H. S., and Stewart, J. E. *J. Res. Natl. Bur. Std.* **62**, 257 (1959).

Tipson, R. S., Layne, W. S., and Cohen, A. unpublished work (1962).

Tipson, R. S., Cerezo, A. S., Deulofeu, V., and Cohen, A. *J. Res. Natl. Bur. Std.* **71A**, 53 (1967).

Tsuboi, M. *J. Polymer Sci.* **25**, 159 (1957).

Turvey, J. R., Bowker, D. M., and Harris, M. J. *Chem. Ind.* (*London*) **1967**, 2081.

Urbański, T., Hofman, W., and Witanowski, M. *Bull. Acad. Polon. Sci. Sér. Sci., Chim. Géol. Géograph.* **7**, 619 (1959).

Verstraeten, L. M. J. *Anal. Chem.* **36**, 1040 (1964).

Verstraeten, L. M. J. *Carbohyd. Res.* **1**, 481 (1966).

Wallenfels, K., Bechtler, G., Kuhn, R., Trischmann, H., and Egge, H. *Angew. Chem.* **75**, 1014 (1963); *Angew. Chem. Intern.* (*ed. Eng.*) **2**, 515 (1963).

Whiffen, D. H. *Chem. Ind.* (*London*) **1957**, 129.

Wolfrom, M. L. and Hanessian, S. *J. Org. Chem.* **27**, 1800 (1962).

Wolfrom, M. L., Vercellotti, J. R., Tomomatsu, H., and Horton, D. *Biochem. Biophys. Res. Commun.* **12**, 8 (1963).

Wood, H. B., Jr. and Fletcher, H. G., Jr., *J. Am. Chem. Soc.* **79**, 3234 (1957).

Yovanovitch, O. *Compt. Rend.* **252**, 2884 (1961).

Zhbankov, R. G., Ivanova, N. V., and Komar, V. P. *Vysokomol. Soedin.* **8**, 1778 (1966).

Chapter 7

LIPIDS

Introduction

Infrared spectra have been very useful in deducing the structure of lipids. This topic has been reviewed by Wheeler (1954) and subsequently O'Connor (1956) and Kauffman (1964). Table 7.1 gives absorption bands employed in the application of infrared spectroscopy to lipid chemistry.

Table 7.1. Some Absorption Bands Employed in the Applications of Infrared Spectroscopy to Fatty Acid Chemistry[a] (Adapted from O'Connor, 1956)

Wave number position of observed absorption band, cm^{-1}	Vibrating group giving rise to observed absorption band

I. O—H, C—H, N—H, C—D, P—OH, and P—H stretching vibrations. Region 5000 to 2000 cm^{-1}.

A. O—H stretching

3636–3571	Free —O—H
3546–3448	Bonded —O—H$\cdots$O of single-bridged dimer
3390–3077	Bonded H$\cdots$O—H$\cdots$O— of double-bridged polymer or cyclic

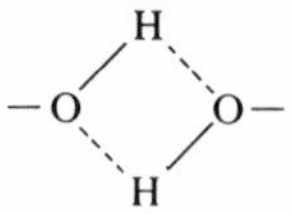

dimer

B. C—H stretching

3333–3279	R≡C—H
3106–3077	R=CH$_2$
3049–3012	R=CHR'
2941–2899	R—CH$_3$
2924–2857	R$_2$—CH$_2$
2899–2874	R$_3$—CH
2857–2703	R—C(=O)—H

Table 7.1 (*Continued*)

Wave number position of observed absorption band, cm^{-1}	Vibrating group giving rise to observed absorption band
2703	C—H and bonded O—H···O combination band
	C. N—H stretching
3509 and 3390	Free N—H primary amide
3333 and 3175	Bonded N—H··· primary amide
3448–3390	Free N—H secondary amide
3333–3279	Bonded N—H··· secondary amide, single bridge (*trans*)
3175–3145	Bonded N—H··· secondary amide, single bridge (*cis*)
3106–3077	Bonded N—H··· secondary amide cyclic dimer
3509 and 3311	N—H primary amine
3509–3311	N—H secondary amine
3390–3205	N—H imines
3125–3030	NH$_3^+$ amino acids
	D. C—D stretching
2155	C—D
2066	C—D
	E. P—H, P—OH stretchings
2469–2353	P—H
2703–2564	P—OH

II. C=O and C=C, C≡C stretching vibrations. Region 3333 to 1667 cm^{-1}.

A. Aldehydes

1739–1724	RC=O, saturated (with H)
1715–1695	PhC=O, aryl (with H)
1709–1681	R—CH=CH—C=O, α, β unsaturated (with H)

B. Ketones

1724–1709	RCH$_2$—C(=O)—CH$_2$R, saturated
1695–1681	Ph—C(=O)—CH$_3$, aryl-alkyl
1667–1661	Ph—C(=O)—Ph, diaryl
1667–1653	R—CH=CH—C(=O)—R, α, β unsaturated
1776	C=O, 4-membered, saturated ring

Table 7.1 (*Continued*)

Wave number position of observed absorption band, cm^{-1}	Vibrating group giving rise to observed absorption band
1745	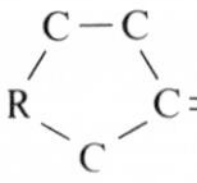 C=O, 5-membered, saturated ring
1721	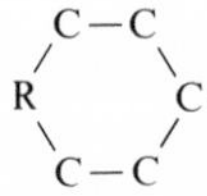 C=O, 6-membered, saturated ring, or
	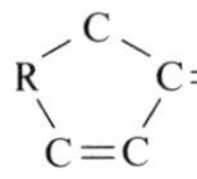 C=O, 5-membered, α, β unsaturated ring
1681	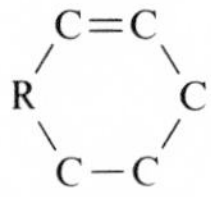 C=O, 6- (or 7-) membered, α, β unsaturated ring
1695–1667	O=Ph=O, quinone, 2 C=O's on 1 ring
1653–1639	O=Ph—Ph=O, quinone, 2 C=O's on 2 rings

C. Acids

1761	R—C(=O)—OH, saturated monomer
1724–1701	R—C(=O)—OH···O=C—R, saturated dimer
1695–1689	R—C=C—C(=O)OH, α, β unsaturated
1695–1681	Ph—C(=O)OH, aryl
1667–1653	Chelated hydroxy-acids, some dicarboxylic acids

D. Esters

1770	H₂C=C—C(=O)OCH₃, vinyl ester
1739	R—C(=O)OCH₃, saturated
1724–1718	R—C=C—C(=O)OCH₃, α, β unsaturated, or R—COOPh, aryl

Table 7.1 (*Continued*)

Wave number position of observed absorption band, cm^{-1}	Vibrating group giving rise to observed absorption band

E. Lactones

1818

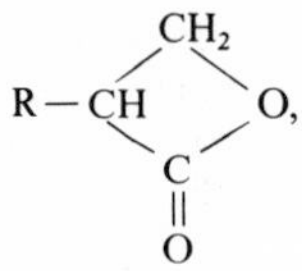

$$R-\underset{\underset{\underset{O}{\parallel}}{\overset{}{C}}}{\overset{\overset{CH_2}{\diagdown}}{CH}}\diagup O,\ \beta \text{ or 4-membered saturated ring}$$

1770

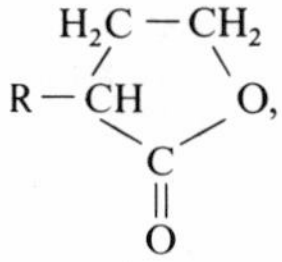

$$R-\underset{\underset{\underset{O}{\parallel}}{\overset{}{C}}}{\overset{\overset{H_2C-CH_2}{\diagdown}}{CH}}\diagup O,\ \gamma,\ \text{or 5-membered saturated ring}$$

1748

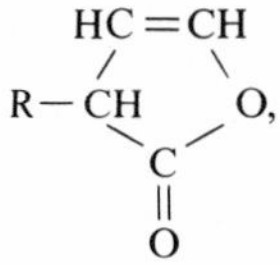

$$R-\underset{\underset{\underset{O}{\parallel}}{\overset{}{C}}}{\overset{\overset{HC=CH}{\diagdown}}{CH}}\diagup O,\ \gamma,\ \text{or 5-membered } \beta,\ \gamma \text{ unsaturated ring}$$

1739

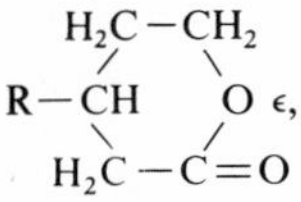

$$R-\underset{\underset{H_2C-C=O}{\diagdown}}{\overset{\overset{H_2C-CH_2}{\diagdown}}{CH}}\diagup O\ \epsilon,\ \text{or 6-membered saturated ring}$$

F. C=C, C≡C stretching

1667–1639	C=C *cis* only (weak when internal in symmetrical molecules)
3300	HC≡CH
2141–2101	RC≡CH
2252–2183	RC≡CR
1946 and 1058	C=C=C

III. C—H deformations, saturated groups. Region 1667 to 1429 cm^{-1}

1493–1449	—CH$_2$— group
1471–1429	—C—CH$_3$ group, asymmetrical deformation
1399–1389	—C(CH$_3$)$_3$ group
1389–1379	—C(CH$_3$)$_2$ group
1379–1370	C—CH$_3$ group symmetrical deformation
1370–1361	—C(CH$_3$)$_2$ group
1342–1333	—C—H group

IV. C—O stretching and C—OH bending. Region 1299 to 1000 cm^{-1}

A. Alcohols

1389–1163	Phenols
1205–1124	Tertiary open-chain saturated
1124–1087	Secondary open-chain saturated

Table 7.1 *(Continued)*

Wave number position of observed absorption band, cm^{-1}	Vibrating group giving rise to observed absorption band
1087–1053	Primary open-chain saturated
1205–1124	Highly symmetrically-branched secondary
1124–1087	α-Unsaturated or cyclic tertiary
1099–1087	Secondary with branching on one α-carbon
1087–1053	Secondary, α-unsaturated or alicyclic 5- or 6-membered ring
1053–1000	Secondary: di-unsaturated, α-branched and unsaturated, or 7- or 8-membered ring
	Primary: α-branched and/or unsaturated
	Tertiary: highly unsaturated
	B. Acids
1290–1282	C—O
1190–1183	C—O
	C. Esters
1266–1250	C—O
1190–1176	C—O
	D. Ethers
1149–1064	CH$_2$—O—CH$_2$, alkyl
1282–1235	Ph—O—Ph, aryl or =C—O, unsaturated
	E. Anhydrides
1299–1205	Cyclic
1176–1053	Open chain
	F. Phosphorus
1449 and 1000	P—O—Ph, aromatic
1050	P—O—CH$_3$, aliphatic

V. C—H deformation about a C=C and skeletal and "breathing" vibrations. Region 1000 to 667 cm^{-1}.

A. C—H bending

995–985

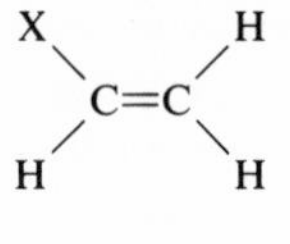

980–965 *(trans only[b])*

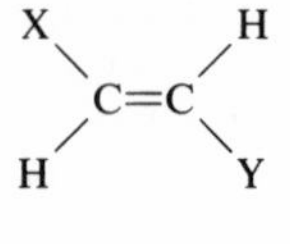

917–905

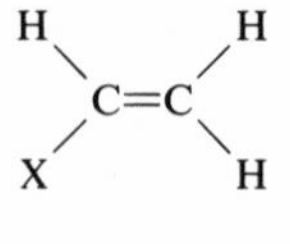

Table 7.1 (*Continued*)

Wave number position of observed absorption band, cm^{-1}	Vibrating group giving rise to observed absorption band
895–885	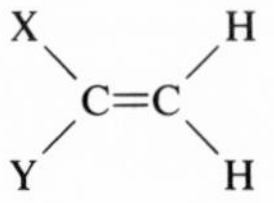 $\begin{smallmatrix} X & & H \\ & C=C & \\ Y & & H \end{smallmatrix}$
840–800	$\begin{smallmatrix} X & & H \\ & C=C & \\ Y & & Z \end{smallmatrix}$
769 > 667	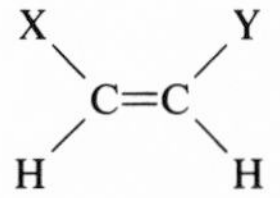 $\begin{smallmatrix} X & & Y \\ & C=C & \\ H & & H \end{smallmatrix}$ (*cis* only[b])

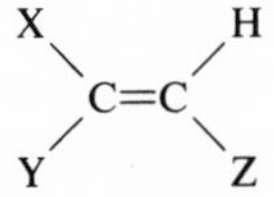

	B. Skeletal and "breathing"
1026 and 866	Cyclopropane
917 and 885	Cyclobutane
970 and 896	Cyclopentane
1038, 1014, 905 and 862	Cyclohexane
893	Epoxy-oxirane ring derived from internal R—C=C—R (*trans* only)
833	Epoxy-oxirane ring derived from internal R—C=C—R (*cis* only)
847, 775, and 680	Benzene ring
833	Hydroperoxide
772	Ethyl
769	CH$_2$ rocking on long carbon chain
741	*n*-Propyl
725	Hydroperoxide
1333–1176	Progression of bands in solid state spectra, probably due to wagging and/or bending mode of vibration of the C—H bonds of methylene groups. The number of bands in the progression is indicative of chain length.

[a]The exact position of maximum absorption depends upon whether the measurements were made on the pure liquid, solid, mull, or solution and on the nature of the particular solvent. Several band positions are also critically dependent upon neighboring groups. The value and range given in this table are from O'Connor's collection of these bands in fatty acid materials mostly from original reports in technical journals. They represent average values of ranges of the various data that have been reported for the specific absorption.
[b]See text for table of wave number positions of various combinations of conjugations involving these two internal groups.

The collection of infrared curves of Barnes *et al.* (1944) contains spectra of several fatty acid materials. Barceló and Bellanato (1953) have reported the spectra of some vegetable oils and compared them with the spectra of long-chain fatty acids and glycerides. Bands displayed by the oils, but not the fatty acids, were found at ∼1095, 1115, 1140, and 1271 cm^{-1}. Shreve *et al.* (1950a) have studied long-chain fatty acids, esters, and alcohols. One of the assignments they made was that of the 2703 cm^{-1} band as a bonded O—H$\cdots$O fused with the 2941 cm^{-1} band coming from C—H stretching. O'Connor *et al.* (1951) have recorded quantitative spectra of the fatty acids and their methyl and ethyl esters and noted that a band at 1111 cm^{-1} may be used to distinguish ethyl from methyl esters.

Sinclair *et al.* (1952*a*,*b*) have discussed the spectra of saturated and unsaturated fatty acids and esters. The intensity of a 769 cm^{-1} band (CH$_2$ rocking vibration) in the spectra of solutions of saturated fatty acids and esters increased progressively with chain length. In the spectra of solutions of unsaturated fatty acids the intensity of the 3030 cm^{-1} band (C—H stretching of the C=C—H) increased with increasing unsaturation, whereas the intensity of the 2915 and 2849 cm^{-1} bands (CH$_2$ stretching) decreased in the same series. The crystalline acids display more detail in their spectra. In the spectra of the even-carbon acids the carbonyl stretching band comes at 1701 to 1698 cm^{-1} but at 1704 cm^{-1} for the odd-carbon acid spectra. This difference distinguishes between odd and even numbers of carbons. In the spectra of the solids there is a progression of bands (CH$_2$ wagging and/or twisting) spaced fairly evenly between 1351 and 1176 cm^{-1}, although the solutions give only broad diffuse bands. Jones *et al.* (1952) have found a regular increase in the number of bands in the progressions in the C$_{12}$ to C$_{21}$ series, with no alteration between odd or even numbered carbon chains. A similar study was done by Meiklejohn *et al.* (1957) with long-chain fatty acids, in which details were given on band progressions.

Freeman (1952) has studied the position of branching, length of branches, and number of branches in long-chain fatty acids containing such structures. Branching on the α carbon or near the carboxyl group is detected by comparing the 1285 and 1250 cm^{-1} bands. In straight chain compounds the former band is the more intense. Substitution on the α carbon makes the 1250 cm^{-1} band the more intense one. Substitution within five carbons of the carboxyl group causes shifts in frequency positions of the 1250 cm^{-1} band. A band at 772 cm^{-1} signifies an ethyl group branch; 741 cm^{-1} signifies an *n*-propyl group branch; and isopropyl and *t*-butyl group branches cause splitting of the band at $\sim$1379 cm^{-1} into two components. The number of branches can be found by determination of terminal methyl groups. Guertin *et al.* (1956) have shown that use of the 1285 and 1250 cm^{-1} band intensities could identify α substitution in branched-chain hexanoic acids.

Sobotka and Stynler (1950) have discussed the spectra of *iso, anteiso*, and *neo* acids. (The definitions of these terms are, respectively: —CH$_3$ group on the next-to-the-end carbon; —CH$_3$ group on the second-from-the-end carbon; and, two —CH$_3$ groups on the next-to-the-end carbon.) *Iso* acids are characterized by a splitting of the 1379 cm^{-1} band into two parts of about equal intensity. *Neo* acids show the same splitting, but the smaller frequency component has the greater intensity, characteristic of *t*-butyl splitting. No splitting occurs in *anteiso* acids, but the 1379 cm^{-1} band is much stronger than in the spectra of normal acids.

O'Connor (1956) has given many references and a detailed discussion about *cis* and *trans* unsaturation, the ability to distinguish between them and to determine *trans* bonds in the presence of *cis* bonds. Ahlers *et al.* (1953) have given the following data for unsaturated systems (Table 7.2):

Table 7.2. Characteristic Absorption Bands of Unsaturated Systems (Adapted from Ahlers, 1953, cited in O'Connor, 1956)

Unsaturated system	Wave number maxima, cm^{-1}
Isolated *trans*	967
Cis, trans-conjugated	983
Trans-trans-conjugated	988
Cis-cis-trans-conjugated	989
Cis-trans-trans-conjugated	991
Trans-trans-trans-conjugated	994
Isolated *cis*	913
Cis, trans-conjugated	950

O'Connor (1956) has discussed the use of *cis–trans* infrared data in hydrogenation and in oxidation studies, and has given details concerning studies on autoxidation and rancidity.

Fatty Acids and Esters

Kaneda (1963) has described the isolation and characterization of the bacterial fatty acids found in a strain of *B. subtilis* (ATCC 7059), during the course of work on the microbial synthesis of branched-chain fatty acids. Spectra were given for a number of fatty acids. The terminal carbon atoms showed characteristic absorption in the range between 1380 and 1360 cm^{-1}. Normal long-chain fatty acids showed a weak band at 1380 cm^{-1}, *iso* fatty acids showed a doublet in the range from 1380 to 1360 cm^{-1}, and *anteiso* fatty acids showed a stronger band in the region of 1380 cm^{-1} than that given by normal fatty acids. In this region, the absorptions of the bacterial fatty acids agree with those given by Sobotka and Stynler (1950). The doublet in the range from 1380 to 1360 cm^{-1} was shown not only by *iso* acids but also by *neo* acids.

In the region from 1300 to 1100 cm^{-1}, a regular band progression is observed in the spectra of the solid acids (i.e., a series of bands either increasing or decreasing progressively in intensity with decrease in frequency). The number of progression bands is directly related to the chain length of the straight-chain fatty acids (Jones *et al.*, 1952; Jones, 1962).

A regular decline of progression band intensities toward lower frequencies was found in the spectra of isopentadecanoic and isoheptadecanoic acids, but isomyristic, isopalmitic, and 14-methylhexadecanoic acids displayed an increase in progression band intensities toward lower frequencies. No band was found at about 1020 cm^{-1} in the fatty acids of *B. subtilis*, thus indicating the absence of cyclopropane compounds (known to occur in some lactic acid bacteria).

Fischmeister (1967) has studied the infrared spectra of all the seventeen isomeric ketooctadecanoic acids from 4000 to 400 cm^{-1}. She has found a characteristic band sequence in the region 1380 to 1160 cm^{-1}, corresponding to CH$_2$ wagging and twisting vibrations as well as rocking and skeletal bands from 1160 to 700 cm^{-1}, to be suitable for the identification of individual isomers in the solid state. Progressing from the 17- to the 5-keto acid the stronger bands of the sequence are regularly displaced to higher frequencies. As their number decreases the number of weak bands or inflections increases. The strong bands of the sequence were assigned to the CH_2 groups between the carbonyl and the carboxyl groups. In the 2-keto acid regularly spaced weak bands overlap the strong carboxyl band at 1288 cm^{-1}. The absorptions of the residual CH_2 groups are as weak as in ketones.

The incubation of ethanol-1-^{14}C with rat liver or kidney homogenates produced a previously unrecognized metabolite, which has been isolated and identified as 5-hydroxy-4-ketohexanoic acid (Bloom and Westerfeld, 1966). Infrared and NMR spectral data were given for the compound and its dehydroabietylamine salt.

Lactobacillic acid, $C_{19}H_{36}O_2$, isolated by Hofmann et al. (1952) is a saturated acid with two less hydrogen atoms than a normal C_{19} saturated acid. These facts suggest a cyclic structure. A band found at 1020 cm^{-1} is characteristic of alkyl-substituted cyclopropanes. The structure of lactobacillic acid (XXVIII) is, therefore

$$\overset{\displaystyle CH_2}{\overset{\displaystyle \diagup \diagdown}{CH_3(CH_2)_x-CH-CH-(CH_2)_yCOOH}}$$

XXVIII

where $x + y = 14$.

Golmohammadi (1966) has investigated the effect on the infrared spectrum of the increase in the number of selenium atoms incorporated in a fatty acid chain. The intensity of the bands near 1250 cm^{-1} was selenium dependent. A progressive increase in the intensity of the bands in this region together with a displacement of the bands toward lower frequency was observed when the number of selenium atoms in the chain was increased progressively from one to three.

Infrared spectra have been used by Freeman (1953) and by Cason et al. (1951) in determining the structure of the C_{27}-phthienoic acid isolated from crude methyl phthioate. Ultraviolet spectra and chemical data showed the acid to be a 2-alkyl-2-alkenoic acid having a second substituent in the 4 or 5 position and a third substituent on a carbon further from the carboxyl. A band in the spectrum at 1645 cm^{-1} indicated a C=C group. The carbonyl band appearing at 1692 cm^{-1}, a frequency somewhat lower than the usual C=O stretching of saturated esters, indicated that the C=O was conjugated with the C=C group. Bands at 994, 800, 758, and 708 cm^{-1} were attributed to C—H deformations about C=C groups, indicating that the C=C group had at least one hydrogen atom attached to it. No iso or neo configurations were present. The possibility of a quaternary carbon atom further away from the carboxyl than the α position was removed by the lack of a band at 1136 cm^{-1}, and absence of bands at ∼772 and 741 cm^{-1} eliminated both ethyl and propyl groups. Bands at 1282 and 1235 cm^{-1} and a knowledge of the spectra of branched-chain fatty acids (Freeman, 1952) indicated a methyl group α to the carboxyl and a second methyl group no further away than the δ position. Four methyl groups were indicated from the intensity of the methyl band at 1370 cm^{-1}. The above considerations and chemical evidence led to the following structure (XXIX) for the C_{27}-phthienoic acid:

$$C_4H_9-(CH_2)_{15}-\underset{\underset{CH_3}{|}}{CH}-CH_2-\underset{\underset{H}{|}}{C}=\underset{\underset{CH_3}{|}}{C}-COOH$$

XXIX

The infrared and ultraviolet spectra of a product of the biohydrogenation of linolenic acid in *Butyrivibrio fibrisolvens* displayed a conjugated *cis,trans*-diene system and the product was shown to be *cis*-9,*trans*-11,*cis*-15-octadecatrienoic acid (Kepler and Tove, 1967). Wilde and Dawson (1966) had found that linolenic acid was first converted to a trienoic acid in which two of the double bonds were conjugated. This was tentatively identified as either $\Delta^{9,11,15}$- or $\Delta^{9,13,15}$-octadecatrienoic acid. Since no *trans* peaks occurred around 969 cm^{-1} in the infrared spectrum, a *cis,cis,cis* configuration was assigned to this acid.

During the determination (van den Oord *et al.*, 1965) of the structure of emulsifiers (fatty acylsarcosyltaurines) in gastric juice from the crab, *Cancer pagurus L.*, it was necessary to identify dodecenoic acid. The acid had no prominent band at 965 cm^{-1}, but it did at 720 cm^{-1}, indicating the presence of a *cis* double bond (Ahlers *et al.*, 1953).

Jart (1960) has recorded spectra of many different fatty acids and fatty acid esters in carbon disulfide solution. Stress was laid on the study of the *trans* absorption at 962 cm^{-1}, and absorptivities of the pure substances at the *trans* maximum were calculated. Beer's law applied for concentrations up to 200 g of substance per liter of solution. This worker used a baseline method for the quantitative determination of *trans* monoene fatty acids in mixtures with saturated acids and *cis* forms of unsaturated fatty acids and fatty acid esters. Unlike the *cis* compounds and the saturated compounds, *trans* compounds display a considerable absorption at 962 cm^{-1} (CH bending about the *trans* C=C group).

Isolated (nonconjugated) double bonds in the *trans* configuration present in long-chain triglycerides, acids, and esters can be determined by infrared spectrophotometry. These *trans* double bonds display an absorption band at about 971 cm^{-1}, whereas *cis* double bonds do not absorb here. Unsaturated constituents of most vegetable fats and oils contain only nonconjugated double bonds in the *cis* configuration; these may isomerize to the *trans* form during extraction and processing due to oxidation or partial hydrogenation. Animal and marine fats may naturally contain some *trans* isomers.

Long-chain fatty acids also show a band at about 943 cm^{-1}, which is due to carboxyl. Correction for this band and any "background" has been made by the baseline technique applied by Firestone and LaBouliere (1965) to margarines and shortenings. If the isolated *trans* content is small, correction may greatly affect the absorption at 971 cm^{-1} and quantitation is not possible. Therefore, long-chain fatty acids containing less than 15 % isolated *trans* isomers must be converted to their methyl esters before analysis by the infrared method. The method applies only to materials containing less than 5 % of *total* conjugates, since triglycerides produce isolated *trans* values which are about 2–3% high, whereas methyl esters produce isolated *trans* values which are about 1.5–3% low.

The method should not be applied without specific precautions to samples containing greater than 5% conjugated unsaturation (tung oil), to materials containing functional groups that modify absorption of the C—H deformation about the *trans* double bond (castor oil containing ricinoleic or ricinelaidic acids), to mixed glycerides with long- and short-chain moieties (diacetostearin), or to any materials where specific bands may appear close to 971 cm^{-1}.

Infrared analysis for *trans*-octadecenoic acid is a standard method (Report of the Spectroscopy Committee, 1959). Good quantitative results for the *trans*-isomers are possible by infrared analyses, and such analyses are done routinely by many laboratories. Kauffman and Lee (1960) found good agreement between the infrared analysis and the gas–liquid partition chromatographic analysis of methyl elaidate in hydrogenated vegetable oil.

Fatty Amides

Infrared spectroscopy has been used for the analysis of mixtures of fatty amides (Kauffman, 1964). In dilute chloroform solutions, the amides $CH_3(CH_2)_nCONH_2$ absorb consistently at ~1681 cm^{-1} and do not display apparent association or enolization. The concentration of unsubstituted amides was quantitatively related to the intensity of an amide I band throughout the range from 1 to 100%. With scale expansion the sensitivity of the method may be extended to 0.03%.

Ethers, Esters, and Ether Esters of Glycerol and Various Diols

Baumann and Ulshöfer (1968) have recorded and interpreted spectra of long-chain ethers, esters, and ether esters of glycerol, 1,2-ethanediol, and propanediols. They have discussed the influence of the environment of alkoxy, acyloxy, and hydroxy groups on the absorption frequencies of these functional groups.

Mayers and Haines (1967) have isolated a sulfolipid, 1,14-docosyl disulfate, from the phytoflagellate, *Ochromonas danica*. Acid hydrolysis of the sulfatide yielded a diol. The structures of the sulfatide (lipoid sulfate ester) and the diol were determined by infrared, NMR, and mass spectroscopy. The infrared spectrum of the sulfatide diol had bands indicative of a long-chain diol: 1060 cm^{-1} (primary alcohol CO), 1120 (secondary alcohol CO), and 718 cm^{-1} (long-chain aliphatic CH_2rocking).

Glycerides and Cholesterol Esters

Szonyi *et al.* (1962) have described a differential infrared spectrophotometric method for the determination of *trans* unsaturation in fats. The method utilizes absorption at 965 cm^{-1}, which is due to C—H out-of-plane deformation vibrations of *trans* unsaturated compounds. The method is rapid, accurate, and directly applicable to the determination of *trans* unsaturation in triglycerides. It is applicable to samples which contain low concentrations of *trans* acids (down to 2%) and also to samples with fatty acids of mixed chain length. The absorptivities used in the calculations for this method are those established by Shreve *et al.* (1950*b*).

Some useful papers concerning glycerides and their infrared spectra are those by Kuhrt *et al.* (1952*a,b*) and O'Connor *et al.* (1955). The latter authors have compared the infrared spectra of mono-, di-, and triglycerides. The O—H stretching region at about 3333 cm^{-1} can be used to confirm the absence of mono- or diglycerides in a preparation of triglycerides by the complete absence of the O—H stretching vibration band. Stretching of the C—O in the α-substituted secondary alcoholic groups appears in monoglycerides at 1053 cm^{-1}, and in diglycerides at 1042 cm^{-1}. Triglycerides have no bands at these frequencies.

Freeman (1964) has analyzed mixtures of cholesterol esters (e.g., cholesteryl oleate) and triglycerides (e.g., triolein) by using the difference in the positions of the ester carbonyl bands. To analyze serum lipids he first removed phospholipids (1735-cm^{-1} ester groups) by adsorbing them on silicic acid. Measurements were then made on carbon tetrachloride solutions of the unadsorbed lipids at 1745 and 1730 cm^{-1}, the approximate band positions of triglycerides and cholesterol esters, respectively. Minor lipid constituents produced only negligibly small errors and the over-all accuracy was about $\pm 5\%$. Freeman said that with minor modifications in procedure as little as 0.4 ml of serum could be analyzed and with microcells as little as 0.05 ml.

Freeman *et al.* (1967) have developed an instrument which uses infrared absorption for the semiautomatic analysis of mixtures of triglycerides and cholesteryl esters. The method of analysis is based on the carbonyl absorption bands of triolein and cholesteryl oleate at 1745 and 1730 cm^{-1}, respectively. Sample preparation consists of an extraction of lipids from serum in such a way as to exclude phospholipids. With adequate resolution and good precision of frequency setting, measurements at the two positions can be used successfully in a standard two-component spectrophotometric analysis. The nonautomatic part of the system is the handling of samples.

Kataura and Kataura (1967*a*) have taken the lipid fraction from dry and wet types of cerumen collected from Japanese children and have separated it into several kinds of lipid component by silicic acid column chromatography. These authors (1967*b*) have identified the various lipid components by thin-layer chromatography and infrared spectrometry. Fatty acid components of the purified lipids were determined by gas chromatography. Larger amounts of stearic and linoleic acids were found in triglyceride from the dry type. In the wet type, palmitic acid was found instead. The contents of the other lipid fractions such as cholesterol esters, diglyceride, and monoglyceride were almost equal in both types of cerumen.

Substitution of a methyl group at the α-methylene carbon of methyl esters of laurate, myristate, and palmitate results in characteristic infrared spectra that readily distinguish them from the straight-chain homologs and facilitate their unequivocal identification when separated from complex lipid mixtures by gas–liquid partition chromatography (Napier, 1966). Spectral changes observed with the methyl esters are also evident in the cholesterol esters of 2-methylalkanoates and also serve in their identification.

Characteristic spectra of methyl and cholesterol esters of 2-methyllauric, 2-methylmyristic, and 2-methylpalmitic acids have been given by Napier (1966). Alkyl substitution at the 2-position is readily identified (Freeman, 1952) by a reversal in intensity of the 1290 and 1236 cm^{-1} bands, as compared to straight-chain acids. Reversal of intensity of the 1464 and 1414 cm^{-1} bands, which is associated with alkyl substitution at the α-methylene carbon (Guertin et al., 1956), is also observed.

Lecithins, Cephalins, and Phospholipids

Baer (1953) has recorded the spectra of pure dimyristoyl, dipalmitoyl, and distearoyl L-α-lecithins. Baer et al. (1952) have also recorded spectra of L-α-cephalins, namely, the same kinds of fatty acid derivatives. The spectra have differences, although this is not true for the corresponding straight-chain fatty acids. In the region from 870 to 714 cm^{-1} several large differences occur.

The introduction of methyl groups into the carbon skeleton of ethanolamine causes little or no change in the infrared spectra of cephalins (Baer and Blackwell, 1963). This may be seen by comparing the spectra of alkyl-substituted ethanolamines, phosphatidyl ethanolamines (Baer and Gróf, 1960), and phosphatidyl cholines (Baer, 1953) containing the same fatty acid substituents. Baer and Blackwell (1963) noted, however, that the stepwise replacement of the two hydrogen atoms of the amino group by methyl groups leads successively to the disappearance of the bands at 1538 and 1626 cm^{-1}. At the same time, the broad region of absorption above the peaks at 2899 and 2817 cm^{-1}, which is characteristic for phosphatidyl ethanolamines, changes into a narrow one with the introduction of the second methyl group. The spectra of phosphatidyl-N,N-dimethylethanolamines (Baer and Pavanaram, 1961) thus resemble much more closely those of lecithins (Baer, 1953), except that the latter have a strong band at 966 cm^{-1}, which is not shown by cephalins nor by N-methylcephalins or N,N-dimethylcephalins. The differences in the spectra in the Baer and Blackwell paper (1963) for phosphatidyl ethanolamine compounds and their N-alkyl-substituted derivatives are sufficiently characteristic to permit their group identification by infrared spectroscopy.

Marinetti and Stotz (1954a,b) have studied the structures of phospholipids. They found bands at 971 cm^{-1} due to the P—O—C group in saturated lecithin and the trans double bond and the P—O—C group in unsaturated phospholipids. The band at 1087 cm^{-1} was thought to be due to the P—O—C vibration and the one at 1220 cm^{-1} to a C—O—C linkage. All phospholipids displayed a band near 725 cm^{-1}, arising from long carbon-to-carbon chains, and a band at 694 cm^{-1}, not assigned. These workers also studied the double-bond configuration of sphingomyelin and related lipids by infrared spectroscopy. Glycerophosphides displayed a very strong ester carbonyl stretching band at 1754 to 1730 cm^{-1}, and sphingolipids exhibited a strong carbonyl band at 1639 cm^{-1}. The latter compounds also showed O—H and N—H stretching bands in the region 3333 to 2941 cm^{-1}.

Stanacev et al. (1964) have reported the synthesis of the first diether analog of a lecithin, L-α-(dioctadecyl) lecithin, and have given infrared spectra of the diether and L-α-(dioctadecanoyl) lecithin, the diester. Fontanges et al. (1964) have used thin-layer chromatography and infrared spectroscopy to identify samples of lecithin down to 22 μg.

The effect of a dc potential applied across brain cephalin and lecithin films has been studied with infrared methods (May and Kamble, 1968). When an electric field was applied (0 to 10,000 V/mm) certain bands in the infrared spectrum were altered in intensity. The bands whose intensities varied with the field were related to the phosphate group and the $-(CH_2)_n-$ group in the fatty acid constituent of the lipid. No spectral changes were found with synthetic phosphatidyl ethanolamine, phosphatidyl serine, and phosphatidyl choline.

Abramson *et al.* (1964*a*) have found that the acid and salt (ionic) forms of the phosphate and of the carboxyl groups of natural lipids can be detected by differences in their infrared spectra. Lipids were prepared in several ways: Nujol mulls, KBr pellets, and CCl_4 and C_2Cl_4 solutions. Phosphatidyl serine and phosphatidic acid were prepared in the acid form by acid dialysis, and salt forms of these lipids and of stearic acid were prepared by titration to the desired pH, the cation concentration being confirmed by analysis. The 1739-cm^{-1} band (esters plus COOH) diminished as cation concentration increased, the COO$^-$ band then appearing (at 1587 to 1563 cm^{-1} for stearic acid and 1639 cm^{-1} for phosphatidyl serine). Ratios of the heights of these bands to a CH band could be used to approximate the extent of ionization of the carboxylic group. Phosphatidic acid had a band at 1020 cm^{-1} that diminished as the acid form was neutralized, forming the phosphate ion. This could then be used to identify the ion form in other phospholipids. Even at low pH these workers found that the phosphate group is ionized in lecithin, sphingomyelin, and phosphatidyl serine, indicating the presence of a dipolar ion. With phosphatidyl serine this point was confirmed by electrophoresis of aqueous solutions which show an isoelectric point of pH 1.2.

Abramson *et al.* (1964*b*) have prepared by ultrasonication optically clear aqueous dispersions of phosphatidyl serine (PS). Infrared and titration data indicated that at the isoelectric point the molecule is a dipolar ion of phosphate and amino groups. They also found evidence for the existence of the following three forms: isoelectric or acidic (XXX), HPS; monosodium salt (XXXI), NaPS; and disodium salt (XXXII), Na_2PS.

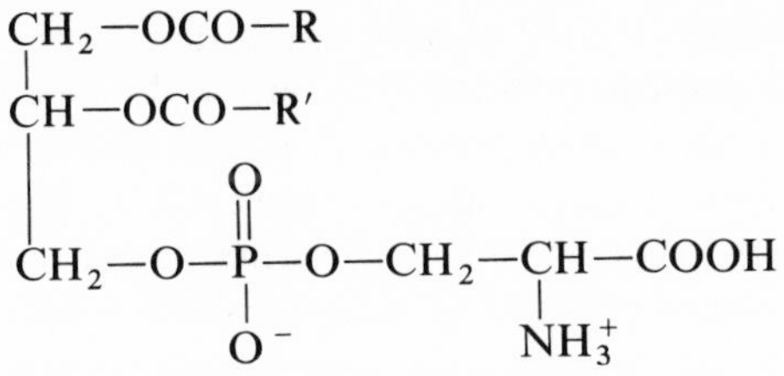

$$\begin{array}{l} CH_2-OCO-R \\ | \\ CH-OCO-R' \\ | \qquad\qquad O \\ | \qquad\qquad \| \\ CH_2-O-P-O-CH_2-CH-COOH \\ \qquad\qquad | \qquad\qquad\quad | \\ \qquad\qquad O^- \qquad\qquad NH_3^+ \end{array}$$

XXX

$$\begin{array}{l} CH_2-OCO-R \\ | \\ CH-OCO-R' \\ | \qquad\qquad O \\ | \qquad\qquad \| \\ CH_2-O-P-O-CH_2-CH-COO^-Na^+ \\ \qquad\qquad | \qquad\qquad\quad | \\ \qquad\qquad O^- \qquad\qquad NH_3^+ \end{array}$$

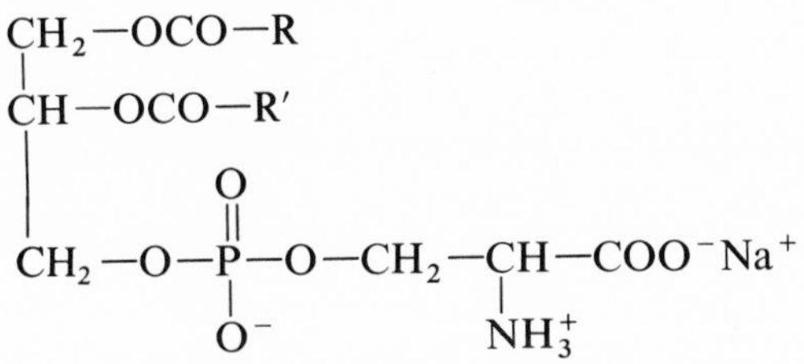

XXXI

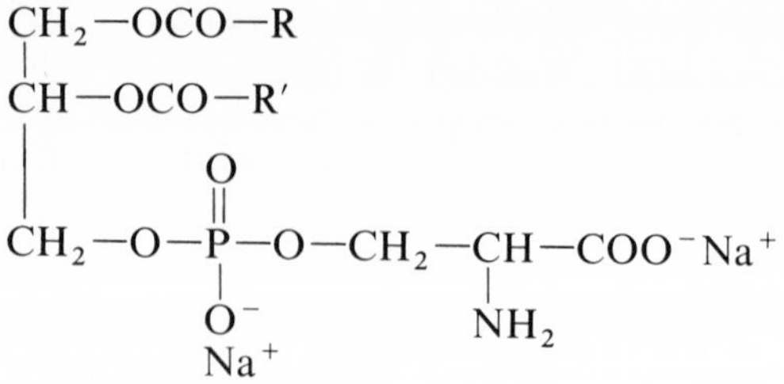

$$CH_2-OCO-R$$
$$CH-OCO-R'$$
$$CH_2-O-\overset{\overset{\displaystyle O}{\|}}{\underset{\underset{\displaystyle Na^+}{O^-}}{P}}-O-CH_2-\underset{\underset{\displaystyle NH_2}{|}}{CH}-COO^-Na^+$$

XXXII

The main differences in the spectra of HPS, NaPS, and Na_2PS appeared in the regions of 1640 and 1740 cm^{-1}. The absorption at 1740 cm^{-1} was attributed to C—O in the ester groups and in the carboxy acid group. The absorption at 1639 cm^{-1} was attributed to COO$^-$. The study was made because of the importance of acidic lipids as tissue ion-exchange agents and because these lipids were thought to have a possible role in transmembrane cation transport.

The spectra of octadecyl dihydrogen phosphate, dioctadecyl hydrogen phosphate, and phosphatidic acid have been compared (Abramson et al., 1965) with those of their sodium salts. The regions showing significant changes in absorption are at 3430 to 3330 cm^{-1}, 2780 to 2630 cm^{-1}, 1230 cm^{-1}, 1100 cm^{-1}, and 1070 to 1030 cm^{-1}. The increased absorption at 3430 to 3330 cm^{-1} upon salt formation was attributed to the OH of water bound to the ionized species and was accompanied by a corresponding decrease in absorption at 2780 to 2630 cm^{-1}, resulting from the decrease in inter-molecular associations of the P—OH groups. The major region of absorption of the phosphate occurs at 1100 to 950 cm^{-1}. Here the absorption maximum of the P—OH in the un-ionized form is at 1030 cm^{-1} while the PO$^-$ group in the salt displays a strong absorption at 1100 cm^{-1}. The P=O group absorbs at 1250 to 1220 cm^{-1}. The spectra of lecithin and sphingomyelin are interpreted as showing a fully ionized phosphate group bonded to H_2O with no association of P—OH groups. The major band at 1100 cm^{-1} is indicative of the PO$^-$ group. Phosphatidyl ethanolamine and phosphatidyl serine in CCl_4 show bonded P—OH groups and a stronger absorption in the region of P—OH (1075 to 1050 cm^{-1}) than of PO$^-$ (1100 cm^{-1}). The following structures were proposed for phosphatidyl serine (at the isoelectric point) (XXXIII), phosphatidyl ethanolamine (XXXIV), and lecithin (XXXV) or sphingomyelin in organic solvents, although in the micelle these molecules would be involved in chains of intermolecular bonds:

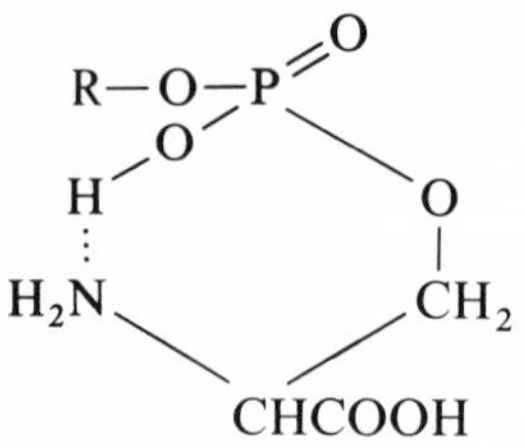

XXXIII 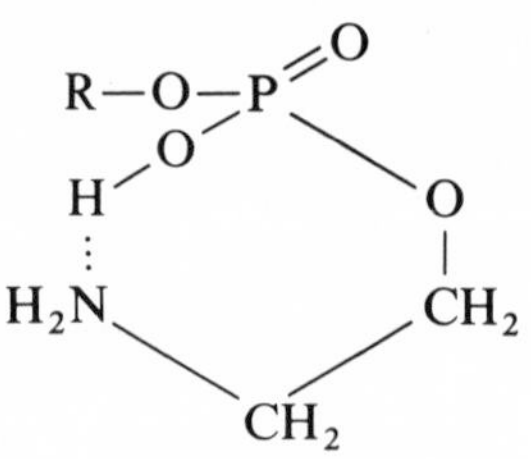   XXXIV 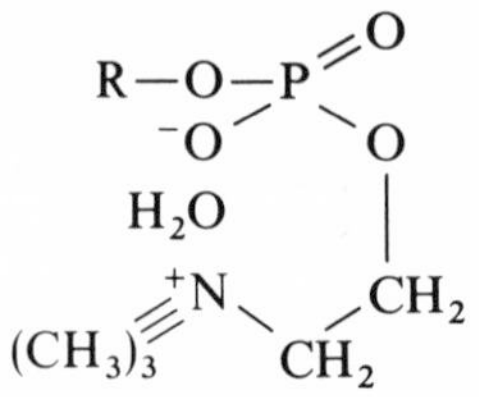   XXXV

Chapman and Morrison (1966) have found NMR evidence favoring a dipolar ionic form for the phosphatidyl ethanolamines. Also, their infrared spectra of chloroform solutions favor a dipolar ionic structure. The evidence was as follows: if dioleoyl-phosphatidyl ethanolamines exist in chloroform in a nonionic form, then intense bands in the 3300 cm^{-1} region should occur because of NH stretching frequencies. Bands were found at 3058, 2710, 2538, and a probable band at 3021 cm^{-1}, which they correlated with vibrations of an NH_3^+ group. A comparison of the spectra of dioleoyl-phosphatidyl ethanolamine and a dipolar ionic amino acid, such as alanine, showed almost identical spectra in the 4000 to 2000 cm^{-1} region. The spectrum of the non-ionized compound, DL-α-alanine methyl ester in chloroform shows intense absorption in the 3300 cm^{-1} region characteristic of a free primary amino group.

Phosphonic Acids, Phosphonates, and Phosphonolipids

The occurrence of organophosphonic acids and their derivatives has recently been reported in a wide variety of biological material. Their presence in lipids (phosphonolipids, as analogs of phospholipids) suggests participation in cell membrane structure. Alam and Bishop (1968) have reported the presence of derived aminoethyl-phosphonate in the phospholipid of human aortas. Choline phosphonate was detected in some of the aorta samples with atherosclerotic plaques and not detected in the other samples. No aminoethylphosphonate was found in the samples.

Baer and Stanacev (1964) have reported the synthesis of dipalmitoyl L-α-glyceryl (2-aminoethyl) phosphonate, the first synthesis of a complex lipid containing phosphonic acid, and have shown spectra of L-α-(dipalmitoyl)cephalin and of dipalmitoyl L-α-phosphono(deoxy)cephalin.

Synthetic analogs of naturally occurring glycerol phospholipids with promise of biological stability possess considerable interest as potential clinical and therapeutic agents. For this reason and in anticipation of the isolation of diether phosphonolipids from natural sources, preparation of such substances was undertaken by Baer and Stanacev (1965a) to provide individual diether phosphonolipids of known structure [dialkyl L-α-glyceryl-(2-aminoethyl)phosphonate (XXXVI)] and configuration for research purposes.

$$CH_3(CH_2)\underset{\text{13, 15, or 17}}{\rule{2cm}{0.4pt}}O-CH_2$$
$$CH_3(CH_2)\underset{\text{13, 15, or 17}}{\rule{2cm}{0.4pt}}O-CH$$
$$H_2C-O-\overset{\overset{O}{\|}}{\underset{\underset{O^-}{|}}{P}}-CH_2-CH_2NH_3^+$$

XXXVI

This report described the synthesis of diether phosphonolipids which resemble cephalins structurally. Infrared spectra were given for dihexadecanoyl L-α-glyceryl-(2-aminoethyl) phosphonate and dihexadecanoyl L-α-glycerylphosphorylethanolamine.

The possibility of the natural existence of phosphonic acid analogs of lecithins prompted Baer and Stanacev (1965b) to synthesize such compounds, which they felt could be useful reference compounds in the elucidation of the structure and configuration of naturally occurring phosphonolipids and valuable substrates for the study of enzymatic reactions. The infrared spectra of lecithins and their phosphonic acid analogs showed several differences in the region of 1333 to 833 cm^{-1}, thus differing from cephalins and their phosphonic acid analogs whose spectra are almost identical, except for one band (Baer and Stanacev, 1964). The phosphonic acid analogs of lecithins differ from lecithins by having a band at 1111 cm^{-1} and no band at 1176 cm^{-1}. Also, their absorption bands at 1212, 1081, and 1058 cm^{-1} are at slightly lower frequencies than the corresponding bands of lecithins.

The occurrence of the phosphonic acid analogs of ethanolamine phosphate, serine phosphate, and choline phosphate, the three major constituents of phospholipids, has been established by Kittredge *et al.* (1967) in the sea anemone, *Anthopleura xanthogrammica*. They isolated and characterized 2-methylaminoethylphosphonic acid (XXXVII) and 2-trimethylaminoethylphosphonic acid (XXXVIII).

$$
\begin{array}{cc}
\overset{\displaystyle CH_3}{\underset{\displaystyle |}{}} \quad \overset{\displaystyle OH}{\underset{\displaystyle |}{}} & \overset{\displaystyle CH_3}{\underset{\displaystyle |}{}} \quad \overset{\displaystyle OH}{\underset{\displaystyle |}{}} \\
HNCH_2CH_2P \rightarrow O & CH_3 - N^+ CH_2CH_2 P \rightarrow O \\
\underset{\displaystyle OH}{|} & \underset{\displaystyle CH_3}{|} \qquad \underset{\displaystyle O^-}{|} \\
\end{array}
$$

XXXVII   XXXVIII

Plasmalogens

Goldfine (1964) has isolated the dimethylacetal derivatives of the aldehyde chains derived from *Clostridium butyricum* plasmalogens, then separated and quantitated them by gas-liquid chromatography. He found the major components to be 16-carbon saturated, 16-carbon mono*un*saturated, and 17- and 19-carbon cyclopropane aldehydes. The aldehydes containing cyclopropane rings were identified by their infrared spectra and by comparison of their NMR spectra with those of model compounds, and by oxidation to the corresponding acids followed by chromatographic comparison with authentic cyclopropane acids. The bands at 1020 cm^{-1} in the spectra of two *C. butyricum* dimethylacetals studied are characteristic of cyclopropane rings. Also, the two unknowns showed weak bands at 1660 to 1665 cm^{-1}, usually seen in compounds containing an isolated double bond. Bands due to acetal structure were found in the region of 1200 to 1040 cm^{-1}.

Prostaglandins

Infrared spectroscopy has been used extensively (Bergström *et al.*, 1963) in establishing the structures of prostaglandins. Ramwell *et al.* (1968) have presented a very interesting review on the prostaglandins including nomenclature, identification, biosynthesis, metabolism, methods for their analysis, and physical data.

The infrared spectra (mineral oil mulls) of six crystalline prostaglandins are shown in Fig. 7.1. (Ramwell *et al.*, 1968). These spectra display characteristic bands consistent with the proposed structures. The figure shows infrared spectra of PGE$_1$, PGA$_1$, PGB$_1$, PGF$_{1\alpha}$, PGF$_2\beta$, and PGE$_2$ (where "PG" means prostaglandin).

These compounds are, respectively: 11α,15(S)-dihydroxy-9-oxo-13-*trans*-prostenoic acid; 15(S)-hydroxy-9-oxo-10,13-*trans*-prostadienoic acid; 15(S)-hydroxy-9-oxo-8(12), 13-*trans*-prostadienoic acid; 9α,11α,15(S)-trihydroxy-13-*trans*-prostenoic acid; 9β, 11α,15(S)-trihydroxy-5-*cis*,13-*trans*-prostadienoic acid; and, 11α,15(S)-dihydroxy-9-oxo-5-*cis*,13-*trans*-prostadienoic acid. NMR and optical rotatory dispersion spectra have been determined for several of these compounds, and ultraviolet data have been given (Ramwell *et al.*, 1968).

Enzymes in the particle-free fraction of guinea pig lung homogenates convert prostaglandin E$_1$ (11α,15-dihydroxy-9-ketoprost-13-enoic acid) into two less polar metabolites, 11α,15-dihydroxy-9-ketoprostanoic acid and 11α-hydroxy-9,15-diketoprostanoic acid. The structures of the metabolites have been determined (Änggård and Samuelsson, 1964) by means of mass and infrared spectra, reversed-phase partition chromatography and chemical procedures. The numbering system is shown in prostaglandin E$_1$ (XXXIX):

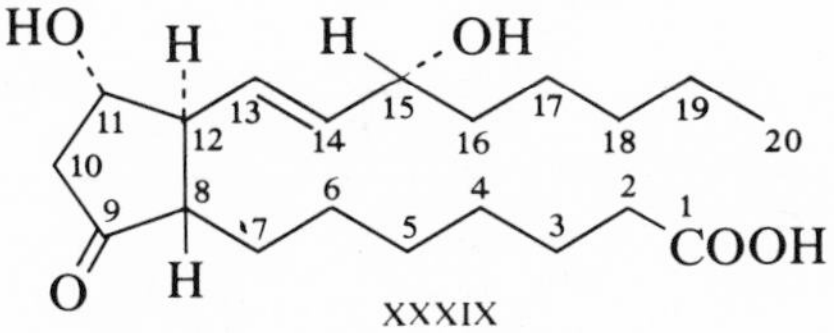

Erythrocyte Composition

Infrared spectra have been used as part of an investigation (Hanahan *et al.*, 1963) to prove that a glyceryl ether phospholipid (XL),

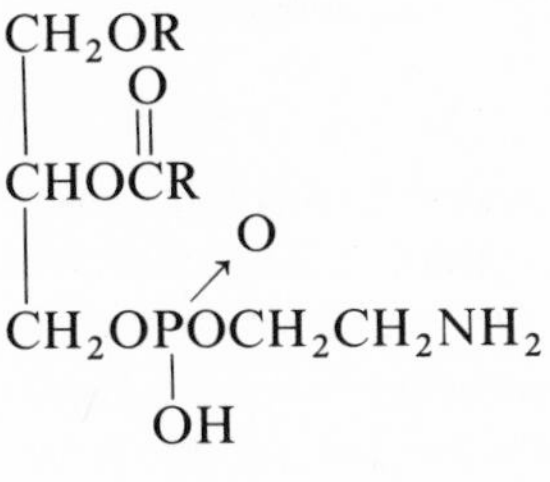

with one mole of a long-chain fatty ether group and one mole of a long-chain fatty acyl group on a glyceryl phosphoryl ethanolamine, is present in bovine erythrocytes.

Nelson (1967) has presented data on the neutral lipid composition of the erythrocytes of several species, namely, the cow, dog, goat, horse, pig, rabbit, rat, and sheep. He determined the cholesterol content by three methods: gas–liquid, thin-layer, and column chromatography, the last in conjunction with infrared spectrophotometry. The results obtained by the three methods were in good agreement. In erythrocytes of the cow, for example, cholesterol comprised 30.2, 28.4, and 27.9 % of the total lipid extract by gas–liquid chromatography, infrared spectrometry, and thin-

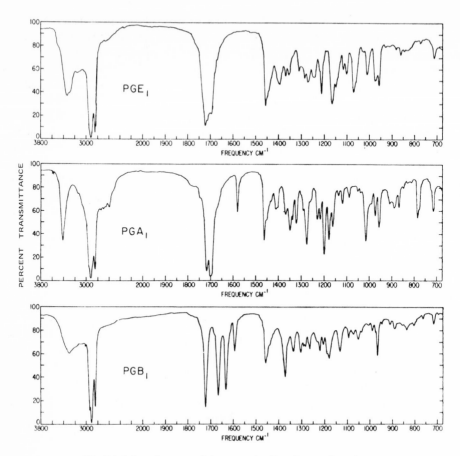

Fig. 7.1. Infrared spectra of six prostaglandins. (Ramwell *et al.*, 1968.)

layer chromatographic analysis, respectively. In general, the results obtained by one method did not differ by more than $\pm 5\%$ from those obtained by the other two. Tables 7.3 and 7.4 (p. 162) give the total erythrocyte lipid of the various mammals and their cholesterol content, respectively.

No evidence was found by Nelson for the "oxycholesterol" reported by Irie *et al.* (1961) in appreciable quantities in dog erythrocytes.

Tissue Analysis

Fatty acid ester content of bovine bone marrow has been determined (Thompson and Hanahan, 1963) by measurement of the infrared peak at 1739 cm^{-1} in extracts of the marrow with methanol and chloroform.

Freeman *et al.* (1953) have compared the spectra of lipoproteins obtained from human serum with similar spectra of such substances as ovalbumin, egg lecithin, vegetable oil, and cholesterol. They reported correlations of the observed spectra with molecules comprising the lipoprotein. The intensity of the ester carbonyl band can be used to estimate lipid content. The band at 1053 cm^{-1} can be used to determine

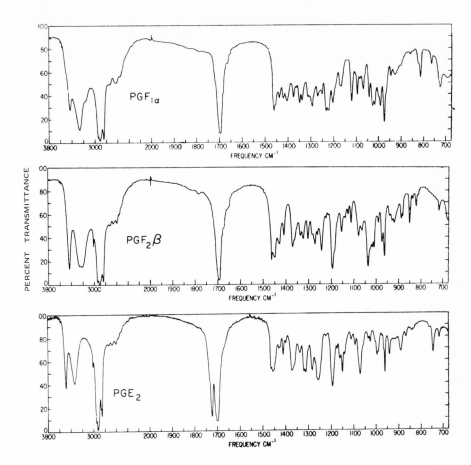

Fig. 7.1 (*continued*).

unesterified cholesterol. The amount of protein can be measured by using the ratio of the intensities of the bands at 1724 and 1639 cm^{-1}.

Hatch *et al.* (1967) have developed an ultracentrifugal method for isolating chylomicron-containing fractions from serum by flotation, with the use of either standard Spinco swinging-bucket rotors or a specially fabricated swinging-bucket rotor. They used an infrared spectrophotometric method for quantitation of the isolated lipoproteins having S_f greater than 400. The method makes use of the fact that the major constituent (about 85–90% by weight) of these macromolecules is triglyceride. The infrared absorption by ester carbonyl groups of the triglycerides at 1742 cm^{-1} provides a convenient measure of this lipid class. Cholesterol esters and phospholipids are present in relatively small amounts, and Hatch *et al.* estimate that the contribution of such compounds to the peak carbonyl absorption is no more than about 5% of the total.

The lipids in various membranes and in mitochondria are discussed in Chapter 19.

Table 7.3. Total Erythrocyte Lipid of Various Common Mammals (Nelson, 1967)

Species	Total lipid[a] mg/ml of packed cells
Cow	4.15
	4.27
Dog	5.16
	5.00
Goat	5.69
	5.89
Horse	4.70
	4.82
Pig	4.19
Rabbit	4.42
Rat	4.86
Sheep	4.44
	4.40

[a]Excluding gangliosides.

Table 7.4. Erythrocyte Cholesterol in Various Common Mammals (Nelson, 1967)

Species	Cholesterol[a]	
	% of total lipid	mg/ml of packed cells
Cow	28.8	1.20
	29.5	1.26
Dog	28.2	1.46
	28.0	1.40
Goat	27.5	1.57
	28.6	1.68
Horse	28.7	1.35
	29.3	1.41
Pig	27.5	1.15
Rabbit	29.9	1.32
Rat	26.1	1.27
Sheep	27.8	1.24
	28.2	1.24

[a]Values given in table are the average of the results obtained by the three different analytical methods.

REFERENCES

Abramson, M. B., Norton, W. T., and Katzman, R. *Federation Proc.* **23**, 222 (1964*a*).

Abramson, M. B., Katzman, R., and Gregor, H. P. *J. Biol. Chem.* **239**, 70 (1964*b*).

Abramson, M. B., Norton, W. T., and Katzman, R. *J. Biol. Chem.* **240**, 2389 (1965).

Ahlers, N. H. E., Brett, R. A., and McTaggart, N. G. *J. Appl. Chem.* (*London*) **3**, 433 (1953).

Alam, A. U. and Bishop, S. H. *Abstracts of the 156th National Meeting of the American Chemical Society*, Atlantic City, New Jersey, Sept. 1968, Paper No. 276.

Änggård, E. and Samuelsson, B. *J. Biol. Chem.* **239**, 4097 (1964).

Baer, E. *J. Am. Chem. Soc.* **75**, 621 (1953).

Baer, E. and Gróf, T. *Can. J. Biochem. Physiol.* **38**, 859 (1960).

Baer, E. and Pavanaram, S. K. *J. Biol. Chem.* **236**, 2410 (1961).

Baer, E. and Blackwell, J. *J. Biol. Chem.* **238**, 3591 (1963).

Baer, E. and Stanacev, N. Z. *J. Biol. Chem.* **239**, 3209 (1964).

Baer, E. and Stanacev, N. Z. *J. Biol. Chem.* **240**, 44 (1965*a*).

Baer, E. and Stanacev, N. Z. *J. Biol Chem.* **240**, 3754 (1965*b*).

Baer, E., Maurukas, J., and Russell, M. *J. Am. Chem. Soc.* **74**, 152 (1952).

Barceló Matutano, J. and Bellanato, J. *An. Real Soc. Españ. Fis. Quim* (*Madrid*), *Ser. B*, **49**, 557 (1953).

Barnes, R. B., Gore, R. C., Liddel, U., and Williams, V. Z. *Infrared Spectroscopy, Industrial Applications and Bibliography*, Reinhold, New York, 1944.

Baumann, W. J. and Ulshöfer, H. W. *Chem. Phys. Lipids* **2**, 114 (1968).

Bergström, S., Ryhage, R., Samuelsson, B., and Sjövall, J. *J. Biol. Chem.* **238**, 3555 (1963).

Bloom, R. J. and Westerfeld, W. W. *Biochemistry* **5**, 3204 (1966).

Cason, J., Freeman, N. K., and Sumrell, G. *J. Biol. Chem.* **192**, 415 (1951).

Chapman, D. and Morrison, A. *J. Biol. Chem.* **241**, 5044 (1966).

Firestone, D. and LaBouliere, P. *J. Ass. Offic. Agr. Chemists* **48**, 437 (1965).

Fischmeister, I. *Ark. Kemi* **26**, 453 (1967).

Fontanges, R., Heritier, P., and Coeur, P. *Bull. Soc. Chim. Biol.* **46**, 1223 (1964).

Freeman, N. K. *J. Am. Chem. Soc.* **74**, 2523 (1952).

Freeman, N. K. *J. Am. Chem. Soc.* **75**, 1859 (1953).

Freeman, N. K. *J. Lipid Res.* **5**, 236 (1964).

Freeman, N. K., Lindgren, F. T., Ng, Y. C., and Nichols, A. V. *J. Biol. Chem.* **203**, 293 (1953).

Freeman, N. K., Lampo, E., and Windsor, A. A. *J. Am. Oil Chemists' Soc.* **44**, 1 (1967).

Goldfine, H. *J. Biol. Chem.* **239**, 2130 (1964).

Golmohammadi, R. *Acta Chem. Scand.* **20**, 563 (1966).

Guertin, D. L., Wiberley, S. E., Bauer, W. H., and Goldenson, J. *Anal. Chem.* **28**, 1194 (1956).

Hanahan, D. J., Ekholm, J., and Jackson, C. M. *Biochemistry* **2**, 630 (1963).

Hatch, F. T., Freeman, N. K., Jensen, L. C., Stevens, G. R., and Lindgren, F. T. *Lipids* **2**, 183 (1967).

Hofmann, K., Lucas, R. A., and Sax, S. M. *J. Biol. Chem.* **195**, 473 (1952).

Irie, R., Iwanaga, M., and Yamakawa, T. *J. Biochem.* (*Japan*) **50**, 122 (1961).

Jart, A. *Acta Chem. Scand.* **14**, 1867 (1960).

Jones, R. N. *Can. J. Chem.* **40**, 321 (1962).

Jones, R. N., McKay, A. F., and Sinclair, R. G. *J. Am. Chem. Soc.* **74**, 2575 (1952).

Kaneda, T. *J. Biol. Chem.* **238**, 1222 (1963).

Kataura, A. and Kataura, K. *Tohoku J. Exp. Med.* **91**, 215 (1967*a*).

Kataura, A. and Kataura, K. *Tohoku J. Exp. Med.* **91**, 227 (1967*b*).

Kauffman, F. L. and Lee, G. D. *J. Am Oil Chemists' Soc.* **37**, 385 (1960).

Kauffman, F. L. *J. Am Oil Chemists' Soc.* **41**, 4 (1964).

Kepler, C. R. and Tove, S. B. *J. Biol. Chem.* **242**, 5686 (1967).

Kittredge, J. S., Isbell, A. F., and Hughes, R. R. *Biochemistry* **6**, 289 (1967).

Kuhrt, N. H., Welch, E. A., Blum, W. P., Perry, E. S., and Weber, W. H. *J. Am. Oil Chemists' Soc.* **29**, 261 (1952*a*).

Kuhrt, N. H., Welch, E. A., Blum, W. P., Perry, E. S., Weber, W. H., and Nasset, E. S. *J. Am. Oil Chemists' Soc.* **29**, 271 (1952*b*).

Marinetti, G. and Stotz, E. *J. Am. Chem. Soc.* **76**, 1345 (1954a).

Marinetti, G. and Stotz, E. *J. Am. Chem. Soc.* **76**, 1347 (1954b).

May, L. and Kamble, A. B. *Abstracts of the 156th National Meeting of the American Chemical Society,* Atlantic City, New Jersey. Sept., 1968, Biol. Chem. Div., Paper No. 203.

Mayers, G. L. and Haines, T. H. *Biochemistry* **6**, 1665 (1967).

Meiklejohn, R. A., Meyer, R. J., Aronovic, S. M., Schuette, H. A., and Meloch, V. W. *Anal. Chem.* **29**, 329 (1957).

Napier, E. A., Jr. *Biochemistry* **5**, 1279 (1966).

Nelson, G. J. *J. Lipid Res.* **8**, 374 (1967).

O'Connor, R. T., Field, E. T., and Singleton, W. S. *J. Am. Oil Chemists' Soc.* **28**, 154 (1951).

O'Connor, R. T., DuPré, E. F., and Feuge, R. O. *J. Am. Oil Chemists' Soc.* **32**, 88 (1955).

O'Connor, R. T. *J. Am. Oil Chemists' Soc.* **33**, 1 (1956).

Ramwell, P. W., Shaw, J. E., Clarke, G. B., Grostic, M. F., Kaiser, D. G., and Pike, J. E. "Prostaglandins," in *Progress in the Chemistry of Fats and Other Lipids, Vol. 9, Part 2,* Pergamon Press, Ltd., Oxford, 1968, p. 233.

Report of the Spectroscopy Committee, 1958–59, *J. Am. Oil Chemists' Soc.* **36**, 627 (1959).

Shreve, O. D., Heether, M. R., Knight, H. B., and Swern, D. *Anal. Chem.* **22**, 1498 (1950a).

Shreve, O. D., Heether, M. R., Knight, H. B., and Swern, D. *Anal. Chem.* **22**, 1261 (1950b).

Sinclair, R. G., McKay, A. F., and Jones, R. N. *J. Am. Chem. Soc.* **74**, 2570 (1952a).

Sinclair, R. G., McKay, A. F., Myers, G. S., and Jones, R. N. *J. Am. Chem. Soc.* **74**, 2578 (1952b).

Sobotka, H. and Stynler, F. E. *J. Am. Chem. Soc.* **72**, 5139 (1950).

Stanacev, N. Z., Baer, E., and Kates, M. *J. Biol. Chem.* **239**, 410 (1964).

Szonyi, C., Tait, R. S., and Craske, J. D. *J. Am. Oil. Chemists' Soc.* **39**, 276 (1962).

Thompson, G. A., Jr. and Hanahan, D. J. *Biochemistry* **2**, 641 (1963).

van den Oord, A., Danielsson, H., and Ryhage, R. *J. Biol. Chem.* **240**, 2242 (1965).

Wheeler, D. H. in *Progress in the Chemistry of Fats and Other Lipids, Vol. 2* (R. T. Holman, W. O. Lundberg, and T. Malkin, eds.), Pergamon Press, Ltd., London, 1954, p. 268.

Wilde, P. F. and Dawson, R. M. C. *Biochem. J.* **98**, 469 (1966).

Chapter 8

AMIDES AND AMINES

The spectra of amides are of importance especially because they are related to proteins and polypeptides. Table 8.1 (Bellamy, 1958) shows the important group

Table 8.1. Group Frequencies of Amides (Bellamy, 1958)

Type	Band,[a] cm^{-1}
I. NH Stretching Modes	
Primary amides	
Free NH	near 3500 and 3400 (m)
Bonded NH	near 3350 and 3180 (m)
Secondary amides	
Free NH—*trans*	3460–3400 (m)
Free NH—*cis*	3440–3420 (m)
Bonded NH—*trans*	3320–3270 (m)
Bonded NH—*cis*	3180–3140 (m)
Bonded NH—*cis* and *trans*	3100–3070 (w)
II. CO Absorption (Amide I)	
Primary amides	
Solid	near 1650
Dilute solution	near 1690 (s)
Secondary amides	
Solid	1680–1630
Dilute solution	1700–1670 (s)
Tertiary amides	
Solid	1670–1630 (s)
Dilute solution	1670–1630 (s)
Cyclic amides	
Large rings (in solution)	near 1680 (s)
Unfused γ-lactams	near 1700 (s)
Fused γ-lactams	1750–1700
Unfused β-lactams (in solution)	1760–1730 (s)
Fused β-lactams[b] (in solution)	1780–1770 (s)

Table 8.1 (*Continued*)

Type	Band,[a] cm^{-1}
III. NH$_2$ Deformation (Amide II)	
Primary amides only	
Solid	1650–1620 (s)
Solution	1620–1590 (s)
IV. Amide II	
Secondary noncyclic amides only	
Solid	1570–1515 (s)
Solution	1550–1510 (s)
V. Amide III	
Secondary amides only	near 1290 (m)
VI. NH Deformation (Amide V)	
Bonded secondary amides only	near 720 (m broad)
VII. Other Correlations (Amide IV and Amide VI Absorptions)	
Secondary amides only	near 620 and 600
Primary amides	1420–1400 (m)

[a]Key: m, medium; s, strong; w, weak.
[b]Fused to thiazolidine rings.

frequencies of various amides. The absorption bands displayed by the amides are due to N—H and C=O stretching vibrations, N—H deformations, and to mixed vibrations known as "amide bands," e.g., amide I, amide II, etc. The amide I band is caused by stretching of the C=O group. The amide II band comes from NH$_2$ deformation in primary amides and from a mixed vibration of N—H bending and C—N stretching in secondary amides.* The various bands displayed by amides are discussed again in greater detail in Chapter 10. Bentley *et al.* (1968) have summarized the characteristic frequencies of amides in the region ~ 700–300 cm^{-1}.

N—H Stretching Modes

Simple primary amides in dilute solutions display two bands due to free N—H stretching near 3500 and 3400 cm^{-1}. In the solid state two bands due to bonded N—H lie near 3350 and 3180 cm^{-1} (Randall *et al.*, 1949; Clarke *et al.*, 1949; Richards and Thompson, 1947; Darmon and Sutherland, 1949; Badger and Rubalcava, 1954). A detailed study (Cleverley, 1956) of free and associated N—H band positions in fourteen amides in chloroform solution showed that neither the position nor the intensity of the free N—H band is influenced by the length or nature of the alkyl chain. However, N—H stretching vibrations are concentration dependent. For example, *n*-valeramide

*See Hallam (1969) in Chapter 10 for an alternate assignment for the amide II absorption, and for a discussion of amide III, etc.

has associated N—H bands at 3493, 3344, 3295, and 3182 cm^{-1} at relatively high concentrations ($\sim 0.2M$).

Secondary amides have a free N—H stretching band in the range 3470–3400 cm^{-1} in dilute solutions. The frequency of bonded NH absorptions depends on the nature of the solvent and upon the concentration. With increasing concentration, two bonded NH bands are found at 3340–3140 and 3100–3060 cm^{-1} (Randall *et al.*, 1949; Clarke *et al.*, 1949; Richards and Thompson, 1947; Darmon and Sutherland, 1949). The spectra of polypeptides and proteins also have these bands. Two bands that appear as a doublet in secondary amides are due to *cis*- and *trans*-rotational isomers containing the free N—H stretching band (Russell and Thompson, 1956).

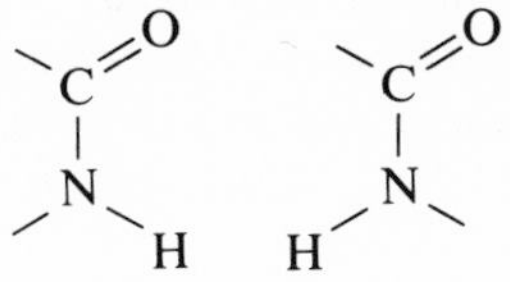

cis   *trans*

N-Methylacetamide and other simple secondary amides are found predominantly in the *trans* configuration. *N-t*-Butylphenylacetamide and other sterically hindered amides are found predominantly in the *cis* form. In open-chain secondary amides, the principal N—H stretching band lies near 3270 cm^{-1} in the solid state (Richards and Thompson, 1947; Darmon and Sutherland, 1949; Letaw and Gropp, 1953).

Secondary amides have N—H stretching bands at lower frequencies that are due to hydrogen bonding between the carbonyl and the N—H group (C=O$\cdots$H—N) (Richards and Thompson, 1947). The *cis* and *trans* forms are also associated with particular regions for the band in the 3340–3140 cm^{-1} range. Besides the free N—H stretching mode, at 3420 cm^{-1}, cyclic amides have a band in the range 3220–3170 cm^{-1} and one near 3080 cm^{-1} (Darmon and Sutherland, 1949; Tsuboi, 1949; Mizushima *et al.*, 1950; Klemperer *et al.*, 1954). The band in the range 3220–3140 cm^{-1} in secondary amides was assigned to the associated N—H stretching of a *cis*-bonded complex, and a band in the range 3340–3270 cm^{-1} was assigned to a *trans*-bonded structure.

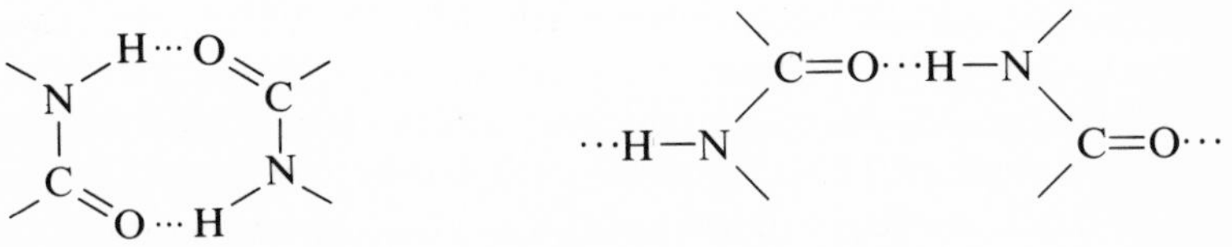

cis   *trans*

The Amide I Band: C=O Stretching

All primary amides display the amide I band, usually between 1715 and 1675 cm^{-1} in solution and around 1650 cm^{-1} in the solid state. For *n*-alkyl amides $CH_3(CH_2)_nCONH_2$ (where $n = 1$ to 10) the band is observed at 1679 cm^{-1} in dilute chloroform and the apparent molar absorptivity is between 600 and 700 (Jones and Cleverley, 1956). The amide I band of acetamide is unusual. It is resolved into two bands at 1714 and 1695 cm^{-1} in carbon tetrachloride, and similar bands in acetonitrile and chloroform. As the concentration increases, the lower frequency band becomes more intense while the higher one decreases, probably due to the upsetting of the monomer $\rightleftarrows$ trimer equilibrium. No doublet appears for the amide I band in the higher homologs, but as concentration increases the band broadens and the maximum moves to a lower frequency (Richards and Thompson, 1947). The lower position of the C=O band in amides compared to alkyl esters and dialkyl ketones can be attributed to the fact that the mesomeric effect (NH_2^+=CR—O$^-$) is much greater than in similar structures in the ester or ketone, and this may cause the amide I band to be more sensitive to intra- and intermolecular interactions than other kinds of carbonyl bands. The amide I band position is affected by substitution on the nitrogen atom and by ring strain. The band is also shifted to higher frequency on α-halogenation, for example, to 1732 cm^{-1} for a dilute solution of trichloroacetamide (Jones and Cleverley, 1956).

Lactam rings containing six or more members display the carbonyl band around 1680 cm^{-1} in solution as do normal amides. However, ring strain produces higher carbonyl frequencies in smaller rings, such as β- and γ-lactams (Clarke et al., 1949; Witkop et al., 1951; Mecke and Mecke, 1956; Brügel (quoted in Bellamy, 1958). Six-membered secondary and tertiary lactams have bands around 1665 and 1637 cm^{-1}. The amide I band occurs in the range 1643–1624 cm^{-1} in N-substituted δ-lactams, but when α,β-unsaturation is introduced (Edwards and Singh, 1954) bands occur at 1675 and 1615 cm^{-1}, and at 1666, 1634, and 1609 cm^{-1}.

Five-membered lactams, like γ-butyrolactam (2-pyrrolidone), which absorbs around 1700 cm^{-1}, show high carbonyl frequencies in solution and in the liquid state (Klemperer et al., 1954; Mecke and Mecke, 1956). γ-Lactams in which the lactam ring is fused to another ring absorb between 1750 and 1700 cm^{-1} (Clarke et al., 1949). β-Lactams absorb between 1760 and 1730 cm^{-1} (Clarke et al., 1949; Sheehan and Bose, 1950, 1951; Sheehan and Ryan, 1951; Sheehan et al., 1951; Sheehan and Laubach, 1951; Sheehan and Corey, 1951). When the α-carbon to the amide nitrogen of β-lactams has substitution of two ester groups, the carbonyl band is shifted up to 1770 cm^{-1}. The β-lactam carbonyl group in benzylpenicillin methyl ester (an example of a lactam ring fused to a second ring) has its C=O frequency at 1780 cm^{-1} (Thompson et al., 1949). This is much higher than the C=O frequency for monocyclic β-lactams (1760–1730 cm^{-1}).

Open-chain secondary N-monosubstituted amides display the C=O stretching band from 1680 to 1630 cm^{-1} in the solid state, and from 1700 to 1650 cm^{-1} in dilute solution. This band has been studied extensively, but the correlations are mainly applicable to solids (Randall et al., 1949; Thompson et al., in Clarke et al., 1949). The groups substituted on the nitrogen produce inductive and conjugation effects which influence the C=O frequency. For example, N-chloroacetamide has its carbonyl band at 1705 cm^{-1} (Gierer, 1953), while N-aryl substituted amides absorb near 1700 cm^{-1} (Richards and Thompson, 1947). Jones and Sandorfy (1956) have discussed the amide I band in N,N-disubstituted amides (tertiary amides, see Table

8.1). This band is usually found near 1650 cm^{-1}, but substitution of a phenyl group,

as in $CH_3-\overset{\overset{\displaystyle O}{\|}}{C}-\overset{\overset{\displaystyle CH_3}{|}}{N}-\langle\rangle$, brings the band up to 1692 cm^{-1}.

The Amide II Band

The amide II band of primary amides comes mainly from the scissoring motion of NH_2. The band is located at $1650–1620 \text{ cm}^{-1}$ in solids, and at $1620–1585 \text{ cm}^{-1}$ in dilute solutions (Randall *et al.*, 1949; Clarke *et al.*, 1949; Richards and Thompson, 1947; Jones and Cleverley, 1956). The band is weaker in most cases than the amide I band. Tertiary amides have no band in the 1620 cm^{-1} region since they have no NH_2 group.

The amide II band of secondary amides is found at $1570–1510 \text{ cm}^{-1}$ in solids and at $1550–1500 \text{ cm}^{-1}$ in dilute solutions (Randall *et al.*, 1949; Clarke *et al.*, 1949; Richards and Thompson, 1947). The physical state affects band locations very strongly. For example, *N*-methylacetamide in liquid, solution, and gaseous states displays the amide II band at 1565, 1534, and 1490 cm^{-1}, respectively (Mizushima *et al.*, 1950; Jones and Cleverley, 1956; Gierer, 1953; Davies *et al.*, 1955). In secondary cyclic amides (lactams) having a ring of less than nine members there is no amide II band. Nine-membered and larger rings do have the amide II absorption (Schiedt, 1954).

The amide II band in secondary amides appears to be due to a mixed vibration involving the $N-H$ in-plane bending and the $C-N$ stretching vibration (Fraser and Price, 1952, 1953; Miyazawa *et al.*, 1956, 1958). (For further discussion on this point see Chapter 10.) The amide I band also appears to be produced by a mixed vibration, the major contribution coming from $C=O$ stretching.

Other Correlations

Randall *et al.* (1949) reported the presence of a band in the range $1418–1399 \text{ cm}^{-1}$ in primary unsubstituted amides. This band does not appear in *N*-substituted amides. Since absorption in this region is not specific, use of this band for identification purposes is of small value (Bellamy, 1958).

The amide III band appears in secondary amides between 1310 and 1200 cm^{-1} and comes from a mixed vibration in which $C-N$ stretching and $N-H$ bending participate (Fraser and Price, 1952, 1953; Miyazawa *et al.*, 1956, 1958). The amide III band occurring at 1299 cm^{-1} in *N*-methylacetamide was characterized as due to 40% $C-N$ stretching, 30% $N-H$ bending, and 20% CH_3-C stretching (Miyazawa *et al.*, 1956, 1958).

The amide IV band is characteristic of secondary amides and appears near 620 cm^{-1}. It is due mainly to $O=C-N$ bending. The amide VI band, which appears near 600 cm^{-1}, is due to $C=O$ out-of-plane bending (Miyazawa *et al.*, 1956, 1958; Miyazawa, 1955, 1956).

The amide V band of secondary amides appears near 700 cm^{-1} and is characteristic of $N-H$ out-of-plane deformation (Miyazawa *et al.*, 1956, 1958; Kessler and

Table 8.2. Vibrations for Amines[a] (Alpert *et al.*, 1970)

Vibration	Frequency, cm^{-1}	Remarks
	A. Primary Amines	
N—H sym. stretch ⎱	3400 ± 100	Asymmetric band the higher of two bands
N—H asym. stretch ⎰		
NH$_2$ deformation	1620 ± 30	
NH$_2$ torsion	290	
Overtone of NH$_2$ torsion	496 ± 24	
C—N stretch		
a. Primary α carbon	1079 ± 11	
b. Secondary α carbon	1040 ± 3	
c. Tertiary α carbon	1030 ± 8	
	B. Secondary Amines	
N—H stretch	3400 ± 100	One band
C—N—C bend	427 ± 14	
C—N asym. stretch		
a. Primary α carbon	1139 ± 7	
b. Secondary α carbon	1181 ± 10	
N—H asym. bend		
a. Primary α carbon	739 ± 11	
b. Secondary α carbon	718 ± 18	
	C. Aromatic Amines	
C—N vibrations		
a. Primary	1350–1250	
b. Secondary (mono-aryl)	1260 ± 4	On deuteration of NH group band is found at 1344 ± 22
c. Secondary (diamyl)	1241	On deuteration of NH the band is above 1370
d. Tertiary	1360–1310	
	D. Ethylenediamine Derivatives	
N—C—N vibrations		
a. Ethylenediamine	1096, 1052	
b. Sym. dimethyl	1147, 1118	
ethylenediamine	1107, 1093	
c. Asym. dimethyl	1023, 1099	
ethylenediamine	1042	

[a]For the series of methyl, dimethyl, and trimethyl amines, vibrations associated with the CN structure are presented in Table 8.3. The bands for the corresponding hydrochlorides are also listed for comparison.

Sutherland, 1953). Secondary amides have a characteristic band near 3100 cm^{-1}. Miyazawa (1960) describes this band as the result of Fermi resonance* of N—H stretching with the combination band of C=O stretching and N—H in-plane bending in *cis*-amides, and the result of Fermi resonance of the N—H stretching with the overtone of the amide II band in *trans*-amides. (When an overtone or combination band is located near a fundamental frequency, the band intensity of the former may

*See Chapter 10 (*Proteins and Polypeptides*) for other comments on Fermi resonance.

be anomalously enhanced or bands may be split. This coupling between an overtone or combination band and a fundamental band is called Fermi resonance; the two levels also have to be of the same symmetry (Nakanishi, 1962).)

Amines

Table 8.2 (Alpert *et al.*, 1970) gives a summary of the important group frequencies of amines. These have also been discussed in Rao (1963) and Bellamy (1958). Table 8.3 (Alpert *et al.*, 1970) presents vibrations associated with the CN structure for the series methyl, dimethyl, and trimethyl amines. The bands for the corresponding hydrochlorides are also given. The infrared spectra of water solutions of the hydrochlorides of several biologically important amines have been recorded in the range 1550–909 cm^{-1} (Kirschenbaum and Parker, 1961). Bentley *et al.* (1968) have discussed alkyl and aromatic amine spectra recorded in the range $\sim 700 \sim 300$ cm^{-1}.

Table 8.3. Vibrations for Amines and Their Hydrochlorides[a] (Alpert *et al.*, 1970)

Compound	C—N symmetric stretch	C—N asymmetric stretch	C—N bend
Methylamine	1044		
Methylamine hydrochloride	995		
Dimethylamine	930	1024	
Dimethylamine hydrochloride	895		
Trimethylamine	826	1043	425, 365
Trimethylamine hydrochloride	817	985	465, 406

[a]In some heterocyclic ring compounds having the nitrogen in a ring with a CH$_3$ group attached to it, a band near 1050 cm^{-1} appears, indicative of the CH$_3$—N structure.

REFERENCES

Alpert, N. L., Keiser, W. E., and Szymanski, H. A. *IR—Theory and Practice of Infrared Spectroscopy*, 2nd Ed. Plenum Press, New York, 1970.
Badger, R. M. and Rubalcava, H. *Proc. Natl. Acad. Sci. U.S.* **40**, 12 (1954).
Bellamy, L. J. *The Infrared Spectra of Complex Molecules*, 2nd Ed., Wiley, New York, 1958.
Bentley, F. F., Smithson, L. D., and Rozek, A. L. *Infrared Spectra and Characteristic Frequencies* $\sim$700–300 cm^{-1}, Interscience, New York, 1968.
Clarke, H. T., Johnson, J. R., and Robinson, R., eds. *The Chemistry of Penicillin*, Princeton University Press, Princeton, New Jersey, 1949.
Cleverley, B., cited in Jones and Sandorfy, 1956.
Darmon, S. E. and Sutherland, G. B. B. M. *Nature* **164**, 440 (1949).
Davies, M., Evans, J. C., and Jones, R. L. *Trans. Faraday Soc.* **51**, 761 (1955).
Edwards, O. E. and Singh, T. *Can. J. Chem.* **32**, 683 (1954).
Fraser, R. D. B. and Price, W. C. *Nature* **170**, 490 (1952).
Fraser, R. D. B. and Price, W. C. *Proc. Roy. Soc.* **B141**, 66 (1953).
Gierer, A. *Z. Naturforsch.* **8b**, 644, 654 (1953).

Jones, R. N. and Cleverley, B., cited in Jones and Sandorfy, 1956.

Jones, R. N. and Sandorfy, C. in *Technique of Organic Chemistry* (A. Weissberger, ed.) *Vol. 9, Chemical Applications of Spectroscopy* (W. West, ed.), Interscience, New York, 1956, p. 513.

Kessler, H. K. and Sutherland, G. B. B. M. *J. Chem. Phys.* **21**, 570 (1953).

Kirschenbaum, D. M. and Parker, F. S. *Spectrochim. Acta* **17**, 785 (1961).

Klemperer, W. Cronyn, M. W., Maki, A. H., and Pimentel, G. C. *J. Am. Chem. Soc.* **76**, 5846 (1954).

Letaw, H., Jr. and Gropp, A. H. *J. Chem. Phys.* **21**, 1621 (1953).

Mecke, R., Jr. and Mecke, R. *Chem. Ber.* **89**, 343 (1956).

Miyazawa, T. *J. Chem. Soc. Japan, Pure Chem. Sect.* **76**, 341, 1018 (1955).

Miyazawa, T. *J. Chem. Soc. Japan, Pure Chem. Sect.* **77**, 171, 321, 526, 619 (1956).

Miyazawa, T. *J. Mol. Spectroscopy* **4**, 168 (1960).

Miyazawa, T., Shimanouchi, T., and Mizushima, S. *J. Chem. Phys.* **24**, 408 (1956).

Miyazawa, T., Shimanouchi, T., and Mizushima, S. *J. Chem. Phys.* **29**, 611 (1958).

Mizushima, S. Simanouti, T., Nagakura, S., Kuratani, K., Tsuboi, M., Baba, H., and Fujioka, O. *J. Am. Chem. Soc.* **72**, 3490 (1950).

Nakanishi, K. *Infrared Absorption Spectroscopy, Practical*, Holden–Day, San Francisco, 1962.

Randall, H. M., Fowler, R. G., Fuson, N., and Dangl, J. R. *Infrared Determination of Organic Structures*, Van Nostrand, Princeton, New Jersey, 1949.

Rao, C. N. R. *Chemical Applications of Infrared Spectroscopy*, Academic Press, New York, 1963.

Richards, R. E. and Thompson, H. W. *J. Chem. Soc.* **1947**, 1248.

Russell, R. A. and Thompson, H. W. *Spectrochim. Acta* **8**, 138 (1956).

Schiedt, U. *Angew. Chem.* **66**, 609 (1954).

Sheehan, J. C. and Bose, A. K. *J. Am. Chem. Soc.* **72**, 5158 (1950).

Sheehan, J. C. and Bose, A. K. *J. Am. Chem. Soc.* **73**, 1761 (1951).

Sheehan, J. C. and Corey, E. J. *J. Am. Chem. Soc.* **73**, 4756 (1951).

Sheehan, J. C. and Laubach, G. D. *J. Am. Chem. Soc.* **7**, 4752 (1951).

Sheehan, J. C. and Ryan, J. J. *J. Am. Chem. Soc.* **73**, 4367 (1951).

Sheehan, J. C., Hill, H. W., Jr., and Buhle, E. L. *J. Am. Chem. Soc.* **73**, 4373 (1951).

Thompson, H. W., Brattain, R. R., Randall, H. M., and Rasmussen, R. S. in *The Chemistry of Penicillin* (H. T. Clarke, J. R. Johnson, and R. Robinson, eds.), Princeton University Press, Princeton, New Jersey, 1949.

Tsuboi, M. *Bull. Chem. Soc. Japan* **22**, 215, 255 (1949).

Witkop, B., Patrick, J. B. and Rosenblum, M. *J. Am. Chem. Soc.* **73**, 2641 (1951).

Chapter 9

AMINO ACIDS, RELATED COMPOUNDS, AND PEPTIDES

Introduction

Many investigations have been made of the infrared spectra of amino acids (Hathway and Flett, 1949; Lenormant, 1946; Wright, 1937, 1939; Sutherland, 1950; Koegel *et al.*, 1955; Brockmann and Musso, 1956; Ehrhart and Hennig, 1956; Freymann and Freymann, 1938; Randall *et al.*, 1949; Fuson *et al.*, 1952; Thompson *et al.*, 1950; Klotz and Gruen, 1948; Clarke *et al.*, 1949; Kuratani, 1949; Larsson, 1950; Ellenbogen, 1956; and Leifer and Lippincott, 1957). Almost twenty years ago Sutherland (1952) reviewed the subject of infrared analysis of the structure of amino acids, polypeptides, and proteins. Gore (1954), Jones and Sandorfy (1956), Bellamy (1958), and Rao (1963) have more recently reviewed work on amino acids and related substances. Bentley *et al.* (1968) have discussed the spectra of amino acids, and many types of compounds related to them, in the region between 700 and 300 cm^{-1}. Most of the above-mentioned studies were conducted on amino acids in the solid state. Gore *et al.* (1949) recorded the absorption spectra and state of ionization of several acids and amino acids as a function of pH and examined hydrogen–deuterium exchange in deuterium oxide solution. Blout and Lenormant showed that infrared spectrophotometry can be used in aqueous solutions to study the structure and ionic state of amino acids, peptides, and proteins (Lenormant, 1952*a,b*; Lenormant and Chouteau, 1952; Lenormant and Blout, 1953; Blout and Lenormant, 1953; Doty *et al.*, 1957). Lenormant (1952*b*) discussed the methodology of the combined use of H_2O and D_2O as solvents.

Duval (1956) has studied the spectra of amino acids recorded from a single drop of aqueous solution. Parker and Kirschenbaum (1960) have recorded the spectra of several α- and non-α-amino acids in water. The spectra of aqueous solutions of glycine, *N*-methylglycine (sarcosine), *N,N*-dimethylglycine, and *N,N,N*-trimethylglycine (betaine) in different states of ionization have also been recorded (Kirschenbaum, 1963). Another solvent which has been suggested for infrared spectral determination of amino acids is antimony trichloride (Lacher *et al.*, 1954).

The α-amino acids in the solid state or at their isoelectric points in aqueous solution exist almost totally as dipolar ions. Edsall's work in Raman spectroscopy of amino acids demonstrated conclusively their dipolar ionic structure (Edsall, 1936,

1937, 1943; Edsall *et al.*, 1950). Because of the dipolar ionic structure many amino acids have a characteristic band at ~ 1587 cm^{-1} which is related to the $-COO^-$ group, as well as a rather weak absorption at ~ 2128 cm^{-1} which comes from NH frequencies in the $-NH_3^+$ ion (Thompson *et al.*, 1950; Klotz and Gruen, 1948).

Koegel *et al.* (1957) have reviewed the infrared absorption of the optically active and racemic straight-chain α-amino acids after presenting an earlier comprehensive paper (1955) on the spectra of about 50 such compounds and their derivatives. In every case these workers found that the spectrum of an L-amino acid was identical with that of the D-isomer in the range 5000 to 667 cm^{-1}.

Spectra–Structure Correlations and Other Structural Considerations

Some spectra–structure correlations have been given for amino acids and amino acid hydrochlorides [Bellamy (1958) and the references cited above in the first paragraph of the *Introduction* (Lenormant, 1946 through Leifer and Lippincott, 1957)], which are summarized in Table 9.1. The general characteristics of the spectra of these types of compounds are given below.

Table 9.1. Important Group Frequencies, in cm^{-1}, of Amino Acids and Their Hydrochlorides (Rao, 1963)

NH_3^+ vibrations (salts do not show these bands)	
Stretching	3130–3030 (m)
Deformation	1660–1590 (w) amino acid band I
	1550–1480 (w) amino acid band II
NH_2^+ vibrations (*N*-monosubstituted amino acids)	
Stretching	2800–2600 (m)
Deformation	1620–1560 (m)
Carboxyl absorption	
Ionized	
Asymmetric stretching	1600–1560 (s)
Symmetric stretching	~ 1410 (w) not easily identified
Un-ionized	
α-amino acid hydrochlorides	1755–1720 (s)
β, γ, and lower amino acid hydrochlorides	1730–1700 (s)
The 3000–2000 region	
Amino acids	2760–2530 (w), 2140–2080 (w)
	not observed in all cases
Amino acid hydrochlorides	3030–2500 (w) a series of bands.
Other correlations	
Amino acids	1300 (m), 880 (w)
Amino acid hydrochlorides	2000 (w)

Key: m, medium; s, strong; w, weak.

N—H Stretching Region

No absorption is displayed in the normal N—H stretching region (3500–3300 cm^{-1}). NH_3^+ stretching in acids causes a band to appear in the 3130–3030 cm^{-1}

region. Although NH_3^+ stretching causes absorption in this region in some amino acid hydrochlorides, such absorption does not appear in all of them. There is no such band for amino acid salts. Sarcosine and proline have no NH_3^+ stretching band since they are *N*-substituted, but show NH_2^+ stretching bands around 2700 cm^{-1}.

NH_3^+ and NH_2^+ Deformation Bands

Two bands probably caused by NH_3^+ deformations appear in all amino acids and their hydrochlorides in the regions 1660–1590 and 1550–1480 cm^{-1}. These bands are known as the amino acid I and II bands, respectively, the latter usually being stronger than the former. Isovaline does not have the amino acid II band. Some amino acids display splitting of this band. In hydroxy acids the II band moves to higher frequencies as the OH and NH_3^+ groups are positioned nearer to each other. For example, the deformation motion at 1504 cm^{-1} in ϵ-hydroxynorleucine shifts to 1522 cm^{-1} in δ-hydroxynorvaline, to 1538 cm^{-1} in homoserine, and does not appear in serine, since it is obscured by the 1600 cm^{-1} carboxyl band. Amino acid salts do not have these bands. The NH_2^+ deformation of *N*-substituted amino acids lies near 1600 cm^{-1}.

Carboxyl Absorption Bands

The asymmetric stretching of the COO$^-$ group causes all amino acids to have a strong band in the 1600–1560 cm^{-1} region. Symmetric stretching of this group causes a weak band to appear near 1410 cm^{-1}, but this band is often not easily identified. Since, as noted above, an NH_3^+ deformation band lies in the region 1660–1590 cm^{-1}, it often interferes with the asymmetric stretching band of the COO$^-$ group. The α-amino acid hydrochlorides display absorption of the un-ionized COOH group in the region 1755–1720 cm^{-1}. The β- and γ-amino acid hydrochlorides absorb in the range 1730–1700 cm^{-1}. Absorption characteristic of the COOH group also appears in the spectra of *N*-substituted amino acid hydrochlorides. Monoaminodicarboxylic acids have a COOH group absorption as well as bands from the NH_3^+ and COO$^-$ groups. The diaminomonocarboxylic acids, e.g., ornithine, have typical amino acid bands and, in addition, absorption bands due to free NH_2. Dicarboxylic amino acids and amino acid hydrochlorides display a C—O stretching vibration band near 1220 cm^{-1} due to the carboxyl group.

Bands in the 3000–2000 cm^{-1} Region

Bands in the 3000–2000 cm^{-1} region are produced by the NH_3^+ group, probably due to its stretching vibrations. Amino acids generally display an absorption band or two in the region 2760–2530 cm^{-1} and another one in the 2140–2080 cm^{-1} region, although exceptions are known. For example, ornithine lacks the 2130 cm^{-1} band and aminocaproic acid does not have the 2600 cm^{-1} band. Aminoadipic acid has a band at 1920 cm^{-1}, and aspartic acid has bands at 2062 and 1905 cm^{-1}. Amino acid hydrochlorides generally display a succession of bands from 3030 to 2500 cm^{-1}.

Other Band Correlations

Amino acid hydrochlorides with the NH_3^+ group have a band near 2000 cm^{-1}. Amino acids have a band near 1300 cm^{-1}. There has been no explanation concerning the structural origins of these characteristic bands. A band appearing near 880 cm^{-1} in many amino acids may come from the rocking motion of the NH_3^+ group.

Other Structural Considerations

None of the correlations referred to in the above discussion changes with chain length. The introduction of the sulfur atom does not appear to cause any striking effects. The dipolar ionic form prevails, even with long-chain molecules such as $NH_2(CH_2)_{10}COOH$. If the amino acid contains an aromatic group, the spectrum will have characteristics different from the usual cases in the region 1600–1500 cm^{-1}.

Among the amino acids, threonine, hydroxylysine, cystine, isoleucine, the two hydroxyprolines, and others possess two optically active centers. Therefore, the synthetic compounds are mixtures of four diastereoisomers: the L- and D- forms, and the L-allo- and D-allo- forms, respectively. For example, threonine can have these four forms: L-threonine (XLI), D-threonine (XLII), L-allothreonine (XLIII), and D-allothreonine (XLIV).

$$
\begin{array}{cccc}
\text{COOH} & \text{COOH} & \text{COOH} & \text{COOH} \\
H_2N-\overset{|}{\underset{|}{C}}-H & H-\overset{|}{\underset{|}{C}}-NH_2 & H_2N-\overset{|}{\underset{|}{C}}-H & H-\overset{|}{\underset{|}{C}}-NH_2 \\
H-\overset{|}{\underset{|}{C}}-OH & HO-\overset{|}{\underset{|}{C}}-H & HO-\overset{|}{\underset{|}{C}}-H & H-\overset{|}{\underset{|}{C}}-OH \\
CH_3 & CH_3 & CH_3 & CH_3 \\
\text{XLI} & \text{XLII} & \text{XLIII} & \text{XLIV}
\end{array}
$$

Examples of cases in which the amino acid and its allo form have spectra which differ from each other are: threonine and allothreonine, hydroxyproline and allo-hydroxyproline, isoleucine and alloisoleucine, phenylserine and allophenylserine.

A special case of compounds having two centers of symmetry includes substances in which the groups about the two centers are identical but the configuration about one of the two centers is a mirror image of the other. Compounds of this type are internally compensated and therefore are optically inactive "meso" compounds. Examples are mesocystine (XLV) and α,ϵ-diaminopimelic acid (XLVI).

$$
\begin{array}{cc}
\begin{array}{cc}
\text{COOH} & \text{COOH} \\
H_2N-\overset{|}{\underset{|}{C}}-H & H-\overset{|}{\underset{|}{C}}-NH_2 \\
CH_2-S & S-CH_2
\end{array}
&
\begin{array}{cc}
\text{COOH} & \text{COOH} \\
H_2N-\overset{|}{\underset{|}{C}}-H & H-\overset{|}{\underset{|}{C}}-NH_2 \\
CH_2-CH_2 & CH_2
\end{array} \\
\text{XLV} & \text{XLVI}
\end{array}
$$

The spectra of L-, D-, and *meso*cystine have been recorded and are almost indistinguishable (Marshall *et al.*, 1957), although earlier work (Wright, 1937, 1939) appears to disagree with these results. The methods of sample preparation may partially explain the discrepancies.

The spectrum of the *meso* form of α,ϵ-diaminopimelic acid displays some sharp bands from 1667 to 1250 cm^{-1}. The L form has a more diffuse spectrum in this region. Between 1250 and 667 cm^{-1} there is no resemblance between the spectra.

L-, D-, and DL-*Amino Acids*

Wright (1937, 1939) had studied infrared spectra of L- and D-amino acids and found them to be identical, but also found that the spectra of L and of D differed from that of the DL-racemate, for example, in the case of cystine, alanine, leucine, valine, and phenylalanine. Glutamic acid, on the other hand, had the same spectrum for the L or DL forms. Acetylleucine exhibits different spectra for the L- and DL-isomers, as is true of acetylisoleucine also (Darmon *et al.*, 1948). Brockmann and Musso (1956) examined L- and DL-alanine and found little difference between their spectra. However, the spectra of L- and DL-serine were quite different.

Koegel *et al.* (1957) also found that the spectrum of DL-alanine from 5000 to 667 cm^{-1} differs very little from that of the L form. The racemate has a band at 1028 cm^{-1}, which is missing for the L form. Also, for the racemate the amino acid I band at 1647 cm^{-1} is more intense and better resolved from the 1595 cm^{-1} band, which is assigned to the antisymmetric COO$^-$. The spectra appear to be nearly identical in all other areas.

The spectra of L- and DL-norleucine have been compared, along with butyrine, norvaline, and others in a homologous series. For the norleucine racemate (Greenstein and Winitz, 1961) there was a sharp, intense band at 1658 cm^{-1} (not present in the L form). The racemate spectrum showed a rather large difference between the intensities of the amino acid II band (1515 cm^{-1}) and the antisymmetric band at 1580 cm^{-1}. (In the L form these two bands had approximately equal intensities.) There were also distinctive differences between the L and DL forms in the methylene and methyl group bending regions. For butyrine, norvaline, and norleucine there are minor differences in the 4000 to 1667 cm^{-1} region, but from 1667 to 667 cm^{-1} the differences between the spectra for L and DL forms are quite pronounced.

The Use of Isotopes for Spectra–Structure Correlation

The study by Laulicht *et al.* (1966) is an example of the introduction of an isotope into an amino acid for the purpose of correlating spectra with structure. They recorded the spectrum of powdered ^{18}O-labeled α-glycine and compared it with the spectrum of normal glycine. (Glycine exists in three crystalline modifications, designated as α-, β-, and γ-glycine, which differ slightly in their vibrational spectra.) These workers showed that the 1334–1324 cm^{-1} band, previously attributed mainly to a CH$_2$ wagging, has a pronounced C—C stretching character. The 894 cm^{-1} band of normal glycine, which had been assigned earlier to C—C stretching, was suggested to be due to the COO$^-$ scissoring. The 700–685 cm^{-1} band was assigned to the COO$^-$ rocking, while that at about 502 cm^{-1} was attributed to an NH$_3^+$ group deformation.

Suzuki *et al.* (1963) studied the spectra of normal and various deuterated glycines and also made band assignments for α-glycine. Some of the assignments made by Laulicht *et al.* (1966) differ from those made by Suzuki *et al.* (1963). For example, the 1334 cm^{-1} band was assigned by the latter group to CH_2 wagging, the 893 cm^{-1} band to C—C stretching, and the 503 cm^{-1} band to COO$^-$ rocking.

Chelate Compounds

Nakamoto (1968) has discussed the theory and experimental results of infrared studies of metal chelate compounds of several types. In one example concerning amino acid complexes, he gave the N—H and CO stretching frequencies of sodium glycinate and *trans*-bis (glycino) complexes. The group of complexes was comprised of nickel, copper, palladium, and platinum compounds. He concluded from the infrared data that the strength of the metal–nitrogen and metal–oxygen coordinate bonds increases as the metal is changed in the order Ni(II) < Cu(II) < Pd(II) < Pt(II).

Ogiso (1967) obtained two kinds of crystals from the reaction of histidine and cupric ion. The ratio of histidine and Cu was found to be 1:1 and 2:1, and the structures of the Cu–histidine chelates were deduced from infrared data. In the 2:1 chelate, coordination from the α-amino N and the N of imidazole to Cu was presumed. The carbonyl–oxygen approaches the Cu closely enough to be considered as loosely bound. In the 1:1 chelate, coordination through the carbonyl group and the N of the imidazole ring was presumed to form a 7-membered ring.

Sarkar and Wigfield (1967) have disputed the view of other investigators that the imidazole group is involved in the Cu(II)–histidine (1:2 molar ratio) chelate and

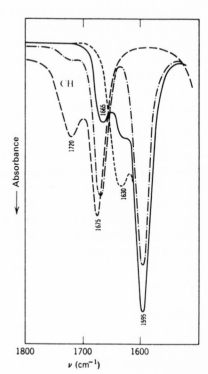

Fig. 9.1. Infrared spectra of glycylglycine in D_2O solution at 0.288M concentration and ionic strength 1.0, adjusted with KCl. Key: (– – –) pD 1.75; (– · – · –) pD 4.31; (——) pD 8.77; (- - - -) pD 10.29. (Kim and Martell, 1964.)

proposed that in Cu(II)–histidine the copper is bound to the amino and the carboxyl groups. Sarkar and Wigfield used titration data and ultraviolet and infrared spectroscopy to support their view. The infrared spectrum of Cu(II)–histidine in D_2O solution displayed a pronounced shift in the antisymmetric carboxyl stretching frequency. Free histidine shows the antisymmetric COOH stretching frequency at 1615 cm^{-1}, and Cu(II)–histidine exhibits the same frequency at 1593 cm^{-1}. This shift supports the concept of coordination of the carboxyl group. Nakamoto *et al.* (1961) have observed a similar shift by other Cu(II)–amino acid complexes.

The effects of intermolecular interaction which are characteristic of different crystal lattices of amino acid complexes can be avoided by comparing the spectra of aqueous solutions (see also, Nakamoto, 1963).

Kim and Martell (1964) have studied solution equilibria involving the formation of glycylglycino–Cu(II) complexes; their study combined aqueous infrared spectroscopy with potentiometric titrations and visible spectroscopy. Figure 9.1 shows the infrared spectra of glycylglycine in D_2O solution as a function of the *p*D. The spectral changes observed were interpreted in terms of the following solution equilibria (Kim and Martell, 1963):

$$
\underset{\text{XLVIII}}{\overset{\displaystyle \overset{O}{\underset{}{\parallel}} \;(1665\,\text{cm}^{-1}) \quad (1595\,\text{cm}^{-1})}{H_3N^+ - CH_2 - C - \underset{H}{\overset{|}{N}} - CH_2COO^-}}
$$

$pK_1 = 3.21$ $pK_2 = 8.12$

$$
\underset{\text{XLVII}}{\overset{\displaystyle \overset{O}{\parallel}\;(1675\,\text{cm}^{-1})}{H_3N^+ - CH_2 - C - \underset{\underset{(1720\,\text{cm}^{-1})}{H}}{\overset{|}{N}} - CH_2COOH}}
\qquad
\underset{\text{XLIX}}{\overset{\displaystyle \overset{O}{\parallel}\;(1630\,\text{cm}^{-1})}{H_2N - CH_2 - C - \underset{\underset{(1595\,\text{cm}^{-1})}{H}}{\overset{|}{N}} - CH_2COO^-}}
$$

At low *p*D values (1.75) two bands are seen at 1720 and 1675 cm^{-1}; the former is due to the un-ionized carboxyl groups and the latter is due to the peptide carbonyl group adjacent to the terminal ammonium group of species XLVII. As the *p*D is raised (4.31) a new band appears at 1595 cm^{-1}, which is due to the ionized carboxyl group. Thus the spectrum at this *p*D value was interpreted as being due to a mixture of species XLVII and XLVIII. At high *p*D (8.77) three bands are seen at 1665, 1630, and 1595 cm^{-1}. The 1665 and 1630 cm^{-1} bands are attributed to the peptide carbonyl groups of species XLVIII and XLIX, respectively. At very high *p*D (10.29) only the two bands (1630 and 1595 cm^{-1}) characteristic of the latter species are observed.

The infrared spectra of glycylglycine mixed with copper chloride in a 1:1 molar ratio in D_2O solution (total concentration of the ligand and the metal is 0.2333*M*) (Kim and Martell, 1964) are shown in Fig. 9.2. As shown above, glycylglycine displays three bands at 1720, 1675, and 1595 cm^{-1} at *p*D 3.58. However, at the same *p*D value, the mixture displays one extra band at 1625 cm^{-1}. This band was attributed to the peptide carbonyl of the metal complex (L) which was formed by the reaction

$$\text{XLVII} \atop \text{and} \quad + \text{Cu}^{2+} \longrightarrow \quad \left[\begin{array}{c} \text{O} \\ \text{H}_2\text{C}-\text{C}^{\,1625\,\text{cm}^{-1}} \\ \text{H}_2\text{N} \qquad \text{NH} \\ \text{Cu} \qquad \text{CH}_2 \\ \text{H}_2\text{O} \qquad \text{O} \atop 1598\,\text{cm}^{-1} \text{O} \end{array} \right]^{+} + x\text{H}^{+}$$

$$\text{XLVIII}$$

L

The same solution displays one broad band at about $1610\,\text{cm}^{-1}$, if the pD is raised to 5.18. This result was interpreted as an indication that the equilibrium is shifted almost completely to the right-hand side, and that the $1610\,\text{cm}^{-1}$ band is due to the overlap of two bands of the carbonyl groups of species LI. The shift of the peptide CO stretching band from 1625 (L) to $1610\,\text{cm}^{-1}$ (LI) was interpreted as direct evidence for the ionization of the peptide NH hydrogen, since such an ionization results in the

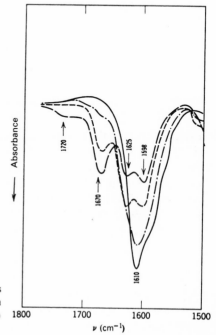

Fig. 9.2. Infrared spectra of Cu(II) glycylglycine complexes in D$_2$O solutions (1:1); ionic strength 1.0, adjusted with KCl. Key: (– – – –) pD 3.58; (- - - -) pD 4.24; (– · – · –) pD 5.18; (——) pD 10.65. (Kim and Martell, 1964.)

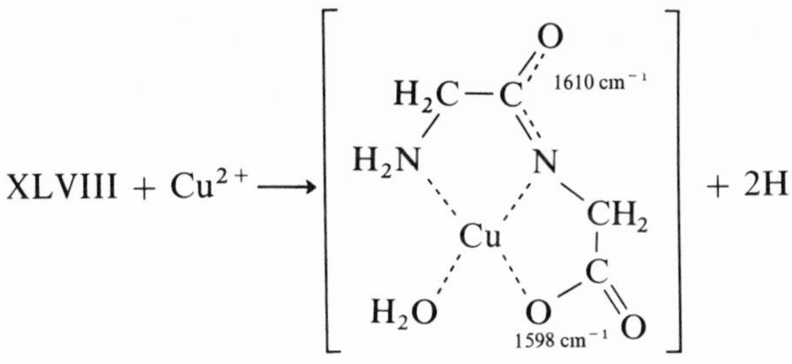

$$\text{XLVIII} + \text{Cu}^{2+} \longrightarrow \left[\begin{array}{c} \text{H}_2\text{C} - \text{C}^{\displaystyle O} \ ^{1610\,\text{cm}^{-1}} \\ \text{H}_2\text{N} \qquad \text{N} \\ \qquad \text{Cu} \qquad \text{CH}_2 \\ \text{H}_2\text{O} \qquad \text{O} - \text{C} \\ \qquad ^{1598\,\text{cm}^{-1}} \ \ \text{O} \end{array} \right] + 2\text{H}^+$$

LI

resonance of the O—C—N system as indicated in structure (LI). The same workers (Kim and Martell, 1966) also extended their investigations to the triglycine– and tetraglycine–Cu(II) systems.

Determination of Ionization Constants

Ionization constants have been determined for D_2O solutions of glycine, glycylglycine, and DL-alanine (Orlov *et al.*, 1967). The constants were calculated from the peak intensities and were found to be 2.7 and 10.0 for glycine; 3.5 and 8.5 for glycylglycine; and 2.6 and 10.4 for DL-alanine, for the amino and carboxyl groups, respectively.

Amino Acid Derivatives

DNP and PTH Derivatives

Kimmel and Saifer (1964) have recorded the infrared spectra of the 2,4-dinitrophenyl (DNP) and 3-phenyl-2-thiohydantoin (PTH) derivatives of 27 amino acids of importance in the study of biological materials, e.g., plasma, tissues, and urine. These workers have also made a quantitative study of the intensities of the major absorption bands according to the method of Flett (1962), and showed that the infrared spectra of DNP- and PTH-amino acids are a useful means for the determination of the N-terminal residues of various peptides and proteins, especially when used in conjunction with thin-layer chromatography.

Murray and Smith (1968) have recorded the spectra of microgram amounts of amino acid phenylthiohydantoins eluted from thin layers of silica gel. Their analyses were performed upon fibrin samples, separated from both autopsy and surgical thrombi, and other protein specimens.

Derivatives from Ribonuclease

In a paper discussing the alkylation and identification of the histidine residues at the active site of ribonuclease, Jaffe (1963) has analyzed infrared spectra of the 1- and 3-carboxymethylhistidines and of the 1- and 3-methylhistidines, and has confirmed spectroscopically the positions assigned to the carboxymethyl groups in the

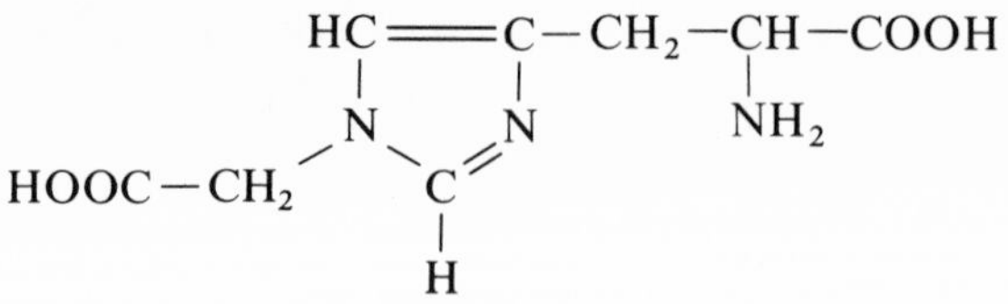

$$HC = C - CH_2 - CH - COOH$$

(Structure LII: imidazole ring with HC=C bearing CH₂–CH(NH₂)–COOH side chain; ring nitrogens N and N with central C–H; HOOC–CH₂ attached to ring N)

HOOC — CH₂

H

LII

$$HC = C - CH_2 - CH - COOH$$

(Structure LIII: imidazole ring with HC=C bearing CH₂–CH(NH₂)–COOH side chain; ring nitrogens N=C and N with central C–H)

H

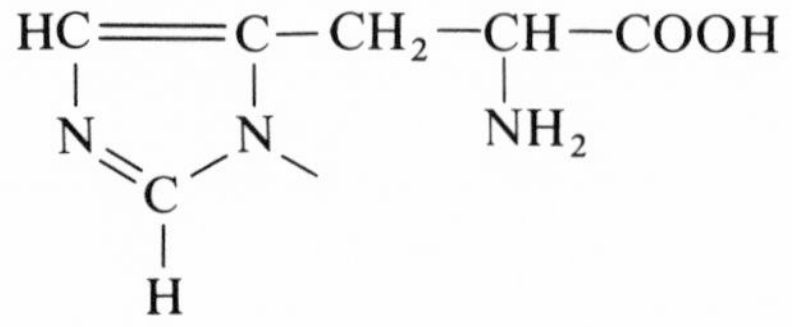

LIII

isomeric carboxymethylhistidines: 3-carboxymethylhistidine (LII) and 1-carboxy-methylhistidine (LIII). Crestfield *et al.* (1963) had demonstrated that when ribo-nuclease A is alkylated by iodoacetate at *p*H 5.5, two inactive products are formed, 1-carboxymethylhistidine-*119*-ribonuclease and 3-carboxymethylhistidine-*12*-ribo-nuclease. Jaffe found a definite and useful correlation of the infrared spectral patterns in the 833 cm^{-1} region with the position of substituents on the imidazole ring. The spectra of both carboxymethylhistidines showed a doubling of the band at about 2222 cm^{-1}, which suggested that these compounds may exist as dizwitterions, the unsubstituted ring nitrogen presumably assuming the second positive charge (NH^{+}), and giving rise to the band at 2000 cm^{-1}.

Compounds Related to Histidine

Assignments have been made for absorption bands of imidazole and several of its derivatives (e.g., tribromo, triiodo, and deuterated). This study, by Cordes and Walter (1968), is a good example for demonstrating how such assignments are made. For example, absorption bands are ascribed to CH, NH, CD, ND, CC, and CX stretchings; NH, ND, CH, and CD bending; overtones; combinations (e.g., 1493 cm^{-1} assigned to 757 + 736 cm^{-1}); ring stretching; torsion; and lattice vibrations.

Reaction of Amino Acids and Amines with N-Substituted Maleimides

Sharpless and Flavin (1966) have described the isolation and determination of the structures of the products arising from the reaction of certain primary and secondary amines and amino acids with *N*-substituted maleimides. Examples of the spectral properties are given in Table 9.2.

Infrared and NMR spectra were consistent with addition of an amino group to the double bond of the maleimide (LIV). R$_1$ and R$_2$ may be carbon atoms in a ring or

Table 9.2. Spectral Properties of Maleimide Adducts
(Sharpless and Flavin, 1966)

Compounds	Carbonyl bands, cm^{-1}		
	Amide I	Amide II	Cyclic imide
Pyrrolidine–maleimide adduct	1715	—	1771
Piperidine–maleimide adduct	1712	—	1770
Benzylamine-N-phenyl maleimide adduct	1709	—	1779
Glycylamide-N-ethyl maleimide adduct	1706⎱ 1672⎰	1634	1779
Glycylamide-N-phenyl maleimide adduct	1718⎱ 1681⎰	1603	1786

additional hydrogen atoms, and R_3 may be H, Et, or phenyl.

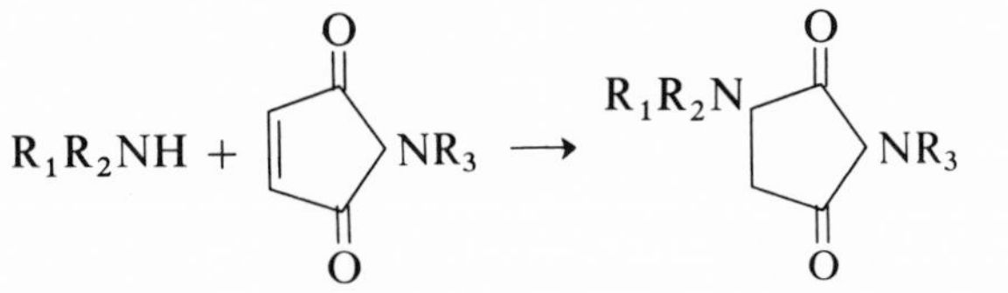

$$R_1R_2NH + \text{(maleimide)} NR_3 \longrightarrow \text{(adduct)} NR_3$$

LIV

Simple Peptides

Several workers have studied the infrared spectra of simple peptides (Thompson *et al.*, 1950; Blout and Linsley, 1952; Ellenbogen, 1956; Berger *et al.*, 1954; Elliott and Malcolm, 1956; Elliott, 1954; Asai *et al.*, 1955). Dipeptides have a dipolar ionic structure. Glycylglycine has bands at 3300, 3080 (NH stretching), 1655 (amide I), 1630 (amino acid I band), 1575 (amino acid II band), 1540 (amide II), and 1240 cm^{-1} (amide III) (Thompson *et al.*, 1950; Blout and Linsley, 1952), and COO$^-$ bands at 1608 and 1400 cm^{-1}. Bellamy (1958) has discussed the effect of the number of glycyl residues in the chain on the influence of the dipolar ionic structure on spectra. He has also discussed other types of polypeptides, including those composed of only one type of amino acid and those composed of mixed amino acids. (See also, Chapter 10.)

Aqueous Solutions of Dipeptides

In a study of several dipeptides in isoionic, basic, and acidic H$_2$O solutions in the range 1538–952 cm^{-1} some properties of the COO$^-$ and COOH groups have also been discussed (Parker, 1967). The spectra of the acidic forms of the peptides (where the COOH group is not dissociated and the amino group is present as NH$_3^+$) are usually quite different from those of the basic forms (the NH$_2$ group is uncharged and the carboxylate has a negative charge) and from those of the isoionic forms (NH_3^+ and COO^- both present), although even between the isoionic and basic forms minor differences exist. The dipeptides studied (about 20) were mainly pairs of those in which there is a reversal of the positions of amino acid residues, e.g., glycyl-L-serine and L-serylglycine, and glycyl-L-proline and L-prolylglycine. All the dipeptides were shown to be distinguishable from each other by means of their spectra.

Low-Temperature Studies with Amino Acids and Simple Polypeptides

Feairheller *et al.* (1969) have used low-temperature spectroscopy to characterize some amino acids and simple polypeptides. The results on the simple amino acids showed that the greatest improvement in spectra can be expected for compounds in the dipolar ionic form. On the other hand, very little improvement was noted for the hydrochlorides of amino acids.

Figure 9.3 shows spectra of glycyl-L-threonine at room temperature ($298°K$; top) and at about $110°K$ (bottom). The region below $1000 \, cm^{-1}$ is definitely improved at liquid-nitrogen temperature. Several bands observed in this region were not visible in the spectrum recorded at room temperature.

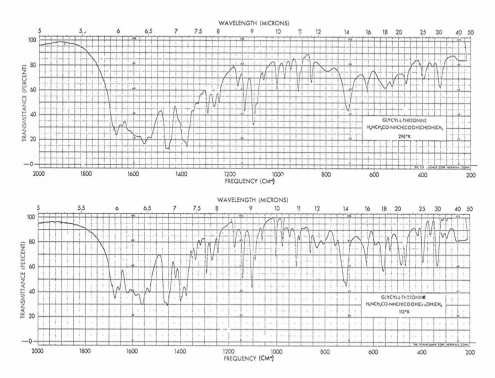

Fig. 9.3. The partial infrared spectrum of a Nujol mull of glycyl-L-threonine at room temperature (upper) and at about $110°K$ (lower). (Feairheller *et al.*, 1969.)

Table 9.3 lists the frequencies that were observed for a number of amino acids in the 400 to 250 cm^{-1} region. Just as the region between 700 and 450 cm^{-1} has been pointed out by Warren *et al.* (1966) to be very useful in the identification of "unknown" amino acids, so the region 400 to 250 cm^{-1} could be quite useful also for this purpose. Feairheller and Miller (1971) have extended studies of these compounds and simple peptides to 33 cm^{-1}.

Table 9.3. Observed Absorption Bands of Amino Acids (400–250 cm^{-1}) (Feairheller et al., 1969)

Compound	Temperature, °K	Band,[a] cm^{-1}					
DL-Alanine	298	376(w),	295(m),	280(vw),			
	113	352(w),	327(w),	297(m),	280(vw)		
DL-Arginine (mono HCl)	298	390(w),	330(m),	310(vw)			
	113	393(m),	333(m),	311(w)			
DL-Asparagine	298	389(s),	351(m),	294(m),	276(vw)		
	113	393(s),	355(m),	306(vw),	301(m),	293(vw),	280(vw)
DL-Aspartic Acid	298	392(s),	348(s)				
	113	397(s),	349(s)				
DL-Cystine	298	384(m),	321(w),	265(m)			
	113	391(m),	327(w),	285(vw),	270(w)		
Glutamic Acid	298	395(w),	380(vw),	325(vw)			
	113	400(m),	382(w),	345(vw),	325(vw),	313(vw)	
DL-Histidine (HCl) mono hydrate	298	377(m),	325(w),	294(w),	271(vw),		
	113	384(s),	343(m),	330(vw),	297(m),	279(w)	
DL-Isoleucine	298	370(m),	339(vw)				
	113	402(w),	372(s),	341(w),	315(vw)		
DL-Leucine	298	400(s),	365(s),	350(m)			
	113	400(s),	369(s),	350(m),	292(w)		
Lysine (mono HCl)	298	355(w),	311(w)				
	113	361(m),	351(w),	313(m)			
DL-Lysine (di HCl)	298	380(vw),	354(s),	295(s)			
	113	395(vw),	387(m),	356(s),	299(s)		
DL-Methionine	298	380(m),	367(vw),	348(vw),	324(s),	302(vw),	280(vw)
	113	378(m),	356(vw),	347(vw),	333(vw),	325(m),	310(w),	290(vw)
DL-Phenylalanine	298	374(s),	337(m)				
	113	378(s),	361(vw),	340(m),	318(vw)		
DL-Proline	298	385(s)	250(vw)				
	113	398(vs),	295(w),	281(vw),	255(vw)		
DL-Tyrosine	298	385(s),	337(w),	325(vw),	306(m)		
	113	387(s),	340(m),	325(w),	311(s)		

[a]Key: m, medium; s, strong; v, very; w, weak.

REFERENCES

Asai, M., Tsuboi, M., Shimanouchi, T., and Mizushima, S. *J. Phys. Chem.* **59**, 322 (1955).

Bellamy, L. J. *The Infrared Spectra of Complex Molecules, 2nd Ed.*, Wiley, New York, 1958.

Bentley, F. F., Smithson, L. D., and Rozek, A. L. *Infrared Spectra and Characteristic Frequencies*, ∼700–300 cm⁻¹, Interscience, New York, 1968.

Berger, A., Kurtz, J., and Katchalski, E. *J. Am. Chem. Soc.* **76**, 5552 (1954).

Blout, E. R. and Linsley, S. G. *J. Am. Chem. Soc.* **74**, 1946 (1952).

Blout, E. R. and Lenormant, H. *J. Opt. Soc. Am.* **43**, 1093 (1953).

Brockmann, H. and Musso, H. *Chem. Ber.* **89**, 241 (1956).

Clarke, H. T., Johnson, J. R., and Robinson, R., eds. *The Chemistry of Penicillin*, Princeton University Press, Princeton, New Jersey, 1949.

Cordes, M. and Walter, J. L. *Spectrochim. Acta* **24A**, 237 (1968).

Crestfield, A. M., Stein, W. H., and Moore, S. *J. Biol. Chem.* **238**, 2413 (1963).

Darmon, S. E., Sutherland, G. B. B. M., and Tristram, G. R. *Biochem. J.* **42**, 508 (1948).

Doty, P., Wada, A., Yang, J. T., and Blout, E. R. *J. Polymer Sci.* **23**, 851 (1957).

Duval, C. *Mikrochim. Acta* **1956**, 741.

Edsall, J. T. *J. Chem. Phys.* **4**, 1 (1936).

Edsall, J. T. *J. Chem. Phys.* **5**, 225, 508 (1937).

Edsall, J. T. *J. Am. Chem. Soc.* **65**, 1767 (1943).

Edsall, J. T., Otvos, J. W., and Rich, A. *J. Am. Chem. Soc.* **72**, 474 (1950).

Ehrhart, G., and Hennig, I. *Chem. Ber.* **89**, 2124 (1956).

Ellenbogen, E. *J. Am. Chem. Soc.* **78**, 363, 366, 369 (1956).

Elliott, A. *Proc. Roy. Soc.* **A221**, 104 (1954).

Elliot, A. and Malcolm, B. R. *Trans. Faraday Soc.* **52**, 528 (1956).

Feairheller, W. R., Jr. and Miller, J. T., Jr. *Appl. Spectrosc.* **25**, 175 (1971).

Feairheller, W. R., Jr., Miller, J. E., Jr., Katon, J. E., Bentley, F. F., and Parker, F. S., Pittsburgh Conference on Analytical Chemistry and Applied Spectroscopy, Cleveland, Ohio, March, 1969, Paper 147.

Flett, M. St. C. *Spectrochim. Acta* **18**, 1537 (1962).

Freymann, M. and Freymann, R. *Proc. Indian Acad. Sci.* **8**, 301 (1938).

Fuson, N., Josien, M. -L., and Powell, R. L. *J. Am. Chem. Soc.* **74**, 1 (1952).

Gore, R. C. *Anal. Chem.* **26**, 11 (1954).

Gore, R. C., Barnes, R. B., and Petersen, E. *Anal. Chem.* **21**, 382 (1949).

Greenstein, J. P. and Winitz, M. *Chemistry of the Amino Acids*, Wiley, New York, 1961.

Hathway, D. E. and Flett, M. St. C. *Trans. Faraday Soc.* **45**, 818 (1949).

Jaffe, H. cited in Crestfield *et al.*, 1963.

Jones, R. N. and Sandorfy, C. *Infrared and Raman Spectrometry: Applications, Technique of Organic Chemistry, Vol. 9* (W. West, ed.) Interscience, New York, 1956, p. 509.

Kim, M. K. and Martell, A. E. *J. Am. Chem. Soc.* **85**, 3080 (1963).

Kim, M. K. and Martell, A. E. *Biochemistry* **3**, 1169 (1964).

Kim, M. K. and Martell, A. E. *J. Am. Chem. Soc.* **88**, 914 (1966).

Kimmel, H. S. and Saifer, A. *Anal. Biochem.* **9**, 316 (1964).

Kirschenbaum, D. M. *Appl. Spectrosc.* **17**, 149 (1963).

Klotz, I. M. and Gruen, D. M. *J. Phys. Colloid Chem.* **52**, 961 (1948).

Koegel, R. J., Greenstein, J. P., Winitz, M., Birnbaum, S. M., and McCallum, R. A. *J. Am. Chem. Soc.* **77**, 5708 (1955).

Koegel, R. J., McCallum, R. A., Greenstein, J. P., Winitz, M., and Birnbaum, S. M. *Ann. N.Y. Acad. Sci.* **69**, 94 (1957).

Kuratani, K. *J. Chem. Soc. Japan, Pure Chem. Sect.* **70**, 453 (1949).

Lacher, J. R., Croy, V. D., Kianpour, A., and Park, J. D. *J. Phys. Chem.* **58**, 206 (1954).

Larsson, L. *Acta Chem. Scand.* **4**, 27 (1950).

Laulicht, I., Pinchas, S., Samuel, D., and Wasserman, I. *J. Phys. Chem.* **70**, 2719 (1966).

Leifer, A. and Lippincott, E. R. *J. Am. Chem. Soc.* **79**, 5098 (1957).

Lenormant, H. *J. Chim. Phys.* **43**, 327 (1946).

Lenormant, H. *Compt. Rend.* **234**, 1959 (1952*a*).

Lenormant, H. *J. Chim. Phys.* **49**, 635 (1952*b*).

Lenormant, H., and Blout, E. R. *Nature* **172**, 770 (1953).

Lenormant, H., and Chouteau, J. *Compt. Rend.* **234**, 2057 (1952).

Marshall, R., Winitz, M., Birnbaum, S. M., and Greenstein, J. P. *J. Am. Chem. Soc.* **79**, 4538 (1957).

Murray, M. and Smith, G. F. *Anal. Chem.* **40**, 440 (1968).

Nakamoto, K. *Infrared Spectra of Inorganic and Coordination Compounds*, Wiley, New York, 1963, p. 204.

Nakamoto, K., in K. Nakamoto and P. J. McCarthy, *Spectroscopy and Structure of Metal Chelate Compounds*, Wiley, New York, 1968, p. 216.

Nakamoto, K., Morimoto, Y., and Martell, A. E. *J. Am. Chem. Soc.* **83**, 4528 (1961).

Ogiso, T. *Yakugaku Zasshi J. Pharm. Soc. Japan* **87**, 617 (1967). (English summary.)

Orlov, I. G., Markin, V. S., Moiseev, Yu. V., and Khurgin, U. I. *Khim. Prir. Soedin.* **3**, 197 (1967); *CA*, **67**, 78918h.

Parker, F. S. in *Progress in Infrared Spectroscopy, Vol. 3* (H. A. Szymanski, ed.) Plenum Press, New York, 1967, p. 89.

Parker, F. S. and Kirschenbaum, D. M. *Spectrochim. Acta* **16**, 910 (1960).

Randall, H. M., Fowler, R. G., Fuson, N., and Dangl, J. R. *Infrared Determination of Organic Structures*, Van Nostrand, Princeton, New Jersey, 1949.

Rao, C. N. R. *Chemical Applications of Infrared Spectroscopy*, Academic Press, New York, 1963.

Sarkar, B. and Wigfield, Y. *J. Biol. Chem.* **242**, 5572 (1967).

Sharpless, N. E. and Flavin, M. *Biochemistry* **5**, 2963 (1966).

Sutherland, G. B. B. M. *Discussions Faraday Soc.* **9**, 319 (1950).

Sutherland, G. B. B. M. *Advan. Protein Chem.* **7**, 291 (1952).

Suzuki, S., Shimanouchi, T., and Tsuboi, M. *Spectrochim. Acta* **19**, 1195 (1963).

Thompson, H. W., Nicholson, D. L., and Short, L. N. *Discussions Faraday Soc.* **9**, 222 (1950).

Warren, R. J., Thompson, W. E., Zarembo, J. E., and Eisdorfer, I. B. *J. Ass. Offic. Anal. Chemists* **49**, 1083 (1966).

Wright, N. *J. Biol. Chem.* **120**, 641 (1937).

Wright, N. *J. Biol. Chem.* **127**, 137 (1939).

Chapter 10

PROTEINS AND POLYPEPTIDES

Conformational Studies of Proteins and Polypeptides

Protein spectra tend to be diffuse owing to the presence of a large variety of side-chain bands. However, infrared spectroscopy has been useful in the study of proteins and particularly as an approach to polypeptide structure. The modern history of the application of infrared spectroscopy to peptide and polypeptide structure determination began in the late 1940's. Following the availability of approximate assignments for certain major absorption bands of the peptide group, workers applied these band frequency positions, and their dichroic properties, to making correlations with the conformations of the peptide and the formation of hydrogen bonds. Mizushima and his colleagues mainly used model amide and peptide compounds as a start. The results they found by using these model compounds in various solvents as a function of concentration allowed them to set up criteria for detecting hydrogen bonds and determining bond strengths. The thorough investigations by Mizushima (1954) and Mizushima and Shimanouchi (1961) of the infrared spectra of amides and peptides showed that their conformation was always *trans*. Tsuboi (1949) had studied the spectra of lactams and diketopiperazines which contained the *cis*-amide group and found that the position of the NH stretching mode of the *cis*-amide group was somewhat below 3250 cm^{-1}, but that of the *trans* conformation was above this frequency. Tsuboi concluded that the peptide groups in most proteins that had been studied had the *trans* conformation. It has been pointed out (Schellman and Schellman, 1964) that *cis*-amide structures rarely occur in polypeptide conformation except for peptide links in which proline and hydroxyproline are present.

In the late 1940's Darmon and Sutherland (1947), Astbury *et al.* (1948), Sutherland (1950), and later the Courtaulds workers obtained data on the spectra of synthetic polypeptides.

The bands most frequently considered in the early studies were the following: ~ 3450 cm^{-1} (N—H stretching); ~ 1680 cm^{-1} (amide I, C=O stretching); and near 1500 cm^{-1} [amide II, N—H deformation (Bamford *et al.*, 1956)]. These frequencies are related to amide groups that are not hydrogen-bonded. When hydrogen-bonded the N—H stretching vibration shifts to ~ 3300 cm^{-1}, a value which is relatively independent of conformation. As a result, workers have frequently used this band shift as a sign of hydrogen bonding in the absence of conformational effects. The amide I and II bands also shift when hydrogen bonds are formed—however, these bands *are* dependent on conformation. Table 10.1 shows the positions of these three bands in the α- and β-forms and in structures not involving hydrogen bonds.

Table 10.1. Characteristic Frequencies of Three Principal Peptide Bands, in cm^{-1} (Kauzmann, 1957)

Type of vibration	α-Form	β-Form	Not hydrogen-bonded
N—H stretch	3290–3300	3280–3300	~3400
C=O stretch (amide I)	1650–1660	1630	1680–1700
N—H deformation (amide II)	1540–1550	1520–1525	<1520?

Ambrose and Elliott (1951) investigated the dichroism of the three bands above and obtained the following information about qualitative dichroic properties (Table 10.2). If we assume that the C=O and N—H stretching vibrations have transition

Table 10.2. Dichroism of the Three Principal Absorption Bands of the Peptide Group[a] (Ambrose and Elliott, 1951)

Assignment	Frequency, cm^{-1}		Dichroism	
	α	β	α	β
N—H stretch	3300	3300	‖	⊥
C=O stretch (amide I)	1660	1640	‖	⊥
N—H deformation (amide II)	1545	1525	⊥	‖

[a]From Schellman and Schellman (1964).

moments parallel to the direction of stretching and that the N—H distortion is perpendicular to the direction of the bond, these findings are completely in line qualitatively with proposals that the β-forms have polypeptide chains connected by transverse hydrogen bonds, and with Pauling and Corey's model for the α-helix (Fig. 10.1) in which the N—H and C=O directions are virtually identical with the direction of the helical axis (Schellman and Schellman, 1964). Figure 10.2 shows a partly idealized schematic drawing of the α-helix.

Elliott and Malcolm (1956a) and Downie *et al.* (1957) showed that the procedure of identification of α- or β-structure by correlation with frequency ranges listed in Table 10.1 can lead to erroneous conclusions in special cases. The Miyazawa theory (to be discussed later) has made possible an analysis more sophisticated than the empirical method mentioned here.

If we consider that the three major absorption bands of the amide group given above actually represent normal modes of motion of the whole amide linkage, along

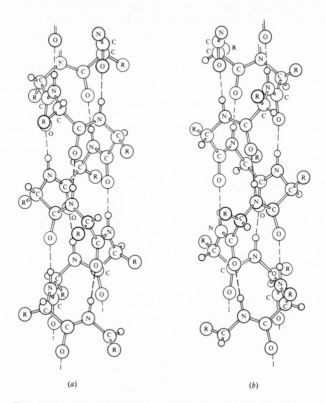

Fig. 10.1. Drawings of the (a) left-handed and the (b) right-handed α-helical forms of a polypeptide chain containing L-amino acids. The side chains, R, and hydrogen atoms attached to the α-carbons in the main chain have positions corresponding to the known configuration of L-amino acids (Pauling *et al.*, 1951). (From Mahler and Cordes, 1966.)

with those of neighboring atoms, we do not expect that the directions of the transition moments characteristic of the bands will have a very simple geometric relationship with particular bonds in the amide group. Sandeman (1955) studied the infrared dichroism of crystalline N,N'-diacetylhexamethylenediamine and thereby determined the directions of the three transition moments (Fig. 10.3). Abbott and Elliott (1956) obtained values of 12°, 20°, and 72°, respectively, for the transition moment directions of the N—H stretch, amide I, and amide II bands, thus demonstrating that these transition moments are not very dependent on the kinds of groups attached at either end of the peptide linkage.

Miyazawa's Theory and Related Considerations

Miyazawa's work set up systems for analyzing the vibronic excited states of a polypeptide molecule. The work began with a normal coordinate treatment of the

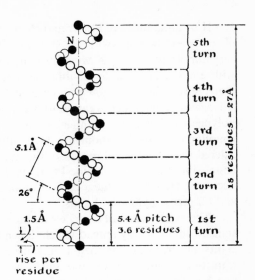

Fig. 10.2. A schematic representation of the α-helix. The indicated dimensions are characteristic (Corey and Pauling, 1955). (From Pimentel and McClellan, 1960.)

trans isomer of *N*-methylacetamide and its deuterated form (Miyazawa *et al.*, 1958). The character of the motions associated with the amide I and II bands was denoted by calculations of the percentage of the total potential energy of the vibration involved in the stretching and bending of single bonds. For the amide I vibration, 80% of the potential energy is associated with C—O stretching, 10% with C—N stretching, and 10% with an in-plane bending motion of the N—H bond. For the amide II vibration, 40% of the potential energy is associated with the C—N stretching and 60% with the

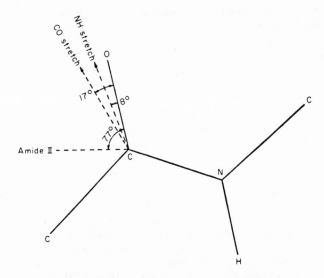

Fig. 10.3. Transition moment directions in the peptide link of *N,N'*-diacetylhexamethylenediamine by Sandeman (1955). (From Schellman and Schellman, 1964.)

N—H bending motion. Fraser and Price (1952) had previously suggested the mixing of the N—H bending with the stretching modes of the N—C—O group.* Another very important finding of Miyazawa and his co-workers was that motions associated with amide I and amide II vibrations are isolated in the peptide linkage itself and there is very little motion produced in the methyl groups of N-methylacetamide. Miyazawa (1960c) then went on to investigate the infrared absorption of an ordered polypeptide chain in a manner similar in many ways to Moffitt's treatment of the coupling of transition moments in the ultraviolet spectra of polypeptides. He considered that the normal modes of the macromolecular system would consist of the coupled motions of the amide I and amide II vibrations. He considered coupling as occurring (1) through the α-carbon atom linking peptide groups with one another, and (2) across hydrogen bonds.

The excited state of a single extended polypeptide chain is made up of a band of states with an energy spread which depends on the magnitude of the coupling constants. The transition moment to any one of these states is the sum of all the individual transition moments in the molecule, each multiplied by a phase factor (Schellman and Schellman, 1964). In the important special situation of the polypeptide chain with a twofold screw axis (sheet structures), a group of two peptide groups in sequence can be considered as a unit cell. In this situation (Davydov, 1962) the allowed motions can be described in terms of transition moments that are either in phase with one another or 180° out of phase (π radians). The former case produces a parallel absorption band, the latter a perpendicular band; thus, the component of the transition moments in the direction of the molecular axis contributes only to the parallel band and the component perpendicular to this axis contributes to the perpendicular band.

The two absorptions occur at slightly different energies, the magnitude of the difference being proportional to the coupling constants between the adjacent peptide vibrations. The frequencies may be written as follows (Miyazawa, 1960c):

$$v_{\parallel}(0) = v_0 + D_1$$

$$v_{\perp}(\pi) = v_0 - D_1$$

---

*Hallam (1969) has offered an alternative assignment for the amide II absorption which is based on the Fraser and Price (1952, 1953) model. According to the calculations of Fraser and Price (1952, 1953) the amide II band is the upper member of a pair of modes arising from the interaction of the symmetric vibration of the $\begin{smallmatrix} O \\ \diagdown \\ C-N \\ \diagup \end{smallmatrix}$ group and the in-plane bending motion of the N—H bond, the lower member being amide III. According to Fraser and Price the frequency of the amide II mode is primarily determined by the O=C—N group but its activity is due mainly to the $\delta_i(N—H)$ contribution. Hallam (1969) considers $\delta_i ND$ rather than $\delta_i NH$ to give rise to the coupling phenomena. He suggests that amide II is primarily a localized $\delta_i NH$ mode and that $\delta_i ND$ couples with $v_s NCO$ (amide III) to give rise to amide II' (as a result of $v_s NCO$ being uncoupled when NH becomes deuterated) at $\sim 1450\,cm^{-1}$ and amide III' at $\sim 900\,cm^{-1}$. Hallam (1969) suggests that in most primary and secondary amides and urethanes, peptides and proteins, amide II is essentially $\delta_i NH_2$ or $\delta_i NH$, and amide III, vCN (strictly speaking $v_s OCN$). To account for the behavior of these types of molecules upon deuteration, Hallam suggests that $\delta_i NH$ shifts by $\sim \sqrt{2}$ to $\sim 1200\,cm^{-1}$ and that this $\delta_i ND$ interacts strongly with vCN (which also shifts slightly in going from OC·NH to OC·ND) to give rise to amide II' and amide III'.

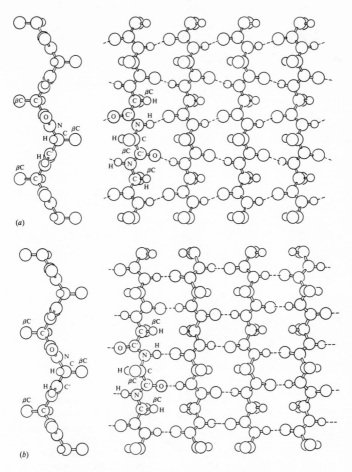

Fig. 10.4. Representation of (a) the parallel and (b) the antiparallel pleated-sheet structures for polypeptides (Pauling and Corey, 1951). (From Mahler and Cordes, 1966.)

where the (0) and (π) indicate the relative phases of adjacent transition moments and D_1 is the coupling constant for intrachain coupling.

The unit cell in a polypeptide sheet structure (Fig. 10.4) contains four peptide groups (Fig. 10.5), giving four phase possibilities for the transition moments corresponding to the two choices which come from each of the chains. The four frequencies related to these transitions may be written as follows:

$$v(0, 0) = v_0 + D_1 + D'_1$$

$$v(\pi, 0) = v_0 - D_1 + D'_1$$

$$v(0, \pi) = v_0 + D_1 - D'_1$$

$$v(\pi, \pi) = v_0 - D_1 - D'_1$$

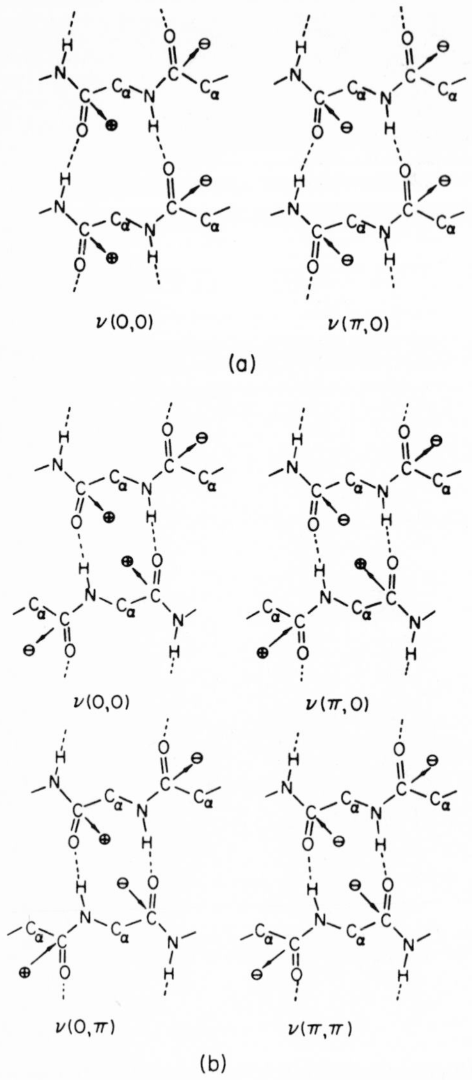

$\nu(0,0)$ $\nu(\pi,0)$

(a)

$\nu(0,0)$ $\nu(\pi,0)$

$\nu(0,\pi)$ $\nu(\pi,\pi)$

(b)

Fig. 10.5. (a) A schematic representation of the vibrational modes of the parallel-chain pleated sheet; the arrows represent the components of the transition moments of peptide groups in the plane of the paper. The plus and minus signs represent the components of the transition moments perpendicular to the plane of the paper, the former pointing upward, and the latter pointing downward. (b) A schematic representation of the vibrational modes of the antiparallel-chain pleated sheet (Miyazawa, 1960c). (From Schellman and Schellman, 1964.)

The factors in parentheses are explained as follows: $(\pi, 0)$, for example, means that there is a 180° phase change as one moves along the polypeptide chain and there is no phase change in going from one amide group to another connected by a hydrogen bond; similarly, $(0, \pi)$ denotes no phase change as we proceed along the polypeptide chain, and a 180° change as we move from one amide group to another hydrogen-bonded to it. Further, D_1 is the intrachain coupling constant, and D_1' is the interchain coupling constant across a hydrogen bond.

We see in Fig. 10.5 that for antiparallel chains $\nu(0, 0)$ leads to complete infrared inactivity as a result of transition-moment cancellations. For antiparallel chains, then, the other three band frequencies are

$$v_\|(0, \pi) = v_0 + D_1 - D_1'$$

$$v_\perp(\pi, 0) = v_0 - D_1 + D_1'$$

$$v_\perp(\pi, \pi) = v_0 - D_1 - D_1'$$

and the two bands of the parallel chain arrangement are designated by

$$v_\|(0, 0) = v_0 + D_1 + D_1'$$

$$v_\perp(\pi, 0) = v_0 - D_1 + D_1'$$

Miyazawa and Blout (1961) have calculated the interaction constants D_1 and D_1' from data on several materials (particularly polyglycine and nylon 66), assuming that the interaction constants are the same in these substances. (This gives the approximate magnitude of the constants, but small changes in v_0 cause disproportionately large changes in D.) The frequency of the motion when unaffected by coupling, v_0, was obtained by examining the spectrum of incompletely deuterated poly-γ-benzyl-L-glutamate, since deuteration was presumed to destroy the resonance interaction of groups by changing their frequencies, with the result that a CONH link surrounded by COND links should have the "intrinsic" value for v_0.

For an α-helix the frequencies are designated by

$$v_\|(0) = v_0 + D_1 + D_3$$

$$v_\perp(\theta) = v_0 + D_1 \cos\theta + D_3 \cos 3\theta$$

where $\theta = 100°$ for an α-helix and D_1, D_3, and v_0 were found by reference to compounds of known structure (see Table 10.3).

Table 10.3 presents a comparison between observed and calculated frequencies of the amide I and amide II bands of polypeptides in various conformations (Miyazawa and Blout, 1961; Miyazawa, 1962). Note the excellent agreement in values. The theory has been very useful to explain some infrared spectral features not previously understood and has been able to make predictions which were later found by experiment to be correct.

Fraser and Suzuki (1970) have described a quantitative analysis of the infrared spectrum of β-keratin, using a method of interpreting the observed dichroism, which allowed all three infrared active components of amide I associated with the antiparallel chain pleated sheet to be detected and estimated. Fraser and Suzuki found the following amide I frequencies for β-keratin: $v_\|(0, \pi) = 1697 \text{ cm}^{-1}$; $v_\perp(\pi, 0) = 1629 \text{ cm}^{-1}$; and $v_\perp(\pi, \pi) = 1683 \text{ cm}^{-1}$. These frequencies were calculated from the following equations:

$$v_\|(0, \pi) = v_0 + \delta v - \delta v'$$

$$v_\perp(\pi, 0) = v_0 - \delta v + \delta v'$$

$$v_\perp(\pi, \pi) = v_0 - \delta v - \delta v'$$

where v_0 is the unperturbed frequency, δv is due to interaction between adjacent

Table 10.3. Observed and Calculated Frequencies, in cm^{-1}, of the
Amide I and II Bands of Polypeptides in Various Conformations[a]
(Schellman and Schellman, 1964)

Conformation	Designation	Theoretical frequency	Amide I frequencies			Amide II frequencies		
			ν_{obs}[c]	ν_{cal}[d]		ν_{obs}[c]	ν_{cal}[d]	
Random coil		ν_0	1656 s	1658		1535 s	1535	
α-Helix	$\nu\|(0)$	$\nu_0 + D_1 + D_3$	1650 s	(1650)	$D_1 = 10$	1516 w	(1516)	$D_1 = -21$
	$\nu\perp(2\pi/n)$[b]	$\nu_0 - 0.17D_1 + 0.50D_3$	1652 m	1647	$D_3 = -18$	1546 s	1540	$D_3 = 2$
Parallel-chain pleated sheet	$\nu\|(0,0)$	$\nu_0 + D_1 + D_1'$	1645 w	1648	$D_1 = 10$	1530 s	1530	$D_1 = -10$
	$\nu\perp(\pi,0)$	$\nu_0 - D_1 + D_1'$	1630 s	1632		1550 w	1550	$D_1' = 5$
Antiparallel-chain pleated sheet	$\nu\|(0,\pi)$	$\nu_0 + D_1 - D_1'$	1685 w	(1685)	$D_1 = 8$	1530 s	(1530)	$D_1 = -10$
	$\nu\perp(\pi,0)$	$\nu_0 + D_1 + D_1'$	1632 s	(1632)	$D_1' = -18$		1540	$D_1' = -5$
	$\nu\perp(\pi,\pi)$	$\nu_0 - D_1 - D_1'$		1668			1550	
Nylon 66		$\nu_0 + D_1'$	1640 s	(1640)		1540 s	(1540)	$D_1' = 5$

[a]From Miyazawa and Blout (1961) and Miyazawa (1962).
[b]The letter n is the number of peptide groups per turn of the helix.
[c]The letter symbols in parentheses indicate observed intensities: s, strong; m, medium; w, weak.
[d]The values in parentheses were used in the calculations of D_1, D_1', and D_3.

residues in the same chain, and $\delta v'$ is due to interaction between residues in adjacent chains. Use of the observed experimental values for dichroic spectra yielded $v_0 = 1663 \, \text{cm}^{-1}$, $\delta v = 7 \, \text{cm}^{-1}$, and $\delta v' = -27 \, \text{cm}^{-1}$.

Monosubstituted amides and polypeptides display two characteristic bands in the $3000 \, \text{cm}^{-1}$ region, namely the strong band at $\sim 3300 \, \text{cm}^{-1}$ (the amide A band) and the weaker band at $\sim 3100 \, \text{cm}^{-1}$ (the amide B band) (Miyazawa, 1962). These bands disappear on N-deuteration and are associated with the N—H bond. The origin of the amide B band has been in question for some time. Badger and Pullin (1954) proposed the explanation that the amide A and B bands arise from the Fermi resonance* between the fundamental N—H stretching vibration and the first over-tone of the amide II vibration. Miyazawa's (1960b) work with amides in the gaseous, liquid, and solid phase substantially supports this assignment. For the case of the cis-amide group characteristic bands are seen at $\sim 3200 \, \text{cm}^{-1}$ and $\sim 3100 \, \text{cm}^{-1}$ and these bands are believed to arise from the Fermi resonance between the fundamental N—H stretching vibration and the combination tone of the C=O stretching vibra-tion ($\sim 1650 \, \text{cm}^{-1}$) and the N—H in-plane bending vibration ($\sim 1450 \, \text{cm}^{-1}$) (Miya-zawa, 1960$a$).

In addition to the bands discussed above, the amide bands III, IV, V, VI, and VII have also been analyzed by Miyazawa (1962). For the amide III vibration ($\sim 1300 \, \text{cm}^{-1}$) the amplitude of the N—H bending displacement is the largest one as in the case of the amide II vibration. Of the total vibrational potential energy, 30% is associated with each of the N—H bending and C—N stretching modes, and 10% with each of the C=O stretching and O=C—N bending modes. There are other vibrational interactions to be taken into account, for example, 20% of the potential energy is associated with the C_α—C stretching mode. This vibration is therefore not isolated within the peptide group.

The amide IV, V, and VI bands of the CONH group are observed in the 800–600 cm^{-1} region. The amide IV band observed at $\sim 620 \, \text{cm}^{-1}$ is primarily due to the O=C—N bending mode. The amide V band observed at $\sim 700 \, \text{cm}^{-1}$ is due to the N—H out-of-plane bending mode (Kessler and Sutherland, 1953). The amide VI band observed at $\sim 600 \, \text{cm}^{-1}$ for liquid N-methylacetamide is considered to be due to the C=O out-of-plane bending mode (Fig. 10.6). The amide VII band, due essen-tially to the torsional vibration about the C—N bond of the peptide link, is observed at $206 \, \text{cm}^{-1}$ (in the Raman effect) in liquid N-methylacetamide. Miyazawa (1961) has estimated the energy barrier for cis–$trans$ isomerization at about $14 \, \text{kcal/mole}$ using observations of this band in N-methylformamide and N-methylacetamide. Only limited application has been made of the amide III–VII bands in conformational studies. The amide V bands are found in the following positions in polypeptides (Miyazawa, 1967): at about $700 \, \text{cm}^{-1}$ for the extended form (of poly-γ-methyl-L-glutamate, poly-L-alanine, and sodium polyglutamate); at about $650 \, \text{cm}^{-1}$ for the disordered form (of polyserine and sodium polyglutamate); and at about $620 \, \text{cm}^{-1}$

*Fermi resonance is a coupling between a fundamental band and an overtone or combination band. Coupling occurs when the two bands are at similar frequencies and symmetry and can result in a sub-stantial increase in intensity of the weaker band and the appearance of a split absorption (Cross and Jones, 1969).

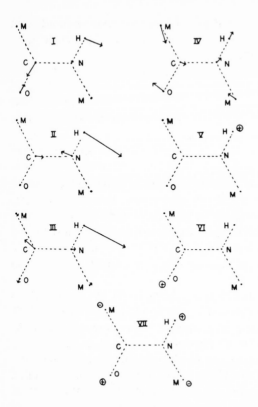

Fig. 10.6. Calculated normal modes of the amide I, II, III, and IV vibrations of N-methylacetamide and approximate representations of the amide V, VI, and VII vibrations. (Miyazawa, 1962.)

for the α-helical form (of poly-γ-methyl-L-glutamate, poly-γ-benzyl-L-glutamate, and poly-L-alanine). The amide V bands are found also in proteins: 600 cm^{-1} for the band of the α-helix of α-keratin (Beer *et al.*, 1959); and 690, 650, and 600 cm^{-1}, respectively, for the extended form, the disordered form, and the α-helical form of lysozyme (Fukushima and Miyazawa, 1964).

A band at ∼410 cm^{-1} (observed for perpendicular orientation in polarized spectra) is characteristic of the α-helical form of polypeptides, e.g., high-molecular-weight poly-γ-methyl-L-glutamate, and is not seen for the extended form. High-molecular-weight poly-γ-benzyl-L-glutamate has an α-helical band at 408 cm^{-1}, and poly-γ-ethyl-L-glutamate has one at 405 cm^{-1}. The band at ∼410 cm^{-1} is thus useful for characterizing conformation and can be used to estimate the fraction of the α-helical form (Miyazawa, 1967).

Miyazawa's theory gave a theoretical basis for our understanding of polypeptide spectra, although Elliott and Bradbury (1962) and Tsuboi (1963) have questioned certain details. The British workers expressed doubt that unique values for v_0 and for the interaction constants exist which are applicable to a variety of polypeptides in different conformations. They studied intrachain coupling in a series of polymers of the form $[-CO-NH-(CH_2)_m-]_n$ where m is an odd integer progressing from 1 to 9. They found v_0 to have small variations and D_1 to be larger than D_1', and they felt that the more important interaction of adjacent amide groups is the intrachain one, noting that it may be necessary to take account of next nearest as well as nearest neighbors (Bradbury and Elliott, 1963). Tsuboi (1963) claimed that Fermi resonance is not an adequate explanation for the $3100\ cm^{-1}$ band.

Besides the bands characteristic of the peptide link there are many bands associated with the primary chain itself and with side chains. Some assignments of prominent bands have been made even though the side-chain bands depend on the individual polypeptide examined. Beer et al. (1959) have discussed the characteristic bands of several fibrous proteins.

Tsuboi (1962) studied the infrared characteristics of poly-γ-benzyl-L-glutamate. The spectra of this compound are presented in Fig. 10.7a–d. Frequencies, dichroic ratios, and assignments for several of these bands are given in Table 10.4. The angle θ

Table 10.4. Wave Numbers, Transition Moment Directions, and Assignments of Some of the Absorption Bands of Poly-γ-benzyl- L-glutamate[a] (Schellman and Schellman, 1964)

Wave number, cm^{-1}		Dichroic ratio $R(\|/\perp)$	Angle θ between transition moment and fiber axis, in degrees	Assignment
$\|$	$\perp$			
3292	3294	$\begin{cases}7.3\ (A)^{[b]}\\4.2\ (B)\end{cases}$	27 29	N—H stretch
1734	1733	1.1 (A)	53	C=O stretch (ester)
1652	1655	2.9 (A)	39	Amide I
1518	1549	$\begin{cases}0.18(A)\\0.25(B)\end{cases}$	74 76	Amide II
1498	1498	1.7	~46	C—C stretch (phenyl A_1)
1453	1453	0.67(B)	61	C—C stretch (phenyl B_1)
1328	1314	2.3 (B)	~40	Amide III
1168	1168	1.0 (B)	54	C—O stretch (ester)
697	697	0.65(B)	62	Phenyl B_2
—	613	~0.1 (B)	>80	Amide V
563	—	$\sim\infty$ (B)	~ 0	Skeletal deformation in long helix

[a]From Tsuboi (1962).
[b]Thinner or thicker films (A) or (B) are used to determine dichroic ratios.

(column 4 of the table) is obtained from the observed dichroic ratio by means of the formula (Fraser, 1953; Beer, 1956)

$$R\left(\frac{\|}{\perp}\right) = \frac{2\cos^2\theta + g}{\sin^2\theta + g}$$

where g is an empirical parameter (equal to Fraser's $\frac{2}{3}(1 - f)/f$; see Fraser, 1953; Zbinden, 1964) which is a measure of deviation from the perfect orientation of the fibers in the sample film. Tsuboi's conclusions, which were based on these results and the known or assumed transition moment directions in the groups concerned, are as follows: (a) The orientation of the peptide group with respect to the fiber axis is

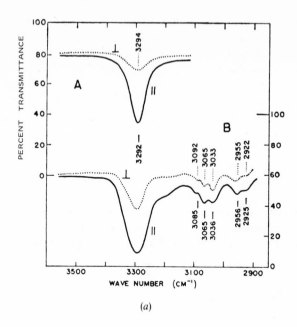

(a)

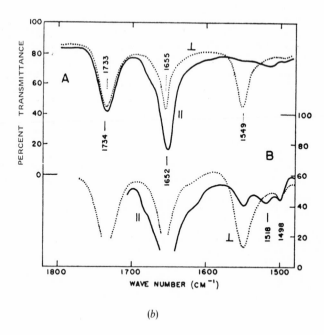

(b)

Fig. 10.7a, b, c, and d. Infrared absorption curves of oriented films of poly-γ-benzyl-L-glutamate: (———) electrical vector parallel to the fiber axis; (· · · ·) electrical vector perpendicular to the fiber axis. Film A is thinner than film B (Tsuboi, 1962). (From Schellman and Schellman, 1964.)

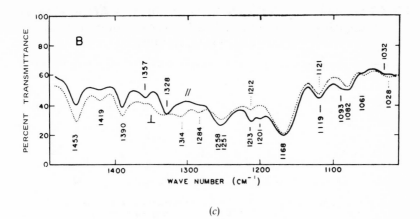

(c)

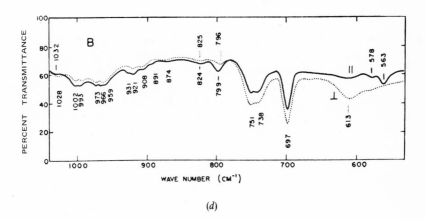

(d)

Fig. 10.7. (*Continued*)

almost the same as that for the α-helix, since there is general agreement between the calculated angle θ and the observed angle θ between transition moments of the peptide group and the fiber axis (see Table 10.5). If the amide group is revolved about its own plane about 10° to bring the C—N bond nearer to the equatorial position, the θ values calculated (column 3) agree somewhat better with the observed θ values but a significant change then occurs in the pitch of the α-helix. (b) Using the dichroic properties shown by the side-chain bands and other available information, Tsuboi proposed a provisional model to take care of side chains in poly-γ-benzyl-L-glutamate (Fig. 10.8).

Amide-rich polypeptides have been prepared by Krull *et al.* (1965) for the purpose of employing model systems for studies of the contribution of side-chain amides to protein properties. Such glutamine-containing polypeptides may be regarded as models for prolamine proteins such as gliadin and zein, and also for glutelin proteins,

Table 10.5. Observed and Calculated Values of Angle θ between Transition Moments of the Peptide Group and the Fiber Axis (Tsuboi, 1962)[c]

Mode	θ calc [a]	θ calc [b]	θ obs
N—H Stretch	19	29	28
Amide I	28	38	39
Amide II	87	83	75
Amide III	63	53	40
Amide V	83	83	> 80

[a]From atomic coordinates of Trotter and Brown cited in Bamford *et al.* (1956), p. 124.
[b]Based on a modified α-helical model of Tsuboi which improves agreement of theoretical and observed infrared spectra.
[c]Table from Schellman and Schellman (1964).

such as wheat glutenin. The polypeptides prepared were poly-γ-benzyl-L-glutamate, poly-γ-ethyl-L-glutamate, and poly-γ-methyl-L-glutamate. The infrared spectra of polymers containing different amounts of glutamine residues were examined in solutions of 25 % dichloroacetic acid–75 % ethylene dichloride. The absorbance at 1650 cm^{-1} was associated with random or helical conformation, whereas the 1610 cm^{-1} band was considered a measure of β-structure (see Blout and Asadourian, 1956). In all the polymers the peak at 1650 cm^{-1} was highly predominant, the 1610 cm^{-1}

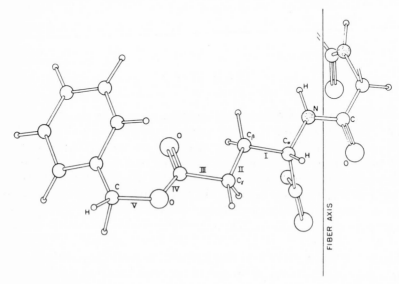

Fig. 10.8. A model for the molecular conformation of the α-form of poly-γ-benzyl-L-glutamate. (Tsuboi, 1962.)

band appearing only as a small shoulder. Therefore, it was felt that most of the molecules occur as random coils or α-helical structures, and the β-form is not stable.

Optical rotatory dispersion (ORD) measurements in the same solvent showed poly-γ-methyl-L-glutamate had a b_0 of −630, corresponding to 98% helix content; the polymer containing 60% L-glutamine and 40% γ-methyl-L-glutamate had a b_0 of −484, corresponding to 76% helix content; and poly-L-glutamine had a b_0 of −261, corresponding to 41% helix content. A b_0 value of zero is assumed for the completely random conformation and −640 for the completely helical structure of polyglutamyl esters.

Dichroism of the polymers was studied by Krull *et al.* to learn about certain properties of the molecular structures. By stroking highly concentrated solutions of the polymer in chloroform on a silver chloride plate as described by Ambrose and Elliott (1951), these workers obtained oriented films of poly-γ-benzyl-L-glutamate, poly-γ-ethyl-L-glutamate, and poly-γ-methyl-L-glutamate. Copolymers containing L-glutamine residues were oriented by stretching dried films on a slide.

Krull *et al.*, compared infrared spectra of oriented films of poly-γ-benzyl-L-glutamate, a polymer consisting of 60% glutamine–40% γ-ethyl-L-glutamate, and poly-L-glutamine. The polymers were inspected with polarized light from a silver chloride polarizer (Makas and Shurcliff, 1955). Going from left to right in Fig. 10.9 (i.e., with increasing numbers of amide groups introduced) the absorbance at 1745 cm⁻¹ (ester group) diminishes, while the amide I band at 1658 cm⁻¹ increases in size.

The poly-γ-benzyl-L-glutamate molecules in the film were well oriented and were predominantly in the α-helical form, since there was marked dichroism of the amide I band at 1658 cm⁻¹ and of the amide II band at 1550 cm⁻¹ in polarized light (Miyazawa and Blout, 1961). The amide I band was most pronounced when the polymer

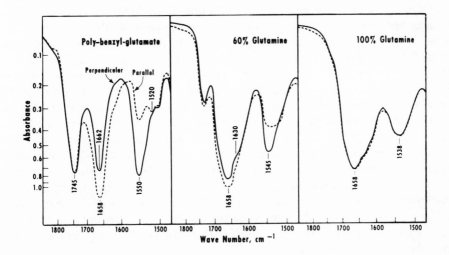

Fig. 10.9. Infrared spectra of oriented polypeptide films obtained with polarized light. (Krull *et al.*, 1965.)

orientation was parallel to the electrical vector of the radiation, but the amide II band was greatest when the polymer orientation was perpendicular to the electrical vector.

As the number of amide groups was increased in the polymers there was diminished dichroism (see figures), since the polymers became less subject to molecular orientation as a result of increased interactions between polymer chains. The infrared spectra displayed characteristics of a shift from α- to β-conformation of the peptide chain as glutamic ester residues were converted to glutamine. A shoulder is present at $1630\ cm^{-1}$ in the amide I band in the 60% and the 100% glutamine polymers. This shoulder is most clearly defined in the 60% glutamine–40% γ-ethyl-L-glutamate film when the polymer orientation is perpendicular to the electrical vector of radiation of the polarized light, which is characteristic of the parallel β-structure (Ambrose and Elliott, 1951). Also there is an increase in the magnitude of the amide II band in the parallel orientation of this polymer. This band shifts to lower frequency as the glutamine content of the polymer increases ($1550 \rightarrow 1545 \rightarrow 1538\ cm^{-1}$).

Little dichroism was shown at $1658\ cm^{-1}$ in the polymers containing more amide groups. The magnitude of this band was probably due to the additional absorbance of hydrogen-bonded side-chain amide groups that absorb in the region of $1650\ cm^{-1}$ (Bellamy, 1956). The position of this band lends support to the presence of side-chain amide associations in the solid amidated polymers.

In aqueous solution, the high local concentration in poly-L-glutamine of amide groups was thought to be possibly sufficient to favor interamide hydrogen bonds over amide–water hydrogen bonds. A hydrophobic local environment, due to the numerous aliphatic side-chain methylenes, may further stabilize the hydrogen bonds between side-chain amide groups.

Fraser et al. (1967) used optical rotatory dispersion, X-ray diffraction, and infrared spectrometry to study the stability and geometry of the α-helix in a series of polypeptides related to poly-γ-benzyl-L-glutamate. The inner portion ($-CH_2-CH_2COOCH_2-$) of the side chain was identical throughout the series, so that differences between the properties of the polymers could be attributed to the outer portions of the various side chains. In addition to the parent benzyl ester, polymers derived from various nitrobenzyl, iodobenzyl, trimethylbenzyl, and naphthylmethyl esters were also examined. In all cases the dichroisms and frequencies of the amide bands elucidated by infrared measurements indicated that the polypeptide chain was in the α-helical conformation. Fraser et al. estimated values of ΔR (in Å), the difference between the $N \cdots O$ hydrogen-bond distance in each polymer and that in poly-γ-benzyl-L-glutamate. Variations in the hydrogen-bond lengths among the polymers were estimated by measuring the frequencies v_A, v_B, and v_{II} of the amide A, B, and II vibrations and applying specific formulas to derive v_{NH} (the unperturbed NH stretching frequency for each polymer) and ΔR. The ΔR values varied between -0.009 and $+0.044$ for the various polymers. These data, and evidence obtained from the other methods used, emphasized the importance of the outer portions of the side chains in determining the stability and precise conformation of the α-helix both in the solid state and in solution.

Krimm (1962) analyzed the infrared spectra of various proteins in terms of the conformations of their component polypeptide chains. He used the general method of Miyazawa and Blout (1961) and made several modifications in their method, which he claimed resulted in better agreement between observed and calculated frequencies [see below (Elliott and Bradbury, 1962) for a discussion of some of Krimm's deductions]. In addition to the standard structures treated previously, namely, the α-helix and the parallel-chain and antiparallel-chain pleated sheets, he extended his calculations to the polar-chain pleated sheet and the polyglycine II structures. He showed from an analysis of the infrared spectrum that a polar-chain structure was favored for feather keratin. An analysis of tobacco mosaic virus protein showed that only a fraction of this protein was in the α-helical form, and that the axes of the α-helices deviated by about 30° from being perpendicular to the axis of the virus.

Elliott and Bradbury (1962) published a short criticism of the interpretations of Krimm (1962) concerning the elucidation of polypeptide and protein conformations. Although the work of Miyazawa (1960c) and Miyazawa and Blout (1961) on the interpretation of amide bands was a big advance and has led to the correct assignment of the weak parallel band at ~ 1690 cm^{-1} in extended chain conformation, Elliott and Bradbury emphasized the need for caution in the application of this work. They had found, from a study of a homologous series of poly-ω-acids (nylons 2, 4, 6, 8, and 10) and of different polypeptides, that there was no unique value for v_0 nor a unique set of interaction constants, as had been assumed (Miyazawa and Blout, 1961; Krimm, 1962), from which the frequencies of amide bands for various conformations of different materials can be accurately predicted. The values of v_0 for the amide I band varied within 7 cm^{-1} for the homologous series and 10 cm^{-1} for different polypeptides in substantially the same extended conformation. The variations in the values of these interaction constants for amide I were small enough to allow the cautious use of the frequencies and splitting (if any) of this band for diagnostic purposes. Such is not the case for the amide II band, where Elliott and Bradbury believed v_0 to be even more conformation sensitive and variable by at least 20 cm^{-1}. Also, their failure to observe the predicted splitting of the amide II band in the study of the homologous series of simple model compounds suggested that the detailed structure of that band was not understood. For these reasons they felt that the application of Miyazawa's theory to a detailed analysis of the amide bands in various proteins, as published by Krimm (1962), was unjustified: The constants used in this analysis were not felt to be reliable because they were based on the invalid assumption of the constancy of v_0, an assumption which leads to large errors in the interaction constants. Thus the conclusion that β-keratin is a parallel-chain pleated structure (Miyazawa and Blout, 1961; Krimm, 1962) was due to the use of inaccurate interaction constants, and to what Elliott and Bradbury (1962) felt had been an uncritical use of the published data.

Elliott and Bradbury (1962) disagreed with some of the deductions made by Krimm (1962) from spectra published by their laboratory. For instance, the enhanced intensity at 1520 cm^{-1} in the spectrum of water-soluble silk, to which he referred and attributed to a random coil form, could be seen in their spectra only when the water-soluble silk had been altered by treatment which produces the β-form (Ambrose et al., 1951). In the spectrum of lithium bromide soluble silk (where there was no β-band or shoulder on the amide I band) the amide II band was single and at ~ 1538 cm^{-1} and this was evidently v_0 for water-soluble silk (*Bombyx mori* silk). Also, Krimm (1962) had stated that in the polyglycine II spectrum (Elliott and Malcolm, 1956) there was a weak band at 1531 cm^{-1}, but this band was not found by Elliott and Bradbury (1962). It was not possible to denote a frequency to a very slight asymmetry of the amide II band found on the low-frequency side and in any case they felt that the presence of a small amount of polyglycine I could cause this asymmetry. Amide I, which is less symmetric than amide II, was taken by Krimm as a single band to fit his calculations. Elliott and Bradbury thought that Miyazawa's treatment of the molecular vibrations in amide groups in polypeptides and proteins was the correct one, but recommended careful use of it.

Other Studies on Polypeptides

Polyglycines I and II

Suzuki *et al.* (1966) have obtained a nearly complete set of assignments of the bands observed for polyglycines by comparing the spectra of the I and II forms of the following five isotopic polyglycines: ordinary polyglycine ($-\text{NHCH}_2\text{CO}-)_n$, N-deuterated polyglycine ($-\text{NDCH}_2\text{CO}-)_n$, C-deuterated polyglycine ($-\text{NHCD}_2\text{-CO}-)_n$, completely deuterated polyglycine ($-\text{NDCD}_2\text{CO}-)_n$, and ^{15}N-substituted polyglycine ($-^{15}\text{NHCH}_2\text{CO}-)_n$. Polyglycine I is the β-form (nearly extended zigzag chain); polyglycine II is a helical form. Tables 10.6 and 10.7 contain the observed frequencies and the assignments made (last column).

Tables 10.8 and 10.9 show the isotope shifts attributed to ^{15}N observed in the frequencies of polyglycine I and polyglycine II, respectively. The amounts of the isotope shifts are given as Δv and as $(\Delta\lambda)/\lambda^0$,

$$\frac{\Delta\lambda}{\lambda^0} = \frac{\lambda(^{15}\text{N}) - \lambda(^{14}\text{N})}{\lambda(^{14}\text{N})}$$

where $\lambda = 4\pi^2 c^2 v^2$, c being the velocity of light and v the observed frequency, in cm^{-1}, at the absorption maximum. Figure 10.10 shows characteristic spectra for ^{14}N-polyglycine II and ^{15}N-polyglycine II in the 3600–2800 cm^{-1} region, demonstrating the effects of the two isotopes. Table 10.8 shows, for example, that in undeuterated polyglycine I there is a skeletal stretching band at 1016 cm^{-1}, which shows a marked ^{15}N isotope shift. The shift is -0.012 in $(\Delta\lambda)/\lambda^0$. In this frequency region only a normal vibration to which the contribution of the $C-N$ stretching mode is great can give rise to such a great ^{15}N isotope shift. The 1016 cm^{-1} band is assigned to the $C-C_\alpha$ and $C_\alpha-N$ stretching modes with 180° phase difference. In undeuterated polyglycine II (Table 10.9) the equivalent band comes at 1028 cm^{-1}.

Krimm (1966) has called attention to experimental evidence which requires that antiparallel chains be present in polyglycine II. On the basis of the spacings in the powder diffraction pattern (Bamford *et al.*, 1955), Crick and Rich (1955) had proposed a model for polyglycine II consisting of parallel polypeptide chains, each with a threefold screw axis, packed in a hexagonal array. Ramachandran *et al.* (1966)

Table 10.6. Observed Frequencies of Polyglycine I and Their Assignments[a] (Susuki et al., 1966)

Frequency, cm^{-1}				Assignment
$(NHCH_2CO)_n$	$(NDCH_2CO)_n$	$(NHCD_2CO)_n$	$(NDCD_2CO)_n$	
3308 s		3297 s		NH str.
3088 m		3055 m		Amide II × 2
2978 w	2976 w			
2929 w	2929 w			CH_2 antisym. str.
2869 vw	2870 vw			CH_2 sym. str.
		2850 vw		
	2462 s		2464 s	ND str.
	2419 m		2416 m	(a combination tone)
		2240 w	2240 w	
		2165 w	2161 w	CD_2 str.
		2118 w	2109 w	
1685 m	1680 m	1684 m	1681 m	Amide I (C=O str.)
1636 s	1629 s	1627 s	1630 s	
1517 s		1498 s		Amide II (CN str. + NH in-plane def.)
	1475 s		1460 s	Amide II' (CN str.)
1432 s	1432 s			CH_2 bend.
1408 w	1352 m			CH_2 wag.
		1297 m		Amide III (CN str. + NH in-plane def.)
	1276 m			CH_2 twist.
	1237 m			
1236 m				Amide III (CN str. + NH in-plane def.)
1214 w				CH_2 twist.
		1189 m	1183 m	CD_2 bend.
		1099 w	1101 w	
		1073 w	1075 w	CD_2 wag.
1057 w				
1016 m	1015 m	1015 w	1016 w	Skel. str.
	950 w		950 w	Amide III' (ND in-plane def.)
		928 w	923 w	CD_2 twist.
			901 w	
		866 w	866 w	CD_2 rock.
708 s		700 m		Amide V (NH out-of-plane def.)
628 w	625 w	610 m	605 m	Amide IV (C=O in-plane def.)
614 m	614 w	564 m	554 m	Skel. def.
589 m		534 m		Amide VI (C=O out-of-plane def.)
	572 m			Amide VI' (C=O out-of-plane def.)
	504 s		493 s	Amide V' (ND out-of-plane def.)

[a]Key: m, medium; s, strong; w, weak.

Table 10.7. Observed Frequencies of Polyglycine II and Their Assignments[a] (Suzuki *et al.*, 1966)

Frequency, cm^{-1}				
$(NHCH_2CO)_n$	$(NDCH_2CO)_n$	$(NHCD_2CO)_n$	$(NDCD_2CO)_n$	Assignment
3303 s		3290 s		NH str.
3086 m		3072 m		amide II × 2
2983 w	2975 w			
2944 w	2940 w			CH₂ antisym. str.
2848 w	2850 w			CH₂ sym. str.
		2857 w		
	2464 w		2485 w	ND str.
			2451 w	
	2416 s		2417 s	(a combination tone)
		2244 w	2246 w	
		2172 w	2173 w	CD₂ str.
		2117 w	2108 w	
1644 s	1639 s	1639 s	1632 s	Amide I (C=O str.)
1554 s		1551 s		Amide II (CN str. + NH in-plane def.)
	1476 s		1462 s	Amide II' (CN str.)
1420 m	1420 m			CH₂ bend.
1377 m	1350 m			CH₂ wag.
		1309 m		Amide III (CN str. + NH in-plane def.)
1283 m, 1249 m				Amide III (CN str. + NH in-plane def.) CH₂ twist.
	1277 m, 1262 m			CH₂ twist.
		1205 m	1222 m	CD₂ bend.
		1057 w	1065 w	
1028 m	1034 m	1016 m	1034 w	Skel. str.
	987 m		993 m	Amide III' (ND in-plane def.)
		943 m	935 w	CD₂ wag.
		920 m	917 w	CD₂ twist.
901 m	886 w			CH₂ rock.
		866 w	866 w	
		805 m	794 m	CD₂ rock.
740 m		715 m		Amide V (NH out-of-plane def.)
698 s	693 s	670 s	667 s	Amide IV (C=O in-plane def.)
573 s		502 m		Amide VI (C=O out-of-plane def.)
	520 s		534 m	Amide V' (ND out-of-plane def.)
			497 s	Amide VI' (C=O out-of-plane def.)
363 s	356 s	347 s	345 s	Amide VII (CO—NH torsion)

[a] Key: m, medium; s, strong; w, weak.

Table 10.8. ^{15}N Isotope Shifts in the Vibrational Frequencies of Polyglycine I (Suzuki *et al.*, 1966)

Frequency v, cm^{-1}		Shift		
^{14}N species	^{15}N species	Δv, cm^{-1}	$\Delta\lambda/\lambda^0$	Assignment
3302.6	3293.1	-9.5	-0.0057	NH str.
3088.0	3064.1	-23.9	-0.0154	Amide II $\times$ 2
1684.6	1685.4	$+0.8$	$+0.0010$ ⎱	Amide I
1635.8	1634.9	-0.9	-0.0011 ⎰	
1517.3	1502.9	-14.4	-0.0189	Amide II
1432	1431	-1	-0.001_4	CH$_2$ bend.
1236	1235	-1	-0.001_6 ⎱	CH$_2$ twist.
1214	1210	-4	-0.006 ⎰	Amide III
1016	1010	-6	-0.012	Skel. str.
628	626	-2	-0.006 ⎱	Amide IV
614	611	-3	-0.009 ⎰	Skel. def.

Table 10.9. ^{15}N Isotope Shifts in the Vibrational Frequencies of Polyglycine II (Suzuki *et al.*, 1966)

Frequency v, cm^{-1}		Shift		
^{14}N species	^{15}N species	Δv, cm^{-1}	$\Delta\lambda/\lambda^0$	Assignment
3303.4	3290.0	-13.4	-0.0081	NH str.
3086.4	3060.7	-25.7	-0.0166	Amide II $\times$ 2
1643.8	1642.7	-1.1	-0.0013	Amide I
1553.7	1540.7	-13.0	-0.0167	Amide II
1420	1419	-1	-0.001_4	CH$_2$ bend.
1283	1280	-3	-0.005	CH$_2$ twist.
1249	1245	-4	-0.006	Amide III
1028	1023	-5	-0.010	Skel. str.
901	900	-1	-0.002	CH$_2$ rock.
698	695	-3	-0.009	Amide IV
573	571	-2	-0.007	Amide VI

have suggested a slight modification of this structure, although the parallel-chain characteristic was retained. Krimm has found that mild stroking of a film of polyglycine II induces a transformation to the β-form. A similar result had been obtained previously by Elliott and Malcolm (1956), and this behavior was consistent with the inability to produce an oriented fiber of polyglycine II (Bamford *et al.*, 1955). The infrared spectrum of the β-form of polyglycine has a characteristic parallel dichroic band at 1685 cm^{-1} (Elliott and Malcolm, 1956) which is almost certainly associated with the presence of an antiparallel extended chain structure (Miyazawa, 1960*c*; Miyazawa and Blout, 1961; Krimm, 1962; Bradbury and Elliott, 1963). Since the mechanical treatment required to transform polyglycine II to the β-form is most

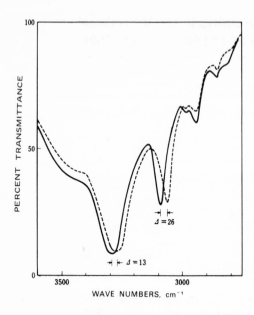

Fig. 10.10. Infrared spectra of ^{14}N-polyglycine II (—) and ^{15}N-polyglycine II (···) in the 3600–2800 cm^{-1} region. (Suzuki *et al.*, 1966.)

unlikely to produce an inversion of the direction of a chain with respect to its neighbors, Krimm concluded that antiparallel chains are also present in polyglycine II. The transformation of polyglycine II to the β-form by mechanical treatment would then simply involve the rotation of peptide groups from the hydrogen-bonding positions of the II structure to those of the antiparallel chain β-form, which is readily feasible. The original structure proposed for polyglycine II can accommodate antiparallel chains without any significant change in the nature of the hydrogen bonds (Crick and Rich, 1955), so that an antiparallel-chain structure would be comparable in stability with one with parallel chains.

Studies of crystal morphology (Padden and Keith, 1965) and infrared studies (Krimm *et al.*, 1967; Krimm and Kuroiwa, 1968) of the possible existence of C—H···O═C hydrogen bonds in polyglycine II gave further evidence for antiparallel chains. For example, although the spectrum of polyglycine I showed two unperturbed CH_2 frequencies at 2850 and 2923 cm^{-1} (C—H···O═C hydrogen bonding being impossible in its structure), the spectrum of polyglycine II displayed four CH_2 frequencies, two near the above values and two others, one at 2800 and one at 2980 cm^{-1} (Fig. 10.11). At $-170°$C, the latter bands shift in a manner which indicates that these perturbed frequencies arise from an attractive interaction rather than a repulsive one. The presence of unperturbed modes is consistent with the antiparallel-chain structure of polyglycine II (Krimm, 1966).

At normal temperatures, bands in the infrared spectrum of polyglycine II which can be associated with bonded CH_2 groups are consistent with the presence of suggested C—H···O═C hydrogen bonds in a structure with parallel chains (Ramachandran *et al.*, 1966), whereas bands associated with nonbonded CH_2 groups require that some chains be antiparallel to each other. Krimm suggested the desirability of determining whether a three-stranded collagen-type helix can be constructed with one strand antiparallel to the other two, since the chain arrangement in polyglycine II is closely related to that of collagen, for which only parallel-chain structures had been proposed (Rich and Crick, 1961; Ramachandran and Sasisekharan, 1965; and earlier papers).

An examination of the structure of myoglobin (Krimm and Watson, 1967) indicated that $C—H\cdots O=C$ hydrogen bonds are present in its α-helical segments. Therefore, although this type of bond is weaker than the $N—H\cdots O=C$ bond, it was suggested that the possibility of an attractive $C—H\cdots O=C$ interaction should be incorporated into the principles underlying the structure of polypeptides and proteins.

Poly-L-proline

Poly-L-proline exists in two different conformations: form I, which assumes a right-handed helix and has each peptide bond in the *cis* configuration (Traub and Shmueli, 1963); and form II, which is a left-handed helix and has its peptide bonds in the *trans* configuration (Sasisekharan, 1959; Cowan and McGavin, 1955). Isemura *et al.* (1968) have differentiated these two forms of poly-L-proline by characteristic far-infrared bands. Form I has two broad bands near 280 and 160 cm^{-1} and form II

Fig. 10.11. Expanded infrared spectra of polyglycine II films in the CH_2 stretching region: top curve, at room temperature, and bottom curve, at about $-170°C$. (Krimm and Kuroiwa, 1968.)

PERCENT TRANSMITTANCE

3050   3000   2950   2900   2850   2800   2750

WAVE NUMBER, cm^{-1}

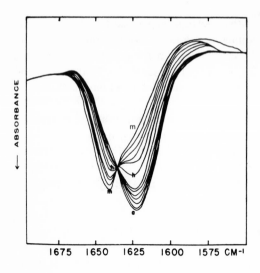

Fig. 10.12. Spectra of the carbonyl region as a function of temperature: curve a, 45°C; curve h, 53°C; curve m, 65°C. (Swenson and Formanek, 1967.)

shows two bands at 400 and 670 cm^{-1}. Form II also has three broad bands at about 250, 200, and 100 cm^{-1}. In the far-infrared region, the bands of the pentamer, hexamer, and octamer of t-amyloxycarbonyl-L-proline were in very good agreement with that of poly-L-proline II, additional evidence to that found in the rest of the 1800–75 cm^{-1} region, that they have a helical structure of a left-handed threefold screw axis. Blout and Fasman (1958) had earlier differentiated poly-L-prolines I and II by means of infrared spectra in the 1800–800 cm^{-1} region. Isemura $et\ al.$ concluded that the molecules of pentamer, hexamer, and octamer had a helical structure of a left-handed threefold screw axis. They also concluded that the tetrapeptide of t-amyloxy-carbonyl-L-proline might also have a left-handed helix, probably one turn, since the tetramer clearly showed an absorption band at about 400 cm^{-1}, characteristic of poly-L-proline II.

Swenson and Formanek (1967) have recorded spectra for H_2O and D_2O solutions of form II poly-L-proline as a function of temperature. The temperature range

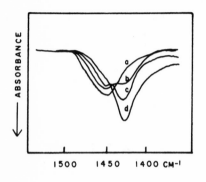

Fig. 10.13. Spectra of the C—H bending region as a function of temperature: curve a, 45°C; curve b, 52°C; curve c, 57°C; curve d, 67°C. (Swenson and Formanek, 1967.)

included the region where a reversible phase transition is known to occur. [Crystallographic analysis has shown that form II poly-L-proline is a left-handed helix which has all *trans*-imide linkages and a repeat distance of 3.1 Å. Form I is a right-handed entirely *cis* helix with a repeat distance of 1.9 Å. Forms I and II can be interconverted in solution by simply changing the solvent, as described by Steinberg *et al.* (1958).] The spectral changes in the carbonyl absorption as this temperature range was approached (Fig. 10.12) were interpreted as a breakdown of solute–solvent interactions which resulted in destabilization of the solute in the solution phase. Simultaneously, a change occurred in the C—H bending absorption (Fig. 10.13), which was interpreted as a small conformational change. Both absorptions showed apparent isosbestic behavior as a function of temperature. Swenson and Formanek estimated a ΔH^0 of 60 ± 10 kcal/mole (residue) for the overall process from the temperature dependence of the carbonyl absorption.

Poly-L-lysine

Davidson and Fasman (1967) have studied the conversion by heating of the α-form of poly-L-lysine to β-poly-L-lysine in aqueous solution by means of infrared spectroscopy and optical rotatory dispersion. A D_2O solution of deuterated poly-L-lysine (pD 12.3) was prepared and infrared spectra were obtained over the region of the amide I and II absorption bands (1300 to 1950 cm^{-1}) at 15, 34, and 47°C (Fig. 10.14). At 15°C, the amide I and II region showed a band at 1628 cm^{-1} (helical) with a shoulder at 1585 cm^{-1} and a second band at 1465 cm^{-1}. At 34°C, there was

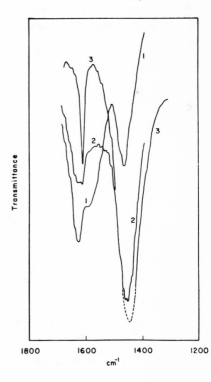

Fig. 10.14. The transition of the α- to β-poly-L-lysine, infrared spectra. Poly-L-lysine in D_2O (0.24 %) at pD 12.3. Spectra were recorded at 15°C (spectrum 1), 34°C (spectrum 2), and 47°C (spectrum 3). (The dashed portion of curve 3 extended beyond the ordinate scale and has been estimated.) A jacketed CaF$_2$ cell of 0.05-mm optical path length was used. (Davidson and Fasman, 1967.)

a broad band (1612 to 1632 cm^{-1}) in the amide I region and a second band at 1455 cm^{-1} in the amide II region. At 47°C, the solution exhibited a very sharp band at 1612 cm^{-1}, a much smaller band at 1680 cm^{-1} (β), and an amide II band which appeared to be centered at about 1445 cm^{-1} (see dashed curve). These results confirmed the ORD finding that poly-L-lysine undergoes a thermal conversion from the α- to the β-form. It should be noted that the amide I band position of the helical form reported in this paper is about 10 cm^{-1} lower than that reported by Townend *et al.* (1966) and by Rosenheck and Doty (1961), but is in agreement with that reported by Sarkar and Doty (1966) and Applequist and Doty (1962). The sample used in Davidson and Fasman's work had been previously deuterated and possibly accounted for the observed differences.

Poly-L-tyrosine

Poly-L-tyrosine absorbs strongly at 1515 cm^{-1}. A band at this frequency has been found in a number of proteins (e.g., α-keratin, ribonuclease, and insulin) and was considered to be due to the tyrosine residue (Bendit, 1967). This assignment was confirmed by examination of the spectra of deuterated proteins which usually exhibit this band at 1513 cm^{-1}.

In order to examine whether the tyrosine content, indicated by the residual band at 1513 cm^{-1}, agreed with the amino acid analyses of the various proteins tested, the intensity of the 1513 cm^{-1} band was plotted against known tyrosine content. The best linear correlation between residual intensity at 1513 cm^{-1} (D_T, for the "tyrosine band" in deuterated specimens) and tyrosine content was obtained with the peak intensity of the amide I band (D_I in undeuterated specimens), i.e., when D_T/D_I was plotted *vs* percentage of tyrosine content (Fig. 10.15). When this ratio

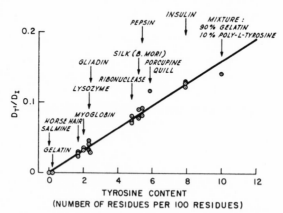

Fig. 10.15. Ratio of absorbance at 1513 cm^{-1} of deuterated proteins (D_T) to absorbance of the amide I band (D_I) plotted against tyrosine content from amino acid analysis. Values of D_T/D_I for silk, gliadin, and pepsin are based on spectra of partially deuterated specimens; all the other proteins were fully or almost fully deuterated. (Bendit, 1967.)

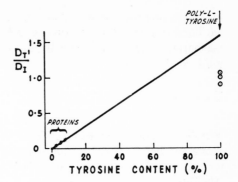

Fig. 10.16. Comparison of the ratio $D_{T'}/D_I$ for poly-
L-tyrosine cast from dimethylformamide with the
extrapolated straight line determined for the proteins
in Fig. 10.15. (Bendit, 1967.)

was plotted for poly-L-tyrosine, the intensity of the tyrosine band was not sufficiently
high to make the experimental points fall on the straight line determined by extrapolation of the data for the proteins (Fig. 10.16). This finding suggested that the interactions
of the tyrosine residues in poly-L-tyrosine are different from those in proteins.

Far-Infrared Spectra of α-Helix and β-Structures

Itoh *et al.* (1969) have measured the far-infrared spectra (700 to 60 cm^{-1}) of the
α-helix structures of poly(L-α-amino-*n*-butyric acid), poly-L-norvaline, poly-L-nor-
leucine, and poly-L-leucine, as well as the spectra of the β-form structures of
poly(L-α-amino-*n*-butyric acid), poly-L-valine, poly(DL-α-amino-*n*-butyric acid), poly-
DL-norvaline, and poly-DL-norleucine. The α-helix has characteristic bands near 690,
650, 610, 380, 150, and 100 cm^{-1}. The β-form has characteristic bands near 700,
240, and 120 cm^{-1}. The vibrations of the main chain are strongly coupled with the
deformation vibrations of the side chains. An earlier study (Itoh *et al.*, 1968) involved
various polyalanines and correlations of spectra with α- and β-structures.

Copolymers of O-Carbobenzoxy-L-tyrosine and Benzyl L-(or D-)glutamate

Infrared spectroscopy and X-ray diffraction data have shown that the copolymers
of *O*-carbobenzoxy-L-tyrosine and benzyl L-(or D-)glutamate as well as benzyl L-
aspartate possess a helical conformation in the solid state, even when they are very
rich in carbobenzoxy-L-tyrosine residues (Vollmer and Spach, 1967). The ordering
of the molecules of poly-*O*-carbobenzoxy-L-tyrosine in a purely helical structure
appeared to be favored by the insertion of a small number of foreign residues in the
polypeptide chain.

β-Benzyl L-aspartate (Transformation from α- to ω-Helix)

Bradbury *et al.* (1968) have used infrared spectroscopy to study the factors
involved in the formation of the ω-helix from β-benzyl L-aspartate in the α-helix form.
The presence of ω-helix* (a 4.0_{13} helix) (Bradbury *et al.*, 1962a; Bragg *et al.*, 1950)
in a polypeptide can be inferred from the characteristic position of the amide bands

*Four residues per turn of the helix. The 13 represents the number of atoms in one hydrogen-bonded
ring (Dickerson, 1964).

for this form; the amide I is at $1675\,cm^{-1}$ for the ω-helix and $1665\,cm^{-1}$ for the left-handed α-helix form according to Bradbury *et al.* (1968). Films of poly-β-benzyl L-aspartate, a left-handed α-helix, are transformed into the ω-helix form when they are heated *in vacuo*.

Gelatin Polypeptide Chains

Milch (1964) obtained infrared evidence in accord with the long-postulated hypothesis concerning the reversible formation of a systematic set of hydrogen bonds along and between the crystalline portions of gelatin polypeptide chains during the temperature-dependent mutarotation processes of the gelatin sol–gel transition (Harrington and von Hippel, 1961). He obtained spectra of gelatin in D_2O sols at 50°C and gels at 15°C. Changes which were characteristic only of the collagen-fold configuration were shown in the N—H frequencies (characterized by appreciable broadening and downward shift of the stretching frequency maximum), suggestive of the occurrence of cold-dependent N—H···O=C forms.

Globular Proteins

Early Investigations

The amide I band has been examined by Elliott *et al.* (1950) in native and denatured insulin, by Elliott *et al.* (1957) in lysozyme, and by Ambrose *et al.* (1951) in water-soluble silk. The band at $\sim 3200\,cm^{-1}$ has also been investigated. Beer *et al.* (1959) have given a comprehensive list of proteins studied up to 1959, along with characteristic absorption bands. Bamford *et al.* (1956) have reviewed work done up to 1956 in the region between 5000 and $4500\,cm^{-1}$ (combination band of the N—H stretching frequency and that of the amide I or amide II band). The infrared dichroic properties of crystals of hemoglobin and ribonuclease have been observed in this region (Elliott and Ambrose, 1951; Elliott, 1952).

β-Lactoglobulins

β-Lactoglobulin A has been observed (Timasheff and Susi, 1966) in the regions of the amide I and amide II absorptions and has been compared with myoglobin and α_s-casein. The spectrum of a β-lactoglobulin A film is shown in Fig. 10.17. The amide I band is a sharp band at $1632\,cm^{-1}$ which is skewed on the high-frequency side, with weaker shoulders near 1685 and $1650\,cm^{-1}$. The amide II maximum lies at $1530\,cm^{-1}$. Krimm (1962) [see also Miyazawa (1960) and Miyazawa and Blout (1961)] has attributed strong bands at $1632\,cm^{-1}$ and $1530\,cm^{-1}$ to characteristic pleated sheet or β structure of polypeptides and proteins. The weak shoulders seen in Fig. 10.17 near $1650\,cm^{-1}$ and $1685\,cm^{-1}$ could be associated with a small amount of α-helical structure and with antiparallel-chain pleated sheets, respectively (Krimm, 1962; Doty *et al.*, 1958). In the amide II region, weak bands at 1516 and $1546\,cm^{-1}$ were compatible with the presence of a small amount of α-helical structure, but the β-structure itself gave rise to weak bands very near to the 1516 and $1546\,cm^{-1}$ bands (Timasheff and Susi, 1966).

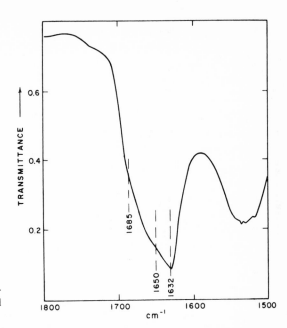

Fig. 10.17. Infrared spectrum of β-lacto-globulin A as a solid film. (Timasheff and Susi, 1966.)

Figure 10.18 shows the spectra of native and denatured β-lactoglobulin A (β-LG), native myoglobin, and α_s-casein in D_2O solution. In the spectra at pD 1.0 and 7.5 (actually these are pD's 1.4 and 7.9 respectively after the proper corrections are made), the amide I band is like the one observed in the dry film (1632 cm^{-1}, with weaker shoulders near 1650 and 1685 cm^{-1}). In the denatured form (pD 11.5, corrected to 11.9), a single amide I band is displayed at 1643 cm^{-1}. Randomly folded α_s-casein (Herskovits and Mescanti, 1965) is known to give an almost identical band.

Myoglobin in D_2O at pD 6.6 (corrected to 7.0) has an amide I band at 1650 cm^{-1}, characteristic of α-helical structure. Myoglobin is known to have at least 77% α-helical structure (Kendrew et al., 1960). Timasheff and Susi's (1966) data indicate that in D_2O solution the β-conformation and the α-helical conformation produce amide I bands at the same frequencies as in dry films. Randomly folded proteins (α_s-casein and denatured β-lactoglobulin) give rise to a sharp band centering at 1643 cm^{-1}.

Slow hydrogen–deuterium exchange (see Chapter 11) of native β-lactoglobulin in D_2O, pD 1.0 (corrected to 1.4) allowed the amide II band (~ 1550 cm^{-1}) to remain and be seen in one sample of the lactoglobulin (Fig. 10.18, top curve) which was run immediately after preparation. The presence of this band shows that the native protein must be quite compact. No such bands are observed in denatured lacto-globulin or in α_s-casein, which have loose randomly folded structures.

The spectra of other genetic variants, β-lactoglobulins B and C, were also found by Timasheff and Susi (1966) to be identical with the data for lactoglobulin A, suggesting very similar secondary structures.

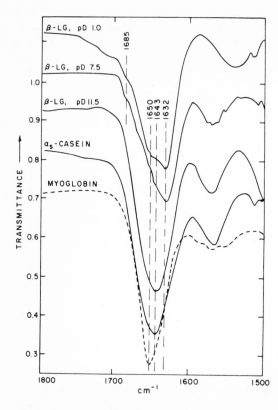

Fig. 10.18. Infrared spectra of native and denatured β-lactoglobulin (LG) A, native myoglobin, and α_s-casein in D_2O solution. The pD values are those read on a pH meter in D_2O solution. Consecutive spectra are linearly displaced by 0.1 scale unit. Concentration, 20 mg/ml; path length, approximately 0.05 mm. (The observed intensities are approximate because of uncertainties in path length. Peak absorptivity values for the amide I band are in the range 3 to 4 liters/g cm, if absorptivity at 1800 cm^{-1} is taken as reference.) (Timasheff and Susi, 1966.)

Synthetic Polypeptides and Nontransferability of Conformation Frequencies to Corresponding Conformations of Globular Proteins

Susi *et al.* (1967) have observed spectra of poly-L-lysine, poly-L-glutamic acid, β-lactoglobulin, myoglobin, and α_s-casein in the region of absorption of the amide I band in H_2O and D_2O solutions, and in the solid state. Their results indicated that characteristic frequencies exhibited by specific conformations of the synthetic polypeptides studied are not transferable to corresponding conformations of globular proteins.

The frequencies obtained for different conformations of globular proteins in H_2O and D_2O solution are internally consistent, in general agreement with corresponding values of fibrous proteins and with the limited data available in the literature concerning deuterated proteins in D_2O solution. Dissolution in aqueous environment by itself does not noticeably alter the amide I frequencies. A tentative set of characteristic frequencies and interaction constants was obtained for the amide I' modes of N-deuterated proteins.* These modes were easily observed in D_2O solution and showed sufficient variations in frequency to permit a distinction between the α-helical, the antiparallel-chain pleated sheet, and the solvated random configurations of globular proteins (Table 10.10).

*The characteristic frequencies of N-deuterated secondary amides, polypeptides, and proteins are conventionally labeled amide I', amide II', etc. (Miyazawa, 1962).

Table 10.10. Observed Amide I and Amide I' Frequencies of Proteins, cm^{-1} (Susi *et al.*, 1967)

Conformation	Mode	State			Sample
		D$_2$O solution	H$_2$O solution	Solid	
Antiparallel-chain pleated sheet	$v(\pi, 0)_A$	1632	1632	1632 1630–1634[a]	β-Lactoglobulin Fibrous proteins
	$v(0, \pi)_A$	1675	1690[b]	1690 1695–1697[a]	β-Lactoglobulin Fibrous proteins
α-Helix	$v(0)\alpha$	1649[c] 1650	1652	1652 1650–1653[a]	β-Lactoglobulin Myoglobin Fibrous proteins
Unordered	$v(u)^d$	1643 1643	1656 1656	1660–1664[a]	β-Lactoglobulin α_s-Casein Fibrous proteins

[a]From Krimm (1962), for comparison. Nomenclature is from this reference.
[b]Unexchanged protein in D$_2$O solution.
[c]In D$_2$O–CH$_3$OD mixed solvent.
[d]Observed value for unordered form; not necessarily equal to v_0.

Timasheff *et al.* (1967a) have recorded infrared spectra in the region of the amide I' band for several other proteins in D$_2$O solution: myoglobin, bovine serum albumin, carboxypeptidase A, lysozyme, insulin, glucose oxidase, deoxyribonuclease, bovine carbonic anhydrase, α-chymotrypsin, chymotrypsinogen, ribonuclease, soybean trypsin inhibitor, γ-globulin, rennin, β-lactoglobulin A, β-lactoglobulin B tryptic core, α_s-casein C, and phosvitin. Some spectra were also recorded on H$_2$O solutions, mineral oil suspensions, and under other conditions. (See Table 10.11 for amide I' frequencies.) Comparison of the observed band positions with structural analyses based on the use of ORD and circular dichroism (CD) data, as well as with structural details based on crystallography, indicated that the technique should be useful for studies of protein conformations in solution. The authors indicated that results obtained by a single technique are frequently difficult to interpret, whereas the simultaneous application of several methods can often resolve such difficulties.

A review has appeared recently (Timasheff and Gorbunoff, 1967) which discusses applications of IR, ORD, and CD spectroscopy in the examination of protein conformations. [See also, Timasheff and Bernardi (1970) concerning nucleases.]

γ-Globulin

Imahori (1963) prepared an oriented film of γ-globulin fibrils by precipitating human γ-globulin with antihuman horse-serum γ-globulin. Infrared dichroism studies showed a pronounced amide I band near 1635 cm^{-1} with light polarized parallel to the fibril axis. The appearance of this band implied the presence of intrachain cross-β structure.

Table 10.11. Amide I Frequencies of Various Proteins
(Timasheff *et al.*, 1967*a*)

Protein	Conditions[a]	Frequency, cm^{-1}
Myoglobin	*p*D 7.4	1650
Bovine serum albumin	Isoionic	1652
	Nujol suspension	1652
	*p*D 2.0	1648
Carboxypeptidase A	2.4 M LiCl	1650, 1637 shoulder
Lysozyme	Unspecified[d]	1650, 1632 shoulder
Insulin	*p*D 2.4	1654
	Nujol suspension	1654
	*p*D 12.5	1644
	*p*D 2.4 film, heated	1633, 1658
	Acidic CH$_3$OD	1654
Glucose oxidase		1643, 1648, 1654
Deoxyribonuclease		1637, 1643, 1650
Bovine carbonic anhydrase	*p*D 7.4, 12.0	1637
	Nujol suspension	1637
	*p*D 1.8	1646
α-Chymotrypsin	*p*D 4.2	1637, 1685 shoulder
Chymotrypsinogen	*p*D 8.7	1637
Ribonuclease	*p*D 4.8	1640, 1685 shoulder
Soybean trypsin inhibitor	*p*D 7.5 to 10.5	1641
γ-Globulin	Oriented film[e]	
	(unpolarized light)	1650, 1635 shoulder
	*p*D 1.6	1637–1643 (broad)
	*p*D 10.0 to 12.0	1637
Rennin	*p*D 1.8, 10.8	1639
β-Lactoglobulin A	*p*D 1.5–8.0[b]	1632, 1675
	Nujol suspension	1632, 1690 shoulder
	*p*D 12.4	1643
	*p*H 12.3 (in H$_2$O)	1656
	90% MeOD	1649, 1620 shoulder[c]
β-Lactoglobulin B tryptic core		1615, 1643
α$_s$-Casein C	*p*D 9–11[f]	1643
	*p*H 9 (in H$_2$O)	1656
Phosvitin	*p*D 3.4; 6.6[g]	1650

[a]Dilute D$_2$O solution, unless specified otherwise.
[b]Genetic variants B and C gave spectra identical with A.
[c]The shoulder is observed at 4 g/liter of protein; it is absent at 0.5 g/liter.
[d]Hamaguchi (1964). [e]Imahori (1963).
[f]Susi, Timasheff, and Stevens (1967).
[g]Timasheff, Townend, and Perlmann (1967*b*).

Glucagon

Gratzer *et al.* (1967) have investigated the conformation of glucagon as a function of its aggregation states. Ultraviolet, ORD, and infrared spectroscopy were used to characterize the conformations. Figure 10.19 shows spectra of glucagon in different states. The acid-aggregated states are substantially in the β-conformation. In D$_2$O

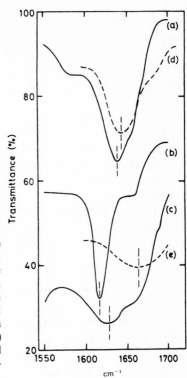

Fig. 10.19. Infrared spectra of glucagon in different states: (a) freshly prepared solution in D_2O/DCl, pD 2; (b) gel formed from (a) on standing (uncompensated); (c) dried film of precipitate from acid solution after standing (hydrogen form); (d) solution in phosphate buffer, pD 10.4; (e) deposited crystals, dried down from phosphate mother liquor (hydrogen form). Note that especially in (e) there is a high scattering background and a very broad band. The shoulder near 1660 cm^{-1} in (a) and (b) arises from the presence of some hydrogen form, since not all the D_2O can be removed from the sample without the possibility of adventitious changes in state. (Gratzer et al., 1967.)

the amide I band is at 1613 cm^{-1}, and in the undeuterated fibrils, which form from the gel on standing, the amide I is at ~ 1630 cm^{-1}, both positions being characteristic of β-conformations (Bamford et al., 1956). There is a shoulder at 1685 cm^{-1} in Fig. 10.19c, which is associated with the $\nu(0, \pi)$ mode (Miyazawa, 1960c; and Miyazawa and Blout, 1961) characteristic of the antiparallel β-form. In films of undeuterated crystals formed at pH 10.2, this effect does not occur, and the amide I band is at 1665 cm^{-1}. In suspensions of deuterated crystals in their mother liquor at pD 10.2 the amide I is at 1635 cm^{-1}. Thus, the β-form does not occur in the crystalline material or in solution at the alkaline pH.

Other Studies on Proteins and Related Substances

Human Hair Keratin

Baddiel (1968) used the ATR method (FMIR) to obtain high-quality spectra of human hair keratin in the range 4000–600 cm^{-1}. The results were generally consistent with those from conventional transmission spectra, but differed sufficiently from them to provide information which is not usually obtained without damaging the protein. The data obtained suggested that the hair cuticle has a mixed protein configuration composed largely of the α-helix with distinct contributions from the β or extended form. The central cortical material was thought to be mainly a mixture of α-helix and random coil or amorphous form. Baddiel described in detail the method

of sample preparation for use of the ATR equipment and gave a diagram of the rig
used for holding the hair in place.

One of the features of ATR spectroscopy is that the depth of penetration d of
the radiation incident on the sample is a function of the wavelength according to the
following equation (Harrick, 1963):

$$d = \frac{0.693\lambda}{2\pi(\sin^2 \theta - n_{21}^2)^{\frac{1}{2}}}$$

where d is the depth of radiation penetration in order to have its intensity drop to
half the original value, θ is the angle of incidence, λ is the wavelength, and n_{21} is the ratio
of the refractive index of the sample to that of the KRS-5 (TlBr–TlI) reflecting crystal,
i.e., n_2/n_1.

Baddiel calculated that for $\lambda = 6\,\mu$ ($\sim 1660\,\text{cm}^{-1}$), d is $3.48\,\mu$ (based on an
average refractive index of hair keratin of 1.555 and on an n_1 value of TlBr–TlI of
2.374). Thus, the incident radiation intensity falls off about 80% at a depth of $7\,\mu$
at $1650\,\text{cm}^{-1}$. Therefore, the main part of the sample signal comes largely from the
cuticle (about 2–$3\,\mu$ thick) of the hair keratin (at lower wavelengths the penetration
decreases, e.g., $1.9\,\mu$ at $3300\,\text{cm}^{-1}$). For this reason spectra had to be recorded both
on normal untreated hair and cuticle-stripped hair. Rubbing with two grades of
emery paper (coarse and fine) followed by washing and drying were the most effective
for removing the hair cuticle.

It should be noted that all the frequencies reported by Baddiel are "corrected"
frequencies, that is, corrected to correspond with transmission data, since differences
were constant and empirical adjustments could be made (ATR frequencies were
always lower by 3–$4\,\text{cm}^{-1}$ in the $3333\,\text{cm}^{-1}$ region and 2–$3\,\text{cm}^{-1}$ in the 1667 to
$1250\,\text{cm}^{-1}$ region).

Bovine Serum Albumin

Scrimshaw et al. (1966) have separated bovine serum albumin into fractions,
which, when analyzed by analytical ultracentrifuge, in one case contained chiefly the
monomeric form, and in other cases consisted of mixtures of monomer, dimer, and
trimer. These investigators have reported that the monomeric species also seems to
be distinguishable from the polymeric forms by infrared spectroscopy. The spectra
of the monomeric and polymeric forms are shown in Fig. 10.20. Small but noticeable
differences can be seen between the monomer and the polymer: (a) there is a
definite band at $845\,\text{cm}^{-1}$ in the polymer, whereas in the monomer the corresponding
absorption appears as a shoulder at $830\,\text{cm}^{-1}$; (b) there is slight but definite
increased absorption at $1085\,\text{cm}^{-1}$ relative to $1110\,\text{cm}^{-1}$ in the polymer; and (c) there
is a marked increase in absorption at $\sim 1100\,\text{cm}^{-1}$ relative to $\sim 700\,\text{cm}^{-1}$ in the
polymer. This appears to have resulted from both an increase at $\sim 1100\,\text{cm}^{-1}$ and
a decrease at $\sim 700\,\text{cm}^{-1}$. A change similar to the one in the $1100\,\text{cm}^{-1}$ region is
also observed in the 1200 to $1100\,\text{cm}^{-1}$ region—the appearance of an apparent
trough at $1195\,\text{cm}^{-1}$ in the polymer. Spectra of monomer–polymer mixtures were
said to be intermediate between the two in the areas mentioned above, as expected.
These authors claimed that their spectra were reproducible and they recognized the

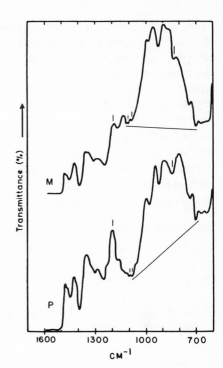

Fig. 10.20. Infrared spectra of eluant fractions from the Sephadex column (potassium bromide disks). (Scrimshaw *et al.*, 1966.) Key: M, BSA monomer; P, BSA polymer.

inherent difficulties with the KBr-disk technique. Perhaps solution spectra would give confirming experimental data.

Difference spectra were used by Vijai and Foster (1967) to study the hydrogen ion titration behavior of bovine plasma albumin in the *p*H range which is acid to the isoionic *p*H. They felt that difference spectra could give better precision (than in previous studies) in defining the number of carboxyl groups already protonated at the isoionic *p*H. Differential spectroscopy had been successfully used earlier by Susi *et al.* (1959) to demonstrate the presence of abnormal carboxyl groups in β-lacto-globulin. The number of protonated carboxyl groups existing in bovine plasma albumin at the isoionic *p*H was found by infrared difference spectroscopy to be 2.5 ± 0.5.

Some details should be noted regarding the carrying out of the infrared observations. Such details are essential for obtaining the necessary precision: (1) Prepare a 10% stock solution of deuterated albumin by dissolving a weighed sample of albumin in D_2O. (2) Since the same concentration is needed in all samples, use the following procedure: to 3.50 ml of stock solution add 0.50 ml of 0.5N KCl (in D_2O) + x ml of DCl + $(1 - x)$ ml of D_2O, where x is varied from 0 to 1 to change the *p*D. (3) Match the CaF_2 cells very carefully (0.05-mm fixed thickness). Matching is obtained when a flat baseline is produced, over the range 2100 to 1400 cm^{-1}, in a difference spectrum obtained with the same protein solution in sample and reference cells under conditions where the total absorption is approximately 50%. Trial and error use of polyethylene spacers and variation of the compression on the spacers by adjustment of the screws attains matching. (4) Into the reference cell inject a solution of protein at pD 8.9 (all carboxyls are assumed to be ionized). (5) In the sample cell place protein solutions of varying pD and scan over the desired region of the spectrum. (6) Rinse the sample cell three or four times with fresh solution and fill finally with the fresh solution to ensure that the pD of the solution in the cell is the same as in the flask. (7) After scanning of a sample is finished, thoroughly flush the cell with H_2O and dry with dry nitrogen gas.

Absorbance at the 1710 cm^{-1} absorption maximum was calculated by the peak-height method. In this case, the peak-height method had a distinct advantage over the peak-area method because near the carboxyl peak there would have been a negative difference spectrum, due to carboxylate ions, at 1550 cm^{-1}, and the area of the peak could have been complicated by this factor.

Determination of Molecular Weight of Insoluble Polypeptides

The determination of N-terminal amino acids of polypeptides or proteins by means of reaction with 2,4-dinitrofluorobenzene is used as a chemical method for the estimation of molecular weight. This operation is rather complex because of the need for hydrolysis, separation of DNP-amino acids, and colorimetric comparison with a standard curve. Schiedt and Restle (1954) have estimated the degree of polymerization of oligopeptides using dinitrophenylation and infrared spectroscopic

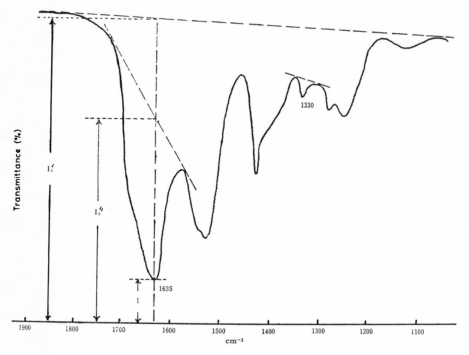

Fig. 10.21. Method of calculating absorbances of DNP-polypeptide; % transmittance *vs.* cm^{-1}. (Okamoto and Hamamoto, 1964.)

examination, but their method was not applicable to high molecular weight polypeptides. Okamoto and Hamamoto (1964) have modified this method, and the modification is presented here: (1) A sample of polypeptide (0.1 g) is dinitrophenylated by DNFB by the usual method. The DNP-polypeptide is obtained as a yellow precipitate. (2) The 1–2 mg precipitate is made into a KBr disk. The infrared spectrum shows a special band of NO which is attributed to the DNP group near 1330 cm^{-1} besides the normal absorption curve of the polypeptide containing the amide I, amide II, etc. (3) Absorptions of amide I near 1635 cm^{-1} and NO near 1330 cm^{-1} are measured with a ruler in mm units. As shown in Fig. 10.21, Okamoto and Hamamoto used I_0'' as I_0, whereas the German workers used I_0'. The value of the absorption ratio shows the length of the chain. Okamoto and Hamamoto have used both chemical and infrared methods to determine the molecular weights of polyglycines or copoly-(gly, ala).

The Japanese workers proposed an equation relating Z (degree of polymerization) to q, $Z = 30q/7$, and Fig. 10.22 shows the degree of correlation. The equation was found to be applicable to a wide range of molecular weights, not only to polypeptides but to copolymers of caprolactam and glycine. The method was reported to be a comparatively rapid and easy one for estimation of molecular weight because the degree of polymerization is obtained by simple measurements of absorption ratios. However, the method cannot apply to a polypeptide having side chains with free amino groups such as lysine or arginine residues.

The determination of molecular weight by measurement of the integrated absorbance of functional groups introduced by forming chemical derivatives has been described by O'Brien and Swindlehurst (1967).

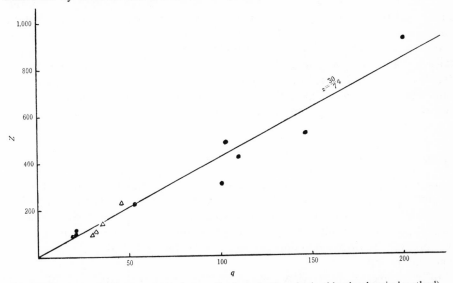

Fig. 10.22. Relationship between Z (the degree of polymerization obtained by the chemical method) and q (the absorption ratio of NO-band of DNP-polypeptide to amide I band), where q is defined by $q = [\log(I_{0_{co}}/I_{co})]/[\log(I_{0_{DNP}}/I_{DNP})]$; $\bullet$, polyglycine; $\triangle$, copoly (Gly, ala). (Okamoto and Hamamoto, 1964.)

Protein–Dye Interactions

Karyakin *et al.* (1967) have used infrared spectroscopy to study the interaction of proteins with dyes and chlorophyllin. When less than 10^{-5} g/ml (final concentration) of eosin or chlorophyllin k was added to glycinin (a globulin from soybean seeds) in aqueous and D_2O solutions, the absorption bands of the glycinin at 1660, 1535, and 1455 were intensified. Plant albumin showed similar spectral changes at 1455, 1410, 1312, 1180, and 1135 cm^{-1}. A new band appeared for albumin at 1400 cm^{-1} upon addition of acridine orange or chlorophyllin k. Interaction with eosin (above) involved NH groups, and interaction with acridine orange involved carboxyl groups of the protein. The interaction of albumin with chlorophyllin k involved both NH and carboxyl groups.

Oxidation of Wool

Strasheim and Buijs (1961) have obtained evidence from infrared spectroscopy that when wool is oxidized by either peroxyacetic acid or hydrogen peroxide, the disulfide bonds are changed to sulfonic acid groups via intermediate sulfenic acid (R—SOH) groups and sulfinic acid (RSO$_2$H) groups. They also found that samples of weathered wool contained sulfonic and sulfinic acid groups. Sulfonic acid groups were detected by bands at 1040 and 1175 cm^{-1}, the sulfenic group at 1059 cm^{-1}, and the sulfinic group at 1090 cm^{-1}.

Hydrogen Bonding of Ammonium Ion to Proteins

Barker (1968) has detected hydrogen bonding of ammonium ion to bovine serum albumin and bovine and equine α-globulins. Ammonium ion caused perturbations of the NH stretching mode at 3300 cm^{-1} of films of proteins cast from solutions of ammonium chloride. The interaction of NH$_4^+$ with α-globulins caused them to lose stability to heating. The extent of tryptic digestion of bovine serum albumin was decreased in ammonium chloride solutions but increased in $6M$ urea solutions, in comparison with solutions in distilled water. Barker (1968) thought that NH$_4^+$ ions may alter protein structure by means of hydrogen bonding in ways which may be of biological significance.

Hydrogen Bonding in Peptides

To investigate the hypothesis that dimerization of tripeptides occurs because they form hydrogen-bonded complexes (Schwyzer, 1958), Shields (1966) synthesized a peptide which consists of two tripeptide sequences joined together through the side chains of the middle residue of each. The properties of this peptide indicated that intramolecular hydrogen bonding can occur in peptides containing as few as six amino acid residues.

In a protected form of N^{α}-glycyl-N^{ϵ}-[β-(glycyl-α-L-aspartylglycine)]-L-lysyl-glycine the 3440 cm^{-1} band was assigned to free NH, and the 3360–3330 cm^{-1} concentration-dependent (2.5–5% and greater) band was assigned to hydrogen-bonded NH. Nuclear magnetic resonance studies confirmed these data.

Glycopeptides and Glycoproteins

Since ester linkages had been postulated as the bonds joining protein and poly-saccharide in chondroitin sulfate (Muir, 1958), submaxillary mucin (Gottschalk and Murphy, 1961), and ovomucoid (Hartley and Jevons, 1962), Rothfus and Smith (1963) investigated the possible occurrence of ester linkages in human γ-globulin glycopep-tides. As a result of their peptide nature and high content of N-acylglucosamine, the glycopeptides from γ-globulin may contain as many as 16 amide bonds per molecule. Consequently, strong absorption was found at an infrared frequency which is charac-teristic of carboxamide groups (1665 cm^{-1}). Also, substantial CO absorption occurred near 1100 cm^{-1}, a feature characteristic of substances containing carbohydrate. Significantly, the spectrum had no absorption band characteristic of normal carboxylic esters. The difference between the spectrum of an $18:1$ mixture (molar ratio) of acet-amide and gluconic acid–δ-lactone and that of glycopeptides, particularly in the region of ester carbonyl absorption (1740 cm^{-1}), indicated that not even one ester bond was present in the glycopeptides from γ-globulin.

Native bovine submaxillary mucin (BSM) showed a distinct absorption band near 1725 cm^{-1}, whereas the spectrum of porcine submaxillary mucin was devoid of this band (Hashimoto *et al.*, 1964). Since there were two possible types of ester bonds (O-glycosidic ester to the protein core or the O-acetyl groups of the sialic acid residues, or both) in the BSM molecule, a sample of hydroxylamine-treated (de-O-acetylated) BSM was analyzed by infrared spectroscopy (Bertolini and Pigman, 1967) in order to observe whether any changes in the spectrum had resulted from the treatment. No absorption in the 1725 cm^{-1} region was present in the case of the de-O-acetylated BSM. The absence of such a band appeared to be a measure of the lack of O-acetyl content of the mucins. Johnson and Chilton (1966) have also interpreted the absence of a band near 1730 cm^{-1} to signify absence of O-acetyl groups in a galactosaminoglycan of myxobacterium *Chondrococcus columnaris*.

Marshall and Porath (1965) have prepared a glycopeptide from α_1-acidic glyco-protein of human serum. A comparison is given in Fig. 10.23 of the infrared spectra of glycopeptide III (designated IIIα_1) from the parent glycoprotein, and the parent glycoprotein itself (α_1-acidic glycoprotein). The figure suggests that the broad band from 1200 to 1000 cm^{-1} is due to the oligosaccharide side chains. The ratio of the intensities, $I_{3400 \text{ cm}^{-1}}/I_{1650-1500 \text{ cm}^{-1}}$, is greater for the glycopeptide IIIα_1 than for the acidic glycoprotein. This increase in the ratio in the case of the glycopeptide was interpreted as reflecting the decrease in the number of peptide bonds (amide I and amide II) in the molecule.

Nucleoproteins

Bradbury *et al.* (1962*b*) have done polarized infrared studies on oriented sheets of deoxyribonucleoprotamine and of the deuterated nucleoprotein over a range of relative humidities between 0 and 100%. At high humidities the DNA was in the B-configuration with the bases perpendicular to the helix axis and this form persisted down to 76% relative humidity. They found the deuteration rate to be very fast, less than three minutes for the complete deuteration of a film exposed to D_2O vapor, and

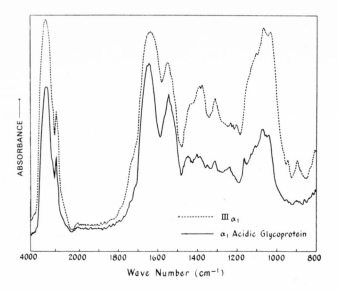

Fig. 10.23. Comparison of infrared spectra of glycopeptide III from its parent protein, α_1-acid glycoprotein, in KBr pellets. (Marshall and Porath, 1965.)

interpreted the data to mean that the protein was not in a helical configuration but in some extended form where the exchangeable hydrogens were accessible to D_2O. The location of the amide I band at $1640\,\text{cm}^{-1}$ was characteristic of neither the α-helix nor the fully extended form in β-sheets.

Infrared, deuteration, and optical rotation studies on films of histone have shown (Bradbury et al., 1962c) that in fresh material about half of the protein was in the α-helical form. As the specimen aged this proportion diminished. For freshly prepared nucleohistone, measurement of the residual N—H of partly deuterated films showed that over half (58%) of the protein in the nucleoprotein complex was of the slowly deuterating type and probably in the α-helical form. Polarized infrared studies were done (Bradbury et al., 1962c) on oriented sheets of nucleohistone, in hydrogenated and deuterated states, over a range of relative humidity between 0 and 95%. At high humidities the DNA was thought to be in the B-configuration with the bases perpendicular to the helix axis, and this form persisted down to about 80% relative humidity. (See also, Chapter 12.)

REFERENCES

Abbott, N. B. and Elliott, A. Proc. Roy. Soc. (London), **A234**, 247 (1956).

Ambrose, E. J. and Elliott, A. Proc. Roy. Soc. (London) **A205**, 47 (1951).

Ambrose, E. J., Bamford, C. H., Elliott, A., and Hanby, W. E. Nature **167**, 264 (1951).

Applequist, J. and Doty, P. in Polyamino Acids, Polypeptides, and Proteins (M. Stahmann, ed.) University of Wisconsin Press, Madison, 1962, p. 161.

Astbury, W. T., Dalgliesh, C. E., Darmon, S. E., and Sutherland, G. B. B. M. Nature **162**, 596 (1948).

Baddiel, C. B. J. Mol. Biol. **38**, 181 (1968).

Badger, R. M. and Pullin, A. D. E. *J. Chem. Phys.* **22**, 1142 (1954).

Bamford, C. H., Brown, L., Cant, E. M., Elliott, A., Hanby, W. E., and Malcolm, B. R. *Nature* **176**, 396 (1955).

Bamford, C. H., Elliott, A., and Hanby, W. E. *Synthetic Polypeptides*, Academic Press, New York, 1956.

Barker, A. V. *Biochim. Biophys. Acta* **168**, 447 (1968).

Beer, M. *Proc. Roy. Soc. (London)* **A236**, 136 (1956).

Beer, M., Sutherland, G. B. B. M., Tanner, K. N., and Wood, D. L. *Proc. Roy. Soc. (London)* **A249**, 147 (1959).

Bellamy, L. J. *The Infrared Spectra of Complex Molecules*, Methuen, London, 1956, p. 176.

Bendit, E. G. *Biopolymers* **5**, 525 (1967).

Bertolini, M. and Pigman, W. *J. Biol. Chem.* **242**, 3776 (1967).

Blout, E. R. and Asadourian, A. *J. Am. Chem. Soc.* **78**, 955 (1956).

Blout, E. R. and Fasman, G. D. (cited in Isemura *et al.*, 1968) in *Recent Advances in Gelatin and Glue Research*, Pergamon Press, London, 1958, p. 122.

Bradbury, E. M. and Elliott, A. *Polymer* **4**, 47 (1963).

Bradbury, E. M., Brown, L., Downie, A. R., Elliott, A., Fraser, R. D. B., and Hanby, W. E. *J. Mol. Biol.* **5**, 230 (1962*a*).

Bradbury, E. M., Price, W. C., and Wilkinson, G. R. *J. Mol. Biol.* **4**, 39 (1962*b*).

Bradbury, E. M., Price, W. C., Wilkinson, G. R., and Zubay, G. *J. Mol. Biol.* **4**, 50 (1962*c*).

Bradbury, E. M., Carpenter, B. G., and Stephens, R. M. *Biopolymers* **6**, 905 (1968).

Bragg, W. L., Kendrew, J. C., and Perutz, M. F. *Proc. Roy. Soc. (London)* **A203**, 321 (1950).

Corey, R. B. and Pauling, L. *Proc. International Wool Textile Res. Conf. Australia*, **1955**, Vol. B. 249.

Cowan, P. M. and McGavin, S. *Nature* **176**, 501 (1955).

Crick, F. H. C. and Rich, A. *Nature* **176**, 780 (1955).

Cross, A. D. and Jones, R. A. *Introduction to Practical Infra-red Spectroscopy*, 3rd Ed., Plenum Press, New York, 1969.

Darmon, S. E. and Sutherland, G. B. B. M. *J. Am. Chem. Soc.* **69**, 2074 (1947).

Davidson, B. and Fasman, G. D. *Biochemistry* **6**, 1616 (1967).

Davydov, A. S. *Theory of Molecular Excitons* (translated by M. Kasha and M. Oppenheimer, Jr.) McGraw-Hill, New York, 1962.

Dickerson, R. E. in *The Proteins, Vol. II, 2nd Ed.* (H. Neurath, ed.) Academic Press, New York, 1964, p. 603.

Doty, P., Imahori, K., and Klemperer, E. *Proc. Natl. Acad. Sci. U.S.* **44**, 424 (1958).

Downie, A. R., Elliott, A., Hanby, W. E., and Malcolm, B. R. *Proc. Roy. Soc. (London)* **A242**, 325 (1957).

Elliott, A. *Proc. Roy. Soc. (London)* **A211**, 490 (1952).

Elliott, A., Ambrose, E. J., and Robinson, C. *Nature* **166**, 194 (1950).

Elliott, A. and Ambrose, E. J. *Discussions Faraday Soc.* **9**, 246 (1951).

Elliott, A. and Bradbury, E. M. *J. Mol. Biol.* **5**, 574 (1962).

Elliott, A. and Malcolm, B. R. *Biochim. Biophys. Acta* **21**, 466 (1956*a*).

Elliott, A. and Malcolm, B. R. *Trans. Faraday Soc.* **52**, 528 (1956*b*).

Elliott, A., Hanby, W. E., and Malcolm, B. R. *Nature* **180**, 1340 (1957).

Fraser, R. D. B. *J. Chem. Phys.* **21**, 1511 (1953).

Fraser, R. D. B. and Price, W. C. *Nature* **170**, 490 (1952).

Fraser, R. D. B. and Price, W. C. *Proc. Roy. Soc. (London)* **141B**, 66 (1953).

Fraser, R. D. B. and Suzuki, E. *Spectrochim. Acta* **26A**, 423 (1970).

Fraser, R. D. B., Harrap, B. S., Ledger, R., MacRae, T. P., Stewart, F. H. C., and Suzuki, E. *Biopolymers* **5**, 797 (1967).

Fukushima, K. and Miyazawa, T. Annual Meeting of the Chemical Society of Japan, Tokyo, April, 1964 (cited in Miyazawa, 1967).

Gottschalk, A. and Murphy, W. H. *Biochim. Biophys. Acta* **46**, 81 (1961).

Gratzer, W. B., Bailey, E., and Beaven, G. H. *Biochem. Biophys. Res. Commun.* **28**, 914 (1967).

Hallam, H. E. *Spectrochim. Acta* **25A**, 1785 (1969).

Hamaguchi, K. *J. Biochem. (Tokyo)* **56**, 441 (1964).

Harrick, N. J. *Ann. N.Y. Acad. Sci.* **101**, Art. 3, 928 (1963).

Harrington, W. F. and von Hippel, P. H. *Advan. Protein Chem.* **16**, 1 (1961).

Hartley, F. K. and Jevons, F. R. *Biochem. J.* **84**, 134 (1962).

Hashimoto, Y., Hashimoto, S., and Pigman, W. *Arch. Biochem. Biophys.* **104**, 282 (1964).

Herskovits, T. T. and Mescanti, L. *J. Biol. Chem.* **240**, 639 (1965).

Imahori, K. *Biopolymers* **1**, 563 (1963).

Isemura, T., Okabayashi, H., and Sakakibara, S. *Biopolymers* **6**, 307 (1968).

Itoh, K., Nakahara, T., Shimanouchi, T., Oya, M., Uno, K., and Iwakura, Y. *Biopolymers* **6**, 1759 (1968).

Itoh, K., Shimanouchi, T., and Oya, M. *Biopolymers* **7**, 649 (1969).

Johnson, J. L. and Chilton, W. S. *Science* **152**, 1247 (1966).

Karyakin, A. V., Nikitina, A. A., and Chmutina, L. A. *Biofizika* **12**, 344 (1967) (in Russian); *CA*, **67**, 17856q (1967).

Kauzmann, W. *Ann. Rev. Phys. Chem.* **8**, 413 (1957).

Kendrew, J. C., Dickerson, R. E., Strandberg, B. E., Hart, R. G., Davies, D. R., Phillips, D. C., and Shore, V. C. *Nature* **185**, 422 (1960).

Kessler, H. K. and Sutherland, G. B. B. M. *J. Chem. Phys.* **21**, 570 (1953).

Krimm, S. *J. Mol. Biol.* **4**, 528 (1962).

Krimm, S. *Nature* **212**, 1482 (1966).

Krimm, S. and Kuroiwa, K. *Biopolymers* **6**, 401 (1968).

Krimm, S. and Watson, H. C. cited in *Science* **158**, 530 (1967), from abstract given at Natl. Acad. Sci. Meeting, Oct. 23–25, 1967.

Krimm, S., Kuroiwa, K., and Rebane, T. in *Conformation of Biopolymers, Vol. 2* (G. N. Ramachandran, ed.) Academic Press, New York, 1967, p. 439.

Krull, L. H., Wall, J. S., Zobel, H., and Dimler, R. J. *Biochemistry* **4**, 626 (1965).

Mahler, H. R. and Cordes, E. H. *Biological Chemistry*, Harper and Row, New York, 1966.

Makas, A. S. and Shurcliff, W. A. *J. Opt. Soc. Am.* **45**, 998 (1955).

Marshall, W. E. and Porath, J. *J. Biol. Chem.* **240**, 209 (1965).

Milch, R. A. *Nature* **202**, 84 (1964).

Miyazawa, T. *J. Mol. Spectrosc.* **4**, 155 (1960a).

Miyazawa, T. *J. Mol. Spectrosc.* **4**, 168 (1960b).

Miyazawa, T. *J. Chem. Phys.* **32**, 1647 (1960c).

Miyazawa, T. *Bull. Chem. Soc. Japan* **34**, 691 (1961)

Miyazawa, T. in *Polyamino Acids, Polypeptides, and Proteins* (M. A. Stahmann, ed.) University of Wisconsin Press, Madison, Wisconsin, 1962, p. 201.

Miyazawa, T. in *Poly-α-Amino Acids* (G. D. Fasman, ed.) Marcel Dekker, New York, 1967.

Miyazawa, T. and Blout, E. R. *J. Am. Chem. Soc.* **83**, 712 (1961).

Miyazawa, T., Shimanouchi, T., and Mizushima, S. *J. Chem. Phys.* **29**, 611 (1958).

Mizushima, S. *Structure of Molecules and Internal Rotation*, Academic Press, New York, 1954.

Mizushima, S. and Shimanouchi, T. *Adv. in Enzymology* **23**, 1 (1961).

Muir, H. *Biochem. J.* **69**, 195 (1958).

O'Brien, R. N. and Swindlehurst, R. N. *Can. J. Chem.* **45**, 2856 (1967).

Okamoto, S. and Hamamoto, M. *Agr. Biol. Chem.* (*Tokyo*) **38**, 55 (1964).

Padden, F. J. and Keith, H. D. *J. Appl. Phys.* **36**, 2987 (1965).

Pauling, L. and Corey, R. B. *Proc. Natl. Acad. Sci. U.S.* **37**, 729 (1951).

Pauling, L., Corey, R. B., and Branson, H. R. *Proc. Natl. Acad. Sci. U.S.* **37**, 205 (1951).

Pimentel, G. C. and McClellan, A. L. *The Hydrogen Bond*, W. H. Freeman, San Francisco, 1960.

Ramachandran, G. N. and Sasisekharan, V. *Biochim. Biophys. Acta* **109**, 314 (1965).

Ramachandran, G. N., Sasisekharan, V., and Ramakrishnan, C. *Biochim. Biophys. Acta* **112**, 168 (1966).

Rich, A. and Crick, F. H. C. *J. Mol. Biol.* **3**, 483 (1961).

Rosenheck, K. and Doty, P. *Proc. Natl. Acad. Sci. U.S.* **47**, 1775 (1961).

Rothfus, J. A. and Smith, E. L. *J. Biol. Chem.* **238**, 1402 (1963).

Sandeman, I. *Proc. Roy. Soc.* (*London*) **A232**, 105 (1955).

Sarkar, P. K. and Doty, P. *Proc. Natl. Acad. Sci. U.S.* **55**, 981 (1966).

Sasisekharan, V. *Acta Cryst.* **12**, 897 (1959).

Schiedt, U. and Restle, H. *Z. Naturforsch.* **9b**, 182 (1954).

Schellman, J. A. and Schellman, C. in *The Proteins, Vol. II, 2nd Ed.* (H. Neurath, ed.) Academic Press, New York, 1964, p. 1.

Schwyzer, R. in *Ciba Foundation Symposium on Amino Acids and Peptides with Antimetabolic Activity* (G. E. W. Wolstenholme and C. M. O'Connor, eds.), Little, Brown, and Co., Boston, 1958, p. 171.

Scrimshaw, G. F., Oberhauser, D. F., and Hess, E. L. *Biochem. Biophys. Res. Commun.* **24**, 290 (1966).

Shields, J. E. *Biochemistry* **5**, 1041 (1966).

Steinberg, I. Z., Berger, A., and Katchalski, E. *Biochim. Biophys. Acta* **28**, 647 (1958).

Strasheim, A. and Buijs, K. *Biochim. Biophys. Acta* **47**, 538 (1961).

Susi, H., Zell, T., and Timasheff, S. N. *Arch. Biochem. Biophys.* **85**, 437 (1959).

Susi, H., Timasheff, S. N., and Stevens, L. *J. Biol. Chem.* **242**, 5460 (1967).

Sutherland, G. B. B. M. *Discussions Faraday Soc.* **9**, 222 (1950).

Suzuki, S., Iwashita, Y., Shimanouchi, T., and Tsuboi, M. *Biopolymers* **4**, 337 (1966).

Swenson, C. A. and Formanek, R. J. *Phys. Chem.* **71**, 4073 (1967).

Timasheff, S. N. and Bernardi, G. *Arch. Biochem. Biophys.* **141**, 53 (1970).

Timasheff, S. N. and Gorbunoff, M. J. *Ann. Rev. Biochemistry* **36**, 13, Part I (1967).

Timasheff, S. N. and Susi, H. *J. Biol. Chem.* **241**, 249 (1966).

Timasheff, S. N., Susi, H., and Stevens, L. *J. Biol. Chem.* **242**, 5467 (1967a).

Timasheff, S. N., Townend, R., and Perlmann, G. E. *J. Biol. Chem.* **242**, 2290 (1967b).

Townend, R., Kumosinski, T. F., Timasheff, S. N., Fasman, G. D., and Davidson, B. *Biochem. Biophys. Res. Commun.* **23**, 163 (1966).

Traub, W. and Shmueli, U. in *Aspects of Protein Structure* (G. N. Ramachandran, ed.) Academic Press, New York, 1963, p. 81.

Tsuboi, M. *Bull. Chem. Soc. Japan* **22**, 215, 255 (1949).

Tsuboi, M. *J. Polymer Sci.* **59**, 139 (1962).

Tsuboi, M. Symposium on Quantum Aspects of Polypeptides and Polynucleotides, Stanford University, March, 1963 (cited in Schellman and Schellman, 1964).

Vijai, K. K. and Foster, J. F. *Biochemistry* **6**, 1152 (1967).

Vollmer, J. -P. and Spach, G. *Biopolymers* **5**, 337 (1967).

Zbinden, R. *Infrared Spectroscopy of High Polymers*, Academic Press, New York, 1964, p. 166.

Chapter 11

HYDROGEN–DEUTERIUM EXCHANGE*

In the field of protein chemistry, it has become more and more important to be able to determine tertiary as well as secondary and primary structure in order to decide what properties of a given protein determine its functions. One of the techniques introduced in recent years for such purposes is hydrogen–deuterium exchange.

Qualitative observations by infrared spectroscopy of hydrogen–deuterium exchange between several proteins and their surrounding solvent water were introduced into protein chemistry by Lenormant and Blout (1953). Quantitative measurement of the rates of hydrogen–deuterium exchange between a dissolved protein and its solvent water was carried out by Hvidt and Linderstrøm-Lang (1954), who employed a density gradient method. Subsequently, studies of hydrogen exchange using the three isotopes of hydrogen (H, D, and T) have been conducted on many proteins and related compounds with a variety of experimental conditions. Several informative reviews concerning such research are available in the literature (Linderstrøm-Lang, 1955; Linderstrøm-Lang, 1958; Kauzmann, 1957; Leach, 1959; Hvidt *et al.*, 1960; Perlmann and Diringer, 1960; Scheraga, 1961; Schellman and Schellman, 1964; Englander, 1967; Harrington *et al.*, 1966; Bryan, 1970). An important finding has been that in aqueous solution simple peptides and randomly coiled polypeptides exchange their labile hydrogen atoms (those bound to oxygen, nitrogen, and sulfur) with solvent water in a few minutes. However, native proteins exchange at least some of their peptide group hydrogens at a much slower rate, the exchange often taking much longer than 24 hours to go to completion. The slow rate of hydrogen exchange in a protein, compared with the exchange rate found for simple peptides under the same experimental conditions, is closely related to the conformation of the protein molecule in aqueous solution (sum of the secondary and tertiary structures). By quantitatively measuring the rate of hydrogen exchange in a given protein under specific conditions, one can obtain a characterization of the protein conformation (or distribution of conformations) present under these conditions.

The tertiary structure of the macromolecule and its molecular weight can be ascertained from measurements of intrinsic viscosity and diffusion, from ultracentrifugation studies, and from light scattering determinations. These methods can

*Adapted in part from Parker and Bhaskar (1969).

differentiate spheres, prolate or oblate ellipsoids, rigid rods, and flexible coils. To make use of the hydrogen exchange rate data in connection with the above type of information and that obtained from optical rotatory dispersion (ORD) and circular dichroism (CD) measurements (e.g., percentage helix content), a detailed knowledge of the molecular mechanism of hydrogen exchange is needed.

Hvidt and Nielsen (1966) have reviewed the results obtained up to 1966 from measurements of hydrogen exchange (H–H, H–D, H–T) in proteins and related compounds and have discussed the experimental techniques that have been used to measure quantitatively the rate of hydrogen exchange. Hydrogen isotope exchange has an advantage (compared to other "group reagents" used for studying the reactivity of functional groups such as SH and tyrosine hydroxyl groups) in that the hydrogen isotope has a negligible space requirement as well as properties very similar to those of the hydrogen atoms it replaces in the native protein. Another advantage is that *all* proteins have "reactive" groups, namely, the peptide groups, in *all* regions of the protein molecule. Thus, a complete kinetic exchange curve (degree of exchange *vs* time) for a protein reflects the conformation of the whole molecule and not only of parts of it. On the other hand, this advantage is partly counterbalanced by the need to determine exactly which parts of the protein molecule exchange their labile hydrogen atoms with a given measured first-order rate constant (see Hvidt and Nielsen, 1966).

Methods

The various methods of studying hydrogen exchange reactions have been reviewed recently (DiSabato and Ottesen, 1967). A brief description of the methods available for hydrogen–deuterium exchange follows.

Linderstrøm-Lang Method

The first quantitative measurement of hydrogen–deuterium exchange was made by Hvidt and Linderstrøm-Lang (1954), who employed a density gradient method. In this method, the protein is dissolved in D_2O and allowed to exchange all its hydrogens for deuterium, after which it is lyophilized. This deuterated protein is then redissolved in H_2O and the back exchange is measured as follows: Aliquots of the solution are removed at various time intervals, frozen quickly in a mixture of acetone and dry ice and lyophilized. During the lyophilization a representative sample of the solvent water is collected (cryosublimation) and its deuterium content is determined by measuring its density relative to that of nonenriched water. A few variations of the method have been developed but the basic features are essentially the same as described above (Hvidt and Nielsen, 1966; DiSabato and Ottesen, 1967).

A criticism has been made of the assumption in this method that exchange ceases on freezing the sample. Hallaway and Benson (1965) have found that some exchange does take place between protein and ice during the cryosublimation step. Attempts have been made to slow down the exchange during the cryosublimation step by decreasing the temperature of the protein solution to $-25°C$ (Hvidt and Nielsen, 1966) but it is not known if this procedure prevents all exchange. Byrne and Bryan (1970) have introduced an improved freeze-drying method, which involves the use of tritiated water, a freezing temperature of $-60°C$, and liquid scintillation counting. The method has demonstrated that the problem concerning cryosublimation has been corrected.

Infrared Spectroscopic Method

While the Linderstrøm-Lang method measures the exchange of all labile hydrogens (those attached to O, N, and S atoms) the infrared method can be used to study the exchange of hydrogens belonging to a specific group. Since measurable exchange rates of most practical utility in biological systems are those involving peptide hydrogens, application of the infrared spectroscopic method would depend upon the changes, on deuteration, in the absorptions involving NH vibrations.

The first observation of such a change was made by Lenormant (1950), who found that deuteration of proteins decreases the intensity of the amide absorption band at $\sim 1550\,cm^{-1}$ and results in a new band at $\sim 1450\,cm^{-1}$. This observation was confirmed by the studies of Lenormant and Blout (1953) on the infrared spectra of bovine plasma albumin in D_2O. Practical application of this observation was not made however until almost a decade later when simultaneous reports appeared in the literature from different laboratories (Blout *et al.*, 1961; Nielsen, 1960; Nielsen *et al.*, 1960; and Bryan and Nielsen, 1960).

Some of the main features of this method are described below and for greater detail reference may be made to the papers by Blout *et al.* (1961) and Hvidt (1963).

Secondary amides exhibit a strong absorption band at $\sim 1650\,cm^{-1}$ (amide I), due mainly to the C=O stretching vibration, and a weaker band at $\sim 1550\,cm^{-1}$ (amide II), which is ascribed to a mixed vibration involving N—H in-plane bending and the C—N stretching modes (Miyazawa, 1963). The amide II band, which has a major contribution from the NH bending vibration is very sensitive to hydrogen–deuterium exchange and has the advantage of occurring in the window region of D_2O absorption. This makes possible measurements on D_2O solutions of reasonable concentration.

The infrared absorption bands of an exchanging α-chymotrypsin solution (Bhaskar and Parker, 1969) are shown in Fig. 11.1. As the deuteration proceeds, the amide II band at $\sim 1540\,cm^{-1}$ is progressively replaced by the deuterated amide II band at $\sim 1450\,cm^{-1}$, while the intensity of the amide I band remains practically constant. The extent and rate of deuteration can be determined by measuring either the disappearance of the amide II band or the appearance of the deuterated amide II band. Interference from HOD absorption, however, makes this latter band less suitable than the undeuterated amide II band for exchange studies. A practical problem which arises in measuring the intensity of the amide II band is the difficulty in determining the baseline of this band. It can be seen in Fig. 11.1 that this does not coincide with the baseline for the amide I band, which is usually drawn parallel to the 100% transmittance line from the transmittance at $1800\,cm^{-1}$. Following the suggestion of Blout *et al.* (1961) the baseline of the amide II band is taken as the absorption of a solution of the completely deuterated protein. Complete deuteration is usually achieved by heating the D_2O solution of the exchanging compound. The temperature and duration of heating varies from compound to compound and is

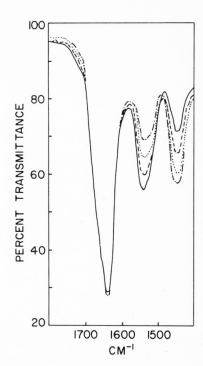

Fig. 11.1. Amide absorption bands of α-chymotrypsin (pH, 3.3) at different time intervals after dissolution in D_2O: (——) 3.8 min; (– – – –) 12 min; (·····) 38 min; (·—·—·) 209 min. (Bhaskar and Parker, 1969.)

arrived at by noting the stage beyond which further heating causes no reduction in amide II absorption. (Precipitation of the compound is also a factor which limits the temperature and duration of heating.) Although Blout *et al.* recognized that this is not a very accurate method they used it as the most practical method and it has found wide acceptance. Complete deuteration has also been achieved in some cases by addition of sodium dodecyl sulfate (Fig. 11.2). (See also DiSabato and Ottesen, 1965.)

Another suggestion of Blout *et al.* (1961), which has also been accepted as a standard procedure, is the use of the ratio $A_{\text{amide II}}/A_{\text{amide I}}$ (the A's are absorbances) for exchange measurements (see below). Besides being concentration independent, this ratio also corrects for any instrumental drift and provides a more accurate measure of exchange than does the absorbance of the amide II band.

In a typical exchange experiment, a measured volume of an H_2O solution of the compound of known concentration ($\sim 3\%$) is lyophilized. The lyophilized sample is then dissolved in enough D_2O to bring the volume (and hence the concentration) to the original value. The zero time of the reaction is usually taken as the time when D_2O is introduced into the sample tube. As soon as dissolution is complete the solution is introduced into a calcium fluoride cell and its spectrum between 1800 and 1300 cm^{-1} is recorded as a function of time. The time of each reading is taken as that at which the ν_{max} of the amide II band (~ 1550 cm^{-1}) is scanned.

From the measured absorbances, the ratio $A_{\text{amide II}}/A_{\text{amide I}}$ is plotted as a function of time. A typical exchange curve obtained in the case of lysozyme (Hvidt, 1963) is

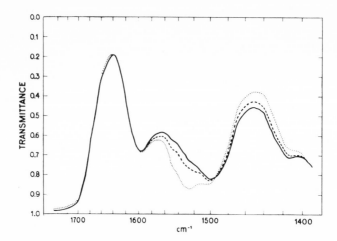

Fig. 11.2. Infrared absorption spectra of ovalbumin at pH 8.7
after 9 (———) and 156 min (————) of exchange at 21°C, and after
$2\frac{1}{2}$ hr at 80°C in the presence of $0.02M$ sodium dodecyl sulfate ($\cdots$).
(Willumsen, 1967.)

shown in Fig. 11.3. From the absorbance ratio at any time during the reaction and
that at zero time (for the completely nondeuterated species) one can estimate the
percentage of unexchanged hydrogens. However, the absorbance ratio for the completely nondeuterated sample is difficult to measure because some of the peptide
hydrogens will have exchanged with D_2O before the absorbance can be measured.
The strong absorption of water in this region does not permit its use for this purpose.
Blout *et al.* (1961) therefore determined the absorbance ratio $A_{\text{amide II}}/A_{\text{amide I}}$ for
films of several proteins and found it to be reasonably constant (0.4 to 0.5). Hvidt
(1963) also obtained a similar value (0.46) for lysozyme dissolved in D_2O at pH 3.2 by
extrapolation of the exchange curve to zero time. This extrapolation procedure is
only possible when the exchange rate is slow. Where the exchange rates are too fast
to permit a reliable extrapolation, the values obtained by Hvidt (0.46 ± 0.01) and
Blout *et al.* (0.45 ± 0.05) are used.

First-order rate constants for the exchange reaction have been obtained from
the absorbance measurements, and by graphical and computer analyses different
classes of peptide hydrogens with different rate constants of exchange have been
identified.

While the speed and ease with which this process can be performed have made it
one of the most frequently employed methods for exchange studies, some of the
assumptions and approximations involved in analyzing the experimental data make
it a little less than a quantitative method. The problem involved in measuring the
amide II absorbance accurately has already been mentioned. Use of the completely
deuterated sample to determine the baseline assumes that the native and denatured
protein have identical background absorption. There is no experimental evidence
to justify this assumption. The calculation of the number of hydrogen atoms

exchanged is based on the assumption that all peptide hydrogens have the same infrared oscillator strength. The dependency of the oscillator strength of a group on its molecular environment suggests that this assumption cannot always be valid. Use of absorbance at v_{max} as an index of absorption intensity introduces additional errors because the broad amide II band is made up of several bands associated with different classes of peptide groups. This problem can be overcome by taking the band area as a measure of the intensity but in doing this one has to take into account the absorption due to ionized carboxyl groups occurring in the same region (1575 cm^{-1}). Exposure to infrared radiation results in heating of the exchanging solutions but this effect is usually minimized by removing the cells from the cell compartment between readings. Use of a temperature-controlled chamber could possibly avoid this problem altogether.

It may be said, then, that this method can only be considered as semiquantitative at the present time. When comparative studies on the same protein or closely related proteins are made, some of these errors are eliminated and the method can yield more precise data.

Changes in other NH absorptions have also been employed for exchange studies. The N–H stretching vibration in the 3300 cm^{-1} region has limited utility for solution studies because of interference from other absorptions due to stretching involving hydrogen. However, it has been employed with reasonable success in the solid state (Haggis, 1957).

Hermans and Scheraga (1960) have found it advantageous to use the overtone region ($\sim 1.5 \mu$) rather than the fundamental region ($\sim 3 \mu$) and employed the first overtone of the N–H stretching vibration to study the exchange behavior of

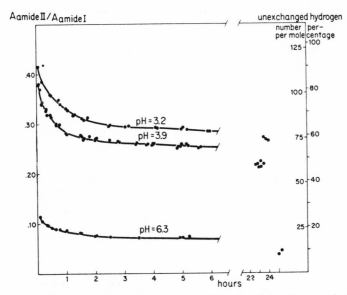

Fig. 11.3. $A_{\text{amide II}}/A_{\text{amide I}}$ as a function of time for three different values of pH at 21°C. (Hvidt, 1963.)

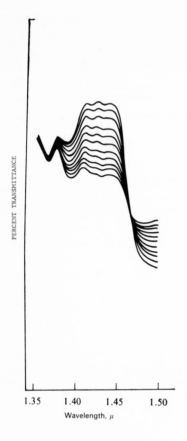

PERCENT TRANSMITTANCE

1.35   1.40   1.45   1.50
            Wavelength, μ

Fig. 11.4. Absorption spectra in the near-infrared region at different times after mixing N-methylacetamide and D_2O in dioxane. (Klotz and Frank, 1965.)

ribonuclease. Klotz and co-workers have also made use of this band for exchange studies of model amides and peptides (Fig. 11.4). (See Chapter 2.)

Miscellaneous Methods

Increase in weight upon deuteration (Morris, 1961) and line broadening in nuclear magnetic resonance spectra (Wishnia and Saunders, 1962) have also been employed to study hydrogen–deuterium exchange. Englander (1967) has described both the use of Sephadex and rapid dialysis to study hydrogen exchange. The previous practice of extensive deuteration is replaced by trace labeling with the radioisotope tritium. The Sephadex column techniques have been described in most detail in Englander (1963). The rapid dialysis technique has been described in Englander and Crowe (1965).

Theory

The foundation of a theoretical framework for the interpretation of hydrogen exchange behavior of biological macromolecules was laid by Linderstrøm-Lang (1955). It has since been developed by other workers, including some of his co-workers

(Englander, 1967; Hvidt and Nielsen, 1966). These authors have also reviewed the present status of the theory of hydrogen exchange recently. In the following paragraphs only the basic outlines of the theory, necessary for an understanding of its applications, are given.

Hydrogen atoms occurring in biological molecules can be classified as labile and nonlabile on the basis of the ease with which they exchange with aqueous solvent. Hydrogen atoms bound directly to carbon are nonlabile and exchange at negligible rates under ordinary experimental conditions. Hydrogen atoms bound to oxygen, nitrogen, or sulfur, on the other hand, are labile and exchange readily when freely exposed to aqueous solvent. In proteins, the labile hydrogen atoms of the side-chains are fully exposed and therefore exchange at rates too fast to be measured by the ordinary methods. Peptide hydrogens, however, exchange at considerably slower rates and are, therefore, of the greatest interest and utility in exchange studies. To explain the exchange of peptide hydrogens buried deep inside the protein matrix, Linderstrøm-Lang (1958) introduced the concept of transconformational reactions in a protein.

According to this concept, a protein molecule can exist in solution in a number of different conformations, which are undergoing rapid interconversion. If we consider any one labile peptide hydrogen atom, in some of these, termed the N conformations, the labile hydrogen is in a folded region and its exchange proceeds at an immeasurably slow (almost zero) rate. In the others, designated I conformations, the labile hydrogen is in an open conformation and exchange proceeds at rates characteristic of small peptides. The exchange of the peptide hydrogens of the protein molecule can then be described schematically by Eq. (1) (Hvidt and Nielsen, 1966).

$$N \underset{k_2}{\overset{k_1}{\rightleftharpoons}} I \xrightarrow{k_3} \text{exchange} \tag{1}$$

A complete solution of the system of rate equations describing an exchange reaction proceeding according to this scheme has been described by Hvidt (1964).

The first-order rate constant of exchange of a peptide hydrogen atom in a protein is given by Eq. (2).

$$k = \frac{k_1 k_3}{k_1 + k_2 + k_3} \tag{2}$$

A few special cases of this equation were considered by Hvidt (1964):

(i) The protein is present in solution mainly in the rapidly exchanging I conformation, i.e., $k_1 \gg k_2$ and $k_2 \gg k_3$. The observed rate constant in this case is equal to k_3, which is almost identical with the rate constant of exchange of small peptides under the same conditions.

(ii) The protein is present mainly in the nonexchanging N conformation, i.e., $k_2 \gg k_1$.

Hvidt has considered two possibilities in such a situation:

(a) The rate of exchange is fast compared to the rate of closing, i.e., $k_3 \gg k_2$. The rate constant, k, in such a case is given by $k = k_1$. Hvidt and Nielsen (1966) have referred to this exchange behavior as "EX_1," designating a unimolecular type of exchange.

(b) The other case arises when the rate of closing is fast compared to the rate of exchange, i.e., $k_2 \gg k_3$. The rate constant, k, in this case is given by Eq. (3).

$$k = \frac{k_1}{k_2} \cdot k_3 \qquad (3)$$

This is called an "EX$_2$" mechanism, designating a bimolecular type of exchange.

Hydrogen exchange data can, therefore, enable one to obtain thermodynamic parameters of protein structure if one can determine the mechanism by which exchange proceeds. In the case of the EX$_1$ mechanism, the observed rate is equal to the rate of opening, and thus gives the free energy of activation of the opening reaction. In an EX$_2$ mechanism, use of k_3 values obtained from exchange studies on model compounds can help in determining the equilibrium constant for the opening reaction (and hence the free energy difference between the open and closed states) from the observed exchange rates.

For the exchange reaction in proteins, a choice between the two mechanisms, EX$_1$ and EX$_2$, has been made based on the pH dependence of exchange rates. The exchange rates of most proteins studied increase sharply with pH, even when no conformational transition is observed in the pH range. For the EX$_1$ mechanism this would imply an increase of k_1 with pH, while for the EX$_2$ mechanism k_3 should increase with pH, the stability of the protein structure (k_1/k_2) being insensitive to pH (as indicated by absence of any conformational transition in the pH range). Studies on model peptide compounds have shown that the rate constant of exchange (k_3) increases with pH (Nielsen, 1960; Nielsen *et al.*, 1960; Bryan and Nielsen, 1960) and Hvidt (1964) has therefore suggested that exchange in proteins proceeds by the EX$_2$ mechanism. Although it is not improbable that k_1 increases with pH, the EX$_2$ mechanism for protein hydrogen exchange has met with general approval.

The rate equations (2 and 3) apply to a group of labile hydrogen atoms which are exposed to the solvent by the same transconformational reaction. In a protein there can be other classes which require different transconformational reactions involving larger or smaller free energy changes and these will therefore exchange at different rates. Linderstrøm-Lang (1958) found that hydrogen exchange curves can be analyzed as a sum of first-order reactions, as in Eq. (4):

$$n_0 - n_t = \sum_i n_i \exp(-k_i t) \qquad (4)$$

In Eq. (4), the terms k_i are the first-order rate constants for each kinetic class containing n_i hydrogens, n_0 is the total number of exchangeable hydrogens (sum of n_i), and n_t is the number of hydrogens which have exchanged at time t. It has generally been found that three or four kinetically distinct classes of peptide hydrogens can explain the observed exchange curves. Both graphical and computer analyses have been used to determine n_i and k_i values from the exchange data.

Factors Affecting Hydrogen–Deuterium Exchange of Biological Molecules

When exchange studies of insulin were first published (Hvidt and Linderstrøm-Lang, 1954), it was suggested that the peptide hydrogens which did not exchange with the solvent were those involved in the N—H···O=C bonding of the peptide backbone (α-helix). This suggestion received support from the work of Elliott and Hanby (1958) on the exchange behavior of poly-γ-benzyl-L-glutamate. Optical rotatory dispersion studies had shown that this polypeptide exists as a random coil in pure dichloroacetic acid, while in 10% (v/v) dichloroacetic acid in chloroform, it is entirely in the α-helical form. The polypeptide dissolved in O-deuterated dichloroacetic acid, to which chloroform was added to give a 10% solution of the acid, did not show the N—H absorption at 3300 cm^{-1}, indicating rapid exchange. This finding is as would be expected for a random coil structure. When the polypeptide was first dissolved in chloroform and then O-deuterated dichloroacetic acid was added to give a 10% solution of the acid, the resulting solution exhibited a strong N–H absorption band which did not diminish appreciably in a few days. This result is also consistent with the known helical structure of the polypeptide in this solvent. Exchange studies on poly-α-L-glutamic acid in a D_2O–dioxane solvent system (Blout et al., 1961; Bryan and Nielsen, 1960) also gave evidence in support of the suggestion that peptide hydrogens in an α-helix exchange slowly and the percentage of hard-to-exchange peptide hydrogens came to be considered an index of helical content of proteins. It was realized, however, that while involvement in helical structure does slow the exchange, slowly exchanging hydrogens cannot always be equated with helical content. Hydrogen atoms could be shielded from exchange by their location in the hydrophobic regions of the protein (Blout et al., 1961). It was also found experimentally that the percentage of hard-to-exchange hydrogens is not always in agreement with the helical content of proteins determined by other methods like ORD and X-ray crystallographic analysis (Beychok and Blout, 1961; Kendrew et al., 1961). It is thus clear that a more thorough understanding of the factors affecting exchange is essential for a complete and useful interpretation of hydrogen–deuterium exchange results.

The simplest approach to a study of the factors affecting hydrogen exchange in biological molecules is to examine the exchange behavior of small peptides since most hydrogens with measurable exchange rates come from peptide groups. The simplest of model peptide compounds, N-methylacetamide, was the subject of the earliest investigations. Nielsen (1960) studied the exchange reaction employing a "stopped flow" infrared spectroscopic method at different pD values ranging from 3.23 to 5.97. The back exchange of a deuterated sample of N-methylacetamide was also studied at different pH values by measuring the increase in absorbance at 2600 cm^{-1} due to the formation of HOD. Both reactions followed first-order kinetics and the rates varied with pH or pD as the case may be, indicating acid and base catalysis. The rate constants were at a minimum at pD 5.4 (pH 5.1 for the back exchange). Nielsen's results agreed with the mechanism of exchange [Eqs. (5) and (6)] proposed by Berger et al. (1959) based on NMR studies.

$$CH_3-CO-NH^*-CH_3 + H_3O^+ \rightarrow CH_3-CO-\overset{+}{N}H\overset{*}{H}-CH_3 + H_2O \tag{5}$$
$$CH_3-CO-\overset{+}{N}H\overset{*}{H}-CH_3 + H_2O \rightarrow CH_3-CO-NH-CH_3 + \overset{*}{H}H_2O^+$$

$$CH_3-CO-N\overset{*}{H}-CH_3 + OH^- \rightarrow CH_3-CO-\overset{-}{N}-CH_3 + HO\overset{*}{H} \tag{6}$$
$$CH_3-CO-\overset{-}{N}-CH_3 + H_2O \rightarrow CH_3-CO-NH-CH_3 + OH^-$$

Exchange studies with di- and tripeptides (Nielsen *et al.*, 1960) gave results similar to those obtained in the case of N-methylacetamide. The rates varied with pD, indicative of acid and base catalysis, and the minimum rates (~ 0.6 min^{-1} for glycylglycine and ~ 0.3 min^{-1} for the N-terminal peptide group of alanylglycylglycine) were about the same as found in the case of N-methylacetamide (~ 0.5 min^{-1}). The pD minimum (pD at k_{min}), however, occurred at lower values (2.4 for glycylglycine and 2.2 for alanylglycylglycine) compared to that for N-methylacetamide (5.4), which can be explained by the presence of the electron-withdrawing groups on the amide (see below).

The polypeptide, poly-DL-alanine, also gave exchange results (Bryan and Nielsen, 1960) similar to those obtained for the smaller peptides. The exchange reaction followed first-order kinetics, the rate constants (0.14 to 1.2 min^{-1}) were of the same order of magnitude as for the simpler peptides and the exchange was subject to H$^+$ and OH$^-$ (D$^+$ and OD$^-$) catalysis. The pH (pD) dependence of the exchange rate has been summarized in two approximate equations (Bryan and Nielsen, 1960)

$$k_{H_2O} \simeq 50(10^{-pH} + 10^{pH-6})10^{0.05(t-20)} \text{ min}^{-1} \tag{7}$$

$$k_{D_2O} \simeq 50(10^{0.3-pD} + 10^{pD-6.3})10^{0.05(t-20)} \text{ min}^{-1} \tag{8}$$

where t is the temperature in °C.

Klotz and his co-workers (Klotz and Frank, 1965; Leichtling and Klotz, 1966; Scarpa *et al.*, 1967) have also carried out detailed studies of the exchange reaction in several model compounds and examined the effects of solvent, temperature, and acid–base catalysts on the exchange rates. The hydrogen–deuterium exchange of N-methylacetamide in a dioxane–water solvent system was found to be markedly dependent on pD, varying parabolically with it, with a minimum at $\sim$ pD 7.5. While this was in agreement with the D$^+$ and OD$^-$ catalysis reported by Nielsen (1960), Klotz and Frank (1965) found that the exchange is also subject to general acid–base catalysis. The catalysis studied, imidazole, acetic acid, and formic acid, however, had smaller catalytic constants than those found for the proton and hydroxyl ion. The exchange reaction of N-methylacetamide in pure D$_2$O was also found to be influenced by general acid–base catalysis, in addition to the D$^+$ and OD$^-$ catalysis reported earlier. Figure 11.5 shows the parabolic dependence of the exchange rate on pD and the influence of catalysts on the rates.

Leichtling and Klotz (1966) extended such studies to other model compounds and examined the three features characteristic of exchange reactions, namely, pD_{min}, k_{min}, and E_D, the activation energy of the exchange reaction. They derived mathematical expressions relating these features to the factors under consideration and analyzed the experimental data in terms of these expressions. The rate constant for the D$^+$ and OD$^-$ catalyzed exchange reaction can be expressed as Eq. (9)

$$k = k_0 + k_D(D^+) + k_{OD} \cdot \frac{K_w}{(D^+)} \tag{9}$$

where k is the observed rate constant, k_0 is that for the spontaneous reaction, k_D is that for the acid-catalyzed reaction, k_{OD} is that for the base-catalyzed reaction, and K_w is the self-dissociation constant for water.

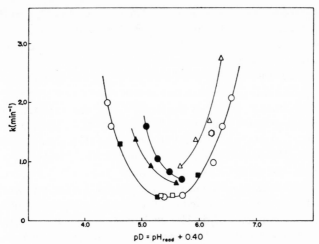

Fig. 11.5. Rates of hydrogen–deuterium exchange for N-methylacetamide ($1M$) in D_2O at 24°C: ○, 0.02M sodium acetate buffer; ●, 0.9M acetic acid-acetate; □, 0.8M sodium chloride plus 0.02M sodium acetate; ■, Nielsen's results (1960); △, 0.4M hydroxylamine; ▲, 0.7M methoxyamine; ◌, 0.6M imidazole. (Klotz and Frank, 1965.)

Mathematical analysis of Eq. (9) yielded the following expressions for the minimum rate constant and the corresponding D^+ concentration (and pD_{min}):

$$(D_{min}^+)^2 = \frac{k_{OD}}{k_D} \cdot K_w \quad \text{or} \quad pD_{min} = \tfrac{1}{2}pK_w - \tfrac{1}{2}\log\frac{k_{OD}}{k_D} \tag{10}$$

and

$$k_{min} = k_0 + 2k_D(D_{min}^+) \tag{11}$$

Leichtling and Klotz also showed that the rate constants of the acid- and base-catalyzed exchange reactions of an amide or peptide following the mechanism given in Eqs. (5) and (6) are equal to k_D and k_{OD}, respectively. With these expressions it is possible to predict the influence of inductive effect on the exchange rates. If an electron-withdrawing substituent is attached to the amide group, it will reduce the basicity of the nitrogen and hence its rate of protonation. At the same time the acidity of the amide group is increased, resulting in a higher rate for the base-catalyzed reaction. From the expression for pD_{min} [Eq. (10)], it will be clear that these two effects act in unison to lower pD_{min}. The expression for pD_{min} indicates that it can be affected by solvent character also. The presence of apolar substances can raise pK_w and hence shift the pD_{min} to higher values. Leichtling and Klotz also showed that inductive effects would not, to a first approximation, affect the k_{min}, although the nature of the solvent would.

The experimental data were found to be in good agreement with the above predictions. The rate constants of exchange of amide hydrogens in both acetylglycine ethyl ester and N-chloroacetylglycine ethyl ester showed parabolic dependence on

pD, with a pD$_{min}$ of 5 and 3.6, respectively. The decrease in pD$_{min}$ in the case of the haloester can be understood in terms of the electron-withdrawing effect of chlorine. The minimum rate constants for the two esters (0.016 min^{-1} and 0.015 min^{-1}) are, however, the same, again in accordance with the predictions. The results of hydrogen–deuterium exchange studies on poly-L-lysine were also found to follow the predicted behavior. The k_{min} for this polypeptide (0.013 min^{-1}) was close to that for the smaller model compounds but the pD$_{min}$ was lower, 2.55. The near-infrared spectrum of this compound did not give any evidence of N—H$\cdots$O=C bonding and ORD studies indicated a nonhelical configuration for the molecule. This may account for the k_{min} while the low pD$_{min}$ has been attributed to the presence of the charged ϵ NH$_3^+$ groups which would favor the base-catalyzed reaction.

Hydrogen exchange studies, under various conditions of temperature, pD, and presence of catalysts, were also carried out (Leichtling and Klotz, 1966) on the helical polypeptide, poly-L-glutamic acid. The exchange raction was studied in a mixed dioxane–D$_2$O (1:1) solvent at pD values below 6 where viscosity measurements, acid–base titrations, and ORD studies indicated that the polypeptide had a helical conformation. Near-infrared spectra also showed a doublet in the 1.53–1.57 μ region, characteristic of N—H$\cdots$O=C bonding. (A sharp rise in the exchange rate was observed at a pD ~ 6, corresponding to the helix–coil transition.)

The rate constants of exchange were found to vary parabolically with pD as in the case of simpler amides, with a pD$_{min}$ at ~ 3.2. The inductive effect of amino acid residues on either side of each peptide group was suggested as the factor responsible for the low pD$_{min}$ (Leichtling and Klotz, 1966). As in the case of N-methylacetamide, added salts did not produce any significant changes in the exchange rates even at concentrations far in excess of those present in the solution at the extremes of pD. The rate constants were linearly dependent on [D$^+$] and [OD$^-$]. The exchange was also catalyzed by acid and base forms of trifluoroacetic acid, trichloroacetic acid, and dichloroacetic acid. Trichloroacetic acid had the highest catalytic rate constant and produced marked increases in the observed rate constant of exchange at a concentration of $\sim 0.5M$. The catalytic effects were similar to those found for N-methylacetamide.

The activation energy of the exchange reaction of poly-L-glutamic acid (27 kcal/mole) was not significantly higher than the value obtained for the simpler amides (23 kcal/mole for N-chloroacetylglycine ethyl ester, 17 kcal/mole for the acid-catalyzed, and 23 kcal/mole for the base-catalyzed exchange reaction of N-methylacetamide).

The major difference between the hydrogen exchange behavior of the helical polypeptide and the simpler amides is in the k_{min}, which for the former is a thousand times smaller than for the latter. Leichtling and Klotz cautioned, however, against attributing the entire decrease, as is very often done, to the helicity of the molecule. While the locking in of the peptide hydrogens in N—H$\cdots$O=C bonds can produce a decrease in k_{min}, other factors found to produce similar effects in the simpler amides must also be effective. Thus, the presence of apolar groups in the side chain can cause a decrease in K_w in the vicinity of the molecule as compared with that of the bulk solvent thereby lowering k_{min}.

It has already been mentioned that the slow exchange of peptide hydrogen can result from factors other than the helical structure of proteins. Scarpa *et al.* (1967) examined the influence of side-chain interactions, interactions of side chains with solvent, etc., by studying the exchange characteristics of poly-isopropylacrylamide (LV).

$$\left[\begin{array}{c} -CH_2-CH- \\ | \\ C=O \\ | \\ H-N \\ | \\ CH \\ H_3C \quad\quad CH_3 \end{array}\right]_n$$

LV

Viscosity measurements indicated a random conformation for this macromolecule. Infrared studies showed presence of N—H$\cdots$O=C bonding but from models of the polymer segment it was concluded that only intramacromolecular hydrogen bonding was possible, steric interference from the bulky isopropyl groups precluding formation of intermacromolecular hydrogen bonding. The absence of α-helical conformations and the presence of large pendant apolar groups thus made this polymeric amide an ideal case for the study of the effect of side-chain interactions on H–D exchange. The rate constants of exchange, determined from infrared absorption measurements in the overtone region, were found to vary parabolically with pD, as in the case of other amide molecules, indicating D$^+$ and OD$^-$ catalysis. The pD_{min} (~ 5) and the activation energy for the exchange reaction (20 kcal/mole) were also close to the values found for other amides and peptides, and Scarpa *et al.* (1967) therefore assumed that the exchange must proceed by the same mechanism as proposed for the model amide and peptide compounds. The minimum rate constant k_{min} ($\sim 4 \times 10^{-3}$ min^{-1}) was, however, very much smaller than in the other cases.

Scarpa *et al.* also studied the exchange reaction of N-isopropylpropionamide, which corresponds in structure to a single residue unit of the polymer. The rate constants showed the same kind of parabolic variation with pD, with a pD_{min} of 5.38 and a k_{min} of $\sim 1.8 \times 10^{-1}$ min^{-1}. The lower pD_{min} for the polymer of N-isopropylpropionamide must be caused by the inductive effect of neighboring amide groups. The k_{min} for the polymer is, however, very much smaller, but this could not be attributed to hydrogen bonding for the following reason. Infrared spectra of the polymer indicated the presence of free N—H groups to the extent of one third of the total of N—H groups. The kinetic data, however, indicated only one class of exchanging hydrogens and the rate constants were the same whether computed from the disappearance of the infrared absorption of free N—H, bonded N—H$\cdots$O=C, or from the appearance of the O–H absorption (due to formation of HOD). Scarpa *et al.* therefore suggested that the bonded and free groups are in rapid equilibrium with each other and explained the exchange kinetics by the usual "motility" mechanism proposed for the exchange of peptide hydrogens buried in the protein matrix (see *Theory*). The bonded form was treated as the nonexchanging state N and the free form as the state I where the amide groups are fully exposed to the solvent:

$$N-H\cdots O{=}C \overset{k_1}{\underset{k_2}{\rightleftharpoons}} N-H + O{=}C \xrightarrow[\text{in D}_2\text{O}]{k_3} N-D + O{=}C \tag{12}$$

where k_3 is the rate-determining step.
Therefore,

$$\frac{d(N-D)}{dt} = k_3(N-H) \tag{13}$$

Also,

$$\frac{(N-H)}{(N-H\cdots O{=}C)} = \frac{k_1}{k_2} = K \tag{14}$$

Therefore,

$$(N-H) = (NH)_{\text{total}} \cdot \frac{K}{K+1} \tag{15}$$

$$\frac{-d(NH)_{\text{total}}}{dt} = \frac{d(N-D)}{dt} = k_3 \cdot (NH)_{\text{total}} \cdot \frac{K}{K+1} \tag{16}$$

and, the observed rate of exchange,

$$k = k_3 \cdot \frac{K}{K+1} \tag{17}$$

Based on the infrared absorbances of the free and bonded $N-H$ groups, Scarpa et al. obtained a value of ~ 0.5 for the equilibrium constant, K, but assuming k_3 to be the same for monomer and polymer, they could not account for the hundred-fold decrease in k_{min} for the polymer. Scarpa et al. concluded that the reduced rate of exchange is not entirely due to the hydrogen-bonded groups but that the exchange of the free group itself is slowed down by environmental effects, which should be quite different for the two species. From the expression for k_{min} (Eq. 11) it can be seen that a change in k_D, k_{OD}, or K_w can affect the k_{min}. Increase in viscosity of the solution of the polymer was considered unlikely to cause the change since previous studies had shown that a ten-fold increase in viscosity produced only a 20% drop in the rate of exchange of N-methylacetamide. Scarpa et al. therefore attributed the slow exchange in the polymer to polymer–solvent interactions. The marked solvent effect on the infrared absorption of the free $N-H$ group indicates that this group is exposed to solvent. Thus, in chloroform the polymer shows a single peak at 1.487 μ, corresponding to the free $N-H$ group, while in water and dioxane two additional bands appear in the region 1.53–1.57 μ, which is characteristic of hydrogen-bonded $N-H$ groups. The apolar isopropyl groups attached to the amide groups must also, therefore, be exposed to the solvent and could decrease K_w in the neighborhood of the polymer. This change in the solvent character would also reduce the concentration of the charged intermediates involved in the rate-controlling step of the acid–base catalyzed exchange reaction. Scarpa et al. therefore suggested that the reduced hydrogen exchange rates in the polymer are mainly due to the difference in the nature of the solvent.

The studies on model peptide compounds have been particularly helpful in explaining the pH dependence of the exchange reaction of proteins (see *Theory*). Hvidt (1964) has attributed the pronounced increase in exchange observed in several proteins with increasing pH above pH 3 to hydroxyl ion catalysis similar to that observed in model peptides. Willumsen (1967) has found support for Hvidt's suggestion from his exchange studies on ovalbumin at different pH values. Using the relation between the exchange rate and pH for the model peptides [Eqs. (7) and (8)], Willumsen obtained the following relation (Eq. 18) between the fraction of unexchanged peptide hydrogens p remaining at time t and the pH (for pH > 4) for a protein (containing h peptide hydrogens) exchanging by the EX_2 mechanism:

$$p = \frac{1}{h} \sum_m^h \exp\left(-\frac{k_{1,m}}{k_{2,m}} \times 5 \times 10^9 [OH^-]t\right) \tag{18}$$

Since ORD studies did not indicate any conformational changes, the ratio $k_{1,m}/k_{2,m}$ for each peptide hydrogen was considered to be constant in the pH range studied (2.4 to 10.5). The number of unexchanged hydrogens would therefore depend only on the product of $[OH^-] \times t$, and a graphical plot of the two quantities resulted in a continuous curve in the pH region 6.3–10.5, supporting Willumsen's suggestion that exchange in ovalbumin follows the EX_2 mechanism.

While increasing pH has been found to increase the extent of exchange in most of the proteins studied (Hvidt and Nielsen, 1966; Hvidt, 1964; Willumsen, 1966; Kägi and Ulmer, 1968), there have been a few exceptions (Hvidt, 1964; Parker and Stryker, 1969). Even in the case of ovalbumin, the exchange data at pH 4.1 and 4.8 deviated from the curve and Willumsen has suggested that general acid–base catalysis by the side chains in the proteins might be operative in this region. Bovine submaxillary mucin (Parker and Stryker, 1969) was found to be another exception to the general dependence of exchange on pH. Both the extent and rate constant of exchange of this glycoprotein decreased with increase in pH in the range 3.7 to 7.2. Since ORD data on the glycoprotein are not available, the possibility of a conformational change in the pH range 3.7 to 7.2 cannot be excluded but it is not improbable that general acid–base catalysis might be operative in this case too. The main opposition to Klotz's theory of general acid–base catalysis has stemmed from the fact that most of his work was carried out in nonaqueous solvents and it is argued that the same factors may not be operative in aqueous medium. The low values of the catalytic constants in aqueous medium (Klotz and Frank, 1965) do seem to lend support to this argument against the importance of general acid–base catalysis. Klotz, however, has suggested that the high concentration of acidic and basic groups in the side chains of proteins might to some extent compensate for their low catalytic activity, making general acid–base catalysis important in the exchange reaction of these molecules.

The investigations of Klotz and his co-workers (Klotz and Frank, 1965; Leichtling and Klotz, 1966; Scarpa *et al.*, 1967) have also raised doubts about the validity of some of the earlier interpretation of exchange data. The exchange studies on a nonhelical polyamide have shown that slow exchange can result from factors other than involvement in helical structures. The attribution of the high activation energy of ~ 20 kcal/mole for the slowly exchanging hydrogens in proteins to their involvement in bridges in helical segments is not necessarily correct, since exchange of free N—H groups was found to require comparable activation energies of 17–23 kcal/mole (Klotz and Frank, 1965). The investigations on model compounds have shown that factors like inductive effects of adjacent groups and presence of apolar groups can affect the exchange characteristics. It would seem, therefore, that the exchange characteristics of a polypeptide reflect the environment of the macromolecule in addition to its conformation.

More investigations on polypeptides with known structural features will have to be made before exchange measurements can provide fairly specific information about macromolecular structure. Used in conjunction with other physical studies like ORD, CD, X-ray analysis, etc., exchange studies can provide useful information. In the detection of conformational changes, however, exchange studies can play a very useful role and most investigations of exchange reactions have been directed toward this end.

Applications

Polypeptides

Synthetic Polypeptides. The structure and properties of synthetic polypeptide monolayers spread at the air–water interface have been investigated by Malcolm (1968). Two series of high molecular weight polymers were examined—esters of polyglutamic acid and polymers with hydrocarbon side chains. The structure was investigated by measurement of force–area relations and surface potentials with a Langmuir trough, and by measurement of H–D exchange rates. These direct methods were supplemented by observations of infrared spectroscopy and electron diffraction on collapsed films removed from the surface. In all cases the properties of the monolayer were consistent with a structure consisting of condensed ordered arrays of α-helices.

The exchange measurements show that exchange can take place in an intact α-helix, and show how the exchange rate is influenced by the accessibility of the peptide group to water, the pH of the substrate, and the hydrophobic nature of the side chain.

Two models of collagen structure have been developed since 1954 which are similar, except for a difference in the number of hydrogen bonds. Rich and Crick (1961) have assumed one hydrogen bond for each tripeptide unit (one-bonded structure), while Ramachandran and Sasisekharan (1961) have assumed two hydrogen bonds per unit (two-bonded structure). In Rich and Crick's collagen structure II, position 1 of the tripeptide unit is always occupied by a glycine unit; the positions 2 and 3 may be occupied by any other amino acid including proline and hydroxyproline. In the structure proposed by Ramachandran and Sasisekharan an imino acid unit can only occupy position 3 in order to avoid changes in the collagen structure. Ramachandran's structure has been supported by the data of Harrington (1964) and Bensusan and Nielsen (1964).

Heidemann and Bernhardt (1967) have synthesized and polymerized a series of tripeptides corresponding to both of the proposed collagen structures. They used the H–D exchange properties of these polymers to compare their stabilities. The extent of H–D exchange in the solid state in the sequence (gly-ala-pro)$_n$ and (pro-gly-ala)$_n$ was higher than in (gly-pro-ala)$_n$, and therefore these workers concluded that the structure (gly-pro-ala)$_n$ is more stable than the other sequences. The polypeptides containing the sequence (gly-ala-pro), which corresponds to the Ramachandran two-bonded collagen model, were less stable than polypeptides having the sequence (gly-pro-ala), which are capable of forming a one-bonded structure according to Rich and Crick's proposal.

Angiotensin II. Paiva *et al.* (1963) have studied the kinetics of H–D exchange of angiotensinamide. The first-order rate constants obtained were of the same order of magnitude and had similar pH dependence as those for poly-DL-alanine and simple peptides. Evidence against a helical model for angiotensinamide was obtained in the H–D experiments showing that all the hydrogens exchanged at the same rate, and the first-order rate constants were of the same order of magnitude as those for simple amides and peptides and of randomly coiled poly-DL-alanine. A slower rate of exchange would be expected for half the NH groups of angiotensinamide if three intrachain hydrogen bonds were present, as proposed by Smeby *et al.* (1962). Other evidence, such as titration data and thermodynamic calculations, supported the conclusion of Paiva *et al.* All the evidence supported a random conformation for angiotensin octa-peptides in aqueous solution, and a predetermined ordered spatial arrangement of the molecule was not necessary for biologic activity.

Gramicidin S–A. Laiken *et al.* (1969) have performed tritium–hydrogen exchange experiments on the cyclic decapeptide gramicidin S–A by means of a new rapid countercurrent dialysis technique. The molecule has eight exchangeable peptide hydrogens, all of which may be observed in the pH_{min} region. These workers found slowly exchanging peptide hydrogens. In the acid-catalyzed region, there are four slow hydrogens, found in one kinetic class. In the base-catalyzed region, there are four rapidly exchanging hydrogens and two groups of two slow hydrogens each, the slowest of which exchange more than an order of magnitude more slowly than those of poly-DL-alanine. The data obtained fit an EX_2 mechanism of exchange, and con-stitute important support for the validity of this type of mechanism.

Proteins

Muscle Proteins. Willumsen (1966) has made measurements of hydrogen ex-change rates to provide information about the conformation of the heavy mero-myosin fragment of the muscle protein myosin. Optical rotatory dispersion studies indicate an α-helical content of about 50 % for this fragment (Lowey and Cohen, 1962).

In this protein 30 % of the peptide hydrogens remained unexchanged after 2.5 hr at pH 6.3. Results indicated that heavy meromyosin contains a relatively high number of peptide hydrogens that are hard to exchange, more than are found in lysozyme, insulin, yeast alcohol dehydrogenase, or bovine plasma albumin. Willumsen con-cluded that the peptide hydrogens in heavy meromyosin are of many different energy levels, similar to the case in globular proteins. No large group of peptide hydrogens with identical free energy of unfolding was suggested by the data. Heavy meromyosin contains a relatively large number of slowly exchanging peptide hydrogens which decreases with increasing pH in the range pH 5.5 to 10.8.

Gabelova and Kobyakov (1965) have used deuterium exchange methods to investigate the simple muscle contraction system of actomyosin (myosin B) and its property of syneresis during interaction with adenosine triphosphate (ATP). The purpose of the work was to look for a molecular basis of muscle contraction and to see if there is any change in protein structure in a contracted muscle. The initial spectrum of actomyosin (the control) and that of actomyosin after syneresis (i.e., with ATP present) showed no essential difference. However, the usual spectral changes produced by partial deuteration of proteins appeared in the spectrum of actomyosin and of actomyosin plus ATP. Instead of a broad band near $3300 \, cm^{-1}$ a much narrower and less intense band appeared in the actomyosin control and a $3297 \, cm^{-1}$ band in the control-plus-ATP films. There was no evidence in the control or after syneresis for the presence or absence of the β-form of polypeptide chains. No band of un-ionized carboxyl near $1730 \, cm^{-1}$ was observed in the control or control-plus-ATP, although most of the polar side groups in actomyosin are acidic.

The increase in intensity of amide bands after syneresis suggested to these authors that arginine residues were as available to H–D exchange after syneresis as before, and that the increase was really due to increase in the fraction of hard-to-exchange peptide NH groups. Syneresis was accompanied by the formation of new α-helices out of formerly disordered polypeptide chains. With contraction in the presence of ATP, the structural equilibrium was displaced towards increased α-helical content.

Kobyakov and Gabelova (1965) have also studied the kinetics of H–D exchange in muscle proteins with "native" structure during the process of structural transition and after its completion.

Hartshorne and Stracher (1965) have found that at the apparent pH value of 6.3 (apparent pH in D_2O, not pD) the percentage of hard-to-exchange peptide hydrogens for certain proteins bore little apparent relationship to their reported helical contents (Cohen and Szent-Györgyi, 1957; Shechter and Blout, 1964; Kay, 1958). These proteins were: light meromyosin fraction I, tropomyosin, actin, myosin, and heavy meromyosin. Increasing pH increased the exchange rate for all these proteins. The effect of pH on the exchange rates of the highly helical light meromyosin fraction I and tropomyosin was more marked than for heavy meromyosin and myosin. Because of this finding Hartshorne and Stracher proposed that in heavy meromyosin some conformation other than the α-helix is partly responsible for the higher percentage of unexchanged peptide hydrogens. The exchange characteristics of light meromyosin fraction I and of tropomyosin would seem to represent the exchange of peptide hydrogens involved in hydrogen bonding.

Collagen. Bensusan and Nielsen (1964) have obtained information on the mechanism of the gelatin → collagen-fold transition by a direct experimental approach of determining the rate of formation of hydrogen bonds between peptide groups in the gelatin chains following quenching (sudden cooling), and comparing this rate with those of a slow increase in levorotation and still slower increase in viscosity (von Hippel and Harrington, 1959; 1960). Purified calfskin tropocollagen was dissolved in $6.7 \times 10^{-3}M$ citrate buffer of pH 3.2 containing 0.8% tetramethylammonium chloride, was heated for 10 minutes at 55–60°C and the completely gelated solution cooled quickly to 14°C. The gelatin → collagen-fold transition was observed polarimetrically, viscometrically, and by infrared studies of the rate of exchange of hydrogen atoms of peptide groups in aliquots which were removed from the reaction mixture, quickly lyophilized, and then dissolved in D_2O at 20°C. The average rate constant for the rapidly exchanging peptide hydrogens was $0.16 \pm 0.01 \, \text{min}^{-1}$. In the quenched gelatin solutions investigated at pD 3.75 the peptide hydrogens fell into two classes: one in which the exchange proceeded with a uniform rate constant, $0.16 \, \text{min}^{-1}$, and the other in which the exchange was much slower, with estimated rate constants less than 1 per day. The kinetic exchange results could be accounted for by varying the number of peptide groups in these two classes. No indication was found of peptide groups exchanging with intermediate rate constants. The rate of the conversion for all the rapidly exchanging peptide hydrogens to slowly exchanging hydrogens was found to be closely proportional to the rate of increase of levorotation, in agreement with the data of Englander and von Hippel (1962).

Bensusan and Nielsen have given the following interpretation to the exchange data. The slowly exchanging hydrogens observed by infrared measurements of the amide II band can be bound only in peptide groups. The rate of hydrogen exchange in randomly coiled gelatin should be essentially the same as that predicted for a gelatin peptide chain having the conformation of a single-stranded poly-L-proline II type helix (essentially a fully extended peptide chain). Experimental evidence in the gelatin–collagen system exists for only three peptide-chain conformations: random coil, single-stranded poly-L-proline II type helix, and triple-stranded collagen type helix (Harrington and von Hippel, 1961). The slowly exchanging hydrogens are, therefore, most likely the ones involved in cooperative interchain hydrogen bonds in the collagen-fold triple helix. The H–D exchange results were interpreted as indicating that the interchain hydrogen bonds of the collagen triple helix are formed rapidly after the helical coiling of the individual peptide chains. This interpretation is in agreement with the mechanism proposed by Flory and Weaver (1960) for the gelatin → collagen-fold transition. The type of mechanism of hydrogen exchange in the collagen triple helix is thought most likely to be of the EX_2 type (see earlier discussions). Hvidt and Nielsen (1966) think the intermediate I in the process

$$N \underset{k_2}{\overset{k_1}{\rightleftharpoons}} I \overset{k_3}{\longrightarrow} \text{exchange}$$

may be a collagen triple helix in which a section of the triple helix has all its hydrogen bonds between two of the chains broken, while all other interchain hydrogen bonds are intact.

Heidemann and Srinivasan (1967) have used H–D exchange to measure orderliness of collagen, gelatin, and poly-DL-alanine after treatment of these substances with phenolic compounds. They treated films of these substances with phenol, 2,4-dihydroxyphenylmethane, and special phenolic tannin extracts, after which they measured the H–D exchange rate. Poly-DL-alanine, a protein model with an α-helix, was more firmly linked by hydrogen bonds than was collagen. The collagen used was regenerated from acid-soluble material and had ∼50% unordered regions. In gelatin the unordered fraction was much greater. The data from poly-DL-alanine show that polyvalent macromolecular tannins split hydrogen bonds and then renew them. Heidemann and Srinivasan think that this reaction probably occurs also with collagen and explains why large tannin molecules can penetrate into the fibril network of structured collagen. 2,4-Dihydroxyphenylmethane was a good model of a bivalent tannin but, being a smaller molecule than the macromolecular tannins, formed less stable hydrogen bonds. This result indicated that hydrogen bond stability between peptide and OH groups increases with size and polyvalence of the phenol.

Plasma Albumin. Foster *et al.* (1965) have used H–D exchange studies in conjunction with solubility and zone-electrophoresis measurements to examine the reversible conformational changes of plasma albumin in the pH 4–5 region. Electrophoretic studies have shown the presence of two forms of plasma albumin in this pH range, which have been designated as the N and F forms. At pH 4 a single moving boundary was seen in electrophoresis which has been attributed to the F form, and the single band seen at pH 5 is attributed to the N form. Density gradient studies had already shown (Benson *et al.*, 1964) that a large number of hydrogens in plasma albumin which are nonexchangeable at pH 5 are readily exchanged when the pH is lowered to 4. Benson *et al.* (1964) had confirmed by ORD measurements that this finding is due to a conformational change. Foster and his co-workers studied the hydrogen exchange at several pH values in the N → F transformation region and found that the exposure of the nonexchangeable peptide hydrogens accompanies the N → F transformation. The variation with pH of the residual unexchanged hydrogens as a fraction of the total which exchange in the F form but not in the N form closely followed the variation of the fraction of N form with pH. This close parallel between the two also provided confirmation of the existence, at intermediate pH values, of stable noninterconvertible isomers. The argument used was as follows: If N and F were to exist in dynamic equilibrium, then at intermediate pH values all the albumin molecules would exist as the F isomer for part of the time. Since the unexchangeable hydrogens of the N isomer become exchangeable in the F isomer, the extent of exchange at the intermediate pH values should reflect this condition. Thus, at the midpoint of the transition all the molecules would exist as the F isomer for about half the time and since the exchange is measured after a considerable amount of time, all the nonexchangeable hydrogens of the N isomer should have exchanged. It was found, however, that the extent of exchange corresponded only to the fraction of F isomer as assayed by optical activity measurement.

From this, as well as from the other physical studies, Foster *et al.* concluded that while the pH-dependent N → F transformation is rapidly and completely reversible, at intermediate pH values, bovine plasma albumin consists of stable noninterconvertible N and F isomers. To explain this apparent contradiction, they proposed the "microheterogeneity" model for plasma albumin. This model postulates the existence in plasma albumin (of both N and F forms) of stable noninterconverting closely related species. Each of these species has a narrow pH range in which rapid $N \rightarrow F$ transformation occurs. The different transition points of these species result in the continuous distribution of N and F forms between pH 4 and 5.

Ovalbumin. Willumsen (1967) has studied the H–D exchange reaction in ovalbumin and has found that this protein has a large number of slowly exchanging peptide hydrogen atoms. Approximately 45% of the peptide hydrogens remained unexchanged after 2.5 hr at pH 6.3 and 21°C. He examined the H–D exchange over a wide pH range from 2.4 to 10.5 in which ovalbumin is stable and discussed the data in relation to a suggested mechanism of exchange. The estimated percentage of peptide hydrogens that exchanged between 9 and 66 minutes after dissolution decreased gradually in the pH range 2.4–10.5 from 19% to 5%. Between 66 and 156 minutes after dissolution a further exchange took place which was 6% at pH 2.4 and 0.5% at pH 10.5. In the intermediate pH interval, 4.1–8.7, the further exchange amounted to 3.5% and was independent of pH. The extent of exchange increased with increasing pH in agreement with Hvidt's mechanism (1964) for the exchange reaction in proteins (see *Factors Affecting Hydrogen–Deuterium Exchange of Biological Molecules*).

γ-Globulin. Conflicting reports have appeared on the rate of H–D exchange of normal γ-globulin. Blout *et al.* (1961) have reported less than 10% slowly exchanging hydrogens in γ-globulin at pD 4 whereas Imahori and Mommoi (1962) have reported 50% slowly exchanging hydrogens. Gould *et al.* (1964*a*) have found that the rapid exchange in normal rabbit γ-globulin was completed within 30 min and at neutral and acid conditions about 30% of the hydrogens remained unexchanged. The hydrogens remaining unexchanged at the end of 24 hr (21% at pD 7 and 13% at pD 1) did not exchange even on heating at 55°C for 12 to 24 hr. At pD 12 to 13, all the hydrogens exchanged within the first few minutes of dissolution in D_2O. These results, indicating greater lability of the protein in alkali than in acid, were in agreement with the decreased helical content found in alkaline solution from ORD measurements. Gould *et al.* (1964*b*) found that the loss of antibody activity of γ-globulin was also more rapid and complete in alkaline solution, while it was gradual and incomplete for a comparably large change in pH in acid solution.

Imahori and Mommoi (1962) have employed H–D exchange studies for the structural characterization of γ-globulin and myeloma protein. The extent of hydrogen exchange after 30 min at 20°C was measured by the infrared method. Myeloma protein underwent an almost complete exchange, while in the case of γ-globulin the exchange was only half complete. Imahori and Mommoi have suggested that H–D exchange could be employed to differentiate between the two proteins. The presence of a higher helical content in γ-globulin was confirmed by ORD studies.

Cytochrome c. The usefulness of H–D exchange studies in detecting subtle conformational changes in proteins has been very ably demonstrated in the recent papers of Kägi and Ulmer (1968) and Ulmer and Kägi (1968) on horse heart cytochrome c. The remarkable differences between the physicochemical properties of the oxidized and reduced forms of cytochrome c had been attributed to conformational differences, and by their studies on the H–D exchange characteristics of ferrocyto-chrome and ferricytochrome, Ulmer and Kägi were able to establish that conforma-tional differences are induced by oxidoreduction.

The rates of exchange of peptide hydrogens under the same conditions of pD and temperature were quite different for the two states. At pD 7.4 and 20°C, ferro-cytochrome exchanged 59% of its peptide hydrogens within 5 min of dissolution in D_2O and another 21% slowly within the next 24 hr, while the remaining 20% were unexchanged. Ferricytochrome had, under the same conditions, 68% rapidly ex-changing hydrogens and 21% slowly exchanging ones, while 11% remained un-exchanged. Ferricytochrome thus had 9 or 10 more exchangeable hydrogens (based on 9% of the 99 potentially exchangeable peptide hydrogens) than ferrocytochrome, implying a more compact conformation for the latter which is in accordance with earlier findings (Margoliash and Schejter, 1966).

The exchange curves for the oxidized and reduced forms follow a parallel course, with a displacement corresponding to 9% of exchanged hydrogens. A simple explanation would be that 9 of the most slowly exchangeable hydrogens in the reduced

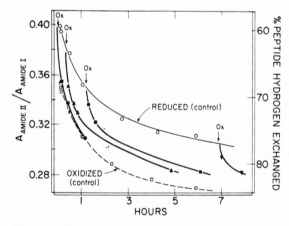

Fig. 11.6. Effect on ferrocytochrome of oxidation at intervals during the time course of exchange of hydrogen for deuterium. The ratio of absorbances of the amide bands, $A_{amide\,II}/A_{amide\,I}$ (left ordinate), and the corresponding percent peptide hydro-gen exchanged (right ordinate) are plotted against time. The kinetic exchange curves of unmodulated reduced (○—○) and oxidized (□— —□) cytochromes are shown as controls. Oxidation (Ox) of ferrocytochrome with potassium ferri-cyanide after a 1.5-min exposure to deuterium rapidly acceler-ates exchange to that of the control oxidized protein (●—●). Oxidation of ferrocytochrome at later times [21 min (▲—▲), 74 min (■—■), and 415 min (▼—▼)] also accelerates ex-change but there is a progressive decrease in the rate and extent to which the exchange curves approach that of the oxi-dized control. The concentrations were protein, $4.9 \times 10^{-3} M$ and potassium ferricyanide, $1 \times 10^{-2} M$ in $0.11 M$ sodium phosphate (pD 7.4). (Ulmer and Kägi, 1968.)

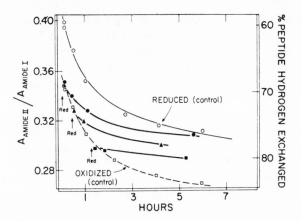

Fig. 11.7. Effect on ferricytochrome of reduction at intervals during the time course of exchange of hydrogen for deuterium. The ratio of absorbances of the amide bands, $A_{\text{amide II}}/A_{\text{amide I}}$ (left ordinate), and the corresponding percent peptide hydrogen exchanged (right ordinate) are plotted against time. The kinetic exchange curves of unmodulated reduced (○—○) and oxidized (□ ---□) cytochrome are shown as controls. Reduction (Red) of ferricytochrome with ascorbate after 1.5-min exposure to deuterium retards exchange and, after 6 hr the extent of exchange approaches that of the reduced control (●—●). Reduction of ferricytochrome at later times [28 min (▲—▲) and 78 min (■—■)] also retards exchange although the value for the reduced control is no longer completely attained. The concentrations were protein, $4.9 \times 10^{-3} M$ and ascorbate, $1 \times 10^{-2} M$, in $0.11 M$ sodium phosphate (pD 7.4). Ulmer and Kägi, 1968.)

state become rapidly exchangeable in the oxidized state. By inducing oxidoreduction during the exchange, Ulmer and Kägi were able to study the exchange characteristics of the two states in greater detail and show that the number of hydrogens affected by the conformational change was considerably larger than the 9 indicated by the exchange curves.

Dissolving one of the forms in D_2O containing ferricyanide or ascorbate in sufficient concentration to reverse the oxidation state instantaneously on dissolution in D_2O was found to result in complete reversal of the exchange pattern. Reversal of oxidation state during the exchange also produced reversal of exchange pattern but the extent and rate of reversal decreased with increasing duration of exchange (Figs. 11.6 and 11.7). Thus, when a ferrocytochrome solution was oxidized after 1.5 min of exchange, acceleration of exchange was observed. Within 5 min the total exchange had reached the level of exchange of a ferricytochrome control solution at the same time interval, and after this point was reached the exchange curve of the oxidized solution closely paralleled that of the control. When the oxidation was carried out at later intervals the exchange pattern still moved toward that of the control but the move was slower and less complete. The change of the exchange pattern of ferricytochrome to that of ferrocytochrome produced by reduction during exchange also showed a similar dependence on the duration of exchange before reduction. The reversal of exchange pattern throughout the course of exchange indicated that oxidoreduction affects both rapidly and slowly exchanging hydrogens. If the difference in the exchange behavior was due to a fixed number of the most slowly exchanging hydrogens in the reduced state becoming rapidly exchangeable in the oxidized state, reduction of the latter after the period of rapid exchange should not produce any further retardation of exchange. Similarly, oxidation of an exchanging ferrocytochrome should result in an immediate increase in exchange as a result of the formation of the rapidly exchanging hydrogens. Ulmer and Kägi therefore concluded that the observed difference of nine to ten hydrogens between the two states is an average of difference between a much larger number of hydrogens of varying rates of exchange in the two states.

Kägi and Ulmer (1968) examined the exchange behavior over a wide range of pD from 1.9 to 12.4 and found that the difference in the number of exchangeable hydrogens between the two forms was maintained except at extremes of pD, indicating stability of the two structures over a wide range of pD. The percentage of exchanged hydrogens was at a minimum at a pD between 4 and 5 for both forms and increased with increasing and decreasing pD. However, the kinetics of exchange at alkaline pD values was different from that at acid pD values. At the extreme alkaline pD (12.4), the exchange is almost complete within the first 5 min. At the acid pD of 1.9, the percentage of hydrogens exchanged at the start of the reaction is even lower than at neutral pD (7.4) but within 15 min it passes the neutral value and the exchange reaches virtual completion within 2 hr. This anomalous exchange behavior at the acid pD seemed to indicate occurrence of a conformational transition and this was confirmed by ORD studies. However, no conformational transition was indicated by ORD results at the alkaline pH, and the increase of exchange in this region should be due to hydroxyl ion catalysis. Following Willumsen's (1967) treatment of base-catalyzed reactions by the EX_2 mechanism, the percentage of unexchanged hydrogens was plotted against $log[OH^-]t$ for both the redox forms. The resulting monotonically decreasing curves confirmed that the exchange followed the EX_2 mechanism and that the increase in the alkaline pH region was due to OH^- catalysis.

The curves for the two forms were, however, displaced from each other, indicating difference in their bimolecular rate constant, $(k_1/k_2) \times c \times 10^{T/20}$. Since c is a pH-independent constant presumed to be the same for all proteins, the difference must be in the ratio k_1/k_2, implying variation in the $N \rightarrow I$ equilibrium for the two states. The relative positions of the two curves indicated a more open structure for the oxidized form as revealed by the percentage of exchange. The two curves were closely parallel to each other and from the displacement a value of 2 kcal/mole was estimated to be the difference in energy for all the $N \rightarrow I$ equilibria of the two redox states.

An interesting parallel was drawn between the $N \rightarrow I$ equilibrium and the open-crevice–closed-crevice equilibrium of cytochrome c. Spectroscopic evidence had indicated that the heme iron was buried in a crevice and to account for its reactivity with ligands, conformational motility was proposed between a nonreactive closed crevice and a reactive open crevice. The effect of oxidoreduction on this equilibrium is similar to that found for the $N \rightarrow I$ equilibrium: reduction shifted the equilibrium to the nonreactive closed crevice. Ulmer and Kägi have therefore suggested that both $N \rightarrow I$ and crevice equilibria represent the same dynamic alterations in the structure of cytochrome c. Proceeding on this assumption, they calculated a value of -5 to -6 kcal/mole for the crevice-closing reaction in ferrocytochrome from the known value for ferricytochrome (-3 to -4 kcal/mole) and the free energy difference between the $N \rightarrow I$ equilibria derived from the H–D exchange measurements (2 kcal/mole). They also estimated a value of $+0.170$ eV for the redox potential of the open-crevice conformation in cytochrome c and found it close to the value (0.120 eV) for myoglobin. This would seem to indicate that the alterations in the $N \rightarrow I$ and crevice equilibria are parallel effects of the conformational transition of cytochrome c on oxidoreduction.

Growth Hormone. Squire and Ottesen (1968) have studied the H–D exchange of human pituitary growth hormone at different pH values and compared the exchange data at three arbitrarily selected time intervals. During the first interval of about 15 min, before the first scan was made, approximately 50% of the secondary amide hydrogens exchanged at pH 2.8 and 3.6, while virtually complete exchange had taken place at pH 9.3. During the second period, extending from 15 min to 3 hr, an additional 20% of the hydrogens exchanged at the low pH values. During the third period, extending from 3 to 24 hr, little or no further exchange occurred at pH 3.6, but an additional 15% of the hydrogens exchanged at pH 2.8.

Bewley and Li's (1967) optical rotatory data on human growth hormone were used to calculate α-helix contents of 58%, 51%, 49%, and 43% at pH values of 1.3, 7.4, 10.1, and 13, respectively. Thus, the helical content appears to be remarkably independent of pH. The data of Blout et al. (1961) on the rate of deuteration of poly-glutamic acid at pD 3.5, under conditions where this polypeptide is known to be helical, show negligible H–D exchange within the first three hours. Since the optical rotatory data indicate that human growth hormone is at least 50% helical below pH 7 and the H–D exchange data show 70% H–D exchange within the first 3 hr, Squire and Ottesen (1968) state that it appears that at least 40% of the α-helical portion of growth hormone exchanges more rapidly than does polyglutamic acid at the same pH.

Miscellaneous. To determine whether the polypeptide chains of soybean 7S and 11S protein molecules (Svedberg units) are rigid or flexible, Fukushima (1967) has measured the rates of H–D exchange of these proteins. The difference spectrum calculated from $A_{\text{amide II}}/A_{\text{amide I}}$ at pD 7.6 and 12.5 (those at 12.5 were the background absorbances) gave a value of approximately 0.2 for this ratio at 1525 cm^{-1}. Since this value had been found earlier to be about 0.40 to 0.45 when the peptide groups are completely undeuterated (Blout et al., 1961), Fukushima estimated that about half of the peptide groups in the 7S and 11S proteins were not exchanged at pD 7.6 and reached the conclusion that both proteins are very rigid and compact, even in their randomly folded parts. Evidence from ORD data also supported the conclusion concerning the compactness.

Jeffries (1964) has studied the H–D exchange reaction in ten proteins and nylons 6, 11, and 66 which were immersed in saturated D_2O vapor. With all the samples an initial rapid exchange was followed by a slower exchange, which gradually leveled off to give equilibrium values. The extent of H–D exchange was discussed in relation to order and disorder in the polymer structures. The extent of exchange was related to the proportions of large side chains in the samples and to the H_2O sorption values. The H–D exchange was measured by determining the increase in dry weight of the sample after exposure to D_2O and also by measuring the decrease in absorbance of the N–H band at 3300 cm^{-1}. The infrared measurements were made on films of the material. The decrease in N—H band absorption was shown to be a reasonable measure of the extent of exchange, provided the changes in band widths are taken into account.

Beer *et al.* (1958) have estimated from the relative intensities of the 3300 cm^{-1} N—H band and the 2410 cm^{-1} N—D band the approximate degree of exchange in silkworm gut, keratin, gelatin, and other proteins. Fraser and Macrae (1958) have calculated exchange values for keratin from infrared spectra in the overtone region.

The exchange of deuterium for the amide-bound hydrogen atoms of human β-lipoproteins has been studied (Crolla and McDonald, 1969) by means of the infrared band at 1449 cm^{-1}. The reaction mixture was comprised of 1:1 D_2O-dioxane and the lipoprotein solution. At the pD_{min} value of 2.4 the rate constant was 1.20×10^{-3} sec^{-1}, and probably characterizes the apoprotein instead of the intact molecule because of the type of solvent.

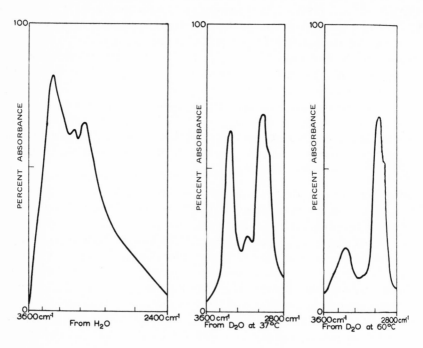

Fig. 11.8. Ribonuclease from solution (dry film). (Haggis, 1957.)

Enzymes

Ribonuclease. Hydrogen–deuterium exchange of ribonuclease has been studied by several workers. Haggis (1957) studied the exchange by examining the infrared spectra of dry films from H_2O and D_2O solutions in the fundamental stretching region (Fig. 11.8). At 37°C, only $70 \pm 5\%$ of the peptide hydrogens exchanged, while at 60°C the exchange was almost complete, $(90 \pm 3\%)$. This observation was in contrast to the finding of Hvidt and Linderstrøm-Lang (1955) that exchange is complete at 37°C. Haggis proposed that the discrepancy might be due to deuteration occurring during the drying process at 60°C involved in the Linderstrøm-Lang method.

Since ribonuclease undergoes a transition above 50°C (Harrington and Schellman, 1956), Schildkraut and Scheraga (1960) undertook a detailed study on the effect of temperature on the exchange reaction, employing the Linderstrøm-Lang method. They also found that complete deuteration occurs only at temperatures above the transition point and suggested that the 20 hydrogens, which are not exchanged at the lower temperatures, form part of the hydrogen-bonded peptide backbone. The studies of Hermans and Scheraga (1960) in the overtone region of NH stretching vibration confirmed this hypothesis.

The first overtone of NH stretching vibration occurs at $1.5\,\mu$ for the free group and at $1.54\,\mu$ for the hydrogen-bonded group. By deuterating ribonuclease at temperatures below and above the transition temperature (38 and 65°C, respectively), Hermans and Scheraga were able to obtain samples which had been deuterated to different extents. The sample deuterated at the lower temperature had all but 20 of the hydrogens replaced by deuterium (designated as $RNase-H_{20}$) while the other sample had undergone complete deuteration. A difference spectrum between the partially deuterated sample and the completely deuterated sample showed a double peak with maximum intensity at $1.54\,\mu$. The shape, position, and intensity of this difference spectrum were very similar to the NH overtone bands of poly-γ-benzyl-glutamate in 1,2-dichloroethane, in which solvent it exists predominantly as an α-helix (Fig. 11.9). The difference spectrum is due only to the 20 hydrogens in $RNase-H_{20}$ and Hermans and Scheraga were, therefore, able to support the earlier hypothesis that the protected hydrogens are involved in the $N—H\cdots O{=}C$ hydrogen bonding of the peptide backbone.

Glyceraldehyde 3-Phosphate Dehydrogenase. Kinetics of H–D exchange between peptide amino groups of glyceraldehyde 3-phosphate dehydrogenase and D_2O has been studied (Abaturov *et al.*, 1968) within the pH range 2.5–10.5 at 20°C by infrared methods. The results indicate the presence in this enzyme of four classes of peptide amino groups which differ markedly in their H–D exchange rates, expressed by hydrogen atom half-lives of 1 min, 2 min, 17–25 min, and 50–60 hr. The exchange rates increased and the fraction of slowly exchangeable hydrogen atoms decreased with increasing pH from 5.3 to 9.2. Above pH 9.2 the rates increased sharply, while below pH 4.5 the exchange rates could not be measured because of their very high values and the aggregation and low solubility of the enzyme. At pH 2.5 the exchange was almost instantaneous. Addition of sodium dodecyl sulfate resulted in a lower intensity of the amide II band; the exchange rate increased without an accompanying

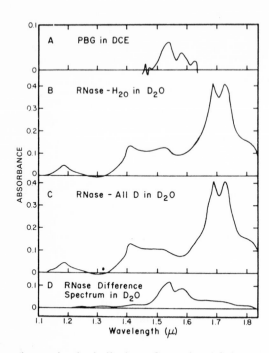

Fig. 11.9. Near-infrared spectra of poly-γ-benzyl-glutamate (PBG) in 1,2-dichloroethane (DCE) and of ribonuclease in D_2O. The light path was 10 cm in curve A, 1 cm in curves B and C, and 2 cm in curve D. In curve D, RNase-all D in D_2O is the reference for the measurement of RNase-H_{20} in D_2O. (Hermans and Scheraga, 1960.)

change in the helical configuration of the enzyme. A comparison study of the H–D exchange kinetics between protamine sulfate and D_2O showed that the rate reached its minimum value at pH 2–3. The results were interpreted in terms of the mechanism of Hvidt and her colleagues, according to which the protein molecules contain energetically heterogeneous regions with varying degree of availability of peptide groups and with high conformational mobility. The pH dependence of the H–D exchange kinetics suggested that the rate of conformational transitions in glyceraldehyde 3-phosphate dehydrogenase exceeds the rate of H–D exchange of peptide groups completely available to D_2O molecules.

Aspergillopeptidase. Ichishima and Yoshida (1966) have shown the deuterium-exchanged enzyme aspergillopeptidase A to have strong amide I absorption at 1632 cm^{-1} and amide II absorption at 1540 cm^{-1}, and have also found a weak band at around 1685 cm^{-1}. They have assigned the 1632 and 1685 cm^{-1} bands to antiparallel β-structure (Miyazawa and Blout, 1961; Imahori, 1963; Timasheff and Susi, 1960) for this enzyme.

Trypsin. Lenz and Bryan (1969) have studied the H–D exchange of trypsin and two trypsin derivatives, DIP-trypsin (diisopropylphosphoryl-trypsin) and TLCK-trypsin (trypsin inhibited by reaction with 1-chloro-3-tosylamido-7-amino-2-heptanone or N-tosyl-L-lysine chloromethyl ketone). They made measurements at pD values from 2 to 8 and found four classes of exchanging hydrogen: (1) a fast class with a pD-dependent rate constant for exchange, (2) a medium class with an essentially pD-indpendent rate constant for exchange, (3) a slow class with a pD-independent rate constant for exchange, and (4) a "core" class that does not exchange within 24 hr.

They could only observe the fast class at pD values around 2 and this class consists of completely exposed peptide hydrogens. The medium and slow classes appear to exchange by a mechanism in which the rate is determined by transconformational exposure of unexposed peptide groups. The rate constants for exchange of the medium or slow classes are essentially the same for trypsin and the two derivatives. The numbers of hydrogens per class are essentially the same for the medium, slow, or core classes for all three substances. These results imply that the medium, slow, or core peptide hydrogens are essentially structurally identical.

α-Chymotrypsin. α-Chymotrypsin was among the compounds used by Blout *et al.* (1961) in demonstrating the applicability of the amide I and II infrared absorption bands to the study of H–D exchange reactions. At a pD of 4.5, 43 % of the amide hydrogens remained unexchanged after the initial rapid exchange (occurring in the first 10 min of the reaction). As an index of helical content, this value was, however, higher than that based on ORD measurements (Doty, 1959), and Blout *et al.* (1961) suggested that a part of the slowly exchanging hydrogens may come from hydrophobic regions of the protein rather than α-helix. Fifteen percent of the amide hydrogens, which remained unexchanged after 24 hr, but exchanged on heating, were also attributed to those in hydrophobic regions.

More recently, Willumsen (1968) has done H–D exchange studies on chymotrypsinogen A, α-chymotrypsin, and diisopropylphosphoryl-α-chymotrypsin. Over a broad pH and pD range, chymotrypsinogen showed at all times of incubation about 5 % more unexchanged peptide hydrogens than α-chymotrypsin. Diisopropylphosphoryl-α-chymotrypsin had nearly the same number of unexchanged peptide hydrogens after 2.5 hr as chymotrypsinogen. The hydrogen isotope exchange fitted a model in which rapid equilibrium between a closed and an open molecular conformation is assumed and in which only the open form is expected to exchange. Bhaskar and Parker (1970) have found that the extent of hydrogen–deuterium exchange in α-chymotrypsin varied parabolically with pH, with a minimum of exchange at pH 3.5. The exchange data of diisopropylphosphoryl chymotrypsin showed significant variations from those of the pure enzyme, particularly in the region of pH 5.

Stevens *et al.* (1964) have studied the effect of radiation on the enzymatic activity and hydrogen-exchange characteristics of α-chymotrypsin, and related activity to the conformation of the enzyme. The enzyme was irradiated by γ-rays from a ^{137}Cs source and the resulting sample was subjected to amino acid analysis, chromatographic examination, and hydrogen-exchange measurements. In α-chymotrypsin, hard-to-exchange amide hydrogens and activity are both lost (as a function of γ-ray dosage) at about the same rate. The heat perturbation of the ultraviolet absorption of α-chymotrypsin (decrease in absorbance at 292 nm occurring between 30 and 50°C), which has been attributed to the removal of tryptophyl residues from the hydrophobic regions, is also lost upon irradiation at the same rate. This result is taken as further evidence of the close relation between loss of conformation and of enzymatic activity.

The parallel decrease in the percentage of remaining ΔA_{292} (ΔA at 292 nm) between 30 and 50°C and the percentage of remaining hard-to-exchange amide hydrogens after irradiation also conforms with the suggestion of Blout *et al.* (1961) that the hydrophobic regions of α-chymotrypsin account for a significant portion of the hard-to-exchange amide hydrogens.

Lysozyme. Hvidt (1963) has used lysozyme for a detailed investigation of the infrared method of studying H–D exchange reactions employing the amide I and II absorption bands. (See *Methods* and Fig. 11.3.) The exchange was studied at 21°C in solutions of *p*H 3.2, 3.9, and 6.3. There was a pronounced increase in the number of exchanging peptide hydrogen atoms with increasing *p*H. At *p*H 3.2, ~60% of the peptide hydrogens remained unexchanged after 24 hr, while at *p*H 6.3 only 10% of the peptide hydrogens remained unexchanged after 24 hr. Density-gradient measurements had shown (Hvidt and Kanarek, 1963) that the total number of exchanging hydrogens also increased with *p*H. By comparison between the two measurements, it was found (Hvidt, 1963) that almost all the hydrogens remaining unexchanged belonged to peptide groups. Since other physical studies (Glazer, 1959) had shown that the structure of lysozyme is insensitive to changes in *p*H in the range 2–7, Hvidt (1964) attributed the increase in exchange with *p*H to hydroxyl ion catalysis.

Stevens *et al.* (1964) have studied the effect of radiation on the enzymatic activity and hydrogen-exchange characteristics of lysozyme. Upon irradiation lysozyme lost its enzymatic activity at half the rate of loss of hard-to-exchange amide hydrogens, indicating that only half the native conformation of lysozyme is essential for activity.

Pepsin and Pepsinogen. Abaturov *et al.* (1967) have studied the H–D exchange in pepsin and pepsinogen by following the changes of the 1549 cm^{-1} band. The lyophilized protein was treated with 99% D_2O and the rate of deuteration of the peptide NH bonds was observed at 25°C. The *p*H ranges used were 3.6–7.2 for pepsin, 5–7.3 for pepsin inactivated with urea, and 6–9.1 for pepsinogen. The exchange was nearly instantaneous outside these *p*H ranges. All the curves for pepsin and pepsinogen could be approximated by three straight-line sections, each of which shows the exchange of one group of hydrogen atoms. Thus, besides the very rapidly exchanging hydrogen atoms, there are in both proteins three other groups of hydrogen atoms differing in the facility for deuterium exchange. The slowest acting of these showed a large *p*H dependence with a maximum rate at *p*H 3.6–4.2. This group accounted for about 60% of total activity. Inactivated pepsin had but 3% of such hydrogen atoms.

Carbonic Anhydrase. Rickli *et al.* (1964) have made some observations of infrared absorption of carbonic anhydrase B in D_2O solution in the region 1800–1300 cm^{-1} in order to determine the hard-to-exchange amide hydrogens. Analysis of their data led to the estimate that 20% of the peptide hydrogens in native carbonic anhydrase B, and 30% in the acid-denatured enzyme, were difficult to exchange.

Effect of Coenzymes and Other Compounds on Enzyme Conformations

Yeast Alcohol Dehydrogenase. Conformational changes are known to play an important role in the mechanism of enzyme action. The existence of conformational changes in enzymes bound to coenzymes or substrates has been postulated by several investigators (Nozaki *et al.*, 1957; Sekuzu *et al.*, 1957; Yonetani and Theorell, 1962; DiSabato and Kaplan, 1964; DiSabato and Kaplan, 1965). Hvidt and her co-workers (Hvidt and Kägi, 1963; Hvidt *et al.*, 1963) were the first to study the effect of coenzyme binding on the exchange characteristics of an enzyme. They have studied the hydrogen exchange of yeast alcohol dehydrogenase (YADH) and the effects on it of the presence of the oxidized and reduced forms of the coenzyme nicotinamide adenine dinucleotide (NAD^+ and NADH). Yeast alcohol dehydrogenase exchanged 80% of all its peptide hydrogens within 10 min of dissolution in D_2O. Of those remaining, 9% exchanged slowly between 10 and 180 min while the other 11% exchanged only on exposure to 0.2% sodium dodecyl sulfate.

Addition of $0.05M$ NAD^+ shifted the exchange curve of $2.1 \times 10^{-4}M$ yeast alcohol dehydrogenase upwards, indicating a decrease in the number of exchangeable hydrogens. The resulting exchange curve was, however, parallel to that of the free enzyme, suggesting that the coenzyme did not affect the rate of the slowly exchanging hydrogens. Time difference spectra of the exchanging solutions between 27 and 152 min of exchange were similar for the enzyme and the enzyme-plus-coenzyme mixture. Both had an absorption band at $1545 \, cm^{-1}$, which is the position of the N–H absorption band (amide II), and a shoulder at $1525 \, cm^{-1}$. These investigators therefore concluded that the hydrogens affected by the coenzyme belong to the rapidly exchanging class. From the difference spectra between the exchanging enzyme solutions with and without the coenzyme, the spectral characteristics of the hydrogens protected from exchange by NAD^+ were obtained (Fig. 11.10). These had a broad band extending from 1500 to $1590 \, cm^{-1}$ and were different from the time difference spectra, implying structural differences in the environments of the protected hydrogens and those exchanging slowly. From measurements of the band area it was calculated that about 4% of all peptide hydrogens of the enzyme are protected by NAD^+.

NADH also had a protective effect on some peptide hydrogens of the enzyme although its effect was different from that of NAD^+. The main difference was that NADH produced a time-dependent change in the amide absorption of an exchanging enzyme solution (Fig. 11.11). The difference between the two decreased by about 40% between 28 and 154 min of exchange. It would thus seem that NADH slows the rate of exchange of some of the hydrogens while NAD^+ prevents their exchange. From measurements of band areas, it was estimated that about 4% of the peptide hydrogens are protected at 28 min of exchange, while at 154 min the percentage protected drops to about 2.3%. The constituent difference spectra reveal that the effect of NADH on the NH absorption band is different at the different time intervals. Thus, between 28 and 154 min of exchange a considerable drop occurs in the intensity at $1540 \, cm^{-1}$ while the intensity of the band at $\sim 1525 \, cm^{-1}$ remains constant (Fig. 11.10). Hvidt and Kägi (1963) suggested that the broad amide II band of yeast alcohol dehydrogenase is a composite band of several closely spaced bands arising from different conformations of the enzyme and that the coenzyme protects the hydrogens in these to varying degrees. The hydrogens belonging to the structure with the $1544 \, cm^{-1}$ band seem to exchange at measurable rates while those with the $1525 \, cm^{-1}$ band do not seem to exchange at all. The two bands were also seen in the time difference spectra.

The effect of coenzyme on an enzyme solution that had been rendered catalytically inactive by storage was found to be less by a factor corresponding to the loss of

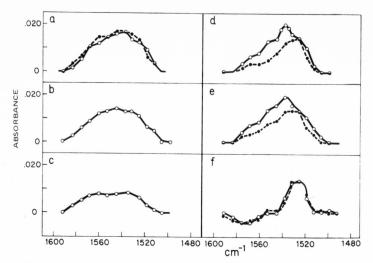

Fig. 11.10. Effect of NAD$^+$ and NADH on the infrared spectra of YADH; constituent difference spectra of $2.1 \times 10^{-4}M$ YADH in D_2O ($0.1M$ sodium phosphate, pD 8). The various conditions were in the presence of (a) $5 \times 10^{-2}M$ NAD$^+$: solid line, 27 minutes of exchange; broken line, 152 minutes of exchange, (b) $1 \times 10^{-2}M$ NAD$^+$: 156 min of exchange, (c) partly inactivated YADH in $5 \times 10^{-2}M$ NAD$^+$, (d) $2 \times 10^{-2}M$ NADH: solid line, 28 min of exchange; broken line, 154 min of exchange, (e) $5 \times 10^{-3}M$ NADH: solid line, 28 min of exchange; broken line, 153 min of exchange, and (f) $5 \times 10^{-2}M$ NADH: solid line, 152 min of exchange; broken line, $0.5M$ ethanol, 153 min of exchange. All difference spectra were obtained by subtracting the spectrum of a control sample which was exposed to D_2O for the same length of time but did not contain coenzyme. (Hvidt and Kägi, 1963.)

activity. Hvidt and Kägi, therefore, concluded that protection of exchangeable hydrogens of yeast alcohol dehydrogenase cannot be due to an unspecific effect of the coenzyme and have offered an explanation for the observed results in terms of restriction of "motility" of the protein. They suggested that the coenzyme might stabilize some conformations by preferential binding and the observed hydrogen exchange would be characteristic of the structures so stabilized. They have also suggested an alternative, though less likely, possibility. Here the coenzyme is pictured as binding tightly to the enzyme surface and preventing exposure of the peptide hydrogens to the solvent. The diminished effect of the reduced coenzyme could, in this case, be due to a less tight fit on the active surface of the enzyme. It is not possible, however, to state definitely which mechanism is operative and it is even likely that the observed results come from a combination of the different effects.

An important outcome of this investigation was the realization that H–D exchange can be successfully employed to study subtle conformational changes in enzymes brought about by coenzymes.

It is interesting to note that the effect of the two forms of the coenzyme on the H–D exchange of yeast alcohol dehydrogenase followed the order of their effectiveness in protecting the enzyme from denaturation by urea (Sekuzu et al., 1957).

Lactic Dehydrogenase. The H–D exchange of the peptide groups of chicken heart lactic dehydrogenase (CHLDH) has been studied (DiSabato and Ottesen, 1965) at *p*H 6.9 in the absence and in the presence of various coenzymes. The technique used was similar to that of Hvidt and Kägi (1963). Native CHLDH had about 50% fast-exchangeable and about 35% hard-to-exchange peptide hydrogens. All the coenzymes tested decreased the hydrogen exchange of the native enzyme. NADH gives protection as shown by a decrease in the rate of exchange of the slowly exchangeable hydrogens. NAD$^+$ decreases the rate of exchange of fast-exchangeable hydrogens. Reduced and oxidized 3-acetylpyridine-adenine dinucleotides decrease the rate of exchange of the fast-exchangeable hydrogens. Sodium dodecyl sulfate greatly enhances the hydrogen exchange of the enzyme.

The coenzymes tested partially counteract the increase in hydrogen exchange brought about by the sodium dodecyl sulfate; stronger protection effects were noted for the reduced coenzymes and the 3-acetylpyridine analogs than for the oxidized and natural coenzymes. The results suggest that conformational changes are induced by the coenzymes of CHLDH.

Concerning the mechanism by which the coenzymes influence the hydrogen exchange of CHLDH, three hypotheses were considered:

(1) The conformation of the peptide chain in the protein molecule might be of different "tightness" of folding in different sections. If this is the case, the hydrogen exchange rates could reflect these structural differences. The coenzyme could then act on the protein molecule by increasing the tightness of some sections. Conformational changes caused by the binding of coenzyme

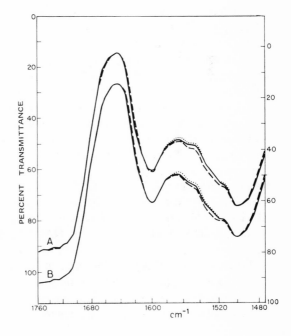

Fig. 11.11. Effect of NADH on the infrared spectrum of $(2.1 \times 10^{-4}M)$ YADH in D_2O $(0.1M$ sodium phosphate, *p*D 8). Solid line: in the presence of $2 \times 10^{-2}M$ NADH; broken line: control; and dotted line: in the presence of $5 \times 10^{-2}M$ NAD$^+$. The ordinates are: (a) left ordinate, 27 min of exchange; and (b) right ordinate, 153 min of exchange. (Hvidt and Kägi, 1963.)

to enzyme might be due to variations of electrostatic factors, breaking and forming of hydrogen bonds, or modifications in the water structure around the protein molecule.

(2) The coenzyme could modify the "motility" of the enzyme molecule by stabilizing some conformational forms which would be favored over the others, with consequent modifications of the rate and extent of hydrogen exchange. Motility indicates continuous structural changes in parts of the protein molecule (Linderstrøm-Lang and Schellman, 1959).

(3) The coenzymes decrease the hydrogen exchange of the CHLDH simply by "shielding" some parts of the enzyme molecule from contact with the solvent. This possibility is unlikely if one considers that the coenzymes protect up to 200 peptide hydrogens of the n_1 (very fast exchangeable) and n_2 group (those exchangeable at a measurable rate) and up to 200 of the n_3 group (hard-to-exchange). Therefore the coenzymes might decrease the rate of exchange in about one-third of the total number of peptide hydrogens. It is hard to visualize how a relatively small molecule like a coenzyme could "shield" the exchange of such a large number of peptide hydrogens unless some special structural features are present in the enzyme molecule. For instance, the decrease in rate of hydrogen exchange might have been caused by the coenzymes shielding an area of the molecule from which the denaturation by sodium dodecyl sulfate or other agents had to start, and from there proceed to the rest of the molecule. This possibility has been discussed elsewhere (DiSabato and Kaplan, 1965) and deemed unlikely from the experimental data.

L-*Glutamate Dehydrogenase.* L-Glutamate dehydrogenase (GDH) has been extensively studied as a model allosteric enzyme (Tomkins et al., 1963; Yielding et al., 1964; Frieden, 1965; Henderson et al., 1969). The theoretical basis of allosteric control is that small molecules (effectors) can bind to regulatory sites on an enzyme, triggering a conformational change which affects the activity of the enzyme. Many small molecules have been shown to influence the activity of bovine liver GDH. Some studies with fluorescence techniques (Dodd and Radda, 1969; Thompson and Yielding, 1968) and immunochemical methods (Talal et al., 1964) have shown a correlation between the presence of effectors and conformational changes in the enzyme. Stryker and Parker (1970) have studied bovine liver GDH with the infrared hydrogen–deuterium exchange method. They measured the effects of NAD, NADH, NADP, NADPH, ADP, ATP, GDP, GTP, diethylstilbestrol, ethanol, L-glutamate, L-leucine, and L-methionine on the hydrogen–deuterium exchange characteristics of GDH. GTP, NADP, and L-methionine increased the extent of exchange, while NAD, GDP, GTP plus NADH, GDP plus NADP, and L-leucine decreased the extent of exchange. The effect of NADH varied with enzyme concentration. Diethylstilbestrol, NADPH, ADP, ATP, and L-glutamate had very little effect on the exchange.

Sund et al. (1968, 1969) have employed sedimentation, light scattering, and X-ray small-angle scattering to establish that bovine liver GDH undergoes a reversible concentration-dependent linear aggregation. Stryker and Parker (1970) have found that the extent of exchange in the absence of modifiers was markedly decreased by an increase of the GDH concentration. They attributed this finding to a change in conformation or motility in conjunction with the known linear aggregation of GDH.

Creatine Phosphokinase. The change in optical rotation of creatine phospho-kinase (CPK) in the presence of a mixture of its substrates has been attributed by Samuels *et al.* (1961) to a conformational change of the enzyme. The protection of the enzyme from inhibition by iodoacetamide or *p*-nitrophenyl acetate by the presence of substrate mixture was ascribed to this conformational change. Lui and Cunning-ham (1965) have studied the effect of substrates and substrate analogs on the inhibition of CPK and also on other physicochemical properties such as sedimentation constant, reduced viscosity, and deuterium exchange. These studies indicated only small structural alterations in the enzyme to be related to the binding of the combined substrates: $MgATP^{2-}$, creatine, $MgADP^-$, and phosphocreatine.

When infrared difference spectra were recorded for CPK in D_2O, in tris-HCl buffer (*pH* 7.1) with and without the presence of various additional substances, the transmittances of the two samples at 1552 cm^{-1} (amide II) almost coincided. (Sample 1 contained CPK + Mg^{2+}; sample 2 contained CPK + Mg^{2+} + ADP + creatine.) In fact, the two spectra were almost identical. The second combination did produce a marked protection against iodoacetamide and *p*-nitrophenyl acetate inhibition of CPK. Between 60 and 65% of the amide hydrogens exchanged instantaneously, while only 70 to 74% exchanged after four hours. Only a small decrease in the rate of exchange was detected when ADP was present, but the addition of creatine had no further effect upon the exchange. ADP interaction with the enzyme slowed the exchange of about 3–4% of the peptide hydrogens which correspond to about 11–18 groups per active site. This effect of nucleotide binding on the H–D exchange of creatine phosphokinase is of comparable magnitude to that found in the case of yeast alcohol dehydrogenase and chicken heart lactic dehydrogenase on addition of NAD. Lui and Cunningham have suggested that the protective action of the mixture of substrates (as against the presence of one of the substrates) is not due to a con-formational change in the enzyme.

Aspartate Aminotransferase. The conformation of pig heart aspartate amino-transferase in the presence of pyridoxal phosphate has been investigated (Torchinskii *et al.*, 1967) by determination of the rate of H–D exchange, ORD, and an immuno-chemical method. The change of optical rotation at 228 nm between the apoenzyme and the holoenzyme was not large and it was possibly due to inductive Cotton effects in longer wavelength regions of the spectrum. Therefore, other independent methods for the conformation study were used. The rate of H–D exchange was followed by measurement of the infrared spectra of the solution of apoenzyme and holoenzyme in D_2O. In the presence of pyridoxal phosphate retardation of exchange was observed.

The immunochemical method was based on the precipitation of apoenzyme or holoenzyme with antibodies produced after immunization by holoenzyme. The amount of precipitated protein was higher in the case of holoenzyme than in the case of apoenzyme. These workers concluded that the binding of pyridoxal phosphate had induced changes in the secondary and tertiary structures of aspartate aminotransferase.

REFERENCES

Abaturov, L. V., Ginodman, L. M., and Varshavskii, Ya. M. *Dokl. Akad. Nauk S.S.S.R.* **172**, 475 (1967); *CA*, **66**, 72774*r* (1967).

Abaturov, L. V., Zavodsky, P., and Varshavskii, Ya. M. *Mol. Biol.* **2**, 136 (1968); *CA*, **69**, 16476*v* (1968).

Beer, M., Sutherland, G. B. B. M., Tanner, K. N., and Woods, D. L. *Proc. Roy. Soc. (London)* **A249**, 147 (1958).

Benson, E. S., Hallaway, B. E., and Lumry, R. W. *J. Biol. Chem.* **239**, 122 (1964).

Bensusan, H. B. and Nielsen, S. O. *Biochemistry* **3**, 1367 (1964).

Berger, A., Loewenstein, A., and Meiboom, S. *J. Am. Chem. Soc.* **81**, 62 (1959).

Bewley, T. A. and Li, C. H. *Biochim. Biophys. Acta* **140**, 201 (1967).

Beychok, S. and Blout, E. R. *J. Mol. Biol.* **3**, 769 (1961).

Bhaskar, K. R. and Parker, F. S. *Abstracts of 158th National Meeting of the American Chemical Society*, New York, Sept. 1969, Paper 247, Div. Biol. Chem.

Bhaskar, K. R. and Parker, F. S. *J. Biol. Chem.* **245**, 3302 (1970).

Blout, E. R., de Lozé, C., and Asadourian, A. *J. Am. Chem. Soc.* **83**, 1895 (1961).

Bryan, W. P. *Recent Prog. Surface Sci.* **3**, 101 (1970).

Bryan, W. P. and Nielsen, S. O. *Biochim. Biophys. Acta* **42**, 552 (1960).

Byrne, R. H. and Bryan, W. P. *Anal. Biochem.* **33**, 414 (1970).

Cohen, C. and Szent-Györgyi, A. *J. Am. Chem. Soc.* **79**, 248 (1957).

Crolla, L. J. and McDonald, H. J. *Federation Proc.* **28**, 845 (1969).

DiSabato, G. and Kaplan, N. O. *J. Biol. Chem.* **239**, 438 (1964).

DiSabato, G. and Kaplan, N. O. *J. Biol. Chem.* **240**, 1072 (1965).

DiSabato, G. and Ottesen, M. *Biochemistry* **4**, 422 (1965).

DiSabato, G. and Ottesen, M. in *Methods in Enzymology, Vol. XI* (C. H. W. Hirs, ed.) Academic Press, New York, 1967, p. 734.

Dodd, G. H. and Radda, G. K. *Biochem. J.* **114**, 407 (1969).

Doty, P. *Rev. Mod. Phys.* **31**, 107 (1959).

Elliott, A. and Hanby, W. E. *Nature* **182**, 654 (1958).

Englander, S. W. *Biochemistry* **2**, 798 (1963).

Englander, S. W. in *Poly-α-Amino Acids* (G. D. Fasman, ed.), Marcel Dekker, New York, 1967, Chap. 8.

Englander, S. W. and Crowe, D. *Anal. Biochem.* **12**, 579 (1965).

Englander, S. W. and von Hippel, P. H. *Am. Chem. Soc. Meeting Abst. Papers*, 56 C, 1962, Atlantic City, New Jersey.

Flory, P. J. and Weaver, E. S. *J. Am. Chem. Soc.* **82**, 4518 (1960).

Foster, J. F., Sogami, M., Petersen, H. A., and Leonard, W. J., Jr. *J. Biol. Chem.* **240**, 2495 (1965).

Fraser, R. D. B. and MacRae, T. P. *J. Chem. Phys.* **28**, 1120 (1958).

Frieden, C. in *Developmental and Metabolic Control Mechanisms and Neoplasia*, Williams and Wilkins, Baltimore, 1965, p. 392.

Fukushima, D. *Agr. Biol. Chem. (Tokyo)* **31**, 130 (1967).

Gabelova, N. A. and Kobyakov, V. V. *12th Colloq. Spectrosc. Int.*, Exeter, England, 559 (1965).

Glazer, A. N. *Australian J. Chem.* **12**, 304 (1959).

Gould, H. J., Gill, T. J., and Doty, P. *J. Biol. Chem.* **239**, 2842 (1964*a*).

Gould, H. J., Gill, T. J., and Kunz, J. H. *J. Biol. Chem.* **239**, 3071 (1964*b*).

Haggis, G. H. *Biochim. Biophys. Acta* **23**, 494 (1957).

Hallaway, B. E. and Benson, E. S. *Biochim. Biophys. Acta* **107**, 157 (1965).

Hanlon, S. and Klotz, I. M. in *Developments in Applied Spectroscopy, Vol. 6* (W. K. Baer, A. J. Perkins, and E. L. Grove, eds.) Plenum Press, New York, 1968, p. 219.

Harrington, W. F. *J. Mol. Biol.* **9**, 613 (1964).

Harrington, W. F. and Schellman, J. A. *Compt. Rend. Trav. Lab. Carlsberg. Sér. Chim.* **30**, 21 (1956).

Harrington, W. F. and von Hippel, P. H. *Advan. Protein Chem.* **16**, 1 (1961).

Harrington, W. F., Josephs, R., and Segal, D. M. *Ann. Rev. Biochem.* **35**, 633 (1966).

Hartshorne, D. J. and Stracher, A. *Biochemistry* **4**, 1917 (1965).

Heidemann, E. and Bernhardt, H. W. *Nature* **216**, 263 (1967).

Heidemann, E. and Srinivasan, S. R. *Leder* **18**, 145 (1967); *CA*, **67**, 101038g (1967).

Henderson, T. R., Henderson, R. F., and Johnson, G. E. *Arch. Biochem. Biophys.* **132**, 242 (1969).

Hermans, J. J., Jr. and Scheraga, H. A. *J. Am. Chem. Soc.* **82**, 5156 (1960).

Hvidt, A. *Compt. Rend. Trav. Lab. Carlsberg* **33**, 475 (1963).

Hvidt, A. *Compt. Rend. Trav. Lab. Carlsberg* **34**, 299 (1964).

Hvidt, A. and Kägi, J. H. R. *Compt. Rend. Trav. Lab. Carlsberg* **33**, 497 (1963).

Hvidt, A. and Kanarek, L. *Compt. Rend. Trav. Lab. Carlsberg* **33**, 463 (1963).

Hvidt, A. and Linderstrøm-Lang, K. *Biochim. Biophys. Acta* **14**, 574 (1954).

Hvidt, A. and Linderstrøm-Lang, K. *Biochim. Biophys. Acta* **16**, 168 (1955).

Hvidt, A. and Nielsen, S. O. *Advan. Protein Chem.* **21**, 287 (1966).

Hvidt, A., Johansen, G., and Linderstrøm-Lang, K. in *A Laboratory Manual of Analytical Methods of Protein Chemistry, Vol. 2* (P. Alexander and R. J. Block, eds.) Pergamon Press, New York, 1960, p. 101.

Hvidt, A., Kägi, J. H. R., and Ottesen, M. *Biochim. Biophys. Acta* **75**, 290 (1963).

Ichishima, E. and Yoshida, F. *Biochim. Biophys. Acta* **128**, 130 (1966).

Imahori, K. *Biopolymers* **1**, 563 (1963).

Imahori, K. and Mommoi, H. *Arch. Biochem. Biophys.* **97**, 236 (1962).

Jeffries, R. *J. Polymer Sci., Pt. A* **2**, 5161 (1964).

Kägi, J. H. R. and Ulmer, D. D. *Biochemistry* **7**, 2718 (1968).

Kauzmann, W. *Ann. Rev. Phys. Chem.* **8**, 413 (1957).

Kay, C. M. *Biochim. Biophys. Acta* **43**, 259 (1958).

Kendrew, J. C., Watson, H. C., Strandberg, B. E., Dickerson, R. E., Philips, D. C., and Shore, V. C. *Nature* **190**, 666 (1961).

Klotz, I. M. and Frank, B. H. *J. Am. Chem. Soc.* **87**, 2721 (1965).

Kobyakov, V. V. and Gabelova, N. A. *Tr. Komis. Spectroskopii Akad. Nauk S.S.S.R.* **3**, 631 (1965); *CA*, **64**, 17902h (1966).

Laiken, S. L., Printz, M. P., and Craig, L. C. *Biochemistry* **8**, 519 (1969).

Leach, S. J. *Rev. Pure Appl. Chem.* **9**, 33 (1959).

Leichtling, B. H. and Klotz, I. M. *Biochemistry* **5**, 4026 (1966).

Lenormant, H. *Ann. Chim. (Rome)* **5**, 459 (1950).

Lenormant, H. and Blout, E. R. *Nature* **172**, 770 (1953).

Lenz, D. E. and Bryan, W. P. *Biochemistry* **8**, 1123 (1969).

Linderstrøm-Lang, K. *Chem. Soc. (London) Spec. Publ.* **2**, 1 (1955).

Linderstrøm-Lang, K. in *Symposium on Protein Structure* (A. Neuberger, ed.) Methuen, London, 1958, p. 23.

Linderstrøm-Lang, K. and Schellman, J. A. in *The Enzymes, Vol. 1* (P. D. Boyer, H. Lardy, and K. Myrbäck, eds.) Academic Press, New York, 1959, p. 443.

Lowey, S. and Cohen, C. *J. Mol. Biol.* **4**, 293 (1962).

Lui, N. S. T. and Cunningham, L. *Biochemistry* **5**, 144 (1965).

Malcolm, B. R. *Proc. Roy. Soc.* **A305**, 363 (1968).

Margoliash, E. and Schejter, A. *Advan. Protein Chem.* **21**, 113 (1966).

Miyazawa, T. in *Aspects of Protein Structure* (G. N. Ramachandran, ed.) Academic Press, New York, 1963, p. 258.

Miyazawa, T. and Blout, E. R. *J. Am. Chem. Soc.* **83**, 712 (1961).

Morris, J. L. *Biochim. Biophys. Acta* **47**, 606 (1961).

Nielsen, S. O. *Biochim. Biophys. Acta* **37**, 146 (1960).

Nielsen, S. O., Bryan, W. P., and Mikkelsen, K. *Biochim. Biophys. Acta* **42**, 550 (1960).

Nozaki, M., Sekuzu, I., Yamashita, J., Hagihara, B., Yonetani, T., and Okunuki, U. *J. Biochem. (Japan)* **44**, 595 (1957).

Paiva, T. B., Paiva, A. C. M., and Scheraga, H. A. *Biochemistry* **2**, 1327 (1963).

Parker, F. S. and Bhaskar, K. R. *Appl. Spectrosc. Rev.* **3**, 91 (1969).

Parker, F. S. and Stryker, M. H. *Appl. Spectrosc.* **23**, 245 (1969).

Perlmann, G. E. and Diringer, R. *Ann. Rev. Biochem.* **29**, 151 (1960).

Ramachandran, G. N. and Sasisekharan, V. *Nature* **190**, 1004 (1961).

Rich, A. and Crick, F. H. C. *J. Mol. Biol.* **3**, 483 (1961).

Rickli, E. E., Ghazanfar, S. A. S., Gibbons, B. H., and Edsall, J. T. *J. Biol. Chem.* **239**, 1065 (1964).

Samuels, A. J., Nihei, T., and Noda, L. *Proc. Natl. Acad. Sci. U.S.* **47**, 1992 (1961).

Scarpa, J. S., Mueller, D. D., and Klotz, I. M. *J. Am. Chem. Soc.* **89**, 6024 (1967).

Schellman, J. A. and Schellman, C. in *The Proteins*, Vol. *II*, 2nd Ed. (H. Neurath, ed.) Academic Press, New York, 1964, p. 1.

Scheraga, H. A. *Protein Structure*, Academic Press, New York, 1961.

Schildkraut, C. L. and Scheraga, H. A. *J. Am. Chem. Soc.* **82**, 58 (1960).

Sekuzu, I., Yamashita, J., Nozaki, M., Hagihara, B., Yonetani, T., and Okunuki, K. *J. Biochem. (Japan)* **44**, 601 (1957).

Shechter, E. and Blout, E. R. *Proc. Natl. Acad. Sci. U.S.* **51**, 695 (1964).

Smeby, R. R., Arakawa, K., Bumpus, F. M., and Marsh, M. M. *Biochim. Biophys. Acta* **58**, 550 (1962).

Squire, P. G. and Ottesen, M. *Biochim. Biophys. Acta* **154**, 226 (1968).

Stevens, C. O., Henderson, L. E., and Tolbert, B. M. *Arch. Biochem. Biophys.* **107**, 367 (1964).

Stryker, M. H. and Parker, F. S. *Arch. Biochem. Biophys.* **141**, 313 (1970).

Sund, H. and Burchard, W. *Eur. J. Biochem.* **6**, 202 (1968).

Sund, H., Pilz, I., and Herbst, M. *Eur. J. Biochem.* **7**, 517 (1969).

Talal, N., Tomkins, G. M., Mushinski, J. F., and Yielding, K. L. *J. Mol. Biol.* **8**, 46 (1964).

Thompson, W. and Yielding, K. L. *Arch. Biochem. Biophys.* **126**, 399 (1968).

Timasheff, S. N. and Susi, H. *J. Biol. Chem.* **241**, 249 (1960).

Tomkins, G. M., Yielding, K. L., Talal, N., and Curran, J. F. *Cold Spring Harbor Symp. Quant. Biol.* **28**, 461 (1963).

Torchinskii, Yu. M., Abaturov, L. V., Varshavskii, Ya. M., and Nezlin, R. S. *Mol. Biol.* **1**, 603 (1967); *CA*, **67**, 105467a (1967).

Ulmer, D. D. and Kägi, J. H. R. *Biochemistry* **7**, 2710 (1968).

von Hippel, P. H. and Harrington, W. F. *Biochim. Biophys. Acta* **36**, 427 (1959).

von Hippel, P. H. and Harrington, W. F. *Brookhaven Symp. Biol.*, *BNL 608*, (C22), 213 (1960).

Willumsen, L. *Biochim. Biophys. Acta* **126**, 382 (1966).

Willumsen, L. *Compt. Rend. Trav. Lab. Carlsberg* **36**, 247 (1967).

Willumsen, L. *Compt. Rend. Trav. Lab. Carlsberg* **36**, 327 (1968).

Wishnia, A. and Saunders, M. *J. Am. Chem. Soc.* **84**, 4235 (1962).

Yielding, K. L., Tomkins, G. M., Bitensky, M. W., and Talal, N. *Can. J. Biochem.* **42**, 727 (1964).

Yonetani, T. and Theorell, H. *Arch. Biochem. Biophys.* **99**, 433 (1962).

Chapter 12

NUCLEIC ACIDS AND RELATED COMPOUNDS

Nucleic acids are an important class of compounds, which have been studied in great detail. Pauling and Corey (1953) proposed a helical model composed of three chains, each having ribofuranose, phosphate ion, and pyrimidine or purine moieties. The chains wound around the axis to form a rope-like molecule, the core of which was formed by the phosphates with hydrogen bonds and ribose rings connecting them. The pyrimidine and purine rings were connected to the sugars along the outer periphery of the polymer. The Watson–Crick (1953) model for DNA was a hydrogen-bonded one which involved two chains. Each chain had a backbone of sugar–phosphate–sugar–phosphate–sugar, etc. In this model the pyrimidine and purine rings were attached to the sugar molecules in such a way that the bases projected inward toward the core of the polymer. The hydrogen bonds formed in such a manner that two hydrogen bonds linked each purine to a pyrimidine on the opposite chain, and *vice versa*. Pauling and Corey (1956) then proposed a model in which instead of two hydrogen bonds between guanine and cytosine there were three (Fig. 12.1). The original Watson–Crick model has since been revised by Wilkins and his co-workers (see Mahler and Cordes, 1966). Figure 12.2a shows the space-filling model (with van der Waals radii indicated) of the revised structure of DNA in the

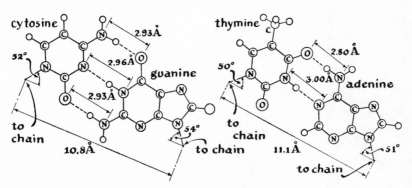

Fig. 12.1. H-bonding of purine and pyrimidine bases in nucleic acids. Left, cytosine and guanine; right, thymine and adenine (Pauling and Corey, 1956). (From Pimentel and McClellan, 1960.)

271

B-configuration. Figure 12.2b shows a projection of the model. The B-configuration of DNA, which is the one stable at greater than 66% relative humidity, is made up

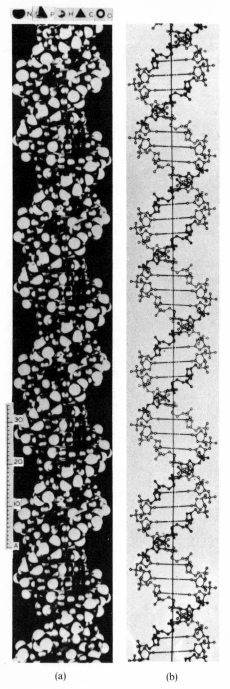

(a)   (b)

Fig. 12.2. (a) A space-filled model of B-configuration DNA (van der Waals radii indicated). (b) A projection of the space-filled model (M. H. F. Wilkins). (From Mahler and Cordes, 1966.)

of two right-handed helical polynucleotide chains in which the internucleotide linkage in one strand is $3' \rightarrow 5'$ and in the other is $5' \rightarrow 3'$, that is, the strands are of opposite polarity. The strands are wound around the axis of the molecule so that the double helix they form cannot be separated into two strands unless unwinding occurs. Complementarity of the organic bases is still maintained, that is, the allowed base pairs are adenine and thymine, and guanine and cytosine. Figure 12.3 shows these base pairs hydrogen-bonded. Unusual bases, such as substituted cytosines (for

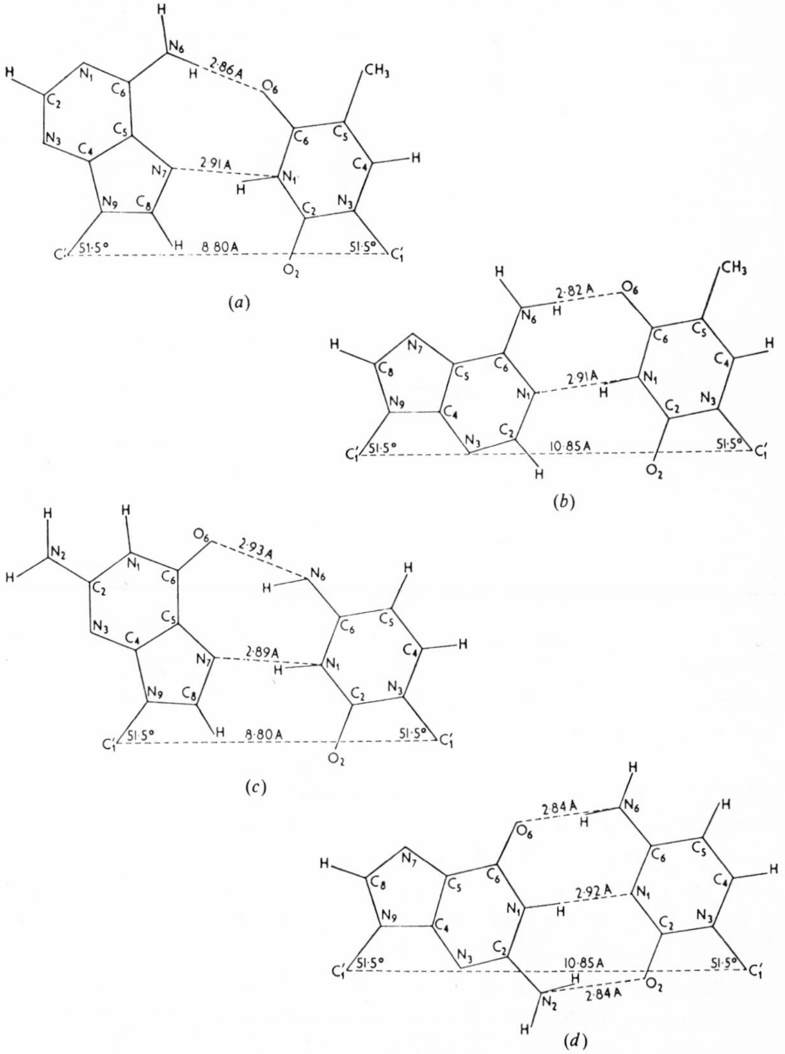

Fig. 12.3. Adenine–thymine hydrogen-bonded pairs: (a) found by Hoogsteen in a 1:1 complex of 3-N-methylthymine and 9-N-methyladenine; (b) postulated by Watson and Crick (current version refined by Arnott) for part of the structure of DNA. Guanine–cytosine pairs: (c) Hoogsteen type; (d) Watson–Crick type (Wilkins and Arnott, 1965). (From Mahler and Cordes, 1966.)

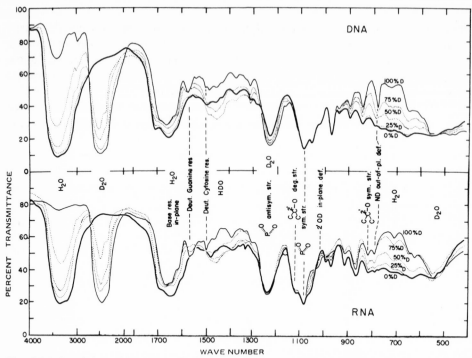

Fig. 12.4. Infrared absorption spectra of calf thymus DNA and yeast tRNA (bulk) films placed in air of 92% relative humidity. The spectra were observed in the undeuterated, partly deuterated, and nearly completely deuterated states. Partial deuteration was achieved by placing the films under partially deuterated water vapor. (Tsuboi, 1969.)

example, 5-methylcytosine) can replace cytosines wherever the former occur. The planes of the aromatic rings are at right angles to the helix axis and the hydrogen-bonded base pairs are stacked on top of one another 3.4 Å apart. The phosphodiester groups are on the outside of the cylinder-like molecule. Since each turn of the double helix contains ten bases and each turn has a pitch of 34 Å, the helix has an exact ten-fold screw axis in a right-handed screw.

There are several lines of experimental evidence to support the Watson–Crick structure (Mahler and Cordes, 1966). We are concerned here only with the type of evidence obtainable from infrared spectroscopy, an example of which is the following: Thin sheets of sodium or lithium DNA, equilibrated with D_2O, show characteristic bands at 1685, 1665, and 1644 cm^{-1} (weak bands at 1620 and 1575 cm^{-1}) that have been assigned to C=O, C=N, and C=C stretching and NH$_2$ deformation vibrations of the bases in the helical pattern. The band intensity at ~ 1685 cm^{-1} of DNA in D_2O is reduced significantly, while those at 1665 and 1620 cm^{-1} are increased when the molecular structure is denatured by heat, alkali, formamide, or deoxyribonuclease [see below] (Bradbury et al., 1961; Kyogoku et al., 1961).

Absorption Band Assignments

Figure 12.4 presents spectra of films of deoxyribonucleic acid (DNA) and ribonucleic acid (RNA) in the undeuterated, partly deuterated, and deuterated forms.

Both DNA and RNA have a broad intense band at $3400 \, \text{cm}^{-1}$ (or at $2500 \, \text{cm}^{-1}$ when deuterated), which is assigned to the OH (or OD) stretching vibration of adsorbed water molecules. In the range $1800–1500 \, \text{cm}^{-1}$ the group of strong bands are due to the $C=O$ stretching, skeletal stretching, and NH bending vibrations of the organic base residues (Sutherland and Tsuboi, 1957). The PO_2^- antisymmetric stretching vibration is found at $1220 \, \text{cm}^{-1}$ (Sutherland and Tsuboi, 1957; Tsuboi, 1957). Several intense bands in the $1100–1000 \, \text{cm}^{-1}$ region are assigned to the PO_2^- symmetric stretching vibration of the phosphate group and the $C—O$ stretching vibrations of the ribose or deoxyribose moiety (Sutherland and Tsuboi, 1957; Tsuboi, 1957; Shimanouchi et al., 1964). The $1000–700 \, \text{cm}^{-1}$ region contains several medium or weak bands, which come from the $P—O$ stretching, $C—O$ stretching, and NH out-of-plane bending vibrations, and also from the librations of adsorbed water molecules. The $600–400 \, \text{cm}^{-1}$ region should contain the librations of D_2O molecules and the ND out-of-plane vibrations. Table 12.1 (Tsuboi, 1969) presents the frequency, relative intensity, and probable vibrational mode for the bands seen in nucleic acids.

Table 12.1. Band Frequencies of Nucleic Acids[a] (Tsuboi, 1969)

Residue	Band, cm^{-1}	Comments
	4000–2000 cm^{-1} Region	
	3300 m	NH_2 antisymmetric and symmetric stretching vibrations of
	3100 m	adenine, guanine, and/or cytosine residues.[b,c]
	2500 w	ND_2 stretching vibrations of adenine, guanine, and/or cytosine
	2360 w	residues.[b,c]
	2920 w	CH stretching vibrations in the ribose group.
	3380 s, br	Stretching vibrations of the adsorbed H_2O and D_2O molecules.
	2500 s, br	
	1750–1550 cm^{-1} Region	
	1640 s, br	Scissoring vibration of the adsorbed H_2O molecule. This vibration can be eliminated by deuteration.
Undeuterated base residues		
Adenine	1665 s	
	1605 s	NH_2 bending and $C=N$ stretching vibrations.[d]
	1573 w	
Uracil	1700 s	$C=O$ stretching vibration.
	1680 s	NH in-plane deformation vibration.[d]
	1650 s	$C=O$ and $C=C$ stretching vibration.
	1620 w	$C=O$ and $C=C$ stretching vibration.
Thymine	1720 s	$C2=O$ stretching vibration.[c]
	1660 s	$C4=O$ stretching vibration.[c]
	1575 w	Ring stretching vibration.[c]
Guanine	1690 s	$C=O$ stretching and NH_2 scissoring vibration.[e,f]
	1630 w	

Table 12.1 *(Continued)*

Residue	Band, cm^{-1}	Comments
Hypoxanthine	1680 s, br	In-plane ring vibration.
Cytosine	1700 w	
	1670 sh	NH$_2$ scissoring vibration.[d]
	1647 s	In-plane ring vibrations.
	1603 s	In-plane ring vibrations.
	1585 w	In-plane ring vibrations.
Deuterated base residues		
Adenine	1627 s	C=N and C=C stretching vibrations.[d]
	1579 w	
Uracil	1692 m	C2=O stretching vibration.[g,d]
	1657 s	C4=O and C=C stretching vibrations.[g,d]
	1618 w	C4=O and C=C stretching vibrations.[g,d]
Pseudouridine[f]	1690 m	C2=O stretching vibration.
	1653 s	C4=O and C=C stretching vibration.
	1631 m	C4=O and C=C stretching vibration.
Guanine	1665 s	C=O stretching vibration.[h]
	1581 s	C=N stretching vibrations.[i]
	1568 m	C=N stretching vibrations.[i]
Hypoxanthine [j]	1672 s	C=O stretching vibration.[h]
Cytosine	1652 s ⎫	
	1616 m ⎬	In-plane ring vibrations.
	1585 w ⎭	

1550–1300 cm^{-1} Region

Deuterated		There are several weak bands in this region. They are assigned
adenine residue	1483 w	to in-plane vibrations of the base residues involving the
		N—H and C—H in-plane deformation modes. Some
Undeuterated		examples are given for both deuterated and undeuterated
cytosine residue	1530 w	residues.
	1500 m	For RNA, a band should appear near 1400 cm^{-1}, which is
Deuterated		due to a vibration involving the in-plane C—O—H
cytosine residue	1525 m	deformation mode at the 2′ position of the ribose ring.[k]
	1503 s	

1300–1000 cm^{-1} Region

Undeuterated		
cytosine residue[k,l,i]	1296 w	
Undeuterated		
thymine residue[l,i]	1292 w	
	1276 w	
	1225 s	Antisymmetric stretching vibration of the $\begin{bmatrix} & O \\ P & \\ & O \end{bmatrix}^-$ group.[b,m]

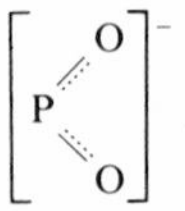

An absorption of the adsorbed D$_2$O molecules is superimposed here.

Table 12.1 *(Continued)*

Residue	Band, cm^{-1}	Comments
	1135 s	Nearly degenerate stretching vibrations of the $\begin{array}{c} \text{C} \\ \diagdown \\ \text{C—O} \\ \diagup \\ \text{C} \end{array}$
	1116 s	structure in RNA at the 2′ position of the ribose residue.[n]
	1084 s	Symmetric stretching vibration of the $\left[\begin{array}{c} \text{O} \\ \diagup \\ \text{P} \\ \diagdown \\ \text{O} \end{array}\right]^{-}$ group.[b,m,o]
	1050 s	Some C—O stretching vibration of the ribose or the deoxyribose residue.
	1020 m, br	C—O—D in-plane vibration in deuterated RNA at the 2′ position of the ribose residue.[k]

1000–700 cm^{-1} Region

Residue	Band, cm^{-1}	Comments
	800–760 w	A few weak bands are present. These are due to out-of-plane vibrations of base residues of undeuterated DNA's and RNA's.[i,p,c]
	800–760	Several bands which are stronger than the above are present. These are due to out-of-plane vibrations of base residues of deuterated DNA's and RNA's.[i,p,c]
	967 s 890 m 830 w	Every DNA, both deuterated and not. The 967 cm^{-1} band is similar in position and intensity to the 950 cm^{-1} band of diethylphosphate anion, which is assigned to the C—C stretching vibration.[o]
	995 m 970 m 915 s 880 m 870 s 815 s	Many of these appear in all the synthetic polyribonucleotides and in all natural RNA. The 995 and 970 cm^{-1} bands disappear on deuteration; the others remain almost unchanged. All are due to ribose–phosphate chain-backbone vibrations. The 815 cm^{-1} band may be assigned to a nearly symmetrical stretching of the $\begin{array}{c} \text{C} \\ \diagdown \\ \text{C—O} \\ \diagup \\ \text{C} \end{array}$ structure at the 2′ position of the ribose residue.[n]

700–400 cm^{-1} Region

Residue	Band, cm^{-1}	Comments
	700–600 vbr	Due to adsorbed H$_2$O molecules.
	550–450 br	Due to D$_2$O molecules.
	650 m 645 m 622 m 598 w 580 w 568 w 548 s 518 m 510 w 485 w	These bands are superimposed upon the above broad bands, and are found in deuterated RNA of rice dwarf virus. The 650, 580, and 548 cm^{-1} bands are strongly polarized along the fiber axis. Those at 645, 598, 518, and 485 cm^{-1} are polarized perpendicular to the fiber axis.

[a]Key: br, broad; m, medium; s, strong; sh, shoulder; v, very; w, weak. [b]Sutherland and Tsuboi (1957). [c]Kyogoku *et al.* (1967*b*).[d]Tsuboi *et al.* (1962). [e]Angell (1955, 1961). [f]Tsuboi and Kyogoku (1969). [g]Miles (1964). [h]Howard and Miles (1965). [i]Tsuboi *et al.* (1968). [j]Miles (1961). [k]Tsuboi (1964). [l]Tsuboi and Shuto (1966). [m]Tsuboi (1957). [n]Tsuboi *et al.* (1963). [o]Shimanouchi *et al.* (1964). [p]Sato *et al.* (1966).

Studies of DNA

It is essential for the recording of infrared spectra of nucleic acid fibers to maintain them in an environment of known relative humidity if structural studies are to be done, since the crystallinity is known to vary with the humidity of the air surrounding the fibers. In order to achieve these conditions one can take the oriented nucleic acid film which is on a plate of CaF_2, BaF_2, or AgCl, and seal it with another plate and a spacer over a chamber holding a saturated solution of the required salt with excess solid (Sutherland and Tsuboi, 1957). Spencer (1926) describes solutions for obtaining the proper constant humidities. One can almost completely deuterate a nucleic acid film by vapor-phase exchange with a D_2O solution of the right salt in the chamber of the sealed cell described above. When the nucleic acid film is placed in the cell and a D_2O salt solution is placed in the bottom of the cell, virtually all the hydrogen atoms in the film (except CH hydrogens) are replaced by deuterium atoms. Figure 12.4 shows that spectra of deuterated and undeuterated films of identical thickness can be compared in this way.

Sutherland and Tsuboi (1957) have studied the polarized spectra of oriented films of sodium DNA at twelve different humidities, and have correlated their data with Crick and Watson's (1954) proposed structure and that of Feughelman *et al.* (1955). One example of the type of work done on films is that by Bradbury *et al.* (1961).

Bradbury *et al.* (1961) have described in detail the experimental techniques used in a study of films of DNA polymers and have recorded polarized infrared spectra of oriented sheets of the sodium and lithium salts of DNA, in both undeuterated and deuterated states, over a range of humidities between 0 and 94% relative humidity (r.h.). They found two forms of sodium DNA, one at humidities greater than 90% r.h. where the bases are perpendicular to the helix axis (B-form) and the other at humidities between 70 and 80% r.h. where the bases have tilted by an angle not less than 13° to the normal to the helix axis (A-form). Lithium DNA exists in the B-form for humidities greater than 66% r.h., and in a second form, different from the A-form of sodium DNA, between 44 and 56% r.h. where the bases are tilted by about 4° from the perpendicular (C-form). These workers have discussed the orientation of the phosphate groups, which they inferred from dichroic effects.

The dichroic ratio may be defined as

$$R = \int_{band} \alpha_\perp \, dv \bigg/ \int_{band} \alpha_\| \, dv$$

where $\alpha_\perp = \log(I_0/I)_\perp$ and $\alpha_\| = \log(I_0/I)_\|$. The ratio I_0/I is the reciprocal transmission at frequency v, measured either perpendicular or parallel to the fiber axis, the integration being over the whole absorption band. If the absorption is symmetrical, the maximum absorption α^{max} is directly proportional to $\int \alpha \, dv$ and therefore the dichroic ratios may be calculated from the simpler expression $\alpha_\perp^{max}/\alpha_\|^{max}$, which was used by Bradbury *et al.* (1961) in this study.

When the dichroic ratio ($R = \alpha_\perp^{max}/\alpha_\|^{max}$) of a specific band (1672 cm^{-1}) for the deuterated film of sodium DNA was plotted against the water content of the film, the curve shown in Fig. 12.5 was obtained. The band at 1672 cm^{-1} was used for the

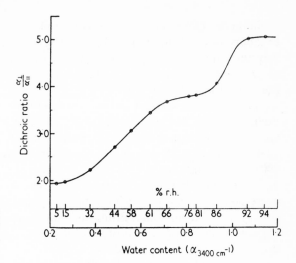

Fig. 12.5. Plot of the dichroic ratio of in-plane base vibrations for NaDNA against water content (D_2O) of the specimen. (Bradbury *et al.*, 1961.)

calculation of the dichroic ratio of the in-plane vibrations of the bases because it gave the maximum polarization ratio and because it lay close to the centroid of the base absorptions. Figure 12.5 shows two plateaus, one for humidities greater than 90% r.h. and the other for humidities between 70% and 80% r.h. (the A-form at 75% r.h. and the B-form at 92% r.h.). In the X-ray analysis of the sodium DNA structure, A- and B-conformations have been found; A is obtained at 75% r.h. and B at 92% r.h. (Franklin and Gosling, 1953).

Falk *et al.* (1963*a*) have studied the infrared spectrum of DNA as a function of relative humidity. From frequency and intensity changes of infrared bands they concluded that the $PO_2^- Na^+$ groups of DNA become hydrated in the range of 0 to 65% r.h., while the hydration of the bases begins above this range. The strength of hydrogen bonding is greatest for the first water molecules to adsorb and decreases thereafter, approaching the strength of hydrogen bonding of liquid water.

The same workers (1963*b*) have investigated the polarized infrared and ultraviolet spectra of oriented films of the sodium salt of DNA as a function of relative humidity. They presented quantitative data for the behavior of the dichroic ratios of the infrared band at 1660 cm^{-1}, and the ultraviolet band at 260 nm. Between 75% and 55% r.h. a sharp increase occurs in the dichroic ratios and the absorbance at 260 nm. Falk *et al.* concluded that DNA films are stable in the B-configuration at relative humidities as low as 75% and that at still lower humidities a reversible transition occurs to a disordered form in which the bases are no longer stacked one above another and are no longer perpendicular to the axis of the helix. The loss of base stacking upon drying suggested that the B-configuration of DNA is stabilized by the stacking of the bases in the presence of water.

As seen in Chapter 11 (*Hydrogen–Deuterium Exchange*), infrared spectroscopy can be used conveniently for measuring the deuteration rate of biological polymers, e.g., polypeptides and proteins. Hydrogen isotopic exchange of DNA has been studied by infrared spectroscopy (Sutherland and Tsuboi, 1957; Bradbury *et al.*, 1961). Bradbury *et al.* (1961) found that hydrogen exchange took place in DNA in less than three minutes. They used solid films of DNA and followed the deuteration by D_2O vapor. Similar results of infrared studies on DNA films were obtained by Sutherland and Tsuboi (1957). Printz and von Hippel (1965), using the hydrogen–tritium exchange method of Englander (1963), studied the kinetics of hydrogen exchange in the range 120 to 2100 sec, and observed an instantaneous exchange of amino hydrogens bonded to solvent and of hydroxyl hydrogens of sugar moieties as well as a fast exchange of interchain bonded hydrogens with a half-time of about 300 sec. "Instantaneous exchange" is completed in a fraction of a second; "fast exchange" has a half-time of greater than 1 sec but less than 3 to 4 hr; "slow exchange" corresponds to a half-time of hours or days [these classifications are those of Osterman *et al.* (1966)]. Furthermore, Printz and von Hippel (1965) found that some hydrogens underwent slow exchange in their experiments, but the effects were small and their origin could not be explained.

Printz and von Hippel (1968) later studied the hydrogen-exchange kinetics of native calf thymus DNA as a function of pH and salt concentration. The DNA hydrogen exchange proceeds at a minimum rate at 0°C and pH values close to neutrality (pH_{min}); increasing or decreasing the pH from pH_{min} increases the rate of exchange. The rate of exchange at any given pH, as well as the value of pH_{min}, depends on the salt concentration.

Osterman *et al.* (1966) studied slow hydrogen exchange of DNA by Englander's method. They did not observe any slow hydrogen exchange at room temperature. At 55°C the half-time of exchange was about 10 days. These investigators attributed the slow exchange to the CH hydrogens in purine and pyrimidine base residues of DNA. Fritzsche (1967a) studied H–D exchange of DNA by the infrared method. After incubation in D_2O solutions at elevated temperatures, changes in spectra of DNA are similar for native and heat-denatured DNA. A new absorption band appears at 2325 cm^{-1}, while the band at 3120 cm^{-1} disappears. These bands were tentatively assigned to the C—D and C—H stretching vibrations, respectively, of adenine and guanine. The tentative assignments were confirmed by comparison with spectra of model compounds.

The C—D stretching vibrations absorb in the region between 2000 and 2500 cm^{-1}, corresponding to the C—H stretching vibrations between 2800 cm^{-1} and 3350 cm^{-1}. In the DNA molecule there are aliphatic C—H groups and C—H groups contained in the base residues with C—H stretching vibrations absorbing at about 2800 to 3000 cm^{-1} and 3000 to 3150 cm^{-1}, respectively. The corresponding C—D stretching absorptions are expected at about 2050–2250 cm^{-1} and 2250–2350 cm^{-1}, respectively. The aliphatic C—H stretching vibration bands are caused by sugar residues and the methyl group of thymine residues.

The new band of DNA at 2325 cm^{-1}, observed after the H–D exchange, falls into the category corresponding to the C—D groups of the heterocyclic purine and pyrimidine base residues. This statement is confirmed by the disappearance of the band at 3120 cm^{-1} after incubation of DNA in D_2O. These observations exclude hydrogen of aliphatic C—H groups as sites of slow hydrogen exchange and leave the hydrogens attached to the carbon of the heterocyclic base residues as the only possible slowly exchanging groups.

The purine bases contain the pyrimidine and imidazole moieties. Hydrogen exchange of the C—H group between the two nitrogens of imidazole was observed for imidazole (Bellocqu et al., 1965) and benzimidazole (Fritzsche et al., 1967a). The position of the corresponding C—D stretching vibration is 2335 cm^{-1} in both cases. The same position is to be expected for C—D stretching vibrations of adenine and guanine moieties of DNA.

Thymine and cytosine residues have one and two C—H groups, respectively, besides the methyl group of thymine. Párkányi and Šorm (1963) synthesized thymine containing a deuterium atom at position 6; the corresponding C—D stretching vibration lies at about 2285 cm^{-1}. The position of the C—D stretching vibration was estimated from the corresponding C—H stretching values, by means of the ratio $v_{C-H}/v_{C-D} = 1.343$. The following C—D stretching vibration values result for the nucleosides when C—H stretching values from Fritzsche (1967a) are used: adenosine, 2331 cm^{-1}; guanosine, 2328 cm^{-1}; cytidine, 2296 cm^{-1}; and thymidine, 2295 cm^{-1}. These values are in agreement with observed positions of band maxima after incubation at 95°C in D$_2$O solutions of nucleosides.

After the DNA films have been exposed to D$_2$O vapor, C—H stretching vibration bands can be observed at 3090 and 3120 cm^{-1}. The 3090 cm^{-1} band is assigned to C—H stretching of pyrimidine bases in agreement with bands of cytidine at 3083 cm^{-1} and thymidine at 3082 cm^{-1}. The 3120 cm^{-1} band is assigned to C—H stretching of purine bases at position 8 in accordance with the band of adenosine at 3131 cm^{-1} and that of guanosine at 3127 cm^{-1}.

The observed decrease or disappearance of the 3120 cm^{-1} band of DNA and the appearance of a new absorption band at 2335 cm^{-1} after incubation with D$_2$O was interpreted (Fritzsche, 1967a) to be the result of H–D exchange of hydrogen at position C-8 of adenine and guanine. No evidence was found of exchange of hydrogen attached to carbon of pyrimidine base residues (cytosine and thymine) of DNA. Fritzsche's data support those of Osterman et al. (1966), who found that the exchange rates for purine nucleoside diphosphates are higher by a factor of about ten than that for pyrimidine nucleoside disphosphates. Thus, the lower exchange rate of C—H groups in pyrimidine nucleosides and related nucleoside diphosphates is preserved in DNA. In contrast, exposure of DNA to tritium gas results in the preferential exchange of hydrogens of the thymine residues (Borenfreund et al., 1959) but the exchange occurs in this case in the thymine methyl groups. Fritzsche emphasized that denatured DNA at 95°C exchanges hydrogens in the same manner as native DNA at 55°C and therefore, as pointed out by Osterman et al. (1966), no connection between the conformation of DNA and slow isotopic hydrogen exchange of DNA could be deduced.

The exchangeable hydrogens in DNA could be divided into three groups: (1) Those OH and NH hydrogens which are not involved in the interstrand base pairing by hydrogen bonding. Exchange is instantaneous. (2) Those hydrogens of base moieties which are engaged in interstrand base pairing by hydrogen bonds. Interaction with solvent is restricted. Exchange is fast but the kinetics of exchange can be measured by Englander's method (1963). (3) Those hydrogens attached to C-8 of purine base moieties. Hydrogens on C-4 in thymine, and on C-4 and C-5 in cytosine may belong to this group, too. Exchange is slow and is observable only at elevated temperatures.

Studies of Deoxyribonucleoproteins

The infrared spectra of nucleoprotamine (Bradbury *et al.*, 1962*a*) and nucleo-histone (Shimanouchi *et al.*, 1964; Bradbury *et al.*, 1962*b*; Bradbury and Crane-Robinson, 1964) display strong bands at 1710, 1225, and 1084 cm^{-1} at 75–94% r.h. The 1710 cm^{-1} band is removed by treatment with DNase or by drying, in a manner like that for the 1705 cm^{-1} band of sodium DNA film. (The 1705 cm^{-1} band is considered to be caused by base moieties involved in the pairing of guanine with cytosine and adenine with thymine; it is missing in single-stranded f 1-phage DNA, and also absent or weaker in DNAse-treated double-helical DNA. The 1705 cm^{-1} band is also removed by drying of calf thymus sodium DNA film, a fact considered to mean that hydration is needed for base pairing to take place.) The 1710 cm^{-1}

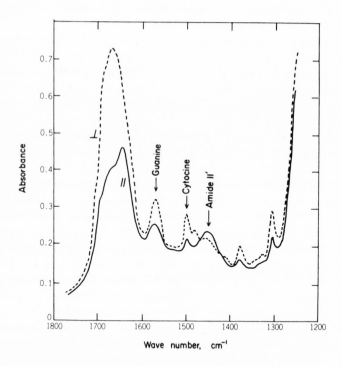

Fig. 12.6. Infrared absorption spectrum of partial nucleoprotein which was obtained by exposing calf thymus nucleohistone to 1.2*M* NaCl. After the film was prepared, it was deuterated by D$_2$O vapor. (——), the electric vector of the incident radiation was placed parallel to the fiber axis; (————), the electric vector of the incident radiation was placed perpendicular to the fiber axis (Bradbury *et al.*, 1967). (From Tsuboi, 1969.)

band of the nucleoprotein has a strong perpendicular dichroism. These observations are evidence that the DNA molecule in nucleohistone and nucleoprotamine has the planes of the adenine–thymine and guanine–cytosine pairs perpendicular to the fiber axis.

Bradbury et al. (1967) have studied partial nucleoproteins produced by removal of histone fractions from nucleohistone. Figure 12.6 shows a polarized infrared spectrum of one of the nucleoproteins oriented by shearing. The spectrum is that of a film deuterated by D_2O vapor. The bands of the in-plane vibrations of cytosine and guanine moieties are highly polarized when the electric vector is placed perpendicular to the fiber axis. The 1452 cm^{-1} absorption is that of the amide II' band (Miyazawa et al., 1956; Miyazawa, 1962) for the readily deuterated fraction of the partial nucleoprotein. As seen in Fig. 12.6, the 1452 cm^{-1} band is polarized considerably in the direction parallel to the fiber axis, giving evidence that the extended polypeptide chain is situated so that its axis lies between the angle of the groove in the DNA double helix and the axis of the DNA helices (Bradbury et al., 1967).

In nucleohistone, there is a component which requires more than three weeks for complete deuteration to take place in D_2O vapor of 94% r.h. (Bradbury et al., 1962b; Kyogoku, 1963), although nucleoprotamine deuterates very fast, the protein and DNA components becoming completely deuterated within a few minutes of exposure of the film in D_2O vapor of 94% r.h. (Bradbury et al., 1962a; Kyogoku, 1963). These results are explained by considering that nucleoprotamine does not contain any α-helical part but nucleohistone does contain the α-helix (Tsuboi, 1969).

Double-Helical RNA

Tsuboi (1969) has reviewed work done on double-helical RNA and has given infrared spectra of oriented films of the sodium salt of double-helical RNA from rice dwarf virus. The RNA was studied in both the undeuterated and deuterated forms at 75% r.h. (Sato et al., 1966). In RNA, the PO_2^- symmetric stretching band at 1083 cm^{-1} shows a parallel dichroism and the PO_2^- antisymmetric stretching band at 1225 cm^{-1} a perpendicular dichroism. However, in DNA the former shows perpendicular dichroism and the latter no dichroism. Sato et al. (1966) concluded from a study of the parameters involved in determining the dichroic ratio that θ, the angle between the helix axis and the transition moment, equals (40 ± 4)° for the 1083 cm^{-1} band and (69 ± 4)° for the 1225 cm^{-1} band. Sato et al. therefore considered that the PO_2^- group of this RNA is oriented so that its O$\cdots$O line makes an angle of about 70° with the molecular axis, and the bisector of angle O—P—O makes an angle of about 40° with it. The orientation of the PO_2^- group mentioned here seems to be a general property among several double-helical RNA molecules from a variety of sources. Arnott et al. (1967) took the PO_2^- orientation calculated from the infrared analysis into account when constructing the detailed model of the RNA structure.

DNA–RNA Hybrid

Higuchi et al. (1969) have used polarized radiation to study the spectrum (Fig. 12.7) of an oriented film of a DNA–RNA hybrid in its undeuterated and deuterated forms. The band at 1225 cm^{-1} displayed perpendicular dichroism and the one at

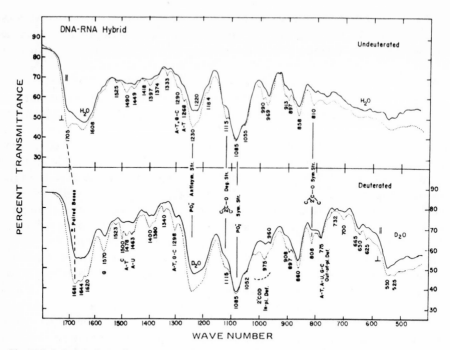

Fig. 12.7. Infrared absorption spectrum of (top) undeuterated and (bottom) deuterated oriented films of DNA–RNA hybrid in the 1800–400 cm^{-1} region at 92 % r.h.; (——) electric vector of the incident radiation parallel to the fiber axis; (————) electric vector of the incident radiation perpendicular to the fiber axis. Some of the absorption bands assignable to vibrations localized in the base residues are indicated by A, T, G, and/or C, which are respectively adenine, thymine, guanine, and cytosine. (Higuchi *et al.*, 1969.)

1085 cm^{-1} showed almost no dichroism. These workers found that the orientation of

the P group with respect to the helix axis in the hybrid structure is not entirely

the same as that in the double-helical sodium DNA at 75 % r.h. The PO_2^- orientation is not the same as that in the double-helical RNA either. Higuchi *et al.* (1969) explained the observed dichroism by considering that the PO_2^- group in the RNA moiety has nearly the same orientation as that in the double-helical RNA, and the PO_2^- group in the DNA moiety has nearly the same orientation as that in the double-helical DNA.

Aqueous Solutions of Nucleosides, Nucleotides, Polynucleotides, and DNA

Infrared spectra of heavy-water solutions have been used to determine the tautomeric forms of nucleosides, nucleotides, and polynucleotides and to study the interactions of polynucleotides in aqueous solution (Miles, 1956; 1958a, b; 1959; and references cited therein). Miles (1959) reported that polyinosinic acid exists in the keto form and polycytidylic acid probably in the amino form both before and after interaction between the two. Uridine and uridylic acid were found to be in the diketo form (Miles, 1956). Thymidine is mainly present in the diketo form but probably exists to some extent also in the enolic form in solution.

Quantitative Measurements

In order to make quantitative measurements, Miles (1958b) has determined integrated infrared absorption intensities for some nucleosides, nucleotides, and poly-nucleotides in D_2O solution (Fig. 12.8 and Table 12.2) by application of Ramsay's "method I" (Ramsay, 1952). When there is a well-resolved band, the application of Ramsay's method encounters no difficulty, but when overlapping bands occur, as in uridine and its derivatives, some uncertainty exists in determining the half-band width of a particular band. (The half-band width is a factor in Ramsay's equation and is defined as, $\Delta v_{1/2}$ = the width of the band in cm^{-1} at half-maximum intensity.) The equation as used by Miles to obtain A, the true integrated absorption intensity

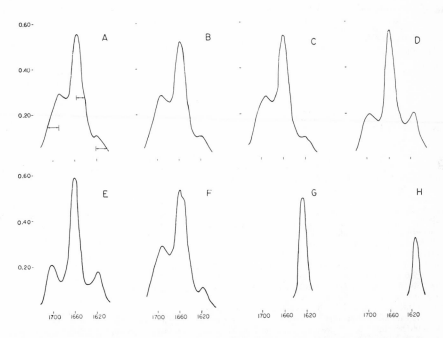

Fig. 12.8. Infrared spectra in D_2O, solvent compensated, 25μ path length. Abscissa is wave number in cm^{-1}, and ordinate is absorbance. (A) uridine ($0.225M$); the arrows represent the value $\Delta v_{1/2}/2$ used to obtain $\Delta v_{1/2}$ in Eq. (1). (B) deoxyuridine ($0.192M$). (C) 1-β-D-glucopyranosyluracil ($0.219M$). (D) 3-methyluridine ($0.208M$). (E) 1-β-D-gluco-pyranosyl-3-methyluracil ($0.225M$). (F) Na_2UMP ($0.195M$). (G) Na_2AMP ($0.195M$). (H) 6-dimethylamino-9-β-D-ribofuranosylpurine ($0.113M$). (Miles, 1958b.)

Table 12.2. Integrated Absorption Intensities of Infrared Spectra in D$_2$O Solution[a] (Miles, 1958)

Compound	Molar concentration[b]	ν_{max}, cm^{-1}	A^c	ν_{max}, cm^{-1}	A	ν_{max}, cm^{-1}	A
Uridine	0.225	1689	3.7	1656	4.3	1622	1.1
	0.075	1689	3.8	1655	4.3	1619	0.7
Deoxyuridine[d]	0.192	1692	3.7	1658	4.4	1620	1.0
	0.096	1691	3.6	1657	4.4	1617	0.9
1-β-D-Glucopyranosyl-uracil	0.219	1691	3.8	1659	3.9	1621	0.8
	0.109	1689	3.7	1657	3.7	1619	1.1
1-β-D-Glucopyranosyl-3-methyluracil	0.225	1703	1.4	1661	3.6	1618	1.2
	0.112	1703	1.5	1660	3.5	1619	1.5
3-Methyluridine	0.208	1695	2.0	1658	4.0	1615	1.5
	0.110	1693	2.1	1654	4.2	1613	1.6
Na$_2$UMP	0.195	1692	3.9	1658	4.7	1618	0.9
	0.088	1692	3.6	1658	4.4	1621	1.0
Na$_2$AMP	0.197					1626	3.0
	0.097					1626	2.8
Na$_2$UMP + Na$_2$AMP[e]	0.066	1694	3.4	1661	—	1628	3.7
	0.033	1692	3.5	1658	—	1626	3.6
6-Dimethylamino-9-β-D-ribofuranosylpurine	0.113					1613	2.9
	0.055					1616	2.7
Poly A[f]	0.084					1630	2.4
	0.062					1631	2.2
	0.042					1632	2.2
Poly U[f]	0.068	1695	3.7	1661	3.9	1623	1.0
Poly A + Poly U[g]	0.060	1693	3.2	1672	—	1635	1.9
	0.045	1695	3.5	1672	—	1634	2.1
	0.026	1695	3.3	1671	—	1632	1.9

[a]The spectra were usually run at gain settings of 7.5; response, 4; resolution, 5; and intensity, 0.3A. A speed of about 30 min/μ and a very slow pen response were used. The calculated spectral slit width varied from 4.1 cm^{-1} at minimum to 5.6 cm^{-1} at maximum, depending on the wavelength and value of $F(S)$ used for the calculation. This variation had no important effect on the value of A. The thickness of the matched CaF$_2$ cells was 25 μ. The 3 μ region was also measured to check on the amount of water present.
[b]Concentrations were determined by measurement of the ultraviolet spectra of aliquots and by use of absorptivities from the literature.
[c]The integrated intensity A is calculated from Eq. (1) and has the units 10^4 liters mole^{-1} cm^{-2}.
[d]This material was "thymine-free" grade from the California Foundation (now Calbiochem).
[e]At pH 7.6 and NaCl concentration 0.1M.
[f]The polymers were dialyzed extensively against distilled water and lyophilized at least twice from D$_2$O before measurement of the spectra.
[g]Stock solutions of the two polymers were prepared separately at about one-tenth the concentration to be used for the infrared spectra. The NaCl concentration was 0.01M, the sodium cacodylate concentration was 0.005M, and the pH was adjusted to 7.5. After the concentrations were determined by ultraviolet absorption, equivalent amounts of the polymers were mixed, and a 15 % decrease in intensity was observed in the solution, which was 0.0032M in each polymer and 0.01M in NaCl. The solution was allowed to stand 5 hr, then lyophilized, redissolved in 0.1 ml D$_2$O, and the pH adjusted to 7.6. The NaCl concentration was then about 0.18M and the sodium cacodylate concentration was about 0.09M. The 0.060M solution of poly A + poly U was made up with 0.1 ml of 0.1M NaCl and 0.1M sodium cacodylate instead of pure D$_2$O. A molar absorptivity of 15,400 was calculated from the values 9800 for poly A (Warner, 1957) and 9430 for poly U (Felsenfeld and Rich, 1957) as reduced by a 20 % hypochromic effect in the more concentrated salt solution. The concentration of an aliquot of the polymer solution was then determined using this value.

of a band, is

$$A = \frac{K}{Cl} \log\left(\frac{T_0}{T}\right)_{v_{max}} \times \Delta v_{1/2} \qquad (1)$$

where K is a constant taken from Table III in Ramsay (1952); C is the molar concentration, l is the path length in cm; and $\log(T_0/T)$ is the absorbance (or optical density). [Wexler (1967) has recently reviewed the topic of integrated intensities.]

Marked changes occurred in the infrared spectra when polyuridylic acid and polyadenylic acid were mixed, and the changes were considered to result from hydrogen-bonding interaction between them. These quantitative measurements supported the conclusion that infrared spectra of such mixtures provide experimental evidence that the uracil units are present in the keto form and that the adenine units are probably present in the amino form, as in the Watson–Crick DNA model. Earlier it had been demonstrated (Miles, 1956; 1958a) by means of infrared spectra of appropriate model compounds that the band at 1692 cm^{-1} is characteristic of the keto form of uridine and would be absent in the enol form. It had also been shown that the 1630 cm^{-1} band is probably indicative of the amino form of adenosine. If interaction of the polymers involved the enol form of uridine, there would be no band at 1695 cm^{-1} in the mixture.

The aqueous method has been used for the determination of: tautomeric structures (Miles, 1961; Miles et al., 1963); the demonstration of helix "strandedness" [i.e., the stoichiometry of interaction of polynucleotides to form two- or three-stranded helices (Miles and Frazier, 1964a)]; the demonstration of a strand disproportionation reaction (Miles and Frazier, 1964b), of a strand displacement reaction (Sigler et al., 1962), and the quantitative analysis of different helical structures (Miles and Frazier, 1964a,b) in complex mixtures. It has been important in most of these applications that there are comparatively large differences of the spectra of the components from one another and there are marked and characteristic changes that the spectra undergo upon interaction to form helical secondary structures.

Infrared observations of 5'-GMP-6-^{18}O established the assignment of the 1665 cm^{-1} band of 5'-GMP-6-^{16}O (in the unassociated nucleotide) as a coupled carbonyl vibration and suggested that in the helical structure formed at low temperature (above 1°C) the carbonyl group is a site of interaction, presumably a hydrogen bond (Howard and Miles, 1965). A solution of guanosine-6-^{18}O in basic D$_2$O ($pD \sim 13$) displayed a spectrum quite similar to that of guanosine-6-^{16}O in the region 1800–1500 cm^{-1}. This fact supported a conclusion that the anion has a major charge localization on O-6. The lack of a carbonyl band in the spectrum of the anion indicated that the C-6—O bond has little double bond character.

Double and Triple Helices

Miles (1960) has shown spectra of the three-stranded helices formed between polyadenylic acid (poly A) and polyuridylic acid (poly U) and between poly U and tetra A (tetraadenylic acid, pApApApA). For the related spectra, the assignment of bands, and the experimental methods used, some of Miles's previous papers (1956; 1958a,b; 1959) should be consulted. The infrared spectra of the double helix formed by interaction of poly A and poly U had shown a decrease in the intensity of the 1632 cm^{-1} band and shifts of this band and of the 1660 cm^{-1} band to higher frequencies. The 1632 cm^{-1} band was believed to be due to the amino form of adenosine, and the observed decrease in its intensity was probably due to the hydrogen bonding occurring during the interaction of the strands.

The most striking observation in the case of the triple helix is seen with the 1630 cm^{-1} band. As mentioned above, this band is decreased in intensity in the double helix. It has disappeared as a separate band in the case of triple-helical poly (A + 2U) and of tetra A + 2 poly U (Fig. 12.9).

Polynucleotide helices are known to dissociate reversibly (Lipsett *et al.*, 1960; Doty *et al.*, 1959) upon heating, the melting temperature depending upon the polymers used and upon the kind and concentration of salt. By using heated solutions, Miles (1960) has followed the thermal dissociation of these three-stranded helices (Fig. 12.9 and Table 12.3) by their infrared spectra. The hot solution of tetra A + 2 poly U

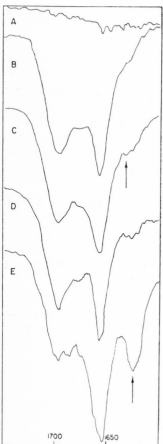

Fig. 12.9. Infrared absorption spectra in D_2O of polynucleotide solutions. Abscissa, frequency of absorption in cm^{-1}. Ordinate, % transmittance on arbitrary scale. Conditions indicated in footnote to Table 12.2. (A) D_2O *vs* D_2O to show compensation obtained. (B) Poly A + 2 poly U at 30°C. (C) Poly A + 2 poly U at about 60°C; the arrow at 1632 cm^{-1} indicates the slight increase in absorption. (D) Tetra A + 2 poly U at 30°C. (E) Tetra A + 2 poly U at ~60°C. The arrow indicates the large increase in absorption at 1625 cm^{-1}. (Miles, 1960.)

Table 12.3. Infrared Spectra in D_2O Solution[a] (Miles, 1960)

Material	Temperature, °C	ν_{max}, cm^{-1}
Poly (A + 2U)	30	1659, 1699
Poly (A + 2U)	~60	1632, 1657, 1698
Tetra A + 2 poly U	30	1654, 1697
Tetra A + 2 poly U	~60	1625, 1659, 1697

[a]The spectra were measured with a Beckman IR-7 spectrophotometer at high scale expansion, using matched CaF_2 cells of 0.025-mm path length. The instrument was purged with dry air. The high-polymer solutions were approximately $0.01M$ in repeating units of AU_2, $0.1M$ NaCl, $0.01M$ sodium cacodylate, and $0.01M$ $MgSO_4$; pH was 7.0. The tetra A + poly U solution was approximately $0.005M$ in repeating units of AU_2, $0.004M$ $MgCl_2$, and $0.01M$ sodium cacodylate; pH was 7.2. No NaCl was added. The high temperature measurements were made by warming the filled cells to about 60°C, measuring the spectra immediately, and then scanning repeatedly as the cells cooled to room temperature.

displays a well-developed band at 1625 cm^{-1} (Fig. 12.9, curve E), evidence of complete dissociation of the helix. With slow cooling of the solution a gradual decrease in intensity of this band occurs until the original spectrum (Fig. 12.9, curve D) is again shown at room temperature. This change in spectrum was much less in the case of poly (A + 2U) (Fig. 12.9, curve C), since the attainable temperature was not high enough to produce extensive dissociation. The above observation with the oligo-nucleotide constituted additional evidence in support of the three-stranded structure for tetra A + 2 poly U (Lipsett et al., 1960).

Miles and Frazier (1964a) have extended the earlier studies (see Miles, 1960) to the determination of the kinds and amounts of the different polymer species and complexes from quantitative infrared spectra in D_2O solution. They showed that spectra of poly (A + U), the double helix, are qualitatively and quantitatively quite different from the spectra of poly (A + 2U), the triple helix, under specific conditions (Fig. 12.10): Poly (A + U) has uridine bands at 1691 cm^{-1} and 1672 cm^{-1}. The adenosine band is at 1631 cm^{-1}. Poly (A + 2U), on the other hand, has uridine bands at 1696 cm^{-1}, ~1677 cm^{-1}, and a very strong band at 1657 cm^{-1}. The intensity of the adenosine band at ~1630 cm^{-1} decreases almost to zero.

Disproportionation Reaction of a Double Helix

Miles and Frazier (1964b) have employed quantitative melting curves measured by infrared spectroscopy and analysis of digitized spectra to demonstrate that the following disproportionation reaction takes place in D_2O solution under certain specified conditions:

$$2 \text{ poly (A + U)} \rightarrow \text{poly (A + 2U)} + \text{poly A} \qquad (2)$$

This reaction occurs when poly (A + U) is heated under appropriate conditions (e.g., $0.14N$ Na$^+$ + $0.02N$ Mg^{2+} or about 0.5 equivalent of Mg^{2+} per phosphate), and was

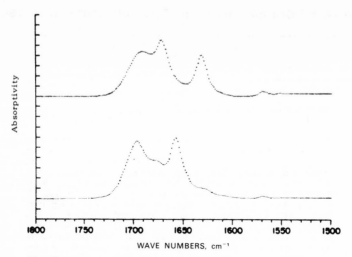

Fig. 12.10. Spectra of poly (A + U) (upper) and poly (A + 2U) (lower) under conditions of $0.14N$ Na^+ and $0.02N$ Mg^{2+} at pD 7.3 ± 0.2 in D_2O. (Miles and Frazier, 1964a.)

first detected by observing the infrared spectra of a 1 : 1 poly A–poly U mixture as a function of temperature. There is a rapid change of spectrum (see Fig. 12.11 for typical spectra) from that at 33°C to a new spectrum upon heating, which subsequently does not change appreciably over a wide temperature range. The change in the absorbance at $1657\ cm^{-1}$ is characteristic of the chemistry involved. When the absorbance at this frequency is plotted from spectra recorded during the melting, a clear biphasic curve is produced (Fig. 12.12, light circles), which suggests that two distinct processes have occurred during the heating. The plateau spectrum (e.g., the spectrum recorded at 56.9°C) is identical with a synthetic curve calculated by adding the spectra of poly (A + 2U) and poly A, and this finding demonstrates that the reaction involving strand disproportionation given above does occur as the first step of the biphasic curve. The second step is due to a simultaneous dissociation of the triple helix according to the following reaction:

$$poly\,(A + 2U)\ \rightarrow\ poly\,A + 2\,poly\,U \tag{3}$$

Dissociation of a Double Helix

In contrast with the results mentioned above, the spectrum of a 1 : 1 poly A–poly U mixture under different salt conditions (e.g., $0.14N$ Na^+, no Mg^{2+}) displays quite different changes with temperature (Fig. 12.13). A discrete $1657\ cm^{-1}$ band appears much later in this case and a continuous change of spectrum takes place all through the heating process rather than the production of a spectrum which remains fixed over a wide temperature range as in Fig. 12.11. The 1 : 1 poly A–poly U mixture with no Mg^{2+} present displays a monophasic melting curve (Fig. 12.12, dark circles). The process observed in Fig. 12.12 (dark circles) is essentially the simple dissociation of the double helix:

$$\text{poly}\,(A + U) \;\rightarrow\; \text{poly}\,A + \text{poly}\,U \tag{4}$$

Included in this paper (Miles and Frazier, 1964b) was a brief investigation of the dependence of the occurrence of reactions (2) or (4) upon Na^+ and Mg^{2+} concentration.

By selecting frequencies of well-resolved peaks definitely assignable to the separate polymers (e.g., the $1657\,cm^{-1}$ band of uridine and the $1625\,cm^{-1}$ band of adenosine) these workers were able to demonstrate through the coincidence of the melting curves (Fig. 12.14) that the spectral changes specifically reflect an interaction between poly A and poly U and not, for example, independent structural changes in the separate polymers.

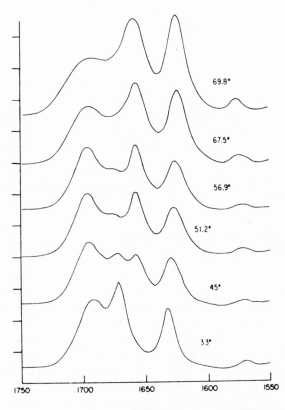

Fig. 12.11. Infrared spectra in D_2O solution of poly $(A + U)$, $0.045M$ in polymer phosphate, $0.14N\ Na^+$ and $0.02N\ Mg^{2+}$; 4.83-fold scale expansion; $55.6\,\mu$ matched CaF_2 cells. Ordinate is absorbance, the index marks being 0.2 unit apart as observed on the original spectra, uncorrected for scale expansion. (Miles and Frazier, 1964b.)

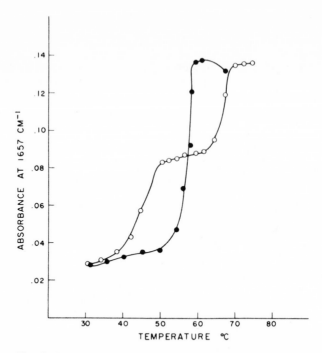

Fig. 12.12. Melting curves for poly $(A + U)$ in presence of Mg^{2+} (conditions of Fig. 12.11; light circles) and in absence of Mg^{2+} (conditions of Fig. 12.13; dark circles). In both cases the total polymer concentration (as phosphate) was $0.045M$, and the ordinate has been corrected for scale expansion to give true absorbance values. (Miles and Frazier, 1964b.)

A Double Helical Polynucleotide Stabilized by Three Interbase Hydrogen Bonds

Howard *et al.* (1966a) have prepared and investigated a new double-stranded polynucleotide helix stabilized by three interbase hydrogen bonds. In nucleic acid helices, it is well known that guanine–cytosine base pairs have higher stability than adenine–thymine or adenine–uracil base pairs (Marmur and Doty, 1959; Inman and Baldwin, 1964). The possibility of forming three hydrogen bonds between guanine and cytosine had been pointed out by Pauling and Corey (1956) and has often been assumed to be the explanation for the greater stability of this pair. The new polynucleotide is made up of a $1:1$ complex of 2-aminoadenine polyribonucleotide and uracil polyribonucleotide (r$\overline{\text{2-AA}}$:rU), where r$\overline{\text{2-AA}}$ is 2-aminoadenine polyribonucleotide (or poly $\overline{\text{2-AA}}$), and rU is uracil polyribonucleotide (or poly U). The bonding scheme is shown in Fig. 12.15.

The spectrum of poly $\overline{\text{2-AA}}$ (at pD 7.9 in $0.1N$ NaCl at 25°C) has ν_{max} at 1617 cm^{-1} and a slight shoulder at about 1600 cm^{-1}. At lower temperature the neutral polymer has bands with ν_{max} at 1626, 1621, and $\sim$1599 cm^{-1} which do not change over the range 1–10°C but show a marked temperature dependence of intensity above 10°C (Fig. 12.16). These spectral changes of the polymer presumably reflect a structural change rather than nonspecific temperature dependence since 2-aminoadenosine does not undergo such changes at low temperature. Poly $\overline{\text{2-AA}}$ also forms an ordered acid "self-structure" with ν_{max} at 1663, 1611, and 1698 cm^{-1}.

The spectrum (Fig. 12.17) of the 1:1 complex between poly U and poly $\overline{\text{2-AA}}$ is very similar to that of rA:rU (Miles and Frazier, 1964a,b), having uridine bands at 1692 and 1672 cm^{-1}. The purine ring vibrations undergo large reductions in intensity because of interaction and are clearly resolved with ν_{max} at 1621 and 1593 cm^{-1}. The parallel temperature dependence of purine (1610 cm^{-1}) and pyrimidine (1657 cm^{-1}) bands (Fig. 12.18) shows that the spectral changes result from a specific base-pairing interaction (Howard *et al.*, 1966b; Miles and Frazier, 1964a,b). The availability of the extra amino group on the purine derivative for hydrogen bonding between poly-2-aminoadenylic acid and polyuridylic acid produces a 25°C increase in T_m over that of rA:rU. (T_m is defined as the midpoint in the steep part of the melting curve. It is analogous to the melting point of a crystal.) The 25°C increase amounts to about half of that attributed to the third hydrogen bond in guanine–cytosine pairs. An estimate

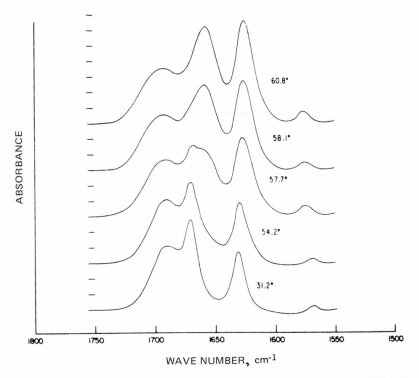

Fig. 12.13. Infrared spectra in D$_2$O solution of poly (A + U), 0.045M, 0.14N Na$^+$, and no Mg^{2+}, 4.50-fold scale expansion, 55.7 μ path length. The index marks are 0.1 unit apart, uncorrected for scale expansion. (Miles and Frazier, 1964b.)

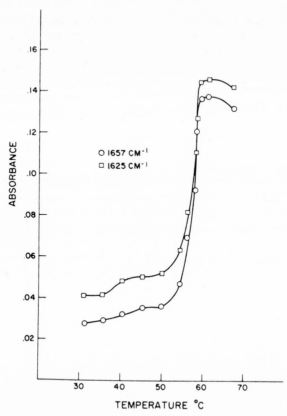

Fig. 12.14. Melting curve in D_2O of poly $(A + U)$ under conditions of Fig. 12.12 (dark circles). The 1657 cm^{-1} band is attributed to the uridine (primarily the C-4 carbonyl), and the 1625 cm^{-1} to the adenosine (ring vibration). The coincidence of the curves indicates that the interaction is specifically between the two different bases. (Miles and Frazier, 1964b.)

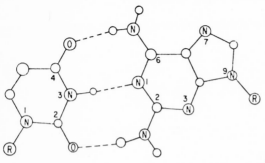

r2-AA: rU

$\vdash \quad 2\text{Å} \quad \dashv$

Fig. 12.15. The bonding scheme for a 1:1 complex of 2-aminoadenine polyribonucleotide and uracil polyribonucleotide. (Howard et al., 1966a)

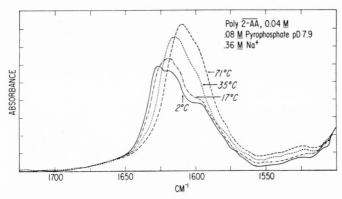

Fig. 12.16. Infrared spectra of poly $\overline{2\text{-AA}}$ in D_2O solution: 24.8 μ path length, 8.1-fold scale expansion. Ordinate index marks are 0.1 absorbance unit apart. The frequency maxima at the various temperatures are (in cm^{-1}): 2°C, 1626, $\sim$ 1621, $\sim$ 1600; 71°C, 1610. The infrared spectrum of the monomer, 2-aminoadenosine [see Howard and Miles (1965), Fig. 3] does not undergo these striking spectral changes but shows a very slight decrease of frequency of the maximum with increasing temperature (e.g., ν_{max} at 2°C = 1615 cm^{-1}; 35°C = 1613 cm^{-1}; and 51°C = 1613 cm^{-1}) with little change of band shape or intensity. It therefore seems probable that the spectral changes shown are the result of structural changes of the polymer. (Howard et al., 1966a.)

is made of approximately 600 calories for the additional hydrogen bond present in the rU :r$\overline{2\text{-AA}}$ complex.

Interaction of Polynucleotides with Nucleosides, Nucleotides, and Purines

Stoichiometric mixtures of polyuridylic acid with either adenosine or 2-amino-adenosine have spectra (Howard et al., 1966b) at low temperatures which closely resemble that of the three-stranded complex (A + 2U) (Figs. 12.19 and 12.20). With heating, the solutions show progressive spectral changes until a temperature is

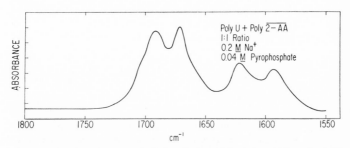

Fig. 12.17. Infrared spectra in D_2O solution of the 1:1 complex of poly U and poly $\overline{2\text{-AA}}$ at 30°C. Peak maxima at 1692, 1672, 1621, and 1593 cm^{-1}; 24.8 μ path length and 4.65-fold scale expansion. Ordinate index marks are 0.1 absorbance unit apart; pD 7.95. (Howard et al., 1966a.)

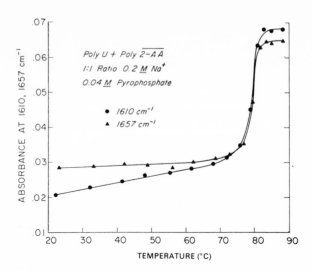

Fig. 12.18. Infrared melting curve of the 1:1 complex formed between poly U and poly $\overline{2\text{-AA}}$ in $0.21M$ Na$^+$ and $0.04M$ pyrophosphate, pD 7.95. Concentrations were: poly U, $0.02M$ and poly $\overline{2\text{-AA}}$, $0.02M$. The $1610\ \text{cm}^{-1}$ absorption corresponds to the maximum of a purine ring vibration at higher temperature (*cf.* Fig. 12.16). The interpretation of curves similar to this one has been discussed in Howard *et al.* (1966*b*) and Miles and Frazier (1964*a,b*). (Howard and Miles, 1966*a*.)

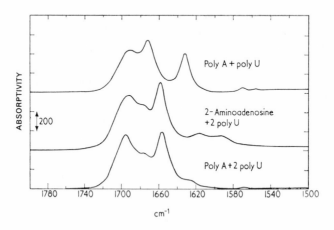

Fig. 12.19. Infrared spectra in D$_2$O solution. Top and bottom spectra from Miles and Frazier (1964*a*). The middle spectrum was measured at 2°C, poly U ($0.03M$), nucleoside ($0.015M$), and Na$^+$ ($0.15M$); pD 7. All the infrared spectra of this complex observed under varying concentrations of reactant and of sodium ion had the same appearance as the one shown here. (Howard *et al.*, 1966*b*.)

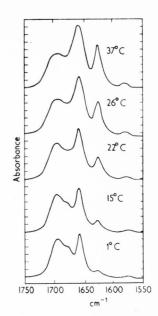

Fig. 12.20. Infrared spectrum of a 2:1 mixture of poly U (0.10*M*) and adenosine (0.05*M*) in 0.15*M* NaCl, 0.015*M* sodium cacodylate buffer (*p*D 7). Path length 23.5 *μ*, 3.0-fold scale expansion. Ordinate index marks are 0.1 absorbance unit apart in this figure. (Howard *et al.*, 1966*b*.)

reached at which no further changes take place, the spectra then being characteristic of the two noninteracted components. The melting point curves are sigmoid, indicative of cooperative melting of ordered structures (Fig. 12.21).

Interaction Between Polyuridylic Acid and Purine Compounds

Figure 12.22 shows a comparison between the melting curves of the complexes formed between polyuridylic acid and adenine, and between polyuridylic acid and 2-aminoadenine. The T_m of the latter complex (right part of figure) shows that it is more stable than that with adenine ($\Delta T_m \cong 20°C$), presumably because of the possibility of forming a third hydrogen bond in the case of the extra amino group (Howard *et al.*, 1966*b*).

Miles (1968) has presented data for the interaction between 5'-GMP and polycytidylic acid, and has discussed the use of 5'-GMP-6-^{18}O to assign a band at 1685 cm^{-1} to a guanine carbonyl in 5'-GMP-6-^{16}O, since the 1685 cm^{-1} band was the only one to decrease in frequency upon 6-^{18}O substitution of GMP. He also discussed the melting curves of a poly U-6-methylamino-9-methylpurine complex, and a poly UC (30/70)–adenosine complex. In this paper he has described cells, temperature controls, spectroscopic instrumentation and conditions, and experimental procedures used throughout most of the work discussed in this section on *Aqueous Solutions*.

Determination of Melting Point of DNA

Fritzsche (1966) has studied the helix-coil transition of calf thymus DNA in D_2O. A plot of the ratio of absorbances at 1662 and 1682 cm^{-1} during an examination of the thermal denaturation of the DNA gave a T_m of 89.6°C for a 10% DNA solution.

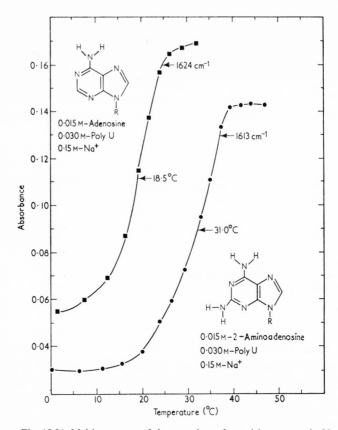

Fig. 12.21. Melting curves of the complexes formed between poly U
and adenosine (left) and 2-aminoadenosine (right), respectively. In this
and the following infrared melting curves, the points have been normalized for scale expansion to give true absorbance for the stated
concentration and path length. (Howard *et al.*, 1966*b*.)

By ultraviolet measurement this value was found to be quite similar, 87°C, for a
0.004 % solution.

Pairing of Bases in Various Nucleic Acids

Thomas (1969*a*) has described a method for determining the fractions of adenine–
uracil and guanine–cytosine base pairs of partly double-helical ribonucleic acids in
aqueous solution. The method is based upon the ability to differentiate between paired
and unpaired bases by examining infrared spectra of their D_2O solutions in the
1800–1500 cm^{-1} region. He applied the method to yeast ribosomal RNA and to
crystallizable fragments of ribosomal RNA. At 30°C, ribosomal RNA is $64 \pm 6\%$
paired (35 % G–C, 29 % A–U) and fragmented RNA is $66 \pm 6\%$ paired (34 % G–C,
32 % A–U). At 5°C the proportions of paired bases are increased to $69 \pm 7\%$ and
$76 \pm 7\%$, respectively, for ribosomal RNA and fragments. The method is applicable

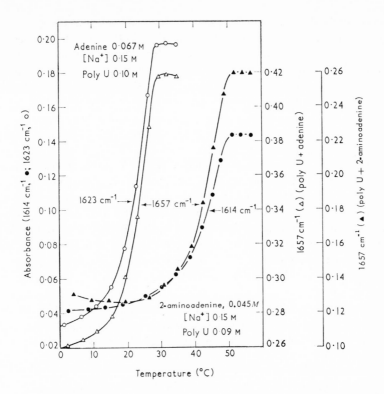

Fig. 12.22. Infrared melting curves of the complexes formed between poly U and adenine (left) and poly U and 2-aminoadenine (right) under the conditions shown; path length 23.8 μ. This figure shows the parallel melting behavior of the uridine band at 1657 cm^{-1} and bands corresponding to ν_{max} of the two purines. (Howard *et al.*, 1966*b*.)

to the determination of the fractions of Watson–Crick base pairs at a given temperature in any RNA containing the four common bases in known ratios [see also, Thomas and Spencer (1969)].

Tsuboi *et al.* (1969) have examined the spectra in the 1750-1450 cm^{-1} region of formyl–methionine transfer ribonucleic acid (tRNA) from *E. coli* and isoleucine-specific tRNA from *Torulopsis utilis* (torula yeast) in D_2O solutions containing 0.2M Na$^+$, 0.5M Na$^+$, or 0.005M Mg^{2+}. The spectrum of formyl-methionine tRNA at 33°C displayed seven absorption bands in this region, and the spectrum of isoleucine-specific tRNA showed nine bands. Using the positions and relative intensities of these bands, Tsuboi *et al.* estimated the amounts of the adenine, uracil, guanine, and cytosine residues in the free and hydrogen-bonded states. They found that a calculated spectrum of 2 A–U + 17 G–C + 13 A + 6 U + 7 G + 8 C was in agreement with the observed spectrum of the formyl-methionine tRNA. A calculated spectrum of 6 A–U + 14 G–C + 9 A + 4 U + 6 G + 6 C was in accord with the observed spectrum of isoleucine tRNA. When the solutions were heated up to 97°C, these workers observed pronounced changes in the positions and intensities of the bands, and explained them to be the results of the disruption of the A–U and G–C hydrogen bonds.

Tsuboi *et al.* gave a similar explanation for the "anticodon loop fragment" of the formyl-methionine tRNA (Fig. 12.23), which contains 19 nucleotides, and is the same fragment as the one reported by Clark *et al.* (1968). The spectrum of this tRNA fragment in D_2O containing $0.5M$ Na^+ indicated that at 50°C it has three G–C base pairs, at 63°C it has two such pairs, at 76°C it has one, and at 80°C none.

The absorption spectra of four tRNA's in the $1750–1400\ cm^{-1}$ region are shown in Fig. 12.24 (Tsuboi, 1969): alanine-specific tRNA from *Torulopsis utilis*, valine-specific tRNA (named valine-I tRNA) from *T. utilis*, methionine-specific tRNA from *E. coli*, and tyrosine-specific tRNA from *E. coli*. The curves should reflect the amounts of A–U and G–C base pairs and the amounts of unpaired A, U, G, and C in each tRNA. The $1690\ cm^{-1}$ band is due to the A–U and G–C base pairs. The strength of this band is much less in each of the tRNA's than in double helical RNA from rice dwarf virus (Tsuboi, 1969), showing that in the tRNA's some base residues are hydrogen-bonded and some are unpaired. Figure 12.24 shows the $1623\ cm^{-1}$ band in the alanine tRNA to be weaker than in the other three tRNA's. Thus, there are fewer unpaired adenine residues in that tRNA. On the basis of the "clover leaf" structure (Takemura *et al.*, 1969; Dube *et al.*, 1968; Goodman *et al.*, 1968), and the standard curves of Thomas (1969b), a calculated spectrum was plotted for each tRNA. These are also given in Fig. 12.24. They agree with the observed curves.

Fritzsche (1967b) has developed a rapid infrared method for the determination of the guanine + cytosine (G–C) content in DNA. He investigated a series of DNA samples from microorganisms, the base composition of which varied from 27 to 72 mole % G–C, and looked for differences among the infrared spectra of the DNA samples which could be related to different base composition. He used the following expressions derived from the Beer–Lambert law:

$$n_{G-C} = (D_{1505}/\epsilon_{1505})c_0 d \tag{5}$$

$$n_{A-T} = (D_{1485}/\epsilon_{1485})c_0 d \tag{6}$$

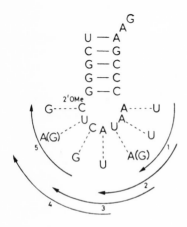

Fig. 12.23. The nucleotide sequence of the anticodon loop fragment is shown with letters representing nucleotides in order from the 5'-phosphate towards the diol end. The numbered arrows indicate possible triplet sequences antiparallel and complementary with the unpaired region. (Clark *et al.*, 1968.)

where n_{G-C} is the mole fraction of G and C bases in the DNA ($n_{G-C} \times 100 =$ G–C content in mole %), n_{A-T} is the mole fraction of A and T bases, D_{1505} and D_{1485} are the measured absorbances at the indicated wave numbers, and ϵ_{1505} and ϵ_{1485} are the corresponding molar absorptivities, related to the total concentration of respective base pairs c_0 and thickness d of the deuterated DNA film. (The assumption is made that in deuterated DNA at 1485 cm^{-1} only A and T bases absorb and at 1505 cm^{-1} only G and C bases absorb.)

For DNA the following expression is valid:

$$n_{G-C} + n_{A-T} = 1 \tag{7}$$

Combining Eqs. (5)–(7) gives

$$D_{1485}/D_{1505} = \epsilon_{1485}/\epsilon_{1505}\,(1/n_{G-C}) - (\epsilon_{1485}/\epsilon_{1505}) \tag{8}$$

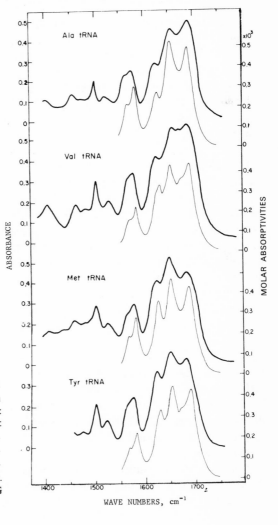

Fig. 12.24. Thick lines: Infrared absorption curves observed of tRNA films placed in air of 92% r.h. (D$_2$O vapor). Thin lines: Curves calculated on the assumption that: Ala. tRNA = 2A–U + 17G–C + 6A + 9U + 8G + 6C; Val. tRNA = 7A–U + 13G–C + 8A + 4U + 4G + 8C; Met. tRNA = 2A–U + 17G–C + 13A + 5U + 6G + 8C; and Tyr. tRNA = 7A–U + 17G–C + 11A + 6U + 4G + 10C. (Tsuboi, 1969.)

As can be seen from Eq. (8), the intensity ratio D_{1485}/D_{1505} depends linearly on $1/n_{G-C}$. A plot of D_{1485}/D_{1505} versus $1/n_{G-C}$ confirmed the expected linear correlation. Least squares treatment of the experimental data led to the following equation:

$$D_{1485}/D_{1505} = 0.429[(1/n_{G-C}) - 0.568] \qquad (9)$$

By using Eq. (9) as a calibration curve Fritzsche was able to determine the G–C content of deuterated DNA films by measuring the infrared spectra in the region 1400 to 1600 cm^{-1}. On the average, he found the difference between the n_{G-C} values determined by conventional methods and those calculated from the infrared absorption by Eq. (9) not to exceed the limits of ± 2 mole % G–C.

Hydrogen Bonding Between Bases in Chloroform

Hamlin et al. (1965) have shown that in concentrated solutions of either 9-ethyladenine or 1-cyclohexyluracil in CDCl$_3$, absorption bands in the infrared spectrum display hydrogen bonding of the adenine and uracil derivatives with themselves. In dilute solutions, very little hydrogen bonding occurs. However, when dilute solutions of 9-ethyladenine and 1-cyclohexyluracil are mixed, a series of bands appear, which indicate that these molecules hydrogen bond with each other much more strongly than with themselves. The purine and pyrimidine derivatives associate as 1:1 hydrogen-bonded pairs in solution.

Infrared studies have yielded information on the equilibrium constants for the pairing of adenine with uracil derivatives (Kyogoku et al., 1966, 1967a) and for the pairing of guanine and inosine with cytosine derivatives. The specificity (i.e., that A interacts only with U, and G only with C) of these interactions has also been demonstrated by infrared spectroscopic methods (Kyogoku et al., 1966). At the solute concentrations employed the infrared evidence shows that only dimers are formed and that the mode of interaction is hydrogen bonding. These data agree with those determined by NMR (Katz and Penman, 1966; Shoup et al., 1966) and by ultraviolet spectroscopy (Thomas and Kyogoku, 1967).

The association constants K for hydrogen-bonded dimer formation in chloroform solution by 9-ethyladenine and by 1-cyclohexyluracil, as well as for the formation of the 1:1 hydrogen-bonded complex of the two bases, were measured by infrared methods at 4 to 58°C. At 25°C, K_{AU} for the mixed dimer was 30 times larger than K_{AA} for the 9-ethyladenine dimer and 15 times larger than K_{UU} for the 1-cyclohexyluracil dimer. From the temperature dependence of association constants, ΔH^0 for association was calculated to be -4.3 ± 0.4 kcal/mole of dimer for cyclohexyluracil, -4.0 ± 0.8 kcal for ethyladenine, and -6.2 ± 0.6 kcal for the mixed dimer. The entropy differences ΔS^0 were -11.0 ± 1, -11.4 ± 2, and -11.8 ± 1.2 e.u., respectively. The method used for the calculations was based on work by Lord and Porro (1960).

A study of hypoxanthine derivatives and uracil derivatives has shown that no selective hydrogen bonding occurs between these molecules in solution. (Kyogoku et al., 1969).

Pitha *et al.* (1966) have studied hydrogen bonding in chloroform between pairs of derivatives of guanine and cytosine and their analogs. The infrared spectra of binary solutions of 2',3',5'-tri-*O*-acetyl-guanosine and 1-methylcytosine; of 2',3',5'-tri-*O*-acetyl-guanosine and compound LVI (an aza-cytosine derivative); and of 2',3',5'-tri-*O*-acetyl-guanosine and compound LVII (another aza-cytosine derivative), each exhibited

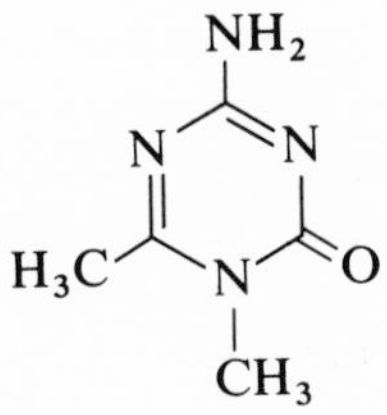

LVI

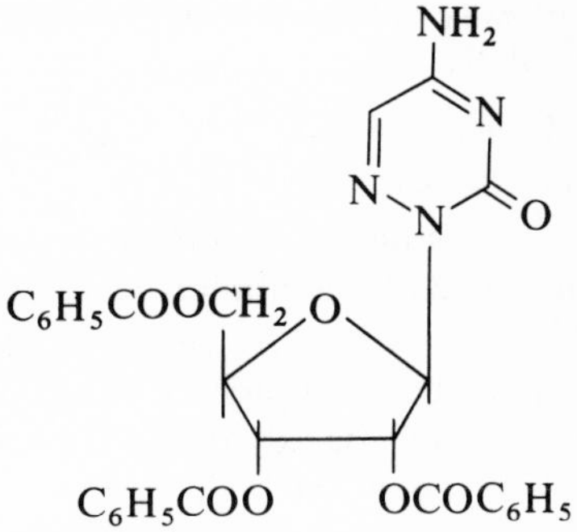

LVII

a prominent new absorption band at 3489, 3488, and 3572 cm^{-1}, respectively, a band attributed to the higher N—H stretching vibration of a hydrogen-bonded primary amine group. The introduction of the additional ring nitrogen in the pharmacologically active 5-azacytosine and 6-azacytosine derivatives does not change the complexing behavior. The positions of the N—H stretching bands in the spectra indicate that these complexes have a structure similar to that of the complexes formed between the corresponding pairs of bases in the naturally occurring DNA chain.

Hydrogen Bonding of Derivatives of Adenine with Barbiturates in Chloroform

In a continuation of their earlier work on purine–pyrimidine base pairing, Kyogoku *et al.* (1968) have shown that barbiturates form strong specific hydrogen bonds with derivatives of adenine. These interactions are more extensive than the hydrogen bonding between thymine or uracil derivatives and adenine derivatives. These workers feel that the highly specific interaction between barbiturates and adenine derivatives may help to explain the different biological effects of barbiturates. Among the compounds studied were: phenobarbital, secobarbital, thiopental, thiamylal, methohexital, mephobarbital, barbital, uracil, and thymine. Self-association constants and association constants for hydrogen bonding were given. For example, phenobarbital has a self-association constant of 8.1 liters/mole, whereas that of mephobarbital is 2.3 liters/mole. In the interactions between 9-ethyladenine and phenobarbital the association constant is 1,200 liters/mole. For 9-ethyladenine and mephobarbital it is only 200 liters/mole. The authors have discussed the roles of *pK* of the barbiturates, substitution by a sulfur atom for an oxygen atom, and methylation of a nitrogen atom, on the association constants of the hydrogen-bonded products.

Hydrogen Bonding of Adenine with Riboflavin Derivatives in Chloroform

Certain derivatives of riboflavin and adenine formed cyclic dimers (1 : 1 proportion) by means of hydrogen bonds through imino and C-2 carbonyl groups of the isoalloxazine ring and the amino group of the adenine moiety (Kyogoku and Yu, 1969). The association constants for this type of reaction ranged from 95 to 130 liters/mole.

Complexes of Nucleosides, Nucleotides, DNA, and Related Compounds with Metal Ions

Tu and Reinosa (1966) have studied complex formation of guanosine, guanosine 3'(2')-monophosphate, inosine, inosine 5'-monophosphate, theophylline, caffeine, uridine 3'(2')-monophosphate, and ribose with silver ion at neutral pH. Guanosine, guanosine 3'(2')-monophosphate, inosine, inosine 5'-monophosphate, and theophylline combined with silver ion in equimolar ratio. The infrared spectra indicated that the carbonyl stretching bands disappeared after complexing, except in the case of theophylline. Based on infrared and ultraviolet data, chemical titrations, conductometric titrations, and hydrogen ion release upon the addition of Ag^+, structures were postulated for the Ag complexes (LVIII), e.g., Ag-guanosine ($R_1 = NH_2$, $R_2 =$ ribosyl group), Ag-GMP ($R_1 = NH_2$, $R_2 =$ ribosyl 3'(2')phosphate), Ag-inosine ($R_1 = H$, $R_2 =$ ribosyl group), and Ag-IMP ($R_1 = H$, $R_2 =$ ribosyl 5'-phosphate). An Ag-theophylline complex (LIX) was also postulated.

LVIII   LIX

In an investigation of the complexing of copper(II) ion with guanosine, guanosine 5'-monophosphate, inosine, inosine 5'-monophosphate, and theophylline, Tu and Friederich (1968) found considerable diminution of the keto stretching vibration band in the region of 1720–1680 cm^{-1}. These data and conductometric and potentiometric titration experiments led to the proposal that the copper atom formed a complex involving enolic oxygen at C-6 and nitrogen at position 7 as in LX, where for guanosine, $R_1 = NH_2$ and $R_2 =$ ribosyl group; for GMP, $R_1 = NH_2$ and $R_2 =$ ribosyl 5'-phosphate; for inosine, $R_1 = H$ and $R_2 =$ ribosyl group; and for IMP, $R_1 = H$ and $R_2 =$ ribosyl 5'-phosphate. No complex formation was observed with similar methods for uridine, uridine 3'(2')-monophosphate, cytidine, cytidine 3'(2')-monophosphate, caffeine, and ribose.

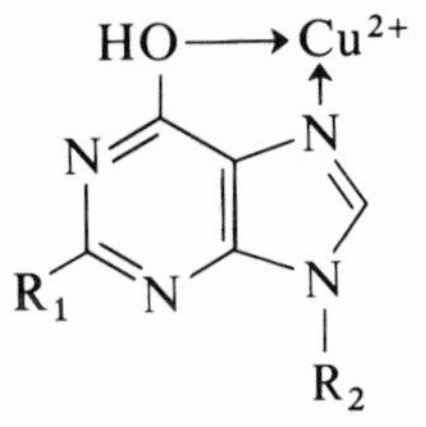

$$\text{LX}$$

In earlier investigations (Blout and Fields, 1950; Tu and Reinosa, 1966) it was found that keto groups in purine and pyrimidine bases give strong stretching vibration bands in the region of 1720–1680 cm^{-1}. The keto group in aromatic rings displays an intense stretching vibration band in the region of 1700–1680 cm^{-1}. Aromatic C=C and C=N stretching bands in purine rings usually appear in the 1630–1555 cm^{-1} region. After theophylline combined with copper ion an intense band at 1600 cm^{-1} was formed, which is within the region of C=C and C=N stretching bands of purine compounds. The proposed copper-theophylline complex is represented in LXI (Tu and Friederich, 1968).

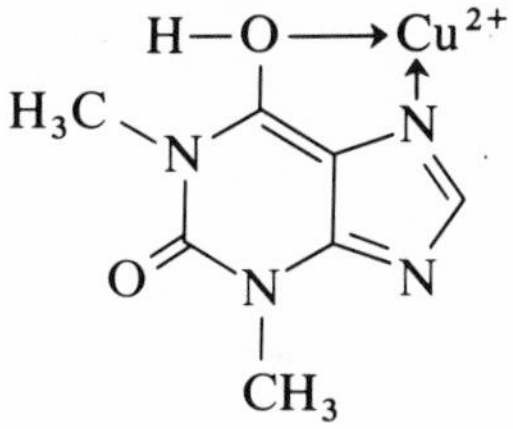

$$\text{LXI}$$

Hartman (1967) has studied the infrared spectra (1800–1500 cm^{-1}) of uridine, cytidine, guanosine monophosphate, and adenosine monophosphate—each complexed in D_2O solution with silver and gold ions, and in certain cases with mercury(II). These compounds all showed spectral changes from the cases in which no metallic ions were present. Salts which did *not* show complex formation with the RNA bases were: $CuCl_2$, $CoCl_2$, $MnCl_2$, $MgCl_2$, $FeCl_2$, and $NiCl_2$ for GMP; $CuCl_2$, $FeCl_2$, $CoCl_2$, $NiCl_2$, and $ZnCl_2$ for cytidine; $FeCl_2$, $CuCl_2$, $NiCl_2$, $CoCl_2$, and $ZnCl_2$ for uridine; $FeCl_2$, $CoCl_2$, and $NiCl_2$ for AMP. Hartman also observed a DNA–$HgCl_2$ complex, which had been observed earlier (Yamane and Davidson, 1961).

Fritzsche and Zimmer (1968) have studied the infrared spectra of thymidine, deoxyadenosine, deoxyguanosine, and deoxycytidine (dT, dA, dG, and dC, respectively) in D_2O solution in the absence and presence of copper sulfate. The spectra of dT and dA do not change upon addition of Cu(II), but new bands appear at 1517 and 1547 cm^{-1} in the spectrum of dC, and at 1590 cm^{-1} in the spectrum of dG after addition of Cu(II) (Figs. 12.25 and 12.26).

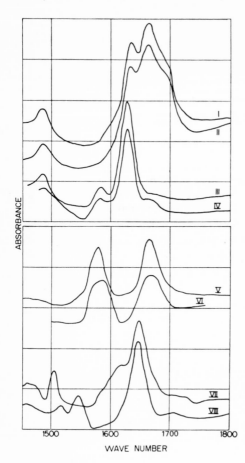

ABSORBANCE

1500   1600   1700   1800
WAVE NUMBER

Fig. 12.25. The infrared spectra of the deoxynucleo-sides with and without addition of copper(II). Deuterium oxide solutions contained 0.015M deoxy-nucleosides, in the case of the less soluble dG, only 0.0075M. Matched cells of fluorite and of a thickness of 0.005 cm were used. The curves I, III, V, and VII are the infrared spectra of the deoxynucleosides dT, dA, dG, and dC, respectively. The curves II, IV, VI, and VIII are the corresponding infrared spectra of the deoxynucleotides with addition of approximately 20 moles $CuSO_4$ per mole of deoxynucleosides. The distance between the index lines on the ordinate corresponds to a change of absorbance of 4%. For greater clarity the spectra are shifted against one another in the ordinate direction. The spectral slit-width was 15 cm^{-1}. (Fritzsche and Zimmer, 1968.)

The changes in the spectrum of DNA from *Streptomyces chrysomallus* (72 mole % guanine plus cytosine) after interaction with Cu(II) are very similar to the observed changes in the nucleoside spectra (Fig. 12.27). New bands at 1518 and 1551 cm^{-1} indicate that the cytosine bases of DNA interact with Cu(II), and the band at 1590 cm^{-1} indicates the interaction of guanine bases of DNA with Cu(II).

Raman Spectroscopic Examination of Base Pairing and of Nucleoside Interaction with Metal Ions

Aqueous solutions of complementary purine and pyrimidine base pairs, when examined by Raman spectroscopy (Lord and Thomas, 1967a), revealed no signs of specific base-pair interactions, although interactions between soluble nucleosides and heavy-metal ions have been detected, e.g., cytidine-$HgCl_2$ complex. Aqueous solutions of RNA derivatives have been examined (Lord and Thomas, 1967b). This work represents an application of growing importance, where further progress will be

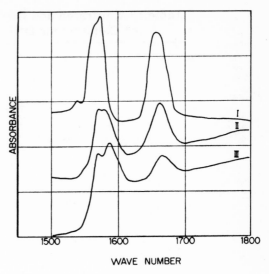

Fig. 12.26. The infrared spectra of dG with and without addition of copper(II) under higher spectral resolution. Deuterium oxide solution of dG (0.0075M) with addition of 0, 2, and 20 moles CuSO$_4$ per mole dG (Curves I, II, and III, respectively). The distance between the index lines on the ordinate corresponds to a change of absorbance of 8%. Spectral resolution of the grating monochromator was estimated to better than 5 cm^{-1}. (Fritzsche and Zimmer, 1968).

made possible by the advances in techniques for obtaining spectra from dilute solutions.

Raman Spectra of DNA

Tobin (1969) has presented Raman spectra of calf thymus DNA and of salmon testes DNA as solids, aqueous gels, D$_2$O gels, and alkaline aqueous gels. The spectra showed features which Tobin ascribed to the disordered helix B-helix transition. In the 1700–1600 cm^{-1} region the spectra of the two DNA's did not agree with each other or with the infrared spectra. Table 12.4 shows data from infrared and Raman spectra of calf thymus DNA.

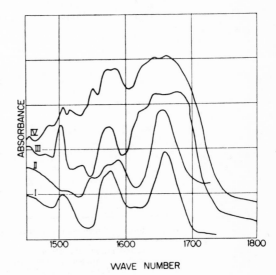

Fig. 12.27. Comparison between the infrared spectra of a 1:1 mixture of dG and dC and of DNA from *Streptomyces chrysomallus* (72 mole % GC) with and without addition of copper(II). Curve I: D$_2$O solution containing 0.0075M dG and 0.0075M dC. Curve II: the same solution, but a 20-fold excess of CuSO$_4$ is added. Curve III: film of DNA in a D$_2$O atmosphere. Curve IV: film of DNA–copper complex under the same conditions. The distance between the index lines on the ordinate corresponds to a change of absorbance of 4% for curves I and II and to a change of absorbance of 20% for curves III and IV. (Fritzsche and Zimmer, 1968.)

Table 12.4. Infrared and Raman Spectra, in cm^{-1}, of Calf Thymus DNA^a (Tobin, 1969)

D$_2$O Gel	Raman^d	Solid	Infrared^b
		113	—
		196	—
		336 wbr	—
		415 wbr	396 w^c
		—	450 w
496	496?	496 s	510 w
		537 w	530 w
		570 w	580 w
		597 w	—
		625 w	—
	676 pp	666 m	650 w
683		683 w	—
		706	698 vw
724	724 pp	731 s	728 w
744	748 pp	746 w	—
778	783 pp	786 vs	784 w
792	∼801	801 m	798 w
838	834 pp	—	819 w
	873	873	883 w
		—	918 w
		963 w	957 m
1017	1014	1013 m	1024 m
1056		1062 m	1069 s
1092	1091 pp	1102 s	1096 s
∼1164		1144 vw	—
1198	1171	1183 m	1188 vs
	1217	1208 m	—
1260	1255 p	1247 s	1241 s
1304	1302 p	1306 s	—
1349	1339 p	1335 s	1322 vs
1377	1373	1372 s	1362 m
1419	1419	1420 w	1410 vw
		1449 w	1440 vw
	1460	1465 w	—
1486	1490 pp	1490 s	1487 m
	1512	1512 w	1500 vw
1534	1533	1537 w	1533 w
1577	1576 p	1586 s	1580 vw
	∼1604?	1612 w	1607 m
		—	1649 s
1673	1662 p	1670 s	1690 s
		—	2800 w
	2900 pp	2900 m	2920 w
	2952 p	2952 s	—
		3050 –	
		3150 vw	3100 m
		—	3300 m

^{a}Key: br, broad; m, medium; p, polarized; pp, strongly polarized; s, strong; v, very; w, weak.
^{b}Sutherland and Tsuboi (1957; p. 452): for undeuterated, dry DNA.
^{c}The line positions below 690 cm^{-1} were eliminated from the infrared spectrum of 0 % r.h. sodium DNA given in Shimanouchi et al. (1964; p. 437).
^{d}In gel form.

Tobin stated that if the "assignment of the $834\,cm^{-1}$ band to the B-helix is correct, even bringing a gel of DNA to pH 12 does not suffice to break the double helix or to change the force constants very much." It should be noted that the spectra of the DNA gel at pH 12 were recorded after allowing the gel to stand overnight at this pH. According to White et al. (1968), ionization of enolic hydroxyl groups occurs near pH 12, preventing the keto–amino group hydrogen bonding of the double-stranded DNA.

The Conformation of Nucleosides in Solution

The conformations of a number of adenosine derivatives in dilute aqueous solution have been investigated with the aid of ORD, infrared, ultraviolet, and dissociation constant measurements by Klee and Mudd (1967). Ehrlich and Sutherland (1954) had shown that both the COO^- and the COOD groups of carboxylic acids give rise to easily measurable absorption bands in the infrared region which may be observed conveniently in D_2O solutions. Klee and Mudd measured the pK' of compounds by infrared spectroscopy in D_2O as a function of pD. The samples were twice equilibrated with D_2O and lyophilized and then dissolved in D_2O at a concentration of about 5%. Spectra were recorded at room temperature in the $1818–1250\,cm^{-1}$ region in CaF_2 cells (0.05 mm), with D_2O in a matched cell in the reference beam. The pD of the sample was adjusted with DCl or NaOD; pD values were estimated by adding 0.40 to the pH meter reading (Mikkelsen and Nielsen, 1960; Lumry et al., 1951). The pK' values for COO^- groups as estimated in D_2O were corrected to the corresponding value in H_2O by subtracting 0.5 (La Mer and Chittum, 1936; and Li et al., 1961).

The reliability of this method of determining pK' was tested with methionine, the pK' value of which is well known. The pK' of 2.2 ± 0.1 obtained by spectrophotometric titration is in good agreement with 2.28 (Emerson et al., 1931).

Preliminary studies (Klee and Mudd, 1967) with models of S-adenosylmethionine and S-adenosylhomocysteine led to the conclusion that there are a large number of conformations which can be assumed by both compounds. A number of conformations of S-adenosylmethionine result in the charged sulfonium center or the adenine ring being in close proximity to either the positive or the negative charge of the aminobutyric side chain. Measurements of pK' and infrared and ultraviolet spectra produced evidence against the predominant existence of conformations which involve such interactions. S-adenosylmethionine is given by LXII. Table 12.5 summarizes the infrared spectroscopic results.

The adenine frequencies of S-adenosylmethionine and S-adenosylhomocysteine in acid solution are identical with one another and with those for adenosine under similar conditions (Tsuboi et al., 1962). The absorption bands due to the COO^- and COOD of these compounds are also very close to one another. The small shift in the COO^- band may reflect a slight perturbation by a relatively far removed sulfonium center. Closely spaced groups lead to a larger change in frequency.

The pK' and infrared data serve to rule out a number of possible conformations of S-adenosylmethionine which might have been proposed (Klee and Mudd, 1967). For example, there are ample possibilities for hydrogen bond formation between the

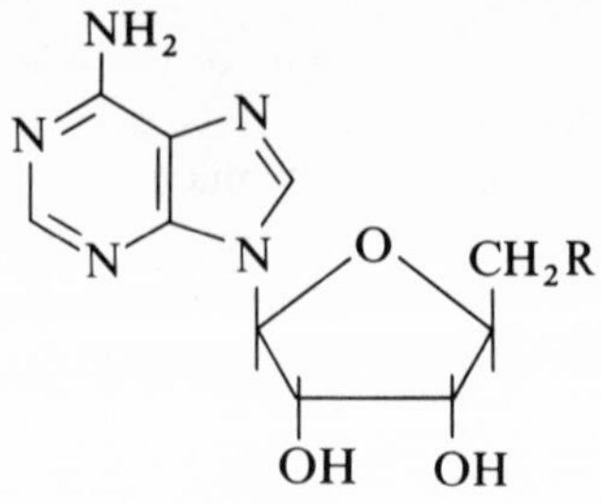

LXII

where R is: $\overset{+}{-S}-CH_2CH_2\overset{COO^-}{\underset{NH_3^+}{\diagup}}CH$

       |
      CH$_3$

For S-adenosylhomocysteine R is: $-SCH_2CH_2\overset{COO^-}{\underset{NH_3^+}{\diagup}}CH$

Table 12.5. Infrared Absorption Bands, in cm^{-1}, of S-adenosyl-methionine and of Some Related Compounds in D$_2$O Solution (Klee and Mudd, 1967)

Compound	COO$^-$	COOD	Adenine
$(-)S$-adenosyl-L-methionine	1626, 1408	1730	1667, 1582, 1511
S-adenosyl-DL-homocysteine	1618, 1412	1742	1667, 1582, 1506
DL-methionine	1616, 1404	1724	
S-methyl-L-methionine	1621, 1404	1727	
Dimethylacetothetin^a	1621, 1379	1718	
Dimethylpropiothetin^b	1592, 1397	1715	

^{a}Dimethylacetothetin is $(CH_3)_2S^+CH_2COO^-$.
^{b}Dimethylpropiothetin is $(CH_3)_2S^+CH_2CH_2COO^-$.

ribose oxygens and the carboxylate group of S-adenosylmethionine. Such hydrogen bonds would have the effect of bringing the negatively charged COO$^-$ group close to the positively charged sulfonium center. However, all structures involving strong coulombic interactions are considered to be highly unlikely because of the relatively normal pK' values found for S-adenosylmethionine. In addition, the infrared and ultraviolet spectral data do not indicate the existence of any strong side-chain–ring interactions.

The following picture of the conformation of S-adenosylmethionine evolves from the data: a reasonably, but not completely, extended molecule which is highly hydrated due to many charged and other polar groups. The adenine ring may be *anti* more often than *syn* (see adenosine LXIII) (Klee and Mudd, 1967), but not for as much of the time as is the adenine of S-adenosylhomocysteine, which molecule has more opportunities for intramolecular bonding due to the absence of a charged sulfur atom.

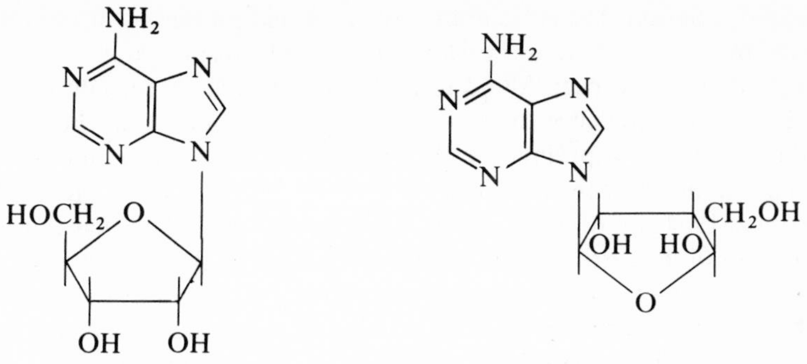

Syn

Anti

LXIII

Photochemistry of Nucleic Acids and Related Compounds

Varghese and Wang (1968) have observed an increase in absorbance at 320 nm in ultraviolet-irradiated DNA, have isolated a compound with the probable structure 6,4'-[pyrimidin-2'-one]-thymine, and have suggested the formation of a photoinduced cytosine–thymine adduct in DNA (Varghese and Wang, 1967; Wang and Varghese, 1967). They later characterized, by IR, UV, NMR, and mass spectrometry, the thymine–thymine adduct (LXIV), 5-hydroxy-6-4'-[5'-methylpyrimidin-2'-one]-dihydrothymine (analogous to the cytosine–thymine adduct) from ultraviolet-irradiated thymine and its dehydration product (LXV), 6-4'-[5'-methyl-pyrimidin-2'-one]-thymine. These characterizations demonstrate that the formation of such adducts between various pyrimidine bases is a common photoreaction and may be of considerable significance in the study of the photochemistry and photobiology of nucleic acids.

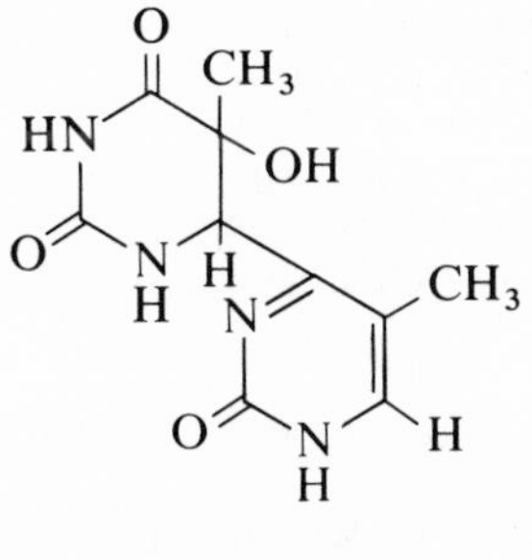

LXIV

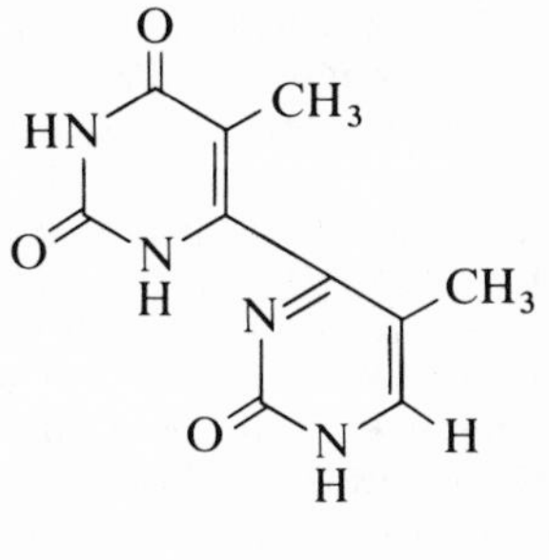

LXV

Smith and Aplin (1966) have irradiated a solution of uracil ($2.8 \times 10^{-3}M$) and cysteine·HCl ($10^{-2}M$) with ultraviolet light (2537 Å). The irradiation produced a heterodimer, 5-S-cysteine-6-hydrouracil, the structure of which was determined by ultraviolet, infrared, NMR, and mass spectroscopy. This compound may serve as a model for the mechanism by which DNA and protein are cross-linked *in vivo* by ultraviolet irradiation. The cross linking of DNA and protein by ultraviolet light was discovered by Smith (1962) and by Alexander and Moroson (1962).

Cysteine had also been found to add photochemically to poly U, poly C, and DNA, thus adding credence to Smith and Aplin's postulate that the formation of a heterodimer between a pyrimidine and a sulfur (or hydroxy) amino acid may constitute a mechanism for the photochemical cross-linking of DNA and protein *in vivo* (Smith, 1962, 1964; Smith *et al.*, 1966). Infrared data suggested that the COOH and amino groups of the cysteine residue were free, implying that the linkage to the uracil skeleton was through the sulfur bond.

Khattak and Wang (1969) have isolated two products from uracil irradiated with ultraviolet light in frozen aqueous solution. Evidence from NMR, UV, IR, and mass spectra suggest that one is a photopolymer, U_3, and the other is probably 6-4'-[pyrimidin-2'-one]-uracil.

REFERENCES

Alexander, P. and Moroson, H. *Nature* **194**, 882 (1962).
Angell, C. L. Ph.D. Thesis, Cambridge, 1955, cited in Tsuboi (1969).
Angell, C. L. *J. Chem. Soc.* **1961**, 504.
Arnott, S., Wilkins, M. H. F., Fuller, W., and Langridge, R. *J. Mol. Biol.* **27**, 535 (1967).
Bellocqu. A. M., Perchard, C., Novak, A., and Josien, M. L. *J. Chim. Phys.* **62**, 1334 (1965).
Blout, E. R. and Fields, M. *J. Am. Chem. Soc.* **72**, 479 (1950).
Borenfreund, E., Rosenkranz, H. S., and Bendich, A. *J. Mol. Biol.* **1**, 195 (1959).
Bradbury, E. M. and Crane-Robinson, C. in *The Nucleohistones* (J. Bonner and P. O. P. Ts'o, eds.) Holden–Day, San Francisco, 1964, p. 117.
Bradbury, E. M., Price, W. C., and Wilkinson, G. R. *J. Mol. Biol.* **3**, 301 (1961).
Bradbury, E. M., Price, W. C., and Wilkinson, G. R. *J. Mol. Biol.* **4**, 39 (1962a).
Bradbury, E. M., Price, W. C., Wilkinson, G. R., and Zubay, G. *J. Mol. Biol.* **4**, 50 (1962b).
Bradbury, E. M., Crane-Robinson, C., Rattle, H. W. E., and Stephens, R. M. in *Conformation of Biopolymers*, Vol. 2 (G. N. Ramachandran, ed.) Academic Press, New York, 1967, p. 583.
Clark, B. F. C., Dube, S. K., and Marcker, K. A. *Nature* **219**, 484 (1968).
Crick, F. H. C. and Watson, J. D. *Proc. Roy. Soc.* (*London*) **A223**, 80 (1954).
Doty, P., Boedtker, H., Fresco, J. R., Haselkorn, R., and Litt, M. *Proc. Natl. Acad. Sci. U.S.* **45**, 482 (1959).
Dube, S. K., Marcker, K. A., Clark, B. F. C., and Cory, S. *Nature* **218**, 232 (1968).
Ehrlich, G. and Sutherland, G. B. B. M. *J. Am. Chem. Soc.* **76**, 5268 (1954).
Emerson, O. H., Kirk, P. L., and Schmidt, C. L. A. *J. Biol. Chem.* **92**, 449 (1931).
Englander, S. W. *Biochemistry* **2**, 798 (1963).
Falk, M., Hartman, K. A., Jr., and Lord, R. C. *J. Am. Chem. Soc.* **85**, 387 (1963a).
Falk, M., Hartman, K. A., Jr., and Lord, R. C. *J. Am. Chem. Soc.* **85**, 391 (1963b).
Felsenfeld, G. and Rich, A. *Biochim. Biophys. Acta* **26**, 457 (1957).
Feughelman, M., Langridge, R., Seeds, W. E., Stokes, A. R., Wilson, H. R., Hooper, C. W., Wilkins, M. H. F., Barclay, R. K., and Hamilton, L. D. *Nature* **175**, 834 (1955).
Franklin, R. E. and Gosling, R. G. *Acta Cryst.* **6**, 673 (1953).
Fritzsche, H. *Biochim. Biophys. Acta* **119**, 645 (1966).
Fritzsche, H. *Biochim. Biophys. Acta* **149**, 173 (1967a).
Fritzsche, H. *Biopolymers* **5**, 863 (1967b).
Fritzsche, H. and Zimmer, C. *Eur. J. Biochem.* **5**, 42 (1968).
Fritzsche, H., Dieu, N. Q., and Te, N. T. unpublished results, cited in (Fritzsche, 1967a).
Goodman, H. M., Abelson, J., Landy, A., Brenner, S., and Smith, J. D. *Nature* **217**, 1019 (1968).
Hamlin, R. M., Jr., Lord, R. C., and Rich, A. *Science* **148**, 1734 (1965).

Hartman, K. A., Jr. *Biochim. Biophys. Acta* **138**, 192 (1967).

Higuchi, S., Tsuboi, M., and Iitaka, Y. *Biopolymers* **7**, 909 (1969).

Howard, F. B. and Miles, H. T. *J. Biol. Chem.* **240**, 801 (1965).

Howard, F. B., Frazier, J., and Miles, H. T. *J. Biol. Chem.* **241**, 4293 (1966*a*).

Howard, F. B., Frazier, J., Singer, M. F., and Miles, H. T. *J. Mol. Biol.* **16**, 415 (1966*b*).

Inman, R. B. and Baldwin, R. L. *J. Mol. Biol.* **8**, 452 (1964).

Katz, L. and Penman, S. *J. Mol. Biol.* **15**, 220 (1966).

Khattak, M. N. and Wang, S. Y. *Science* **163**, 1341 (1969).

Klee, W. A. and Mudd, S. H. *Biochemistry* **6**, 988 (1967).

Kyogoku, Y. Ph.D. Thesis, University of Tokyo, 1963, cited in Tsuboi (1969).

Kyogoku, Y. and Yu. B. S. *Bull Chem. Soc. Japan* **42**, 1387 (1969).

Kyogoku, Y., Tsuboi, M., Shimanouchi, T., and Watanabe, I. *J. Mol. Biol.* **3**, 741 (1961); *Nature* **189**, 120 (1961).

Kyogoku, Y., Lord, R. C., and Rich, A. *Science* **154**, 518 (1966).

Kyogoku, Y., Lord, R. C., and Rich, A. *J. Am. Chem. Soc.* **89**, 496 (1967*a*).

Kyogoku, Y., Higuchi, S., and Tsuboi, M. *Spectrochim. Acta* **23A**, 969 (1967*b*).

Kyogoku, Y., Lord, R. C., and Rich, A. *Nature* **218**, 69 (1968).

Kyogoku, Y., Lord, R. C., and Rich, A. *Biochim. Biophys. Acta* **179**, 10 (1969).

La Mer, V. K. and Chittum, J. P. *J. Am. Chem. Soc.* **58**, 1642 (1936).

Li, N. C., Tang, P., and Mathur, R. *J. Phys. Chem.* **65**, 1074 (1961).

Lipsett, M. N., Heppel, L. A., and Bradley, D. F. *Biochim. Biophys. Acta* **41**, 175 (1960).

Lord, R. C. and Porro, T. J. *Z. Elektrochem.* **64**, 672 (1960).

Lord, R. C. and Thomas, G. J., Jr. *Biochim. Biophys. Acta* **142**, 1 (1967*a*).

Lord, R. C. and Thomas, G. J., Jr. *Spectrochim. Acta* **23A**, 2551 (1967*b*).

Lumry, R., Smith, E. L., and Glantz, R. R. *J. Am. Chem. Soc.* **73**, 4330 (1951).

Mahler, H. R. and Cordes, E. H. *Biological Chemistry*, Harper and Row, New York, 1966.

Marmur, J. and Doty, P. *Nature* **183**, 1427 (1959).

Mikkelsen, K. and Nielsen, S. O. *J. Phys. Chem.* **64**, 632 (1960).

Miles, H. T. *Biochim. Biophys. Acta* **22**, 247 (1956).

Miles, H. T. *Biochim. Biophys. Acta* **27**, 46 (1958*a*).

Miles, H. T. *Biochim. Biophys. Acta* **30**, 324 (1958*b*).

Miles, H. T. *Biochim. Biophys. Acta* **35**, 274 (1959).

Miles, H. T. *Biochim. Biophys. Acta* **45**, 196 (1960).

Miles, H. T. *Proc. Natl. Acad. Sci. U.S.* **47**, 791 (1961).

Miles, H. T. *Proc. Natl. Acad. Sci. U.S.* **51**, 1104 (1964).

Miles, H. T. in *Methods in Enzymology, Vol. XII, Nucleic Acids, Part B* (L. Grossman and K. Moldave, eds.) Academic Press, New York, 1968, p. 256.

Miles, H. T. and Frazier, J. *Biochem. Biophys. Res. Commun.* **14**, 21 (1964*a*).

Miles, H. T. and Frazier, J. *Biochem. Biophys. Res. Commun.* **14**, 129 (1964*b*).

Miles, H. T., Howard, F. B., and Frazier, J. *Science* **142**, 1458 (1963).

Miyazawa, T., Shimanouchi, T., and Mizushima, S. *J. Chem. Phys.* **24**, 408 (1956).

Miyazawa, T. in *Polyamino Acids, Polypeptides, and Proteins* (M. A. Stahmann, ed.) University of Wisconsin Press, 1962, p. 201.

Osterman, L. A., Andler, V. V., Bibilaschvili, R., Savochkina, L. P., and Varshavskii, Ya. M. *Biokhimiya* **31**, 398 (1966); *CA* **65**, 951*b* (1966).

Párkányi, C. and Šorm, F. *Coll. Czech. Chem. Commun.* **28**, 2491 (1963).

Pauling, L. and Corey, R. B. *Proc. Natl. Acad. Sci. U.S.* **39**, 84 (1953).

Pauling, L. and Corey, R. B. *Arch. Biochem. Biophys.* **65**, 164 (1956).

Pitha, J., Jones, R. N., and Pithova, P. *Can. J. Chem.* **44**, 1045 (1966).

Printz, M. P. and von Hippel, P. H. *Proc. Natl. Acad. Sci. U.S.* **53**, 363 (1965).

Printz, M. P. and von Hippel, P. H. *Biochemistry* **7**, 3194 (1968).

Ramsay, D. A. *J. Am. Chem. Soc.* **74**, 72 (1952).

Sato, T., Kyogoku, Y., Higuchi, S., Mitsui, Y., Iitaka, Y., Tsuboi, M., and Miura, K. *J. Mol. Biol.* **16**, 180 (1966).

Shimanouchi, T., Tsuboi, M., and Kyogoku, Y. in *Advances in Chemical Physics, Vol. VII, The Structure and Properties of Biomolecules and Biological Systems* (J. Duchesne, ed.) Interscience, New York, 1964, pp. 435, 437.

Shoup, R. R., Miles, H. T., and Becker, E. D. *Biochem. Biophys. Res. Commun.* **23**, 194 (1966).

Sigler, P. B., Davies, D. R., and Miles, H. T. *J. Mol. Biol.* **5**, 709 (1962).

Smith, K. C. *Biochem. Biophys. Res. Commun.* **8**, 157 (1962).

Smith, K. C. *Photochem. Photobiol.* **3**, 415 (1964).

Smith, K. C. and Aplin, R. T. *Biochemistry* **5**, 2125 (1966).

Smith, K. C., Hodgkins, B.. and O'Leary. M. E. *Biochim. Biophys. Acta* **114**, 1 (1966).

Spencer, H. M. in *International Critical Tables, Vol. I*, McGraw-Hill, New York, 1926, p. 67.

Sutherland, G. B. B. M. and Tsuboi, M. *Proc. Roy. Soc. (London)* **A239**, 446 (1957).

Takemura, S., Mizutani, T., and Miyazaki, M. *J. Biochem. (Tokyo)*, cited in Tsuboi (1969).

Thomas, G. J., Jr. *Abstracts of the Third International Biophysics Congress of the I.U.P.A.B.*, Cambridge, Mass., Aug.–Sept. 1969*a*, p. 171.

Thomas, G. J., Jr., *Biopolymers* **7**, 325 (1969*b*).

Thomas, G. J., Jr. and Kyogoku, Y. *J. Am. Chem. Soc.* **89**, 4170 (1967).

Thomas, G. J., Jr. and Spencer, M. *Biochim. Biophys. Acta* **179**, 360 (1969).

Tobin, M. C. *Spectrochim. Acta* **25A**, 1855 (1969).

Tsuboi, M. *J. Am. Chem. Soc.* **79**, 1351 (1957).

Tsuboi, M. *J. Polymer Sci. Pt. C* No. 7, 125 (1964).

Tsuboi, M. *Appl. Spectrosc. Rev.* **3**, 45 (1969).

Tsuboi, M. and Kyogoku, Y. in *Synthetic Procedures in Nucleic Acid Chemistry, Vol. II* (W. W. Zorbach and R. S. Tipson, eds.) Interscience, New York, 1969.

Tsuboi, M. and Shuto, K. *Chem. Pharm. Bull. (Tokyo)* **14**, 784 (1966).

Tsuboi, M., Kyogoku, Y., and Shimanouchi, T. *Biochim. Biophys. Acta* **55**, 1 (1962).

Tsuboi, M., Matsuo, K., Shimanouchi, T., and Kyogoku, Y. *Spectrochim. Acta* **19**, 1617 (1963).

Tsuboi, M., Shuto, K., and Higuchi, S. *Bull. Chem. Soc. Japan* **41**, 1821 (1968).

Tsuboi, M., Morikawa, K., Higuchi, S., Kyogoku, Y., Nishimura, S., Seno, T., and Takemura, S. *Abstracts of the Third International Biophysics Congress of the I.U.P.A.B.*, Cambridge, Mass., Aug.–Sept. 1969, p. 19.

Tu, A. T. and Friederich, C. G. *Biochemistry* **7**, 4367 (1968).

Tu, A. T. and Reinosa, J. A. *Biochemistry* **5**, 3375 (1966).

Varghese, A. J. and Wang, S. Y. *Science* **156**, 955 (1967).

Varghese, A. J. and Wang, S. Y. *Science* **160**, 186 (1968).

Wang, S. Y. and Varghese, A. J. *Biochem. Biophys. Res. Commun.* **29**, 543 (1967).

Warner, R. C. *J. Biol. Chem.* **229**, 711 (1957).

Watson, J. D. and Crick, F. H. C. *Nature* **171**, 737 (1953).

Wexler, A. S. *Appl. Spectrosc. Rev.* **1**, 29 (1967).

White, A., Handler, P., and Smith, E. L. *Principles of Biochemistry, 4th Ed.*, McGraw-Hill, New York, 1968.

Wilkins, M. H. F. and Arnott, S. *J. Mol. Biol.* **11**, 391 (1965).

Yamane, T. and Davidson, N. *J. Am. Chem. Soc.* **83**, 2599 (1961).

Chapter 13

STEROIDS

The applications of infrared spectroscopy to steroid identification by Herget and Shorr (1941), Furchgott *et al.* (1946), and Dobriner *et al.* (1948) were important advances in this field. Other early work on the infrared absorption of steroids has been reviewed by Jones and Dobriner (1949), Cole (1954; 1956), Rosenkrantz (1955), Jones and Herling (1954), Page (1957), and Jones and Sandorfy (1956). A collection of infrared charts of a large number of steroids and steroidal sapogenins has been published by Dobriner *et al.* (1953) and a second volume by Roberts *et al.* (1958). Studies concerning the steroidal sapogenins have been summarized by Kasturi (1963).

The work by Jones and Dobriner (1949) contains applications of infrared spectrometry to metabolism. A recent textbook on the metabolism of steroid hormones (Dorfman and Ungar, 1965) should be consulted for structures, nomenclature, and for the multifarious pathways in steroid metabolism, and for a discussion of conformational analysis of steroids. A chapter applying conformational analysis to steroids has been published by Eliel *et al.* (1965). A recently published handbook (Sober, 1968) shows the structures of 108 steroids and gives references to infrared data for many of them. The atlases by Neudert and Röpke (1965), Dobriner *et al.* (1953), and Roberts *et al.* (1958), are also very useful. Caspi and Scrimshaw (1967) have given some of the elementary aspects of the infrared spectroscopy of steroids.

Infrared spectroscopy has been applied in identifying and determining the structures of steroids isolated from natural sources, for example, testis, ovary, placenta, adrenal cortex, bile, and blood. The technique has been used to identify compounds along the biosynthetic routes of, for example, cholesterol, pregnenolone, androgens, adrenal hormones, aldosterone, and estrogens.

Steroid Nomenclature

The steroids contain the perhydrocyclopentanophenanthrene nucleus. Among these compounds are such important substances as androgens, estrogens, progesterone, adrenocortical hormones, cholesterol, ergosterol, bile acids, cardiac glycosides, sapogenins, and various metabolites. Figure 13.1 presents the structure of testosterone to illustrate the basic common nucleus. Generally, the steroid structures are written in a shorthand method, which does not include all the carbons and hydrogens, as shown in Figs. 13.1 through 13.3. Figure 13.1 also shows the numbering system and

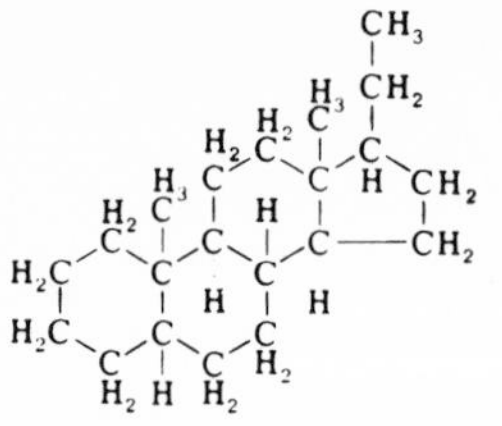

Fig. 13.1. Illustrative steroid structures: (I) 5β-pregnane (showing all carbon and hydrogen atoms), (II) 5β-pregnane (showing conventional representation), and (III) testosterone. (Dorfman and Ungar, 1965.)

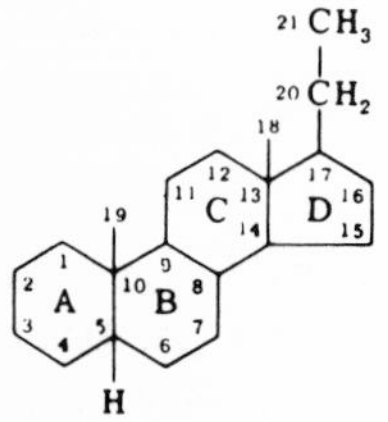

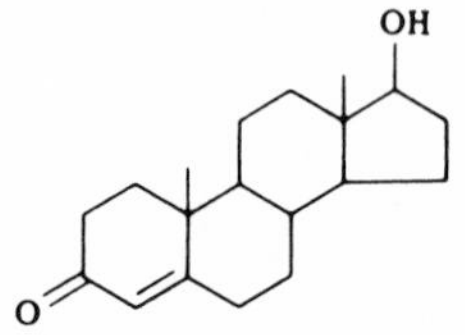

the way of designating the rings. The methyl groups, containing carbon atoms 18 and 19, are called angular methyl groups and are indicated by solid straight lines between rings C and D, and A and B, respectively.

The androgenic, estrogenic, adrenocortical, and progestational series of steroids all have structures that can be thought of as deriving from the eight basic hydrocarbons shown in Fig. 13.2. When the steroid molecule is shown in the usual way, the angular methyl groups lie above the plane of the page. The entire nucleus of the steroid molecule, a relatively flat structure (except for the angular methyl groups), is thought of as lying in the plane of the page. The methyl groups are connected to the nucleus by solid lines indicating a β-stereochemical configuration. The angular methyl groups act as the reference groups for assignment of stereochemical configuration. Whenever a solid line is indicated for a substituent group (where stereoisomerism is possible), that group is above the plane of the molecule, on the same side of the molecule as the two angular methyls, and has a β-configuration. A dotted line, on the other hand, indicates that the group is below the plane of the molecule, on the opposite side with respect to the angular methyls, and has an α-configuration. Two groups, when lying on the same side of the molecule, are called *cis* to each other, and when on opposite sides are said to be *trans* to each other.

The structures for 5α- and 5β-gonane, which contain no angular methyl groups, are given in Fig. 13.2. In the 5α hydrocarbon, the hydrogen atom projects behind or below the plane of the molecule, while in the 5β compound the hydrogen atom at carbon 5 projects before or above the plane of the molecule. The estrane series (5α and 5β) has only one angular methyl group (at C-18) and includes among its naturally occurring members, 19-norandrost-4-ene-3,17-dione and unsaturated derivatives of rings A and B.

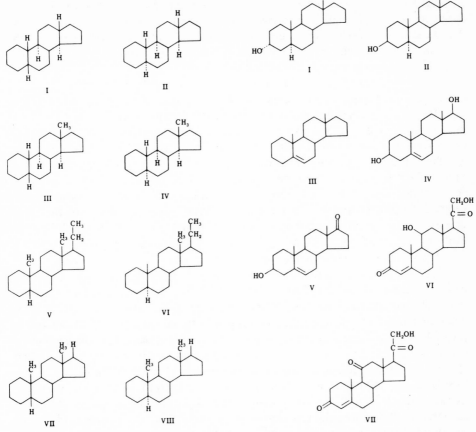

Fig. 13.2. Basic hydrocarbons: (I) 5β-gonane, (II) 5α-gonane, (III) 5β-estrane, (IV) 5α-estrane, (V) 5β-pregnane, (VI) 5α-pregnane, (VII) 5β-androstane, and (VIII) 5α-androstane. (Dorfman and Ungar, 1965.)

Fig. 13.3. Illustration of nomenclature rules: (I) 5α-androstan-3α-ol, (II) 5α-androstan-3β-ol, (III) androst-5-ene, (IV) androst-5-ene-3β,17β-diol, (V) 3β-hydroxyandrost-5-en-17-one (dehydroepiandrosterone), (VI) 11β-21-dihydroxypregn-4-ene-3,20-dione (corticosterone), and (VII) 21-hydroxypregn-4-ene-3,11,20-trione (11-dehydrocorticosterone). (Dorfman and Ungar, 1965.)

Two angular methyl groups are present in 5β-androstane and 5α-androstane (carbons 18 and 19) and the only difference is the spatial configuration of the hydrogen at carbon 5. The 5α-configuration of androstane was called "allo" in the older literature. The 5β-configuration of androstane is sometimes called the "normal" form. The same relationship holds for 5α-pregnane and 5β-pregnane with regard to spatial configuration of the hydrogen at carbon 5.

Comparison of 5α-pregnane and 5α-androstane shows that these compounds differ by an ethyl side chain on the former at carbon 17. The β-compounds,

5β-pregnane and 5β-androstane, are similarly related. The side chains in both 5α-pregnane and 5β-pregnane are connected to carbon 17 by a solid line, which shows that the side chain is *cis* to the angular methyl groups and lies above the plane of the molecule. Such a configuration is labeled β, while side chains *trans* to the angular methyls are called α.

Carbon 3 is a position where the steroids frequently show differences in spatial configurations. Figure 13.3 presents this type of isomerism. In 5α-androstan-3α-ol, the OH group at carbon 3 is *trans* to the methyl group on carbon 10. This *trans* position is labeled α. In 5α-androstan-3β-ol, the OH group lies *cis* to the methyl on carbon 10, and this position is labeled β. Carbon 11 frequently has an OH group attached. In vertebrates, the OH at carbon 11 in all naturally occurring steroids is β, that is, the OH group is *cis* to the angular methyls.

Nomenclature rules were adopted by Commissions on the Nomenclature of Organic Chemistry and the Nomenclature of Biological Chemistry of the International Union of Pure and Applied Chemistry in 1957. They were published by Butterworth's Scientific Publications in 1958 as *Nomenclature of Organic Chemistry*.

Three types of steroid names are commonly used. The trivial name is simple but does not reveal the nature, the position, or the spatial configuration of substituents on the steroid nucleus. The second kind of name starts with a definite common name of a compound, and denotes the substance to be named by changes in the common name. For example, corticosterone is the trivial name for 11β,21-dihydroxy-pregn-4-ene-3,20-dione and the name commonly used, 11-dehydrocorticosterone (or dehydrocorticosterone), tells that the derived compound is the same as corticosterone except that two hydrogens are now missing at carbon 11 (one is lost from the OH group and one from carbon 11, resulting in a ketone group). The third kind of name defines exactly the compound at hand by emphasizing the parent hydrocarbon and designating the changes in the parent substances by the kind and the position of the change, and the spatial configuration.

The presence of a double bond in the steroid nucleus is indicated by changing a name, for example, from androstane to androstene. In positioning double bonds, for example, "-3" means that the double bond lies between carbons 3 and 4, the bond being designated by the lower of the consecutive numbers. When the bond is found between carbons 9 and 11, the numbering "-9(11)" is used. When more than one double bond is present, say between carbons 3 and 4, and 5 and 6, "-3,5" is used. The following method names compounds with a double bond: The first part of the hydrocarbon name, the designation of the double bond(s) position, followed by the suffix "ene." See, for example, androst-5-ene (Fig. 13.3). A common, older designation uses the symbol Δ, e.g., Δ^5-androstene.

An alcohol or OH group is denoted either as the suffix "ol" preceded by the number of the carbon atom to which it is connected or by the term hydroxy preceding the hydrocarbon name. An example of a steroid having one double bond and one OH group would be androst-5-ene-3β-ol (Fig. 13.3). The suffix "one" denotes a ketone group. The term keto or oxo is used when it precedes the hydrocarbon name. Figure 13.3. gives other examples. Table 13.1 shows prefixes and suffixes.

Table 13.1 Nomenclature^a (Klyne, 1957)

Chemical group	Prefix	Suffix
Double bond	—	ene
Triple bond	—	yne
Acetate	Acetoxy	yl acetate
Hydroxyl	Hydroxy	ol
Benzoate	Benzoyloxy	yl benzoate
Carbonyl	Oxo (or keto)	one
Carboxylic acid	Carboxy	oic acid
Carboxylic ester	Methoxycarbonyl (or carbomethoxy)	methyloate
Epoxide	Epoxy	—
Amine	Amino	amine

^{a}Dorfman and Ungar (1965).

The prefix "nor" denotes shortening of the side chain or the elimination of a methyl group. Illustrations of norsteroids are 21-nor-5α-pregnane (LXVI) and 18-nor-5α-pregnane (LXVII). The term "nor" is also used when a ring is decreased,

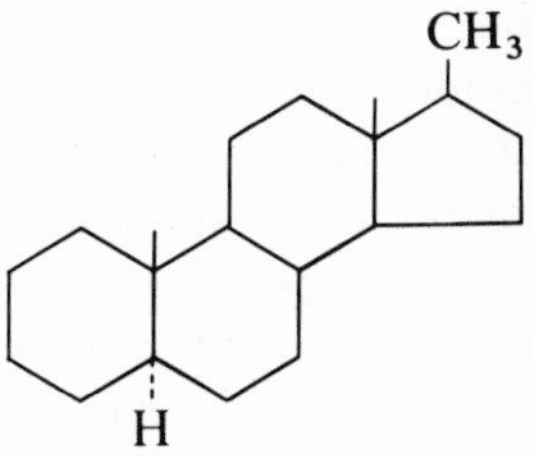

LXVI

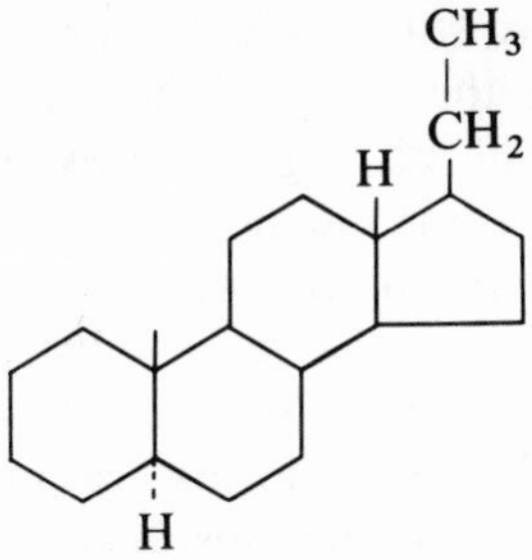

LXVII

as in A-nortestosterone (LXVIII). The prefix "homo" is used when a ring is increased, as in D-homotestosterone (LXIX). The term "des" followed by a ring letter indicates

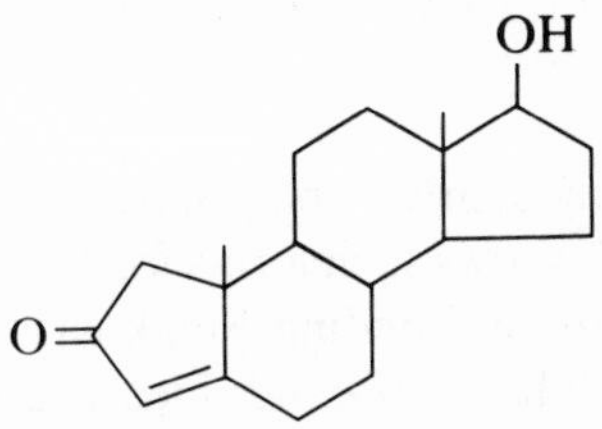

LXVIII

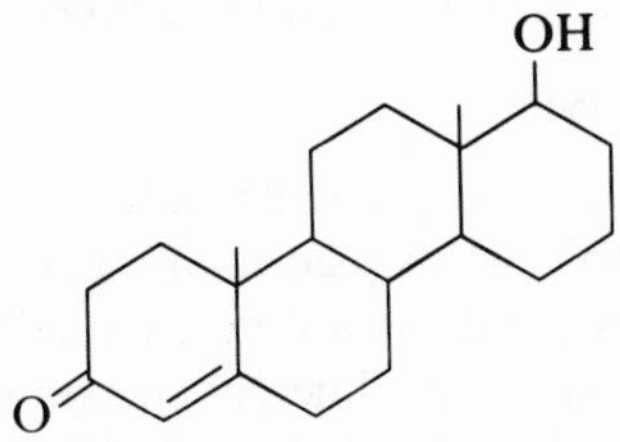

LXIX

the absence of that ring, as in des-D-5α-androstane (LXX). The opening of a ring is indicated by the term "seco." A rupture between carbons 1 and 2 is designated as in the name 1:2-seco-5α-androstane (LXXI).

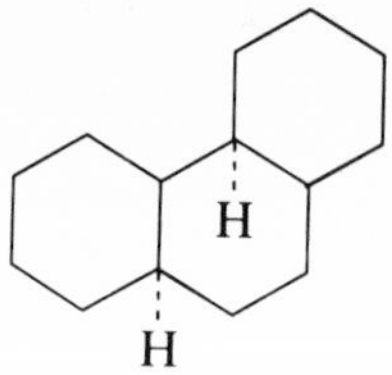

The symbol for an unknown configuration is drawn with a wavy line and is designated as ξ.

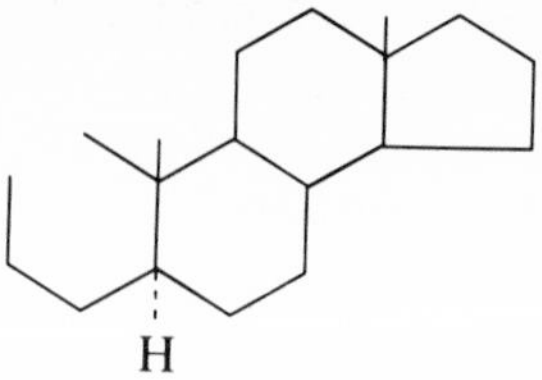

Conformations of Steroids

Because the steroid nucleus is composed of three cyclohexane rings and one cyclopentane ring there are various conformations which the nucleus can assume. The valencies of the carbon atoms of the cyclohexane ring can be found in the general plane of the ring, where they are called equatorial (e); or they are perpendicular to the plane of the ring, and are called axial (a). When drawing these parallel or perpendicular bonds, we still retain the convention and significance of writing the bonds extending above the plane of the ring as a solid line (β-configuration) and those below the plane of the ring as a broken line (α-configuration).

The chair form is the most stable conformation of the cyclohexane ring and of the steroid ring system. Because of the greater interatomic distances, the equatorial substituents can cause greater stability than do the corresponding axial ones, owing to differences in repulsion between atoms. The chair forms of the rings A, B, and C in the 5α- and 5β-steroid forms are presented in Fig. 13.4 showing the equatorial and axial substituents. The substituents on ring D are found at nearly 45° angles, and therefore their bonds are called indeterminate (i), quasiequatorial, or quasiaxial. The conformations for carbons 6 to 17 are the same for steroids having 4- and 5-unsaturations as well as for 5-compounds with 5α- and 5β-orientation. Ring A of the 5β-steroids lies under or below the plane of the rest of the molecule. The conformations in this ring are different from those in the 5α-molecule (and also 4- and 5-compounds). Fieser and Fieser (1959) give a detailed pictorial treatment of the conformations.

Polymorphism

Polymorphic forms of the same compound frequently display marked differences. A good example of this is given in Fig. 13.5, which shows such forms for 17β-estradiol. These forms depend on the different modes of hydrogen bonding (Smakula et al., 1957). Steroids that give such different spectra in the solid state produce identical spectra in solution. Therefore, it is best to use dilute solutions of nonpolar solvents to record the spectra.

When 35 steroids were examined by infrared spectroscopy and compared with authentic specimens, it was found that the spectra were an excellent means of identification, provided that the effects of polymorphism were precluded (Mesley and Johnson, 1965). No polymorphism was evident in 16 of the samples and it was

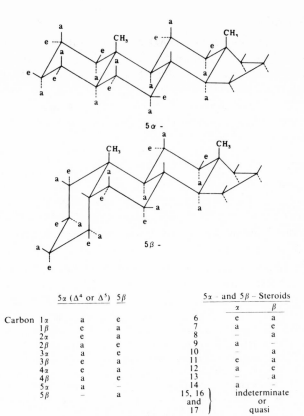

		5α (Δ⁴ or Δ⁵)	5β		5α - and 5β - Steroids	
					α	β
Carbon	1α	a	e	6	e	a
	1β	e	a	7	a	e
	2α	e	a	8	–	a
	2β	a	e	9	a	–
	3α	a	e	10	–	a
	3β	e	a	11	e	a
	4α	e	a	12	a	e
	4β	a	e	13	–	a
	5α	a	–	14	a	–
	5β	–	a	15, 16 and 17	indeterminate or quasi	

Fig. 13.4. The equatorial and axial substituents on the chair forms of steroid rings *A*, *B*, and *C*. (Dorfman and Ungar, 1965.)

possible to examine another 12 in solution. For the remaining seven substances (betamethasone, fludrocortisone acetate, hydrocortisone, methylprednisolone, prednisolone, prednisone, and triamcinolone) specific solvent treatments may be necessary if authentic and sample spectra are not identical on first examination. It should be noted that cortisone acetate was identified in seven different forms and dexamethasone acetate in four, but these compounds were soluble enough to obtain solution spectra. For additional examples of polymorphism among steroids the reader is also referred to Mesley (1966).

Correlations of Spectra with Structures

Much of the work on steroids has been involved with identifying carbonyl and hydroxyl groups, as well as unsaturations in various parts of the steroid nucleus. Jones *et al.* (1955) and Jones and Herling (1956) have given much information on correlations of steroid structures with their spectra. Some of the more important correlations are given here.

C—H Stetching Region

In the C—H stretching region (3000–2850 cm^{-1}) there is much overlap of absorption bands and as a result not much structural information is obtainable from methyl and methylene C—H stretching vibrations of steroids. Deuterated steroids have the C—D stretching band in the region 2275–2105 cm^{-1} (Dobriner et al., 1949; Jones et al., 1952c; Nolin and Jones, 1953). Unsaturated group C—H bonds are easily recognized by their absorption above 3000 cm^{-1} (C—H stretching frequencies are usually lower).

Bending Vibrations of Methyl and Methylene Groups (1480–1350 cm^{-1})

The region of C—H bending in methyl and methylene groups is quite useful when carbonyl bands are contemplated at the same time, especially in locating a keto group. Jones and Cole (1952) and Jones et al. (1952c) studied many compounds and their deuterated derivatives and obtained correlations like those in Table 13.2.

Table 13.2. Methyl and Methylene Bending Vibrations of Steroids in CCl$_4$ Solution[b] (Kasturi, 1963)

Frequency,[a] cm^{-1}	Structure assignment
1468 (1470–1462)	Side-chain methylene groups
1450 (1464–1446)	Ring system methylene groups
1438 (1440–1435)	Always present in spectra of methyl esters
1438 (1445–1432)	Methylene group adjacent to a double bond
1434 (1438–1426)	Ketosteroids with α-methylene group adjacent to a carbonyl group at C-4, C-6, C-7, C-11, or C-12
1422 (1476–1415)	3-ketosteroid with α-methylene group at C-2 or C-4
1408 (1411–1404)	17-ketosteroids with α-methylene at C-16, or quite generally, a methylene adjacent to a carbonyl group on a five-membered ring
1385 (1392–1374)	Angular methyl group between two six-membered rings (C-10)
1380 (1386–1374)	Side-chain methyl group at C-21 and C-28
1377 (1383–1372)	Angular methyl group between a five- and a six-membered ring (C-18)
1375 (1382–1369)	Acetate methyl group
1368 (1374–1360)	Gem-dimethyl group at end of side chain
1365 (1370–1360)	Acetate methyl group
1357 (1359–1356)	Methyl group of —CO—CH$_3$ side chain.

[a]The first figure is the average band position as determined from compounds in which there is no appreciable overlap with neighboring bands. The figures in parentheses indicate the extreme range found and include cases where the band may be reduced to an inflection by overlap.
[b]Data compiled from Jones and Cole (1952) and Cole (1954, 1956).

Differences between a 3- and a 7-ketone are easily found in the intensity of the α-ketomethylene bands around 1430 cm^{-1} (Figs. 13.6 and 13.7) (Jones, 1958). The frequency near 1410 cm^{-1} arising from the scissoring motion of one or more methylene groups next to the carbonyl group of a D-ring ketone can be used to distinguish this ketone from esters or acetates having carbonyl stretching bands that overlap (Jones and Cole, 1952).

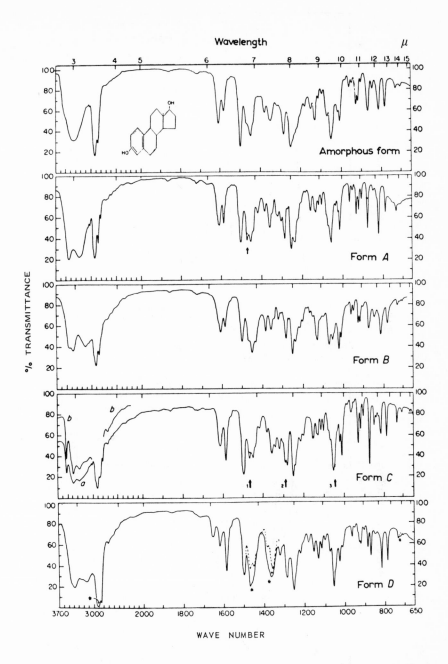

Fig. 13.5. Infrared spectra of the various solid forms of estradiol-17β: Form A, crystallized on a NaCl plate from a glassy melt. Form B, KBr pellet. Form C, (a) crystallized on NaCl plate from Form A and (b) Nujol mull. Form D, Nujol mull. Asterisk indicates Nujol bands; arrows indicate variations in intensity due to orientation or incomplete crystallization (Smakula et al., 1957). (Rao, 1963.)

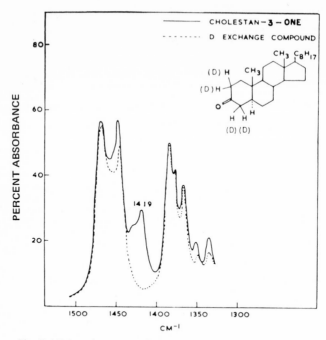

Fig. 13.6. Infrared spectrum of cholestan-3-one showing the α-keto-methylene bands (Jones, 1958). (Kasturi, 1963.)

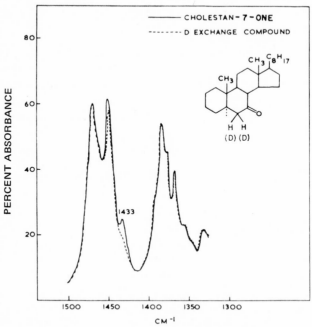

Fig. 13.7. Infrared spectrum of cholestan-7-one showing α-keto-methylene band (Jones, 1958). (Kasturi, 1963.)

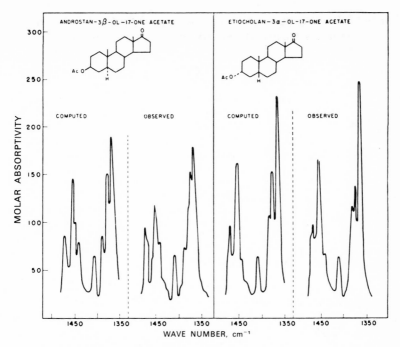

Fig. 13.8. Comparison of observed and computed spectra of 3-acetoxy-17-ketosteroid isomers; CS_2 solution. (Jones *et al.*, 1957.)

Barton *et al.* (1954) devised a method for locating cyclopropane rings in steroids and triterpenoids. They used hydrogen chloride to open the cyclopropane ring, and in a separate operation used deuterium chloride for the same reaction; then, they compared the intensities of the bands for methyl and methylene bending.

Jones *et al.* (1957) applied integrated intensity measurements to predict the spectra of certain steroids. Figure 13.8 shows comparisons of computed and observed spectra for the compounds, androstan-3β-ol-17-one acetate and etiocholan-3α-ol-17-one acetate.

Ethylenic Double Bonds

To characterize ethylenic double bonds, bands from three available regions can be applied: C—H stretching, C=C stretching, and C—H bending vibrations. The C—H stretching vibrations associated with unsaturation were mentioned above. Several research groups have studied the stretching frequencies of the double bonds in a variety of positions in the steroid molecule (Bladon *et al.*, 1951; Henbest *et al.*, 1954; Jones *et al.*, 1953). Cole's (1956) summary of their results are given in Table 13.3. Sometimes the position of the isolated di- and trisubstituted ethylene bonds of the steroid nucleus can be characterized (Henbest *et al.*, 1954; Jones *et al.*, 1953). Slomp and Johnson (1958) developed a quantitative method for the determination of side-chain and nuclear double bonds in 4,22-stigmastadien-3-one. The method involves

Table 13.3. Characteristic Absorption Bands of Ethylenic Double Bonds in Steroids[i] (Kasturi, 1963)

Position of C=C	C=C stretching vibration in CHCl$_3$, cm^{-1}	C=C—H stretching vibration in CCl$_4$, cm^{-1}	C—H bending vibration in CS$_2$, cm^{-1}	Reference
		Isolated Double Bonds		
Δ^1	1644	3021–2995	754–700	a
Δ^2	1657–1653	3034	774–664	a,b
Δ^3	1647	3015	773–671	a
Δ^4	1657	3040	810	c
Δ^5	1672–1667	3030	803–799	b,c,d
			814–812	
			840–830	
Δ^6	1639–1633	3017–3000	772–704	a,e
Δ^7	1666–1664	3040–3013	847, 827	b,c
$\Delta^{9(11)}$	1648–1643	3042	852, 827	c,e
Δ^{11}	1628–1620	3050–3033	839–703	a,b
Δ^{14}	1648–1646	3055	807, 797	c,f
Δ^{16}	1630–1621	—	—	b
Δ^{22}	1666–1664	—	974–970	e,g,h
		Conjugated Double Bonds[b]		
$\Delta^{3,5}$-Diene	1618, 1578			
$\Delta^{3,5}$-Diene-3-ylester	1671–1670, 1639			
Δ^1-3-ketone	1609–1604			
Δ^4-3-ketone	1609–1615			
$\Delta^{9(11)}$-12-ketone	1607			
Δ^{15}-17-ketone	1587			
Δ^{16}-20-ketone	1592–1588			
$\Delta^{1,4}$-Diene-3-ketone	1621, 1606–1603			
$\Delta^{4,6}$-Diene-3-ketone	1619–1616, 1587			
$\Delta^{3,5}$-Diene-7-ketone	1627, 1598			

[a]Henbest et al. (1954); [b]Jones et al. (1950); [c]Bladon et al. (1951); [d]Hirschmann (1952); [e]Jones and Herling (1954); [f]Jones and Dobriner (1949); [g]Turnbull et al. (1950); [h]Jones (1950); [i]Cole (1956).

infrared analysis of ozonized samples. Slomp et al. (1955) also used the infrared method to follow the selective and stereospecific hydrogenation of the double bond in the steroid nucleus of 4,22-stigmastadien-3-one.

Hydroxy Steroids

The O—H stretching region has no value for locating hydroxyl group positions in the ring system, but below 1350 cm^{-1} infrared spectra can be used to locate O—H groups at carbon atoms 5, 17, and 20 (Jones and Roberts, 1958). In 3-hydroxysteroids the predominant band lies between 1056 and 999 cm^{-1} owing to C—O stretching. The ϵ^a_{max} for these steroids is between 160 and 220, whereas C-17 or C-20 hydroxylated steroids have an ϵ_{max}^a ranging from 60 to 120 for the corresponding bands. The location of the band within the range given here is related to the stereochemistry at C-3 and C-5. In the 5α- and 5β-series, however, the 17β-hydroxysteroids are not distinguishable. The 20α- and the 20β-hydroxysteroids have a medium strong band at 1100–1090 cm^{-1} and 1016–1008 cm^{-1}, respectively. Table 13.4 gives a summary of some of the correlations of the frequencies with the orientation of the O—H group relative to the

Table 13.4. Stereochemical Characterization of Hydroxy Steroids[c]
(Kasturi, 1963)

A/B ring fusion	C—O bond configuration and conformation[a]	Main C—O frequency in CS_2 soln, cm^{-1}	Additional bands, cm^{-1}
trans	3β(e)	1040–1035	993–989, 978–976, 957–951, 937–934, 976–970, 962–952
trans	3α(a)	1004–999	935–927, 909–904
cis	3α(e)	1038–1035	1014–1007, 952–943, 920–908, 900–893
cis	3β(a)	1033–1030	986–980, 962–959, 916–908
$\Delta^5 C{=}C$	3β(e)	1056–1047	1025–1019, 989–978, 958–953, 841–838
$\Delta^5 C{=}C$	3α(a)	1034	—
trans	2α(e)	1035–1030[a]	—
trans	2β(a)	1010[b]	—
trans	4α(e)	1040	—
trans	4β(a)	1000	—
trans	17β(e)	1012–1008	—

[a](e), equatorial; (a), axial.
[b]Nujol mull.
[c]Data compiled from Jones and Roberts (1958), Cole et al. (1952), Rosenkrantz and Zablow (1953), Fürst et al. (1952), Jones et al. (1951) and Wiggins and Klyne (1955).

junction of the A and B rings (Wiggins and Klyne, 1955). These correlations are useful for analyzing steroid spectra by the differential method (Jones, 1958; Jones et al., 1957). One can estimate the number of O—H groups in a steroid molecule by measuring the apparent molar absorptivity of the O—H stretching band in solutions of approximately $0.01M$ concentration. The contribution from each O—H group is ~50 ϵ units (Cole, 1956). Weinman (1967) has studied the OH free stretching vibration band in 71 monohydroxylated steroids. The group included the mono- and multifunctional steroids such as 3-, 5-, 7-, 17-, and 17β-hydroxy-17α-methyl compounds.

Hydrogen Bonding

Hydrogen bonding of steroids has been investigated in detail by infrared spectroscopy. One can frequently obtain information concerning the stereochemical relationship between O—H groups that are near each other or between a C=O group

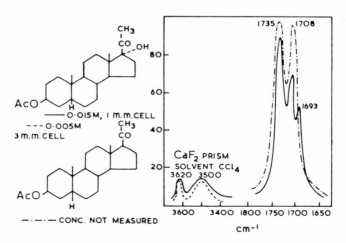

Fig. 13.9. Infrared spectrum illustration of the effect of intramolecular
hydrogen bonding. (Jones *et al.*, 1952*d*.) (Ordinate is % absorbance.)

and an O—H group (Jones *et al.*, 1951; Rothman and Wall, 1955; Wall and Serota, 1960). Not only is the frequency of the O—H stretching lowered by intramolecular hydrogen bonding, but also the frequency of the acceptor group involved in the hydrogen bond. The effect of intramolecular hydrogen bonding is demonstrated in Fig. 13.9 by comparing the spectra of 3β-acetoxypregnan-20-one and its 17α-hydroxy derivative (Jones *et al.*, 1952*d*). The latter compound displays equally intense bands at 3620 and 3500 cm^{-1} in the O—H stretching region. These bands are not changed by further dilution. The hydroxy derivative has three bands in the carbonyl region at 1735, 1708, and 1693 cm^{-1}, resulting from the acetate, 20-ketone (LXXII), and the hydrogen-bonded 20-ketone (LXXIII). The intensity of the band due to the

free 20-ketone is much lower than that of the nonhydroxylated compound. The use of infrared spectra of the epimeric 12-hydroxy-12-methylpregnane-3,20-dione allowed a definite assignment of the configuration at the C-12 position to be made (Nagarajan and Just, 1961). Although the spectrum of compound LXXIV (an α-hydroxy compound) displays a band at 3680 cm^{-1} due to free O—H and one at 1713 cm^{-1} due to the 3,20-dione, compound LXXV (the β-hydroxy form) has a bonded O—H band at 3450 cm^{-1}, a 3-ketone band at 1716 cm^{-1}, and another band

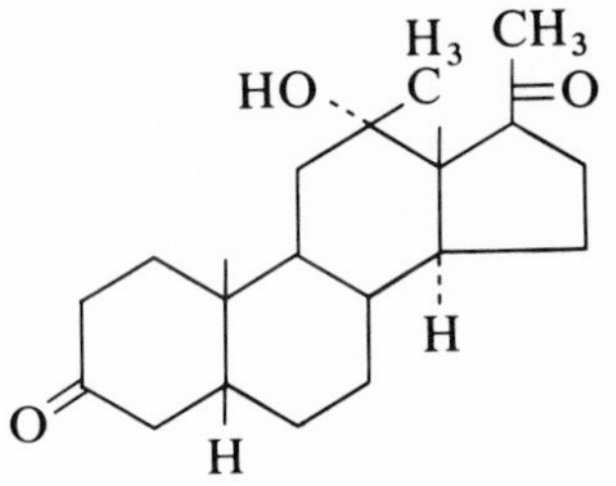

LXXIV 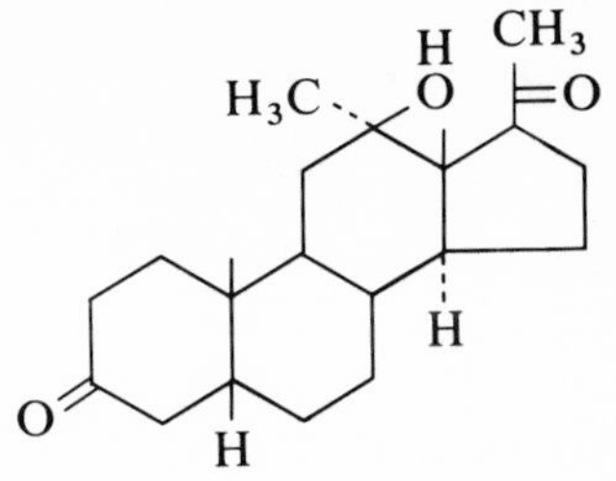   LXXV

at 1695 cm^{-1} ($>$C20$=$O$\cdots$H$-$O$-\overset{|}{C}$12$-$). Inspection of the models of these com-
pounds shows that only a 12β-hydroxy group can interact with the 20-carbonyl
group to yield an intramolecular hydrogen bond.

During the preparation of 3β-acetoxy-5α-Δ^{16}-pregnene-12,20-dione (LXXVI)
from hecogenin, a by-product (LXXVII) was formed, whose structure was determined
with the help of infrared spectra (Rothman and Wall, 1955). The ultraviolet absorption
spectrum of compound LXXVII and optical rotation differences confirmed the

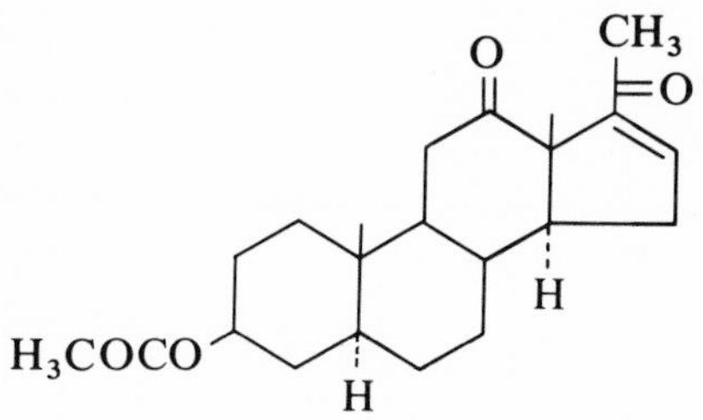

LXXVI   LXXVII

ν_{max} 1732 cm^{-1} (acetate) ν_{max} 1732 cm^{-1} (acetate)
1740 cm^{-1} (12, 20-diketone)   1655 cm^{-1} (C-20 conjugated
1668 cm^{-1} (C-20 conjugated   ketone)
ketone)

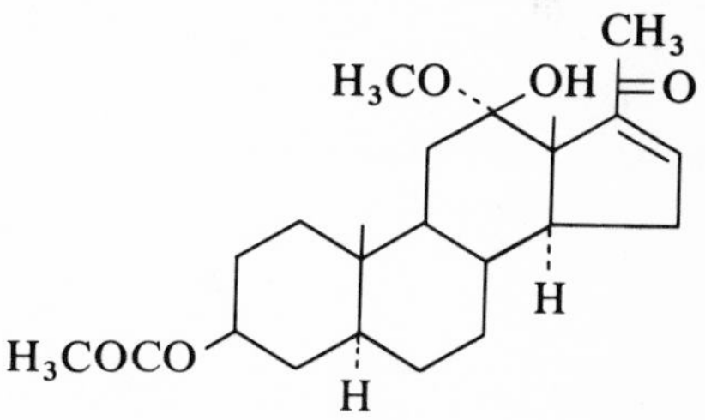

structure. The infrared spectrum of the hemiketal (LXXVII) proved the over-all
structure and stereochemistry at C-12. The band at 3340 cm^{-1} (concentration-
independent) showed that there was an O$-$H group present which was intramolecularly hydrogen bonded. A strong acetate carbonyl band at 1732 cm^{-1} and a strong
conjugated-20-ketone band at 1655 cm^{-1} were found. The latter band showed clearly
that the O$-$H group in compound LXXVII was hydrogen bonded to the conjugated
ketone. The possibility of this hydrogen bond arises only when the O$-$H is in the
equatorial 12β-configuration. The spectrum of compound LXXVII also showed a
strong band at 1082 cm^{-1}, which was absent in the enone (LXXVI); this band arises
from an axial $-$OCH$_3$ group (Page, 1955).

Parker and Bhaskar (1968) have measured quantitatively the self-association by
hydrogen bonding of cholesterol (use of the O$-$H stretching bands), and the intermolecular hydrogen bonding between cholesterol and several triglycerides by infrared

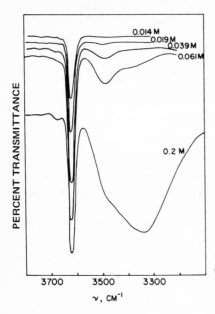

Fig. 13.10. OH-stretching vibration bands of cholesterol at different concentrations. Solvent, CCl_4; temperature, 23°C. The ordinate axis is not drawn to scale. (Parker and Bhaskar, 1968.)

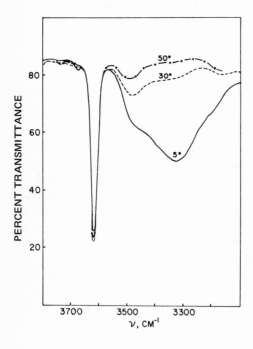

Fig. 13.11. Effect of temperature on the bonded OH peaks of cholesterol (0.06M); solvent, CCl_4. (Parker and Bhaskar, 1968.)

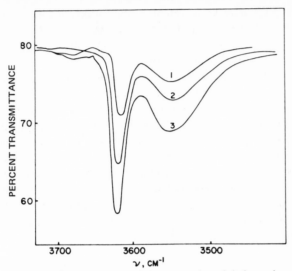

Fig. 13.12. Effect of varying the concentration of cholesterol on the associated OH band of cholesterol–tributyrin complex. Solvent, CCl_4; temperature, 23°C. Curves (1)–(3): cholesterol (0.006, 0.01, and 0.015M) plus tributyrin (0.15M). (Parker and Bhaskar, 1968.)

methods (Liddel and Becker, 1957; Pimentel and McClellan, 1960). At concentrations below 0.014M in carbon tetrachloride, cholesterol exists only as a monomer. As the concentration is increased, cholesterol associates to form a dimer, and at a concentration of ~0.06M a higher aggregate begins to form, which becomes the predominant species at ~0.2M concentration. The dimerization constant was determined (K_d at 23°C = 4.5 liters/mole) at different temperatures, from which the enthalpy was evaluated ($\Delta H = -1.8$ kcal/mole). Infrared spectra of mixed solutions of cholesterol and the triglycerides gave evidence of the formation of a 1:1 hydrogen-bonded complex. The equilibrium constants and enthalpies of formation of the complexes of cholesterol with triacetin, tributyrin, and trilaurin were also reported ($K_{23°C} = 2.4$ to 3.7 liters/mole; $\Delta H = -3.5$ to -5.4 kcal/mole). Figure 13.10 shows the OH-stretching bands of cholesterol at various concentrations; Fig. 13.11 shows the effect of temperature on the bonded O—H peaks of cholesterol at 0.06M concentration; and Fig. 13.12 shows the effect of varying the concentration of cholesterol on the associated O—H band (3540 cm^{-1}) of the cholesterol–tributyrin complex (the spectra in Fig. 13.12 were recorded against a blank solution containing the triglyceride at the same concentration as in the mixed solutions of triglyceride plus cholesterol).

Zull et al. (1968) have used the ATR technique (see Chapter 3) to obtain spectra of solid films of egg lecithin-cholesterol mixtures. In films cast from organic solvents, cholesterol can form a sterol:phospholipid (2:1) complex which involves interaction of the cholesterol —OH with lecithin polar groups. Association with the quaternary nitrogen of lecithin was ruled out, and the formation of a hydrogen bond between cholesterol and some unidentified polarizing group in lecithin was believed to be responsible for the shift of the OH stretching absorption of cholesterol by $\sim 150\,\text{cm}^{-1}$.

In a 1:1 molar mixture of cholesterol and lecithin the OH stretching band of cholesterol lies at $\sim 3250\,\text{cm}^{-1}$, having been shifted from $3420\text{--}3400\,\text{cm}^{-1}$ in the spectrum of cholesterol alone. These workers have cited some of the evidence supporting the idea that cholesterol–lecithin interactions may be important in cellular membranes and in certain other biological systems.

Eger and Norton (1965) have studied androgenic steroid complexes with p-bromophenol in order to circumvent the problem of molecular distortion in steroids (Norton et al., 1963, 1964) by the covalent binding of heavy atoms in the X-ray method (Robertson, 1953) of structure determination. Their infrared measurements supported the assumption that intermolecular complexes between androgenic steroids and p-bromophenol are hydrogen bonded. For example, androsterone by itself had the ν_{CO} stretching frequency at $1744\,\text{cm}^{-1}$ and the ν_{OH} (monomer) at $3628\,\text{cm}^{-1}$. In the steroid–p-bromophenol complex the ν_{CO} band and the ν_{OH} (monomer) band were found at 1728 and $3610\,\text{cm}^{-1}$, respectively.

Carbonyl Bands

Thorough investigations have been made of the C=O stretching region (1900–$1580\,\text{cm}^{-1}$) (Jones and Dobriner, 1949; Jones et al., 1952d; Jones et al., 1949; Jones et al., 1950b; Jones et al., 1948).

About 150 different types of carbonyl functions have been classified by Dobriner et al. (1953) and Roberts et al. (1958). Table 13.5 presents the frequency ranges of typical steroid carbonyl groups.

Table 13.5. Carbonyl Stretching Frequencies in Steroids[b]
(Kasturi, 1963)

A. Saturated Ketones			
Keto group	Type	Series	Frequency in CCl_4 and CS_2 solution, cm^{-1}
1	6-membered ring	5α	1709–1704
2		5α	1710
3		$5\alpha, 5\beta$	1719–1713
4		5α	1712
6		5α	1714–1713
6		5β	1708–1706
7		5α	1718–1713
11		5β	1710–1704
12		5β	1712–1706
15	5-membered ring		1738
16		Unsubstituted at C-17	1749
		Substituted at C-17	1738
17			1745–1741

Table 13.5 *(Continued)*

B. Ester Carbonyl and Dicarbonyl Compounds

Carbonyl type	Frequency in CCl_4 and CS_2 solution, cm^{-1}
Formate	1729–1725
Acetate	1739–1733
Benzoate	1724–1717
Methyl ester	1742–1735
Hexahydrobenzoate	1739–1735
Phenolic-3-benzoate, naphtholic-3-benzoate	1746
Phenolic-3-acetate, propionate	1767–1764
20,21-Diacetate	1749
3-Ketone enolacetate (Δ^2)	1758–1753
Δ^4-3-ketone enolacetate ($\Delta^{3,5}$)	1754
11-Keto enolacetate ($\Delta^{9(11)}$)	1749
20-Ketone enolacetate (Δ^{20})	1755–1752
20-Ketone enolacetate ($\Delta^{17(20)}$)	1756–1749
γ-Lactone	1781–1777
δ-Lactone	1744–1742
11,12-Diketone	1726
3,6-Diketone	1723–1718
11α-Acetoxy-12-*ketone*[a]	1730–1726
11α-*Acetoxy*-12-ketone[a]	1753–1748
11β-Acetoxy-12-*ketone*[a]	1720
11β-*Acetoxy*-12-ketone[a]	1753–1752
12α-Acetoxy-11-*ketone*[a]	1720
12α-*Acetoxy*-11-ketone[a]	1752
12β-Acetoxy-11-*ketone*[a]	1730–1727
12β-*Acetoxy*-11-ketone[a]	1755–1752
17α-Acetoxy-20-*ketone*[a]	1719
17α-*Acetoxy*-20-ketone[a]	1736
17β-Acetoxy-20-*ketone*[a]	1716
17β-*Acetoxy*-20-ketone[a]	1742–1736
21-Acetoxy-20-*ketone*[a]	1736–1724
21-*Acetoxy*-20-ketone[a]	1758–1754
11,*17*-Diketone[a]	1752–1748
11,17-Diketone[a]	1719–1713
Shoulder due to *monomeric* carboxylic acid[a]	1758–1748
Dimeric carboxylic acid[a]	1710–1700

C. α,β-Unsaturated Ketones

Chromophore	Type	Frequency in CCl_4 and CS_2 solution, cm^{-1}
Δ^2-1-Ketone		1682
Δ^1-3-Ketone	6-membered ring	1684–1680
Δ^4-3-Ketone		1681–1677
Δ^4-3,6-Diketone		1692
$\Delta^{1,4}$-Diene-3-ketone		1671–1663
$\Delta^{4,6}$-Diene-3-ketone		1669–1666

[a] These structures give rise to two carbonyl bands. The group causing the absorption is italicized.
[b] Data compiled from Cole (1954; 1956) and Jones and Herling (1954).

Table 13.5 *(Continued)*

C. α,β-Unsaturated Ketones *(continued)*

Chromophore	Type	Frequency in CCl₄ and CS₂ solution, cm⁻¹
Δ^5-7-Ketone		1682–1674
$\Delta^{3,5}$-Diene-7-ketone		1663
Δ^8-7-Ketone		1667
Δ^8-11-Ketone		1660
$\Delta^{9(11)}$-12-Ketone		1684–1680
$\Delta^{8(14)}$-15-Ketone	5-membered ring	1705–1703
Δ^{15}-17-Ketone		1716
Δ^{16}-20-Ketone	No ring	1670–1666

[a]These structures give rise to two carbonyl bands. The group causing the absorption is italicized.
[b]Data compiled from Cole (1954; 1956) and Jones and Herling (1954).

Generally, carbonyl groups at C-3, C-4, C-6, C-7, C-11, C-12, and C-20 cannot be distinguished only by the absorption bands they produce in the region 1900–1580 cm⁻¹. It is necessary to have additional information from methyl and methylene bending frequencies in order to identify some of them. Ketones of the five-membered ring are distinguishable from the remainder of the carbonyls by the greater stretching frequency. The carbonyl stretching frequency is lowered by 20 to 40 cm⁻¹ when the carbonyl is conjugated with an ethylenic double bond. Conjugation also increases the intensities of the C=O and C=C stretching absorptions. However, in enol acetates the carbonyl stretching is increased to 1755 cm⁻¹. The carbonyl of formate, acetate, benzoate, carboxylic acid, γ-lactone, and δ-lactone groups are distinguishable, as shown by the information in Table 13.5. Formates also have strong C—O stretching bands at 1180–1178 cm⁻¹; acetates have such bands at 1256–1236 cm⁻¹; and the benzoate C—O stretching band lies at 1270 cm⁻¹. Benzoates also display the phenyl group C—H out-of-plane deformation bands at ~705 cm⁻¹. When a five-membered ring ketone is present, acetates and esters which have overlapping C=O stretching bands can be distinguished from their C—O stretching bands at lower frequencies. Esters have these absorptions at 1155–1135 cm⁻¹; acetates at 1240 cm⁻¹.

The two spectra in Fig. 13.13 demonstrate how specific carbonyl bands can be. In spectrum *A* there are three bands due to three carbonyl groups. In spectrum *B* there is an extra band near 1710 cm⁻¹, which results from the 11-keto group in the third ring.

Adjacent ketones such as the 11,12-diketones absorb in a higher range (1729–1726 cm⁻¹) than do saturated ketones (1716–1710 cm⁻¹), owing to dicarbonyl inter-actions. (See Table 13.5.) Such interactions take effect again in the case of the 21-acetoxy-20-ketone compounds seen in Fig. 13.13. Absorption is at 1758 and 1732 cm⁻¹ instead of at 1735 cm⁻¹ (acetate) and 1707 cm⁻¹ (20-ketone). Shifts of this type have been used in other structural determinations (Jones and Herling, 1954; Jones *et al.*, 1949; Jones *et al.*, 1950b; Dickson and Page, 1955). Jones *et al.* (1952a) observed that the carbonyl stretching frequency in a six-membered ring is about 20 cm⁻¹ higher when an equatorial bromine atom is attached. This increase comes from interaction between the C=O and C—Br dipoles, which is maximal when these dipoles are coplanar. On the other hand, an axial bromine has essentially no effect (Table 13.6)

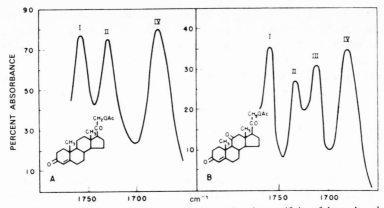

Fig. 13.13. Infrared spectra of two steroids illustrating the specificity of the carbonyl stretching vibration (Jones, 1958). (Rao, 1963.)

Table 13.6. Carbonyl Band Positions and Probable Conformations of Haloketones in CS₂ Solution[c] (Kasturi, 1963)

Compound	Carbonyl position, cm^{-1}	Conformation of C–halogen bond[a]
Cyclohexanone	1712[b]	—
2-Bromocyclohexanone	1716[b]	a
2-Chlorocyclohexanone	1722[b]	a
2,2-Dichlorocyclohexanone	1745[b]	a,e
2-Methylcyclohexanone	1715	—
2-Bromo-2-methylcyclohexanone	1722	a
2,6-Dibromo-2-methylcyclohexanone	1738	a,e
2,6,6-Tribromo-2-methylcyclohexanone	1737	a,a,e
2,6-Dimethylcyclohexanone	1714	—
2-Bromo-2,6-dimethylcyclohexanone	1722	a
Cholestan-3-one	1718	—
2-Bromocholestan-3-one	1733	e
2,2-Dibromocholestan-3-one	1735	e,a
2,4-Dibromocholestan-3-one	1758–56	e,e
2-Iodocholestan-3-one	1724	e
Coprostan-3-one	1716	—
4-Bromocoprostan-3-one	1733	e
2,4-Dibromocoprostan-3-one	1756	e,e
11-Ketosteroids	1713	—
12α-Bromo-11-ketosteroids	1714	a
12-Ketosteroids	1710	—
11α-Bromo-12-ketosteroids	1735	e
11β-Bromo-12-ketosteroids	1706	a
allo-Pregnan-20-one	1710	—
17α-Bromo-*allo*-pregnan-20-one	1705	a

[a] a, axial; e, equatorial.
[b] Carbon tetrachloride solution.
[c] Data compiled from Cole (1956), Corey (1953), Corey and Burke (1955), and Jones *et al.* (1952a).

since in this case the C=O and C—Br dipoles are almost 90° apart and there is very little interaction between them. Jones et al. (1952a) give a range of displacements from 13 to 30 cm^{-1} for equatorial bromine and -5 to $+8$ cm^{-1} for axial bromine. α-Substitution of chlorides produces similar results (Cummins and Page, 1957).

Rosenkrantz and Gual (1959) and Gual et al. (1958) examined the spectra of some lactones of the D ring, in which bands at 1219, 1110, and 989 cm^{-1} appeared

consistently. The first two bands are from the $-\overset{\overset{\displaystyle O}{\|}}{C}-O$ and $-\overset{|}{\underset{|}{C}}-O-$ groups,

respectively. In the acetoxy derivatives of the D-ring lactones the 1219 cm^{-1} band is imperceptible, but the 1110 cm^{-1} band is quite evident. Jones and Gallagher (1959), Jones et al. (1959), and Angell et al. (1960) have examined the infrared spectra of several types of lactones and have confirmed the regions of absorption assigned earlier to the carbonyl stretching bands of saturated γ- and δ-lactones (Bellamy, 1958). Also, other bands characteristic of different types of steroid lactones have been recorded in the range 1450–1350 cm^{-1}.

Measurement of Integrated Absorption Intensities

Jones et al. (1952a,b) and Ramsay (1952) have correlated band intensities with molecular structure. Wexler (1967) has recently reviewed the subject of integrated intensities. Jones et al. (1952a; 1957) have measured the integrated intensities for several steroids.

The integrated intensity A is given by

$$A = \frac{1}{cl} \int \log_e \left(\frac{I_0}{I} \right)_v dv = 2.303 \int \epsilon \, dv$$

and various corrections are made in this equation for errors due to the finite slit width. One unit of integrated absorption intensity is defined as 10^4 mole^{-1} liter cm^{-2}. The use of such integrated intensities for structural determinations is limited to bands free from overlap by other bands. Carbonyl integrated intensities for different carbonyl groups lie within the range 1–5 units. Table 13.7 gives a summary of the values for the integrated intensities of steroid carbonyl stretching bands. The carbonyl stretching band intensities increase in the presence of conjugation (a factor causing the carbonyl stretching frequency to be reduced); they decrease when α-substitution is present (for example, α-halogenation which makes the carbonyl stretching frequency greater). The integrated intensities of the carbonyl stretching bands of steroids change only slightly for a C=O group in a particular molecular arrangement, and this constancy can help to identify structures. In compounds containing several carbonyl groups where the carbonyls are separated by more than two carbon atoms, the same absorption frequency is usually exhibited as for the monocarbonyl compounds. For 3,12-diketo steroids only one band is usually seen, but examination of its intensity indicates that two carbonyls are there. Table 13.8 presents integrated intensity data on some steroids having one, two, or several carbonyl groups. Table 13.7 gives examples of the application of intensity measurements for learning how many

Table 13.7. Integrated Carbonyl Band Intensities of Polycarbonyl Steroidal Derivatives Exhibiting only One Carbonyl Peak (Jones *et al.*, 1957)

Compound	ν_{max}	A_{obs}	A_{calc}
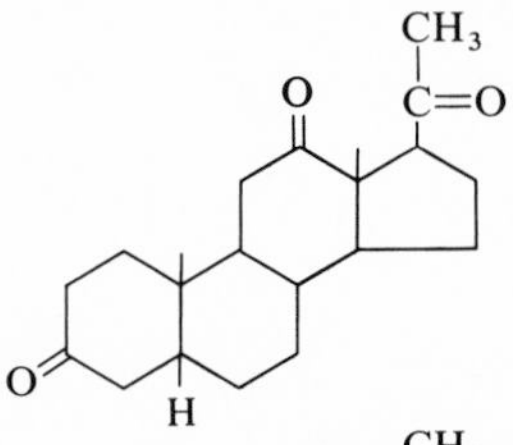	1712	6.93	6.61
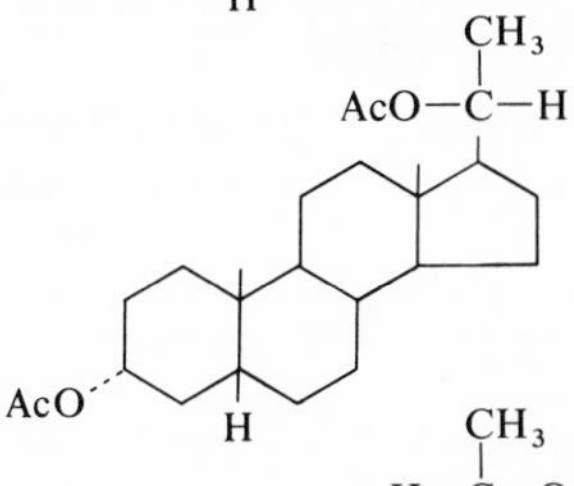	1739	6.21	6.48
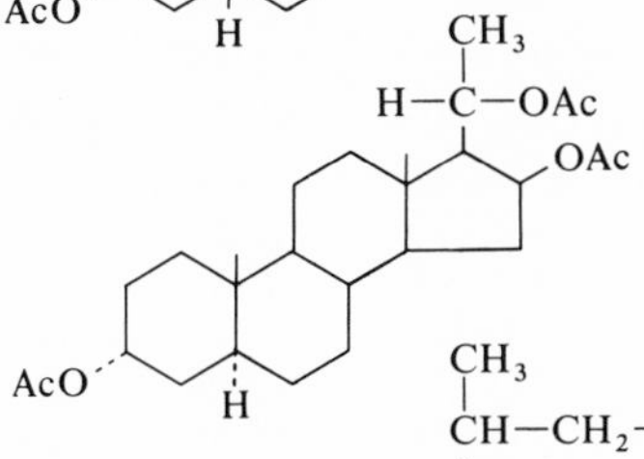	1739	9.05	9.72
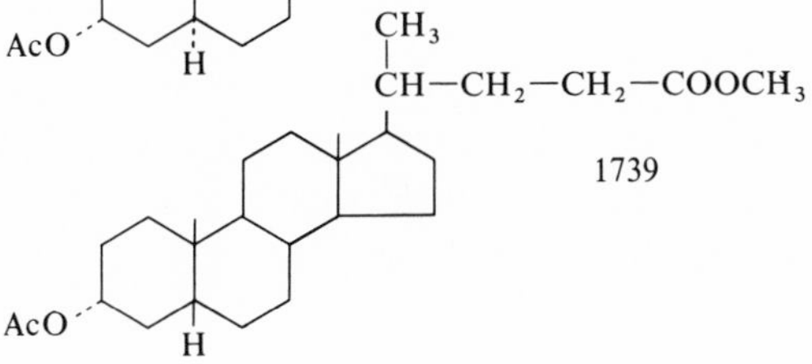	1739	6.21	6.37

carbonyls are present in steroids which display only one carbonyl stretching band. One adds up the values for each carbonyl group of the corresponding compounds with one C=O group. Then, comparing the observed value of the integrated intensity with the calculated value, one learns directly the number of carbonyl groups in the compound.

Another example of the application of integrated intensity measurements is given by the differentiation between such structures as LXXVIII and LXXIX (Jones *et al.*, 1957) by making such measurements at 1740 cm^{-1} and in the 1290–1190 cm^{-1} region.

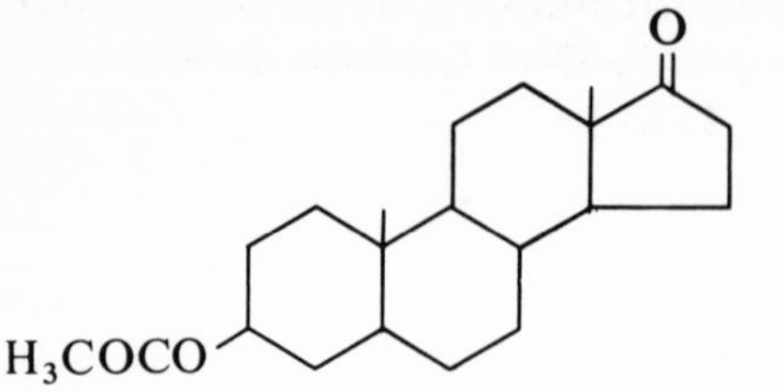

LXXVIII

A_{1740} 5.95
$A_{1290-1190}$ 4.00

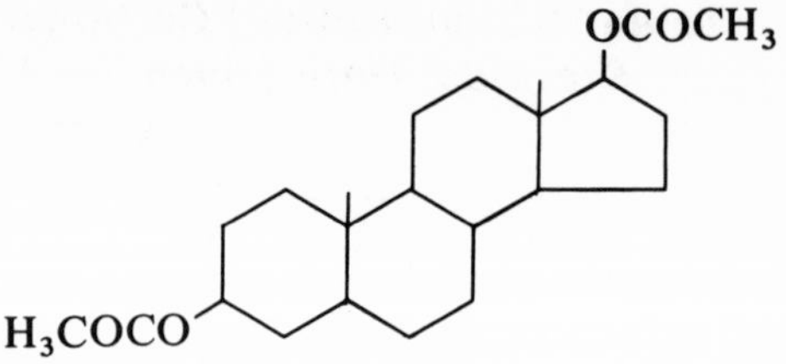

LXXIX

A_{1740} 6.5
$A_{1290-1190}$ 8.0

Table 13.8. Integrated Intensities of the Carbonyl Stretching Frequencies of Steroids[d] (Rao, 1963)

Compound	ν_{max}, cm^{-1}	Integrated absorption intensity[a,b] (10^4 mole^{-1} liter cm^{-2})
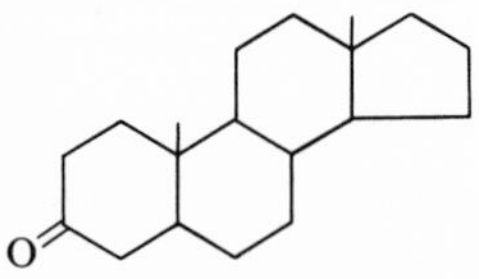	~1716	2.55
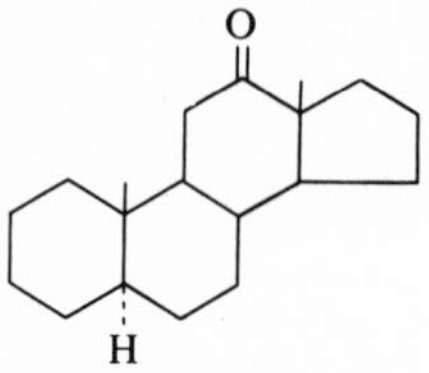	~1710	2.27
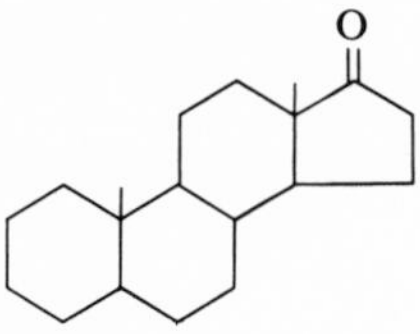	~1745	2.69
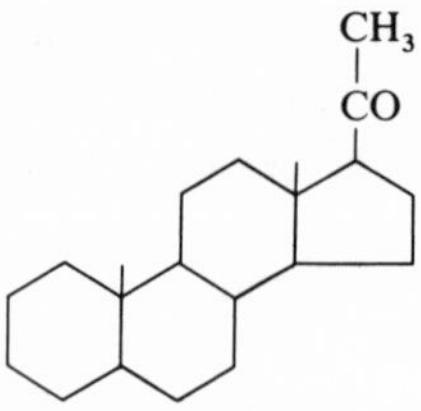	~1710	1.79

Table 13.8 *(Continued)*

Compound	v_{max}, cm^{-1}	Integrated absorption intensity^{a,b} (10^4 mole^{-1} liter cm^{-2})
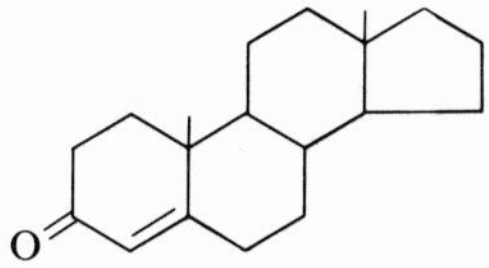	~1678	3.65
Acetate group	~1735	3.24
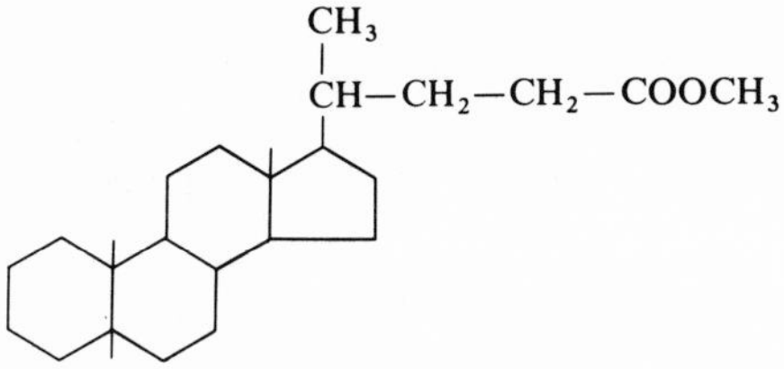	~1735	3.13
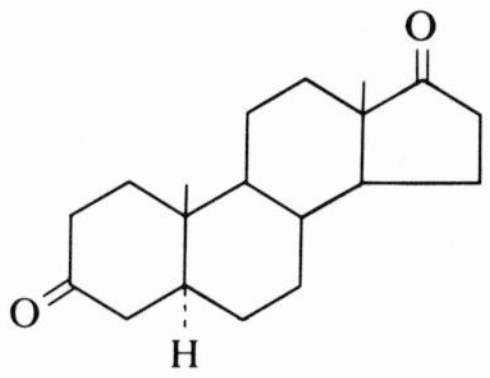	1745 1719	5.19 (5.24)
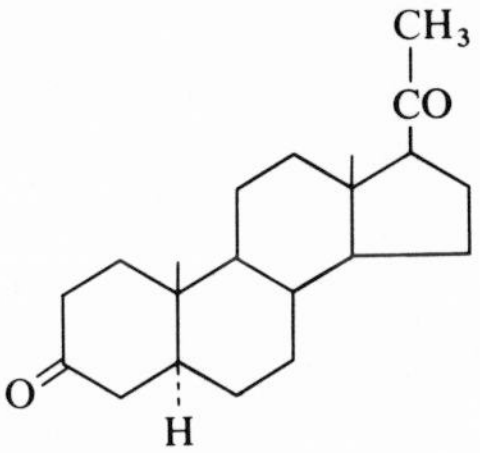	1719 1710	4.45 (4.34)
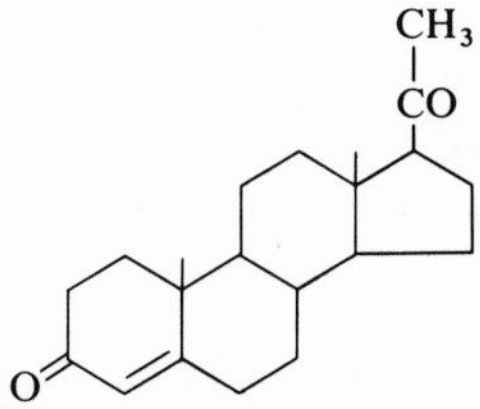	1708 1677	5.90 (5.44)

Table 13.8 *(Continued)*

Compound	v_{max}, cm^{-1}	Integrated absorption intensity[a,b] (10^4 mole^{-1} liter cm^{-2})
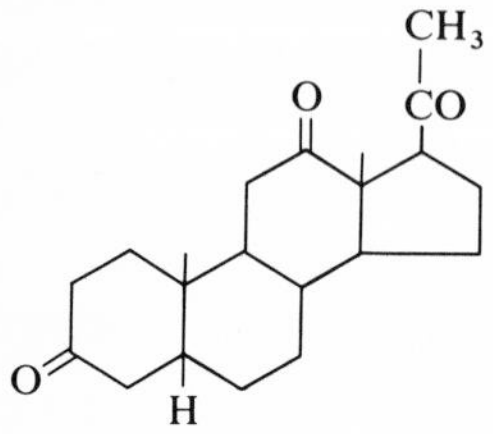	1712[c]	6.93 (6.61)
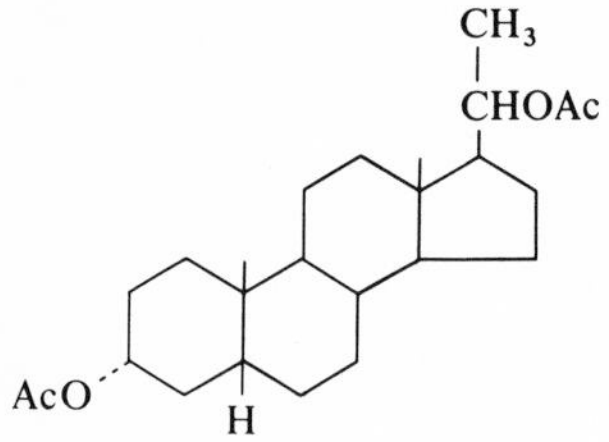	1739[c]	6.21 (6.48)
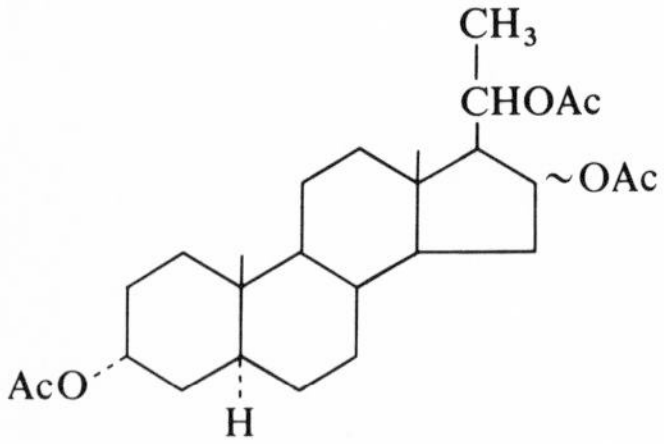	1739[c]	9.05 (9.72)
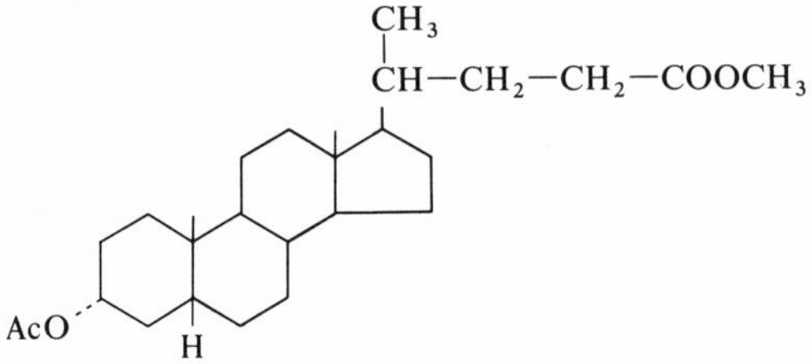	1739[c]	6.21 (6.37)
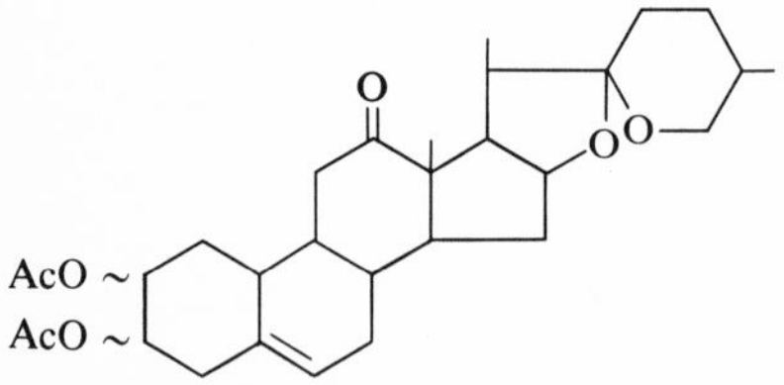	1745[c] 1713	8.47 (8.75)

[a]All measurements in carbon disulfide or carbon tetrachloride solution.
[b]Values in the parentheses are calculated from the data on monocarbonyl steroids.
[c]Composite band from more than one carbonyl peak.
[d]Jones *et al.* (1952*b*).

The "Fingerprint" Region (below 1350 cm⁻¹)

Jones (1958) has shown that large effects are produced in the infrared spectra of stereoisomeric steroids by small structural differences. For example, in the region from 1375 to 900 cm⁻¹, the following isomeric steroids displayed vastly different spectra: (1) 3β-hydroxyandrostan-17-one; (2) 3α-hydroxyandrostan-17-one; (3) 3β-hydroxyetiocholan-17-one; (4) 3β-hydroxyetiocholan-17-one.

Jones (1958) and co-workers have developed a graphical method for obtaining "zone patterns" for particular steroids (see also, Fieser and Fieser, 1959). This method uses a differential technique, and gives characteristic patterns for various functional groups, allowing one to predict what characteristics a spectrum will have. For example, in Fig. 13.14 a combination curve is given which is a summation of the 3β-hydroxy-5β steroid and the 17-keto-5β steroid zone patterns. In Fig. 13.15, a comparison is made of the zone combination pattern from Fig. 13.14 with the observed spectrum of 3β-hydroxyetiocholan-17-one.

Metabolism in Man and Animals

For detailed accounts of the multifarious pathways of steroid metabolism, the reader is referred to Dorfman and Ungar (1965) and Hadd and Blickenstaff (1969).

Calvin and Lieberman (1964) have given evidence that steroid sulfates serve as biosynthetic intermediates. They incubated pregnenolone-³H sulfate-³⁵S (LXXX) with a homogenate of hyperplastic adrenal tissue. The product isolated in greatest yield (> 5%) was 17α-hydroxypregnenolone-³H sulfate-³⁵S (LXXXI), which contained both isotopes in the same ratio as in the substrate. They proved the radiochemical purity of the product by crystallizing the compound with authentic ammonium 17α-hydroxypregnenolone sulfate and by converting to 17α-hydroxypregnenolone, pyridinium dehydroisoandrosterone sulfate, and dehydroisoandrosterone sulfate (LXXXII). Infrared spectroscopy was used for identification and characterization of the pyridinium and ammonium salts of 17α-hydroxypregnenolone

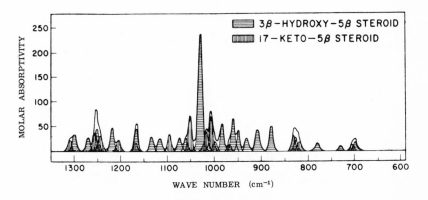

Fig. 13.14. Combination of the characteristic zone pattern of 3β-hydroxy-5β- and 17-keto-5β-steroids (Jones, 1958). (Kasturi, 1963.)

sulfate. Adrenal tissue was shown to be capable of hydroxylating pregnenolone sulfate at C-17. These workers proposed a pathway involving steroid sulfates as intermediates:

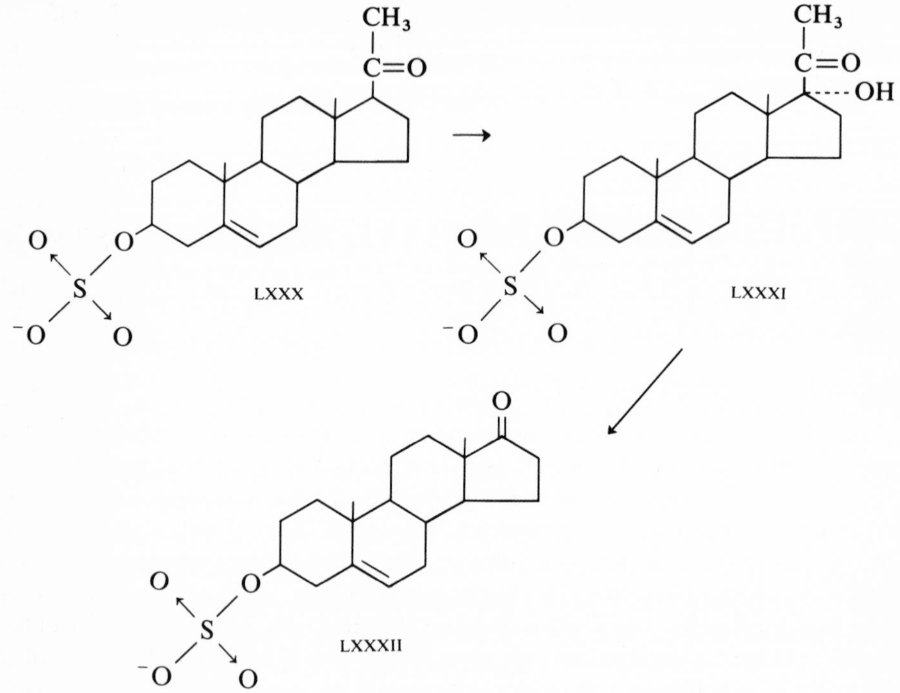

LXXX LXXXI LXXXII

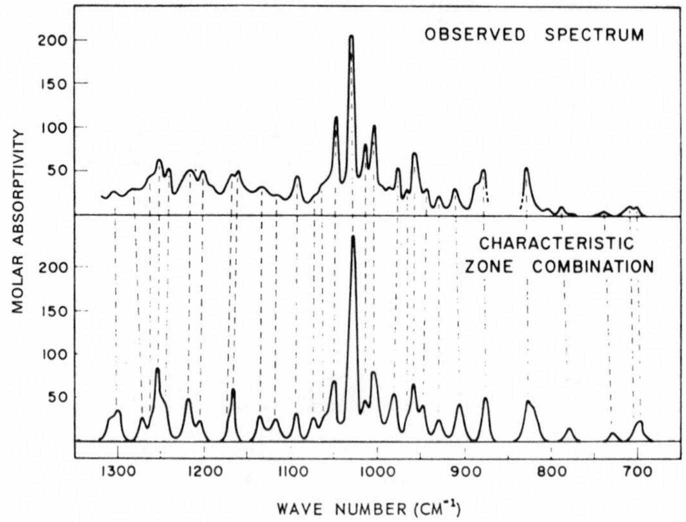

Fig. 13.15. Comparison of the characteristic zone combination pattern from Fig. 13.14 with the observed spectrum of 3β-hydroxyetiocholan-17-one (Jones, 1958). (Kasturi, 1963.)

Using countercurrent distribution Hadd and Dorfman (1963) have isolated 5α-androstan- and crystalline 5β-androstan-[(3α → 1β-oside)-D-glucopyranosuronic acids]-17-one from the urine of a woman with an adrenal adenoma. They characterized these conjugates and their triacetyl methyl esters by infrared absorption, optical rotation, and melting point methods.

YoungLai and Solomon (1967) have characterized the most abundant single urinary product in man of injected 16α-hydroxydehydroisoandrosterone. It was 3α,16α-dihydroxyandrost-5-en-17-one. In the spectrum of 16α-acetoxy-3α-hydroxy-androst-5-en-17-one (the monoacetate derivative) there are two carbonyl bands, at 1730 and 1757 cm^{-1}, the latter being characteristic of a 16α-acetoxy-17-keto group.

Arcos and Lieberman (1967) have isolated two new crystalline conjugates from the urine of normal human subjects to whom pregnenolone had been administered. One of the compounds was 5-pregnene-3β,20α-diol-20-(2′-acetamido-2′-deoxy-α-D-glucoside). Its structure was proved by elemental analysis, infrared and NMR spectra, and the products formed by hydrolysis. The infrared spectrum of 5-pregnene-3β,20α-diol-3-sulfate-20-(2′-acetamido-2′-deoxy-α-D-glucoside) had bands at 1635 and 1550 cm^{-1} compatible with the amide I and II bands and one wide band between 1235 and 1215 cm^{-1}, which is associated with the sulfate group. The isolation of 5-pregnene-3β,20α-diol-3-sulfate-20-(2′-acetamido-2′-deoxy-α-D-glucoside) revealed for the first time the presence of a glucosaminide conjugate of a steroid in human urine.

Hashimoto and Neeman (1963) have obtained pure crystalline estriol 16α-glucosiduronic acid from third trimester human pregnancy urine, completely characterized it and converted it to two derivatives. Infrared spectroscopy was used in the identification of these compounds, methyl(3,17β-dimethoxyestra-1,3,5(10)-trien-16α-yl-2′,3′,4′-tri-O-methyl-β-D-glucopyranosid) uronate, and methyl (3-methoxy-17β-acetoxyestra-1,3,5(10)-trien-16α-yl-2′,3′,4′-tri-O-acetyl-β-D-glucopyranosid)-uronate.

Epitestosterone has been identified in the urine of men and women by Korenman *et al.* (1964). Infrared techniques were used in the characterization. The α,β-unsaturated ketone absorption was found at 1677 cm^{-1}.

Noguchi and Fukushima (1966) have isolated and characterized, by infrared, NMR, and mass spectrometry and chemical techniques, a highly polar metabolite, 3α-ureido-11β-hydroxy-Δ^4-androsten-17-one, from the urine of normal healthy subjects, a substance which is materially increased in patients with myxedema. Infrared spectroscopy helped to eliminate an alternate structure, a pseudourea derivative. The compound found was given the trivial name of ureasterone (LXXXIII).

LXXXIII

Van Lier and Smith (1967) have isolated and characterized 26-hydroxycholesterol from healthy and diseased human aortal tissue by a variety of chromatographic methods, and one method of identification involved comparison of the infrared spectra of the sterol and its $3\beta,26$-diacetate with those of authentic samples.

Oleinick and Koritz (1966) have studied the isomerization of androst-5-ene-3,17-dione in a deuterated medium under several sets of conditions: (a) enzyme catalysis by a rat adrenal microsomal fraction activated by DPN^+; (b) nonenzymatic reaction at pH 7.0; (c) acid catalysis, and (d) base catalysis. They used infrared spectra of the various possible deuterium-labeled androstenediones to determine the positions of D-incorporation into the molecule (see Malhotra and Ringold, 1964). The position of the deuterium label was determined by the $C-D$ stretching frequencies: C-6β, 2140 cm^{-1}; C-6α, 2190 cm^{-1}; C-4, 2255 cm^{-1}; C-2β, 2140 cm^{-1}; and C-2,2, 2140 cm^{-1} and 2220 cm^{-1}.

In all cases, incorporation of deuterium into the 6β-position of the product was found. Incorporation at C-4 of the product was also found for all conditions except acid catalysis.

In a study by Albaum and Staib (1965) 4-androstene-3,17-dione was incubated with reduced nicotinamide adenine dinucleotide phosphate and liver microsomes which had been prepared from normal rats. The following metabolites were demonstrated by infrared spectroscopy: 7α-hydroxy-4-androstene-3,17-dione, 6β-hydroxy-4-androstene-3,17-dione, and androstane-dione. Another metabolite was very likely identical with testosterone. Differences due to sex were observed in the formation of the various metabolites.

In a study (Gaylor, 1963) of the biosynthesis of skin sterols, infrared analysis was used for product identifications and, in addition, for indicating the presence of residual halogen (758 cm^{-1}) in the chemical synthesis of one of the substances, lanosta-7,24-dien-3β-ol.

Frantz *et al.* (1966) have isolated a sterol from the skin of rats treated with triparanol (1-[p-(β-diethylaminoethoxy)phenyl]-1-(p-tolyl)-2-(p-chlorophenyl)ethanol). Infrared and NMR spectroscopy were used to characterize the new compound. The close similarity of the spectra of Δ^7-cholesten-3β-ol and $\Delta^{7,24}$-cholestadien-3β-ol was proof of the presence of the Δ^7 bond and of the identity of the ring system in the two compounds. The diminution in the absorption at 1368 cm^{-1}, prominent in cholesterol, Δ^7-cholesten-3β-ol, and other sterols with a saturated side chain, was evidence for the absence of the isopropyl group and compatible with the presence of a Δ^{24} or Δ^{25} bond. The absence of a band at 887 cm^{-1} eliminated the latter possibility.

The 1β-hydroxylation of 3α,17α,20β,21-tetrahydroxy-5β-pregnan-11-one has been shown by Schneider and Bhacca (1966) to occur *in vivo* in a man and *in vitro* by surviving liver slices of the guinea pig. The position and configuration of the metabolically introduced hydroxyl group were determined by chemical degradation and by NMR spectroscopy. Infrared spectroscopy was also used in the identification of 1β,3α,17α,20β,21-pentahydroxy-5β-pregnan-11-one.

Layne *et al.* (1964) have shown that radioactive 17α-estradiol excreted by the rabbit after the administration of 16-^{14}C-estrone is conjugated at position 17 to

N-acetylglucosamine. By means of ultraviolet and infrared spectroscopy and chemical procedures the conjugate was characterized as estra-1,3,5(10)-trien-3-ol-17α-yl-2'-acetamido-2'-deoxy-β-D-glucopyranoside.

Hofmann and Mosbach (1964) have used chromatography, infrared spectroscopy, and optical rotatory dispersion to identify allodeoxycholic acid as the major component of gallstones induced in the rabbit by 5α-cholestan-3β-ol.

Incubation of estrone with ox adrenal glands leads to the formation of 15α-hydroxyestrone and a second metabolite, which was identified by chemical and physical properties (Knuppen *et al.*, 1967), including the infrared spectrum: 3450 and 3220 cm^{-1} (hydroxyl stretching bands), 1717 cm^{-1} (carbonyl stretching of a C-17 ketone), and 1615, 1581, and 1490 cm^{-1} (carbon–carbon stretching vibrations of an aromatic ring). Ultraviolet and NMR data were also given.

Cholesterol sulfate occurs in bovine adrenal glands (Drayer *et al.*, 1964). This fact raises the possibility that the conjugate may represent an active form of cholesterol in metabolic processes other than those concerned with steroid hormone biosynthesis. Isolation of the sodium and pyridinium salts of cholesterol sulfate and recording of their infrared spectra led to the identification of these compounds.

Besides giving information concerning the infrared spectra of several aldosterone metabolites, Kohler *et al.* (1964) have made the observations that the lactones of the 5α,A:B-*trans* series have maxima of medium intensity at 910 and 880 cm^{-1}, while in the 5β,A:B-*cis* series these maxima are replaced by a maximum of high intensity at 903 cm^{-1}. Inspection of that portion of the infrared spectrum can serve as a guide for the assignment of configuration at C-5.

Saier (1968) has pointed out the advantages of using multiple internal reflection spectroscopy in the study of steroids, where only microgram quantities are available sometimes, and where many of the steroids are relatively insoluble in the usually accepted infrared solvents. She used steroid films containing 20 to 100 μg of compound. Only one MIR spectrum (that of pregnenolone) did not compare favorably with the one in the collection of Neudert and Röpke (1965), who used the KBr technique. With the MIR procedure two carbonyl bands appear in the 1720 cm^{-1} region, whereas KBr pellet spectra show only one band.

Jayle (1965) has presented and interpreted the spectra of some hormonal steroids such as hydrocarbon, acyloxy, and oxo types, in the 1300–450 cm^{-1} region. Jayle's atlas includes 240 spectra of steroids recorded in this spectral region.

A recent catalog by Bernstein *et al.* (1968) gives references to the original literature for the infrared spectra of many steroid conjugates, e.g., glucuronides and sulfates. This collection consists of 147 entries, an entry being composed of all the conjugate derivatives of a parent steroid. Other physical properties recorded are melting point, ultraviolet absorption, and other spectral, chromatographic, or allied information.

Metabolism in Plants

Sterols have been obtained from the seeds of *Xanthium strumarium* Linn., *Helianthus annuus* Linn., and the rhizome of *Smilax glabra* Roxb., *Veratrum grandiflorum* Loesen, *fil.* (Tsukamoto *et al.*, 1963). Gas chromatography and infrared data showed the compounds to be β-sitosterol, stigmasterol, and probably campesterol.

De Souza and Nes (1968) have isolated a crystalline mixture of unsaturated 24-ethylcholesterols from the blue–green alga *Phormidium luridum*. Various techniques were used for identification of the compounds—column and gas–liquid chromatography, as well as UV, mass, NMR, and infrared spectrometry. The positions of unsaturation were corroborated with an infrared band at 1033 cm^{-1} for Δ^7 unsaturation and a shoulder near 1050 cm^{-1} for a Δ^5 unsaturated component. A weak band at 968 cm^{-1} corresponded to a *trans*-Δ^{22} component.

Metabolism in Microorganisms

For a detailed account of steroid transformations by microorganisms the reader is referred to Dorfman and Ungar (1965) and Iizuka and Naito (1968).

Ringold *et al.* (1963) have studied the stereochemistry and mechanism of the bacterial C-1,2 dehydrogenation of steroids in *Bacillus sphaericus* (ATCC 7055) in whole cell and cell-free preparations. The enzymatic dehydrogenation of 2-deuterio-androst-4-ene-3,17-dione with a cell-free preparation in the presence of menadione (externally added electron acceptor) proceeded with a 75 % loss of deuterium, indicating (along with other evidence) a preference for 2β-hydrogen loss. The 2-deuterio compound which was prepared by zinc-deuterioacetic acid reduction of 2α-iodo-androst-4-ene-3,17-dione was very likely primarily 2β-labeled in analogy with the reduction reported by Corey and Sneen (1956) of 2α-bromo-5α-cholestan-3-one and on the basis of the carbon–deuterium stretching band in the infrared, the major band appearing at 2146 cm^{-1}. The shoulder at 2162 cm^{-1} and the weak bands at 2190 and 2220 cm^{-1}, however, indicated the presence of some 2α-deuterio isomer. A sample of 1-deuterioandrost-4-ene-3,17-dione, which according to its infrared spectrum was predominantly the 1β-deuterio compound, underwent only a 22 % loss of deuterium on enzymatic conversion to the 1,4-dienone, showing that the removal of deuterium at C-1 is not a random process.

Exposure of estrone to *Nocardia* sp. (E 110) resulted in the formation of three degradative products. These were characterized (Coombe *et al.*, 1966) as: (a) 3aα-H-4α-[3'-propanoic acid]-5β-[2-ketopropyl]-7aβ-methyl-1-indanone, (b) 3aα-H-4α-[3'-propanoic acid]-5β-[4'-but-3-enoic acid]-7aβ-methyl-1-indanone, and (c) 2-carboxy-7aβ-methyl-7-keto-9aα-H-indano-[5,4f]-5aα,10,10aβ,11-tetrahydroquinoline by melting point, infrared, ultraviolet, NMR, and mass spectra and by chromatography.

Sih *et al.* (1968) have frequently used infrared methods in delineating some of the steps of steroid oxidation by microorganisms of the genus *Nocardia*, specifically in the degradation of the cholesterol side chain, wherein C$_{22}$ acids were shown to be key intermediates.

Two enzymes are responsible for the cleavage of pregnane side chains by microorganisms. A steroid-inducible oxygenase, in the presence of reduced nicotinamide adenine dinucleotide phosphate and molecular oxygen, catalyzes the conversion of progesterone to testosterone acetate (Rahim and Sih, 1966). The latter ester is then hydrolyzed by an esterase to yield testosterone. Infrared spectra were used to prove the identity of testosterone acetate and the progesterone oxygenase reaction product.

REFERENCES

Albaum, G. and Staib, W. *Biochem. Z.* **342**, 120 (1965).

Angell, C. L., Gallagher, B. S., Ito, T., Smith, R. J. D., and Jones, R. N. *The Infrared Spectra of Lactones*, N.R.C. Bull. No. 7, Natl. Research Council, Ottawa, 1960.

Arcos, M. and Lieberman, S. *Biochemistry* **6**, 2032 (1967).

Barton, D. H. R., Page, J. E., and Warnhoff, E. W. *J. Chem. Soc.* **1954**, 2715.

Bellamy, L. J. *The Infrared Spectra of Complex Molecules*, 2nd Ed., Wiley, New York, 1958.

Bernstein, S., Dusza, J. P., and Joseph, J. P. *Physical Properties of Steroid Conjugates*, Springer-Verlag, New York, 1968.

Bladon, P., Fabian, J. M., Henbest, H. B., Koch, H. P., and Wood, G. W. *J. Chem. Soc.* **1951**, 2402.

Calvin, H. I. and Lieberman, S. *Biochemistry* **3**, 259 (1964).

Caspi, E. and Scrimshaw, G. F. in *Steroid Hormone Analysis*, Vol. I (H. Carstensen, ed.) Marcel Dekker, New York, 1967, p. 55.

Cole, A. R. H. *Revs. Pure and Appl. Chem. (Australia)* **4**, 111 (1954).

Cole, A. R. H. *Fortsch. Chem. Org. Naturstoffe* **13**, 27 (1956).

Cole, A. R. H., Jones, R. N., and Dobriner, K. *J. Am. Chem. Soc.* **74**, 5571 (1952).

Coombe, R. G., Tsong, Y. Y., Hamilton, P. B., and Sih, C. J. *J. Biol. Chem.* **241**, 1587 (1966).

Corey, E. J. *J. Am. Chem. Soc.* **75**, 2301 (1953).

Corey, E. J. and Burke, H. J. *J. Am. Chem. Soc.* **77**, 5418 (1955).

Corey, E. J. and Sneen, R. A. *J. Am. Chem. Soc.* **78**, 6269 (1956).

Cummins, E. G. and Page, J. E. *J. Chem. Soc.* **1957**, 3847.

Dickson, D. H. and Page, J. E. *J. Chem. Soc.* **1955**, 447.

Dobriner, K., Lieberman, S., Rhoads, C. P., Jones, N. R., Williams, V. Z., and Barnes, R. B. *J. Biol. Chem.* **172**, 297 (1948).

Dobriner, K., Kritchevsky, T. H., Fukushima, D. K., Lieberman, S., Gallagher, T. F., Hardy, J. D., Jones, R. N., and Cilento, G. *Science* **109**, 260 (1949).

Dobriner, K., Katzenellenbogen, E. R., and Jones, R. N. *Infrared Absorption Spectra of Steroids—An Atlas*, Vol. I, Interscience, New York, 1953.

Dorfman, R. I. and Ungar, F. *Metabolism of Steroid Hormones*, Academic Press, New York, 1965.

Drayer, N. M., Roberts, K. D., Bandi, L., and Lieberman, S. *J. Biol. Chem.* **239**, PC 3112 (1964).

Eger, C. and Norton, D. A. *Nature* **208**, 997 (1965).

Eliel, E. L., Allinger, N. L., Angyal, S. J., and Morrison, G. A. *Conformational Analysis*, Interscience, New York, 1965, p. 256.

Fieser, L. F. and Fieser, M. *Steroids*, Reinhold, New York, 1959.

Frantz, I. D., Jr., Scallen, T. J., Nelson, A. N., and Schroepfer, G. J., Jr. *J. Biol. Chem.* **241**, 3818 1966).

Furchgott, R. F., Rosenkrantz, H., and Shorr, E. *J. Biol. Chem.* **163**, 375 (1946).

Fürst, A., Kuhn, H. H., Scotoni, R., Jr., and Günthard, Hs. H. *Helv. Chim. Acta* **35**, 951 (1952).

Gaylor, J. L. *J. Biol. Chem.* **238**, 1649 (1963).

Gual, C., Dorfman, R. I., and Rosenkrantz, H. *Spectrochim. Acta* **13**, 248 (1958).

Hadd, H. E. and Blickenstaff, R. T. *Conjugates of Steroid Hormones*, Academic Press, New York, 1969.

Hadd, H. E. and Dorfman, R. I. *J. Biol. Chem.* **238**, 907 (1963).

Hashimoto, Y. and Neeman, M. *J. Biol. Chem.* **238**, 1273 (1963).

Henbest, H. B., Meakins, G. D., and Wood., G. W. *J. Chem. Soc.* **1954**, 800.

Herget, C. M. and Shorr, E. *Am. J. Physiol.* **133**, P323 (1941).

Hirschmann, H. *J. Am. Chem. Soc.* **74**, 5357 (1952).

Hofmann, A. F. and Mosbach, E. H. *J. Biol. Chem.* **239**, 2813 (1964).

Iizuka, H. and Naito, A. *Microbial Transformation of Steroids and Alkaloids*, University Park Press, Baltimore, 1968.

Jayle, M. F. *Analysis of Hormonal Steroids, Vol. 3: IR Spectra 1300–450 cm⁻¹*, Masson et Cie, Paris, 1965.

Jones, R. N., *J. Am. Chem. Soc.* **72**, 5322 (1950).

Jones, R. N. *Trans. Roy. Soc. Can. Sect. III* **52**, 9 (1958).

Jones, R. N. and Cole, A. R. H. *J. Am. Chem. Soc.* **74**, 5648 (1952).

Jones, R. N. and Dobriner, K. *Vitamins and Hormones*, Vol. 7 (R. S. Harris and K. V. Thimann, eds.) Academic Press, New York, 1949, p. 293.

Jones, R. N. and Gallagher, B. S. *J. Am. Chem. Soc.* **81**, 5242 (1959).

Jones, R. N. and Herling, F. *J. Org. Chem.* **19**, 1252 (1954).

Jones, R. N. and Herling, F. *J. Am. Chem. Soc.* **78**, 1152 (1956).

Jones, R. N. and Roberts, G. *J. Am. Chem. Soc.* **80**, 6121 (1958).

Jones, R. N. and Sandorfy, C. in *Technique of Organic Chemistry* (A. Weissberger, ed.) *Vol. 9, Chemical Applications of Spectroscopy* (W. West, ed.) Interscience, New York, 1956, p. 247.

Jones, R. N., Williams, V. Z., Whalen, M. J., and Dobriner, K. *J. Am. Chem. Soc.* **70**, 2024 (1948).

Jones, R. N., Humphries, P., and Dobriner, K. *J. Am. Chem. Soc.* **71**, 241 (1949).

Jones, R. N., Humphries, P., Packard, E., and Dobriner, K. *J. Am. Chem. Soc.* **72**, 86 (1950a).

Jones, R. N., Humphries, P., and Dobriner, K. *J. Am. Chem. Soc.* **72**, 956 (1950b).

Jones, R. N., Humphries, P., Herling, F., and Dobriner, K. *J. Am. Chem. Soc.* **73**, 3215 (1951).

Jones, R. N., Ramsay, D. A., Herling, F., and Dobriner, K. *J. Am. Chem. Soc.* **74**, 2828 (1952a).

Jones, R. N., Ramsay, D. A., Keir, D. S., and Dobriner, K. *J. Am. Chem. Soc.* **74**, 80 (1952b).

Jones, R. N., Cole, A. R. H., and Nolin, B. *J. Am. Chem. Soc.* **74**, 5662, 6321 (1952c).

Jones, R. N., Humphries, P., Herling, F., and Dobriner, K. *J. Am. Chem. Soc.* **74**, 2820 (1952d).

Jones, R. N., Katzenellenbogen, E., and Dobriner, K. *J. Am. Chem. Soc.* **75**, 158 (1953).

Jones, R. N., Herling, F., and Katzenellenbogen, E. *J. Am. Chem. Soc.* **77**, 651 (1955).

Jones, R. N., Augdahl, E., Nickon, A., Roberts, G., and Whittingham, D. J. *Ann. N.Y. Acad. Sci.* **69**, Art. 1, 38 (1957).

Jones, R. N., Angell, C. L., Ito, T., and Smith R. J. D. *Can. J. Chem.* **37**, 2007 (1959).

Kasturi, T. R. in *Chemical Applications of Infrared Spectroscopy* by C. N. R. Rao, Academic Press, New York, 1963, Chapter VIII.

Klyne, W. *The Chemistry of the Steroids*, Methuen, London, 1957.

Knuppen, R., Haupt, O., and Breuer, H. *Biochem. J.* **105**, 971 (1967).

Kohler, H., Hesse, R. H., and Pechet, M. M. *J. Biol. Chem.* **239**, 4117 (1964).

Korenman, S. G., Wilson, H., and Lipsett, M. B. *J. Biol. Chem.* **239**, 1004 (1964).

Layne, D. S., Sheth, N. A., and Kirdani, R. Y. *J. Biol. Chem.* **239**, 3221 (1964).

Liddel, U. and Becker, E. D. *Spectrochim. Acta* **10**, 70 (1957).

van Lier, J. E. and Smith, L. L. *Biochemistry* **6**, 3269 (1967).

Malhotra, S. K. and Ringold, H. J. *J. Am. Chem. Soc.* **86**, 1997 (1964).

Mesley, R. J., and Johnson, C. A. *J. Pharm. Pharmacol.* **17**, 329 (1965).

Mesley, R. J. *Spectrochim. Acta* **22**, 889 (1966).

Nagarajan, R. and Just, G. *Can. J. Chem.* **39**, 1274 (1961).

Neudert, W. and Röpke, H. *Atlas of Steroid Spectra*, Springer-Verlag, New York, 1965.

Noguchi, S. and Fukushima, D. K. *J. Biol. Chem.* **241**, 761 (1966).

Nolin, B. and Jones, R. N. *J. Am. Chem. Soc.* **75**, 5626 (1953).

Norton, D. A., Kartha, G., and Lu, C. T., *Acta Cryst.* **16**, 89 (1963).

Norton, D. A., Kartha, G., and Lu, C. T. *Acta Cryst.* **17**, 77 (1964).

Oleinick, N. L. and Koritz, S. B. *Biochemistry* **5**, 3400 (1966).

Page, J. E. *J. Chem. Soc.* **1955**, 2017.

Page, J. E. *Chem. and Ind.* (*London*) **1957**, 58.

Parker, F. S. and Bhaskar, K. R. *Biochemistry* **7**, 1286 (1968).

Pimentel, G. C. and McClellan, A. L. *The Hydrogen Bond*, Freeman, San Francisco, 1960, p. 82.

Rahim, M. A. and Sih, C. J. *J. Biol. Chem.* **241**, 3615 (1966).

Ramsay, D. A. *J. Am. Chem. Soc.* **74**, 72 (1952).

Rao, C. N. R. *Chemical Applications of Infrared Spectroscopy*, Academic Press, New York, 1963.

Ringold, H. J., Hayano, M., and Stefanovic, V. *J. Biol. Chem.* **238**, 1960 (1963).

Roberts, G., Gallagher, B. S., and Jones, R. N. *Infrared Absorption Spectra of Steroids—An Atlas, Vol. II*, Interscience, New York, 1958.

Robertson, J. M. *Organic Crystals and Molecules*, Cornell University Press, Ithaca, 1953, Chapter 6.

Rosenkrantz, H. in *Methods of Biochemical Analysis, Vol. II* (D. Glick, ed.) Interscience, New York, 1955, p. 1.

Rosenkrantz, H. and Gual, C. *Spectrochim. Acta* **13**, 291 (1959).

Rosenkrantz, H. and Zablow, L. *J. Am. Chem. Soc.* **75**, 903 (1953).

Rothman, E. S. and Wall, M. E. *J. Am. Chem. Soc.* **77**, 2229 (1955).

Saier, E. L. *Appl. Spectrosc.* **22**, 445 (1968).

Schneider, J. J. and Bhacca, N. S. *J. Biol. Chem.* **241**, 5313 (1966).

Sih, C. J., Wang, K. C., and Tai, H. H. *Biochemistry* **7**, 796 (1968).

Slomp, G., Jr. and Johnson, J. L. *J. Am. Chem. Soc.* **80**, 915 (1958).

Slomp, G., Jr., Shealy, Y. F., Johnson, J. L., Donia, R. A., Johnson, B. A., Holysz, R. P., Pederson, R. L.,
    Jensen, A. O., and Ott, A. C. *J. Am. Chem. Soc.* **77**, 1216 (1955).

Smakula, E., Gori, A., and Wotiz, H. H. *Spectrochim. Acta* **9**, 346 (1957).

Sober, H. A. *Handbook of Biochemistry*, The Chemical Rubber Co., Cleveland, Ohio, 1968, p. F-3.

de Souza, N. J. and Nes, W. R. *Science* **162**, 363 (1968).

Tsukamoto, T., Yagi, A., and Mihashi, K. *Japanese J. Pharmacognosy*, **17**, 11 (1963). (Syôyakugaku
    Zasshi).

Turnbull, J. H., Whiffen, D. H., and Wilson, W. *Chem. and Ind.* (*London*) **1950**, 626.

Wall, M. E. and Serota, S. *Tetrahedron* **10**, 238 (1960).

Weinman, J. *Bull. Soc. Chim. Fr.* **1967** (11), 4259; *CA* **68**, 82691s. (1968).

Wexler, A. S. *Appl. Spectrosc. Rev.* **1**, 29 (1967).

Wiggins, H. S. and Klyne, W. *Chem. and Ind.* (*London*) **1955**, 1488.

YoungLai, E. and Solomon, S. *Biochemistry* **6**, 2040 (1967).

Zull, J. E., Greanoff, S., and Adam, H. K. *Biochemistry* **7**, 4172 (1968).

Chapter 14

PORPHYRINS AND RELATED COMPOUNDS

Sample Preparation

Since the solubility of porphyrins in the usual infrared solvents (e.g., carbon disulfide and carbon tetrachloride) is limited, spectral studies of porphyrins in such solvents are limited. Bromoform has been used as a solvent for some porphyrins (Alben and Caughey, 1968). Aqueous solutions have been used for porphyrin-containing substances, e.g., metmyoglobin of the sperm whale (McCoy and Caughey, 1967) and various hemoglobins and myoglobins (Caughey *et al.*, 1969). The pressed-disk method is especially suited, although porphyrins can also be studied as films deposited from solution on silver chloride slides, and as mineral oil mulls. Porphyrin spectra are so complex that extraneous bands due to moisture or contaminants must be excluded. Spectra of poorer quality frequently result from hydrates and free acids, as compared with esters. Changes from a crystalline form to an amorphous film may cause changes in shapes or positions of absorption bands. Also, spectral changes frequently become appreciable when recrystallization begins to take place within a plated-out film. Schwartz *et al.* (1960) have reported that large effects of this kind have been found for coproporphyrin III ester and mesoporphyrin IX ester.

Characteristic Group Frequencies

Table 14.1 shows characteristic group frequencies of porphyrins published by Schwartz *et al.* (1960). Table 14.2 shows a comparison of low-frequency assignments for porphin and tetraphenylporphin.

Porphyrins and Metalloporphyrins

Marks (1969) has discussed the physical properties of tetrapyrroles and has presented infrared spectral data for porphyrin and heme esters in the 1800–1500 cm^{-1} region as found by Falk and Willis (1951). (See Table 14.3.)

The biological and chemical importance of metalloporphyrins has caused much interest in the nature of the metal–ligand linkage in such complexes. Fleischer (1970) has discussed certain structural properties of the porphyrin molecule necessary for understanding such physical properties as its solubility, magnetic susceptibility, and visible absorption, electron spin resonance (ESR), and NMR spectra. He has also

Table 14.1. Characteristic Group Frequencies of Porphyrins (Schwartz *et al.*, 1960)

Functional group	Stretching frequency,[a] cm^{-1}	Bending and general deformation
OH		
Alcohols	3340–3330 (m–s)	1100–1050 (s)
Water	3340–3320 (s)	1600 (m)
Acids	3320–2700 (br, m)	910 (m) (tentative)
NH, pyrrolic	3330–3320 (m)	1110 (m) (all porphyrins)
=CH, unsaturated or aromatic	3100–3000 (w)	1440 (m), pyrrolic hydrogens 990 (m) and 913–900 (s), vinyl group 835 (s), methylidyne hydrogen
CH, saturated	2960–2840 (m)	1460–1440, and 1370 (m) (methyl only)
C=O		
Esters	1735–1725 (s)	
Acids	1710–1680 (s)	
Conjugated	1660–1630 (s)	
Carboxylate ion	1600–1560 (s) and 1430–1400 (s)	
C=C		
Olefinic	1680–1630 (w)	
Conjugated or aromatic	1610–1580 (m)	
C=N, pyrrolic	1350 (tentative) (m)	
C—O		
Acids	1270–1230 (m)	
Esters	1200–1150 (s)	
Alcohols	1100–1050 (s)	

[a]Key: br, broad; m, medium; s, strong; w, weak.

related structural data to the clarification of chemical features such as the rates and mechanism of metalloporphyrin formation and decomposition, ligand reactions at the metal center of the metalloporphyrins, substitution reactions on the ring system, and oxidation–reduction reactions of the porphyrin and metalloporphyrin system.

Boucher and Katz (1967*b*) have studied the infrared spectra of complexes of the divalent metals Co, Ni, Cu, Zn, Pd, Ag, Cd, and Mg with protoporphyrin IX dimethyl ester and hematoporphyrin IX dimethyl ester. They found a strong absorption near 350 cm^{-1}, which they attributed to a coupling of the metal–nitrogen stretch with a porphyrin skeletal deformation. The frequencies of the porphyrin skeletal absorptions, particularly those at 970–920 and 530–500 cm^{-1}, are metal-ion dependent. Metal ions, in the order Pd > Co > Ni > Cu > Ag > Zn > Cd > Mg, cause the bands to shift to higher frequency, paralleling the metal–ligand coordinate-bond strength.

Caughey *et al.* (1966) have done infrared studies of a series of substituted deutero-porphyrins IX (LXXXIV) which had been obtained from protoporphyrin IX or its iron(III) chloride (protohemin). These investigations have permitted vibrational

Table 14.2. Comparison of Low Frequency Assignments[a] for Porphin and Tetraphenylporphin (Schwartz et al., 1960)

Vibration	Wave number, cm^{-1}		
	Ref. b	Ref. c	Ref. d
Pyrrole ring deformation	1185		
	1137	1110	
	841 (a)		
	731	770 (d)	
	720 (c)	690	
	650	620	
Pyrrolic CH		1224 (ip)	990
	1057	1184 (ip)	980
	1048	1048 (ip)	965
	773 (d)	853 (op)	
Methylidyne CH	1237 (ip)		
	951 (op)	841 (a) (op)	
C—C stretching	992		
	969 (b)		
N—H bending	694 (d)	970 (b)	
		719 (c)	
C—C—C bending	900 (ip)		
	855 (op)		
	795 (ip)		
	746 (ip)		

[a]Letters a–d in parentheses designate some of the bands to which conflicting assignments have been given. Each of the authors has drawn upon rather different evidence in making assignments. Nelson compared porphin with pyrrole and phthalocyanine. Mason studied spectral changes occurring with tetrasubstitution of methylidyne hydrogens and octasubstitution of pyrrolic hydrogens. Thomas and Martell in their study of tetraphenyl porphins were chiefly interested in phenyl substituents and did not assign many other bands. Key: ip, in-plane; op, out-of-plane.
[b]Nelson (1954).
[c]Mason (1958).
[d]Thomas and Martell (1956).

assignments for most of the peripheral ring substituents as well as more equivocal assignments for bands of the porphyrin ring as a whole. Among the substituted deuteroporphyrins IX were acetyl, formyl, cyano, propionyl, methoxycarbonyl, and other derivatives.

Boucher and Katz (1967b) have made assignments for a number of ligand absorption bands, which are independent of the metal ion, and these assignments are in general agreement with those reported earlier by Caughey et al. (1966) for some metal-free porphyrins. The data of Boucher and Katz suggest that the metal ion is involved in π-bonding with the ligand in the case of the Ni, Co, and Pd metalloporphyrins and that for the other metal atoms, e.g., Cu and Zn, the interaction must be considerably weaker.

Table 14.3. Infrared Spectra of Porphyrin and Heme Esters in the 1800–1500 cm^{-1} Region (after Falk and Willis, 1951). (From Marks, 1969.)

Compound	Absorption frequencies of solid: rock-salt prism (cm^{-1})						Absorption frequencies in chloroform solution: fluorite prism (cm^{-1})					
Mesoporphyrin	1730						1733					
Mesoheme	1730						1734					
Uroporphyrin I	1734						1735					
Uroheme I	1734						1735					
2-Monoformyldeuteroporphyrin	1734		1655				1734		1658			
4-Monoformyldeuteroporphyrin	1736		1654				1734		1659			
Monoformyldeuteroheme	1732		1665				1735		1669			
Diformyldeuteroporphyrin	1736		1673	1658[b]	1584[a]	1533	1734	1674[b]	1657			
Diformyldeuteroheme	1726		1663		1560	1545	1734		1659			
Monoacetyldeuteroporphyrin	1735		1644		1542		[c]					
Monoacetyldeuteroheme	1736		1656				1734		1665			
Diacetyldeuteroporphyrin	1736	1718[b]	1650		1571	1520	1735		1658			
Diacetyldeuteroheme	1732		1656		1558	1543	1735		1662			
Chlorocruoroporphyrin	1734		1654				1733		1658			
Phylloerythrin porphyrin	1735		1690				1732	1691				
Phylloerythrin heme	1732		1696				1735[a]	1694				
Methylpheophorbide a	1736		1698	1620	1584			1699		1636	1620	1584
Methylpheophorbide b	1736		1700	1654	1556	1543	1735	1719	1702	1662	1619 1604	1589

[a] Weak band

[b] Shoulder

[c] Specimen not available for study with the fluorite prism.

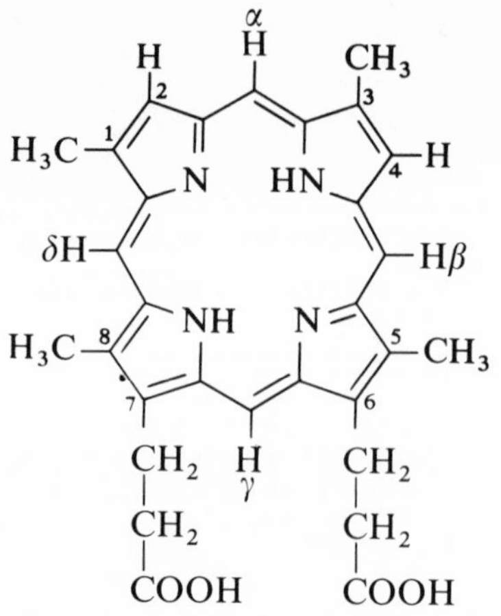

H $\overset{\alpha}{H}$ CH$_3$

H$_3$C—   N   HN   —H

δH—   —Hβ

H$_3$C—   NH   N   —CH$_3$

CH$_2$ $\overset{}{\underset{\gamma}{H}}$ CH$_2$

CH$_2$ CH$_2$

COOH   COOH

LXXXIV

Fuchsman and Caughey (1967), in order to examine the properties of oxygenated hemes, have prepared pure oxo-2,4-diacetyldeuteroporphyrin IX dimethyl ester iron. From ESR and Mössbauer spectroscopy and measurement of the infrared absorption of the acetyl carbonyl, the effective oxidation state of iron in the complex appeared to be +3. Infrared absorption attributable to oxygen was close to the expected frequency for free O$_2$, and was suggestive of binding parallel to the heme plane.

Alben *et al.* (1968) have studied the autoxidation of iron porphyrins, with emphasis on Fe(II) compounds and have discussed mechanisms of autoxidations, which they based on electronic, infrared, and NMR spectra.

Boucher and Katz (1967a) have investigated manganese protoporphyrin IX dimethyl ester complexes with anionic ligands: F$^-$, Cl$^-$, Br$^-$, I$^-$, CN$^-$, and SCN$^-$. Elemental analyses and physical and spectral properties were consistent with the formulation, [Mn(III)porphyrin X]. Infrared spectra of the complexes, in the solid state and in nonpolar solvents, gave evidence for coordination of the anions to an axial position of the metal. Visible spectra also supported this conclusion.

Porphyrin-Containing Proteins

McCoy and Caughey (1967) have observed infrared spectra of aqueous solutions of sperm whale metmyoglobin that had been shown by visible spectra to exist as azido and cyano derivatives. Cyano metmyoglobin had one absorption band at 2040 cm^{-1}. Azido metmyoglobin had bands at 2023 and 2045 cm^{-1}. (See also McCoy and Caughey, 1970.) Azido protoporphyrin IX diethyl ester iron(III), known to be high-spin in CHCl$_3$ and as a solid, exhibited a v_{N-N} at 2062 cm^{-1} in both CHCl$_3$ and KBr. In pyridine (where this hemin exists as a mixture of high- and low-spin forms) infrared bands were found at 2010 and 2045 cm^{-1}. (For a discussion of high- and low-spin forms the reader is referred to Kotani, 1964.) The two v_{N-N} bands of azido metmyoglobin can also arise from species of different spin state, the 2023 cm^{-1}

band corresponding to a low-spin form and the 2045 cm^{-1} band corresponding to a high-spin form. The single v_{C-N} found for cyano metmyoglobin is in accord with only one (low) spin state for this derivative. The half-band widths of NaN_3 and KCN in water and the free hemin derivative in $CHCl_3$ are about twice that found for the proteins. These observations were consistent with a nonpolar environment for the protein-bound ligands.

The Binding of Carbon Monoxide

The Binding of Carbon Monoxide to Hemes

Alben and Caughey (1968) have studied the binding of carbon monoxide to hemes by high-resolution infrared difference spectroscopy. The spectrophotometer was adjusted to resolve water vapor bands at 1792.65 and 1790.95 cm^{-1}. All carbonyl stretching frequencies were calibrated to the water vapor band at 1942.6 cm^{-1}. The NH-stretching frequencies of metal-free porphyrins were calibrated from the 3447.20 cm^{-1} water band. A standard slit-width program, which was just insufficient to resolve water-vapor bands at 1910.1 and 1908.2 cm^{-1} was used, along with a five-fold expanded ordinate.

In a study of the effects of 2,4-substituents on v_{CO} of pyridinecarbonyldeuterohemes and v_{NH} of deuteroporphyrins and of the v_{CO} frequencies for the heme carbonyls in bromoform–pyridine, frequencies increased with changes in the 2,4-substituent in the order ethyl < hydrogen < vinyl < acetyl. This order also represents the order of increasing electron-withdrawing effect of the 2,4-substituents as reflected in nitrogen basicities (pK_3 values) of metal-free porphyrins. The v_{NH} values for metal-free porphyrins had a similar order.

A single sharp band at 1951 cm^{-1} (CO-stretching) for human carboxyhemoglobin (carbonyl hemoglobin) in the intact red blood cell indicated that only a single type of carbon monoxide environment can be present in carbonyl hemoglobin and that differences between the α and β chains do not affect the bound CO. The authors point out that shifts in v_{CO} have been observed with genetically abnormal hemoglobins which contain an amino acid substitution in the vicinity of the heme.

In their studies Alben and Caughey attempted an interesting use of HbCO prepared from carbon monoxide that was 53% enriched in ^{13}C or 86% enriched in ^{18}O. They wanted to identify v_{CO} frequencies by measuring the effects of atomic mass on the shift of position of $v_{12_C16_O}$ to that of $v_{13_C16_O}$ or that of $v_{12_C18_O}$. However, the presumed $^{13}C^{16}O$ and $^{12}C^{18}O$ complexes gave v_{CO} values too near to each other, and another infrared method was finally used (see p. 181 of Alben and Caughey, 1968, Table IV), i.e., they identified CO isotopic mixtures ($^{13}C^{16}O$ + $^{12}C^{16}O$; and $^{12}C^{18}O$ + $^{12}C^{16}O$) by comparing literature values with their own for vibrational–rotational spectra of the respective gas mixtures.

Alben and Caughey also compared mass effects for free gases with those for CO-hemoglobin complexes. For example, they gave the ratio (based on experimental values)

$$\frac{v(\text{Hb }^{13}\text{C }^{16}\text{O})}{v(\text{Hb }^{12}\text{C }^{16}\text{O})} = \frac{1906.88}{1950.3} = 0.97769$$

and compared this ratio for v_{CO} values with values based on mass effects

$$\frac{v^*}{v} = \frac{(M_A^* + M_B^*)/M_A^* M_B^*}{(M_A + M_B)/M_A M_B}$$

where M is the mass of atom A or B and * denotes the isotopic effect. The calculated ratio for $v(^{13}C^{16}O)/v(^{12}C^{16}O)$ was 0.977775. The predicted position from the theoretical isotopic shift due to mass only for $^{13}C(HbCO)$ is (1950.3) (0.977775) = 1906.5 cm^{-1}, which is the same as the value they observed. The prediction for the ^{18}O (HbCO) did not match experimental data so well.

The frequency shifts observed required that CO be directly bound to iron in HbCO. The best fit of their data with data from hemoglobin led to a preference for the structure Hb—Fe—O≡C, rather than the more conventional structure, Hb—Fe—C≡O.

The application of infrared difference spectroscopy as demonstrated in their work provides a new way to study metal–protein complexes with ligands such as carbon monoxide, cyanide, azide, and others, which absorb in the triple-bond region.

The Binding of Carbon Monoxide to Hemoglobins and Myoglobins

Caughey et al. (1969) have determined the infrared stretching frequencies for carbon monoxide (v_{CO}) bound to various hemoglobins and myoglobins. They confirmed the identity of v_{CO} by spectra of $^{12}C^{18}O$ and $^{13}C^{16}O$ derivatives. For human hemoglobins A, F, H, and Chesapeake (Charache et al., 1966) and sheep hemoglobins A, B, C, and lamb, they found one narrow CO absorption band near 1951 cm^{-1}. However, they found two absorption bands of similar area for hemoglobin $M_{Emory}(\alpha_2^A \beta_2^{63\,Tyr})$ (Gerald and Efron, 1961), at 1970 and 1950 cm^{-1} and for hemoglobin Zurich ($\alpha_2^A \beta_2^{63\,Arg}$) (Rieder et al., 1965), at 1958 and 1951 cm^{-1}. Thus, substitution for the amino acid at position 63 of the β chain (the position of the distal histidine in β^A) greatly affected the v_{CO} bound to that chain. However, they found no such effect when the amino acid substitution was in a position more remote from the ligand. The data of Caughey et al. (1969) suggest significant participation of the β^{63} residue in the binding of CO.

Myoglobins from horse heart and sperm whale displayed maximum CO absorptions at about 1944 cm^{-1}, which is consistent with stronger binding of CO than is the case for hemoglobin A.

Caughey et al. (1969) did not find direct correlation between v_{CO} and the affinities for CO or oxygen. Many of the hemoglobins studied by these workers differ markedly in oxygen affinity from hemoglobin A (Charache et al., 1966; van Vliet and Huisman, 1964; Benesch et al., 1961), and yet have essentially the same v_{CO} value. The lack of correlation between affinity and v_{CO} adds further support to the importance of such factors as changes in conformation (Muirhead and Perutz, 1963), α–β chain interactions (e.g., Antonini et al., 1965; Guidotti, 1967), and effects of 2,3-diphosphoglyceric acid (Benesch and Benesch, 1968), in addition to the strengths of ligand–heme binding, in the determination of overall equilibrium constants for ligand binding.

Chlorophyll and Related Compounds

Stair and Coblentz (1933) published the first infrared spectra of chlorophyll and the important related compounds ethyl chlorophyllide, pheophytin, and phytol. The subsequent studies by Weigl and Livingston (1953) and by Holt and Jacobs (1955) demonstrated well the usefulness and applicability of infrared spectroscopy to structural problems in chlorophyll chemistry. Many chlorophylls and related compounds have been studied by infrared spectroscopy. Marks (1969) has discussed chlorophylls and their derivatives and has presented infrared data of Holt and Jacobs (1955) for the region $\sim 3600\text{--}1604$ cm^{-1} (see Table 14.4). Katz *et al.* (1966) have provided a table of references to infrared spectra of the chlorophylls. Their review discusses in detail both infrared and NMR spectroscopy of chlorophylls and related compounds. They analyzed the 1750–1600 cm^{-1} (carbonyl) region of many compounds in the disaggregated monomeric state and in the aggregated state, the disaggregated form being present in polar solvents, and the aggregated form in nonpolar solvents and in the solid state. They distinguished the carbonyls of ester, ketone, aldehyde, aggregated-state ketone, and aggregated-state aldehyde, as well as skeletal $>C=C<$ and $>C=N-$ vibrations. Other regions for which they gave detailed analyses are 4000–2700, 1600–1300, 1300–650, and 650–200 cm^{-1}.

In the carbonyl region the infrared spectra of chlorophyll and its magnesium-containing derivatives show very strong solvent dependence. Although chlorophyll a when dissolved in nonpolar solvents shows four bands in this region (1750–1600 cm^{-1}), in polar solvents the spectra show only three bands. Magnesium-free derivatives display infrared spectra in this region that are not solvent dependent. Thus, the coordination properties of magnesium contribute much to the formation of infrared absorption bands of chlorophyll in this region.

The solvent dependence in the carbonyl region has been interpreted by Holt and Jacobs (1955) in terms of keto–enol tautomerism. However, a more consistent explanation can be offered (Katz *et al.*, 1963) by considering the coordination properties of magnesium and the effects on the ketone carbonyl absorption in ring V (for numbering of rings see LXXXV). According to this view, the coordination number of magnesium in chlorophylls is always greater than four. In solutions of polar solvents, the coordination unsaturation of the magnesium is primarily satisfied by coordination with solvent molecules, and chlorophyll exists in these solutions in monomeric form. In nonpolar solvents the coordination unsaturation of the magnesium is satisfied by coordination to ketone carbonyl oxygen of ring V, or in the case of chlorophyll b, to aldehyde carbonyl oxygen as well. Such coordination leads to aggregation of the chlorophyll. The carbonyl band moves to lower frequencies, and in solutions with nonpolar solvents, where dimers exist, half of the carbonyl oxygens of ring V are free and half are coordinated to magnesium. The carbonyl absorption thus appears as two bands: (a) the normal ketone oxygen absorption; (b) the Mg to $C=O$ coordination absorption, at somewhat lower frequency. The Mg to $C=O$ coordination absorption is called an *aggregation* or association band. Its presence in an infrared spectrum is indicative of chlorophyll aggregates.

The infrared spectrum of chlorophyll (1880–1600 cm^{-1}) gives evidence of aggregation effects in Fig. 14.1. Curve A of this figure is the spectrum of highly aggregated

Table 14.4. Infrared Absorption Bands of Chlorophylls and Derivatives; Frequencies Corrected (cm⁻¹) (after Holt and Jacobs, 1955). (Marks, 1969.)

Compound	Solvent	NH	OH	C—H Phytol	C—H Aldehyde	C=O Ester	C=O Ketone	C=O Carboxyl C-7	C=O Aldehyde	C=O Chelate	C=C
Ethyl chlorophyllide a	$CHCl_3$	....	....	....	....	1733	1678	....	....	....	1610
Chlorophyll a	Crystals in Nujol	3605?	3440+ 3252	2920	....	1727+ 1740	1694	....	....	1640	1604
Chlorophyll a	CCl_4, CS_2	....	3350 (weak)	2916	....	1735	1691	....	....	1652	1610
Chlorophyll a	Ethyl ether	....	....	....	....	1738	1698	....	....	....	1608
Chlorophyll a	Pyridine	....	3400 strong	....	....	1740	1683	....	....	....	....
Ethyl chlorophyllide b	$CHCl_3$	....	....	....	2720	1730	1689	....	1657	....	1610
Chlorophyll b	$CHCl_3$	....	....	2920	2720	1728	1688	....	1655	....	1610
Chlorophyll b	Crystals in Nujol	....	3360	2924	2710	1738	1697	....	1655	....	1607
Chlorophyll b	CCl_4, CS_2	....	3350	2924	2710	1738	1701	....	1664	....	1609
Pheophytin a	CCl_4	3392	....	2930	....	1740	1705	....	....	....	1618
Pheophytin a	Pyridine	3382	3100– 3700	2920	....	1736	1700	....	....	....	1617
Ethyl pheophorbide a	CS_2	3375	....	....	....	1743	1706	....	....	....	1618
Ethyl pheophorbide a	$CHCl_3$	3380	....	....	....	1736	1697	....	....	....	1620
Pheophorbide a	$CHCl_3$	....	2400– 2800	....	....	C-10: 1741	1702	1702	....	....	1622
Pheophorbide a-2,4-dinitrophenylhydrazone	$CHCl_3$	....	....	....	....	C-10: 1728	....	1705	....	....	1613
Pyropheophorbide a	$CHCl_3$	....	....	....	....	....	1685	1704	....	....	1620
Pyropheophorbide a-2,4-dinitrophenylhydrazone	Nujol mull	....	....	....	....	....	....	1725	....	....	1610
Pheophytin b	CCl_4	3410	....	2930	2720	1740	1709	....	1665	....	1618
Ethylpheophorbide b	$CHCl_3$	3380	....	....	2720	1737	1704	....	1663	....	1620
Ethyl pheophorbide b-2,4-dinitrophenylhydrazone	$CHCl_3$	....	....	....	....	1740	1710	....	....	....	1618
Bacteriochlorophyll	CCl_4	....	3350	2916	....	ester 1735	ketone 1683	1683	....	acetyl 1655	1610
Methyl bacteriochlorophyllide	$CHCl_3$	....	....	....	....	1727	1662	1662	....	1662	1610

chlorophyll *a* in carbon tetrachloride. The important characteristic is a very strong aggregation band at about $1640 \, \text{cm}^{-1}$, and an uncoordinated (weak) $C=O$ band at $1695 \, \text{cm}^{-1}$. The composition of the solution is that of a trimer or greater. Curve B is the spectrum of the same solution warmed for 1 minute at 60°C; here, the aggregation band is weakened, and the solution composition is about that of a dimer. Curve C is for the same sample in tetrahydrofuran. The aggregation band is missing, and the chlorophyll exists as the monomer. Molecular weight determinations by Aronoff (1962) and by Katz *et al.* (1963) support these conclusions regarding the relationship of the presence of an aggregation band, the presence of chlorophyll aggregates in nonpolar solvents, and the disaggregating effects of polar solvents.

The greater the frequency difference between the free carbonyl band and the aggregation band, the greater the degree of aggregation. Thus, the aggregation band of chlorophyll *a* in $CHCl_3$ and in CCl_4 lies at $\sim 1650 \, \text{cm}^{-1}$, whereas the aggregation band in highly oriented solid chlorophyll *a* occurs at $1640 \, \text{cm}^{-1}$. The intensity of the aggregation band is also related to the degree of aggregation. The relative intensity of the aggregation band at $\sim 1650 \, \text{cm}^{-1}$ increases as the concentration of chlorophyll

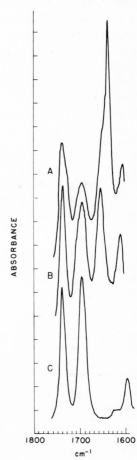

Fig. 14.1. Aggregation and disaggregation in chlorophyll *a*. Curve *A*, sample dissolved in CCl_4; curve *B*, solution warmed for 1 min at 60°C; curve *C*, same material dissolved in tetrahydrofuran. (Katz *et al.*, 1966.)

a in CCl$_4$ increases, presumably a result of higher aggregate concentration with greater chlorophyll concentration. The intensity of the aggregation band is also increased in trichloroethylene or cyclohexane. For chlorophyll *b*, the relative intensity of the band at ~ 1600 cm^{-1} seems to parallel the extent of aggregation, and is greater for the oriented solid than for the solution in CCl$_4$.

Another argument against interpreting the solvent dependence of the infrared spectra of chlorophylls as being due to the enol form (Holt and Jacobs, 1955) is that infrared and NMR data show that chlorophylls *a* and *b* exist in solution predominantly in the keto form (β-keto ester of ring V) (Pennington *et al.*, 1964).

Other chlorophylls also exhibit aggregation bands in nonpolar solvents: pyro-chlorophyll *a*, methyl chlorophyllides, chlorobium chlorophyll (660), and bacteriochlorophyll. The intensity of the 1610 cm^{-1} aggregation band for the last compound can be used to estimate its degree of aggregation. Aggregation is exhibited not only by magnesium-containing compounds, but also by others. The spectra of the divalent metal–pheophytin *a* derivatives can be interpreted in a similar way. Zinc and copper pheophytin *a* spectra and zinc and copper pyropheophytin *a* spectra exhibit aggregation bands at 1650 to 1640 cm^{-1} in nonpolar solvents (Katz *et al.*, 1966), but mercury(II) and cadmium pheophytin *a* and silver pyropheophytin *a* show no evidence of aggregation bands in nonpolar solvents.

The region 650 to 200 cm^{-1} contains absorption bands which originate in metal–ligand fundamental stretching vibrations, as well as some that arise from bending or deformation modes of organic moieties. The spectra of the chlorophylls *a* and *b* (Fig. 14.2) are complex and show several weak and medium bands. Most of these bands are associated with the chlorin ring vibrations and are difficult to assign, but a pair of medium bands at 400 to 370 cm^{-1} are logically assignable to a pyrrole ring deformation, since it is common to chlorins, porphyrins, and pyrroles (Boucher *et al.*,

Table 14.5. Nomenclature and Proton Designations for Structure LXXXV (Katz *et al.*, 1966)

Compound[a]	R	Proton no. of R	R'	Proton no. of R'	R''	Proton no. of R''
Methyl pheophorbide *a*	CH$_3$	3a	CH$_3$	12	CO$_2$CH$_3$	11
Methyl pheophorbide *b*	CHO	3b	CH$_3$	12	CO$_2$CH$_3$	11
Methyl pyropheo-phorbide *a*	CH$_3$	3a	CH$_3$	12	H	10
Methyl bacterio-pheophorbide[b]	CH$_3$	3a	CH$_3$	12	CO$_2$CH$_3$	11
Methyl chlorobium pheophorbide 660[c]	CH$_3$	3a	CH$_3$	12	H	10

[a] In the magnesium-free pheophorbides and pheophytins, the $>$NH protons are designated 13, 14.
[b] In bacteriochlorophyll, the 2 position is occupied by an acetyl, CH$_3$CO-group; these protons are designated as 2; Ring II contains two extra hydrogen atoms, corresponding to protons 7 and 8, and these are designated as 3° and 4°.
[c] Position 2 in chlorobium chlorophyll 660 is presumed to contain a hydroxy ethyl group, CH$_3$CHOH —. These are designated as protons 2 and 2'.

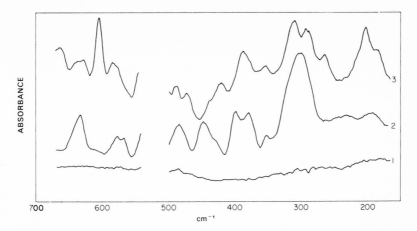

Fig. 14.2. Far-infrared spectra of chlorophylls in cyclohexane. Curve *1*, cyclohexane solvent; curve *2*, chlorophyll *b*, suspension; curve *3*, saturated chlorophyll *a*. (Katz *et al.*, 1966.)

1966). The 306 and 312 cm^{-1} bands come from the magnesium–carbonyl (or aldehyde) oxygen vibration in the aggregate form of chlorophylls *a* and *b*, and serve as evidence of an Mg—O bond in chlorophyll aggregates. The bands at 292 and 195 cm^{-1}, and at 293 and 196 cm^{-1}, for chlorophylls *a* and *b*, respectively, are assignable to Mg—N stretching and bending modes. The spectra of aggregated chlorophylls *a* and *b* in mineral oil mulls, cyclohexane, and benzene are significantly different from the spectra of the monomeric materials measured in methanol or pyridine–cyclohexane solution, particularly in the region 320 to 260 cm^{-1} (Figs. 14.3 and 14.4).

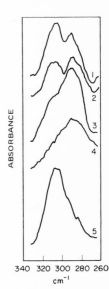

Fig. 14.3. Far-infrared spectra of chlorophyll *a* in various media. Curve *1*, cyclo-hexane; curve *2*, Nujol mull; curve *3*, pyridine–cyclohexane (10 % v/v); curve *4*, methanol–cyclohexane (1 % v/v); curve *5*, benzene. (Katz *et al.*, 1966.)

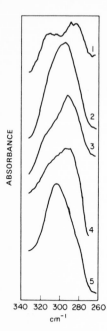

Fig. 14.4. Far-infrared spectra of chlorophyll *b* in various media. Curve *1*, cyclo-hexane; curve *2*, Nujol mull; curve *3*, pyridine–cyclohexane (10% v/v); curve *4*, methanol–cyclohexane (1% v/v); curve *5*, benzene. (Katz *et al.*, 1966.)

Attempts have been made in recent years to determine whether chlorophyll possesses exchangeable hydrogen. Efforts to detect labile hydrogen by tritium studies have been largely inconclusive (Coleman and Vishniac, 1963), as were various infrared studies (Karyakin and Chibisov, 1962; Sidorov, 1962, 1963; Kutyurin *et al.*, 1961). NMR spectroscopy (Dougherty *et al.*, 1965) has established that the δ and the C-10 protons of chlorophyll are labile, and exchange with the deuteron in the hydroxyl group of methanol d_4 in the solvent $CDCl_3$ (see LXXXV and Table 14.5

LXXXV

for proton designations). The C-10 proton undergoes exchange at a rate at least two orders of magnitude greater than the exchange rate at the δ position.

REFERENCES

Alben, J. O. and Caughey, W. S. *Biochemistry* **7**, 175 (1968).
Alben, J. O., Fuchsman, W. H., Beaudreau, C. A., and Caughey, W. S. *Biochemistry* **7**, 624 (1968).
Antonini, E., Bucci, E., Fronticelli, C., Wyman, J., and Rossi-Fanelli, A. *J. Mol. Biol.* **12**, 375 (1965).
Aronoff, S. *Arch. Biochem. Biophys.* **98**, 344 (1962).
Benesch, R. and Benesch, R. E. *Science* **160**, 83 (1968).
Benesch, R. E., Ranney, H. M., Benesch, R., and Smith, G. M. *J. Biol. Chem.* **236**, 2926 (1961).
Boucher, L. J. and Katz, J. J. *Abstracts of the 154th National Meeting of the American Chemical Society*, Chicago, Sept. 1967a, Paper 250 C.
Boucher, L. J. and Katz, J. J. *J. Am. Chem. Soc.* **89**, 1340 (1967b).
Boucher, L. J., Strain, H. H., and Katz, J. J. *J. Am. Chem. Soc.* **88**, 1341 (1966).
Caughey, W. S., Alben, J. O., Fujimoto, W. Y., and York, J. L. *J. Org. Chem.* **31**, 2631 (1966).
Caughey, W. S., Alben, J. O., McCoy, S., Boyer, S. H., Charache, S., and Hathaway, P. *Biochemistry* **8**, 59 (1969).
Charache, S., Weatherall, D. J., and Clegg, J. B. *J. Clin. Invest.* **45**, 813 (1966).
Coleman, B. and Vishniac, W. *Natl. Acad. Sci. Natl. Res. Council, Publ.* **1145**, 213 (1963).
Dougherty, R. C., Strain, H. H., and Katz, J. J. *J. Am. Chem. Soc.* **87**, 104 (1965).
Falk, J. E. and Willis, J. B. *Aust. J. Scient. Res.*, Series **A4**, 590 (1951).
Fleischer, E. B. *Accounts of Chemical Research* **3**, 105 (1970).
Fuchsman, W. H. and Caughey, W. S. *Abstracts of the 154th National Meeting of the American Chemical Society*, Chicago, Sept. 1967, Paper 92 C.
Gerald, P. S. and Efron, M. *Proc. Natl. Acad. Sci. U.S.* **47**, 1758 (1961).
Guidotti, G. *J. Biol. Chem.* **242**, 3673 (1967).
Holt, A. S. and Jacobs, E. E. *Plant Physiol.* **30**, 553 (1955).
Karyakin, A. V. and Chibisov, A. K. *Opt. Spectrosk. (USSR) (English Transl.)* **13**, 209 (1962).
Katz, J. J., Closs, G. L., Pennington, F. C., Thomas, M. R., and Strain, H. H. *J Am. Chem. Soc.* **85**, 3801 (1963).
Katz, J. J., Dougherty, R. C., and Boucher, L. J. in *The Chlorophylls* (L. P. Vernon and G. Seely, eds.) Academic Press, New York, 1966, p. 185.
Kotani, M. "Electronic Structure and Magnetic Properties of Hemoproteins, Particularly of Hemoglobins," in *Advances in Chemical Physics, Vol. 7* (J. Duchesne, ed.) Interscience, New York, 1964, p. 159.
Kutyurin, V. M., Karyakin, A. V., Chibisov, A. K., and Artamkina, I. Yu. *Dokl. Akad. Nauk SSSR* **141**, 744 (1961).
Marks, G. S. *Heme and Chlorophyll*, D. Van Nostrand, London, 1969, p. 94.
Mason, S. F. *J. Chem. Soc.* **1958**, 976.
McCoy, S. and Caughey, W. S. *Abstracts of the 154th National Meeting of the American Chemical Society*, Chicago, Ill., Sept. 1967, Paper 51 C.
McCoy, S. and Caughey, W. S. *Biochemistry* **9**, 2387 (1970).
Muirhead, H. and Perutz, M. F. *Nature* **199**, 633 (1963).
Nelson, H. M. "The Infrared Spectrum of Porphin," Ph.D. Thesis, Oregon State College, 1954.
Pennington, F. C., Strain, H. H., Svec, W. A., and Katz, J. J. *J. Am. Chem. Soc.* **86**, 1418 (1964).
Rieder, R. F., Zinkham, W. H., and Holtzman, N. A. *Am. J. Med.* **39**, 4 (1965).
Schwartz, S., Berg, M. H., Bossenmaier, I., and Dinsmore, H. in *Methods of Biochemical Analysis, Vol. 8* (D. Glick, ed.) Interscience, New York, 1960, p. 221.
Sidorov, A. N. *Opt. Spectrosk. (USSR) (English Transl.)* **13**, 206 (1962); **15**, 454 (1963).
Stair, R. and Coblentz, W. W. *J. Res. Natl. Bur. Std.* **11**, 703 (1933).
Thomas, D. W. and Martell, A. E. *J. Am. Chem. Soc.* **78**, 1338 (1956).
van Vliet, G. and Huisman, T. H. J. *Biochem. J.* **93**, 401 (1964).
Weigl, J. W. and Livingston, R. *J. Am. Chem. Soc.* **75**, 2173 (1953).

Chapter 15

ENZYMOLOGY

Infrared spectroscopy has been used advantageously in studies of enzymatic reactions. Chapter 11 contains a discussion of the application of the infrared method of studying hydrogen–deuterium exchange in various enzyme molecules, and the application of this method for studying the effects of substrates and of inhibitor substances on the conformations of the enzymes.

This chapter gives some examples of how infrared measurements have been applied to studies of mechanisms of enzyme reactions, stereospecificity of reactions, inhibition of reactions, kinetics, properties of substrates, characterization of products of reactions, and identification of organisms by the kinds of enzymatic action displayed by them.

Mechanism of Enzyme Action

Ribonuclease, Trypsin, and Chymotrypsin

Drey and Fruton (1965) have studied infrared spectra of imidazole as a function of concentration to detect intermolecular hydrogen bonding. They were interested in the possible cooperative catalytic effects of two imidazolyl groups with defined spatial orientation, for example, in ribonuclease action (Witzel, 1963) and in chymotrypsin or trypsin action (Walsh *et al.*, 1964). Witzel (1963) had suggested that the imidazolyl groups of histidines 12 and 119 are hydrogen-bonded in active ribonuclease. Recent studies on the amino acid sequence of trypsinogen and chymotrypsinogen have shown that in both enzymes derived from these zymogens there are two histidine residues in the sequences His-Phe-Cys and Ala-His-Cys, the two half-cystine residues being joined by a disulfide bridge. Walsh *et al.* (1964) have suggested that these two histidine residues may participate in the "active site" of trypsin or chymotrypsin.

In dimethyl sulfoxide, intermolecular hydrogen bonding of imidazole was demonstrable (Drey and Fruton, 1965). This was also true for 4,4′(5,5′)bis-imidazolylmethane (BIM). For imidazole ($0.65M$) bands were seen at 3440, 3178, 3030, 2930, 2840, 2690, and 2579 cm^{-1}. Upon progressive dilution to $0.076M$, the intensity of the band at 3178 cm^{-1} decreased and that at 3440 cm^{-1} increased. For BIM ($0.27M$), bands were observed at 3440, 3165, 2940, 2858, 2700, and 2581 cm^{-1}. Upon dilution to $0.016M$

the intensity at 3440 cm^{-1} increased and the intensity at 3165 cm^{-1} (hydrogen-bonded NH) decreased. No evidence was available for *intra*molecular hydrogen bonding when *inter*molecular hydrogen bonding was possible. The favored conformation of the BIM molecule was therefore suggested to be the one in which the two imidazolyl groups are furthest apart.

Carbonic Anhydrase

Carbonic anhydrase catalyzes the hydration of carbon dioxide and the dehydration of HCO_3^-. Azide ion, N_3^-, specifically inhibits this enzyme. In aqueous solutions, CO_2 has a very strong absorption band at 2343.5 cm^{-1}, and N_3^- has a strong band at 2049 cm^{-1}, owing to the asymmetric stretching of these linear molecules. Since both these frequencies fall within the 2000 to 2800 cm^{-1} range, a region within which all aqueous protein solutions have relatively high transmittance, Riepe and Wang (1968) have carried out a systematic investigation of the infrared spectra of CO_2 and N_3^- in carbonic anhydrase solutions to learn about the molecular mechanism of action of this enzyme. The infrared data revealed the catalytic mechanism clearly, without reliance on kinetic results.

In the infrared experiments with carbonic anhydrase most of the work was done at *p*H 5.5. After equilibration of enzyme solutions in flasks with CO_2, the solutions were transferred, by positive pressure from injected inert gas, through polyethylene tubing which fit snugly over 20-gauge needles used as inlet ports for the infrared cells. The partial pressure of CO_2 was maintained at a constant level throughout the transfer procedure so that equilibrium was not disturbed. In a typical experiment the infrared sample cell, with 0.075-mm path length and CaF_2 windows, was filled with carbonic anhydrase solution under equilibrium CO_2 pressure. The matched reference cell was similarly filled with the same carbonic anhydrase solution containing an inhibitor, ethoxzolamide (or sodium azide), equilibrated under the same CO_2 atmosphere. Since the infrared absorptions of water, protein, and dissolved CO_2 in the matched cells cancelled exactly, the difference spectrum was assumed to be due to the CO_2 bound at the active site in the sample cell and the specific inhibitor in the reference cell. By accurately measuring the difference spectrum of CO_2-equilibrated carbonic anhydrase against the ethoxzolamide- or azide-inhibited enzyme, these workers were able to obtain the absorption maximum at 2341 cm^{-1} which was due to the asymmetric stretching of the CO_2 molecule loosely bound to a hydrophobic surface at the active site of the enzyme.

Carbon dioxide in water absorbs at 2343.5 cm^{-1}. The observed small shift for bound CO_2 was attributed to a solvent effect. The absence of a larger spectral shift showed that this substrate is not appreciably distorted upon binding by the enzyme. The data of Riepe and Wang (1968) indicated that at a p_{CO_2} of 1 atm and 25°C the active site of the enzyme was about 25% saturated with CO_2. Consequently, the measurement of peak height from the difference infrared spectrum was much less reliable than the measurement of frequency for the bound CO_2. On the other hand, carbonic anhydrase can be completely saturated by azide at fairly low total concentration of the latter. Therefore, quantitative infrared measurements of the concentration of the enzyme-azide complex were more easily carried out.

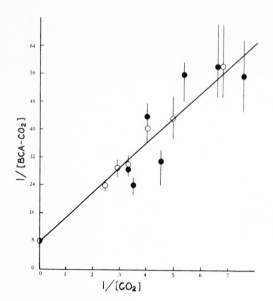

Fig. 15.1. Estimation of K_{CO_2} for 33.3 % carbonic anhydrase solution at pH 5.5 and 22.5°C. The [BCA—CO$_2$] and [CO$_2$] are both expressed in absorbance units. To obtain K_{CO_2} in terms of partial pressure in atmospheres, the x coordinate must be multiplied by a factor of 0.410 (the absorbance of free CO$_2$ at 1 atm). ○, Data for enzyme sample Lot 6JA; ●, Lot 5537. The intercept (◗) is a calculated value, obtained by using the absorptivity of CO$_2$ dissolved in water (at 2343.5 cm^{-1}) as the absorptivity of the CO$_2$ bound at the active site of the enzyme (at 2341 cm^{-1}). (Riepe and Wang, 1968.)

The binding constant of bovine carbonic anhydrase (BCA) for CO_2 was estimated directly from the difference spectra taken at various CO_2 pressures. The height of the difference absorption peak at 2341 cm^{-1} was measured as a function of p_{CO_2}. Values of 1/[BCA–CO$_2$ complex] were plotted vs the corresponding values of $1/p_{CO_2}$ as in Fig. 15.1. The approximate value of the dissociation constant,

$$K_{CO_2} = \frac{[BCA]p_{CO_2}}{[BCA\text{–}CO_2 \text{ complex}]}$$

determined from the slope and intercept of this plot, was 3 ± 0.8 atm, 8 times larger than that determined by Kernohan (1965). Riepe and Wang were not able to determine the cause of this discrepancy.

Azide ion, both free and bound to metal–ion complexes, also absorbs strongly in this range. A comparison between the infrared spectra of azide bound to diethylenetriamine Zn(II) and Co(II) complexes or to the cobalt–enzyme and the corresponding spectrum for the native enzyme showed that the azide ion is coordinate to the Zn(II) atom of the native enzyme. Examination of the difference spectra in the presence of both azide and CO_2 showed that the bound azide sterically interferes with the binding of CO_2 in the hydrophobic cavity adjacent to the Zn(II).

In 33 % inert protein solutions, the azide ion absorbed maximally at 2046 cm^{-1}. When azide was added to a solution of carbonic anhydrase, an absorption band also appeared at 2094 cm^{-1} as shown in spectrum 1 of Fig. 15.2. When an excess of ethoxzolamide was subsequently added to the above solution composed of azide plus enzyme, the absorption band at 2094 cm^{-1} became completely suppressed, as shown in spectrum 2 of Fig. 15.2. Thus, ethoxzolamide and azide compete for binding by the enzyme. Spectrum 1 in Fig. 15.2 represents complete saturation of the enzyme with azide, since the further addition of a large excess of azide caused no further increase in the intensity of the 2094 cm^{-1} band.

Infrared data on the competition of bicarbonate and hydroxide with azide for Zn(II), when combined with the results on CO_2 binding, led to the formulation of a simple molecular mechanism for the catalytic action of carbonic anhydrase, i.e., attack of the bound CO_2 by the OH$^-$ group coordinated to the zinc ion. Both Smith (1949) and Davis (1958) had previously suggested similar mechanisms from kinetic considerations. Coleman (1967) has recently shown that titration data also support such a mechanism.

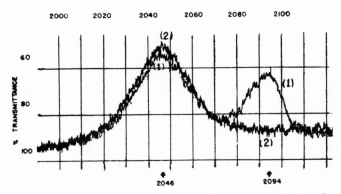

Fig. 15.2. Difference spectra of azide in equilibrium with carbonic anhydrase. Spectrum 1 (lighter trace), [total enzyme] = 0.0075M, [total azide] = 0.035M; spectrum 2 (heavier trace), sample solution contained an excess of ethoxzolamide but was otherwise the same as that in 1. The peak at 2094 cm^{-1} represents azide bound to carbonic anhydrase, and the peak at 2046 cm^{-1} represents free N_3^- in the protein solution. The molar absorptivities of free and bound N_3^- at their respective absorption maxima are different. (Riepe and Wang, 1968.)

Arylsulfatase

The spectra of normal barium sulfate and barium sulfate containing 87% BaS $^{16}O_3$ ^{18}O (Fig. 15.3) have been examined by Spencer (1959), who developed a method of estimating BaS $^{16}O_3$ ^{18}O content in barium sulfate. The method is based on the measurement of relative absorbances at 981 and 961 cm^{-1}, and was used to analyze the barium sulfate prepared from the inorganic sulfate liberated by *Aspergillus nidulans* arylsulfatase acting on potassium *p*-nitrophenyl sulfate in the presence of water containing 8.7 atoms % excess of ^{18}O. The BaS $^{16}O_3$ ^{18}O content of barium sulfate was found to be 7.8%, showing that the type I arylsulfatase of *A. nidulans* splits the O—S bond of arylsulfates. Splitting the C—O bond would result in incorporation of ^{18}O into the hydroxy compound formed as follows:

$$R \!\mid\! O - SO_2 - O^-$$
$$H^{18}O \mid H \qquad \rightarrow R^{18}OH + HOSO_3^-$$

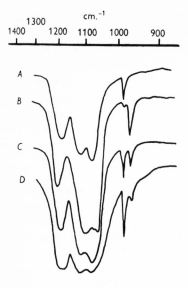

Fig. 15.3. Infrared spectra of Nujol mulls of barium sulfate samples. (*A*) normal barium sulfate, (*B*) Barium sulfate containing 87 % of BaS^{16}O$_3$^{18}O and 11.8 % of BaS^{16}O$_4$, (*C*) 40 % of BaS^{16}O$_3$^{18}O, 59.5 % of BaS^{16}O$_4$, and (*D*) 8.7 % of BaS^{16}O$_3$^{18}O, 91.2 % of BaS^{16}O$_4$. (Spencer, 1959.)

However, in the splitting of the O—S bond the ^{18}O would be found in the liberated sulfate as shown:

$$R-O \!\mid\! SO_2-O^- \longrightarrow ROH + H^{18}OSO_3^-$$
$$H \!\mid\! ^{18}OH$$

β-Methylaspartase

Bright (1964) has studied the mechanism of the β-methylaspartase reaction, in which the enzyme β-methylaspartase catalyzes the reversible conversion of L-β-methylaspartate (LXXXVI) to mesaconate (LXXXVII) and ammonia:

$$
\begin{array}{ccc}
\text{CH}_3 & \text{CH}_3 & \text{COO}^- \\
\text{H}-\text{C}-\text{COO}^- & \rightleftharpoons & \text{C} \\
\text{H}-\text{C}-\text{NH}_3^+ & & \text{C} \\
\text{COO}^- & \text{COO}^- & \text{H}
\end{array}
\quad + \text{H}^+ + \text{NH}_3
$$

LXXXVI   LXXXVII

Bright used infrared spectroscopy to study the β-methylaspartase catalyzed exchange of solvent hydrogen into, and elimination of ammonia from, β-deuterated L-*threo*-β-methylaspartate. The β-deuterium exchange reaction was measured by infrared spectroscopy following product isolation. A standard curve was used to determine HMA (L-*threo*-β-methylaspartate containing ^{1}H at the β-carbon) in (D)MA (L-*threo*-β-methylaspartate containing more than the natural abundance of ^{2}H at the β-carbon atom). This curve was a plot of the ratio of absorbance at 1190 cm^{-1} to that at 1149 cm^{-1} vs percentage HMA in samples of β-methylaspartic acid containing HMA and (D)MA (29 % β-protonated). The 1190 cm^{-1} band is unique to HMA, whereas the 1149 cm^{-1} band is of approximately equal intensity in HMA and (D)MA. The use of a ratio of absorbances was necessitated by the fact that the absolute amount of amino acid in the KBr disk varied somewhat from disk to disk.

An early and obligatory step in the overall reaction was found to be β-hydrogen extraction by a base (sulfhydryl anion) at the active center of the enzyme to form a carbanion intermediate.

Rhodanese—Cleavage of Sulfur–Sulfur Bond

Rhodanese may function as either a thiosulfate:cyanide sulfur transferase or a thiosulfate:lipoate oxidoreductase (Villarejo and Westley, 1963a,b). Kinetic studies comparing methanethiosulfonate and thiosulfate as rhodanese substrates have shown (Mintel and Westley, 1966) that the cleavage of the thiosulfate sulfur–sulfur bond limits the overall maximum velocity of the thiosulfate–cyanide sulfur transfer reaction. A kinetically significant enzyme–thiosulfate complex has thus been shown. The same studies indicated that the Michaelis constant for thiosulfate is a true dissociation constant for this complex. Infrared spectroscopic studies and experiments involving the reaction between thiosulfonates and cyanide in nonenzymic systems showed that the reactivity of the planetary sulfur atom toward cyanide is enhanced by a shift of electrons away from the sulfur–sulfur bond. Studies of the enzyme-catalyzed reaction between the thiosulfonates and cyanide suggested that one aspect of enzymic catalysis in the thiosulfate reaction is such a shift of electrons away from the sulfur–sulfur bond. The authors gave positions of the fundamental S—O stretching band in thiosulfonates: p-bromobenzenethiosulfonate, 1049 cm^{-1}: benzenethiosulfonate. 1063 cm^{-1}: p-toluenethiosulfonate, 1063 cm^{-1}; and methanethiosulfonate, 1092 cm^{-1}. In thiosulfate it was at 1117 cm^{-1}. The pseudo-first-order rate constants for the reaction between thiosulfonates and cyanide were given as: p-bromobenzenethiosulfonate, 0.0042 sec^{-1}; benzenethiosulfonate, 0.0031 sec^{-1}; p-toluenethiosulfonate, 0.0025 sec^{-1}; methanethiosulfonate, 0.0013 sec^{-1}; and ethanethiosulfonate, 0.00063 sec^{-1}. For the reaction between thiosulfate and cyanide, the rate constant was 0.00003 sec^{-1}. The frequency values for the S—O fundamental band showed qualitatively the expected variation with the inductive power of the substituent groups.

Stereospecificity

In several enzyme systems that catalyze the desaturation of stearic acid, oleic acid (cis-Δ^9-octadecenoic acid) is the sole product and the reaction is therefore characterized by positional and geometrical stereospecificity. The desaturating system of Corynebacterium diphtheriae has been found by Schroepfer and Bloch (1965) to have the additional property of selectively removing one particular hydrogen atom from each pair of hydrogens at carbons 9 and 10 of stearic acid. The enzyme can thus distinguish between the two hydrogen atoms attached to a carbon atom of a polymethylene chain, and the system is a notable example of the stereospecificity of enzymatic reactions at meso carbon atoms (Levy et al., 1962). The chemical identity of each of the hydroxystearates used was based on melting points, chromatographic properties, source, and mode of preparation of the compounds and infrared spectra. Spectra were given for 9-D- and 9-L-, 10-D- and 10-L-, and 12-D-hydroxyoctadecanoic acids.

While the infrared spectra of the 9-D- and 10-D-hydroxyoctadecanoic acids are essentially identical with those of their enantiomorphs, there are significant differences in the spectra of the 9-, 10-, and 12-hydroxyoctadecanoic acids in the 1429 to 1000 cm^{-1} region.

The conversion of oleic acid to 10-D-hydroxystearic acid by a pseudomonad has been studied in a medium enriched in D_2O (Schroepfer, 1966). The reaction occurred with stereospecific incorporation of one atom of solvent hydrogen at carbon 9. The location of the deuterium was established by a combination of chemical and mass spectrometric evidence. After extraction with ether and silicic acid column chromatography, crystals were obtained of 10-D-hydroxy-[9-$^2H_1$]-stearic acid. The presence of deuterium in the compound was indicated by a 2128 cm^{-1} C—D stretching band in the infrared spectrum of the free acid and of its methyl ester. The fingerprint region of the spectra differed significantly from that of the spectra of the corresponding undeuterated compounds (Table 15.1) as has been

reported earlier in several cases of deuterium-labeled compounds (Turkevich *et al.*, 1949; Nolin and Jones, 1953; Hall *et al.*, 1963; Childs and Bloch, 1961; Frejaville *et al.*, 1963). Other differences were also noted between the deuterated and non-deuterated stearates.

Table 15.1. The Effects of Deuteration on the Bands, in cm^{-1}, of Methyl 10-Hydroxystearates (Schroepfer, 1966)

Nondeuterated	Deuterated
1319	1319 (reduced in intensity)
1285	1290 and 1272 (shoulders)
1176	1176 and a shoulder at 1185
1112	(absent)
1099	(absent)
998	1105
	998 (weaker than the nondeuterated 998 band)
	990
	912
889 (broad)	881 and 871
	862
	835
	775

Enzyme Inhibition

Inhibition of Cholinesterase

Neely *et al.* (1964) have prepared a series of aryl *N*-methyl methylphosphorami-dates and have examined them for inhibition of fly-head (*Musca domestica* L.) cholin-esterase by a standard Warburg manometric method (Metcalf, 1949). The type formula for *N*-methyl methylphosphoramidates is given by LXXXVIII.

LXXXVIII

The R groups substituted on the benzene ring were p-NO_2, 2,4,5-trichloro, 2,4-dichloro, o-chloro, p-chloro, H, p-methoxy, m-t-butyl, p-t-butyl, and 2-chloro, 4-t-butyl. Neely *et al.* plotted log k for cholinesterase inhibition *vs* frequency of the phosphorus-oxygen-aromatic stretching vibration for the series of compounds.

Fukuto and Metcalf (1956) had found earlier that the shift in frequency of the infrared absorption band attributed to the phosphorus-oxygen-phenyl (P—O—Ph) stretching vibration (see LXXXIX) was a linear function of the ability of the substi-tuted phenyldiethylphosphates to inhibit fly-brain cholinesterase. A similar relation-ship was shown to exist by Neely *et al.* but a puzzling anomaly was evident in the data. It was expected that as the phenyl group became more electrophilic it would draw electrons away from the P—O—Ph bonds, thus weakening the bonds, with resulting reduction in frequency. Ingraham *et al.* (1952) had found this to be the case for the O—H stretching frequency of substituted phenols. Fukuto and Metcalf had found the opposite situation, that is, the more electrophilic the substituted phenol the greater the stretching frequency of the P—O—Ph bond. Neely *et al.* have ex-plained this anomaly in the correlation in the following way: The frequency of the absorption band observed is a function of a Ph—O stretching vibration (Nyquist, 1957) and not a phosphorus—O—aryl stretching vibration. They argued that increasing the electrophilic nature of the substituent group R makes structure XC more important:

LXXXIX   XC

The contribution from XC would make the Ph—O bond stronger owing to increased resonance and thus increase its stretching frequency. The frequencies of the oxygen–phenyl stretching mode are given in Table 15.2.

Table 15.2. Oxygen–Phenyl Stretching Mode for Substituted Aryl-*N*-methyl methylphosphoramidates (Neely *et al.*, 1964)

R group	Frequency, cm^{-1}
p-NO$_2$	1237
2,4,5-Cl$_3$	1259
2,4-Cl$_2$	1245
o-Cl	1235
p-Cl	1228
H	1213
p-OCH$_3$	1207

Renin Inhibitor

A renin inhibitor has been isolated by Sen *et al.* (1967) from canine kidney. By means of infrared and ultraviolet spectroscopy and thin-layer and gas-liquid partition chromatography, it was shown to be a phospholipid similar to bovine phosphatidyl serine but structurally different in fatty acid and amino acid content. The amino acid appeared to be a β-hydroxy-α-amino acid. Assignments of bands were as follows: 3425 (OH), 3226 (ionized NH$_2$), 2924 (CH), 2857 (CH), 2667 (POH), 1730 (COOR), 1639 (COO$^-$), 1515 (CONH), 1449 (CH), 1235 (P=O), 1176–1163 (COR), 1031 (POC), and 719 cm^{-1} ((CH$_2$)$_n$).

Aconitase Inhibition

Fanshier *et al.* (1964) have shown that the aconitase-inhibitory effect of synthetic monofluorocitrate is due to one isomer, synthesized enzymatically from fluoroacetyl coenzyme A and oxaloacetate. They separated diastereoisomers of synthetic monofluorocitric acid, compared the chemical and enzyme-inhibitory properties of these with enzymatically formed isomers, and found that the DL-*erythro* form of fluorocitrate is the diastereoisomer which inhibits aconitase action. Infrared spectroscopy helped to characterize the DL-*erythro* and DL-*threo* isomers.

Enzyme Kinetics

Adenosine Deaminase

Howard and Miles (1964) have carried out the reaction

$$\text{Adenosine} + \text{H}_2\text{O} \xrightarrow{\text{adenosine deaminase}} \text{inosine} + \text{NH}_3$$

in D$_2$ ^{18}O in order to synthesize inosine specifically labeled with ^{18}O for use as an aid in assigning a band at 1673 cm^{-1} for inosine-6-^{16}O. The use of ^{18}O-substitution as an aid in the interpretation of vibrational spectra has been employed by a number of investigators [see the references cited in Howard and Miles (1964)].

Howard and Miles followed the enzymic deamination of adenosine to form inosine-6-^{18}O directly in a CaF_2 cell (Fig. 15.4). Although they found the method to be less convenient than ultraviolet spectroscopy as a method of enzyme assay, it was a clearly more convenient and economical use of the isotope than total chemical synthesis followed by isolation of the compound and measurement of the infrared spectrum. Adequate kinetic data were obtainable (Fig. 15.5). (The authors' claim to be the first users of infrared spectroscopy to follow enzyme kinetics is not valid [see Parker, 1962a,b and Parker and Kirschenbaum, 1959b].) The carbonyl band of inosine-6-^{18}O shifted 4 cm^{-1} to lower frequency in 32.9% D_2 ^{18}O and 13 cm^{-1} in 90.7% D_2 ^{18}O (Fig. 15.4). Digitized spectra of inosine formed in D_2 ^{18}O were normalized to an absorptivity basis by a computer, and the spectrum of ordinary inosine-6-^{16}O (weighted by the fraction of ^{16}O in the water) was subtracted from the spectrum of enzymically formed inosine-6-^{18}O. The difference curves were then renormalized to 100% D_2 ^{18}O. The same normalized frequency shift ($\Delta v = 14$ cm^{-1}) was found for both the 33% and 91% D_2 ^{18}O solutions, although, as would be expected, the quality of the normalized spectrum obtained from the 91% D_2 ^{18}O solution was much

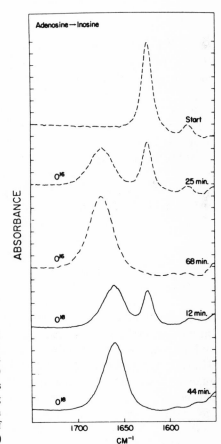

Fig. 15.4. Enzymic conversion of adenosine to inosine in D_2 ^{16}O (broken lines) and D_2 ^{18}O (90.7%, solid lines). The carbonyl absorption band of inosine-6-^{18}O has shifted to a lower frequency by 13 cm^{-1} (1673–1660 cm^{-1}). The faster rate of inosine-6-^{18}O formation was due to a larger amount of adenosine deaminase being present in the reaction mixture. No attempt was made in these experiments to determine an intrinsic effect of isotope on the reaction rate. (Howard and Miles, 1964.)

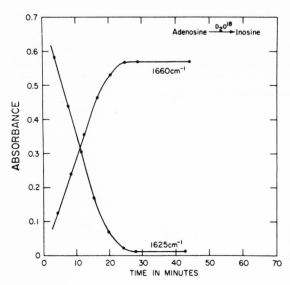

Fig. 15.5. Kinetics of adenosine deamination ($1625\,\text{cm}^{-1}$) and inosine-6-^{18}O formation ($1660\,\text{cm}^{-1}$). (Howard and Miles, 1964.)

higher. The fact that the same normalized frequency shift was obtained from solutions of such different isotope content suggests that there is no significant kinetic isotope effect in the reaction. Pure inosine-6-^{18}O, therefore, would have a $14\,\text{cm}^{-1}$ shift and ϵ_{max} of 1010 at $1659\,\text{cm}^{-1}$. These data allowed the assignment of the $1673\,\text{cm}^{-1}$ band of inosine-6-^{16}O to a strongly coupled carbonyl vibration at position 6, in agreement with previous assignments based on chemical plausibility (Miles, 1961; Brown and Mason, 1957).

Howard and Miles (1965) have described the enzymic synthesis of inosine-6-^{18}O, inosine 5′-phosphate-6-^{18}O, and guanosine-6-^{18}O. In the course of this work they used infrared spectroscopy of $D_2{}^{16}O$ and $D_2{}^{18}O$ solutions to evaluate the conversion of 5′-AMP to 5′-IMP and the conversion of 2,6-diamino-9-β-D-ribofuranosylpurine to guanosine. Kinetic plots of the data for these reactions were made in the same way as already indicated in Fig. 15.5 of Howard and Miles (1964). Figure 15.6 shows spectra for the enzymic conversion of 2,6-diamino-9-β-D-ribofuranosylpurine to guanosine, and Fig. 15.7 shows a plot of the absorbances of certain specific absorption bands during the course of the reaction.

With regard to band assignment, the isotopic frequency shift of $14\,\text{cm}^{-1}$ in D_2O [normalized to 100% ^{18}O content as described in Howard and Miles, (1964)] observed for inosine, IMP, and guanosine, indicates that the $1673\,\text{cm}^{-1}$ band in the first two and the $1665\,\text{cm}^{-1}$ band in the latter are attributable to a coupled carbonyl vibration (Howard and Miles, 1964) (Figs. 15.8 and 15.6). The fact that the 2-amino group in guanosine does not affect the magnitude of the isotopic shift suggests that the bond between the C-2 and N-3 is not strongly coupled to the carbonyl vibration. Guanosine has two ring vibrations in the region observed, one at $1578\,\text{cm}^{-1}$ and one at about $1568\,\text{cm}^{-1}$ (shoulder). In guanosine-6-^{18}O the former band is unchanged in frequency, but the latter is shifted slightly to $1563\,\text{cm}^{-1}$. This observation suggests that the band at about 1568 cm^{-1} results from a normal mode which includes some motion of the carbonyl oxygen.

The infrared spectra of helical aggregates formed by 5'-GMP and 3'-GMP show characteristic frequency and intensity changes for both the lower frequency bands (1578 cm^{-1} and ~1568 cm^{-1} in the unaggregated molecule), different shifts being observed for the high frequency band in the 5' and 3' helices (Miles and Frazier, 1964). The experiments of Howard and Miles (1964), which show a frequency difference of 14 cm^{-1} between the two isotopic 5'-GMP species both in the helical and in the unaggregated state, indicate that the bands above 1650 cm^{-1} are attributable to carbonyl vibrations in both cases and suggest that the carbonyl group is a site of localized interaction, presumably a hydrogen bond. Changes in the ring vibrations

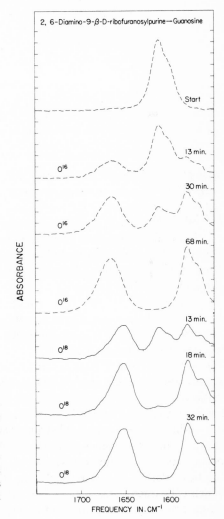

Fig. 15.6. Enzymic conversion of 2,6-diamino-9-β-D-ribofuranosylpurine to guanosine in D$_2$^{16}O (-----) and D$_2$^{18}O [90.7% ^{18}O, (——)] in a cell of 55.7-μ path length. The carbonyl absorption band of guanosine-6-^{18}O has shifted to a lower frequency by 13 cm^{-1} (1665 to 1652 cm^{-1}). The quantity of adenosine deaminase was not determined for either reaction mixture. Index marks on the ordinate are 0.1 absorbance unit apart. (Howard and Miles, 1965.)

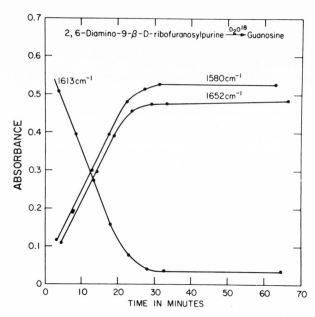

Fig. 15.7. Kinetics of 2,6-diamino-9-β-D-ribofuranosylpurine deamination (1613 cm^{-1}) and guanosine-6-^{18}O formation (1580 and 1652 cm^{-1}). (Howard and Miles, 1965.)

upon gel formation are seen in Fig. 15.9. Figure 15.10 is a plot of the melting curve for 5′-GMP-6-^{16}O under conditions given in Fig. 15.9.

β-Glucosidase

Infrared studies have been done on glucose oxidase (Parker, 1962 *a*), invertase (Parker, 1962*b*), and β-glucosidase (Parker, 1964), all in H$_2$O (*not* D$_2$O) solutions. These studies have been reviewed recently (Parker, 1967). A description of the β-glucosidase study follows:

The system consisted of 0.5 ml of enzyme solution added to 2.0 ml of substrate solution, producing concentrations of 3.2% phenyl β-D-glucoside and 0.2% β-glucosidase (wt/vol) in pH 5.2 acetate buffer (ionic strength 0.1) at the beginning of the reaction, stirred magnetically. Figure 15.11 shows the course of action. The dotted line is the spectrum of water in an IRTRAN-2 cell of 0.025-mm thickness, measured against water in a reference IRTRAN-2 cell. The solid line is the spectrum of the enzymatic system near the beginning of the reaction, recorded against acetate buffer in the reference beam at a rate of 1μ/min. This curve was begun at 1538 cm^{-1}, 2.5 min after the mixing of enzyme and substrate. The enzyme itself does not display a spectrum at 0.2% concentration. As the reaction proceeds, several changes in the spectrum occur. The new band at 1475 cm^{-1} is characteristic of free phenol in aqueous solution (Parker and Kirschenbaum, 1959*a*). The band at 1227 cm^{-1} (solid line), which is characteristic of the β-phenyl group linked to glucose through an oxygen atom, gradually disappears and is replaced by a band at 1241 cm^{-1} (dashed line),

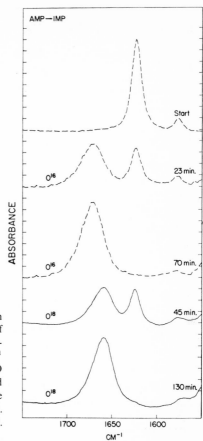

Fig. 15.8. Enzymic conversion of 5'-AMP to 5'-IMP in $D_2{}^{16}O$ (----) and $D_2{}^{18}O$ [90.7% ^{18}O (——)] in a cell of 55.5-μ path length. The carbonyl absorption band of 5'-IMP-6-^{18}O has shifted to a lower frequency by 13 cm^{-1} (1672 to 1659 cm^{-1}). The faster rate of 5'-IMP-6-^{16}O formation was caused by a larger amount of adenylic acid deaminase being present in the reaction mixture. The spectra are plotted as absorbance vs frequency in cm^{-1}. Index marks on the ordinate are 0.1 absorbance unit apart. (Howard and Miles, 1965.)

characteristic of phenol that is no longer bonded to glucose (Parker and Kirschenbaum, 1959a). As the reaction proceeds, a new band appears at 1149 cm^{-1}, which does not exist in the spectrum of the original phenyl β-D-glucoside. This band has been shown earlier (Parker, 1960) to be characteristic of the glucose molecule, and its intensity increases gradually as more free glucose appears. The spectrum of the final solution is identical with a solution made up of stoichiometric amounts of phenol and glucose, assuming complete hydrolysis. The IRTRAN-2 (ZnS) did not interfere with the assay of the enzymatic reaction, since it was demonstrated that spectral changes were the same whether they were determined on aliquots removed from the reaction mixture at various time intervals or were determined directly on the reaction mixture contained in the cell. First-order kinetics was calculated from absorbance data of the band at 1149 cm^{-1} from the equation

$$k = \frac{1}{t} \log \frac{A_0 - A_\infty}{A_t - A_\infty}$$

In a typical run A_∞, the absorbance of the band at the end of the reaction, was 0.2864; A_0, the initial absorbance of the band (extrapolated to time zero), was 0.2528; and A_t

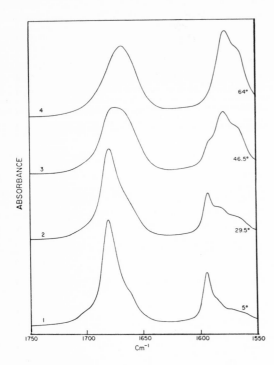

Fig. 15.9. Infrared spectra (curves 1 to 4) of 5'-GMP-6-^{16}O (0.20 M 5'-GMP-6-^{16}O, 0.11 M NaCl, and pD 5.8) in D_2O solution, 25.8-μ path length. The solutions were stored for several hours at 1°C before observation. A series of spectra was then measured as the temperature increased. Curves 1 to 4 were selected from the series for reproduction. (Howard and Miles, 1965.)

was the absorbance of the band at time t, varying from 0.2557 to 0.2857. Some typical examples of k were 0.0106, 0.0096, 0.0134, and 0.0112, with an average k of 0.0112/min.

Fumarase

Rajender and McColloch (1965) have determined equilibrium and rate data in the reaction of the hydration of fumarate by fumarase (L-malate hydrolase) in both H_2O and D_2O. Fumaric acid has a strong band at 965 cm^{-1} attributable to the out-of-plane deformations of the hydrogen atoms on the carbons connected by the double bond, and malate displays strong absorption at 1093 cm^{-1} due to the C—O stretching of the alcohol group. Figure 15.12 presents spectra showing band changes with time due to C—O stretching during the enzymic reaction. Figure 15.13 shows changes at 965 cm^{-1} during the reaction. Another band (1210 cm^{-1}) also undergoes change as the reaction proceeds.

These workers used intensity-change measurements to obtain the apparent equilibrium constant (Bock and Alberty, 1953)

$$K_{app} = \frac{[M]_T}{[F]_T} = \frac{[S]_0 - [F]_T}{[F]_T}$$

where $[M]_T$ and $[F]_T$ are the total concentration of L-malate and fumarate at equilibrium and $[S]_0$ is the initial concentration of fumarate. They obtained a K_{app} of 4.2 $\pm$ 0.4 at 25°C. A van't Hoff plot gave a ΔH for the reaction of -4300 ± 200 cal. In D_2O, K_{app} was found to be 3.0 $\pm$ 0.4.

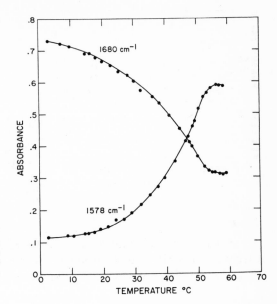

Fig. 15.10. Melting curve for 5'-GMP-6-^{16}O under conditions given in Figure 15.9. The experimental points were taken from the same series of spectra from which curves 1 to 4 of Fig. 15.9 were selected. (Howard and Miles, 1965.)

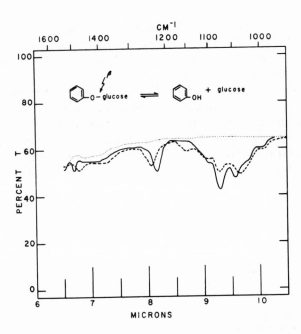

Fig. 15.11. The action of β-glucosidase on 3.2 % (wt/vol) phenyl β-D-glucoside. ($\cdots$) Water in a 0.025-mm thick IRTRAN-2 cell *vs* water in the reference beam, (——) reaction mixture near beginning of the reaction (see text), (————) reaction mixture at end of enzymatic action (170 min). (Parker, 1964.)

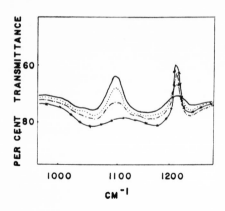

Fig. 15.12. Spectra showing changes in absorption due to C—O stretching during reaction of fumarase on 1.0 M sodium fumarate–fumaric acid (0.04 mg enzyme/ml). Observed absorption maxima are at 1093 and 1220 cm^{-1}: (×) 1 min, (– · –) 10 min, (· ·) 20 min, and (———) 30 min. (Rajender and McColloch, 1965.)

Urease

Jencks (1963) has studied the rates of disappearance and formation of urea (XCI), carbamate (XCII), and bicarbonate in the urease-catalyzed hydrolysis of urea (Figs. 15.14 and 15.15). Urea hydrolysis occurs according to the following scheme:

$$\underset{\text{XCI}}{H_2N\overset{\overset{\displaystyle O}{\|}}{C}NH_2} \rightarrow NH_4^+ + \underset{\text{XCII}}{H_2NCO_2^-} \rightarrow HCO_3^- + NH_3$$

Spectra recorded at intervals after the addition of urease to an unbuffered solution of urea first show the disappearance of urea peaks at 1604 and 1490 cm^{-1} and the appearance of carbamate peaks at 1441 and 1540 cm^{-1}, along with some bicarbonate. At the end of the reaction, the urea and carbamate bands have been replaced by strong bicarbonate absorption at 1363 and 1630 cm^{-1}. Figure 15.15 shows the time course of the appearance and disappearance of the individual bands in a similar experiment; carbamate forms rapidly and then slowly disappears, while bicarbonate forms more slowly and approaches a maximum value. The absorption band near the 1441 band of carbamate does not disappear completely because of the appearance of a band of CO_3^{2-}, which is in equilibrium with HCO_3^-, at 1416 cm^{-1}.

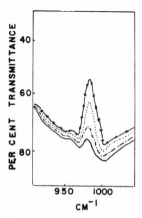

Fig. 15.13. Spectra showing changes in absorption due to out-of-plane deformations of the *trans*-hydrogens during reaction of fumarase on 1.0M sodium fumarate–fumaric acid in D_2O (*p*H 7.3, 0.04 mg enzyme/ml). (×) 2 min, (– – –) 15 min, (– · –) 30 min, and (———) 60 min. (Rajender and McColloch, 1965.)

Penicillinase

Penicillinase attacks penicillin (XCIII) according to the following hydrolytic reaction:

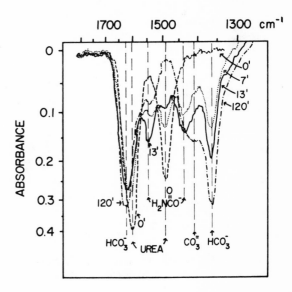

Zugaza and Hidalgo (1966) have studied the kinetics of this reaction by following the disappearance of the stretching vibration of the β-lactam $>C=O$ group frequency around $1780\ \text{cm}^{-1}$ in H_2O and D_2O at pH 7.2. Figure 15.16 shows pen tracings for penicillin $G(K^+)$ at various time intervals, with concomitant changes in the 1780 cm^{-1} band. These workers studied the relative resistance of various penicillins to penicillinase action and found that 3-phenyl-5-methyl-4-isoxazolyl penicillin was much more stable than penicillin V (Na^+). The energy of activation was 4.7 kcal/mole for enzyme action on penicillin V (Na^+), and the Q_{10} was 1.30. Among the other penicillins studied were the phenoxymethyl and phenoxyethyl derivatives.

Fig. 15.14. Infrared spectra recorded during the course of the urease-catalyzed hydrolysis of urea in D_2O. The reaction mixture contained $0.2M$ urea, 2 mg of urease (Sigma, Type V), and 0.4 mg of Diamox (a carbonic anhydrase inhibitor) in 1.0 ml; the reaction was allowed to proceed in a 0.1-ml aliquot in a calcium fluoride cell with a path length of 0.05 mm. The final pH, after dilution with water, was 9.3. (Jencks, 1963.)

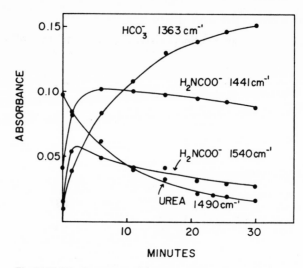

Fig. 15.15. The time course of the appearance and disappearance of urea, carbamate, and bicarbonate absorption during the urease-catalyzed hydrolysis of urea. Experimental conditions were similar to those for the experiment described in Fig. 15.14, except for a urea concentration of $0.1M$. (Jencks, 1963.)

The spectra of many penicillins in KBr pellets were also given by these authors, e.g., Na^+, K^+, Cs^+, and Ca^{2+} salts. The spectra of the diastereoisomers D- and L-α-phenoxyethyl penicillins were also given and shown to be distinctly different from each other.

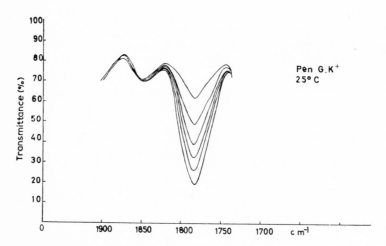

Fig. 15.16. Pen tracings for penicillin $G(K^+)$ showing changes with time of the $1780\ cm^{-1}$ band (see text). (Zugaza and Hidalgo, 1966.)

Properties of a Nitrilase and Its Substrates

Robinson and Hook (1964) have synthesized a number of compounds structurally related to the naturally occurring nitrile ricinine (XCV) and have reported some of the properties of ricinine nitrilase, an enzyme that catalyzes the hydrolysis of ricinine as shown below:

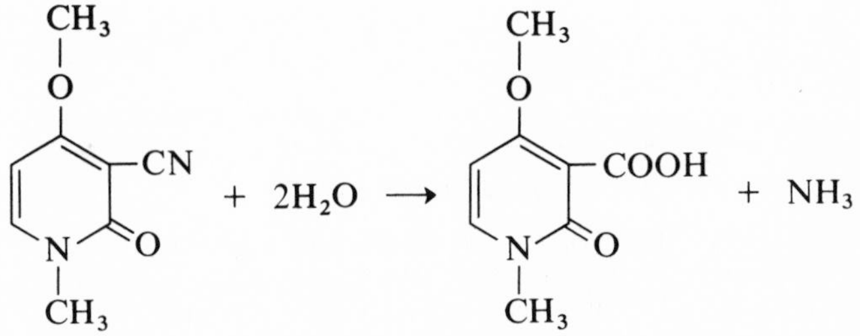

$$+ \; 2H_2O \; \rightarrow \qquad\qquad + \; NH_3$$

XCV   XCVI

Infrared spectra were given for ricinine and for several compounds used in the study of the properties of the enzyme. In the spectrum for ricinine the typical $C{\equiv}N$ stretching band was present at about 2250 cm^{-1}. The cyano group of 3-cyano-2-pyridones was hydrolyzed if position 1 of the 2-pyridone was unsubstituted or substituted with a methyl or ethyl group and position 4 was unsubstituted or substituted with a methoxyl or ethoxyl group.

Characterization of Reaction Products

3-Enolpyruvylshikimate 5-phosphate

In the presence of partially purified extracts of E. coli, shikimate 5-phosphate and enolpyruvate phosphate react to yield 3-enolpyruvylshikimate 5-phosphate and orthophosphate (Levin and Sprinson, 1964). 3-Enolpyruvylshikimate 5-phosphate (ES-5-P) has been isolated as an essentially pure barium salt by anion exchange chromatography of incubation mixtures on Dowex 1 (Cl). The chemical properties of 3-enolpyruvylshikimate 5-phosphate and its infrared spectrum were consistent with the proposed structure (XCVII):

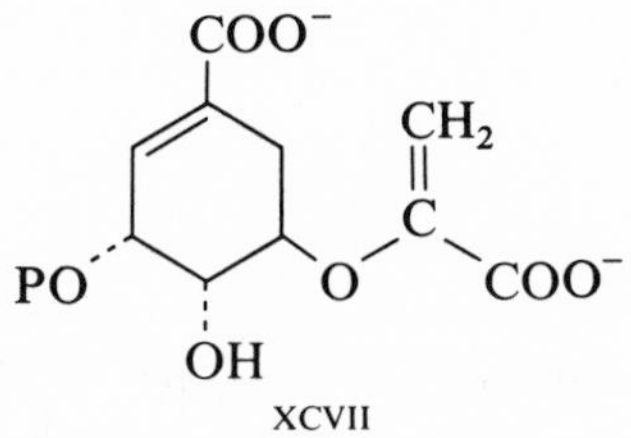

XCVII

ES-5-P, enolpyruvylshikimate, and shikimate 5-phosphate all show two bands resulting from the absorption of carboxylate ion at 1575 and 1403 cm^{-1}. No absorption was observed in the carbonyl region, indicating the absence of a lactone or ester

Wave number, cm⁻¹

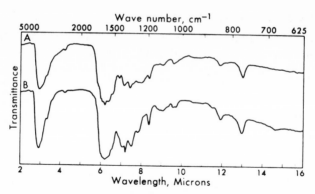

Fig. 15.17. Infrared spectra of tetrahydrofolate derivatives: (A)
formiminotetrahydrofolate, (B) 5-formyltetrahydrofolate. Spectra
were measured in KBr pellets with a Baird-Atomic infrared spectrophotometer. The tetrahydrofolate derivatives (1 mg) were pulverized
in 100 mg of KBr by means of a mortar and pestle. (Uyeda and
Rabinowitz, 1965.)

structure. Only ES-5-P and enolpyruvylshikimate showed a band at 1220 cm^{-1}
characteristic of the C—O vibration of vinyl ethers. Only ES-5-P and shikimate 5-
phosphate showed bands at 1124 to 1075 cm^{-1} (ionic phosphate) and 976 cm^{-1}
(P—O—alkyl).

5-Formiminotetrahydrofolate

Glycine formiminotransferase catalyzes the reaction of formiminoglycine
(XCVIII) and tetrahydrofolate (XCIX) to glycine (C) and 5-formiminotetrahydro-
folate (CI) (R^1 is benzoyl-L-glutamate). Uyeda and Rabinowitz (1965) have purified

this enzyme of *C. cylindrosporum* and have described the properties and the character-istics of the reaction. The formimino group on the tetrahydrofolate was assigned to the N-5 position on the basis of several lines of evidence, including infrared and ultra-violet spectroscopy. The ultraviolet and infrared spectra of formiminotetrahydrofolate closely resembled those of 5-formyltetrahydrofolate. Figure 15.17 shows infrared spectra of the two compounds.

The evidence cited by Uyeda and Rabinowitz did not rule out a saturated cyclic structure (CII) for the formiminotetrahydrofolate, but the infrared spectrum suggested an N-5 substitution rather than the saturated structure, since the absorption bands due to $>N-C=O$ and $>N-C=NH$ are known to be in the same region (Bellamy, 1954).

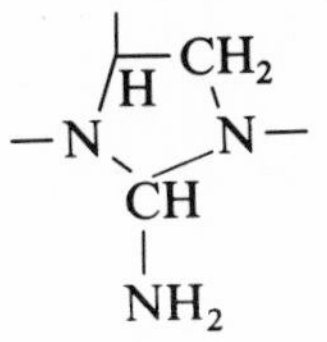

$$-N\underset{\underset{\displaystyle \underset{\displaystyle NH_2}{|}}{CH}}{\overset{\displaystyle \overset{\displaystyle |}{H}\ CH_2}{\diagdown\ \diagup}}N-$$

CII

Identification of Organisms by Their Enzymic Action

β-Lactamase (Penicillinase) and Amidase Actions

Thompson *et al.* (1949) reported a strong band in benzylpenicillin at ~ 1760 cm^{-1}. This band is in the general region of stretched double bond absorption of small unit molecular structure, but is well apart from the main complex of stretched bond absorptions within the range 1500 to 1700 cm^{-1}, given in varying forms by all 6-aminopenicillanic acid derivatives [see the spectra of benzylpenicillin and 6-amino-penicillanic acid before hydrolysis (Figs. 15.18 and 15.20)]. The 1760 cm^{-1} band is characteristic of the carbonyl frequency of the intact β-lactam ring, and has been clearly identified in penicillins as well as in the parent compound. It is also present as a strong, distinct absorption in 7-aminocephalosporanic acid (7-ACA) and its derivatives (Fleming *et al.*, 1963). The presence of the 1760 cm^{-1} band in its isolated position thus ensures a reliable demonstration of the unaltered β-lactam ring struc-ture.

Chapman *et al.* (1964) were able to differentiate between the enzymes β-lactamase and amidase by virtue of the fact that the carbonyl absorption band of the β-lactam of penicillins and cephalosporins at 1760 cm^{-1} is detectable when the antibiotics are mixed with enzymes or bacterial protein. Coliform organisms may produce either of two inactivating enzymes: (1) a β-lactamase which abolishes all the anti-bacterial activity of 6-aminopenicillanic acid, 7-aminocephalosporanic acid, and certain derivatives, by hydrolyzing the lactam ring; (2) an amidase which does not alter the lactam-ring structure but causes a lesser degree of inactivation by deacylating the side chains of some derivatives at the peptide linkage.

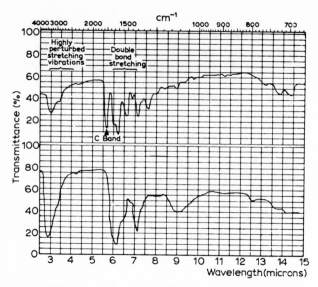

Fig. 15.18. (*top curve*). Benzylpenicillin before hydrolysis. (Holt and Stewart, 1965.)

Fig. 15.19. (*bottom curve*). Benzylpenicillin after β-lactamase hydrolysis. (Holt and Stewart, 1965.)

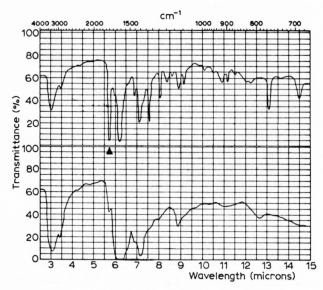

Fig. 15.20. (*top curve*). 6-Aminopenicillanic acid before hydrolysis. (Holt and Stewart, 1965.)

Fig. 15.21. (*bottom curve*). 6-Aminopenicillanic acid after β-lactamase hydrolysis. (Holt and Stewart, 1965.)

The β-lactamase-forming organisms can be identified directly by incubating whole-cell concentrates with substrate for a few hours; loss of the characteristic carbonyl band at 1760 cm^{-1} is conclusive evidence of this. Some organisms, however, inactivate derivatives of 6-aminopenicillanic acid (6-APA) and 7-aminocephalosporanic acid (7-ACA) without altering this band; the substrate residue can then be reactivated by phenylacetylation, proving that enzymes in this category have no action upon the lactam ring and that deacylation is clearly the consequence of a specific amidase present in coliform cells acting at the CONH linkage of the side chain. The substrates studied in this paper (Chapman *et al.*, 1964) were 6-aminopenicillanic acid, benzylpenicillin, methicillin, cloxacillin, ampicillin, quinacillin, 7-aminocephalosporanic acid, 7-thienylacetamido-cephalosporanic acid, and 3-pyridine-7-thienylacetamido-cephalosphoranic acid.

Holt and Stewart (1965) have studied the breakdown of β-lactam antibiotics by infrared spectroscopy. The method is based on observation of the rupture of the integral lactam ring by penicillinase, β-lactamase, with the production of penicilloic acid derivatives totally lacking in antibiotic activity. Figure 15.18 is the spectrum of benzylpenicillin before hydrolysis by a staphylococcal suspension, and Fig. 15.19 is the spectrum after rupture of the lactam ring by β-lactamase of the bacteria. Figure 15.20 is the spectrum of 6-APA before hydrolysis, and Fig. 15.21 after lactam ring

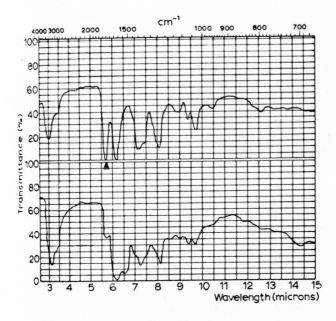

Fig. 15.22. (*top curve*). 7-Aminocephalosporanic acid before hydrolysis. (Holt and Stewart, 1965.)

Fig. 15.23. (*bottom curve*). 7-Aminocephalosphoranic acid after β-lactamase hydrolysis. (Holt and Stewart, 1965.)

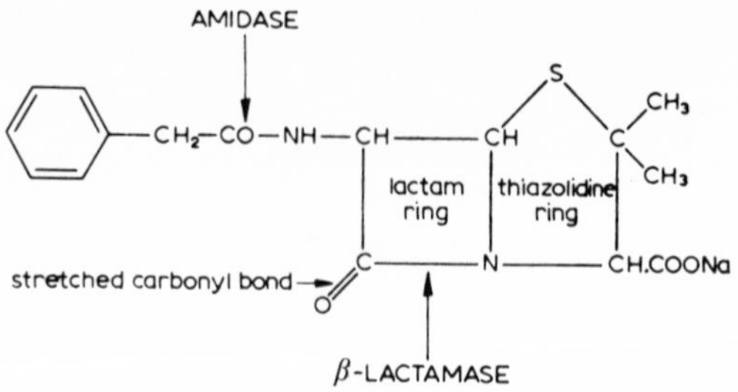

AMIDASE

lactam ring

thiazolidine ring

stretched carbonyl bond →

β-LACTAMASE

Fig. 15.24. Action of β-lactamase on benzylpenicillin. (Holt and Stewart, 1965.)

opening. Figure 15.22 is the spectrum of 7-ACA before hydrolysis, and Fig. 15.23 is the spectrum of that compound after β-lactamase hydrolysis.

The action of the enzyme is represented in Figure 15.24. All these spectra were run on freeze-dried samples from which KBr pellets were prepared. Total inactivation of an antibiotic with abolition of the so-called "c" band (~ 1760 cm^{-1}) is characteristic of the action of this enzyme. Partial inactivation with preservation of this band implies deacylation or other degradation leaving the β-lactam ring intact.

REFERENCES

Bellamy, L. J. *The Infra-red Spectra of Complex Molecules*, Wiley, New York, 1954, pp. 181–183, 227–229.

Bock, R. M. and Alberty, R. A. *J. Am. Chem. Soc.* **75**, 1921 (1953).

Bright, H. *J. Biol. Chem.* **239**, 2307 (1964).

Brown, D. J. and Mason, S. F. *J. Chem. Soc.* **1957**, 682.

Chapman, M. J., Holt, R. J., Mattocks, A. R., and Stewart, G. T. *J. Gen. Microbiol.* **36**, 215 (1964).

Childs, C. R., Jr. and Bloch, K. *J. Org. Chem.* **26**, 1630 (1961).

Coleman, J. E. *J. Biol. Chem.* **242**, 5212 (1967).

Davis, R. P. *J. Am. Chem. Soc.* **80**, 5209 (1958).

Drey, C. N. C. and Fruton, J. S. *Biochemistry* **4**, 1 (1965).

Fanshier, D. W., Gottwald, L. K., and Kun, E. *J. Biol. Chem.* **239**, 425 (1964).

Fleming, P. C., Goldner, M., and Glass, D. G. *Lancet* **1**, 1399 (1963).

Frejaville, G., Gounelle, Y., Jullien, J., and Paillous, A. *Bull. Soc. Chim. France* **1963**, 2171.

Fukuto, T. R. and Metcalf, R. L. *J. Agr. Food Chem.* **4**, 930 (1956).

Hall, G. E., Libby, E. M., and James, E. L. *J. Org. Chem.* **28**, 311 (1963).

Holt, R. J. and Stewart, G. T. *Biochim. Biophys. Acta* **100**, 235 (1965).

Howard, F. B. and Miles, H. T. *Biochem. Biophys. Res. Commun.* **15**, 18 (1964).

Howard, F. B. and Miles, H. T. *J. Biol. Chem.* **240**, 801 (1965).

Ingraham, L. L., Corse, J., Bailey, G. F., and Stitt, F. *J. Am. Chem. Soc.* **74**, 2297 (1952).

Jencks, W. P. in *Methods in Enzymology, Vol. 6* (S. P. Colowick and N. O. Kaplan, eds.) Academic Press New York, 1963, p. 914.

Kernohan, J. C. *Biochim. Biophys. Acta* **96**, 304 (1965).

Levin, J. G. and Sprinson, D. B. *J. Biol. Chem.* **239**, 1142 (1964).

Levy, H. R., Talalay, P., and Vennesland, B. in *Progress in Stereochemistry, Vol. 3* (P. B. D. De la Mare and W. Klyne, eds.) Butterworths, London, 1962, p. 299.

Metcalf, R. L. *J. Econ. Entomol.* **42**, 875 (1949).

Miles, H. T. *Proc. Natl. Acad. Sci. U.S.* **47**, 791 (1961).

Miles, H. T. and Frazier, J. *Biochim. Biophys. Acta* **79**, 216 (1964).

Mintel, R. and Westley, J. *J. Biol. Chem.* **241**, 3381 (1966).

Neely, W. B., Unger, I., Blair, E.H., and Nyquist, R. A. *Biochemistry* **3**, 1477 (1964).

Nolin, B. and Jones, R. N. *J. Am. Chem. Soc.* **75**, 5626 (1953).

Nyquist, R. A. *Appl. Spectrosc.* **11**, 161 (1957).

Parker, F. S. *Biochim. Biophys. Acta* **42**, 513 (1960).

Parker, F. S. *Federation Proc.* **21**, 243 (1962a).

Parker, F. S. *Perkin–Elmer Instrument News* **13**(4), 1 (1962b).

Parker, F. S. *Nature* **203**, 975 (1964).

Parker, F. S. in *Progress in Infrared Spectroscopy Vol. 3* (H. A. Szymanski, ed.) Plenum Press, New York, 1967, p. 75.

Parker, F. S. and Kirschenbaum, D. M. *J. Phys. Chem.* **63**, 1342 (1959a).

Parker, F. S. and Kirschenbaum, D. M. *Federation Proc.* **18**, 299 (1959b).

Rajender, S. and McColloch, R. J. *Anal. Biochem.* **13**, 469 (1965).

Riepe, M. E. and Wang, J. H. *J. Biol. Chem.* **243**, 2779 (1968).

Robinson, W. G. and Hook, R. H. *J. Biol. Chem.* **239**, 4257 (1964).

Schroepfer, G. J., Jr. *J. Biol. Chem.* **241**, 5441 (1966).

Schroepfer, G. J., Jr. and Bloch, K. *J. Biol. Chem.* **240**, 54 (1965).

Sen, S., Smeby, R. R., and Bumpus, F. M., *Biochemistry* **6**, 1572 (1967).

Smith, E. L. *Federation Proc.* **8**, 581 (1949).

Spencer, B. *Biochem. J.* **73**, 442 (1959).

Thompson, H. W., Brattain, R. R., Randall, H. M., and Rasmussen, R. S. in *Chemistry of the Penicillins* (H. T. Clarke, J. R. Johnson, and R. Robinson, eds.) Princeton Univ. Press, 1949, Chap. XIII, p. 382.

Turkevich, J., McKenzie, H. A., Friedman, L., and Spurr, R. *J. Am. Chem. Soc.* **71**, 4045 (1949).

Uyeda, K. and Rabinowitz, J. C. *J. Biol. Chem.* **240**, 1701 (1965).

Villarejo, M. and Westley, J. *J. Biol. Chem.* **238**, PC 1185 (1963a).

Villarejo, M. and Westley, J. *J. Biol. Chem.* **238**, 4016 (1963b).

Walsh, K. A., Kauffman, D. L., Sampath Kumar, K. S. V., and Neurath, H. *Proc. Natl. Acad. Sci. U.S.* **51**, 301 (1964).

Witzel, H. *Progr. Nucleic Acid Res.* **2**, 221 (1963).

Zugaza, A. and Hidalgo, A. *Memorias de la Real Academia de Ciencias, Exactas, Fisicas y Naturales de Madrid, Serie de Fisico-Quimicas, Tomo* **VI**, 7 (1966).

Chapter 16

DRUGS, PHARMACEUTICALS, AND PHARMACOLOGICAL APPLICATIONS

In 1957 Carol published a short review on the application of infrared methods to pharmaceutical analysis. Among the substances for which infrared methods had been developed at that time he mentioned atropine, nitroglycerine, sodium *N*-lauroyl sarcosinate, phenobarbital, testosterone, and progesterone. Other substances for which analytical methods were given were penicillin G in the presence of penicillin F, K, or dihydro-F; ethinyl testosterone; aspirin, phenacetin, and caffeine; and estrogenic substances, e.g., estrone, equilin, and estradiol-17-β (*cis*).

Infrared spectra are now widely used in the examination of pharmaceuticals. The sixteenth revision of *The Pharmacopoeia of the United States* (U.S.P.) and the eleventh edition of the *National Formulary* (N.F.) have presented identification tests which used infrared spectroscopy, whereas no infrared tests were used in U.S.P. XV or N.F. X. Infrared spectra have attained acceptance in legal considerations and are now given in patent applications as characteristics of antibiotics of unknown structure. In the pharmaceutical industry there are many applications for quantitative infrared analyses in research and development work, pharmacy research, and in various phases of pharmaceutical production. For example, infrared data are used to characterize reaction conditions and yields, to assay the purity of intermediate products, to examine such problems as the stability of a drug in the material in which it is suspended, and to maintain quality control in the chemical production of bulk drugs. A recent review (Papendick *et al.*, 1969) has given many references to fractionation and isolation methods for pharmaceutical analysis, such as the various types of chromatography, electrophoresis, countercurrent distribution, and extraction. The authors presented many references to infrared analyses for a wide variety of compounds (Table 16.1).

The techniques of sample preparation of pharmaceuticals are essentially those discussed more generally earlier in this book, such as the preparation of films, mulls, pellets, and solutions, and the use of ATR. Carol (1957) referred to most of these methods for the preparation of pharmaceutical compounds. Warren *et al.* (1967) have discussed pharmaceutical applications of the internal reflection technique. They have given examples of the suitability of the method for the analyses of less

Table 16.1. Compounds (in Their Order of Appearance) for which Papendick *et al.* (1969) Have Given References to Infrared Analyses

Pilocarpine	Urea and derivatives
Brevicolline	Chromones
Morphine	Flavones
Caffeine	Phenolic hydroxyl compounds
Antibiotics (general)	Steroids and hormones (general)
Nystatin	Hydrocortisone
Oxacillin	Prednisolone
Penicillins	Ethinylestradiol
Tetracycline	Estrogens (general)
Amides	Pregnane
Amines (aliphatic)	Benzothiadiazines
Nitrogenous organic bases	Cyclohexysulfamates
Sympathomimetic amines	Mercaptans
Amino acids	Phenothiazines
Amino ethers	Saccharin
Barbiturates	Sulfonamides
Imidazoles	Sulfonic acids
Quaternary ammonium compounds	Thiazine
Quinolines	Thioxanthines
Uracils	Vitamin B_{15}, pangamic acid

than 1 μg of gas–liquid and thin-layer chromatographic samples, tablet coatings, tissue specimens, and vehicles used in ointments.

It should be noted that many polar substances with poor solubilities in the usual solvents can be dissolved in water or in D_2O. Examples of the potential use of aqueous solutions in pharmaceutical research are to be found in the discussions of aqueous solutions (Chap. 3), enzymology (Chap. 15), hydrogen–deuterium exchange (Chap. 11), nucleic acids (Chap. 12), carbohydrates (Chap. 6), amino acids (Chap. 9), amines (Chap. 8), clinical chemistry (Chap. 18), and inorganic ions (Chap. 19).

Spectra Collections

The sources of infrared spectra are quite numerous, some of which contain specific information on drugs and pharmaceuticals and others which have data on compounds indirectly related to them. Sadtler Research Laboratories, Inc. (1969) have compiled collections of pharmaceutical, steroid, and biochemical spectra. The pharmaceutical collection contains 850 infrared and 1500 ultraviolet reference spectra of drugs, medicinals, and pharmaceutical compounds compiled from the latest editions of the U.S.P., *British Pharmacopeia, International Pharmacopeia, National Formulary*, and *New and Nonofficial Drugs*. The steroid collection contains the infrared spectra of 750 steroids and related compounds. The biochemical collection consists of 2000 infrared spectra and 650 ultraviolet spectra. The American Petroleum Institute and the Manufacturing Chemists Association also have collections of spectra available. (The address is given in the References to this Chapter.)

A partial bibliography of spectra of biochemical compounds (Clark and Chianta, 1957) is quite useful for the pharmaceutical field.

In 1962, Hayden *et al.* (1962) published infrared, ultraviolet, and visible spectra of 175 U.S.P. and N.F. reference standards and their derivatives. A list of the compounds recorded is given in Table 16.2.

Table 16.2. Reference Standards for which Hayden *et al.* (1962) Have Given Infrared Spectra

Acetaminophen	Chlortetracycline hydrochloride	Erythromycin lactobionate
Acetazolamide	Cholic acid	Estradiol
Acetyl sulfisoxazole	Choline chloride	Estradiol benzoate
para-Aminobenzoic acid	Cortisone acetate	Estradiol cyclopentylpropionate
Amisometradine	Cyanocobalamin	Estradiol dipropionate
Amodiaquine	Cyclizine	Estrone
Amodiaquine hydrochloride	Cyclizine hydrochloride	Ethinyl estradiol
Anthralin	L-Cystine	Ethisterone
L-Arginine hydrochloride		Evans blue
Ascorbic acid		
Atropine	Activated 7-dehydrocholesterol	Folic acid
Atropine sulfate	Desoxycorticosterone acetate	Furazolidone
Azacyclonol	Desoxycorticosterone tri-	
Azacyclonol hydrochloride	methylacetate	Glutethimide
Azure A	Diaminodiphenylsulfone	Gramicidin
	Dibucaine	
	Dibucaine hydrochloride	
Bacitracin	Dienestrol	Hexachlorophene
Benoxinate	Diethylstilbestrol	Histamine dihydrochloride
Benoxinate hydrochloride	Diethyltoluamide	L-Histidine monohydrochloride
Benzathine penicillin G	Digitoxin	Homatropine methylbromide
Benzestrol	Digoxin	Hydrocortisone
Betazole hydrochloride	Dihydrostreptomycin sulfate	Hydrocortisone acetate
Bishydroxycoumarin	Dihydrotachysterol	Hydrocortisone sodium
	Dihydroxyanthraquinone	succinate
	Dimethisoquin	Hydroxychloroquine
Calciferol	Dimethisoquin hydrochloride	Hydroxychloroquine sulfate
Calcium disodium edetate	Dimethyl tubocurarine iodide	Hydroxystilbamidine isethionate
Calcium pantothenate	Diphenadione	Hydroxyzine
Carbetapentane	Diphenhydramine	Hydroxyzine hydrochloride
Carbetapentane citrate	Diphenhydramine hydrochloride	
Carbinoxamine	Diprotrizoic acid	Iodipamide
Carbinoxamine maleate	Doxylamine	Iodochlorhydroxyquin
Cetylpyridinium chloride	Doxylamine succinate	Iophenoxic acid
Chloramphenicol		L-Isoleucine
Chlorcyclizine		Isoproterenol hydrochloride
Chlorcyclizine Hydrochloride		
Chlorothiazide	Edrophonium chloride	Lanatoside C
Chlorotrianisene	Epinephrine bitartrate	L-Leucine
Chlorpheniramine	Ergonovine maleate	Levallorphan
Chlorpheniramine maleate	Erythromycin	Levallorphan tartrate
Chlorpromazine	Erythromycin ethylcarbonate	Liothyronine
Chlorpromazine hydrochloride	Erythromycin glucoheptonate	L-Lysine monohydrochloride

Table 16.2 (*continued*)

Mecamylamine	Polymixin B sulfate	Testosterone cyclopentyl-propionate
Mecamylamine hydrochloride	Pramoxine	
Meclizine	Pramoxine hydrochloride	Testosterone enanthate
Meclizine hydrochloride	Prednisolone	Testosterone propionate
Menadione	Prednisolone acetate	Tetracaine
Mephenesin	Prednisone	Tetracaine hydrochloride
Mercaptopurine	Primidone	Tetracycline hydrochloride
Metharbital	Probenecid	Thenyldiamine
L-Methionine	Prochlorperazine	Thenyldiamine hydrochloride
Methotrexate	Prochlorperazine maleate	Thiamine hydrochloride
Methyltestosterone	Progesterone	Thiopental
Methyprylon	Promethazine	L-Threonine
	Promethazine hydrochloride	Tolazoline
Neomycin sulfate	Propoxyphene	Tolazoline hydrochloride
Nicotinamide	Propoxyphene hydrochloride	Tolbutamide
Nicotinic acid	Pyridoxine hydrochloride	Triacetyloleandomycin
Nifuroxime	Pyrilamine	Trihexyphenidyl base
Nitrofurantoin	Pyrilamine maleate	Trihexyphenidyl hydrochloride
Nitrofurazone	Pyrimethamine	Trimethaphan camphorsulfonate
Novobiocin		Tripelennamine
Nystatin	Rescinnamine	Tripelennamine citrate
	Reserpine	Tripelennamine hydrochloride
Oleandomycin chloroform	Riboflavin	L-Tryptophan
adduct	Rutin	Tubocurarine chloride
Ouabain		L-Tyrosine
Oxytetracycline	Saccharin	Tyrothricin
	Sitosterols	
Phenindamine	Sodium diatrizoate	Urethan(e)
Phenindamine tartrate	Sodium penicillin G	
Phenoxymethyl penicillin	Streptomycin sulfate	
Phentolamine	Sulfadiazine	L-Valine
Phentolamine methanesulfonate	Sulfamerazine	
L-Phenylalanine	Sulfamethazine	
Phytonadione	Sulfamethoxypyridazine	Warfarin sodium
Pipradrol	Sulfanilamide	
Pipradrol hydrochloride	Sulfisoxazole	Zoxazolamine

Later, Sammul *et al.* (1964) published a supplement to the 1962 paper, and gave infrared spectra of 334 more compounds which were new and nonofficial drugs, U.S.P. and N.F. items, solvents, and reagents. A list of these compounds is given in Table 16.3.

Hayden *et al.* (1966) have compiled 200 infrared spectra of new and unofficial drugs, U.S.P. and N.F. items, related compounds, solvents, and reagents. The substances are listed in Table 16.4. Fazzari *et al.* (1968) have given the infrared and ultraviolet spectra of 42 pharmaceuticals, 12 of which are commonly used tablet excipients. The substances are listed in Table 16.5.

Table 16.3. Compounds Whose Spectra Were Given in a Supplement to the Paper by Hayden *et al.* (1962) (Sammul *et al.,* 1964)

Acacia
Acetic acid
Acetone
Acetophenazine dimaleate
Acetophenetidine
N-Acetyl-d-amphetamine
N-Acetyl-dl-amphetamine
Acetyl salicylic acid
N⁴-Acetylsulfanilamide
Akineton hydrochloride
Aluminum acetate
Alvodine base
dl-Amidon hydrochloride
Aminoacetic acid (glycine)
2-Amino-5-chlorobenzo-
    phenone
4-Amino-6-chloro-m-benzene-
    disulfonamide
Aminophylline
p-Aminosalicylic acid
Ammonium chloride
Amobarbital
d-Amphetamine hydrochloride
n-Amyl alcohol
Amylene hydrate
Amyl metacresol
delta-5-Androstene-3-beta-17-
    beta-diol
delta-5-Androstene-3-beta-ol-
    17-one
delta-4-Androstene-3, 17-dione
Angiotensin amide
Anhalamine hydrochloride
Anhalonine hydrochloride
Ansolysen bitartrate
Anthracene
Anthralin
Anthrone
Antipyrine
Apomorphine hydrochloride
Aprobarbital
Azaphenothiazine
Aspirin anhydride
Atropine

Barbital
Bentyl analog
Benzaldehyde
Benzoic acid
Benzoin

Benzphetamine hydrochloride
Benzthiazide
Benztropine methanesulfonate
Benzyl alcohol, redistilled
Betamethasone
Betazole hydrochloride
d-Brompheniramine maleate
Bunamiodyl
Butabarbital
Butethamine hydrochloride
n-Butyl alcohol
sec-Butyl alcohol
Butylated hydroxytoluene

Caffeine
Calcium Cyanamide
Camphor
Camphor-10-sulfonic acid
Caramiphen ethanedisulfonate
Carbimazole
Carbocaine hydrochloride
Carbon disulfide
Cardrase
Castor oil
Cellulose
Chlordantoin
Chlorhexidine dihydrochloride
Chloroprocaine hydrochloride
Chlorpropamide
Cholesterol
Cinchonidine
Cinchonine
Citric acid
Codeine
Codeine phosphate
Corn oil
meta-Cresol
Cyclothiazide
Cyproheptadine hydrochloride

Dequadin chloride
Dequalinium chloride
Desoxycorticosterone acetate
Desoxycorticosterone trimethyl-
    acetate
Desoxyephedrine hydrochloride
Dexamethasone
Dexamethasone acetate

Dextrochlorpheniramine
    maleate
Dextrose, anhydrous
Diacetylchondocurarine iodide
N,O³-Diacetylphenylephrine
N,O-Diacetyl-p-hydroxy-
    amphetamine
Dibenzothiophene
Dihydrochlorothiazide
1,8-Dihydroxyanthraquinone
Dilabron methanesulfonate
Dimethoxinate hydrochloride
p,p-Dimethoxydiphenylacetylene
Dimethpyrindene maleate
Dimethylaminopropyl-
    d-camphidine dimethyl sulfate
3,5-Dimethyl-2-nitroanisole
3,5-Dimethyl-4-nitroanisole
Diphenylhydantoin
Diphenylpyraline hydrochloride
Dromostanolone propionate
Dynacaine hydrochloride

Ectylurea
Enovid
Ephedrine sulfate
Equilenin
Equilin
Erythritol tetranitrate
Erythromycin propionate
beta-Estradiol
Estradiol dipropionate
Ethanol
Ether
Ethinyl estradiol-3-benzoate
Ethinyl nortestosterone
Ethyl nitrate
2-Ethyl thioisonicotinamide
Ethynyl estradiol-3-methyl ether
Etryptamine acetate
Eugenol

Ferric sodium ethylenediamine
    tetraacetate
Fluocinolone acetonide
5-Fluorouracil
Fluoxymesterone
Fluphenazine dihydrochloride
Flurandrenolone
Formamide

Table 16.3 (*continued*)

Gantanol
Gitogenin
Glycerin
Glycopyrrolate
Griseofulvin
Guaiacol

1,3-Hexachlorobutadiene
Hexachlorophene
Hexadimethrine bromide
Hexestrol
Hexetidine
Hexocyclium
Hexylethyl barbituric acid
Hydrocortamate
Hydrocortisone-21-phosphate
    disodium salt
Hydroflumethiazide
17-Hydroxydesoxycorti-
    costerone
3-Hydroxymercuri-2-methoxy-
    succinimido-propane theo-
    phyllinate
2-Hydroxy-2-phenylethyl-
    carbamate
Hyoscine hydrobromide

Ipodate calcium
Ipodate sodium
Iso-octane
Isopregenone
Isopropamide iodide
Isosorbide dinitrate
Isuprel ethanesulfonate

Kanamycin sulfate

Lactose
Levarterenol bitartrate
Librium base
Librium hydrochloride
Linoleic acid
Lophophorine hydrochloride

Mannitol hexanitrate
Menadione
Menthol
Mephenoxalone
Meprobamate
Mescaline hydrochloride
Mescaline sulfate

Meta-butethamine hydro-
    chloride
Metaxalone
Methalamic acid
Methanol
Methaqualone
Metharbital
Methenamine mandelate
Methenamine undecylenate
Methocarbamol
Methoxyflurane
p-Methoxythiobenzaldehyde
Methsuximide
Methyl acetate
6-alpha-Methyl-17-acetoxy-
    progesterone
Methyldopate hydrochloride
6-Methylhydrocortisone acetate
3-Methyl-4-nitroanisole
3-Methyl-3-pentanol carbamate
6-Methyl-prednisolone
Methyl testosterone
Metopirone
Mikedimide

Neomycin undecylenate
Nialamide
Nicotinic acid
p-Nitrofluorobenzene
Nitroglycerine
Nitromethane
19-Nor-delta-4-androstene-17-
    beta-ol-3-one-beta-phenyl-
    propionate
Norethisterone acetate
Nostal

Olive oil
Orphenadrine citrate
Orphenadrine hydrochloride
Oxyphencyclimine hydro-
    chloride
Oxytetracycline

Palmitic acid
Parabromdylamine maleate
Paraffin oil
Peanut oil
Pentaerythritol tetranitrate
Pentamethylene tetrazole
Pentobarbital
Perphenazine
Persantin

Phenazocine hydrobromide
Phendimetrazine bitartrate
Phenformin hydrochloride
Pheniramine
Phenobarbital
Phenobutiodil
Phentolamine methane-
    sulfonate
Phenylephrine hydrochloride
Phenylpropanolamine hydro-
    chloride
Phenyl-*t*-butylamine hydro-
    chloride
Phenylramidol hydrochloride
Phenylramidol salicylate
Physostigmine
Picric acid
Piperidolate
Placidyl
Poldine methosulfate
Polyethylene glycol
    t-dodecylthioether
Polythiazide
Potassium aluminum sulfate
Potassium aspartate
Potassium bromate
Potassium chlorate
Potassium iodate
Potassium perchlorate
Potassium periodate
Potassium sulfate, anhydrous
Potassium warfarin
Potato starch
Pregnenolone acetate
Preludin
Probarbital
Procaine hydrochloride
Progesterone
Proketazine maleate
1,2-Propanediol acetate
Propanol
1-(4-hydroxyphenyl)-1-pro-
    panone
Propionyl erythromycin lauryl
    sulfate
1-Propoxyphene-2-naphthalene
    sulfonate monohydrate
1-Propoxyphene-*N*-oxide-
    hydrochloride
Prothipendyl hydrochloride
Pyridine

Table 16.3 (*continued*)

Quinethazone	Stigmasterol	Thymol
Quinidine	Succinyl choline picrate	Tofranil
	Sucrose	O^2,O^4-*n*-Triacetyl-l-epine-
	Sulfachloropyridazine	phrine
Salicylic acid	Sulfamethazole	O^3,O^4-*n*-Triacetyl-1-norepine-
Sandoptal	Sulfaphenazole	phrine
Scopolomine	Sulfobromophthalein sodium	Triburon chloride
Secobarbital sodium		Trichlorocarbanilide
Sitosterols		Trichlormethiazide
beta-sitosterol	Taractan	Triethylamine
Sodium diacetate	Tenuate	Triflupromazine
Sodium lauryl sulfate	Testosterone	Triflupromazine hydrochloride
Sodium nitrate	Testosterone propionate	3,5,3'-Triiodothyropropionic acid
d-Sorbitol	Tetrachloroethylene	Triparanol
Soya sitosterols	Thalidomide	
Soybean oil	Theobromine	Uracil mustard
Sparteine sulfate	Thiethylperazine dimaleate	Urea
Spironolactone	Thioguan(ine)	
Stanozolol	Trans-Stilbene	Vanillic acid diethylamide
Starch	Thioridazine hydrochloride	
Stearic acid	Thonzonium bromide	Xanthiol hydrochloride

A recent book edited by E. G. C. Clarke (1969) contains a chapter by Chapman and Moss (1969) on infrared spectrophotometry which should be of great use. Also, in a section called "Analytical and Toxicological Data," infrared characteristic bands for many compounds have been given. Part 3 of the book contains 444 infrared spectra. The compounds for which the spectra are given were thought to be the ones most likely to be encountered in pharmaceutical and toxicological investigations. A very useful feature is the catalogue system listing principal bands in ascending order of wave number (from $650\,\text{cm}^{-1}$) and placing next to these frequencies the compounds which exhibit them, the page number of the compound, and the page number on which the spectrum is to be found.

The system works in the following way: On each of the 444 spectra, the major bands have been labelled "*A*", "*B*", or "*C*" in decreasing order of intensity. With some spectra (for example, that of amphetamine) selection of these bands presents no difficulty; with others (for example, that of brucine) there are a number of bands of almost equal intensity and therefore several bands may carry the same designation, but in practice this does not present much of a problem.

This table of principal bands is then used to determine the identity of "unknowns." In the spectrum of an unknown compound (Fig. 16.1) there are three major bands: (A) $1157\,\text{cm}^{-1}$, (B), $1305\,\text{cm}^{-1}$, and (C) $1342\,\text{cm}^{-1}$. One then turns to the table and lists all bands and compounds (a) between 1156 and $1158\,\text{cm}^{-1}$, (b) between 1304 and $1306\,\text{cm}^{-1}$, and (c) between 1341 and $1343\,\text{cm}^{-1}$. Chlorothiazide is the only compound having bands in all three regions (a), (b), and (c) above, and its spectrum (from the group of 444 spectra) is then compared with that of the

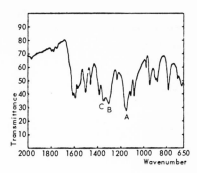

Fig. 16.1. Spectrum of an "unknown" compound. Major peaks are labelled *A*, *B*, and *C*, in order of decreasing intensity. (Clarke, 1969.)

unknown and found to be identical. Of course, not all identifications are so easily made, and lists of other frequencies and possible compounds may be needed.

For other general references the reader is referred to General References and Spectra Collections (Chap. 20).

Table 16.4. New and Unofficial Drugs whose Infrared Spectra Have Been Given by Hayden *et al.* (1966)

Acetanilide	Benzocaine	Clemizole hydrochloride
Acetohexamide	Benzylmethylamine	Colistimethate sodium
Acetyl sulfamethoxypyridazine	Betamethasone acetate	Colistin sulfate
L-Alanine	Betamethasone disodium phos-	Corn phospholipids
Allyl estrenol	phate	Cotton seed oil
2-Amino-6-methylheptane	Bromelain concentrate	Creatine hydrate
Aminopyrine	Brucine	Creatinine
Ammonium sulfate	Butallylonal	Cyclandelate
Amobarbital sodium	Butethal	Cyclobarbital
Amopyroquin dihydrochloride	Butethamine	Cyclopal sodium
d,l-Amphetamine sulfate		Cyproheptadine
Amphomycin	Calcium trisodium pentetate	
Androstane-17β-ol-3-one	Camphor	Desipramine hydrochloride
Anethole	Camphoric acid	*l*-Desoxyephedrine
Anisotropine methylbromide	Cantharidin	Dexamethasone sodium phos-
Antazoline hydrochloride	Carbocaine	phate
Anthrone	Chloral hydrate	Dextrin
Apoatropine hydrochloride	*d,l*-Chloramphenicol	Dextromethorphan hydro-
Aribine	Chloranil	bromide
Atabrine dihydrochloride	Chlormadinone acetate	Diacetyltubocurarine nitrate[b]
Atmul 80	Chlorobutanol, hydrous	Dichlorisone
	Chloroprocaine	Dichlorphenamide
Barbital sodium	Chlorothen citrate	Didexamethasone-21-phosphate
Bendroflumethiazide	8-Chlorotheophylline	*N,N*-Diethyl-D-lysergamide[c]
Benzalkonium chloride	Chlorthalidone	Diisopromine hydrochloride
Benzene	Chondocurarine iodide	Dimenhydrinate
Benzene hexachloride, gamma	Citanest hydrochloride	Dimethisterone
Benzilic acid	Clemizole bisulfate	Dimethylformamide

Table 16.4 (*continued*)

Dimethyl sulfoxide
Dipropylbarbituric acid
Disodium edetate
Dithiazanine iodide
Domiphen bromide
Doxapram hydrochloride
Doxylamine succinate

Epinephrine bitartrate
Estradiol benzoate
Ethosuximide
Ethyl acetate
1-Ethyl hydantoin
Ethyl laurate
Etryptamine

Fludrocortisone acetate
Fluprednisolone

Glyceryl guaiacolate
Glycine anhydride
Guanethidine sulfate
Guanidino acetic acid

Hexethal sodium
Hexylcaine hydrochloride
Hydantoin
Hydrochlorothiazide[a]
p-Hydroxyacetanilide
p-Hydroxybenzoic acid
2-Hydroxyethyl benzyl-
  carbamate
p-Hydroxy-α-(methylamino-
  methyl)-benzyl alcohol tartrate[b]
8-Hydroxyquinoline sulfate

Idoxuridine
Indomethacin
Inositol niacinate
Iso-*N*,*N*-diethyl-D-lysergamide[c]
Isopropyl alcohol
Isopropyl meprobamate[c]

Lucanthone hydrochloride

Magnesium aspartate
Mannitol
Mebutamate
Mefenamic acid
d,l-Menthone
Methandrostenolone
Methapyrilene hydrochloride
Methdilazine hydrochloride
Methenamine
Methyl 3-benzoylpropionate
1-Methyl hydantoin
3-Methyl hydantoin
5-Methyl hydantoin
3-Methyl-2-phenylmorpholine[d]
Methyprylon
Methysergide bimaleate
Metronidazole
Metyrapone

l-Nordefrin

Oxethazaine

Pargyline hydrochloride
Paromomycin sulfate
Pentapiperide methylsulfate
Phenanthrene
Phenelzine sulfate
Phenolsulfonphthalein
Phenothiazine
Phensuximide
Phenylbutazone
Phenylephrine
4-Phenylhexanoic acid
Phenylpropanolamine
Phenyramidol
Pilocarpine hydrochloride
Pipazethate hydrochloride
Piperidolate hydrochloride
Pipradrol
Polyvinylpyrrolidone
Probarbital calcium
Procaine
Propenzolate hydrochloride
Prophenal
Protoveratrine A

d-Pseudoephedrine hydrochloride
d-Pseudoephedrine sulfate
3-Pyridinemethanol tartrate
Pyridostigmine bromide

Quinidine
Quinine

Resorcinol
Ricinoleic acid

Salicylamide
Santonin
Sarcosine
Sesame oil
Sodium *p*-aminosalicylate
Sodium diethyldithiocarba-
  mate trihydrate
Sodium tequinol hydrate
Stovaine
Succinic acid
Sulfaguanidine
Sulfamethizole
Sulfamethoxydiazine
Sulfamic acid
Sulfapyridine

Tartaric acid
Tenuate
Terpin hydrate
Thihexinol methylbromide
Thimerosal
Thioridazine
Triamcinolone
Triamcinolone acetonide
Trimethobenzamide hydro-
  chloride
Tybamate
Tylosin

Valethamate bromide
Vinblastine sulfate

Yohimbine hydrochloride

---

[a]Dihydrochlorothiazide.   [b]Synephrin tartrate.   [c]Carisprodol.   [d]Preludin.

Table 16.5. Pharmaceuticals, including 12 Commonly Used Tablet Excipients, whose Infrared Spectra Have Been Given by Fazzari *et al.* (1968)

Allopurinol	Magnesium hydroxide
Anagestone acetate	Mepenzolate methyl bromide
Beeswax	Methaqualone
Calcium carbonate	Methazolamide
Carnauba wax	Methetoin
Cetyl alcohol	Methopholine
Chlorphentermine HCl	Methoxylin nicotinate
Clamoxyquin HCl	Methyclothiazide
Clopamide	Nalidixic acid
Clorprenaline HCl	Nylidrin HCl
Cyclamic acid	Opipramol HCl
Cyclopenthiazide	Oxazepam
Demecarium bromide	Phenol
Ethambutol HCl	Piposulfan
Furosemide	Polyethylene glycol 400
Gelatin	Polyethylene glycol 4000
Haloperidol	Sodium benzoate
Heptolamide	Sodium sulfate
Hexacosanol	Sulfadimethoxine
Hydroxydione succinic acid	Talcum
Isoxsuprine HCl	Tragacanth

Antibiotics

Evans (1965) has presented details of the chemistry, synthesis, biosynthesis, and clinical applications of many of the antibiotics used in medicine. Although no details of infrared spectra were given, in some cases information was presented concerning the type of information obtained from infrared evidence. Table 16.6 contains some of the information available from Evans' book. If Evans has given a direct reference to the infrared evidence, that reference and its page number is given in column 3. If no direct reference is given, the page containing the infrared information is stated.

Griseofulvin

Eglinton (1964) discussed the total synthesis of the important oral antibiotic griseofulvin and demonstrated how infrared measurements can still be of considerable use even when the compounds under study are polyfunctional and of such complexity that important bands overlap. The structure of this antibiotic is given in Fig. 16.2. The conjugated double bond and the aromatic ring vibrations were all found in the 1600 cm^{-1} region. The value of 1720 cm^{-1} for $v_{C=O}$ of the five-membered coumaranone ring was explained approximately in terms of ring strain ($v_{C=O}$ rises, 5-membered ring ketone), the electronegative oxygen atom in the ring (v rises), and vinylogous conjugation (v falls; aromatic ring with three o and p oxygen substituents).

Table 16.6. Infrared Information on Antibiotics Available from Evans (1965) (See Text)

Antibiotic	Type of Information Gained from Infrared Spectra	Reference and Page Number in Evans Book	
D-Cycloserine (Oxamycin)	Dipolar ionic structure	Ref. (2) on page (18)	
Penicillins	β-Lactam-thiazolidine structure	(2)	(39)
Cephalosporin C	β-Lactam carbonyl	—	(33)
Polymyxin B₁-isomers (α and γ)	Qualitative spectra data	—	(66)
Kanamycin A	Two α-glycosidic linkages	(55)	(101)
Erythromycin	Polyhydroxy-keto-lactone system	—	(136)
Cladinose moiety	Two hydroxyl groups	—	(137)
Erythronolide (dihydro)	Polyhydroxylactone	—	(139)
Erythralosamine	Ketone group position located	(6) (8)	(166)
Carbomycin (Magnamycin)	α,β-unsaturated ketone and a saturated aldehyde group	—	(145)
Substituted glutaric anhydride of a dicarboxylic acid derived from the antibiotic	Determined relative positions of two carboxyl groups	—	(150)
Polyene types e.g., pimaricin and lagosin	Lactone group	—	(169)
Nystatin (Fungicidin)	Carboxyl Carbonyl (lactone?) Hydroxyl groups	(8)	(175)
Amphotericin B	Lactone- or ester-carbonyl	—	(173)
Fumagillin	Lack of carbonyl group	(17)	(176)
Novobiocin			
Methyl glycoside moiety	Carbamate	—	(178)
The substituted benzoic acid moiety	Ester carbonyl Ethylenic linkages C—O linkages	—	(180)
Ristocetins A and B	Qualitative spectra	—	(187)
Vancomycin	OH or NH₂, amide, and aromatic groups	(19) (21)	(191)
Puromycin	Carboxamide group	—	(188)
Purine moiety	6-Dimethylamino purine	—	(188)
Amino pentose moiety	3-Deoxy-3-amino-D-ribose	(28)	(191)

The dienone $v_{C=O}$ at $\sim 1670\ \mathrm{cm}^{-1}$ seemed high, but it was felt that this result might be an effect of measurement in the solid state (and perhaps not strictly reliable).

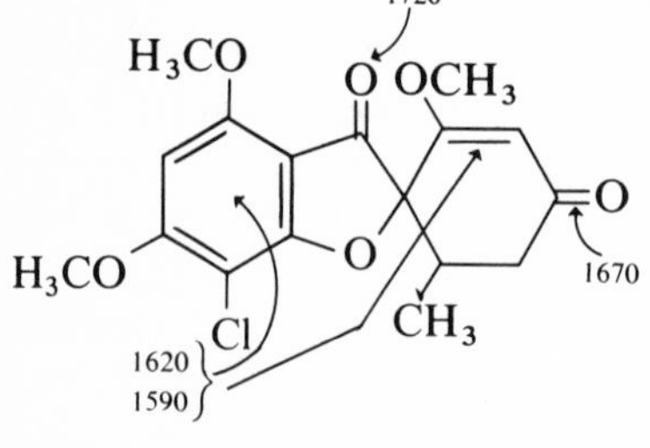

Fig. 16.2. Molecule of griseofulvin showing origin of various bands. (Eglinton, 1964.)

Penicillins

Zugaza and Hidalgo (1965) studied the infrared spectroscopy of a variety of penicillins—natural, biosynthetic, and synthetic—in the form of mulls, pellets, and

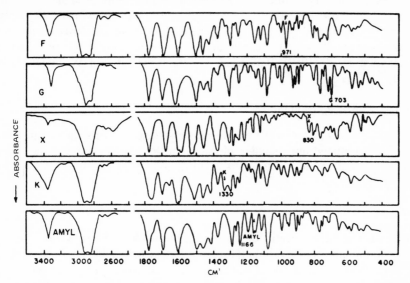

Fig. 16.3. The infrared absorption spectra of five types of penicillin obtained from mineral oil mulls. (Gore and Petersen, 1949.)

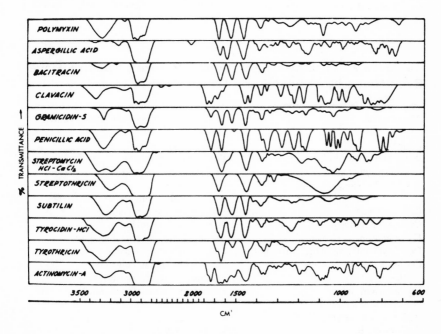

Fig. 16.4. Infrared spectra of antibiotics. (Gore and Petersen, 1949.)

aqueous solutions. The solution spectra were simpler than the solid-state spectra since molecular interactions in the latter caused the occurrence of new absorption band maxima. The spectra obtained from the solid penicillins were influenced by the type of cation and the crystalline form, effects which were suppressed in the solutions.

Figure 16.3 shows the spectra of five types of penicillin which differ only in their side chains (Gore and Petersen, 1949). Zugaza and Hidalgo (1966) have published the spectra of about 20 penicillin compounds. The reader should also refer to Evans (1965), Hayden *et al.* (1962), Papendick (1969), Wayland and Weiss (1965), and Chapter 15—*Enzymology*.

Antibiotics from Actinomycetes

Umezawa (1967) has recently published an index of antibiotics from actino-mycetes. This compendium includes physical and chemical data such as, crystal type, melting point, optical rotation and ultraviolet maxima, infrared spectrum, structural diagram, etc.

General

Figure 16.4 presents spectra of twelve antibiotics including several of protein character and high molecular weight (Gore and Petersen, 1949).

In 1965, Wayland and Weiss published infrared spectra of 33 antibiotics not included in the earlier collections in the same journal (Hayden *et al.*, 1962; Sammul *et al.*, 1964). These antibiotics are listed in Table 16.8. Unterman (1965) has published a survey on spectrophotometric methods for antibiotic determination in the ultra-violet and infrared regions.

Table 16.7. Antibiotics for which Infrared Spectra Have Been Given by Wayland and Weiss (1965)

Amphotericin A	Lincomycin hydrochloride monohydrate
Amphotericin B	Neomycin palmitate
Ampicillin trihydrate	Paromomycin sulfate
Benzathine penicillin V	Potassium penicillin O
Calcium amphomycin	Potassium phenethicillin
Chloramphenicol palmitate	Rolitetracycline
Chloroprocaine penicillin O	Sodium cephalothin
Colistin sulfate	Sodium cloxacillin monohydrate
Cycloserine free acid	Sodium colistimethate
Dactinomycin	Sodium dicloxacillin monohydrate
Demethylchlortetracycline hydrochloride	Sodium methicillin monohydrate
Erythromycin ethylsuccinate	Sodium nafcillin monohydrate
Erythromycin stearate	Sodium oxacillin monohydrate
Erythromycin sulfate	Spectinomycin dihydrochloride
Gentamicin sulfate	pentahydrate
Hydrabamine penicillin V	Vancomycin sulfate
Hygromycin B	Viomycin sulfate

Sulfonamides

Sulfonamide Activity   Infrared studies to determine the polarity of the sulfone group in the sulfonamides (Seydel *et al.*, 1960; Wahl, 1965) have challenged Bell and Roblin's (1942) hypothesis that a sulfonamide with a more polar SO_2 group (i.e., more negative) is more active in its action against microbes. Wahl (1965) did not support a hypothesis of Seydel *et al.* (1960; 1961) that the reactivity of the primary amino group in the sulfonamides is most important for sulfonamide action.

Seydel (1966) has shown that there is an approximate linear relationship between Hammett σ-values for various benzeneamines (related to the sulfonamides) and the logarithm of the lowest concentration of a sulfa drug to show inhibition of bacterial growth. The more positive the Hammett σ-constant (Hammett, 1937) of the amine, the more active was the corresponding sulfonamide, with the exception of the ortho-halogenated amine, which proved to give a more active sulfonamide than could be expected from its Hammett σ-value.

Data obtained by Seydel (1966) from the stretching frequencies of the amino group were in excellent agreement with those obtained by Krueger and Thompson (1959; 1957). Seydel tried to find a correlation of either the stretching frequencies, the force constant (k), the band intensities, or the "s" character of the nitrogen hybrid orbitals of the N-H bonds (expressed in the equation below as the coefficient b of the amines), to the MIC (the logarithm of the lowest concentration of a sulfa drug to show inhibition of bacterial growth) of the corresponding sulfonamides. The relation between b and the nitrogen hybrid orbitals is given by

$$\psi_{hybrid} = b\psi_s + \sqrt{1 + b^2}\,\psi_b$$

Each of these parameters gave acceptable correlations to the *in vitro* activity of the sulfonamide. Seydel decided to use the s character, since it had the best correlation and it was possible to interpret the results in a "more fundamental manner." The H—N—H bond angle θ was calculated from the N—H stretching frequencies by means of Eq. (1) (see Barnard *et al.*, 1949) and Eq. (2) (see Mason, 1958; Linnett, 1945),

$$k = \frac{\pi^2 c^2 m'}{\dfrac{1}{M'} + \dfrac{1}{M''}}(v_s + v_a)^2 \text{ dyne cm}^{-1} \tag{1}$$

$$\cos\theta = 1 + \frac{M'}{M''} \times \frac{\left(\dfrac{v_a}{v_s}\right)^2 - 1}{\left(\dfrac{v_a}{v_s}\right)^2 + 1} \tag{2}$$

where k is the force constant; c is the velocity of light, 2.99776×10^{10} cm sec^{-1}; m' is the mass of particle with molecular weight of 1; M' and M'' are the atomic weights of atoms participating (for NH_2, $M' = 14$; $M'' = 1$); v_s is the symmetrical stretching vibration of NH_2; v_a is the antisymmetrical stretching vibration of NH_2.

The H—N—H bond angle θ increases smoothly from 109.4° in *p*-dimethyl-aminoaniline to 111.1° in aniline, to 113.6° in *p*-nitroaniline (Mason, 1958) (see Table 16.8). The increase in the bond angle parallels the change of the nitrogen atom from almost pure sp^3-hybridization toward a state with higher $s:p$ ratio when the lone pair electrons become more and more delocalized over the aromatic ring. The increase in s character is accompanied by an increase in the NH_2 stretching frequencies and in the C—N stretching frequencies as these acquire a progressively more double-bonded character.

The constant b is the coefficient of the $2s$ orbital of nitrogen in the hybrid (Mason, 1958). It can be used as a measure of the s character of the hybrid nitrogen orbitals binding the hydrogen atoms and can be calculated by Eq. (3):

$$b^2 = -\cos\theta/(1 - \cos\theta) \qquad (3)$$

For some *para*-substituted anilines the calculated b values can also be found in the paper by Krueger (1962). Table 16.8 contains the results of Seydel's (1966) calculations plus data of the NH_2 stretching frequencies and the bond angles. The double-bond character expressed by b increases as the substituents become more and more electron-withdrawing. A plot of the b values of the amines *vs* the logarithm of the MIC of the

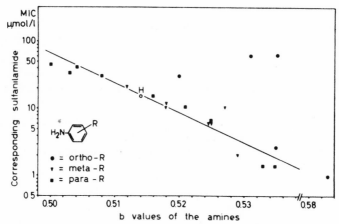

Fig. 16.5. Plot of the b values (Table 16.8, column 6, calculated by Eq. 3) of the amines *vs* the logarithm of the MIC of the corresponding sulfon-amides (Table 16.8. column 7). (Seydel, 1966.)

sulfonamides produces a nearly linear relationship (Fig. 16.5), showing that the sulfon-amide becomes more active with increasing b values, i.e., with increasing double-bond character of the C—N bond. The *ortho*-substituted aniline derivatives deviate from the line in most cases. Seydel explained this behavior to be a result of intra-molecular hydrogen bonding between the protons of the primary amino group and the substituents in the *ortho* position. The hydrogen bonding causes a larger bond

Table 16.8. Structure-Activity Correlation in the Sulfonamide Series[a] (Seydel, 1966)

1	2 σ Hammett	3 v_a (cm^{-1})	3 v_s (cm^{-1})	4 $10^5 \times k$ (dyne cm^{-1})	5 θ	6 b	7 Melting point (°C)	7 MIC μmoles/l E. coli
p-N(CH$_3$)$_2$	−0.600	3453	3377	6.46	109.4	0.499	234–239	45.0
o-OCH$_3$	−0.390[b]	3487	3396	6.56	113.2	**0.531**	203–206	45.0
o-OC$_2$H$_5$	−0.350[b]	3486	3396	6.56	112.9	**0.535**	150–153	45.0
p-OCH$_3$	−0.268	3460	3382	6.48	109.9	0.504	198	34.5
p-OC$_2$H$_5$	−0.250	3459	3381	6.48	109.8	0.503	195	32.0
p-CH$_3$	−0.170	3470	3390	6.51	110.3	0.508	188–195	27.2
o-CH$_3$	−0.170[b]	3482	3398	6.55	113.2	**0.520**	157–158	32.0
m-CH$_3$	−0.069	3475	3393	6.53	110.8	0.512	135–137	22.5
H	0	3479	3396	6.54	111.1	0.514	156–157	16.0
m-OC$_2$H$_5$	+0.115	3482	3397	6.55	111.6	0.518	180	13.3
m-OCH$_3$	+0.115	3483	3398	6.55	111.6	0.518	164–166	11.2
o-Cl	+0.200[b]	3494	3401	6.58	113.7	**0.535**	174–175	2.8
p-Cl	+0.227	3482	3398	6.55	111.3	0.516	202	16
p-Br	+0.232	3485	3399	6.56	111.8	0.521	203–206	11.25
p-I	+0.276	3486	3398	6.55	112.4	0.525	208–216	10.1
m-I	0.352	3487	3399	6.56	112.3	0.525	132–134	8
m-Cl	0.373	3490	3402	6.575	112.4	0.525	129–135	8
m-Br	0.391	3490	3401	6.57	112.6	0.527	138–141	11.25
m-NO$_2$	0.710	3497	3407	6.60	112.9	0.529	180	2
p-COCH$_3$	0.874[c]	3502	3410	6.61	113.4	0.533	195–199	1.4
p-NO$_2$	1.27[c]	3509	3416	6.64	113.6	0.535	169–172	1.4
o-NO$_2$	—	3527	3407	6.66	121.1	**0.583**	179	1.4

Column 7 heading: Sulfonamides synthesized using the amines listed in column 1.

[a] Structures of anilines used in this study (column 1); σ-Hammett constant for the substituents of the anilines (column 2); the position of the symmetric v_s and the antisymmetric v_a stretching vibration of the amino group in anilines (columns 3); the N—H stretching vibration force constant k calculated from v_s and v_a using Eq. (1) (column 4); the H—N—H bond angle θ, calculated from v_s and v_a using Eq. (2) (column 5); the "s" character of the nitrogen hybrid orbitals of the N—H bonds expressed as the coefficient b using Eq. (3) (column 6); and the melting point and minimum inhibitory concentration (MIC) (columns 7) of the sulfonamides synthesized by using the anilines listed in column 1. The *ortho*-substituted compounds are printed in boldface type.

[b] Using Taft's *ortho*-substituent parameter.

[c] Special σ-values for aniline derivatives, Krueger and Thompson, (1957).

angle, calculated from infrared data, than can be attributed to delocalization of the lone electron pair only (Fig. 16.6). Preliminary studies on heterocyclic amines showed that the relationship described above holds in general and is not limited to amines of the aniline type.

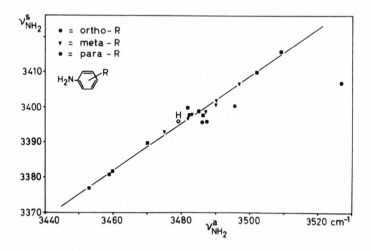

Fig. 16.6. Plot of symmetric-NH_2 vs the antisymmetric-NH_2 stretching frequencies (Table 16.8 column 3) of the anilines used for the synthesis of sulfonamides. (Seydel, 1966.)

Determination of Sulfonamides   Sulfonamides have been determined by infra-red spectroscopy (Oi and Inaba, 1967b). Sulfathiazole and sulfamerazine have been analyzed in mixtures of the two, as have mixtures of sulfathiazole and sulfadiazine. The solvent used was dimethylformamide. For mixtures of sulfathiazole and sulfamer-azine bands at 919 and 958 cm^{-1}, respectively, were used; for mixtures of sulfathiazole and sulfadiazine, 729 and 929 cm^{-1} bands.

Alkaloids

Marion *et al.* (1951) have studied the infrared spectra of 47 alkaloids in chloro-form solution. They paid particular attention to the regions at 3700-3200 cm^{-1}, in which both the hydroxyl and the imino groups absorb, and 1780–1620 cm^{-1}, in which the variously substituted carbonyl groups absorb. Hydroxyl groups were generally detected by the appearance of a sharp band in the 3625–3540 cm^{-1} region with an apparent molar absorptivity of 30–160, although there were exceptions. Imino groups produced a sharp band in the 3480–3440 cm^{-1} region with an apparent molar absorptivity of 100–210 when the imino group was present in an indole nucleus or secondary amido group. When the imino group was present in a piperidine ring, however, the absorption was much weaker but could be observed in increased path lengths. Carbonyl groups were detectable from the appearance of a strong band in the $1780–1620 \text{ cm}^{-1}$ region, with an apparent molar absorptivity from 300–1200. Certain correlations were established by Marion *et al.* between the carbonyl frequency and the type of carbonyl group present. Phenyl rings generally displayed bands in the $1640–1600 \text{ cm}^{-1}$ region and aromatic nitrogen rings generally gave bands between 1600 and 1560 cm^{-1}. The presence of a carbonyl group both in cevine and jervine was indicated by the infrared spectra of these compounds.

The biosynthesis and chemistry of alkaloids have been discussed in a recent book on natural products (Hendrickson, 1965). Hendrickson has pointed out the important contributions made to structural chemistry by instrumental methods of analysis such as UV, IR, NMR, and ORD spectroscopy. One of the alkaloids dis-cussed in which reference to the use of infrared spectroscopy was made was vindoline (CIII) (a compound isolated from periwinkles *Vinca rosea* in a search for anticancer agents). Dihydrovindoline hydrochloride on mild pyrolysis yielded a ketone $C_{21}H_{28}$-N_2O_2. The infrared spectrum of this compound showed absorption at 1709 cm^{-1}, which indicated a cyclohexanone. This result was used, together with other informa-tion, in determining the total structure.

CIII

Kasturi (1963) has presented some infrared data concerning alkaloids. Among the compounds were quinozilidine alkaloids, methyl neoreserpate, 3-epialloyo-himbone, and 3-epialloyohimbine (conformational data). Other compounds men-tioned were the tobacco alkaloid myosmine; the substances histisine, angustifoline, recanescine, and reserpine; other *Rauwolfia* alkaloids; some *Vinca* alkaloids; and powellane. Neuss (1959) has also supplied data on indole and dihydroindole alka-loids.

Holubek and Strouf (1965) have compiled spectral data and physical constants of alkaloids. The collection now consists of 600 cards assembled in book form, each card describing one alkaloid. Structural and molecular formulas, melting points, specific rotations, and apparent pK values in 80% methylcellosolve are given. The spectroscopic data include ultraviolet and infrared absorption spectra.

Clarke (1969) has published infrared spectra of many drugs, pharmaceuticals, and medicinals, and among these are the spectra of numerous alkaloids. Table 16.9 lists the compounds for which spectra are given in the section entitled "Analytical and Toxicological Data" of Clarke's book.

Hubley and Levi (1955), Levi *et al.* (1955), and Pleat *et al.* (1951) have discussed alkaloids from the forensic point of view.

Table 16.9. Compounds Whose Infrared Spectra are Given in Clarke (1969)

Acepifylline	Azacyclonol	Carbamazepine
Acepromazine	Azamethonium bromide	Carbetapentane
Acetanilide	Azapetine	Carbimazole
Acetazolamide		Carbinoxamine
Acetomenaphthone	Bamethan	Carbromal
Acetophenazine	Bamifylline	Carbutamide
Acetylcarbromal	Bamipine	Carphenazine
Acetylcholine chloride	Barbitone	Cetalkonium chloride
Acinitrazole	Barbituric acid	Cetrimide
Aconitine	Bemegride	Chlophedianol
Adiphenine	Benactyzine	Chloral hydrate
Adrenaline	Bendrofluazide	Chloramphenicol
Alcuronium chloride	Benzamine	Chlorbutol
Allantoin	Benzethonium chloride	Chlorcyclizine
Allobarbitone	Benzhexol	Chlordiazepoxide
Allylbarbituric acid	Benzilonium bromide	Chlorhexidine
Alverine	Benzocaine	Chlorisondamine chloride
Ambazone	Benzoic acid	Chlormethiazole
Ambenonium chloride	Benzquinamide	Chlormezanone
Amethocaine	Benzthiazide	Chloroprocaine
Amicarbalide	Benztropine	Chloropyrilene
Amidopyrine	Benzyl nicotinate	Chloroquine
Aminacrine	Bephenium hydroxynaphthoate	Chlorothiazide
Aminobenzoic acid	Berberine	Chlorotrianisene
Aminometradine	Betamethasone	Chlorpheniramine
Aminopentamide	Bethanechol chloride	Chlorphenoxamine hydro-
Aminosalicylic acid	Bibenzonium bromide	chloride
Amiphenazole	Biperiden	Chlorphentermine
Amisometradine	Bisacodyl	Chlorproguanil
Amitriptyline	Bretylium tosylate	Chlorpromazine
Amodiaquine	Bromodiphenhydramine	Chlorpropamide
Amolanone	Bromvaletone	Chlorprothixene
Amopyroquine	Brucine	Chlortetracycline
Amphetamine	Buclizine	Chlorthalidone
Amprolium	Buformin	Chlorzoxazone
Amprotropine	Bufylline	Choline chloride
Amydricaine	Butacaine	Choline theophyllinate
Amylobarbitone	Butanilicaine	Cinchocaine
Amylocaine	Butethamate	Cinchonidine
Anileridine	Butethamine	Cinchonine
Apoatropine	Butobarbitone	Cinnarizine
Apomorphine	Butoxamine	Clamoxyquin
Aprobarbitone	Butriptyline	Clemizole
Apronal	Butyl aminobenzoate	Clomiphene
Arecoline	Butylchloral hydrate	Clopenthixol
Aspirin		Clothiapine
Atropine	Caffeine	Cocaine
Atropine methobromide	Captodiame	Codeine
Azacosterol	Caramiphen	Colchicine

Table 16.9 (*continued*)

Cotarnine
Cropropamide
Crotethamide
Cyclizine
Cyclobarbitone
Cycloguanil
Cyclomethycaine
Cyclopenthiazide
Cycloserine
Cyclothiazide
Cycrimine
Cyproheptadine

Debrisoquine
Decamethonium iodide
Demecolcine
Demethylchlortetracycline
Denatonium benzoate
Dequalinium chloride
Deserpidine
Desferrioxamine
Desipramine
Dexamethasone
Dexamphetamine sulfate
Dexbrompheniramine maleate
Diamthazole
Diaveridine
Diazepam
Dibromopropamidine
Dichloralphenazone
Dichlorophen
Dicyclomine
Dienoestrol
Diethazine
Diethylpropion
Digoxin
Dihydrallazine
Di-iodohydroxyquinoline
Dimefline
Dimercaprol
Dimethisterone
Dimethoxanate
Dioxyline
Diperodon
Diphenhydramine theoclate
Diphenidol
Diprophylline
Disulfamide
Domiphen bromide
Doxapram
Doxylamine

Droperidol

Edrophonium chloride
Embramine
Embutramide
Emetine
Emylcamate
Ephedrine
Etafedrine
Ethchlorvynol
Ethinamate
Ethinylestradiol
Ethionamide
Ethotoin
Ethylnoradrenaline

Fencamfamin
Fenethylline
Fenfluramine
Fenpiprane
Fluanisone
Fluopromazine
Fluphenazine
Folic acid
Frusemide
Furaltadone
Furazolidone

Glutethimide

Haloperidol
Halopyramine
Heteronium bromide
Hexachlorophene
Hexamethonium bromide
Hexetidine
Hexobarbitone
Hexocyclium methylsulfate
Hexylresorcinol
Homatropine
Homidium bromide
Hordenine
Hydrallazine
Hydrastine
Hydrastinine
Hydrochlorothiazide
Hydrocortisone
Hydroflumethiazide
Hydromorphinol
Hydroquinidine
Hydroxyamphetamine

Hydroxychloroquine
Hydroxyquinoline
Hydroxyzine
Hyoscine
Hyoscyamine

Imipramine
Iproniazid
Isobutyl aminobenzoate
Isocarboxazid
Isoniazid
Isopropamide iodide

Leptazol
Lignocaine
Lincomycin
Lobeline
Lucanthone
Lysergic acid

Meclozine
Mefenamic acid
Mepacrine
Mephenesin
Mephentermine
Mepivacaine
Meprobamate
Mepyramine
Meteraminol
Metformin
Methaqualone
Metharbitone
Methimazole
Methixene
Methocarbamol
Methotrimeprazine
Methoxamine
Methoxypromazine maleate
Methsuximide
Methyclothiazide
Methyl nicotinate
Methyl phenidate
Methylamphetamine hydro-
  chloride
Methyldopa
Methyldopate
Methylephedrine
Methylergometrine maleate
Methylpentynol
Methylphenobarbitone
Methyltestosterone

Table 16.9 (*continued*)

Methyprylone
Methysergide maleate
Metronidazole
Metyrapone
Morphine

Naepaine
Nalorphine hydrobromide
Naphazoline
Nealbarbitone
Neostigmine bromide
Nicotinamide
Nicotine
Nicotinic acid
Nicotinyl alcohol
Nikethamide
Nitrazepam
Nitrofurantoin
Noradrenaline acid tartrate
Norbormide
Norbutrine
Norharman
Normetanephrine hydro-
   chloride
Nortriptyline hydrochloride
Noscapine

Orphenadrine citrate
Oxazepam
Oxybuprocaine hydrochloride
Oxycodone
Oxymetazoline hydrochloride
Oxyphenbutazone
Oxyphenonium bromide

Pamaquin
Papaverine
Paracetamol
Paraldehyde
Paraquat
Pargyline
Pecazine hydrochloride
Pemoline
Pempidine tartrate
Pentobarbitone
Pericyazine
Pethidine hydrochloride
Phenacemide
Phenacetin
Phenadoxone hydrochloride
Phenazone

Phencyclidine hydrochloride
Phenelzine
Pheneturide
Phenformin hydrochloride
Phenglutarimide hydrochloride
Phenindamine
Phenindione
Pheniramine maleate
Phenmetrazine hydrochloride
Phenobarbitone
Phenothiazine
Phenoxybenzamine
Phenoxypropazine maleate
Phensuximide
Phenylbutazone
Phenylephrine hydrochloride
Phenylpropanolamine hydro-
   chloride
Phenyltoloxamine citrate
Phenytoin
Pholcodine
Pholedrine sulfate
Physostigmine
Picrotoxin
Pilocarpine
Pipamazine
Pipazethate
Pipenzolate bromide
Piperazine citrate
Piperocaine hydrochloride
Pipradrol hydrochloride
Prednisolone
Prednisone
Prilocaine hydrochloride
Primaquine phosphate
Procainamide hydrochloride
Procaine hydrochloride
Procyclidine hydrochloride
Progesterone
Proguanil hydrochloride
Promazine hydrochloride
Promethazine hydrochloride
Propantheline bromide
Prothionamide
Prothipendyl hydrochloride
Protriptyline hydrochloride
Proxyphylline
Pseudoephedrine hydrochloride
Pyridostigmine bromide
Pyridoxine hydrochloride
Pyrimethamine

Pyrithidium bromide
Pyrrobutamine

Quinalbarbitone sodium
Quinidine
Quinine
Quinuronium sulfate

Reserpine
Riboflavin

Saccharin
Salicylic acid
Serotonin
Sparteine sulfate
Stilbestrol
Strophanthin-K
Strychnine
Styramate
Sulfacetamide sodium
Sulfadiazine
Sulfaguanidine
Sulfamerazine
Sulfamethoxypyridazine
Sulfanilamide
Sulfaphenazole
Sulfasomidine
Suxamethonium chloride

Testosterone
Tetracycline hydrochloride
Tetrahydrozoline hydrochloride
Thebaine hydrochloride
Theobromine
Theophylline
Thiabendazole
Thialbarbitone sodium
Thiethylperazine maleate
Thiopentone sodium
Thiopropazate hydrochloride
Thioridazine hydrochloride
Tolazoline
Tolbutamide
Tolpropamine hydrochloride
Tranylcypromine sulfate
Triamterene
Tridihexethyl chloride
Trifluoperazine hydrochloride
Trimeprazine tartrate
Trimetaphan Camsylate
Trimipramine maleate

Table 16.9 (*continued*)

Tripelennamine hydrochloride	Urethane	Warfarin sodium
Triprolidine hydrochloride		
Troxidone		Xylometazoline hydrochloride
Tubocurarine chloride		
Tybamate	Veratrine	
Tymazoline	Vinbarbitone	Yohimbine hydrochloride

Heroin

Mills (1963) has described a method for the preparation of samples to obtain positive identification of heroin hydrochloride by infrared spectroscopy in material containing as little as 6 to 10 μg. A five-fold scale expansion is required.

Barbiturates and Ureides

The infrared spectra of barbiturates and of barbiturate complexes have been examined by several groups of workers (Umberger and Adams, 1952; Price *et al.*, 1954; Levi and Hubley, 1956; Chatten and Levi, 1957; and Mesley, 1970). The barbiturates are derivatives of barbituric acid (malonylurea) and have substitutions

CIV

in R_1 and R_2 of position 5 (CIV). There can also be substitutions in position 1. In thiobarbiturates the oxygen atom in position 2 is replaced by sulfur. Some barbiturates are produced as sodium salts.

The free barbiturates do not generally absorb infrared radiation in the region above 3300 cm^{-1} (exceptions are barbituric acid and sodium phenobarbital). The ureides, $R-CO-NH-CO-NH_2$, on the other hand, all have three strong N—H stretching bands between 3450 and 3250 cm^{-1}. All the barbiturates have N—H stretching bands near 3200 and 3100 cm^{-1}. In the region 3000 to 2800 cm^{-1} a group of up to four bands of medium to very strong intensity occur, caused by alkyl C—H stretching bonds of the substituents in positions 1 and 5.

Up to three strong carbonyl stretching bands occur between 1765 and 1670 cm^{-1} in the spectra of the barbiturates. The C=O vibrations at positions 4 and 6 perpendicular to the molecular axis of symmetry lead to a band at $\sim$1760 to 1740 cm^{-1}. The next band, $\sim$1720 to 1700 cm^{-1}, is also produced by C=O vibrations at positions 4 and 6, but is affected by the C=O vibration at position 2. The band at $\sim$1700 to 1670 cm^{-1} is produced by the C=O vibration at position 2.

Only two bands occur for the sodium salts of barbiturates between 1700 and 1650 cm^{-1}. A strong broad band also occurs in the region 1600 to 1500 cm^{-1}, while the free bases have almost no absorption in this region. The sodium thiobarbiturates display only the lowest of the three carbonyl bands, which shows up in the 1700 to 1680 cm^{-1} region. These salts also exhibit the strong broad band, which is displayed between 1650 and 1600 cm^{-1}.

Barbiturates generally have several strong bands in the region 1460 to 1250 cm^{-1} due to C—H deformation and C—N stretching vibrations. Barbiturates can be distinguished from thiobarbiturates by the lack of a strong broad band in the former in the region 1500 to 1480 cm^{-1}.

Mesley (1970) has studied the spectra of some 5,5-disubstituted barbituric acids and their salts, including many polymorphic modifications. For most of the disubstituted barbituric acids in the solid phase the most stable structure is that in which both NH groups form intermolecular hydrogen bonds with the 4- and 6-carbonyl groups. Examples of compounds having this type of structure are allobarbitone, amylobarbitone I and II, barbitone III, *cyclo*barbitone, and pentobarbitone I, II, and III. These compounds all have spectra showing two bands of approximately equal intensity at about 3200 and 3080 cm^{-1}. The major exceptions to this type of structure are barbitone and phenobarbitone, in each of which a number of other structures have comparable or greater stability, including some with stronger hydrogen bonds than usual. In these cases the band near 3200 cm^{-1} is significantly stronger than that at 3100 cm^{-1}, the latter being shifted from its usual position of 3080 cm^{-1}. Another feature is that the highest frequency carbonyl band, at 1745 cm^{-1} (due to the 2-carbonyl group) in the compounds mentioned earlier, is missing in phenobarbitone and barbitone. Heptabarbitone, hexabarbitone, and the thiobarbituric acids show strong absorption at 3230 cm^{-1} or above, which is indicative of weaker hydrogen bonds. Mesley (1970) has noted that the groups of compounds separated here according to their spectral properties are also distinguished from one another by the durations of action on the central nervous system, barbitone and phenobarbitone being long acting, whereas heptabarbitone, hexabarbitone and the thiobarbituric acids are all short acting, the other compounds being mainly of intermediate duration. Price *et al.* (1954) felt that fat solubility is the most important factor influencing duration of action, and this in turn will depend on the degree of association of the acid and the ease with which hydrogen bonds can be broken. The pharmacological activities of these compounds are thus correlated to some degree with their structures in the solid state, as deduced from infrared spectra.

Kyogoku *et al.* (1968) have demonstrated that adenine derivatives form strong specific hydrogen bonds with barbiturates in chloroform solution, e.g., phenobarbital, secobarbital, mephobarbital, and several others (see Chap. 5). This highly specific interaction is more extensive than the hydrogen bonding between thymine (or uracil) derivatives and adenine derivatives. These investigators believed that the highly specific interaction between adenine derivatives and barbiturates may help to explain the various biological effects of barbiturates, although the relevance of this observation to the physiological activities of the barbiturates is not yet established. Kyogoku *et al.* pointed out that NMR and Raman spectroscopic studies also clearly indicated hydrogen bonding of barbiturates to adenine derivatives in aqueous solutions.

The biologically active barbiturates have two important characteristics. One is the hydrophobic side chains found on carbon-5 of the barbiturate ring and the other is the ring system containing the CO—NH groupings involved in hydrogen bonding. Changing the hydrophobic side chains on C-5 changes the speed and effectiveness of barbiturate action. It is generally believed that the specificity and necessity for the lipophilic side chains arises because the barbiturates are active in nonpolar regions, perhaps in membranes. Thus, studies in chloroform solution may be related in some way to their physiological activity (Kyogoku et $al.$, 1968).

The ureides, R—CO—NH—CO—NH$_2$, can be divided into three groups, based on the type of acyl group present: (a) aryl or alkyl, (b) alkyl or alkenyl, and (c) halogenated. Three strongly absorbing N—H stretching bands occur in all ureides, in the range 3450 to 3250 cm^{-1}, which characterize them as different from the barbiturates. Three bands of medium intensity generally occur between 3000 and 2800 cm^{-1} and they are caused by a variety of alkyl C—H stretching vibrations. Two strong bands caused by C=O stretching vibrations are displayed in the area 1720 to 1660 cm^{-1}. These bands occur in positions different from those in the barbiturate free bases and the intensities also differ, but they are similar to the ones present in the sodium salt spectra. All the ureides possess a broad medium band or a doublet in the 1400 cm^{-1} region. This band may be produced by the amide group.

Amphetamines

Heagy (1970) has developed a method by which three distinctive infrared spectra can be produced for d-, dl-, and l-amphetamine as the d-mandelate salts. The greatest differences in the spectra are found in the 800–600 cm^{-1} region. The method has been used successfully on samples of illicit amphetamine tablets.

Marijuana

Various types of $Cannabis$ resin exhibit marked and characteristic differences in their infrared spectra in the region 1300–700 cm^{-1} (Grlic, 1965). The observed characteristics have been explained as being caused by differences in the content of cannabidiolic acid, cannabidiol, tetrahydrocannabinols, and cannabinol in samples that have been examined. The absorption spectra and particularly the value $(T_{890 cm^{-1}}) - (T_{1130 cm^{-1}})$ offered possibilities for examining the progress of the phytochemical interconversion of cannabinols (the "ripening" process) as well as estimating the type and chemical composition of the resin. Two bands at 1130 and 1160 cm^{-1} are characteristic for ripe and overripe samples.

Caffeine and Theobromine

The infrared spectra of caffeine and its salts indicate that when protonation occurs it is the N-9 position to which the proton attaches (Cook and Regnier, 1967a). From the spectra of theobromine salts Cook and Regnier (1967b) have concluded that the salts are probably arranged in hydrogen-bonded centrosymmetric pairs (CV) involving N1 H$\cdots$O=C6 interactions, as for example, in the hydrogen-bonded structures of 1-methyluracil (CVI, R = H) and 1-methylthymine (CVI, R = CH₃), which contain the same pyrimidine-2,4-dione structure as theobromine. A^- represents the anion of the salt in (CV).

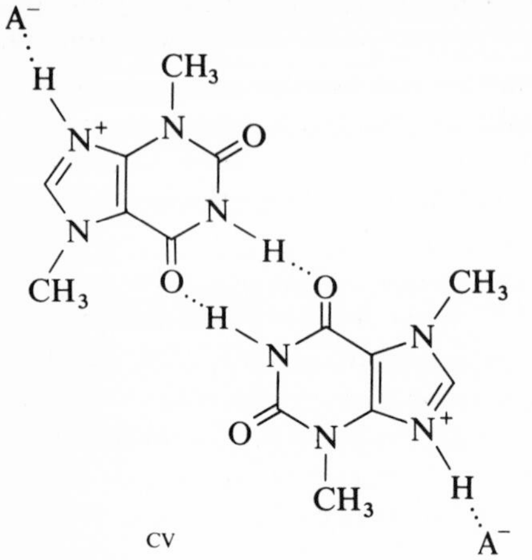

CV 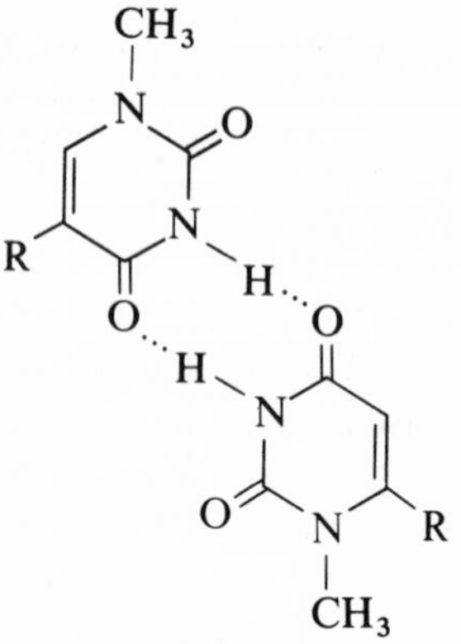   CVI

Choleretics

Chihara *et al.* (1961) have reported that infrared spectroscopy is effective for the study of the pharmacological action of choleretics. They administered ten kinds of choleretics to laparotomized dogs, collected the liver bile every 30 min for 5 to 8 hr, and besides measuring volume of flow, amount of solid components, and quantity of bilirubin, followed the changes in bile composition by examination of infrared spectra. After administration of sodium ursodeoxycholate, 1-phenyl-1-propanol, or 1-(5-norbornen-2-yl) ethyl hydrogensuccinate, almost no changes in the above characteristics were observed. In the case of the ursodeoxycholate, transient intensification of the lipid absorption (1738 cm⁻¹) occurred. Sodium dehydrocholate caused increased secretion of bile markedly 30 to 60 min after its injection but the amount of solid components increased only slightly. The infrared absorption of sodium taurocholate (982, 951, and 917 cm⁻¹) at this stage was not proportional to the quantity of solid components and actually its absorption became extremely weak. A substance with absorption at ~ 1700 cm⁻¹ was secreted in large amounts instead.

The compounds α-4-dimethylbenzyl camphorate diethanolamine, magnesium bis (4-methoxy-γ-oxo-1-naphthalenebutyrate), and 5-(*p*-methoxyphenyl)-1,2-dithiol-3(3H)-thione, which were supposed to be cholaneretics, did not increase the secretion of natural bile. The infrared spectra indicated that the constituion of the bile components was different after administration of each of these choleretics. The common feature of each was the near disappearance of bands of sodium taurocholate (mentioned earlier). No change in the absorption of lecithin (1738 cm⁻¹) was caused by α-4-dimethylbenzyl camphorate diethanolamine, but this band disappeared in the bile after administration of the magnesium and thione compounds above. Some bands not observed in natural bile appeared instead, notably near 1420 and 800 cm⁻¹, after the administration of the diethanolamine compound; others appeared at 1455, 877, and 838 cm^{-1} after administration of the magnesium compound, and at 836 and 749 cm^{-1} after administration of the thione compound.

Marked changes in the infrared spectrum of bile were found after the administration of the following compounds: 2-(1-hydroxycyclohexyl) butyric acid, 1,4-O-dicaffeylquinic acid, or N-(p-hydroxyphenyl) salicylamide. All these compounds caused new bands to occur. The first compound caused a new band at 830 cm^{-1}, the second compound caused new ones at 1396, 1261, and 1170 cm^{-1}, and the third produced new bands at 1512, 1175, 832, and 754 cm^{-1}. The absorption of bile acid weakened in intensity, but did not disappear, as was the case with some of the choleretics mentioned earlier. After the administration of the quinic acid compound, the absorption of lecithin characteristically disappeared.

Chihara et al. (1961) have found the changes in the infrared spectra of bile after the administration of each of the choleretics to be characteristic of these compounds and were able to confirm their reproducibility. The method was an effective means for assaying the potency of choleretics.

Cyclamate, Dulcin, and Saccharin

Oi and Inaba (1967a) have determined sodium saccharin, dulcin, and sodium cyclamate in mixtures of these compounds. In methanol solution the bands at 1149 and 1512 cm^{-1} are used for the determination of saccharin and dulcin, respectively, and in aqueous solution the band at 1194 cm^{-1} is used for the determination of cyclamate. These methods have been found to be quite useful for routine analyses.

Isopropamide

Oi and Miyazaki (1967) have described an infrared compensation method for the rapid determination of small amounts of isopropamide in pharmaceutical preparations containing aminopyrine, phenacetin, caffeine, and isopropamide. The key band used for isopropamide was 702 cm^{-1} in an acetone solvent.

Thyroid Drug Preparations

Gas–liquid chromatographic spectra have been prepared by Hansen (1968) for thyroxine, 3,5,3'-triiodothyronine, 3,5-diiodothyronine, and 3,5-diiodotyrosine in order to develop a method for quantitation of the physiologically active components of thyroid drug preparations. The tris(trimethylsilyl) derivatives of thyronine, tyrosine, and their halogen analogs are formed rapidly and are stable. The infrared spectra of the free amino acids and tris(trimethylsilyl) derivatives have also been given. The silylation of the phenolic groups was demonstrated by the disappearance of the strong phenolic absorbance at $\sim$3250 cm^{-1} and the appearance of the Si—O—Ar stretching vibration band at 930 cm^{-1}. The silylation of the nitrogen was shown by the disappearance of the weak overtone band at 2100 cm^{-1}, which is distinctive for amino acids and which is absent if the nitrogen atom is substituted (Silverstein and Bassler, 1967).

Analysis of Dicumarol in Blood Serum

Stubbs *et al.* (1962) have devised a simple infrared method for the analysis of Dicumarol, 3,3'-methylene bis (4-hydroxycoumarin), in blood. The method is sensitive to concentrations as low as 0.1 mg% (1 ppm). The Dicumarol in acidified blood serum is extracted with carbon tetrachloride, and the solvent layer, containing the Dicumarol, is scanned in a 10-mm infrared microcell. The absorption at 1667 cm^{-1} is measured and the concentration of the drug calculated. Compounds known to interfere with the determination of the drug in blood by other methods (e.g., vitamin K, salicylic acid, aspirin, warfarin, and others) do not interfere with the infrared method.

Addendum

Sutherland *et al.* (1971) have recently reviewed the literature of analytical methods for pharmaceuticals and related drugs. These authors have given numerous references to fractionation and isolation methods, and have presented many references to infrared analyses. Infrared spectrophotometric methods were listed for the following substances (in their order of appearance): ephedrine, alkaloids (general), noscapine, pilocarpine, ajmaline, brucine, lysergic acid diethylamide, strychnine, codeine, diamorphine, dihydrocodeine, dihydrocodeinone, morphine, narcotine (noscapine), papaverine, thebaine, cinchonine, quinine ethyl carbonate, atropine, chloramphenicol, ristomycins, thiamphenicol, inorganic halogenated compounds, water, amides, procaine and analogs, sympathomimetic amines, amino acids, barbiturates, benzodiazepines, carbamates, nucleosides (internal reflection, i.r.), quinolines, alcohols (i.r.), carboxylic aliphatic acids, aromatic esters, salicylic acid and derivatives, digitalis, polyhydric alcohols, aldosterone (i.r.), dexamethasone, pregnenolone (i.r.), testosterone, steroid phosphates, steroids (general), sulfonyl ureas, thiamine intermediates, pantothenic acid, antiseptics (i.r.), dyes (i.r.), general techniques (including i.r.), aspirin (i.r.), *Cannabis*, and salicylic acid (i.r.).

REFERENCES

American Petroleum Institute, Research Project 44 and the Manufacturing Chemists Association Research Project, located at Department of Chemistry of the Agricultural and Mechanical College of Texas, College Station, Texas.

Barnard, D., Fabian, J. M., and Koch, H. P. *J. Chem. Soc.* **1949**, 2442.

Bell, P. H. and Roblin, R. O., Jr. *J. Am. Chem. Soc.* **64**, 2905 (1942).

Carol, J. *Ann. N.Y. Acad. Sci.* **69**, Art. 1, 190 (1957).

Chapman, D. I. and Moss, M. S. in *Isolation and Identification of Drugs* (E. G. C. Clarke, ed.) The Pharmaceutical Press, London, 1969, p. 103.

Chatten, L. G. and Levi, L. *Appl. Spectrosc.* **11**, 177 (1957).

Chihara, G., Matsuo, K., Arimoto, K., Sugano, S., Shimizu, K., and Mashimo, K. *Chem. Pharm. Bull. (Tokyo)* **9**, 939 (1961).

Clark, C. and Chianta, M. *Ann. N. Y. Acad. Sci.* **69**, Art. 1, 205 (1957).

Clarke, E. G. C. (ed.) *Isolation and Identification of Drugs*, The Pharmaceutical Press, London, 1969.

Cook, D. and Regnier, Z. R. *Can. J. Chem.* **45**, 2895 (1967*a*).

Cook, D. and Regnier, Z. R. *Can. J. Chem.* **45**, 2899 (1967*b*).

Eglinton, G. in *Physical Methods in Organic Chemistry* (J. C. P. Schwarz, ed.) Holden–Day, Inc., San Francisco, 1964, p. 114.

Evans, R. M. *The Chemistry of the Antibiotics Used in Medicine*, Pergamon Press, London, 1965.

Fazzari, F. R., Sharkey, M. F., Yaciw, C. A., and Brannon, W. L. *J. Ass. Offic. Anal. Chem.* **51**, 1154 (1968).

Gore, R. C. and Petersen, E. M. *Ann. N. Y. Acad. Sci.* **51**, Art. 5, 924 (1949).

Grlic, L. *Planta Med.* **13**, 291 (1965).

Hammett, L. P. *J. Am. Chem. Soc.* **59**, 96 (1937).

Hansen, L. B. *Anal. Chem.* **40**, 1587 (1968).

Hayden, A. L., Sammul, O. R., Selzer, G. B., and Carol, J. *J. Ass. Offic. Agr. Chem.* **45**, 797 (1962).

Hayden, A. L., Brannon, W. L., and Yaciw, C. A. *J. Ass. Offic. Anal. Chem.* **49**, 1109 (1966).

Heagy, J. A. *Anal. Chem.* **42**, 1459 (1970).

Hendrickson, J. B. *The Molecules of Nature*, W. A. Benjamin, Inc., New York, 1965.

Holubek, J. and Strouf, O. *Spectral Data and Physical Constants of Alkaloids, Issues 1 and 2* (in 2 volumes) Heyden and Son, Ltd., London, 1965. *Issues 3 and 4* have recently been published, the latter in 1969.

Hubley, C. E. and Levi, L. *Bull. Narcotics, U. N. Dept. Social Affairs* **7**, No. 1, 20 (1955).

Kasturi, T. R. in *Chemical Applications of Infrared Spectroscopy* (by C. N. R. Rao) Academic Press, New York, 1963, p. 451.

Krueger, P. J. *Z. Naturforsch.* **17a**, 692 (1962).

Krueger, P. J. and Thompson, H. W. *Proc. Roy. Soc.* **A243**, 143 (1957).

Krueger, P. J. and Thompson, H. W. *Proc. Roy. Soc.* **A250**, 22 (1959).

Kyogoku, Y., Lord, R. C., and Rich, A. *Nature* **218**, 69 (1968).

Levi, L. and Hubley, C. E. *Anal. Chem.* **28**, 1591 (1956).

Levi, L., Hubley, C. E., and Hinge, R. A. *Bull. Narcotics, U. N. Dept. Social Affairs* **7**, No. 1, 42 (1955).

Linnett, J. W. *Trans. Faraday Soc.* **41**, 223 (1945).

Marion, L., Ramsay, D. A., and Jones, R. N. *J. Am. Chem. Soc.* **73**, 305 (1951).

Mason, S. F. *J. Chem. Soc.* **1958**, 3619.

Mesley, R. J. *Spectrochim. Acta* **26A**, 1427 (1970).

Mills, A. L. *Anal. Chem.* **35**, 416 (1963).

Neuss, N. *Physical Data of Indole and Dihydroindole Alkaloids, 3rd Ed.*, Eli Lilly and Co., Indianapolis, 1959.

Oi, N. and Inaba, E. *Yakugaku Zasshi* **87**, 640 (1967a) (Japanese).

Oi, N. and Inaba, E. *Bunseki Kagaku* **16**, (1967b) (Japanese).

Oi, N. and Miyazaki, K. *Yakugaku Zasshi* **87**, 739 (1967) (Japanese).

Papendick, V. E., Sutherland, J. W., and Williamson, D. E. *Anal. Chem.* **41**, 190R (1969).

Pleat, G. B., Harley, J. H., and Wiberley, S. E. *J. Am. Pharm. Assoc., Sci. Ed.* **40**, 107 (1951).

Price, W. C., Bradley, J. E. S., Fraser, R. D. B., and Quilliam, J. P. *J. Pharm. Pharmacol.* **6**, 522 (1954).

Sadtler Research Laboratories, Inc., 3316 Spring Garden Street, Philadelphia, Pa. (1969).

Sammul, O. R., Brannon, W. L., and Hayden, A. L. *J. Ass. Offic. Agr. Chem.* **47**, 918 (1964).

Seydel, J. K. *Mol. Pharmacol.* **2**, 259 (1966).

Seydel, J. K., Krüger-Thiemer, E., and Wempe, E. *Z. Naturforschung*, **15b**, 628 (1960).

Seydel, J. K., Krüger-Thiemer, E., and Wempe, E. *Tuberk.—Forschungsinst. Borstel Jahresber.* **5**, 651 (1961).

Silverstein, R. M. and Bassler, G. C. *Spectrometric Identification of Organic Compounds, 2nd Ed.*, Wiley, New York, 1967, pp. 97, 102.

Stubbs, B. T., Stewart, R. D., and Boettner, E. A. *J. Lab. Clin. Med.* **59**, 667 (1962).

Sutherland, J. W., Williamson, D. E., and Theivagt, J. G. *Anal. Chem., Ann. Rev.* **43** (5), 206R (1971).

Umberger, C. J. and Adams, G. *Anal. Chem.* **24**, 1309 (1952).

Umezawa, H. (ed.) *Index of Antibiotics from Actinomycetes*, University Park Press, Baltimore, 1967.

Unterman, H. W. *Antibiotiki* **10**, 867 (1965) (Russian).

Wahl, M. "Spectroscopic Investigations of Several Sulfonamides," Thesis, The University of Munich, 1965.

Warren, R. J., Eisdorfer, I. B., Thompson, W. E., and Zarembo, J. E. *Microchem. J.* **12**, 555 (1967).

Wayland, L. and Weiss, P. J. *J. Ass. Offic. Agr. Chem.* **48**, 965 (1965).

Zugaza, A. and Hidalgo, A. *Rev. Real Acad. Cienc. Exact., Fis. Nat. Madrid* **59**, 221 (1965) (Spanish).

Zugaza, A. and Hidalgo, A. *Memorias Real Acad. Ciencias Exactas, Fisicas y Naturales de Madrid, Ser. Fis.-Quim.* **6**, No. 1, 1966.

Chapter 17

MICROBIOLOGY

Bacterial species can be differentiated by spectral analysis of organic solvent extracts, and by typing spectra of crude capsular polysaccharides. Infrared spectra can be used as a control over processes of inoculation, growth, harvesting, and in the production of fractions to be tested for biologic activity. However, it must be pointed out that rigid standards must be maintained with respect to the use of the spectrophotometer, the incubation time and temperature, culture media, harvest procedures, and the preparation of a dried film on an optical window.

Parent strains of certain bacteria, e.g., antibiotic-resistant or antitubercular-drug-resistant strains, can be differentiated from nonresistant substrains. The spores of phytopathogenic fungi can be differentiated by means of their spectra. Viruses can also be differentiated.

Isolated bacterial capsular and cell-wall polysaccharides, peptides, and other compounds can be identified with the help of infrared spectra. Cell-wall materials of fungi may be extracted and analyzed for taxonomic purposes. Certainly, one of the most obvious uses of infrared spectroscopy is to characterize isolated metabolic products and to establish metabolic pathways.

Identification of Organisms

The application of infrared spectrophotometry to the study of bacteria and fungi has been investigated by many workers. Stevenson and Bolduan (1952) proposed the use of this method for the identification of bacteria. They found that infrared spectra do not always group species as in the Bergey classification. For example, *Pseudomonas* (Family II) organisms and *Escherichia* (Family X) had similar spectra, whereas *Micrococcus rosaceous* and *M. pyogenes* var. *aureus* (Genus I, Family V) had very different spectra.

Thomas and Greenstreet (1954) used infrared spectrophotometry for the differentiation of bacterial species. Levine *et al.* (1953*a*) have applied it to distinguish pneumococcal polysaccharides and to follow glycogen production in *Aerobacter aerogenes* (Levine *et al.*, 1953*b*). Levine *et al.* (1953*a*) also presented over 60 infrared spectra of pneumococcal polysaccharides and divided them into four groups depending on whether the spectra exhibited amide bands, carboxylate bands, both, or neither.

In these studies they found that O-acetyl and N-acetyl groups could be clearly distinguished.

Randall *et al.* (1951) recorded reproducible spectra of organic solvent extracts of *Mycobacterium tuberculosis*. They established that with sufficient care in preparation, the different fractions (chloroform followed by acetone and alcohol precipitation) can be reproduced to the extent that they give constant spectra. The spectrum of each fraction was characteristic of that fraction. These workers felt that infrared spectra might be used as a control over the processes of inoculation, growth, harvesting, and fractionation in the production of fractions to be tested for biologic activity. In a series of papers Randall and Smith (1953) related certain biological properties of *Mycobacterium tuberculosis* H37Rv strains to various wavelengths of the spectrum. Levine *et al.* (1953c) used infrared spectroscopy to study enteric bacteria and to type *Klebsiella* (Levine *et al.*, 1955) from the spectra of crude capsular polysaccharides. Kull and Grimm (1954) in a preliminary report indicated that infrared spectroscopy might be useful in the study of bacterial resistance. Kull and Grimm (1956a) felt that although the infrared spectra of purified extracts or of relatively simple components of closely related species might be used as a criterion for separation or distinguishing studies, the application of infrared techniques as a qualitative differentiating tool was limited if crude extracts, or whole or intact cells were used. The same workers [Kull and Grimm (1956b)] compared the infrared spectra of several parent bacterial strains with the spectra of substrains resistant to various antibiotics and antituberculous compounds. The spectrum of a bacitracin-resistant strain of *B. megaterium*, for example, differed (Fig. 17.1) at 1735, 1230, and 835 cm^{-1} from that of the spectrum of the parent strain sensitive to 0.2 unit/ml bacitracin. The increased intensity of the 1230 cm^{-1} band and the new bands at 1735 and 835 cm^{-1} were not only noted in strains resistant to 225 units/ml but also in mutants resistant to 37.5, 50, 87.5, and 150 units/ml of the antibiotic. Another example was found in the spectrum of a substrain of *B. megaterium* resistant to the soluble antituberculous thiourea, Su 2079. This spectrum differed from that of the sensitive parent strain at 1735 and 1176 cm^{-1}, and the intensity of the 1310 cm^{-1} band was increased (see Fig. 17.1).

Comparison of the two mutant strains of *B. megaterium*, resistant to bacitracin and Su 2079, respectively, revealed that the 1735 cm^{-1} band was present in both cases, while the 1176 cm^{-1} band seen in the strain resistant to the thiourea compound was absent in the curve produced by the cells resistant to bacitracin. The strain resistant to the Su 2079 had a spectrum without a band at 835 cm^{-1}. The 835 cm^{-1} band was distinct in the spectrum of the bacitracin-resistant cells. Thus, infrared spectroscopy differentiated between sensitive and resistant strains.

Figure 17.2 shows spectra of *M. tuberculosis BCG*, sensitive and resistant to isoniazid, as well as spectra of *BCG* cells sensitive and resistant to the antitubercular diphenylthiourea derivative, namely Su 1906. The shoulder at 1735 cm^{-1} and the bands at 1409 and 893 cm^{-1} became exaggerated with the development of resistance.

The results of the study by Kull and Grimm (1956a) with different strains of bacteria showed that the application of infrared analysis to whole bacterial cells for taxonomic purposes was possible only under certain well-defined and specific conditions.

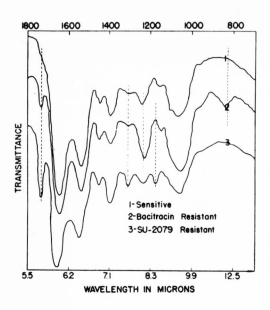

Fig. 17.1. Infrared spectra of *Bacillus mega-terium* sensitive and resistant to bacitracin and to Su 2079. (Kull and Grimm, 1956*b*.)

Smith *et al.* (1960) made comparisons of the lipids of 72 strains of mycobacteria including virulent and avirulent strains of *Mycobacterium tuberculosis*, and *M. bovis*. They also studied *M. avium*, *M. phlei*, *M. smegmatis*, and representatives of the atypical acid-fast group. The cultures were grown on synthetic medium and were extracted with ether–ethanol. The lipids were separated by chromatography on magnesium silicate-Celite and were identified by means of their infrared spectra.

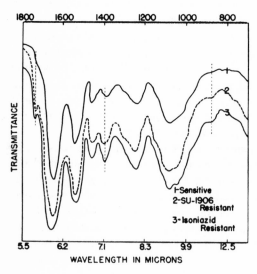

Fig. 17.2. Infrared spectra of *BCG* which are sensitive and resistant to isoniazid and Su 1906. (Kull and Grimm, 1956*b*.)

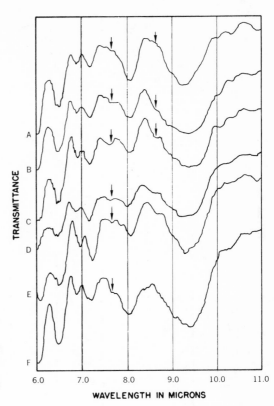

Fig. 17.3. Spectral differences between 6.0–10.0 μ due to the following growth conditions: (A), (B), and (C), P. fluorescens after 24, 72, and 120 hr incubation on TGE; (D), (E), and (F), B. subtilis after 24 hr incubation on nutrient agar, TGE, and TGE followed by a saline solution wash. (Reynolds et al, 1967.)

This study permitted the recognition of five lipid substances which are characteristic for the type of organism from which they were isolated. Two of these substances were glycolipids and two were glycolipid peptides.

Infrared spectrophotometry has been used for accelerated determination of sanitary-indicator microbes (Bugrova, 1965). A difference in the absorption bands in the range 2000 to 833 cm^{-1} was established for three different families of bacteria: Enterobacteriaceae (Escherichia coli), Enterococci, and Staphylococci. Studies of the spectra of both entire microbe cells and of acetone extracts were conducted. The method must be used in conjunction with certain classical methods of bacteriological research to confirm the identity of the microorganisms.

Reynolds et al. (1967) have investigated the use of infrared spectroscopy as a rapid means of identifying microorganisms which are viable in hydrocarbon fuels. As stated previously by others, these workers noted that rigid standards must be maintained with respect to use of the instrument, incubation time and temperature, culture media, harvest procedures, and dry film preparation on the optical material. In Fig. 17.3 are shown spectral differences due to growth conditions. The upper tracings show Pseudomonas fluorescens after incubation periods of 24, 72, and 120 hr on tryptone–glucose extract agar (TGE). The lower curves show B. subtilis after

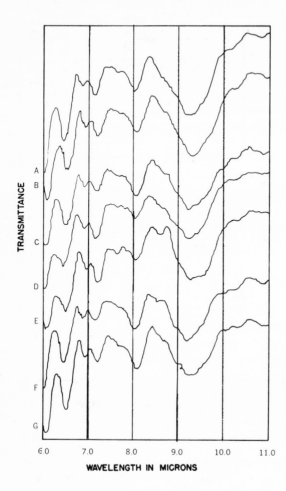

Fig. 17.4. Spectrogram of characteristic spectra of selected bacteria: (A) A. aerogenes (strain 1B21); (B) A. aerogenes (strain B-7); (C) E. coli; (D) B. subtilis; (E) B. cereus; (F) P. aeruginosa; (G) P. fluorescens. (Reynolds et al., 1967.)

24-hr incubation at 28°–30°C on nutrient agar, TGE, and TGE medium followed by a saline wash.

Figure 17.4 shows spectra of several kinds of bacteria grown under rigid conditions. These spectra possess the same major bands but these bands vary in relative intensity, a property which provides "the most efficient basis for differentiation of whole bacterial cells thus far determined by infrared spectroscopy" (Reynolds et al., 1967). Figure 17.5 shows the differences in the relative intensities of bands in spectra of B. subtilis and P. aeruginosa. The ratios of the intensities at 1460 cm^{-1} (6.85 μ), 1399 cm^{-1} (7.15 μ), 1242 cm^{-1} (8.05 μ), and 1087 cm^{-1} (9.2 μ) readily distinguish these two organisms. Figure 17.6 shows differences in the spectra of A. aerogenes, P. fluorescens, and E. coli, all gram-negative rods. A. aerogenes is identified by the fact that the 1242 cm^{-1} (8.05 μ) band is less than the 1399 cm^{-1} (7.15 μ) one; and the difference between the 1242 and the 1087 cm^{-1} bands is much greater for this organism than for the others. P. fluorescens and E. coli differ mainly (1) in the presence of an 1149 cm^{-1} (8.7 μ) shoulder in the spectrum of E. coli; (2) greater relative intensity

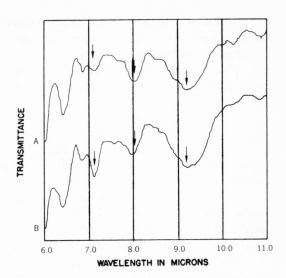

Fig. 17.5. Absorption band differences of
(A) P. aeruginosa and (B) B. subtilis.
(Reynolds et al., 1967.)

of the 1242 and 1087 cm^{-1} bands for *P. fluorescens*; and (3) more pointed general shape of the nucleic acid band at 1087 cm^{-1} for *E. coli*.

Parnas *et al.* (1966) have confirmed the value of infrared spectrophotometry for taxonomic determinations of bacteria. They compared biotypes, serotypes, chemotypes, and infrared-spectrophotometric types of *Brucellae*, *Klebsiellae*, *Listeriae*, *Erysipelothrix*, and *Leptospirae*. They found only one infrared type in various serotypes of *Leptospirae*, *Brucellae*, *Listeriae*, and *Erysipelothrix*. They observed only one serotype and one infrared type in *Klebsiella rhinoscleromatis*, and one chemotype characteristic for every genus in *Brucellae*, *Listeriae*, *Erysipelothrix*, and *Klebsiellae*. They

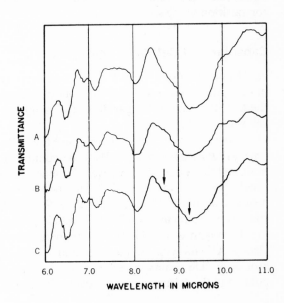

Fig. 17.6. Spectral differences of three
gram-negative bacteria: (A) A. aerogenes,
(B) P. fluorescens, and (C) E. coli. (Reynolds
et al., 1967.)

found infrared spectra to be generically common and to characterize the following genera: *Leptospira, Brucella, Listeria, Erysipelothrix, Klebsiella rhinoscleromatis*, and *Aerobacter aerogenes.*

The infrared spectra of forty strains of sulfate-reducing bacteria have been recorded in the range 1800 to 800 cm^{-1} (Booth *et al.*, 1966). *Desulfovibrio gigas* and *Desulfotomaculum nigrificans* were identified positively. A broad differentiation was possible between salt-water and fresh-water forms of *Desulfovibrio desulfuricans.*

Scopes (1962) has found that the infrared spectra of twenty-two strains of the genus *Acetobacter* were significantly different from those of nine strains of the genus *Acetomonas*, with one exception. Within each genus, however, the spectra formed smooth graduated series which displayed no sudden differences that could be correlated with species boundaries, and exhibited no features that were traceable to the presence or absence of characteristic biochemical properties. These findings supported the division of the acetic acid bacteria into two genera—*Acetobacter* and *Acetomonas*—and made the existence of distinct species within the genera unlikely.

Preobrazhenskaya *et al.* (1966) have studied the infrared spectra of spores of 34 species of actinomycetes. According to their spectral data they divided all the cultures studied into seven groups. Most of the cultures belonged to a group with absorption bands at 3344, 2924, 1650, 1550, 1414, 1250, and 1053 cm^{-1}. Actinomycetes of the *A. coerulescens* series differed from all the other cultures in the shape of the absorption curve, which was more intense at 1550, 1420, and 1250 cm^{-1} and had a characteristic maximum at 880 cm^{-1}.

Peruanskii and Topolovskii (1964) have studied the spores of some phytopathogenic fungi and gave infrared spectra of the conidia of *Phytophthora infestans*, *Erysiphe graminia*, and *Verticillium alboatrum*, of the teliospores of *Puccinia phragmitis*, and of the urediospores of *P. triticina*. The differences in the number of absorption bands and their intensities were related to the differences in the chemical composition of the spores.

Capsular and Cell-Wall Materials

Liu and Gotschlich (1963) have used infrared methods to identify D-galactosamine 6-phosphate, a major constituent of pneumococcal C-polysaccharide. This was the first time that this sugar derivative had been isolated and identified in bacterial cell walls.

Tyler and Heidelberger (1968) have found the specific capsular polysaccharide of type VII *Pneumococcus* (*S* VII) to be composed of D-galactose, D-glucose, L-rhamnose, and the *N*-acetyl derivatives of D-glucosamine and galactosamine (probably D-), roughly in the ratios 4:2:3:2:2. The spectrum of *S* VII allowed the following assignments to be made: 3333 cm^{-1}, OH; 2899 cm^{-1}, CH; and 1724, 1653, 1550, and 1389 cm^{-1} were characteristic of the acetamido group.

Heymann *et al.* (1963) have prepared the group-specific C-polysaccharide of Group A hemolytic streptococci and found (1 →3)-linked rhamnose units. Side groups or side chains were linked to the 2-position of rhamnose. Infrared spectra pointed to the existence of α-glycosidic linkages between the rhamnose units.

In a study (Heymann *et al.*, 1964) involving the structure of deacetylated C-polysaccharide of Group A hemolytic streptococci, infrared spectroscopy has been used to observe the disappearance of amide absorption as a result of treatment at 100°C with 0.49N barium hydroxide.

In a study of the type-specific substance of *H. influenzae*, Type *d* (Williamson and Zamenhof, 1963), infrared spectra have been given for its dissociated and undissociated forms. The substance *d* contains N-acetylglucosamine and N-acetylglucosamine uronic acid. Infrared spectroscopy on the capsular substance has identified the acetamido group and the P=O group in a study of the type-specific substances of *Hemophilus parasuis* (Williamson and Zamenhof, 1964). Williamson and Zamenhof (1964) have identified a nonreducing polymer of disaccharide units of α-galactosyl-α-N-acetylglucosaminide connected through phosphodiester linkages.

Infrared spectroscopy has been useful in identifying two disaccharides as products of the hydrolysis of the cell wall of *S. aureus* by an acetylmuramidase and an amidase from *Streptomyces albus G* (Ghuysen and Strominger, 1963). These disaccharides were β-1,6-N-acetylglucosaminyl-N-acetylmuramic acid and β-1,6-N-acetylglucosaminyl-N,4-O-diacetylmuramic acid.

Eremin *et al.* (1965) have precipitated carbohydrate material with ethanol after alkaline hydrolysis of cultures of Whitmore's bacillus (*Pseudomonas pseudomaller*), *Pasteurella pestis*, and *Vibrio comma*, and have subjected these polyoses to infrared spectroscopy. All spectra had strong absorption at 1660 and 1550 cm^{-1}; the former was related to double-bond vibrations and the latter was associated with stretching vibrations of C—N. The latter absorption was almost completely absent in the spectrum of a complex from *V. comma*. Absorption at 970 cm^{-1} (the C=C double bond in the *trans* position) and traces of absorption at 790 cm^{-1} characteristic of the $1 \rightarrow 3$ bond were always present. A polysaccharide from the cell wall of *P. pestis* had a wide band at 1170–1000 cm^{-1}; the low intensity bands at 1190 and 1160 cm^{-1} indicated the presence of P—O—Me and P—O—Et groups. The spectrum of a complex from Whitmore's bacillus differed from the others by the presence of a band at 1735 cm^{-1} due to esters of fatty acids.

Linker and Jones (1966) have isolated polysaccharides from slime-producing *Pseudomonas* organisms. The polysaccharides contained mannuronic and guluronic acid only, and resembled alginic acid, a polysaccharide in seaweed, as shown by composition, infrared spectra, and digestion by alginase. The polysaccharides appeared to contain O-acetyl groups, which are not found in alginates. Strong negative optical rotation and an infrared band at 893 cm^{-1} indicated the presence of the β-linkage in the polysaccharides.

Taurine has been isolated and characterized as a unique constituent of the cell wall of a variant of *Bacillus subtilis*, designated "opaque" (Kelly and Weed, 1965). Infrared methods were used in the identification. The reader should take note of a correction to the spectra in the paper.

In an investigation of the peptide of the cell wall of a strain of *S. faecalis*, N^{α}-(L-alanyl-D-isoglutaminyl)-N^{ϵ}-(D-isoasparaginyl)-L-lysyl-D-alanine, the location of the D-isoasparaginyl residue as a substituent of the N^{ϵ}-lysine in the disaccharide pentapeptide monomer has emerged from several lines of investigation (Ghuysen *et al.*,

1967), among which was the finding that the infrared spectrum showed no amino-succinimide ring. The structure of the pentapeptide portion was given as CVII.

$$\text{L-ala-D-glu}^{\alpha}\text{—CONH}_2$$
$$|$$
$$\text{L-lys-D-ala}$$
$$\epsilon\,|\quad^{\alpha}$$
$$\text{D-asp}^{\alpha}\text{—CONH}_2$$

CVII

Michell and Scurfield (1967) have studied the composition of extracted fungal cell walls of *Polyporus myllitae* Cke. et Mass. by means of infrared spectroscopy to determine the value of the spectroscopic method for taxonomic purposes. The principal constituent was α-chitin, and possibly β-1,3-glucan was present. The hyphae of fungi of species from diverse taxonomic groups were extracted successively with potassium hydroxide and glacial acetic acid–hydrogen peroxide and the infrared spectra of the residues were recorded. Two distinguishable groups were found, those having cellulose and those having α-chitin as a significant constituent. The presence of undissociated and dissociated carboxyl groups at 1730 and 1550 cm^{-1}, respectively, in fungal walls was believed to indicate the presence of uronic acids (Fig. 17.7).

Michell and Scurfield (1970) have shown that infrared spectra afford a relatively rapid and easy means of obtaining an indication of the nature of the major components of fungal cell walls. Their usefulness extends beyond merely distinguishing between species with chitinous and nonchitinous walls. These authors reached several conclusions: (1) Spectra can be used to indicate the presence of mannans in admixture with β-glucans, such as cellulose I or laminaran, or with α-chitin or protein, and also in heteropolymers such as galactomannans. The spectra cannot be used to determine the type of linkage present or the degree of branching of the mannans. (2) Spectra can be used to indicate the presence of α-glucans in admixture with α-chitin and, with less certainty, cellulose I. (3) Spectra can be used as aids for characterizing β-1,2-, β-1,3-, β-1,4-, and β-1,6-glucans. (4) Spectra can be used to distinguish between α- and β-chitin, and to detect α-chitin in admixture with mannans, α- or β-glucans, protein, or lipid. (5) Spectra can be used to recognize the presence of protein in admixture with either mannan or β-glucan. It would be wise to read the original paper before trying to apply these principles, since these workers have made several qualifying statements regarding them.

Viruses and Viral Infection

Pollard *et al.* (1952) have recorded spectra of meningopneumonitis (Francis strain), ornithosis (P4 strain), mumps (Enders strain) and Newcastle disease (California strain) viruses (Fig. 17.8). There is some similarity between the spectra of meningopneumonitis and ornithosis viruses, whereas the spectra of mumps and of Newcastle disease viruses were distinctly differentiated. Since the viral preparations were relatively crude, however, the curves in Fig. 17.8 were not felt to be definitive.

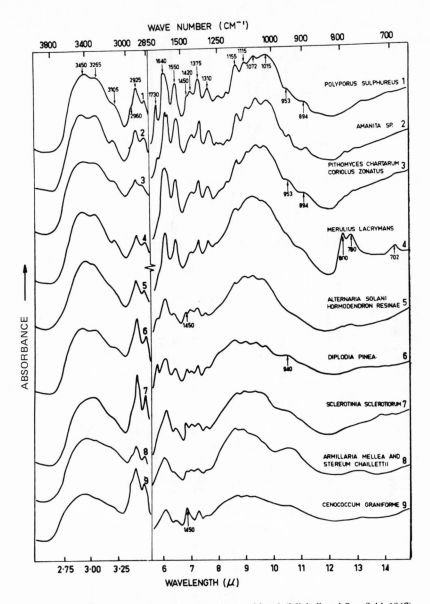

Fig. 17.7. Infrared spectra of hyphal wall residues of fungi. (Michell and Scurfield, 1967).

Benedict (1955) has shown that partial purification of viruses under standard conditions allows them to be grouped according to their infrared spectra. The viruses studied were influenza A (PR 8), influenza A' (FM 1), influenza B (Lee), Newcastle disease (California), mumps, psittacosis, lymphogranuloma venereum, meningo-pneumonitis, feline pneumonitis, mouse pneumonitis, vaccinia, and fowl-pox.

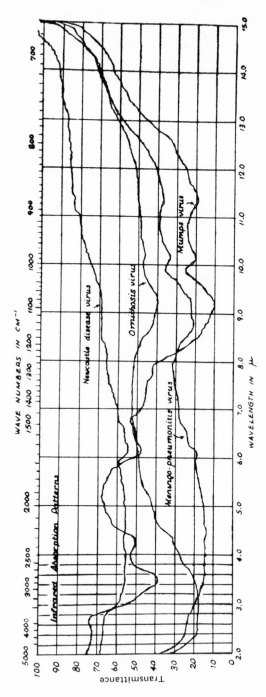

Fig. 17.8. Spectra of meningopneumonitis, ornithosis, mumps, and Newcastle disease viruses. (Pollard *et al*, 1952.)

Figure 17.9 presents the spectra of the viruses and normal host components. All the viruses studied fell into four different spectral groups. The spectra of the influenza strains, mumps virus, and the normal host constituents were alike as seen in curve A. Curve B is similar to curve A except for the shape at about $8.8\,\mu$ ($1136\,\text{cm}^{-1}$). All the numbers of the LGV-psittacosis group gave the same infrared spectra (curve C), which was identified by the missing shallow band at $7.7\,\mu$ ($1299\,\text{cm}^{-1}$), the double hump at 8.3–$8.8\,\mu$ (1205–$1136\,\text{cm}^{-1}$), the reduced absorption of the broad band at 9.0–$10.0\,\mu$ (1111–$1000\,\text{cm}^{-1}$), and the deepest band at $10.4\,\mu$ ($962\,\text{cm}^{-1}$). The shape of the curve at 8.3–$8.8\,\mu$ (1205–$1136\,\text{cm}^{-1}$) resembled NDV, but the peaks of the humps were always at the same height or slightly below that at $7.4\,\mu$ ($1351\,\text{cm}^{-1}$). In the case of NDV, the region of 8.3–$8.8\,\mu$ (1205–$1136\,\text{cm}^{-1}$) was always above the portion of the curve at $7.4\,\mu$ ($1351\,\text{cm}^{-1}$). Vaccinia and fowl-pox preparations varied at $7.8\,\mu$ ($1282\,\text{cm}^{-1}$), but the remainder of the spectrum was reproducible.

Kull and Grimm (1956c) have studied the spectra of bacteriophages active against *Bacillus megaterium*. They compared the spectra of six serologically different M bacteriophages with those of the *B. megaterium* S host cell (Table 17.1). The terminology of the M phages is given in (Friedman and Cowles, 1953). Distinction of the phage spectra was made possible by the exaggerated absorption at $1735\,\text{cm}^{-1}$, which was not observed with purified tobacco mosaic virus or *E. coli* T phages (Fig. 17.10). Figure 17.11 shows spectra obtained with M2, M4, and M6 bacteriophages as well as the host cell *B. megaterium* S. The three phages produced similar spectra.

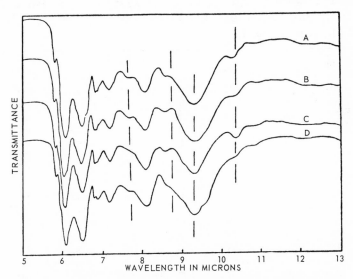

Fig. 17.9. Spectra of virus preparations and normal host components processed by 3 cycles of differential centrifugation: (A) influenza A, A′, B, mumps, and normal host components; (B) Newcastle disease; (C) psittacosis, lymphogranuloma venereum, meningopneumonitis, feline pneumonitis, and mouse pneumonitis; (D) vaccinia and fowl-pox. (Benedict, 1955.)

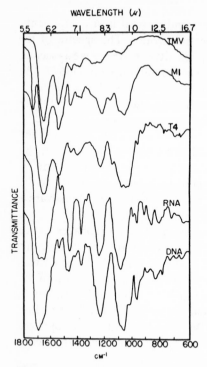

WAVELENGTH (μ)

Fig. 17.10. Infrared spectra of tobacco mosaic virus, M1 and T4 bacteriophage, and deoxyribonucleic and ribonucleic acids. (Kull and Grimm, 1956c.)

Table 17.1. Differences between *B. megaterium* S Host Cell and M1–M6 Bacteriophages at Wave Numbers of Major Absorbance (Kull and Grimm, 1956c)

Wave number, cm^{-1}		% Absorbance[a]						
Bacteria	Phage	M1	M2	M3	M4	M5	M6	S
—	1735	65	80	68	54	74	86	—
1650	1650	100	100	100	100	100	100	100
1545	1545	82	93	83	91	89	92	76
—	1465	55	70	57	60	58	69	—
1450	—	—	—	—	—	—	—	35
1400	—	—	—	—	—	—	—	43
—	1380	46	59	46	50	49	56	—
1310	—	—	—	—	—	—	—	28
—	1295	—	59	52	47	49	—	—
1230	1230	59	72	61	59	63	72	39
—	1175	43	56	48	42	50	55	—
1160	—	—	—	—	—	—	—	30
1060	1060	57	64	62	51	56	64	24
—	830	9	13	13	9	12	18	—
—	740	7	10	4	11	10	7	—
—	700	13	14	8	17	14	13	—
660	660	15	12	11	15	13	13	11

[a] The 1650 cm^{-1} band is arbitrarily assigned 100 % absorbance.

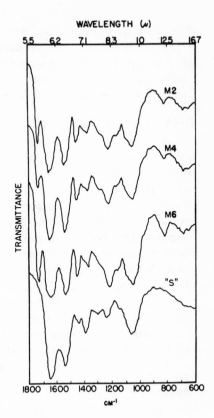

WAVELENGTH (ω)

Fig. 17.11. Infrared spectra of M2, M4, and M6 bacterio-phage and host cell *B. megaterium* S. (Kull and Grimm, 1956c.)

The spectra of M1, M3, and M5, not shown in Fig. 17.11, yielded major bands at corresponding wavelengths. The relative band intensities obtained from the various M phages were different (Table 17.1). As shown in this table and in Fig. 17.11, the major bands of the bacterial cell at 1650, 1545, 1230, and 1060 cm^{-1} were prominently present in the phage spectra. Absent from the bacterial spectrum were bands at 1735, 1465, 1380, 1295, and 1175 cm^{-1}, which were present in all spectra of the M phages. The prominent band of the M phage at 1735 cm^{-1} is absent in the host cell and serves to identify the M phage. This band also did not appear in spectra from tobacco mosaic virus, T3, and T4 *E. coli* bacteriophage, all gram-negative, and most gram-positive bacteria tested.

Cochran *et al.* (1960) have used differential spectroscopy to study changes in the biochemistry of cucumber seedlings as a result of cucumber mosaic virus infection. Figure 17.12 shows a spectrum obtained on dried normal cucumber cotyledon lower epidermis. Figure 17.13 shows virus-induced changes with time. Normal epidermal specimens were used in the reference beam and virus-infected specimens were placed in the sample beam. Of course, the curves are complicated by the variety of chemical changes taking place and the large numbers of substances present. The authors related changes in intensities of bands at 1639 cm^{-1} (6.1 μ) and 1538 cm^{-1} (6.5 μ) to protein increase, and those in the 3333–1667 cm^{-1} (3–6 μ) range and the 1111–1053 cm^{-1}

(9–9.5 μ) range to nucleic acid increase. They noted that the absorption in the 1111–1053 cm^{-1} (9–9.5 μ) region appeared to vary inversely with the absorption at 1538 cm^{-1} (6.5 μ). Their results suggested the occurrence of a cyclic variation in which nucleic acids built up and then fell during a period in which proteins were increased. Then, when protein levels fell on the sixth and seventh days, nucleic acid levels rose.

DNA of Viruses and Cells

An infrared study has been made of the influence of growth media and myoinositol on structural changes in DNA induced by dehydration and ultraviolet light (Webb and Dumasia, 1968). Films of DNA, RNA, and synthetic polynucleotides were exposed to varying degrees of relative humidity in the presence and absence of myoinositol. As other researchers have shown, when the relative humidity was lowered, shifts in the P=O and C=O absorption frequencies occurred, which seemed to be associated with the removal of approximately 12 molecules of water per nucleotide and all shifts were prevented by the presence of two molecules of inositol per nucleotide during desiccation. The irradiation of DNA at 75 % relative humidity with ultraviolet light (2537 Å) also produced spectral shifts which appeared to arise as a result of bound water molecules moving from P=O and C=O groups. These investigators felt that the ability of inositol to prevent spectral shifts in DNA caused by desiccation and irradiation tended to substantiate the suggestion that it preserved the biological integrity of cells and viruses during stress by combining with DNA.

The reader is referred to Chapter 12.

Metabolic Products

Levy and McNutt (1962) have studied the utilization of the naturally occurring pteridine, xanthopterin (CVIII) by a bacterium isolated from soil, hoping to relate some of the intermediates produced to metabolites formed in the normal biogenesis of pteridines. They isolated from cultures of this organism, and identified by ultraviolet

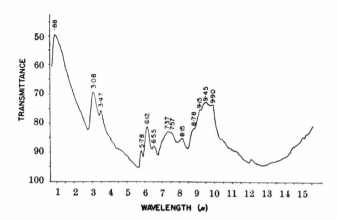

Fig. 17.12. Infrared transmittance through dried normal cucumber cotyledon lower epidermis. Wavelengths of principal peaks are identified. (Cochran et al., 1960.)

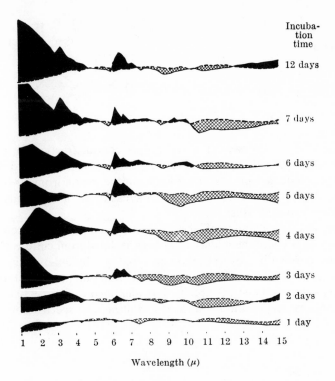

Fig. 17.13. Progressive changes in infrared absorbance of lower epidermis of cucumber cotyledons induced by the cucumber mosaic virus infection process (averaged values of two experiments). (———) graphically determined median value for comparisons of three virus-infected with three normal epidermal strips; (– – – –) I_0 (baseline of the instrument); dark shading, greater than normal absorption; light shading, less than normal absorption. (Cochran et al., 1960.)

and infrared spectra and paper chromatography, 6-oxylumazine (CIX), 6,7-dioxy-lumazine (CX), and leucopterin (CXI). They also obtained enzymes from this organism which catalyze the following reactions: (a) the deamination of xanthopterin and iso-xanthopterin to 6-oxylumazine and 7-oxylumazine, respectively; (b) the oxidation of 6-oxylumazine to 6,7-dioxylumazine; and (c) the transformation of 6,7-dioxy-lumazine to another compound, which was not identified.

CVIII

CIX

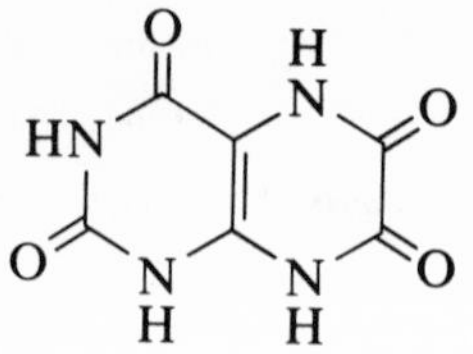

Iwahara *et al.* (1960) have prepared polysaccharide fractions of the endotoxin from nine strains of *E. coli* and six strains of *S. flexneri*, and have recorded the infrared spectra of the fractions. For *E. coli* the bands in the region from 1000 to 650 cm^{-1} differed considerably from strain to strain, whereas those of six strains of *S. flexneri* were very similar to each other.

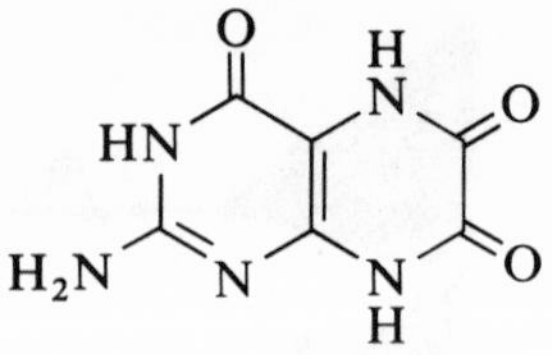

A water-soluble metabolite of indole-3-acetic acid is formed by certain strains of *Pseudomonas savastanoi*. The material has been isolated from culture filtrates of this organism and purified by adsorption on charcoal and by ion-exchange chromatography on sulfonic acid resins and quaternary ammonium resins (Hutzinger and Kosuge, 1968). The isolated metabolite was identified as indole-3-acetyl-ϵ-L-lysine (CXII) by chemical means and by infrared and ultraviolet spectroscopy.

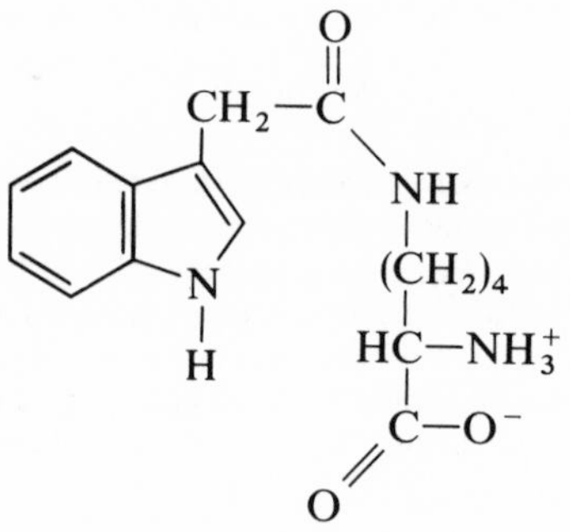

Infrared spectroscopy was one of the tools used (Levy and Frost, 1966) for a definitive identification of 2,3-dihydroxyphenylpropionic acid, the product from the reaction between the enzyme, melilotate hydroxylase (from extracts of the *Arthrobacter* species), and melilotic acid, NADH, and oxygen.

Loos *et al.* (1967*a, b*) have used gas chromatography, and ultraviolet and infrared spectroscopy to identify products of 2,4-dichlorophenoxyacetate (2,4D) and 4-hydroxyphenoxyacetate degradation by a soil arthrobacter. These products were phenol and hydroquinone. The results indicated that the arthrobacter degraded 2,4D via 2,4-dichlorophenol.

Pseudomonas putida has been grown with toluene as the sole source of carbon and was found to oxidize chloro-, bromo-, iodo-, and fluorobenzenes to their respective 3-halogenated catechol derivatives (Gibson *et al.*, 1968). The dihydroxy compounds from the first three substrates were identified by isolation and comparison with synthetic compounds by means of infrared and NMR spectroscopy.

6-Hydroxynicotinic acid has been shown (Tsai *et al.*, 1966) to be the first intermediate in the degradation of nicotinic acid by a clostridium organism. In the presence of pyruvate other intermediates accumulated, of which two were identified: 1,4,5,6-tetrahydro-6-oxonicotinic acid and α-methyleneglutaric acid. Infrared and NMR methods were used in defining the structures.

Riboflavin-catalyzed photooxidation products of indole-3-acetic acid have an inhibitory effect on the growth of *Schizosaccharomyces pombe*. These products also affect *E. coli* and other bacteria, and inhibit the growth of tomato-root tips and the germination of pea seeds. Infrared spectroscopy has aided in the identification of these photooxidation products (Fukuyama and Moyed, 1964) as 3-hydroxymethyloxindole and 3-methyleneoxindole. Both are converted to 3-methyloxindole, which was also identified by infrared spectroscopy, and is a nontoxic substance.

Holowczak *et al.* (1966) have isolated four new glycosides, hygromycin C, and compounds D, E, and F from the fermentation broths of *Streptomyces hygroscopicus* and partially characterized them. Ultraviolet and infrared spectra and isolation of components after acid and basic hydrolysis indicated that the glycosides have a similar aglycon, composed of 3,4-dihydroxy-α-methylcinnamic acid or an isomer of this compound and neoinosamine-2, linked via an amide bond. The carbohydrate component of compound F was D-glucose; of hygromycin C, L-fucose. The carbohydrate component of compounds D and E was 5-keto-6-deoxyarabohexose.

Schubert *et al.* (1967) have isolated cholesterol from *Streptomyces olivaceus*. This is a species of the order Actinomycetales. These workers used chromatography, mass spectrometric, and infrared spectroscopic methods for the isolation and identification. *S. olivaceus* can degrade as well as synthesize cholesterol.

Martin and Perun (1968) have isolated a polyhydroxydioxolactone with the empirical formula $C_{21}H_{36}O_7$ from the fermentation broth of a blocked mutant of *Streptomyces erythreus*. By chemical and physical means they showed the compound to be 5-deoxy-5-oxoerythronolide B (CXIII). Infrared, ultraviolet, NMR and mass spectrometry were among the methods used.

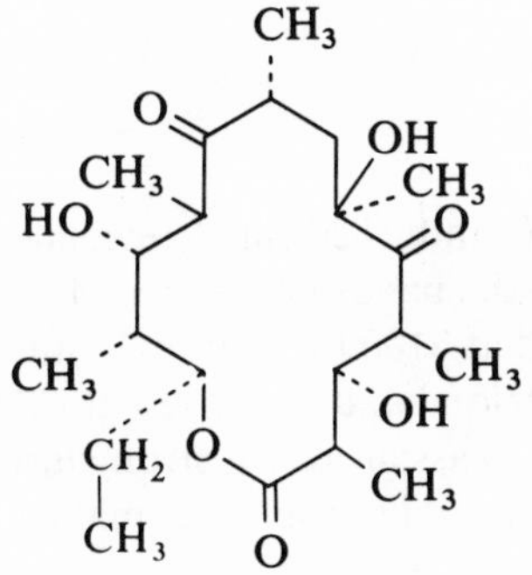

CXIII

Goren *et al.* (1971) have elucidated the structures of the acyl functions of a sulfolipid of *Mycobacterium tuberculosis*, strain H37Rv. In previous work Goren (1970) had characterized the sulfolipid as 2,3,6,6'-tetraacyl-trehalose 2'-sulfate. Goren *et al.* (1971) found three principal (and related) series of carboxylic acids: palmitic and stearic acids with minor amounts of other homologs; a multibranched series, the "average" member of which was 2,4,6,8,10,12,14-heptamethyltriacontanoic acid; and a second, related oxygenated multibranched group consisting chiefly of 17-hydroxy-2,4,6,8,10,12,14,16-octamethyldotriacontanoic acid. The structures were elucidated largely by mass spectrometry. Infrared spectroscopy was also used to examine esters prepared during part of the work. Methyl branches were found abundant in all three substances as indicated by a deep absorption band at 1380 cm^{-1} (confirmed subsequently in other studies). Intermolecular hydrogen bonding in films of hydroxyphthioceranates was responsible for the relative intensities of a split carbonyl band peaking at 1725 and 1745 cm^{-1}, the latter band having about twice the intensity of the former. The C_{40}-hydroxyphthioceranic acid examined in this work was 17-hydroxy-2,4,6,8,10,12,14,16-octamethyldotriacontanoic acid.

Among the tropolone compounds found in nature are four tropolone–carboxylic acids, obtained from mold cultures. Two of these, stipitatonic acid (CXIV, R = COOH) and stipitatic acid (CXIV, R = H) have been obtained from *Penicillium stipitatum* (Bentley and Thiessen, 1963). Puberulonic acid (CXV, R = COOH) and puberulic acid (CXV, R = H) were obtained from *P. aurantio-virens*, *P. puberulum* and *P. johannioli* as well as from strains of the *P. cyclopium-viridicatum* series (sources quoted in this paper). The infrared spectrum of the sodium salt of the anhydride of stipitatonic

CXIV   CXV

acid showed bands at 1750 and 1820 cm^{-1}, evidence for the anhydride ring. In puberulonic acid anhydride the bands occurred at 1770 and 1820 cm^{-1}. In stipitatic acid there was a broad band at 3100 to 3000 cm^{-1} and a sharp band at 3270 cm^{-1}; in stipitatonic acid in this region bands occurred at 3270 and 3140 cm^{-1}. For tropolone itself, this region has been assigned to a strong intramolecular association of the hydroxyl group in tropolone, reflecting presumably hydrogen bonding with the adjacent carbonyl oxygen.

Two new β-unsaturated amino acids have been identified in the mushroom, *Bankera fuligineoalba*, by means of chemical and spectral criteria, including infrared and NMR spectra (Doyle and Levenberg, 1968). These substances are L-2-amino-3-hydroxymethyl-3-pentenoic acid and L-2-amino-3-formyl-3-pentenoic acid.

Substances which accumulate in the culture medium and inhibit or significantly retard the growth and development of the organisms producing them have been called autoantibiotics. Two such compounds produced by the yeastlike fungus *Candida albicans* have been identified (Lingappa *et al.*, 1969), with the aid of mass and infrared spectrometry, as 2-phenylethanol (CXVI) and 3-β-hydroxyethylindole (CXVII).

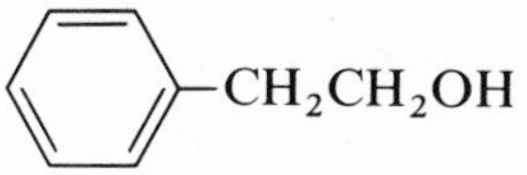

CXVI 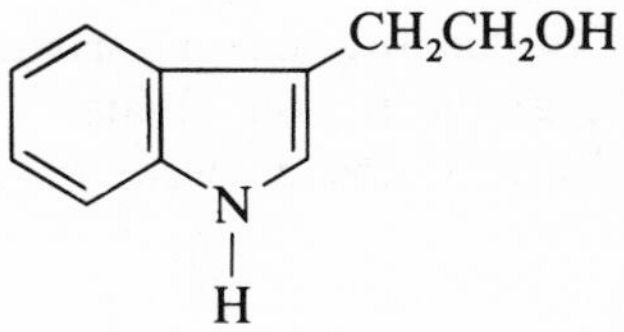   CXVII

The infrared spectrum of 2-phenylethanol displayed bands at 3600, 1600, and 1495 cm^{-1}, indicating the presence of both a hydroxyl group and an aromatic ring. The 3-β-hydroxyethylindole exhibited a spectrum with bands at 3600 cm^{-1} (OH), 3490 cm^{-1} (NH), 1600 cm^{-1}, and 1500 cm^{-1} (aromatic ring).

An iron(III)-binding compound with the properties of a secondary hydroxamic acid has been isolated from supernatant solutions of iron-deficient cultures of a red yeast, *Rhodotorula pilimanae* (Atkin and Neilands, 1968). The compound rhodotorulic acid (CXVIII) was characterized as LL-3,6-bis (*N*-acetyl-3-hydroxyaminopropyl)-2,5-piperazinedione, i.e., the diketopiperazine of δ-*N*-acetyl-L-δ-*N*-hydroxyornithine. The latter is an amino acid which is a constituent of ferrichromes, albomycins, and fusarinines. In the characterization of rhodotorulic acid, IR, UV, NMR, ORD, and mass spectroscopy were applied.

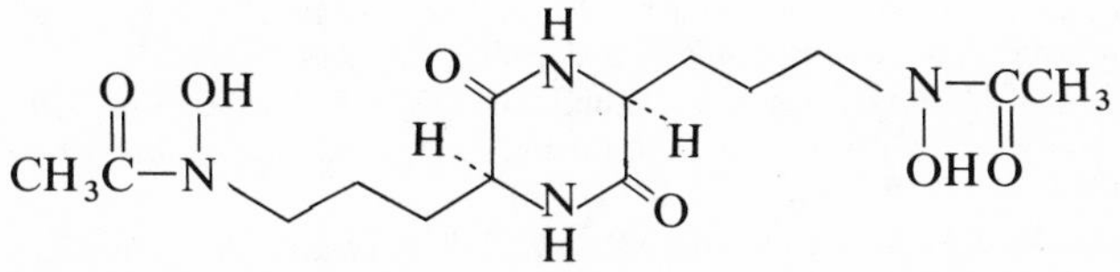

CXVIII

The principal infrared bands of rhodotorulic acid were at 3190, 3095, 2870, 1682, 1594, 1467, 1447, 1430, 1339, 1216, 1164, 969, 827, 797, and 777 cm^{-1}. The 1682 cm^{-1} band was assigned to amide I absorption. Characteristics of the *cis*-peptide bonds of the ring included a lack of N—H stretching bands higher than 3250 cm^{-1}, lack of any bands in the amide II region, and presence of the in-plane-bending vibration near

$1450\ cm^{-1}$, the latter being prominent in all the diketopiperazines studied. A strong band at $1594\ cm^{-1}$ was assigned to the hydroxamic acid carbonyls, although this is a very low frequency for C=O stretch (perhaps due to hydrogen bonding).

The accumulation of β-carboxy-β-hydroxyadipic acid (homocitric acid) in the culture medium of the lysine-requiring yeast mutant, Ly4, grown on limiting amounts of lysine, has been demonstrated by Maragoudakis and Strassman (1966). Infrared spectra were used to affirm identification of the biologically formed acid, as were R_f's, optical rotations, and melting points.

The paralytic poison produced by the dinoflagellate *Gonyaulax catenella* in axenic culture has been isolated in pure form by Schantz *et al.* (1966). A study of its chemical, physical, and biological properties established that it is identical in chemical structure to saxitoxin, the poison isolated from toxic Alaska butter clams (*Saxidomas giganteus*) and to the poison isolated from toxic California sea mussels (*Mytilus californiaus*). Infrared spectra were given for *G. catenella* poison and the mussel poison.

REFERENCES

Atkin, C. L. and Neilands, J. B. *Biochemistry* **7**, 3734 (1968).
Benedict, A. A. *J. Bacteriol.* **69**, 264 (1955).
Bentley, R. and Thiessen, C. P. *J. Biol. Chem.* **238**, 1880 (1963).
Booth, G. H., Miller, J. D. A., Paisley, H. M., and Saleh, A. M. *J. Gen. Microbiol.* **44**, 83 (1966).
Bugrova, V. I. in *Problems of Sanitary Bacteriology and Virology*, Meditsina (Moscow), 1965, p. 61. From *Ref. Zh. Biol.* 1966, No. 6B255 (Translation).
Cochran, G. W., Welkie, G. W., and Chidester, J. L. *Nature* **187**, 1049 (1960).
Doyle, R. R. and Levenberg, B. *Biochemistry* **7**, 2457 (1968).
Eremin, Yu. G., Efimtseva, E. P., and Kuklinskii, A. Ya. *Mater. Nauch. Konf. Sovnarkhoz Nizhnevolzh Ekon Raiona Volgograd Politekh. Inst. Volgograd* **2**, 181 (1965).
Friedman, M. and Cowles, P. B. *J. Bacteriol.* **66**, 379 (1953).
Fukuyama, T. T. and Moyed, H. S. *J. Biol. Chem.* **239**, 2392 (1964).
Ghuysen, J.-M. and Strominger, J. L. *Biochemistry* **2**, 1119 (1963).
Ghuysen, J.-M., Bricas, E., Leyh-Bouille, M., Lache, M., and Shockman, G. D. *Biochemistry* **6**, 2607 (1967).
Gibson, D. T., Koch, J. R., Schuld, C. L., and Kallio, R. E. *Biochemistry* **7**, 3795 (1968).
Goren, M. B. *Biochim. Biophys. Acta* **210**, 116, 127 (1970).
Goren, M. B., Brokl, O., Das, B. C., and Lederer, E. *Biochemistry* **10**, 72 (1971).
Heymann, H., Manniello, J. M., and Barkulis, S. S. *J. Biol. Chem.* **238**, 502 (1963).
Heymann, H., Manniello, J. M., Zeleznick, L. D., and Barkulis, S. S. *J. Biol. Chem.* **239**, 1656 (1964).
Holowczak, J. A., Koffler, H., Garner, H. R., and Elbein, A. D. *J. Biol. Chem.* **241**, 3270 (1966).
Hutzinger, O. and Kosuge, T. *Biochemistry* **7**, 601 (1968).
Iwahara, S., Ishizeki, R., Koshinuma, K., and Oba, T. *Bull. Natl. Inst. Hyg. Sci.* (*Japan*) **78**, 77 (1960), (Japanese, with English summary).
Kelly, A. P. and Weed, L. L. *J. Biol. Chem.* **240**, 2519 (1965).
Kull, F. C. and Grimm, M. R. *Bacteriol. Proc.* **1954**, 26.
Kull, F. C. and Grimm, M. R. *Bacteriol. Proc.* **1956***a*, M 137.
Kull, F. C. and Grimm, M. R. *J. Bacteriol.* **71**, 342 (1956*b*).
Kull, F. C. and Grimm, M. R. *Virology* **2**, 131 (1956*c*).
Levine, S., Stevenson, H. J. R., and Kabler, P. W. *Arch. Biochem. Biophys.* **45**, 65 (1953*a*).
Levine, S., Stevenson, H. J. R., and Bordner, R. H. *Science* **118**, 141 (1953*b*).
Levine, S., Stevenson, H. J. R., Chambers, L. A., and Kenner, B. A. *J. Bacteriol.* **65**, 10 (1953*c*).
Levine, S., Stevenson, H. J. R., Bordner, R. H., and Edwards, P. R. *J. Infectious Diseases* **96**, 193 (1955).

Levy, C. C. and Frost, P. *J. Biol. Chem.* **241**, 997 (1966).

Levy, C. C. and McNutt, W. S. *Biochemistry* **1**, 1161 (1962).

Lingappa, B. T., Prasad, M., Lingappa, Y., Hunt, D. F., and Biemann, K. *Science* **163**, 192 (1969).

Linker, A. and Jones, R. S. *J. Biol. Chem.* **241**, 3845 (1966).

Liu, T.-Y. and Gotschlich, E. C. *J. Biol. Chem.* **238**, 1928 (1963).

Loos, M. A., Roberts, R. N., and Alexander, M. *Can. J. Microbiol.* **13**, 679 (1967*a*).

Loos, M. A., Roberts, R. N., and Alexander, M. *Can. J. Microbiol.* **13**, 691 (1967*b*).

Maragoudakis, M. E. and Strassman, M. *J. Biol. Chem.* **241**, 695 (1966).

Martin, J. R. and Perun, T. J., *Biochemistry* **7**, 1728 (1968).

Michell, A. J. and Scurfield, G. *Arch. Biochem. Biophys.* **120**, 628 (1967).

Michell, A. J. and Scurfield, G. *Aust. J. Biol. Sci.* **23**, 345 (1970).

Parnas, J., Hencner, Z., Pleszczynska, E., Poplawski, S., and Cybulska, M. *Acad. Pol. Sci. Ser. Sci. Biol.* **14**, 475 (1966).

Peruanskii, Yu. V., and Topolovskii, V. A. *Vestnik Sel'skokhoz Nauk Kazakhstana* **1**, 24, (1964) (from *Ref. Zh. Biol.* 1964, No. 20v28, translation).

Pollard, M., Engley, F. B., Jr., Redmond, R. F., Chinn, H. I., and Mitchell, R. B. *Proc. Soc. Exp. Biol. Med.* **81**, 10 (1952).

Preobrazhenskaya, T. P., Altukhova, L. B., Dorozhinskii, V. B., and Ivanovna, L. V. *Mikrobiologiya* **35**, 96 (1966).

Randall, H. M. and Smith, D. W. *J. Optical Soc. Am.* **11**, 1086 (1953).

Randall, H. M., Smith, D. W., Colm, A. C., and Nungester, W. J., *Am. Rev. Tuberc.* **63**, 372 (1951).

Reynolds, R. J., Hedrick, H. G., and Crum, M. G. *Develop. Ind. Microbiol.* **8**, 246 (1967).

Schantz, E. J., Lynch, J. M., Vayvada, G., Matsumoto, K., and Rapoport, H. *Biochemistry* **5**, 1191 (1966).

Schubert, K., Rose, G. and Hörhold, C. *Biochim. Biophys. Acta* **137**, 168 (1967).

Scopes, A. W. *J. Gen. Microbiol.* **28**, 69 (1962).

Smith, D. W., Randall, H. M., MacLennan, A. P., Putney, R. K., and Rao, S. V., *J. Bacteriol.* **79**, 217 (1960).

Stevenson, H. J. R. and Bolduan, O. E. A. *Science* **116**, 111 (1952).

Thomas, L. C. and Greenstreet, J. E. S. *Spectrochim. Acta* **6**, 302 (1954).

Tsai, L., Pastan, I., and Stadtman, E. R., *J. Biol. Chem.* **241**, 1807 (1966).

Tyler, J. M. and Heidelberger, M. *Biochemistry* **7**, 1384 (1968).

Webb, S. J. and Dumasia, M. D., *Can. J. Microbiol.* **14**, 841 (1968).

Williamson, A. R. and Zamenhof, S. *J. Biol. Chem.* **238**, 2255 (1963).

Williamson, A. R. and Zamenhof, S. *J. Biol. Chem.* **239**, 963 (1964).

Chapter 18

APPLICATIONS IN MEDICINE AND RELATED FIELDS

Infrared spectroscopy has been used in many ways in clinical laboratories. The infrared analysis of serum and other fluids from healthy individuals and from patients with various diseases has been an aid in the diagnosis of those diseases. The adsorption of plasma proteins on the surface of polymers which are to be exposed to the bloodstream has been and is actively being investigated by means of internal reflection techniques. Normal and diseased skin surfaces have been examined by the ATR method.

Urines of normal and of diabetic individuals have been examined by infrared spectroscopy for analysis of the components therein, for example, reducing substances such as sugars. In conjunction with other techniques, infrared spectra can be used to examine the urine of patients with phenylketonuria. Also, the serums of blisters formed during certain diseases have been analyzed for protein. Thus, various metabolic products can be identified.

Cholesterol esters and triglycerides have been determined in serums by infrared methods. Biliary and renal calculi are quite readily analyzed from infrared spectra. Sometimes it is necessary to determine body water content, and this can be done by the use of D_2O, and quantitating the deuterium by means of infrared spectroscopy.

Other important uses of infrared techniques are found in the field of toxicology. Infrared spectra are useful for determining toxic gases and volatile organic compounds in human beings after their exposure to such substances, e.g., carbon tetrachloride. Toxicologists can determine such substances as tranquilizers, sedatives, pesticides, etc., in the organs of cadavers by using infrared methods.

An important application in anesthesiology is the use of infrared instruments for regulating the content of vapors to be inhaled by the patient. Carbon monoxide analyzers are used in clinics where lung-function testing is done for estimating lung diffusing capacity. Alveolar carbon dioxide can be determined, as can CO_2 production in human beings, by infrared measurements. The carbon dioxide content and carboxyhemoglobin (bound carbon monoxide) in the blood have also been measured. Biologists have measured CO_2 production in locusts and hens by means of infrared instrumentation. Another application of the infrared method is the determination of alcohol vapor in the breath.

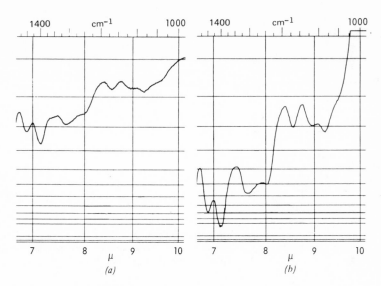

Fig. 18.1. Differential spectra of human blood. (*a*) Specimen: human blood serum; reference cell: distilled water. (*b*) Specimen: human blood serum concentrated 5 times; reference cell: distilled water. (Stewart *et al.*, 1959*a*.)

Infrared spectroscopy has important applications in the areas of air and water pollution, which are problems for everyone. With the well-being of people all over the world at stake, it appears that pollution of all kinds is a public health problem, an area where the influence of the medical and related professions must be brought to bear. It is with this idea in mind that these topics have been included in this particular chapter.

Diagnostic Use of Spectra of Blood Serum and Other Biologic Fluids

Stewart *et al.* (1959*a*) have described a rapid method for the infrared examination of blood serum, living cells, and biologic fluids which requires only two drops of sample. They presented spectra of red blood cells and a variety of normal and pathologic fluids in the natural state, along with spectra showing significant differences between these normal and pathologic specimens. Stewart *et al.* investigated the following materials without the use of any sample preparation: whole blood; human blood serum; rat blood serum; semen; urine; spinal, ascitic, and pleural fluids; milk; saliva; and, gastric juice. They concentrated saliva and spinal fluid about fivefold by vacuum drying at room temperature.

Figures 18.1a and 18.1b are differential spectra of natural human blood serum and human blood serum concentrated five times. Figure 18.2 is a differential spectrum of normal human blood serum in which the reference cell also contained the same fluid. The figure shows a relatively flat baseline with which to compare abnormal specimens. Figure 18.3 shows marked differences in the spectrum of the blood serum from a patient with the disease lupus erythematosus disseminatus. Figure 18.4

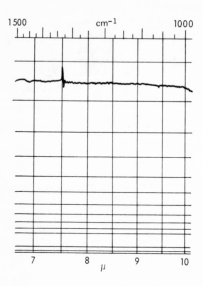

Fig. 18.2. Differential spectra of human blood. Specimen: normal human blood serum; reference cell: normal human blood serum. (Stewart *et al.*, 1959*a*.)

shows the differential spectrum of the serum of a patient with multiple myeloma, and again shows marked differences from the differential spectrum of the normal blood serum.

Figures 18.5 and 18.6 represent differential spectra of serum from patients with monomyelocytic leukemia and cystic fibrosis of the pancreas, respectively. Among the other blood sera studied were those from patients with aplastic anemia, panhypopituitarism, salivary gland carcinoma, and acute myocardial infarction.

Spectra were also given for serum albumin from a patient with Hodgkin's disease, gamma globulin from a patient with multiple myeloma, pleural fluid from a patient with chronic lymphocytic leukemia, and ascitic fluid from a patient with

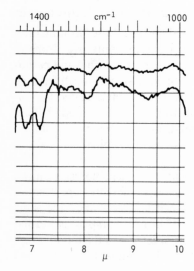

Fig. 18.3. Differential spectra of human blood. Specimen: blood serum from a patient with lupus erythematosus disseminatus; reference cell: normal serum. (Stewart *et al.*, 1959*a*.)

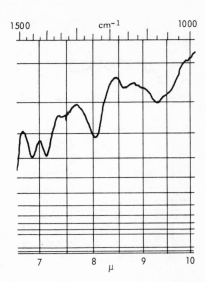

Fig. 18.4. Differential spectra of human blood. Specimen: blood serum from a patient with multiple myeloma; reference cell: normal serum. (Stewart *et al.*, 1959*a*.)

carcinomatosis. A consistent difference was found between the adult red blood cell and the fetal red blood cell (see also, Stewart and Erley, 1961.) In addition, small differences in the structure of the fetal red cells themselves were noted, suggesting that structural differences in the red blood cells could be investigated by this technique, and the method adapted to the study of many cellular elements in blood and body fluids.

Representative infrared spectra of normal blood serum and of serum protein fractions from 105 healthy adults have been presented by Stewart *et al.* (1960*b*) to demonstrate that infrared analysis adds another qualitative dimension to the investigation of the serum proteins. Two drops of blood serum or a serum protein fraction

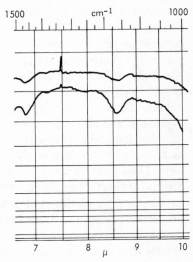

Fig. 18.5. Differential spectra of human blood. Specimen blood serum from a patient with monomyelocytic leukemia: reference cell: normal serum. (Note: Lower tracing obtained with pen expander.) (Stewart *et al.*, 1959*a*.)

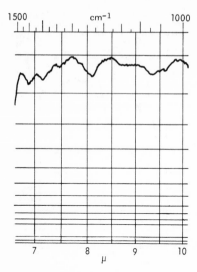

Fig. 18.6. Differential spectra of human blood. Specimen: blood serum from a patient with cystic fibrosis of the pancreas; reference cell: normal serum. (Stewart *et al.*, 1959*a*.)

were analyzed in a double-beam infrared spectrometer. A range of qualitative differences in the various normal albumin fractions and in each of the three globulin fractions was readily apparent. Additional qualitative differences were observed in serum fractions from patients with neoplasms. In contrast, the range of variation in the spectra of normal blood serum was small, with most differences limited to the 1429–1333 cm^{-1} region.

Blood Clotting

The phenomenon of blood clotting has been under study for many years. The consensus now is that a cascade-like sequence of events takes place in which a series of proteinaceous clotting factors, present in the blood as inactive precursors, become activated in a stepwise manner leading, ultimately, to the formation of a fibrin clot (White *et al.*, 1968). The first recognized event in the coagulation of shed blood is believed to be activation of Hageman factor (Factor XII), a sialoglycoprotein. It is generally believed that activation comes about through contact of the factor with a "foreign" surface such as glass or connective tissue. Recent studies by Lyman and Brash (1969), however, cast some doubt on this idea. They adapted the infrared internal reflection technique to study the adsorption of plasma proteins such as albumin, γ-globulin, and fibrinogen on polyethylene, polystyrene, Silastic, and Teflon. In every case they found the behavior to be similar. The polymers rapidly adsorbed the proteins as monomolecular layers. These adsorbed proteins retained their native globular form. These workers say it is conceivable that Hageman factor and the other proteinaceous clotting factors also are adsorbed on the polymer surfaces in a similar fashion to albumin, γ-globulin, and fibrinogen, which they closely resemble in physical properties. Lyman and Brash feel that if this is the case, then it may throw into dispute the theory that contact activation of Hageman factor is the initiating step of clotting, at least in *in vitro* conditions or by the introduction of a man-made surface into the blood circulatory system. These investigators hope that their work will lead to the development of novel functional polymers that do not cause clotting.

Examination of Normal and Diseased Skin

Comaish (1968) has applied the technique of multiple internal reflection (MIR) to intact human skin *in vivo*. Using a germanium reflector plate he obtained spectra (in a Wilks "Skin Analyzer") of normal skin (Fig. 18.7) and various dermatoses (e.g., Fig. 18.8) and found gross changes from the normal in the lesions of psoriasis, eczema, and other disorders. He has also demonstrated abnormalities in the clinically normal skin of psoriatic patients. The technique measures surface changes, so that only the outer layers of stratum corneum, sweat, sebum, and extraneous materials will affect the results obtained. Lactic acid, lactate, and ammonia from sweat probably contribute substantially to the bands at 2935, 1465, 1125, and 1045 cm^{-1}. After the skin was washed and dried these bands were diminished or disappeared entirely. Any contributions the stratum corneum makes to the infrared spectrum are at 2870, 1650, and 1550 cm^{-1}.

In the lesions of psoriasis (Fig. 18.8) the bands at 1650, 1550, 1125, 1045, and 1025 cm^{-1} were grossly diminished and those at 2940, 2870, and 1465 cm^{-1} diminished only slightly. Also, the water bands at 3400 and 1640 cm^{-1} were considerably diminished (from impaired sweating in the lesion). The band at 3400 cm^{-1} was replaced

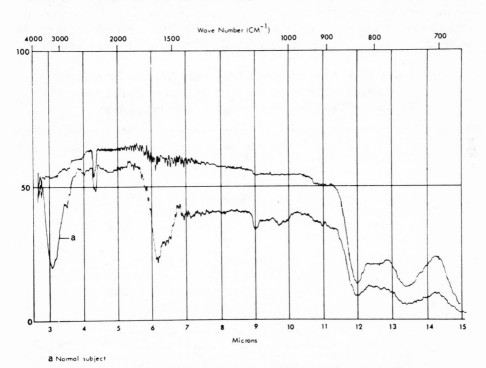

a Normal subject

Fig. 18.7. Multiple internal reflection spectrum of palmar skin in normal subject. Upper curve is that obtained using attachment alone. (Comaish, 1968.)

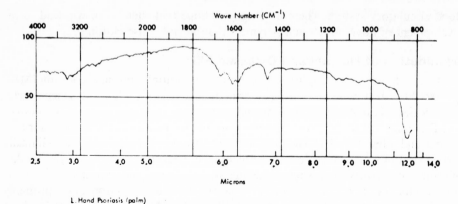

L. Hand Psoriasis (palm)

Fig. 18.8. Spectrum obtained from psoriatic plaque of palm. Note diminution of peaks at 1650 cm^{-1}, 1550 cm^{-1}, 1125 cm^{-1}, and 1045 cm^{-1} compared with normal palm. (Comaish, 1968.)

by two smaller ones at 3450 and 3350 cm^{-1}, and the 1650 cm^{-1} band was replaced by smaller ones at 1675, 1625, and 1595 cm^{-1}. New bands also appeared at 1160 and 1060 cm^{-1} in some lesions (Tables 18.1 and 18.2).

The apparently normal palms of 10 psoriatic patients were studied. As Table 18.1 shows, the findings were significantly different from those in the palmar skin of nonpsoriatic controls as measured by the height of bands recorded by a Hilger and Watts Infrascan spectrophotometer. (A Perkin-Elmer Model 157 instrument did not yield statistically significant differences, although the trend of results was similar.)

Table 18.1. Absorption in Clinically Normal Palms of Nonpsoriatic and Psoriatic Subjects Using an Infrascan Spectrophotometer (Comaish, 1968)

Wave number, cm^{-1}	Absorbance,[a] % Control[b]	Psoriatic[c]	t	P[d]
3400	23.5 ± 4.1	16.3 ± 3.1	1.33	0.2
2940	11.5 ± 2.4	7.5 ± 1.0	1.47	0.2
2870	7.1 ± 1.5	4.4 ± 0.5	4.6	0.001*
1650	32.2 ± 1.8	22.0 ± 3.2	2.7	0.2*
1640	30.9 ± 1.9	21.4 ± 3.1	2.5	0.002*
1550	15.6 ± 1.2	9.2 ± 1.5	3.2	0.01*
1465	10.3 ± 1.1	11.4 ± 1.0	0.7	0.5
1125	9.3 ± 0.7	6.6 ± 0.6	2.8	0.02*
1045	8.2 ± 0.8	5.2 ± 0.7	2.6	0.02*
1025	6.9 ± 0.9	4.4 ± 0.6	2.2	0.05*

[a]Absorbance is expressed as the percentage measured from a baseline.
[b]The control consisted of 11 subjects.
[c]There were 10 psoriatic subjects. Note that the palms of these subjects were *not* involved in the disease.
[d]An asterisk denotes a statistically significant difference.

Table 18.2. Peaks of Absorbance of Normal Palms and Psoriatic Plaques on Palms (Comaish, 1968)

Peaks, cm^{-1}		
Normal[a]	Psoriatic plaques[b]	Interpretation
3400	3450 3350	Water band
2940	2940⎫	
2870	2870⎭	CH stretch, —CH$_2$ groups
1650	1675 1625 1595	C=O stretch, amide I band
1550	—	NH deformation vibrations, amide II band
1465	1465	—CH deformation vibrations
1125	(1160) (1125)	C—O stretch
1045	(1060)	
1025	—	

[a]Control (normal) group consisted of 13 subjects.
[b]There were 7 psoriatic subjects. The peaks given in parentheses are weak.

Comaish states that it is entirely possible that certain diseases may be associated with the deposition of certain chemical compounds in amounts capable of exerting a characteristic effect on the MIR spectrum. Further observation is needed.

Examination of Urine

Rozelle *et al.* (1965) have examined normal and diabetic urines (Fig. 18.9) and have found the major infrared-absorbing components of normal urine to be urea, mono- and dihydrogen phosphates, and sulfates. Glucose absorption bands were readily identified in the spectra of diabetic urine. These workers found it difficult to determine the exact minimum concentration of these urine constituents necessary for visual detection from infrared spectra. The concentrations of components tend to vary from sample to sample and mask some bands, depending upon the relative quantities of the components. However, Rozelle *et al.* made estimates regarding the sensitivity of this infrared technique. From its 6.83 μ (1464 cm^{-1}) band urea can be detected at approximately 0.1 g/100 ml (0.02M). Glucose absorption can be observed easily at 0.7 g/100 ml (0.04M). Below this concentration the glucose peaks are generally masked by the phosphates and sulfates. The mono- and dihydrogen phosphates may be detected at concentrations of 0.1 g/100 ml (0.005M). The sulfates have a recognizable band only at very low phosphate concentrations but since the sulfate absorption is intense as well as broad, it causes a characteristic broadening of the phosphate band at 9.27 μ (1079 cm^{-1}). A trained observer can see the influence of sulfate on the infrared spectrum of urine at levels as low as 0.15 g/100 ml (0.01M).

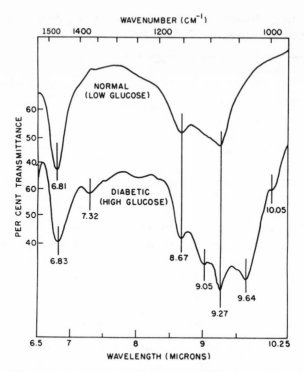

Fig. 18.9. Infrared spectra of normal and diabetic urine samples. (Rozelle *et al.*, 1965.)

The contribution that creatinine, the urates, organic acids, and ammonium ion make to the infrared spectra of urine are masked by sulfate, the phosphates, urea, and glucose, probably because the former group of substances are too low in concentration to absorb very strongly between 6.5 μ (1538 cm^{-1}) and 10.5 μ (952 cm^{-1}). In Fig. 18.9 the following bands can be seen: 6.81 μ (or 6.83) (1468 cm^{-1}) urea; 7.32 μ (1366 cm^{-1}) glucose; 8.67 μ (1153 cm^{-1}) urea, glucose, and $H_2PO_4^-$; 9.05 μ (1105 cm^{-1}) glucose and SO_4^{2-}; 9.27 μ (1079 cm^{-1}) $H_2PO_4^-$, HPO_4^{2-}, and glucose; 9.64 μ (1037 cm^{-1}) glucose; and 10.05 μ (995 cm^{-1}) HPO_4^{2-} and glucose.

The precise identification of urinary reducing substances is a frequent clinical chemical problem. Derivatization of the sugar as the osazone has long been considered a good conclusive distinguishing test. Klein and Weissman (1964) have published the infrared spectra of some clinically important sugars in the form of phenylosazones including: glucosazone, galactosazone, fructosazone (identical with glucosazone), lactosazone, maltosazone, arabosazone, xylosazone, and mannose phenylhydrazone. The infrared spectra make identification of such compounds much easier and more dependable.

By means of gas chromatography and infrared spectroscopy Wells *et al.* (1964) have identified galactitol in the urine of two infants with galactosemia. Excretion of galactitol continued for a prolonged period of time after dietary galactose had been removed.

Clausen *et al.* (1963) have given a method combining paper electrophoresis and infrared analysis for the characterization of the acid mucopolysaccharides in the urine of patients suffering from gargoylism (Hurler's syndrome) and Morquio's disease (atypical chondrodystrophy). In two of three siblings with the latter disease, chondroitin sulfate B and altered hyaluronic acid were excreted in the urine, but not keratosulfate. In a patient with gargoylism the excretion of chondroitin sulfate B was confirmed but no excretion of heparin could be demonstrated. Among the infrared spectra these authors used for clinical evaluation were those of the following: acid mucopolysaccharides, chondroitin sulfates A and B, heparin, and hyaluronic acids treated in various ways.

The diketopiperazine of histidylproline (CXIX) has been found in greatly increased amounts in the urines of patients with phenylketonuria who were being fed a low phenylalanine diet. Perry *et al.* (1965) identified the urinary product by its infrared spectroscopic, chromatographic, electrophoretic, and other chemical properties, e.g., products of hydrolysis. The markedly increased amounts of the product in the urine of phenylketonurics receiving a low phenylalanine diet apparently resulted from ingestion of the diketopiperazine of histidylproline in the low phenylalanine diet itself.

CXIX

Analysis of Blister Serum

Wegmann and Thewes (1956) have used infrared spectroscopy and various clinical tests to examine the serum from blisters of patients with Brocq–Dühring disease (a herpetic dermatitis) and pemphigus *vulgaris* (another skin disease). The infrared spectra were useful for comparing the protein contents of the blisters in the two diseases. In the case of pemphigus *vulgaris* denatured proteins appeared, which were absent in the Brocq–Dühring disease. In the latter disease, the abnormality showed up in the phosphosugar fraction, which was less disturbed than in pemphigus.

Analysis of Lipoprotein in Serum

Hatch *et al.* (1967) have developed an ultracentrifugal method for isolating chylomicron-containing fractions from serum by flotation, with the use of either standard Spinco swinging-bucket rotors or a specially fabricated swinging-bucket rotor. They used an infrared spectrophotometric method for quantitation of the isolated lipoproteins having S_f greater than 400. The method makes use of the fact that the major constituent (about 85–90% by weight) of these macromolecules is triglyceride.

The infrared absorption by ester carbonyl groups of the triglycerides at 1742 cm^{-1} provides a convenient measure of this lipid class. Cholesterol esters and phospholipids are present in relatively small amounts, and Hatch *et al.* estimated that the contribution of such compounds to the peak carbonyl absorption is no more than about 5% of the total.

Analysis of Calculi (Gallstones)

Edwards *et al.* (1958) have used infrared spectroscopy in the qualitative analysis of 30 specimens of human biliary calculi. The spectra of cholesterol, calcium bilirubinate, and calcium carbonate display prominent and characteristic bands that do not overlap in certain areas of the spectrum. Bands at 3380, 2910, and 1055 cm^{-1} indicate the presence of cholesterol, a doublet at 1670 and 1630 cm^{-1} is characteristic of calcium bilirubinate, and a sharp band at 875 cm^{-1} is produced by calcium carbonate. It is thus possible to verify the principal constituents of biliary calculi—whether they are (1) "pure" gallstones that are composed of either cholesterol, calcium bilirubinate, or calcium carbonate, or are (2) mixed gallstones that are composed chiefly of two or three of these components, or are (3) combined gallstones with a nucleus of one kind and a shell of another substance.

Weissman *et al.* (1959) have analyzed renal calculi by the KBr disk method and applied the same technique to prostatic, dental, biliary tract, and salivary gland calculi, and to the identification of alcohols, barbiturates, and salicylates (Klein *et al.*, 1960).

Recent advances in infrared spectroscopy have allowed rapid and simple determination of some gallstone ingredients (Chihara *et al.*, 1958) with a minimum quantity of specimen. Of various absorptions characteristic of calcium bilirubinate, the one at 1624 cm^{-1}, which is due to $\nu_{C=C}$ of pyrrole nuclei (Suzuki and Toyoda, 1966), was chosen as the key band for quantitative analysis. In analyses of 10 gallstone specimens (Toyoda, 1966) the infrared-determined calcium bilirubinate content was found to lie in between values from two kinds of chemical data: those obtained for extracts of the stones with (a) chloroform after HCl treatment, plus a 5% ammonia extraction, and (b) those obtained for extracts with chloroform, but no ammonia extraction. Toyoda claims that the infrared data show satisfactory proportionality with both types of data.

A few situations limit the usefulness of this method. In gallstones containing calcium carbonate beyond a certain level, for instance, the baseline for measuring the absorbance cannot be readily determined because of an obscured absorption minimum at 1485 cm^{-1}. The baseline is usually drawn as a straight line tangential to an absorption minimum around 1800 cm^{-1} and one at 1485 cm^{-1} (see Fig. 18.10). In the variety of bilirubin stones which appear black on the surface and upon transection, the spectrum often is abnormally flat in the range including 1624 cm^{-1}, probably as a result of overlapping of the spectra of other components, and the spectroscopically determined values show a significant discrepancy from the chemical data.

Gallstones occasionally contain black pigments which consist of polymers of bilirubin derivatives. To study whether any bile pigment metal complex is also

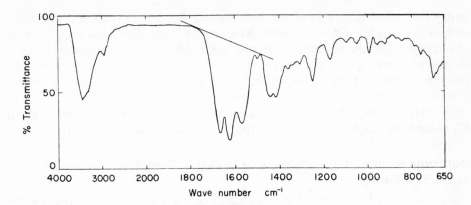

Fig. 18.10. Infrared spectrum of authentic calcium bilirubinate showing the baseline for measuring the absorbance at 1624 cm^{-1}. (Toyoda, 1966.)

concerned with such black pigments of the gallstone, Suzuki (1966) made attempts at synthesis of bilirubin-Cu complexes. When free bilirubin and cupric chloride were mixed in a solvent mixture of chloroform and ethanol, a dark blue or black substance was formed. This pigment was soluble in ethanol and showed a characteristic absorption at 350 nm (Soret band) and two visible absorption bands at 595 nm and 645 nm. Infrared spectroscopy showed that this compound is a complex salt of bilirubin and copper having an analogous structure to metalloporphyrins, in which bilirubin has a ring structure of tetrapyrroles and copper is located in the center of the nitrogen atoms of pyrroles. A comparison between the infrared spectra of the bilirubin-Cu complex and that of free bilirubin shows a number of differences between these two spectra: (1) In free bilirubin, the band at 1692 cm^{-1} was due to the stretching vibration of C=O of COOH, but in the bilirubin-Cu complex this band was shifted to ~ 1710 cm^{-1}. (2) Free bilirubin did not show the 1100 cm^{-1} band that is found in the bilirubin-Cu complex. (3) The band due to N—H stretching at 3400 cm^{-1} was much weaker in the bilirubin-Cu complex than in free bilirubin.

Addition of cupric chloride to bilirubin in sodium hydroxide solution resulted in precipitation of a black substance, which was proved to be a bilirubin–Na–Cu complex formed by coordination of copper to sodium bilirubinate. The black pigments of the gallstone may thus possibly include some metal complexes of bile pigments, although identification of such substances in the bile or in gallstones had not yet been definitely established at the time of this investigation.

Body Water Content

Stansell and Mojica (1968) have described a procedure for the determination of total body water content of human subjects. They used 11- to 12-g doses of D_2O, vacuum sublimation of serum samples, and quantitation of the deuterium by infrared spectroscopy at 2510 cm^{-1}. The chief advantages given for the procedure were (1) a D_2O dose of only 10 ml is required for adult subjects, and (2) the ease of the assay is coupled with satisfactory accuracy and precision. The coefficient of variation based on day-to-day procedure reproducibility is less than 2%. The D_2O dilution procedure compared favorably with the tritium dilution technique, with values obtained on 45 out of 46 individuals falling within the ± 3 combined standard deviation limits for the two methods. A correction for hydrogen–deuterium exchange processes between the ingested deuterium (as D_2O) and serum organic molecules, primarily proteins, is not usually applied when determining changes in the body water within an individual. However, for absolute measurements of water, a correction of about 1% of the body weight is needed to approach the actual value more closely.

Determination of Toxic Gases and Volatile Organic Compounds in Human Beings

Stewart and Erley (1965) have discussed applications of infrared spectroscopy which are of great use to the toxicologist, the medical examiner, and the physician. Rapid methods are needed with which to determine whether human exposure to toxic gases and volatile organic compounds has occurred, and if so, to estimate the magnitude of that exposure. Inherent in the infrared techniques for detecting toxic gases and volatile organic compounds in the expired breath, blood, urine, and tissues are the desired specificity (immediate identification of compound), the sensitivity (detection at or below concentrations effecting physiological changes), the speed, and the simplicity of analysis.

Fortunately, gases such as oxygen and nitrogen have no infrared absorption. Thus, path lengths of 10 and 40 m can be used to detect trace contaminants in air. However, the presence in air of small amounts of carbon dioxide and water vapor, which absorb infrared radiation strongly, precludes extending the path length much beyond 40 m for *general* trace analysis.

Compounds containing either halogen atoms or oxygen atoms can generally be determined with high specificity, whereas saturated aliphatic hydrocarbons, although identifiable as a class, are difficult to distinguish from one another and may be masked by the alkyl groups of other compounds.

Gases

The most commonly used design of a multiple reflection gas cell for the analysis of trace organic vapors is the one shown in Fig. 18.11. Three spherical mirrors of equal radii of curvature are arranged so that the infrared beam is reflected back and forth many times before leaving the cell. Each time the beam passes through the cell, the molecules absorb additional radiation, increasing the sensitivity of detection. A cell of this kind has certain advantages: compactness, relatively small sample volume, and long path length. Cells of 10- and 40-m are sold commercially. The 10-m cell is probably a good compromise between sensitivity and convenience of use.

Liquids

The concentration of the solution to be analyzed and the transmission of the solvent at the wavelength of analysis are factors to be considered in choosing a cell. If the product of the percentage concentration of the sample and the optical path

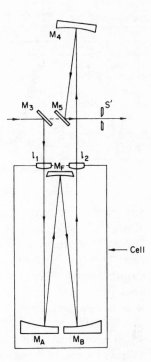

Fig. 18.11. Optical diagram of multiple reflection gas cell set for 1-m path length. The number of beam traversals can be increased by resetting M_A and M_B to give path lengths up to 10 m. (Stewart and Erley, 1965.)

length in millimeters is above 0.1, infrared methods can usually be applied for identification. For example, a cell of at least 10-mm path length is needed to analyze solutions containing 0.01 % of sample. The only solvent with adequate transmission for general analytical applications requiring this sensitivity is carbon disulfide.

Long-path-length cells for liquids may be constructed by clamping two alkali halide crystal windows on either side of a teflon spacer which forms part of the cell chamber. Such cells require about 10 ml of solution, therefore a relatively large amount of sample. An easily assembled cell has been described by Erley *et al.* (1960). A long path microcell requiring only 0.5 ml of solution has been described by Erley (1961). Two silver chloride windows are press-fitted into the ends of a silver cylinder, which forms the sample chamber. The use of silver prevents reaction between AgCl and any other metal higher in the electromotive series, and the silver acts as a heat sink and minimizes the tendency of the solvent to boil in the cell.

Analysis of Expired Breath

The absorption bands of water vapor and carbon dioxide are always present in the infrared spectrum of the expired breath because both these substances strongly absorb infrared radiation. On the other hand, oxygen, nitrogen, and the inert gases do not absorb infrared radiation and therefore cannot be detected—an advantage for the methods to be discussed.

Figure 18.12 is the spectrum of breath normally expired in a gas cell of 10-m path length. The region 2128–1250 cm^{-1} (water vapor) has too many bands to allow most

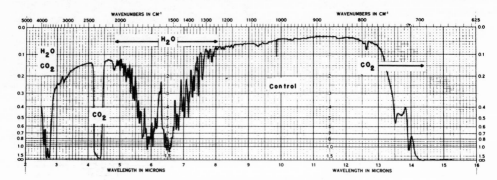

Fig. 18.12. Infrared spectrum of expired breath. (Stewart and Erley, 1965.)

analytical work except where differential techniques are used. There are other bands also: 3922–3636 cm^{-1}, carbon dioxide and water; 2299 cm^{-1}, carbon dioxide; 792 cm^{-1} and 741–625 cm^{-1}, carbon dioxide. The region 1250–769 cm^{-1} is a good "window" area for working with the vapors of many toxic materials, since they absorb strongly here, whereas normal breath vapor offers little interference. (A weak triplet may be noted at 1053 cm^{-1} in the normal control spectrum. It is thought to be caused partly by a trace of ethanol normally present as a metabolite.)

Stewart and Erley (1965) have listed the following groups of compounds for which infrared analysis of the expired breath may be used effectively: (a) halogenated hydrocarbons, such as carbon tetrachloride, trichloroethylene, and methylene chloride; (b) alcohols, such as methanol, ethanol, and isopropanol; (c) ethers and aldehydes, such as ethyl ether and paraldehyde; (d) ketones, such as acetone and methylethylketone; and (e) gases such as carbon monoxide, dioxide, and ammonia. They have given a table of about 160 gases and vapors in which are listed the Threshold Limit Value (ppm), the infrared sensitivity (ppm), and the wavelength at which the infrared sensitivity is measured. For most compounds the detection limit in breath is well below the threshold limit. Examples are carbon tetrachloride, with a threshold limit value of 10 ppm and an infrared sensitivity of 0.5 ppm and dichlorodifluoro-methane, with values of 1000 and 1 for those properties, respectively. Sensitivity is here defined as the minimum concentration of the compound which gives an absorbance of at least 0.01 in a 10-m path length cell.

Sampling of Breath

For the qualitative analysis of most volatile compounds and gases the collection of expired breath samples is quite adequate. For quantitative analysis of breath alveolar air should be collected to minimize the variations arising from individual collection techniques. Collection techniques were described by Stewart and Erley (1965). An apparatus used to desorb vapors from silica gel (used as an alternative technique for sampling) was also described.

Breath Analysis After Exposure to Toxic Gases or Volatile Compounds

Figure 18.13 shows a spectrum of expired breath after exposure to carbon tetrachloride vapor. There is a characteristic band at 795 cm^{-1}. It should be noted

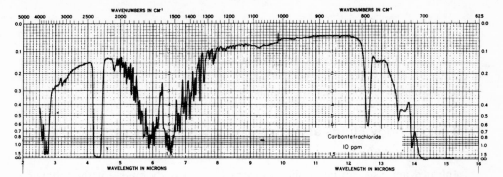

Fig. 18.13. Infrared spectrum of carbon tetrachloride in expired air following vapor exposure. (Stewart and Erley, 1965.)

that the rate of elimination of a volatile compound from the body as reflected by its concentration in the expired air is related to the duration of the exposure as well as to the vapor concentration during the exposure.

Figure 18.14 is a spectrum of a dog's expired air three hours after the forced ingestion of isopropyl alcohol. The alcohol bands and those of its metabolite, acetone, can be specifically and quantitatively identified. Ethyl alcohol can be specifically and quantitatively analyzed in human breath and such infrared analysis permits one to estimate blood alcohol concentration.

In healthy adults the mean concentration of acetone in the breath is 1.1 ± 0.5 μg/liter (Stewart and Boettner, 1964). With a 10-m or 40-m gas cell such concentrations are not detectable. The acetone content of diabetic breath may range from 0.1 μg/liter to more than 2000 μg/liter. Before the onset of significant ketonemia, the acetone content in the diabetic's breath rises to amounts detectable by infrared spectroscopy and therefore analysis of the breath gives a rapid method for determining the degree of ketosis.

Carbon monoxide analysis in the breath by the infrared method is much faster and more sensitive than the standard carboxyhemoglobin technique. Infrared

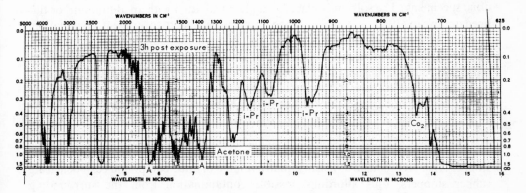

Fig. 18.14. Infrared spectrum of expired air after ingestion of isopropanol. Isopropanol (*i*-Pr) and its metabolite, acetone (A) are quantitatively identified. (Stewart and Erley, 1965.)

spectroscopy can easily detect the increase in CO in the breath after the smoking of one cigarette. The 2174 cm^{-1} band is used for detection.

For diagnostic and forensic purposes it should be remembered that the *absence* of certain infrared bands in a spectrum gives useful information. Since halogenated hydrocarbons have characteristic bands, lack of these bands in a breath sample recently taken would indicate lack of exposure or minimal exposure to the compound.

Analysis of Body Fluids after Extraction with Solvents

Detection of Volatile Organic Compounds

Infrared analysis can be applied to extractions of body fluids for the following classes of compounds: (a) halogenated hydrocarbons; (b) alcohols; (c) ethers and aldehydes; and (d) ketones. Most of the 160 substances (except the gases) given in the table of Stewart and Erley (1965) referred to earlier may be extracted from body fluids and analyzed by infrared methods.

Principles for Detecting Low Concentrations

The presence of proteins in biological fluids tends to mask the presence of most volatile compounds. It is therefore necessary to isolate the volatile organic compounds from the body fluids before infrared analysis. Another problem is that water in the fluids strongly absorbs the infrared radiation and prevents the use of cells longer than 0.05 to 0.1 mm. The sensitivity is thereby limited to the parts-per-hundred range. Of course, the solvent of choice must have the proper optical properties. Among the most useful solvents are carbon disulfide, carbon tetrachloride, methylene chloride, methylene bromide, chloroform, and bromoform. Differential techniques are preferred, particularly for long-path-length cells. It should always be remembered that a solvent which is totally absorbing in certain frequency ranges produces "dead" areas in the spectrum. Such areas cannot be used for analysis.

Collection of Samples and Extraction Procedures

The following technique can be applied to the extraction of aliquots of blood or urine specimens: The solvent is placed in a glass vial fitted with an aluminum- or tin-lined screwcap. If the sample is blood, heparin is added to prevent coagulation. After the blood or urine is added to the vial, the volatile compound is extracted by gentle agitation, and the layers separated by centrifugation. If too vigorous agitation is used an emulsion may form of blood and solvent, which is difficult to separate.

The process of extraction can be enhanced for volatile compounds that are soluble in both water and the solvent by the addition of an anhydrous salt, or by adjustment of the pH of the specimen based on knowledge of the pK of the organic compound.

For sampling purposes, direct injection of blood through a self-sealing rubber stopper is *not* recommended because many volatile organic compounds can penetrate rubber stoppers, thus affording possible contamination from the surrounding atmosphere.

Feldstein (1965a) has also described the analysis of toxic gases in blood. A rapid simple analysis involved aeration of acidified samples of blood into a 10-m path-length infrared cell and measurement at specific wavelengths for carbon tetrachloride, carbon disulfide, ether, ethylene, methane, chloroform, trichloroethylene, cyclopropane, nitrous oxide, benzene, and carbon monoxide. Excess moisture was removed by calcium chloride, which has no effect upon the gases. Normal acidified blood samples show only CO_2 absorption bands. The method has value as a screening procedure, since not only the eleven named compounds but also other volatile organic solvents (petroleum products, turpentine, various halogenated hydrocarbons) also show characteristic absorptions. Feldstein (1965b) pointed out that cyanogen, diazomethane, nitrosyl chloride, and propyne interfere in a determination of blood CO, but these are rarely present in blood and can be ruled out by reference to a complete spectrum. The absorbance at $2174 \, cm^{-1}$ is measured to determine carbon monoxide after it is extracted by sulfuric acid treatment.

Fields of Application

The following disciplines have benefited from infrared analysis of body fluid extracts: (a) toxicology (Stewart *et al.*, 1959b, 1960a, 1961a, 1961b, 1961c, 1962; Stewart and Erley, 1963); (b) industrial and forensic medicine (the papers by Stewart

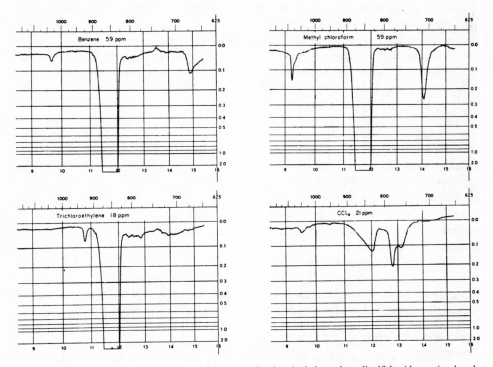

Fig. 18.15. Compensated infrared spectra of four volatile chemicals in carbon disulfide. Absorption bands of benzene are observed at $1036 \, cm^{-1}$ and $671.1 \, cm^{-1}$, methyl chloroform at $1085 \, cm^{-1}$ and $710.2 \, cm^{-1}$, trichloroethylene at $927.6 \, cm^{-1}$, and carbon tetrachloride at $781.3 \, cm^{-1}$ and $762.2 \, cm^{-1}$. Note: the solvent (CS_2) absorbs all the radiation between $877.2 \, cm^{-1}$ and $833.3 \, cm^{-1}$ at this path length (12.5 mm), which renders the instrument "dead" in this region. (Stewart and Erley, 1965.)

et al. already cited, and in addition, Stewart *et al.*, 1963*a*, 1964, 1961*d*; and Erley, 1961); and, (c) pharmacology (Chenoweth *et al.*, 1962).

Figure 18.15 shows four spectra of carbon disulfide extracts of blood samples which contained benzene, 1,1,1-trichloroethane, trichloroethylene, and carbon tetrachloride, respectively. The absorption bands useful for the analyses are: benzene, 1036 cm^{-1}; 1,1,1-trichloroethane, 1085 and 710.2 cm^{-1}; trichloroethylene, 927.6 cm^{-1}; and carbon tetrachloride, 781.3 cm^{-1}.

Tissue Extract Analysis

The method to be described (Robertson and Erley, 1961) has been used to study the content of anesthetics in various organs of anesthetized dogs and monkeys (Chenoweth *et al.*, 1962). Various chlorinated hydrocarbon solvents have also been measured.

The method used is as follows: Five milliliters of carbon disulfide (spectrophotometric purity) is measured into a sealable distillation cup and weighed. About 1 g of freshly cut tissue is placed in this cup, and the combination weighed again to determine sample weight. (This prevents the loss of volatile material during weighing.) The sample is allowed to soak for several hours in a tightly stoppered vessel to achieve quantitative extraction. The cup is attached to a distillation apparatus (see Robertson and Erley, 1961) and warmed in a water bath to 75°C. The receiver is immersed in an ice–brine bath (-10 to $-15°C$) and the distillation continued until no more CS_2 collects. Some water is usually carried over, but is trapped in the receiver. The distillate is drawn out and observed spectrophotometrically in a long-path-length liquid cell. (Use *caution* in distilling CS_2: it has a low flash point of $-30°C$ and an autoignition temperature of 100°C.)

Additional Considerations

A collection of organic vapor spectra determined in one's own laboratory is of particular importance for much of the above work for several reasons: (1) Solid- or liquid-state spectra differ from those of the substances in vapor form. (2) There is no extensive collection of organic vapor spectra. (3) Of the vapor spectra which are available, most have been determined with a 10-cm cell. As a result, if a compound does not have sufficient vapor pressure to give a useful spectrum at this path length, it is not recorded.

One should compare the observed sample spectrum with that of a control sample to ascertain the significance of the absorption bands. Contamination can occur from the sample container, tubing, solvent, etc., and it is therefore wise to run a control sample, especially if one is concerned with trace concentrations.

Analysis of Organs and Organ Contents

Tofranil (a tranquilizer), parathion, endrin, and nicotine (insecticides), amobarbital (a sedative) and disulfiram (used in the treatment of chronic alcoholism) have been detected (Alha *et al.*, 1960) in the stomach, stomach contents, and liver of cadavers by infrared spectroscopy of acetone extracts and by other confirmatory methods.

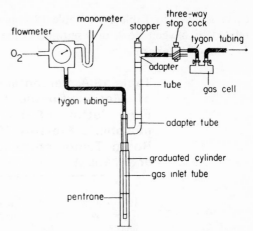

Fig. 18.16. Apparatus used for the standardization procedure. (Paredes *et al.*, 1965.)

Mills *et al.* (1969) have discussed the use of infrared spectroscopy to differentiate brain tissues of patients with Huntington's chorea from those of patients with non-hereditary degenerative disease. (See *Nervous Tissue*, in Chapter 19.)

Use of Gases in Anesthesiology

Paredes *et al.* (1965) have described a method by which it is possible to determine the composition of gaseous mixtures of volatile anesthetics in oxygen flowing from a vaporizer by measuring the absorbance of the mixture at a suitable wavelength. They developed the method for the anesthetic methoxyflurane (2,2-dichloro-1,1-difluoroethyl methyl ether) (Penthrane or Pentrane), but the procedure is of general applicability. Two 10-cm long sodium chloride gas cells were used, the reference-beam cell being filled with oxygen, and the other cell with the sample mixture (anesthetic in oxygen) coming from the anesthesia vaporizer. The apparatus used to standardize the method is depicted in Fig. 18.16 and data used to obtain a standard

Table 18.3. Experimental Data Used to Determine the Standard Curve[a] (Paredes *et al.*, 1965)

Determination number	Oxygen flow, ml/min	Pressure, mm Flow-meter	Pressure, mm Room	Temperature, °K Flow-meter	Temperature, °K Room	Time flow, min	Weight of penthrane vaporized, g	Moles of penthrane	Moles of oxygen	Molar fraction of penthrane, X_m	Average absorbance $\bar{A}$ at 1456 cm^{-1}
1	1,840	675	674	295	295	55.0	4.42	0.0268	3.71	0.00717	0.128
2	1,855	671	671	295	295	33.0	5.39	0.0326	2.24	0.0144	0.237
3	815	673	671	295	295	44.0	6.73	0.0407	1.31	0.0302	0.418
4	407	698	675	296	296	84.3	8.95	0.0542	1.30	0.0402	0.500

[a]For determinations number 1 and 2 the gas inlet tube (Fig. 18.16) was about 1 cm above the surface of the liquid anesthetic. For determination 2 the cylinder containing the anesthetic was submerged in a constant temperature bath at 35°C. For determinations 3 and 4 the inlet tube (Fig. 18.16) was below the surface of the liquid anesthetic and the flowing oxygen bubbled through it. For determination 4 the gas inlet tube had a porous tip to allow the flowing oxygen to come out through the liquid anesthetic in very small bubbles.

curve is given in Table 18.3. Table 18.4 shows absorbances for methoxyflurane as related to oxygen flow, dial setting on the vaporizer, and the percentage molar fraction of the anesthetic.

Table 18.4. Percentage Molar Fraction[a] of Methoxyflurane for Each Oxygen Flow Setting of the Vaporizer at Atmospheric Pressure (674 mm Hg) and Room Temperature (22°C) (Paredes et al., 1965)

Indicated oxygen flow, liters/min	Dial setting	Recorded absorbance 1456 cm^{-1}	Percentage molar fraction of penthrane[a]
6	10	0.275	1.73
6	9	0.275	1.73
6	8	0.260	1.62
6	7	0.232	1.42
6	5	0.107	0.61
6	4	0.055	0.31
6	3	0.013	0.07
4	10	0.300	1.92
4	9	0.280	1.77
4	8	0.290	1.84
4	7	0.261	1.62
4	6	0.150	0.88
4	5	0.099	0.56
4	4	0.034	0.19
2	10	0.345	2.29
2	9	0.330	2.17
2	8	0.340	2.25
2	7	0.285	1.81
2	6	0.110	0.63
2	5	0.010	0.06
2	4	0.054	0.30
0.5	10	0.340	2.25
0.5	9	0.360	2.43
0.5	8	0.360	2.43
0.5	7	0.232	1.42
0.5	6	0.090	0.51
0.5	4	0.038	0.21

[a]The percentage molar fraction is numerically equal to the percentage volumetric composition

Other Medical Applications Using Gas Analyzers

Gas analyzers are used for anesthetic research with N_2O, diethyl ether, halothane ($CF_3CHClBr$), and chloroform (Hill and Powell, 1968). Carbon monoxide analyzers are used in lung-function testing clinics for estimating lung diffusing capacity. Payne *et al.* (1966) used a gas analyzer to determine ethyl alcohol vapor in the breath.

Various conventional medical applications for infrared gas analyzers have been described in the literature: continuous analysis of CO_2 in respired air (Dornhorst *et al.*, 1953); alveolar CO_2 measurement (Collier *et al.*, 1955); measurement of CO_2 in respired gas mixtures (Cullen *et al.*, 1956); measurement of CO_2 in respired gases containing cyclopropane and ether (Linde and Lurie, 1959); and application to anesthesia and respiratory physiology (Powell, 1965).

Determination of Carboxyhemoglobin in Blood

Coburn (1964) has described an infrared method for the measurement of blood carboxyhemoglobin. Bound CO is liberated by treating a hemolyzed blood solution with acid ferricyanide, extracted by means of any oxygen washout technique, collected in a tonometer, and measured in an infrared CO meter. The result is compared with that of a known gas standard containing 0.007% CO. Oxygen is used as the baseline gas. The method is highly specific for CO and can detect ± 0.006 ml of CO per 100 ml blood, or approximately $\pm 0.02\%$ carboxyhemoglobin.

Determination of Carbon Dioxide in Blood

Gimeno Ortega *et al.* (1966) have adapted an infrared analyzer for the determination of carbon dioxide content of blood. They tested the apparatus with various carbon dioxide tensions and compared the results with those obtained with the conventional Van Slyke–Neill technique. The reproducibility of the infrared method was found to be better than the Van Slyke–Neill method under comparable operating conditions. A complete infrared analysis takes only 5 min, but the main disadvantage of the method is its limitation to carbon dioxide.

Figure 18.17 shows the reproducibility of the infrared method. It is a plot of carbon dioxide values for the first and second determinations on 84 blood samples. Figure 18.18 shows a comparison between the carbon dioxide content of 26 blood samples determined by the infrared method and the carbon dioxide content as determined by the Van Slyke–Neill method. Gimeno Ortega *et al.* point out that the infrared method should not be used for samples obtained during surgery because anesthetic gases may interfere.

Other Examples of the Use of an Infrared Gas Analyzer

Many examples of biological applications of the infrared gas analyzer have been mentioned by Hill and Powell (1968). Carbon dioxide production by locusts has been studied by Hamilton (1959, 1964). Romijn and Lokhorst (1961) examined expired air of a hen for CO_2 in studying the hen's heat-regulating mechanism. The CO_2 resulting from the combustion of small amounts of organic carbon in fresh and saline water has been measured by Montgomery and Thom (1962).

Hill and Powell (1968) have also discussed the measurement of CO_2 in human subjects. A CO_2-analyzer using solid-state circuitry, capable of following 30 breaths/ min, was described by Hill and Stone (1964).

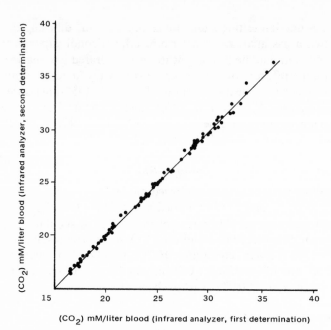

Fig. 18.17. Carbon dioxide content of 84 blood samples determined by the infrared method in duplicate. (Gimeno Ortega *et al.*, 1966.)

Hill and Powell (1968) have recently written a comprehensive text on non-dispersive infrared gas analysis. They have discussed applications and sampling techniques in science, medicine, and industry; instrumentation and detecting systems; and methods for producing calibration gas and vapor mixtures.

In confined spaces it is necessary to keep the concentration of CO from automobile exhaust down to acceptable limits. Therefore, CO infrared gas analyzers have been installed on a large scale in such places (e.g., tunnels, parking areas, etc.). In medicine, respiratory units for lung-function investigations have used infrared analyzers. The use of infrared methods for qualitative and quantitative information on a variety of gases and vapors has been discussed earlier in this chapter in connection with anesthesiology and toxicology applications.

Biologically important industrial uses of infrared gas analyzers are the measurement of air pollution and the monitoring of atmospheres (fruit and crop production and storage). It is interesting to note that the use of a CO_2 blanket inhibits deterioration of a banana cargo and apples stored in an atmosphere containing 1000 ppm v/v CO_2 will keep almost indefinitely.

Monitoring of Environments

Air Pollution and Analysis

A recent comprehensive report has been published by the American Chemical Society (1969) on the widespread problems of air and water pollution, of pesticides in

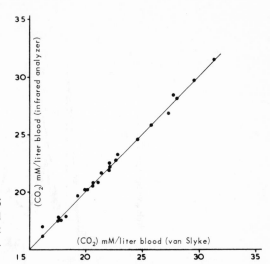

Fig. 18.18. Carbon dioxide content of 26 blood samples determined by the infrared method and compared with the CO_2 content as determined by the Van Slyke–Neill method. (Gimeno Ortega *et al.*, 1966.)

the environment, and of solid waste disposal. The Subcommittee on Environmental Improvement has discussed in detail what some of the problems are, what needs to be done, and how to go about attacking some of the difficulties in which we now find ourselves. The above report is cited because it gives about 500 references dealing with a wide variety of the technical problems, and makes many recommendations for approaching solutions to them.

A recent review (Altshuller, 1969) has discussed papers on the topic of air pollution published through 1968. Among the topics referred to were the following: determination of carbon dioxide by infrared gas analyzers; description of a method for obtaining the infrared spectrum of microgram quantities of atmospheric dust; use of a cryogenic sample technique combined with infrared spectrophotometry for the analysis of multicomponent mixtures of C_1 to C_4 hydrocarbons; the use of a carbon dioxide laser to stimulate aliphatic hydrocarbons to emit infrared radiation (thereby permitting detection); the use of infrared instruments for detection of sulfur dioxide and nitrogen dioxide in exhaust stack gases. In this review (Altshuller, 1969) many references were given to the types of air pollution studies going on all over the world. Another review (Altshuller, 1967) on this topic has appeared earlier.

Low (1967) has characterized air pollutants at a distance from the site of measurement. He coupled a multiple-scan interferometer to a telescope pointed at a smokestack about 600 feet away and analyzed the infrared radiation emitted by the hot smoke of the stack. He readily detected sulfur dioxide (Fig. 18.19), and it should be noted that the measurements were done at 11 P.M., a time when ordinary methods are not possible.

Hanst (1970), in a review of the use of spectroscopic methods in the analysis of the atmosphere, has emphasized the special advantages of infrared vibration–rotation spectra for identification and measurement of air pollutants. He has discussed the capabilities of lasers in atmospheric work, giving particular attention to proposed methods of detecting pollutants by absorption of selected infrared laser lines. Quoting

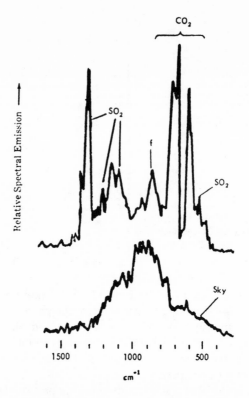

Fig. 18.19. The spectrum of the stack gases is corrected
for sky emission. The peak, f, is probably due to unburnt
or partially burnt fuel. It took 100 sec (for 100 scans)
to make this record. (Low, 1967.)

the work of Scott *et al.* (1957), Hanst has shown the type of spectra that were obtained
for pollutants in Los Angeles air in 1956 by means of cells with folded path lengths of
several hundred meters. Among these pollutants were ozone, carbon monoxide,
ethylene, propylene, and acetylene.

Among the laser lines discussed by Hanst (1970) were the CO_2, He–Ne, I_2,
Ne, Kr, and Xe lines; and the pollutants which can be (or in some cases may in the
future be) detected by absorption of laser radiation were CH_4, C_2H_2, C_2H_4, C_4H_{10},
peroxy acetyl nitrate (PAN), peroxy benzoyl nitrate, peroxy butyl nitrate, peroxy
propyl nitrate, NO, NO_2, CO, O_3, SO_2, NH_3, H_2S, HF, HCl, and D_2O. The sources
of some of these pollutants are as follows: CH_4 comes from marsh gas; C_2H_2, C_2H_4,
C_4H_{10}, NO, NO_2, and CO from auto exhaust and combustion; PAN from atmospheric photochemistry; O_3 from atmospheric photochemistry; SO_2 from the combustion of S-containing fuels; NH_3 from organic wastes and industry; H_2S from
industry; HF and HCl from the burning of plastics; and D_2O from atomic energy
installations.

Hinkley and Kelley (1971) have described the use of tunable $Pb_{1-x}Sn_xTe$ semi-conductor diode lasers in remote or long-range sensing and point sampling of molecular pollutant gases. An example of the use of a $Pb_{0.88}Sn_{0.12}Te$ diode laser for point sampling is its application in the 948 cm^{-1} region to detect ammonia. Another use is the application to the detection of ethylene in automobile exhaust specimens. Specialized infrared techniques may be used with a $Pb_{1-x}Sn_xTe$ diode laser to detect spectral emission lines from pollutants present in smokestack effluent.

Water Pollution

Parker (1971) has discussed various aspects of infrared spectroscopy which are of use in studies of water pollution and in the analysis of water. Rosen (1967) has also discussed water pollution and its control. He has mentioned the problems involved in obtaining samples of polluted water on a routine basis. Mattson *et al.* (1970) have used the ATR technique to identify crude oils and tars contaminating the waters of the Santa Barbara Channel.

Fishman and Erdmann (1971) have recently reviewed the topic of water analysis. Among the infrared methods mentioned were the following: the use of internal reflection spectroscopy for the analysis of optically opaque samples; the determination of total carbon in waste waters; the quantitative determination of total CO_2 in sea water; the determination of the deuterium content in water; the use of spectra of hydrocarbon fractions from the acid part of organic substances in natural waters; and the use of infrared for the identification of organic substances in water.

Analysis of Space-Flight and Submarine Atmospheres

Toliver and Morris (1966) performed chemical analyses of the permanent gases and the trace volatile organic constituents on a 30-day manned space flight experiment with a Mercury spacecraft. The experiment was primarily concerned with the feasibility of providing a suitable atmosphere for three men for a period of 30 days. Infrared and mass spectrometry were used as adjuncts to gas chromatographic isolation of the gaseous materials. Among the 29 volatile atmospheric contaminants recovered and identified were such compounds as carbon dioxide, ethylene, acetylene, propylene, butene-1, ethyl formate, various hydrocarbons (weak infrared absorbers and therefore not specifically identified), ethyl alcohol, ethyl acetate, benzene, toluene, butanol, acetone, butyric acid, and formaldehyde. Many of the compounds identified had been previously found among 59 atmospheric contaminants in the Mercury spacecraft. Also, carbon dioxide partial pressures have been determined in simulated space vehicle atmospheres by one of the major instrument companies (Hill and Powell, 1968). Special purpose infrared gas analyzers have been devised for nuclear submarine and manned spacecraft applications.

REFERENCES

Alha, A. R., Tamminen, V., Mukula, A. -L., Levonen, E., Ruohonen, A., and Salomaa, E. *Arch. Toxikologie* **18**, 347 (1960).

Altshuller, A. P. *Anal. Chem.* **39**, 10R (1967).

Altshuller, A. P. *Anal. Chem.* **41**, 1R (1969).

American Chemical Society, *Cleaning Our Environment, The Chemical Basis for Action*, A Report by the Subcommittee on Environmental Improvement, Committee on Chemistry and Public Affairs, Washington, D. C., 1969.

Chenoweth, M. B., Robertson, D. N., Erley, D. S., and Golhke, R. *Anesthesiology* **23**, 101 (1962).

Chihara, G., Yamamoto, S., and Kameda, H. *Chem. Pharm. Bull.* **6**, 50 (1958).

Clausen, J., Dyggve, H. V., and Melchior, J. C. *Arch. Dis. Childh.* **38**, 364 (1963).

Coburn, R. F. in *Hemoglobin: Its Precursors and Metabolites* (F. W. Sunderman and F. W. Sunderman, Jr., eds.) J. B. Lippincott Company, Philadelphia, Pa., 1964, Section I. 6B, p. 67.

Collier, C. R., Affeldt, J. E., and Farr, A. F. *J. Lab. Clin. Med.* **45**, 526 (1955).

Comaish, S. *Brit. J. Dermatol.* **80**, 522 (1968).

Cullen, W. G., Brindle, G. F., and Griffith, H. R. *Can. Anaesth. Soc. J.* **3**, 81 (1956).

Dornhorst, A. C., Semple, S. J. G., and Young, I. M. *Lancet* **1**, 370 (1953).

Edwards, J. D., Jr., Adams, W. D., and Halpert, B. *Am. J. Clin. Pathol.* **29**, 236 (1958).

Erley, D. S. *Appl. Spectrosc.* **15**, 80 (1961).

Erley, D. S., Blake, B. H., and Potts, W. J. *Appl. Spectrosc.* **14**, 108 (1960).

Feldstein, M. *J. Forensic Sci.* **10**, 207 (1965a).

Feldstein, M. *J. Forensic Sci.* **10**, 43 (1965b).

Fishman, M. J. and Erdmann, D. E. *Anal. Chem., Ann. Rev.* **43** (5), 356R (1971).

Gimeno Ortega, F., Orie, S. A. M., and Tammeling, G. J. *J. Appl. Physiol.* **21**, 1377 (1966).

Hamilton, A. G. *Nature* **184**, 367 (1959).

Hamilton, A. G. *Proc. Roy. Soc.* **160B**, 373 (1964).

Hanst. P. L. *Appl. Spectrosc.* **24**, 161 (1970).

Hatch, F. T., Freeman, N. K., Jensen, L. C., Stevens, G. R., and Lindgren, F. T. *Lipids* **2**, 183 (1967).

Hill, D. W. and Stone, R. N. *J. Sci. Instrum.* **41**, 732 (1964).

Hill, D. W. and Powell, T. *Non-Dispersive Infra-red Gas Analysis in Science, Medicine, and Industry*, Plenum Press, New York, 1968.

Hinkley, E. D. and Kelley, P. L. *Science* **171**, 635 (1971).

Klein, B., Weissman, M., and Berkowitz, J. *Clinical Chem.* **6**, 453 (1960).

Klein, B. and Weissman, M. *Clin. Chem.* **10**, 741 (1964).

Linde, H. W. and Lurie, A. A. *Anesthesiology* **20**, 45 (1959).

Low, M. J. D. *Science and Technology*, February, 1967

Lyman, D. J. and Brash, J. R. cited in *Chem. and Eng. News.* **47**, 37, Jan. 27, 1969.

Mattson, J. S., Mark, H. B., Jr., Kolpack, R. L., and Schutt, C. E. *Anal. Chem.* **42**, 234 (1970).

Mills, J., Orr, A., and Whittier, J. R. *J. Neurochem.* **16**, 1033 (1969).

Montgomery, H. A. C. and Thom, N. S. *Analyst (London)* **87**, 1038 (1962).

Paredes, R., Zapata, A., Palacios, N., and Tejada, V. *Anesthesiology* **26**, 107 (1965).

Parker, F. S. "Infrared Spectroscopy in Water Analysis," in *Water and Water Pollution Handbook, Vol. 4* (L. Ciaccio, ed.) Marcel Dekker, New York, 1971, Chap. 30.

Payne, J. P., Hill, D. W., and King, N. W. *Brit. Med. J.* **1**, 196 (1966).

Perry, T. L., Richardson, K. S. C., Hansen, S., and Friesen, A. J. D. *J. Biol. Chem.* **240**, 4540 (1965).

Powell, T. *Wld. Med. Electron.* **3**, 8 (1965).

Robertson, D. N., and Erley, D. S. *Anal. Biochem.* **2**, 45 (1961).

Romijn, C. and Lokhorst, W. *Tijdschr. Diergeneesk.* **3**, 153 (1961).

Rosen, A. A. *Anal. Chem.* **39**, No. 12, 26A (1967).

Rozelle, L. T., Hallgren, L. J., Bransford, J. E., and Koch, R. B. *Appl. Spectrosc.* **19**, 120 (1965).

Scott, W. E., Stephens, E. R., Hanst, P. L., and Doerr, R. C. *Proc. API* **37**, 171 (1957).

Stansell, M. J. and Mojica, L., Jr. *Clin. Chem.* **14**, 1112 (1968).

Stewart, R. D. and Boettner, E. A. *New England J. Med.* **270**, 1035 (1964).

Stewart, R. D. and Erley, D. S. *Exptl. Cell Research* **23**, 460 (1961).

Stewart, R. D. and Erley, D. S. *J. Forensic Sci.* **8**, 31 (1963).

Stewart, R. D. and Erley, D. S. in *Progress in Chem. Toxicology, Vol. 2* (A. Stolman, ed.) Academic Press, New York, 1965, p. 183.

Stewart, R. D., Erley, D. S., Skelly, N. E., and Wright, N. *J. Lab. Clin. Med.* **54**, 644 (1959a).

Stewart, R. D., Erley, D. S., Torkelson, T. R., and Hake, C. L. *Nature* **184**, 192 (1959b).

Stewart, R. D., Skelly, N. E., and Erley, D. S. *J. Lab. Clin. Med.* **56**, 391 (1960*b*).

Stewart, R. D., Torkelson, T. R., Hake, C. L., and Erley, D. S. *J. Lab. Clin. Med.* **56**, 148 (1960*a*).

Stewart, R. D., Gay, H. H., Erley, D. S., Hake, C. L., and Schaffer, A. W. *Arch. Environ. Health* **2**, 516 (1961*a*).

Stewart, R. D., Gay, H. H., Erley, D. S., Hake, C. L., and Peterson, J. E. *J. Occupational Med.* **3**, 586 (1961*b*).

Stewart, R. D., Gay, H. H., Erley, D. S., Hake, C. L., and Schaffer, A. W. *Am. Ind. Hyg. Assoc. J.* **22**, 252 (1961*c*).

Stewart, R. D., Boettner, E. A., and Stubbs, B. T. *Nature* **191**, 1008 (1961*d*).

Stewart, R. D., Gay, H. H., Erley, D. S., Hake, C. L., and Peterson, J. E. *Am. Ind. Hyg. Assoc. J.* **23**, 167 (1962).

Stewart, R. D., Boettner, E. A., Southworth, R. R., and Cerney, J. C. *J. Amer. Med. Assoc.* **183**, 994 (1963).

Suzuki, N. *Tohoku J. Exp. Med.* **90**, 195 (1966).

Suzuki, N. and Toyoda, M. *Tohoku J. Exp. Med.* **88**, 353 (1966).

Toliver, W. H. and Morris, M. L. *Aerospace Med.* **37**, 233 (1966).

Toyoda, M. *Tohoku J. Exp. Med.* **90**, 303 (1966).

Weissman, M., Klein, B., and Berkowitz, J. *Anal. Chem.* **31**, 1334 (1959).

Wegmann, R. and Thewes, A. *La Presse Médicale* **1956**, 1158.

Wells, W. W., Pittman, T. A., and Egan, T. J. *J. Biol. Chem.* **239**, 3192 (1964).

White, A., Handler, P., and Smith, E. L. *Principles of Biochemistry*, *4th Ed.*, McGraw–Hill, New York, 1968.

Chapter 19

SPECIAL TOPICS OF INTEREST IN BIOCHEMISTRY AND RELATED FIELDS

Up to this point we have attempted to categorize topics under single headings. There are, however, many other areas in which infrared spectroscopy has been used that are of interest. This is not to say that any of these areas is less important or needs less emphasis in research laboratories than do other topics. Several of the topics mentioned here have had books devoted to them. The attempt in this chapter is to bring to the attention of the reader the wide variety of areas in which infrared spectroscopy, along with other tools, can offer information to research scientists.

Water Structure

The absorption bands of water have been tabulated and their origins summarized by Eisenberg and Kauzmann (1969). Falk and Ford (1966) have studied the infrared spectrum of HDO at low concentrations in H_2O and in D_2O between 0 and 130°C. Each of the three fundamentals of HDO (~ 3400, ~ 2500, and ~ 1450 cm^{-1}) showed nearly Gaussian contours, complete absence of shoulders, and a single maximum at a frequency intermediate between those of HDO in ice and vapor. Increase of temperature caused a gradual shift of each band in the direction of the frequency in vapor. These observations were evidence in full support of the "continuum" models of liquid water structure, and in dispute with workers citing evidence for "mixture" models. "Continuum models" have been defined by Falk and Ford as those which describe liquid water as an essentially complete hydrogen-bonded network, with a distribution of hydrogen bond energies and geometries. The average strengths of hydrogen bonds are considered in these models to be weaker than in ice as a result of irregular distortion and elongation, both of which increase with temperature.

"Mixture models" have been defined (Falk and Ford, 1966) as those which describe liquid water as an equilibrium mixture of molecular species with different numbers of hydrogen bonds per molecule. At any temperature there would be a well-defined proportion of hydrogen bonds.

Falk and Ford concluded that hydrogen bonds in liquid water have a broad smooth single-peaked distribution of strengths which gradually shifts with temperature. At one end of this distribution the hydrogen bond strengths are comparable to those in ice, while those at the other end are very weak. These extremes do not represent distinct molecular species, which are characteristic of the "mixture models." Falk and Ford expressed the opinion that Hartman's (1966) data supporting an upper limit of 7% for HDO in D_2O at 25°C should not be taken as spectroscopic evidence of up to 7% of some *distinct molecular species* with non-hydrogen-bonded OH groups. Falk and Ford's data stressed the absence of such species as shown by the lack of any shoulders in the region of 3635 cm^{-1} in their spectra and in Hartman's (the region presumed by Hartman to show absorption for "non-hydrogen-bonded OH groups").

Frequency assignments for "associational bands" found in part of the infrared spectra of water and of D_2O were proposed by Williams (1966). The bands were at 2130, 3950, and 5600 cm^{-1} for H_2O; and 1555, 2900, and 4100 cm^{-1} for D_2O.

Draegert *et al.* (1966) investigated the far-infrared spectrum of liquid water between 30 cm^{-1} and 1200 cm^{-1}. They reported a broad major band with maximum absorption at 685 $\pm$ 15 cm^{-1} together with a second, less intense overlapping band at 170 $\pm$ 15 cm^{-1}. In liquid D_2O the corresponding bands appeared at 505 $\pm$ 15 cm^{-1} and 165 $\pm$ 15 cm^{-1}, respectively.

Polywater

Deryagin and Churayev (1968) reported on a form of water with properties different from those well established for water, and the new form has been referred to as "anomalous water." This water has been prepared by Fedyakin (1962) in a sealed glass capillary 2 to 4 μ in diameter and later by Deryagin *et al.* (1965) by the condensation of water vapor in glass and fused quartz capillaries at relative pressures somewhat less than unity. Among the properties of this water, renamed "polywater" by Lippincott *et al.* (1969), are: (1) low vapor pressure; (2) solidification at -40°C or lower temperatures to a glass-like state with a substantially lower expansion than that of ordinary water when it freezes; and (3) a density of 1.01 to 1.4 g/cm^3 and stability to temperatures of the order of 500°C.

The mid-infrared spectrum of polywater removed from fused quartz capillaries was obtained (Lippincott *et al.*, 1969) by use of a miniaturized diamond cell with a thin diamond platelet and a beam-condensing system. Spectra are given in Fig. 19.1 where the solid curves are the spectrum of polywater and the dashed curve is the spectrum of the diamond background. The spectrum of polywater has several remarkable features and appears to be unique. (Note contrary data on p. 472.) The unusual properties include the complete absence of absorption from 4000 to 2500 cm^{-1}, the presence of a strong band near 1595 cm^{-1}, and an intense doublet in the 1400 cm^{-1} region. The region 2500–1950 cm^{-1} is partially obscured by strong diamond absorption. Close examination of this region, not shown in Fig. 19.1(a), shows that it contains no appreciable absorption bands coming from the polywater. Repeat runs of the spectrum after a few days showed no significant changes.

Although the vibrational spectra of polywater appear to be completely unique and different from those of any known substance (Lippincott *et al.*, 1969), these spectra have several features that are quite similar to those of hydrogen-bonded systems which have very strong symmetric hydrogen bonds such as those occurring in KHF$_2$ and

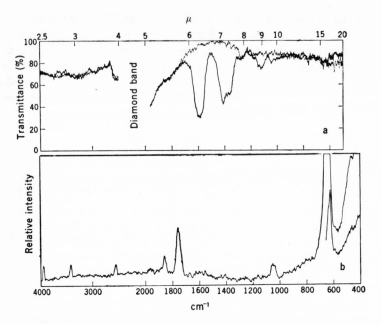

Fig. 19.1. Infrared and Raman spectra of polywater. (a) Infrared spectrum of polywater after transfer from quartz capillary onto diamond platelets. (——) polywater; (————) diamond background. No OH absorption bands between 4000 and 3000 cm^{-1} are observed. New strong bands appear near 1600 and 1400 cm^{-1}. (b) Raman spectrum of polywater obtained in quartz capillary, with 4880-Å argon laser for excitation. Curve 1 obtained at very high gain, curve 2 at a lower gain sufficient to bring band near 620 cm^{-1} on scale. At this gain, all other bands are extremely weak. The weak band at 1050 cm^{-1} is due to the quartz capillary. (Lippincott *et al.*, 1969.)

$HCrO_2$. In KHF_2 the bifluoride ion has a linear configuration with an F—F distance of 2.26 Å and an F—H distance of 1.13 Å. The hydrogen bond energy of the FHF^- system has been given as 58 ± 5 kcal/mole in KHF_2 and as 37 ± 2 kcal/mole in $(CH_3)_4N—HF_2$; it is the strongest known hydrogen-bond system (Waddington, 1958; Harrell and McDaniel, 1964). Lippincott *et al.*, observing that the structural unit O—H—O is isoelectronic with the FHF^- ion, interpreted the major features of the observed spectra on the basis of a structure that would be consistent with the unusual properties of polywater. They believe that the basic structural unit of polywater is an exceedingly strong O—H—O three-center bond, isoelectronic with the bifluoride ion. Whereas the O···O distance in normal water is about 2.8 Å with a hydrogen-bond energy of ~4 kcal/mole, the O···O distance in polywater must be near 2.3 Å, with a hydrogen-bond energy of 30–50 kcal/bond unit or 60–100 kcal/water formula unit. Figure 19.2 shows a possible structure involving the monomer, H_2O, in an extended network system of strong O—H—O bonds.

Donohue (1969) has proposed a structure for polywater which consists of hydrogen-bonded clusters of water molecules lying at the vertices of rhombic dodecahedra.

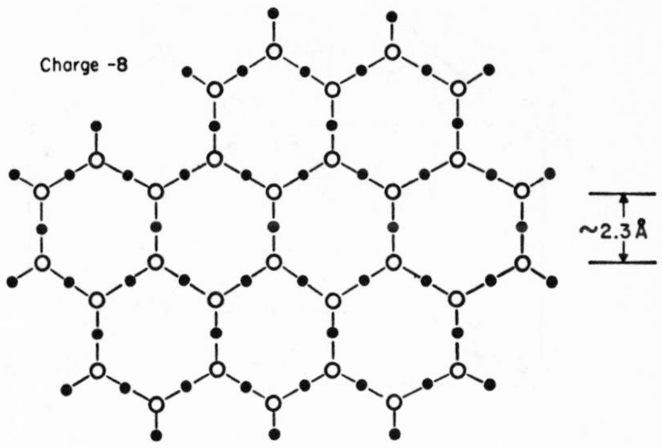

Fig. 19.2. Structural diagram for polywater consisting of a network of hexagonal units. This structure, as drawn, would have a negative charge. An appropriate number of hydronium ions, protons, or tetra-coordinated hydrogen-bonded oxygen atoms would be required to maintain the empirical formula $(H_2O)_n$. (Lippincott *et al.*, 1969.)

Charge −8

~2.3 Å

According to Donohue this structure contains features which are less unattractive than other models proposed earlier by other investigators. Lippincott *et al.* (1969) assumed that the O to O distances in symmetrical $O\cdots H\cdots O$ bonds would be 2.3 Å, and Donohue claims this value is too low. He used 2.44 Å, a figure derived from O to O distances in potassium hydrogen malonate, potassium hydrogen chloromaleate, chromous acid, and potassium hydrogen maleate. The increase in the O to O distance from 2.3 to 2.44 Å reduces the maximum density of 1.40 g/cm^3 calculated by Lippincott *et al.* for the planar structures to 1.24, a value considerably smaller than the maximum value of 1.4 reported by Deryagin and Churayev (1968). Figure 19.3 shows two canonical forms of a hydrogen-bonded rhombic dodecahedral cluster of $(H_2O)_{14}$. Figure 19.4 shows two canonical forms of part of a hydrogen-bonded supercluster, the larger polymeric units.

Page *et al.* (1970) have presented NMR and infrared spectra of polywater. The infrared spectrum of polywater prepared by a new method confirmed the infrared

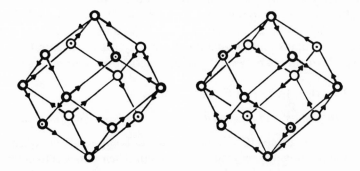

Fig. 19.3. Two canonical forms of a hydrogen-bonded rhombic dodecahedral cluster of $(H_2O)_{14}$. Circles represent oxygen atoms; arrows represent hydrogen atoms. The oxygen atoms which are bonded to the four hydrogen atoms not used in the formation of the cluster are indicated by the dots. (Donohue, 1969.)

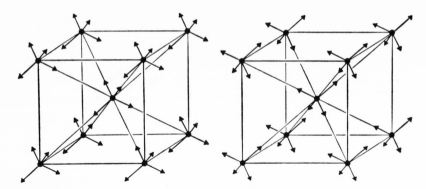

Fig. 19.4. Two canonical forms of part of a hydrogen-bonded supercluster. Filled circles represent rhombic dodecahedral clusters; arrows represent the "unused" hydrogen atoms of Fig. 19.3. (Donohue, 1969.)

spectrum reported earlier (Lippincott *et al.*, 1969). An NMR spectrum from another laboratory (Petsko, 1970) is in agreement with the hexagonal ring structure.

At the recent International Conference on Polywater, Davis *et al.* (1970) have reported that electron spectroscopic chemical analysis (ESCA) showed polywater to be a mixture of hard-to-analyze substances leached out of glass surfaces. Analysis of 15 samples of polywater showed that it contained 95% or more by weight of sodium, potassium, sulfate, carbonate, chloride, nitrate, borates, and silicates with traces of other impurities containing organic carbon. Davis gave evidence showing that poly-H_2O and poly-D_2O have the same infrared spectra. The ESCA studies have, according to Davis, confirmed the views of Rousseau and Porto (1970) and of Kurtin *et al.* (1970) that polywater is nothing more than hydrosols or gels of impurities with very little water present. [See also Davis *et al.* (1971)]. Rousseau (1971) has shown that the infrared spectrum of polywater is remarkably similar to the infrared spectrum of sodium lactate, the primary constituent of sweat. He proposes that this property of polywater, and possibly others, results from accidental contamination. It is interesting to note that Tal'rose [cited by Rousseau (1971)] did mass spectrometric analyses on 25 of Deryagin's samples and all contained organic substances, including lipids and phospholipids, in very high concentrations. The controversy over the reality of polywater as a polymeric form of water is still continuing.

Aqueous Solution Infrared Spectroscopy

Carbonyl Frequencies and Reactivity toward Nucleophilic Reagents

A series of acyl groups (CXX) in a variety of compounds in aqueous solution (D_2O) has been investigated by Jencks *et al.* (1960). Their aim was to determine what correlations might exist between the carbonyl stretching frequency and the reactivity toward nucleophilic reagents, or the stretching frequency and the "energy-rich" nature of activated acyl groups as reflected in a high free energy of hydrolysis. We might expect such correlations in certain instances, because the frequency of the carbonyl absorption is a measure of the length of the C=O bond (Margoshes *et al.*, 1954), and the length of this bond (Hartwell *et al.*, 1948; Freeman, 1958; Kagarise, 1955; Bellamy, 1955), its susceptibility to nucleophilic attack, and its free energy of hydrolysis are all influenced by inductive and resonance effects of neighboring atoms. Among the structures contributing to the carbonyl group, forms CXX*b* and CXX*c* give more single-bond character to the C—O bond and hence a longer bond and smaller carbonyl absorption frequency (Hartwell *et al.*, 1948). Inductive effects

CXX

which favor form CXX*b* by electron donation and resonance effects that increase the single-bond character of the C—O bond by contributions of forms such as CXX*c* decrease reactivity toward nucleophilic reagents by decreasing the positive character of the carbon atom and by stabilizing the carbonyl group through resonance. The same effects decrease the free energy of hydrolysis by stabilizing the reactants (which have a strong carbonyl dipole and a relatively large amount of carbonyl double-bond character) more than the products (such as $RCOO^-$), which have a much smaller dipole and less double-bond character, and accordingly are relatively less subject to such stabilization.

The maximal carbonyl frequencies of the compounds investigated by Jencks *et al.* (1960) are given in Tables 19.1 through 19.6, grouped by chemical class. Other absorption frequencies are also listed, but are not necessarily repeated for compounds that are listed in more than one table. Carbonyl absorption bands have strong intensity; other bands are strong unless indicated otherwise (see footnote *d* of Table 19.1).

The first group of compounds discussed were the esters and acids. The carbonyl band of ethyl acetate, a typical "low-energy" ester, lies at 1710 cm^{-1} in D_2O and is strong (Table 19.1). This was compared to the values of 1765 cm^{-1} reported for ethyl acetate in the vapor state (Hartwell *et al.*, 1948), 1742 cm^{-1} in carbon tetrachloride solution, and 1733 cm^{-1} in carbon tetrachloride solution containing methanol (Searles *et al.*, 1953). Similar shifts to low frequencies in D_2O solution, compared to the vapor state or to solutions in solvents with low hydrogen-bonding capacity, are apparent for several other esters and acids given in Table 19.1. When an electron-withdrawing group is added to either the alcohol moiety, as in acetylcholine (1735 cm^{-1}), *N,O*-diacetylserinamide (1728 cm^{-1}), and 2'(3')-acetyl 5'-adenylate (1737 cm^{-1}) or to the acid moiety, as in ethyl trifluoroacetate (1786 cm^{-1}) and amino acid ester hydrochlorides (1740–1751 cm^{-1}), increases occur in the carbonyl absorption frequency. The carbonyl frequency of isopropenyl acetate, a strong acylating agent, displays an increase to 1728 cm^{-1}, perhaps due partly to inhibition of resonance to form CXX*c* because of competition for the resonance of the free electron pair of oxygen by the carbon–carbon double bond and partly to the electron-withdrawing effect of sp^2 hybridization of the carbon atom adjacent to this oxygen (Jencks *et al.*,

Table 19.1. Absorption Bands, in cm^{-1}, of Esters and Acids, $\overset{\overset{\text{O}}{\|}}{\text{RCOR'}}$ (Jencks *et al.,* 1960)

Compound	Carbonyl frequency	C—O	Other bands[d]
Ethyl acetate	1710	1275b	2970m, 1470w, 1448w,[a] 1397sh, 1380[a]
Acetylcholine chloride	1735	1255b	2960w, 1483, 1460sh, 1425w,[a] 1383[a]
N,O-Diacetylserinamide	1728	1250b	1638b, 1427,[a] 1387,[a] 1370w, 1333w
Ethyl trifluoroacetate	1786		1450m, 1371, 1345
2'(3')-Acetyl 5'-adenylate	1737		2940w, 1700w, 1627, 1562, 1484m, 1419, 1382w,[a] 1342w, 1307w
Alanine ethyl ester hydrochloride	1743		2950w, 1462, 1445, 1393w, 1359w, 1320m
Serine ethyl ester hydrochloride	1740		2950w, 1450b, 1410w, 1380m, 1350w, 1308
Histidine ethyl ester dihydro-chloride	1751		3130w, 1600w, 1449, 1408w, 1300w
Isopropenyl acetate	1728	1271b	1432w, 1680m, 1380[a]
Ethyl carbamate	1681	1347[b]	2950w, 1472m, 1382
O-Carbamylhydroxylamine	1708	1390[b]	
Acetic acid[c]	1710	1318	1430w,[a] 1382[a]
Pyruvic acid	1725	1351m[b]	1440m, 1382m[b]
α-Ketoglutaric acid	1722b	1330m	1397[a]
Glutamine hydrochloride	1729		
Asparagine hydrochloride	1730		

[a] C—H *alpha* to C=O.
[b] Assignment uncertain.
[c] Raman bands at 1720, 1436, 1370, and 1272 cm.$^{-1}$ (Edsall, 1936).
[d] Key to Tables 19.1–19.5: b, broad; m, medium; sh, shoulder; w, weak. The bands indicated are strong (Tables 19.1–19.6) unless indicated by the above code to be otherwise.

1960). A nitrogen atom adjacent to the carbonyl group, as in ethyl carbamate, gives amide character to the bond with a marked decrease of the carbonyl frequency because of resonance to form CXXc (X is N). In *O*-carbamylhydroxylamine this effect is largely offset by the increased electron-withdrawing effect of the second nitrogen atom. Thus, this compound has a band very near to that of ethyl acetate.

The carbonyl band of acetic acid in D$_2$O lies at 1710 cm^{-1}, the same as that in ethyl acetate and similar to the frequency of hydrogen-bonded dimers of carboxylic acid. This 1710 cm^{-1} band is much lower than that of the acetic acid monomer in the vapor state at 1785 cm^{-1} (Hartwell *et al.,* 1948). Pyruvic and α-ketoglutaric acids and glutamine and asparagine hydrochlorides absorb at 12 to 20 cm^{-1} higher, because of the electron-withdrawing effects of the adjacent carbonyl and ammonium groups (Edsall, 1936), effects which also show up in the greater acid strength of these acids. Jencks *et al.* found that in most of the compounds they had investigated, an intense C—O stretching band (coupled with the O—D band in the case of carboxylic acids) was evident, but because of the inherently broad character of this band and the fact that it occurs in a technically poor region of the spectrum under the conditions of measurement, the frequencies given are only approximate. In most cases bands from

C—H bending α to a carbonyl group appear and are usually sharper than the carbonyl bands, which are broader, probably because of their varying amounts of interaction with the solvent.

The second group of compounds studied by Jencks *et al.* were acyl phosphates (Table 19.2). The acetyl phosphate dianion, which is the predominant ionic species at neutral *p*H, has a carbonyl band at 1713 cm^{-1}. Acylated phosphate monoesters, such as acetyl adenylate and acetyl phenyl phosphate absorb at higher frequencies of 1737 to 1747 cm^{-1}, and the structurally similar acetyl phosphate monoanion absorbs

Table 19.2. Absorption Bands, in cm^{-1}, of Acyl Phosphates, $\underset{\text{RCOPO}_3\text{R}'}{\overset{\text{O}}{\overset{\|}{}}}$ (Jencks *et al.,* 1960)

Compound	Carbonyl frequency	$\underset{\text{H—C—C}}{\overset{\text{O}}{\overset{\|}{}}}$	Other bands
Lithium acetyl phosphate, dianion	1713	1374	1425w, 1275
Lithium acetyl phosphate, monoanion	1740	1377	1240
Acetyl adenylate	1737	1372	2922m, 1620, 1577w, 1480m, 1427m, 1337w, 1299w
Acetyl phenyl phosphate	1747	1374	1598m, 1490
Carbamyl phosphate	1670		1405

at 1740 cm^{-1}. Cramer and Gärtner (1958) reported acetyl diethyl phosphate to have a carbonyl band even higher, at 1780 cm^{-1}. Carbamyl phosphate absorbs at a much lower frequency, probably because the carbonyl group has had amide character introduced.

The thiol esters examined by Jencks *et al.,* including *S*-acetylglutathione, all have the C=O band near 1675 cm^{-1} (Table 19.3). Thiol esters also absorb maximally near 1675 cm^{-1} in nonpolar solvents. Hauptschein *et al.* (1952) have reported this absorption to be insensitive to substituent effects.

Table 19.3. Absorption Bands, in cm^{-1}, of Thiol Esters, $\underset{\text{RCSR}'}{\overset{\text{O}}{\overset{\|}{}}}$ (Jencks *et al.,* 1960)

Compound	Carbonyl frequency	Other bands
S-Acetyl mercaptoethanol	1675	
S-Acetyl thiomalate, disodium salt	1677	1573, 1398, 1372, 1303w
S-Acetyl-β-mercaptopropionate sodium salt	1670	1565, 1412, 1398m, 1361m, 1308w
S-Acetylglutathione	1675sh	2900w,[a] 1620b,[a] 1397,[a] 1360m,[a] 1310[a]

[a] Also in glutathione.

Table 19.4. Absorption Bands, in cm⁻¹, of Amides, $R-\overset{\overset{\displaystyle O}{\|}}{C}-N-$ (Jencks *et al.*, 1960)

Compound	Carbonyl frequency	B	Other bands
Acetylimidazole	1740[a]		3130m, 1483m, 1390, 1365sh, 1314m, 1295
O-Carbamylhydroxylamine	1708b	1390	
Ethyl carbamate	1681	1428	2950w, 1472m, 1382, 1347
Glycylglycine	1673[b]	1489	1448m, 1362m, 1315
Carbamyl phosphate	1670	1405	
Glutamine hydrochloride	1639	1448	2950sh, 1729, 1408sh, 1340, 1285sh
Glutamine	1625[c]	1450	2920sh, 1408, 1360w, 1342m
Glutamine, potassium salt	1629	1452	2935w, 1580, 1411, 1336m
Asparagine hydrochloride	1653	1446	2940w, 1730, 1398m, 1346w, 1313w
Asparagine	1626[c]	1440	2940w, 1402, 1353w, 1325w
Asparagine, potassium salt	1635	1440	2920w, 1585, 1408, 1342w, 1315w
Acetamide	1627b	1440b	
Hydroxyurea	1618	1468b	
Urea	1604	1490	
Ammonium carbamate	1540	1441	

[a] Otting (1956) gives a value of 1747 cm⁻¹ in CCl₄.
[b] Lenormant and Chouteau (1952) give a value of 1680 cm⁻¹.
[c] Broad absorption peak includes COO⁻ absorption.

The carbonyl absorption of amides is found at a much lower frequency than that of esters, probably because of the great importance of the resonance contribution of form CXX*c* in the amides (Table 19.4). In D_2O solution the amide I (carbonyl) band of acetamide is at 1627 cm⁻¹, which is much lower than the normal amide frequency of 1710 cm⁻¹ in the vapor state and 1714 cm⁻¹ for the monomer of acetamide in CCl₄ (Davies and Hallam, 1951). Acetamide is monomeric in aqueous solution (same reference). Acetylimidazole, an "energy-rich" compound, absorbs at 1740 cm⁻¹ in D_2O. This high frequency may be due largely to the drawing by the aromatic imidazole ring of the electrons on the nitrogen atom next to the carbonyl group (Otting, 1956; Staab, 1956). This action inhibits form CXX*b* by electron withdrawal and prevents resonance to form CXX*c*. Glycylglycine and other peptide bonds (Lenormant and Chouteau, 1952) also absorb at a wavelength higher than that of acetamide. This fact may be partly attributed to the withdrawal of electrons by NH_3^+ groups and by other peptide bonds. A similar effect is shown by comparison of the amide absorptions of the various ionic forms of glutamine and asparagine. Ethyl carbamate (1681 cm⁻¹), O-carbamylhydroxylamine (1708 cm⁻¹) and carbamyl phosphate (1670 cm⁻¹) display higher absorption frequencies, associated with substitution of an electron-withdrawing oxygen atom for carbon next to the carbonyl group; this indicates that the electron-withdrawing effect is more important than any electron donation by resonance from oxygen to the carbonyl group (Jencks *et al.*, 1960). When a second nitrogen atom is introduced, as in urea and hydroxyurea, a lowering of carbonyl frequency is caused, in spite of the greater electronegativity of nitrogen than carbon. In these compounds, therefore, additional electron donation by resonance from the nitrogen atom appears to be of greater importance than the inductive effect. The carbamate ion, absorbing at 1540 cm^{-1}, is considered to be an extreme example of lengthening of the carbonyl bond due to resonance with the neighboring oxygen anion of the carboxylate group.

Acetone and cyclohexanone in D_2O (Table 19.5) show carbonyl bands at 1698 and 1695 cm^{-1}, respectively. The 1698 cm^{-1} band is much lower than that for acetone vapor (1742 cm^{-1}), liquid acetone, and acetone in organic solvents (1712–1718 cm^{-1}) (Hartwell, 1948). Furfural, in which the carbonyl group is conjugated with an aromatic ring, absorbs at 1665 cm^{-1}. Pyruvic and α-ketoglutaric acids, which have adjacent electron-withdrawing carboxyl groups, absorb at higher frequencies of 1722 to 1725 cm^{-1}. The anions of pyruvate and α-ketoglutarate absorb at 1710 cm^{-1}.

Table 19.5. Absorption Bands, in cm^{-1}, of Aldehydes and Ketones $\overset{O}{\underset{RCR'}{\parallel}}$ **(Jencks et al., 1960)**

Compound	Carbonyl frequency	Other bands
Furfural	1665	3130w, 1570w, 1475m, 1395m, 1368w, 1280w
Cyclohexanone	1695	
Acetone	1698	1420m,[a] 1372[a]
Sodium pyruvate	1710	2980w, 1458sh, 1358m[a]
Pyruvic acid	1725[b]	1440m, 1382m, 1351m
Sodium α-ketoglutarate	1710	1348m, 1310w
α-Ketoglutaric acid	1722[b]	1397, 1330m

[a] C—H *alpha* to C=O.
[b] Includes carboxylic acid band.

The COO$^-$ group in D_2O solution has an asymmetrical stretching frequency, which has some of the character of a C=O stretching vibration, at 1540 to 1630 cm^{-1} and a symmetrical stretching frequency, which is somewhat comparable to the C—O stretching vibration of esters and acids, at 1360 to 1420 cm^{-1} (Table 19.6). The frequency of the asymmetrical stretching, at 1559 cm^{-1}, is progressively increased by the introduction of electron-withdrawing amino, amide (Lenormant and Chouteau, 1952), carbonyl, ammonium (Lenormant, 1952; Edsall, 1937), and hydroxyl groups in glutamine and asparagine anions, glycylglycine, α-ketoglutarate, pyruvate, glutamine, asparagine, and bicarbonate, and is decreased by the introduction of amide character in ammonium carbamate.

The results reported by Jencks et al. (1960) showed that within a group of compounds of a given chemical series a rough correlation exists between the C=O frequency in D_2O solution and the reactivity towards nucleophilic reagents and "energy-rich" character, expressed as free energy of hydrolysis, of a number of carbonyl compounds of biological significance. While there is no general relationship

Table 19.6. Absorption Bands, in cm^{-1},
of Carboxylate Ions, $\overset{\overset{O}{\|}}{RC}-O^-$
(Jencks *et al.*, 1960)

Compound	Carboxylate frequencies
Sodium bicarbonate[a]	1629, 1363
Glutamine	1625,[b] 1408
Asparagine	1626,[b] 1402
Sodium pyruvate	1610, 1410[b]
Sodium α-ketoglutarate	1613, 1408, 1563
Glycylglycine	1597, 1398
Glutamine, potassium salt	1580, 1411
Asparagine, potassium salt	1585, 1408
Sodium acetate[c]	1559, 1419
Ammonium carbamate	1540, 1441[d]
Sodium carbonate[e]	1416[b]

[a] Miller and Wilkins (1952) give 1630 and 1410 cm^{-1} for the solid.
[b] Includes —CONH$_2$ absorption.
[c] Also bands at 2960w and 1351m.
[d] Or amide II.
[e] Miller and Wilkins (1952) give 1440–1450 cm^{-1} for the solid.

between rates and equilibria of reactions, a parallelism may be expected in this case because inductive and resonance effects of substituents affect reactivity toward nucleophilic reagents and free energy of hydrolysis in a similar manner. Factors that stabilize the carbonyl group relative to the transition state will also stabilize it relative to the products of the reaction. Within the series of esters, for example, ethyl acetate (C=O, 1710 cm^{-1}) reacts with hydroxide ion with rate constants of 4.3 and 6.6 liters mole^{-1} min^{-1}, at 19 and 25°C, respectively (Potts and Amis, 1949; Salmi and Leimu, 1947); acetylcholine (absorption at 1735 cm^{-1}) has a rate constant of 57 liters mole^{-1} min^{-1} at 20°C (Butterworth *et al.*, 1953); the ester group of *N,O*-diacetylserinamide (a model for the acylated active site of chymotrypsin) absorbs at 1728 cm^{-1} and has a rate constant of 49 liters mole^{-1} min^{-1} at 25°C (Anderson *et al.*, 1960); and alkyl trifluoroacetates, which absorb at 1786 cm^{-1}, react with hydroxide ion at a rate which is too rapid to be easily measured in water, but which is certainly much greater than that of ethyl acetate (Bunton and Hadwick, 1958). Similarly, *N,O*-diacetyl-serinamide and 2'(3')-acetyl 5'-adenylate (ribose-acetylated AMP, 1737 cm^{-1}) react rapidly with hydroxylamine, whereas similar conditions barely allow reaction with ethyl acetate (Anderson *et al.*, 1960; Jencks, 1957).

The structurally similar amino acid–RNA derivatives also react rapidly with hydroxylamine and are "energy-rich" compounds, since they are in reversible equilibrium with ATP (Hoagland *et al.*, 1958; Mager and Lipmann, 1958; Holley and Goldstein, 1959). Within the acyl phosphate series, acetyl adenylate (1737 cm^{-1}) has a free energy of hydrolysis at *p*H 7 which is similar to that of acetylimidazole and is therefore at least 2800 calories greater than that of acetyl phosphate (Jencks, 1957; Stadtman, 1954). Oxyluciferyl-adenylate, similar in structure, has a free energy of hydrolysis at pH 7.1 of $-13{,}100$ calories (Rhodes and McElroy, 1958). Acetyl diethyl phosphate is a powerful acylating agent which is very quickly hydrolyzed in water (Cramer and Gärtner, 1958).

In the group of amide compounds, acetylimidazole (1740 cm^{-1}) has a much greater free energy of hydrolysis (Stadtman, 1954) and a greater carbonyl frequency than ordinary amides (1627 to 1673 cm^{-1}); the correlations between infrared carbonyl frequency and reactivity for the series acetylpyrrole, acetylimidazole, acetyltriazole, and acetyltetrazole and the analogous acetylbenzimidazole series have been presented by Otting (1956) and Staab (1956, 1957).

Although within a given class of compounds there appears to be some relationship between carbonyl frequency and "energy-rich" nature and/or reactivity toward nucleophilic reagents, no such relationship exists among compounds of widely differing structures. Jencks *et al.* (1960) have suggested that this lack of correlation may be due to a differing relative importance of inductive and resonance effects in affecting carbonyl absorption frequency and reactivity. The markedly lower absorption frequency of carbonyl compounds in D_2O, compared to nonpolar solvents or the vapor state, is attributed to hydrogen bonding of the carbonyl group to the solvent, which, as a form of general acid catalysis, may also increase the susceptibility of carbonyl groups to nucleophilic attack in aqueous solution.

Coordination Compounds

Since metal cyano (Jones and Penneman, 1954, 1956, 1961) and thiocyanato (Fronaeus and Larsson, 1962) coordination compounds display their $C{\equiv}N$ stretching bands between 2200 and 2000 cm^{-1}, it is possible to study equilibria of such complexes with H_2O as the solvent.

One can study solution equilibria of carboxylato complexes in D_2O solution from observation of the CO stretching bands, since the un-ionized, ionized, and coordinated carboxyl groups display relatively strong bands between 1750 and 1550 cm^{-1}. Nakamoto *et al.* (1962, 1963) first studied such complexes in equilibrium studies of iminodiacetic acid, nitrilotriacetic acid, ethylenediaminetetraacetic acid, and related chelating agents by changing the pH (pD) of the solution. They established the following carboxyl group frequencies—type A: un-ionized carboxyl (R_2N-CH_2COOH), 1730–1700 cm^{-1}; type B: α-ammonium carboxylate ($R_2N^+H-CH_2COO^-$), 1630–1620 cm^{-1}; type C: α-aminocarboxylate ($R_2N-CH_2COO^-$), 1595–1575 cm^{-1}.

However, the absorption of the coordinated carboxyl group lies between 1650 and 1590 cm^{-1}, and the kind of metal determines the exact frequency. The frequency

Ionic   Covalent

CXXI

becomes higher as the metal–oxygen bond becomes more covalent (CXXI). The COO group absorbs at 1650–1620 cm^{-1} when coordinated to such metals as Cr(III) and Co(III), and at 1630–1575 cm^{-1} when coordinated with Cu(II) and Zn(II). One must therefore choose the metal ions carefully when trying to distinguish free carboxylate (types B and C, above) from coordinated carboxylate groups (Nakamoto, 1968).

Role of Mg²⁺ and Ca²⁺ in the Hydrolysis of Labile Phosphate Compounds

Oestreich and Jones (1967) have studied the role of magnesium and calcium ions in the hydrolysis of various labile phosphate compounds. Infrared and NMR data on D$_2$O solutions of acetyl phosphate and magnesium sulfate showed no characteristics which could be assigned to chelate structures (in the group state) involving the carbonyl group, as in CXXII:

$$Mg^{2+}$$

$$\begin{array}{ccc} O & & O^- \\ \parallel & & \mid \\ C & -O- & P=O \\ \mid & & \mid \\ R & & O^- \end{array}$$

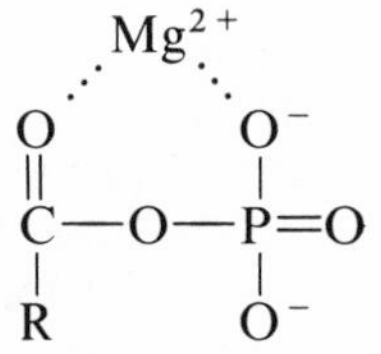

R = CH$_3$, acetyl phosphate
R = NH$_2$, carbamyl phosphate

CXXII

Infrared studies on D$_2$O solutions of carbamyl phosphate led to the same conclusion. Table 19.7 shows infrared bands of acetyl phosphate and carbamyl phosphate in the presence and absence of Mg^{2+} ions. The carbonyl group in acetyl phosphate dianion and carbamyl phosphate dianion absorbs at 1708 and 1666 cm^{-1}, respectively. Table 19.7 shows that the positions of these bands are increased only slightly in the presence of Mg^{2+} ions. These very small increases contrast sharply with large *decreases* in the infrared stretching frequencies of the carbonyl group when it coordinates with metallic ions.

The ground state structure of the magnesium-phosphate complex not involving the carbonyl group as in CXXIII, is favored by the infrared and NMR data (the NMR spectrum of acetyl phosphate dianion in D$_2$O was unchanged by the addition of anhydrous magnesium sulfate).

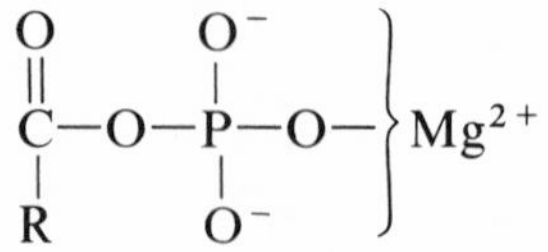

$$\begin{array}{ccc} O & O^- & \\ \parallel & \mid & \\ C-O- & P-O- & \Big\} Mg^{2+} \\ \mid & \mid & \\ R & O^- & \end{array}$$

CXXIII

Table 19.7. Infrared Absorption Bands[a], in cm^{-1}, of Acetyl Phosphate[b] (as the Dianion, AcP^{2-}) and Carbamyl Phosphate[c] (as the Dianion, CAP^{2-}) in the 1200–1800 cm^{-1} Region in D$_2$O Solution (Oestreich and Jones, 1967)

AcP^{2-}		CAP^{2-}	
Mg^{2+} Absent	Mg^{2+} Present	Mg^{2+} Absent	Mg^{2+} Present
1275s	1275s		
1377s	1375m		
1420w	1430sh	1405s	1407sh
1548w	1548m		1429s
1708s	1711s	1666s	1670s

[a] Key: m, medium; s, strong; sh, shoulder; w, weak.
[b] The concentration of AcP^{2-} (as the dilithium salt) was 0.4M; the concentration of MgSO$_4$ was 0.42M.
[c] The concentration of CAP^{2-} (as the dilithium salt) was 0.2M; the concentration of MgSO$_4$ was 0.21M.

Inorganic Ions

Figure 19.5 is a structure correlation chart of common inorganic ions in aqueous solution (Goulden and Manning, 1967). Optimum pH values were calculated from the recorded pK_a's of the appropriate acids and in general were chosen to be about 2 pH units removed from the pK_a value. Where two values are listed for ϵ_m, these apply to the respective absorption bands. It is apparent from Fig. 19.5 that, except for the halide ions, all the more common inorganic anions can be identified readily from bands in this region of the spectrum.

Since the structure of many inorganic ions depends upon pH, it is important to note the pH of a solution whose spectrum is observed. Changes in pH can be quite useful in the assignment of bands appearing in the 1100 cm^{-1} region. For the analysis of "unknown" samples, Goulden and Manning found it convenient to examine spectra of samples in solution at pH 4 and 9, as well as at the original pH. In Fig. 19.6 (Goulden and Manning, 1967) the groups absorbing at frequency intervals of 50 cm^{-1} have been collated for solutions at pH 4 and 9.

Ahlijah and Mooney (1966) used a TlBr–TlI plate with an angle of incidence of 30° to study aqueous solutions of phosphates and phosphites. They showed that HPO$_4^{2-}$, HPO$_3^{2-}$, P$_2$O$_7^{4-}$, H$_2$PO$_4^-$, and other phosphate ion bands fall in fairly definite regions, and the ion present can in some instances be identified. They presented a table of vibrational frequencies for sodium and potassium salts.

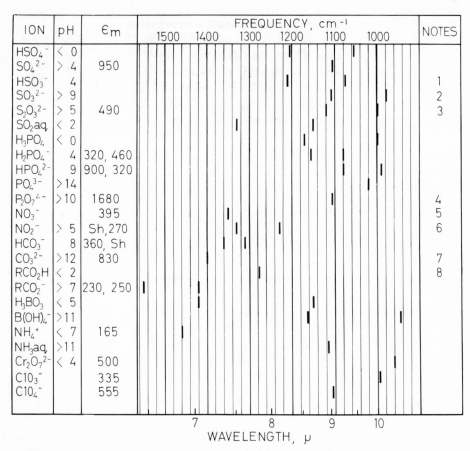

ION	pH	ϵ_m	FREQUENCY, cm⁻¹	NOTES
HSO_4^-	< 0			
SO_4^{2-}	> 4	950		
HSO_3^-	4			1
SO_3^{2-}	> 9			2
$S_2O_3^{2-}$	> 5	490		3
SO_2aq	< 2			
H_3PO_4	< 0			
$H_2PO_4^-$	4	320, 460		
HPO_4^{2-}	9	900, 320		
PO_4^{3-}	>14			
$P_2O_7^{4-}$	>10	1680		4
NO_3^-		395		5
NO_2^-	> 5	Sh, 270		6
HCO_3^-	8	360, Sh		
CO_3^{2-}	>12	830		7
RCO_2H	< 2			8
RCO_2^-	> 7	230, 250		
H_3BO_3	< 5			
$B(OH)_4^-$	>11			
NH_4^+	< 7	165		
NH_3aq	>11			
$Cr_2O_7^{2-}$	< 4	500		
ClO_3^-		335		
ClO_4^-		555		

Fig. 19.5. Infrared absorption of common inorganic ions in aqueous solution. The molar absorptivity (ϵ_m) = absorbance/(molarity × path length, in cm). A shoulder is indicated by sh. Notes: (1) Bands due to solvated SO_2 are also usually present. (2) SO_3^{2-} is readily oxidized in aqueous solution to SO_4^{2-}. Absorption at 1105 cm⁻¹ is due to both v_1 of SO_3^{2-} and v_3 of SO_4^{2-}. (3) Sulfur is precipitated at $pH < 5$. (4) Bands for more acidic forms are not included. Hydrolyzed to orthophosphate by boiling at low pH. (5) Concentrated solutions show band broadening, together with the appearance of a very weak band at 1045 cm⁻¹. (6) The 1330 cm⁻¹ band appears as a shoulder on the side of the 1230 cm⁻¹ band. Nitrites decompose at $pH < 5$. (7) Concentrated solutions show a very weak band at 1060 cm⁻¹. (8) An additional band observed near 1700 cm⁻¹ not included in this figure. (Goulden and Manning, 1967.)

These workers pointed out that recording the ATR spectra in aqueous solution has several advantages over transmission spectra recorded as mulls or in halide disks: (1) the ATR spectra are less complex; (2) no exchange of ions occurs as in halide disks; (3) there is no problem with nonuniform particle size and distribution of the salts. The main difficulty with using water as the solvent is its low transmission of energy over much of the useful frequency range. However, Fahrenfort (1961) showed that the ATR spectrum of water displays excellent "window" properties. Goulden and Manning (1964) raised objections to the preference of ATR over

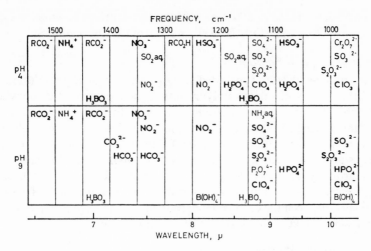

Fig. 19.6. Frequency correlation chart for pH 4 and pH 9. (Goulden and Manning, 1967.)

transmission techniques for aqueous-solution work, since it has been shown that very thin layers of water do transmit energy adequately by transmission methods (Ackermann, 1961).

Ahlijah and Mooney (1969) determined the vibrational frequencies of nitrate, nitrite, cyanide, cyanate, thiocyanate, and ferro- and ferricyanide anions in aqueous solution by ATR. These authors were not entirely in agreement with Goulden and Manning's observation (Goulden and Manning, 1964, 1967) that the ATR method has no advantage over the conventional transmission techniques. A larger "window" is available for studying aqueous solutions by the ATR method than by the normal transmission method. Goulden and Manning (1967) stated that in the transmission spectrum of water there is 90% absorption of available energy at $1250\ cm^{-1}$ and it is necessary to increase the spectral slit width. The increase in spectral slit width results in the loss of resolution which is normally available on the spectrophotometer. Canepa and Mooney (1965) needed such resolution to study the conformation of acetylcholine in mammalian fluid by the use of ATR spectroscopy. Ahlijah and Mooney (1966) showed that, in agreement with data reported by Fahrenfort (1961), there is little absorption of energy by water in ATR. Goulden and Manning (1967) also stated that higher concentrations of salts are required (by a factor of at least five) when the ATR method is used. The work of Ahlijah and Mooney (1969) showed that this statement is not necessarily true.

The use of the ATR method was found to be more reliable (Ahlijah and Mooney, 1969) for analyzing a mixture of sodium nitrate and sodium nitrite than was a standard chemical method (Furman, 1962). Absorption spectra of many inorganic ions have been determined for the solid state in the region $4000–650\ cm^{-1}$ by Miller and Wilkins (1952). Al-Kayssi and Magee (1962) have reported the systematic analysis of the following inorganic anions by infrared spectroscopy: SCN^-, $S_2O_8^{2-}$, ClO_4^-, ClO_3^-, MnO_4^-, IO_4^-, $Fe(CN)_6^{4-}$, $Fe(CN)_6^{3-}$, $S_2O_3^{2-}$, CrO_4^{2-} (or $Cr_2O_7^{2-}$), molybdate, tungstate, and others. Haba and Wilson (1964) have identified arsenite, BrO_3^-, IO_3^-, silicate, PO_4^{3-}, AsO_4^{3-}, and SO_4^{2-}, as well as some of the above anions.

Relationship of pK_a and Frequency

Goulden and Scott (1968) recorded infrared spectra of D_2O solutions of the sodium or lithium salts of various polyuronides as well as several carboxylic acids. The relationship between pK_a and frequency (v) of the antisymmetric stretching vibration of the ionized carboxylate group is shown in Fig. 19.7. The results indicate that the type of acid (aromatic, dibasic, α-oxy) may affect the position of v. Electron-withdrawing substituents can also affect v (Spinner, 1964). Thus, although hexuronic acids in polyuronides all showed slightly higher frequencies than would be expected from their measured pK_a values, factors other than the presence of $CO_2^- \cdots HO$ hydrogen bonds (between the carboxylate group and a hydroxyl group at C-2 or C-3 of the uronic acid) could account for the observed frequencies.

Collagen-reactive Aldehydes

Milch (1964) obtained infrared spectra on aqueous solutions of some aliphatic and aromatic aldehydes. Saturated compounds which are capable of acting as inter-chain cross-linking agents for the polypeptide chains of either gelatin sols or native

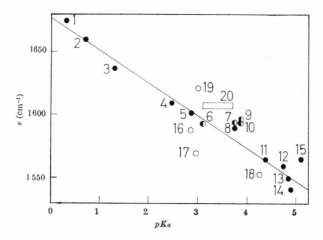

Fig. 19.7. Relationship between acid dissociation constant (pK_a) and frequency (v) of the antisymmetric stretching vibration of salts of carboxylic acids in D_2O solution. 1, Trifluoroacetic; 2, trichloroacetic; 3, dichloroacetic; 4, cyanoacetic; 5, chloroacetic; 6, 2-phenoxypropionic; 7, methoxyacetic; 8, formic; 9, glycollic; 10, lactic; 11, 3-hydroxybutyric; 12, acetic; 13, propionic; 14, cyclohexane carboxylic; 15, polyacrylic; 16, malonic; 17, phthalic; 18, benzoic; 19, salicylic; and 20, polyguluronic, polygalacturonic, galacturonic, alginic, hyaluronic, and chondroitin sulfates A, B, and C. (● ◐) From Goulden and Scott, 1968; (○) from Chapman et al, 1964. The pK_a values are from references given in Goulden and Scott (1968). (Goulden and Scott, 1968.)

and partially denatured collagen tissue preparations appear to exist in aqueous solution either primarily or exclusively in the hydrated, or *gem*-dihydroxy configuration. On the other hand, unsaturated, aliphatic monoaldehydes probably have the β-hydroxycarbonyl or aldol form, and aldehydes which do not act as collagen cross-linking agents have the unhydrated or free carbonyl form. Milch (1964) proposed a "three-carbon" selection rule, which, together with previously recorded structural requirements, provides a basis for predicting the ability of any mono- or dialdehyde to cross link in collagen. He stated that crotonaldehyde appeared to be the only exception to the rule.

Analysis of Milk

The fat, protein, and lactose contents of milk can be determined from measurements of the intensities of the respective absorption bands at 1745, 1548, and 1053 cm^{-1} (Goulden, 1964). The content of solids labelled as "solids-not-fat" (SNF) can be obtained from a single intensity measurement at 1266 cm^{-1} where the fat shows no attenuation and the protein and lactose absorptivities are approximately equal. Photometric errors resulting from natural variations in the size distribution of fat globules can be eliminated by prior homogenization of the milk. Using these observations as a basis, scientists at the National Institute for Research in Dairying (Reading, England) constructed an automatic infrared milk analyzer, which they named IRMA. Preliminary tests on 50 milk samples from individual cows of three different breeds gave standard deviations from the chemical analyses for the percentages of fat, protein, lactose, and SNF of 0.10, 0.07, 0.07, and 0.19, respectively. The standard deviations for the percentages of fat and SNF on 60 farm bulk milk samples were 0.08 and 0.12, respectively. The constructional features of the analyzer have been described by Goulden *et al.* (1964).

Use of D_2O in Studies of Tissue Fluids

Turner *et al.* (1960) devised a rapid infrared method by which D_2O can be determined in biological fluids, including plasma water. The error of this analytical procedure is $\pm 1\%$, and as many as 50 plasma samples may be processed and analyzed in a single day. Neely *et al.* (1962) applied this method to study the possibility of the absorption of water by the body through the skin. A small animal is placed in a specially constructed chamber which encloses the body up to the neck. The chamber is filled with an H_2O–D_2O vapor. Plasma water in blood samples taken from the animal is analyzed in a calcium fluoride cell against a plasma water blank at 2513 cm^{-1}. The results of Turner *et al.* indicated that D_2O moved into the animal (dog) body water at the rate of 45 grams per square meter per hour.

Verzhbinskaya and Sidorova (1966) used spectra of tissue fluids from white rats injected with heavy water to determine the rate of water exchange between blood and tissues. Brain absorbed the greatest amount of D_2O and skeletal muscle the least.

Metal Chelates and Complexes

Infrared spectra have been given for many ethylenediaminetetraacetic acid (EDTA)–metal chelates (Sawyer, 1960) along with a listing of the major absorption bands for the CH_2, COO^- (antisymmetrical and symmetrical vibrations) and CN groups. Many references were given in this paper to much of the background of the chemistry of the metal chelate compounds.

The EDTA chelates studied by infrared were those of the following metals: Mg(II), Ca(II), Sr(II), Ba(II), Mn(II), Co(II), Ni(II), Cu(II), Zn(II), Cd(II), Hg(II), Pb(II), Al(III), Ce(III), Bi(III), V(III), V(IV), Cr(III), Fe(III), Co(III), Ti(IV), Th(IV), Mo(V), and Mo(VI). Spectra were also recorded for the tetrasodium and tetrapotassium salts of EDTA.

The characteristic frequency for the CH_2 groups in chelated EDTA molecules suggests that the COO^- groups are attached directly to the metal ion. The frequency for the CH_2 group decreases generally as the ionic radius of the metal ion increases; this trend is particularly true for closely similar groups of ions, for example, the alkaline earth ions or the divalent ions.

The infrared data for the COO^- bands support the conclusion that the bonding is primarily ionic for the EDTA chelates of Mg(II), Ca(II), Sr(II), Ba(II), Mn(II), Co(II), Ni(II), Cu(II), Zn(II), Cd(II), Hg(II), Pb(II), Al(III), Ce(III), and Bi(III). The chelates of the following metals are primarily covalently bonded: V(III), V(IV), Cr(III), Fe(III), Co(III), Ti(IV), Th(IV), Mo(V), and Mo(VI). Generally, a band for COO^- (antisymmetrical) in the 1610 to 1550 cm^{-1} region is evidence for ionic bonding; a band in the 1660 to 1630 cm^{-1} region is evidence for covalent bonding. The difference in frequency between the major band for the symmetrical vibration (1450 to 1350 cm^{-1}) and the band for the antisymmetrical vibration (1660 to 1570 cm^{-1}) of the COO^- group indicates the degree of covalent bonding for the EDTA chelates. The frequency difference increases as the bonding becomes more covalent.

The infrared spectra of a number of complexes of metals with pyridine and quinoline (Frank and Rogers, 1966) and with isothiocyanate (Clark and Williams, 1966) have been reported. Infrared data have been given for metal—leucine chelates (Jackovitz and Walter, 1966).

Spiro *et al.* (1966) and Allerton *et al.* (1966) have synthesized a polymeric compound, $[Fe_4O_3(OH)_4(NO_3)_2]_n$, which appears to be an analog to the iron-containing core of ferritin. Its diameter of 70 Å is the same, and when the corresponding ferric citrate micelle is placed in a solution of noncrystalline apoferritin, the synthetic core is surrounded by the protein subunits to give a substance whose gross morphology is very similar to that of ferritin (Pape *et al.*, 1968). Brady *et al.* (1968) have studied the structural and magnetic properties of the synthetic iron-core polymer, and have compared them with the properties of the ferritin micelle in the hope that a study of the synthetic compound might yield information relevant to the structure of the natural material. The ferric ions were all found to be tetrahedrally coordinated by O^{2-}, OH^-, and H_2O in a corner-sharing bridged structure, with Fe—O—Fe bond angles of $125 \pm 30°$ (Fig. 19.8). The nitrate ions are not coordinated directly to the iron, but act only as counterions to the positively charged polymer spheres. The infrared spectrum of the synthetic polymer shows a strong band, broad but apparently unsplit, at 1310 cm^{-1} and a weak one at 1030 cm^{-1}, consistent with only a small perturbation away from the ideal D_{3h} symmetry of the free nitrate ion. [The nitrate ion and D_{3h} symmetry are discussed in Herzberg (1945).] In addition to the

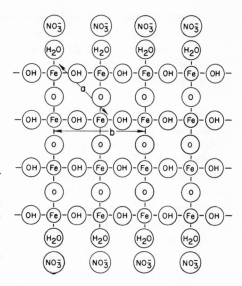

Fig. 19.8. A proposed secondary structure of $Fe_4O_3(OH)_4(NO_3)_2 \cdot 1.5H_2O$ shown as a projection on the plane of the paper. The vectors a and b are the probable Fe–Fe interactions corresponding to the two peaks at 5.6 and 6.4 Å in the radial distribution function. The NO_3^- ions are thought to be hydrogen bonded to the H_2O molecules on the surface. (Brady et al., 1968.)

nitrate ion bands, a moderately strong band occurs at $700 \, cm^{-1}$, a stronger one at $450 \, cm^{-1}$ having a shoulder at $500 \, cm^{-1}$, and a weak band at $270 \, cm^{-1}$. Also, a band at $1600 \, cm^{-1}$ attributable to the bending mode of water is clearly evident in the spectrum, even after prolonged drying of the polymer in vacuo over P_2O_5.

Hewkin and Griffith (1966) have summarized the infrared data on bridged complexes, stating that a sharp band (v_3) is displayed at $\sim 850 \, cm^{-1}$ whenever a linear M—O—M bridge is present. The synthetic polymer contains no such band, and Brady et al. (1968) propose that the features above arise instead from both oxy- and hydroxy-bent bridges.

Membranes

Plasma Membranes, Endoplasmic Reticulum, Mitochondrial Membranes, and Nerve and Muscle Membranes

Plasma membranes of Ehrlich ascites carcinoma have been examined by Wallach and Zahler (1966) by infrared and fluorescence spectroscopy and optical rotatory dispersion. As a result, these workers have suggested that there are important classes of membrane proteins—both structural and functional—whose unique amino acid sequences impose tertiary and quaternary structures in which two hydrophilic pep- tide regions are widely separated by a hydrophobic zone. They envision the two hydrophilic sections to lie at the membrane surfaces, connected by hydrophobic rods penetrating the membrane normal to its surface. The length of the hydrophobic units would equal the width of the apolar membrane core. The hydrophobic units would consist of helical peptide segments, packed amidst the hydrocarbon residues of membrane lipids in at least two types of arrangement: (a) single units, with only nonpolar side chains; (b) aggregates (possibly reversible) in the form of microtubules which, depending on primary sequence, could have a polar interior. They believe that membrane transport could occur largely via such structures, controlled by the conformational state of the protein subunits and/or by subunit aggregation.

The hydrophilic peptide segments, located at the hydrated membrane surfaces, would be in polar interaction with other (including nonpenetrating) peptide chains, as well as with the headgroups of membrane lipids.

The same authors (Wallach and Zahler, 1968) have compared the infrared spectra of plasma membranes of Ehrlich ascites carcinoma with those of endoplasmic reticulum. These spectra are dominated, as one would expect, by the amide I and amide II bands of membrane proteins and by bands associated with the O—H, C—H, C=O, P=O, C—O—C, and P—O—C vibrations of membrane lipids. The two membrane types differed in how much lipid was extractable from them. Extraction with acetone–water (9:1, v/v) caused over 90% of the lipid phosphorus to be removed from endoplasmic reticulum but only 50 to 60% from plasma membrane. However, the same solvent extracted most of the cholesterol from both types of membrane.

The infrared spectra of plasma membrane and endoplasmic reticulum films, extracted at neutral pH, differ noticeably in the region 1750 to 1710 cm^{-1} (C=O stretching). In plasma membrane this region is featureless after lipid extraction (Fig. 19.9), whereas the spectra of endoplasmic reticulum have a pronounced shoulder there, of uncertain origin. Absorbance changes which are pH-dependent at 1720 cm^{-1} and 1590 cm^{-1} (i.e., absorbance at 1720 cm^{-1} increases in $0.1M$ HCl, whereas absorbance at ~ 1590 cm^{-1} decreases) are attributable to various carboxyl groups.

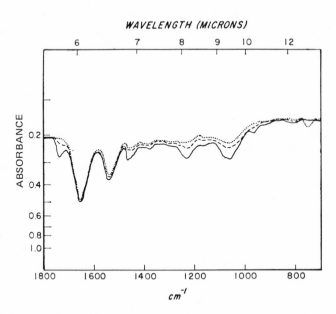

Fig. 19.9. Infrared spectrum of a plasma membrane film. (——) Untreated; (– – – –) after extraction with acetone–water (9:1, v/v); and (· · · · ·) after extraction with chloroform–methanol (2:1, v/v). (Wallach and Zahler, 1968.)

The authors attributed the appearance of a shoulder near $1630\,cm^{-1}$, when both membrane films were exposed to acid pH, to refolding of some of the membrane peptide into a β-conformation. The pH-dependent changes in the infrared spectra of lipid-free membranes suggested that the proteins were highly amidated. This suggestion was also indicated by high yields of ammonia in amino acid analyses.

Many pieces of evidence, including IR, ORD, NMR, and CD examination of isolated plasma membranes, provide reasons for questioning the validity of the traditional structural model of the plasma membrane and other biological membranes (Korn, 1967). Some of the evidence given by Korn (1967) includes: (1) The observation that the electron microscope reveals a globular substructure in most membranes (but not as yet in the plasma membrane) when viewed by various techniques. (2) The fact that the trilaminar image of at least one membrane, the inner mitochondrial membrane, is essentially unchanged after removal of all the lipid. (3) The realization that the chemistry of osmium tetroxide fixation is only poorly understood and may be quite different from that previously supposed. (4) Wide differences exist in the chemical composition of membranes both in the ratio of protein to lipid and in the composition of the lipids and the proteins. (5) Phospholipids can assume many stable forms other than the bimolecular leaflet. (6) There are preliminary suggestions that lipoproteins, not phospholipids, are the biosynthetic precursors of natural membranes. (7) Studies on isolated plasma membranes by infrared spectroscopy, optical rotatory dispersion, nuclear magnetic resonance, and circular dichroism demonstrate that the protein is in the α-helical and random coil configurations and that the protein–lipid bonds are largely hydrophobic. (8) Studies with proteins and lipids isolated from membranes suggest that protein–protein interactions may be the fundamental forces in membrane formation. (9) There are indications that a "structural" protein, which interacts with phospholipids through nonpolar bonds, may be a common feature of all membranes.

No unique structural model for the plasma membrane can yet be deduced from these data (Korn, 1967) but it is now possible to consider that lipoprotein molecules may be the fundamental structural, as well as functional, components of membranes. Individual protein molecules could easily extend through the 75–$100\,\text{Å}$ width of the plasma membrane. This implies that the "carrier" molecule and the "structural" molecule may be identical and that transport across the plasma membrane might be conceived as a conformational change of the membrane substructure. The membrane need no longer be viewed primarily as an inert barrier but rather as a dynamic aggregate of functional polymers (Korn, 1967).

Rat liver mitochondrial membrane proteins, in contrast to other membrane types studied, have considerable β-conformation, according to IR, ORD, and CD spectral studies by Gordon et al. (1969). The membranes and their "structural protein" exhibited an absorption band due to helical and/or "unordered" conformations (near $1650\,cm^{-1}$), and shoulders near 1630 and $1690\,cm^{-1}$, attributed to β-conformations. The inner membranes and "structural protein" showed the highest proportion of β-structure.

The physical mechanism of the permeability change which occurs in nerve and muscle cells during excitation has been the subject of much recent discussion

(Sherebrin, 1969). The electrical field across a cellular membrane in the resting state is $\sim 10^5$ V/cm, which is almost the value required for dielectric breakdown. During excitation this field decreases to zero, reverses sign, and returns to the resting value, all in a few milliseconds. A molecule in the membrane having an unsymmetrical charge distribution and, hence, a net dipole moment could undergo very profound conformational changes and cause the transient changes in permeability to sodium and potassium which are observed. Sherebrin (1969) has modified a Perkin–Elmer model 521 infrared spectrophotometer to examine conformational changes which should be observable by means of infrared spectral shifts. A system of mirrors was used to modify the sampling procedure. A specimen of living nerve or muscle was placed on an internal reflecting prism and the difference spectrum was measured for the same sample between the excited and unexcited states. Sherebrin found that changes were emphasized in the region of the membrane, and that by means of the differential technique only changes with excitation were recorded.

Erythrocyte, Plasma, and Myelin Membranes

Studies on erythrocyte membranes have shown that proton magnetic resonance (Chapman *et al.*, 1968*a*) and infrared spectroscopy (Maddy and Malcolm, 1965; Chapman *et al.*, 1968*b*) can provide useful information about membrane organization. The combination of such studies and data from ORD and CD experiments on erythrocyte membrane and other plasma membranes (Wallach and Zahler, 1966; Lenard and Singer, 1968) suggest the possibility of interaction between lipid hydrocarbon chains and the apolar amino acids of the membrane protein. Proton wide-line nuclear resonance studies of myelin from beef brain in excess ^2H_2O (D_2O) have shown a spectrum very similar to that of the total lipid of myelin swollen in ^2H_2O (Jenkinson *et al.*, 1969). The lipids are known to be in a lamellar arrangement in the latter case and therefore the evidence was consistent with myelin itself having a lamellar organization. Jenkinson *et al.* (1969) have also shown infrared evidence that a considerable degree of planar *trans* (or *anti*) configuration of the carbon skeleton of the hydrocarbon chains of the lipids exists within the membrane structure. [See Corish and Davison (1955) for a discussion of the *trans* configuration of the $(CH_2)_n$ chain in crystalline $COOH-(CH_2)_n-COOH$ molecules, where $n = 4$ to 16.] The infrared spectra of dried films of myelin and its total lipid extract are shown in Figs. 19.10*a,b*. Partially hydrated spectroscopic cells (AgCl windows plus AgCl spacer, yielding a sealed cell of 0.25-mm path length) containing myelin equilibrated with ^2H_2O gave spectra like the one in Fig. 19.10*a* except for a little broadening between 1660 and 1220 cm^{-1} due to $H-O-H$, $H-O-^2H$ (HDO), and $^2H-O-^2H$ (D_2O). The spectrum of myelin is similar to spectra presented earlier (Hulcher, 1963). Amide I and II bands of the protein are displayed at 1656 and 1535 cm^{-1}, respectively, with no band or shoulder at 1628 cm^{-1}. A weak band occurs at 700 cm^{-1}, perhaps from NH out-of-plane deformations (Miyazawa *et al.*, 1962). Equilibration with ^2H_2O did not reduce the 700 cm^{-1} band but there was some reduction of the amide II band.

Figure 19.10*b* shows no absorption at 700 cm^{-1} for the lipid extract, but small bands are present at the amide I and II frequencies due to cerebrosides and sphingomyelin. Other characteristics of lipid material common to both spectra are the $C=O$

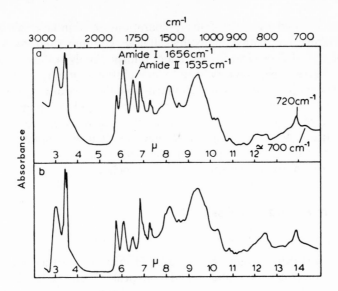

Fig. 19.10. (a) Myelin, room temperature dried AgCl film, (b) total lipid, room temperature AgCl film. (Jenkinson et al., 1969.)

stretch (ester) at 1739 cm^{-1}, P=O stretch at 1226 cm^{-1}, (P)—O—C stretch at 1070 cm^{-1} (broad), P—O—(C) at 970 cm^{-1}, Δ^5 sterol at 839 and 799 cm^{-1} and $(CH_2)_n$ rock at 720 cm^{-1}. Chapman et al. (1967) have given assignments of absorption bands for lecithins in this region.

The band at 720 cm^{-1} is characteristic of the rocking motion of four or more connected CH_2 groups in planar trans (or anti) configuration. (Planarity here refers to the carbon skeleton and not to unsaturation.) The band is present in the spectrum of hydrated myelin (5–50% water) at room temperature, and in the spectra of dried films of myelin as well as its lipid extract over a wide range of temperatures. This fact indicates (Jenkinson et al., 1969) that even though protein is present, a considerable degree of order exists within the lipid chains of myelin. Erythrocyte membranes, on the other hand, only develop this band distinctly at subzero temperatures, although it can be seen in the spectrum of erythrocyte lipids above room temperature (Chapman et al., 1968b). The amide I and II bands in the spectrum of myelin, either completely or partially hydrated, are typical of protein in either α-helical and/or random coil conformation, with no β-configuration present (Maddy and Malcolm, 1965). Steim (1967) has given infrared and ORD data which support the presence of helical content in myelin protein.

The NMR and infrared data of Jenkinson et al. (1969) on myelin and erythrocyte membrane suggests that there are differences in lipid-chain organization and inter-action between lipid and protein in these two membrane systems.

Mitochondria

Fleischer et al. (1967) have studied the lipid composition of highly purified preparations of mitochondria from bovine heart, liver, and kidney. They determined components of lipid extracts of the preparations by thin-layer chromatography, chromatography on diethylaminoethylcellulose columns, and by spectrophotometric procedures. They identified the major phospholipids by their chromatographic behavior, infrared spectrometry, and paper chromatography of their hydrolysis products. The phospholipids of mitochondria from the three organs were qualitatively and quantitatively very similar. The major polar lipid components, cardiolipin, choline glycerophosphatides, and ethanolamine glycerophosphatides, were present in molar ratios of 1:4:4. Fleischer *et al.* suggested that mitochondria from different sources contain characteristic lipids, mainly phospholipids, of which cardiolipin is particularly diagnostic of the source of the mitochondria.

Some Early Studies on Tissues

Normal and Cancerous Tissue

Woernley (1952) has published infrared absorption spectra for various types of normal and cancerous tissues and related biological substances. He drew the following conclusions: (1) The infrared pattern of RNA and DNA, consisting of the 1235, 1070, and 971 cm^{-1} bands was due mainly to the pentose and phosphoric acid components. (2) DNA and RNA were present in many tissues in sufficient concentrations to cause bands of reasonable intensity in the region 1250–909 cm^{-1}. (3) The spectrum of mouse liver and tumor nuclei, and of many tissues both normal and cancerous, were very much like that for DNA and RNA in the region 1250–909 cm^{-1}. (4) Intensities of the bands at $\sim$1235, 1075, and 971 cm^{-1} could be correlated with the concentrations of nucleic acids in tissues. (5) Muscular tissue generally displayed less marked absorption bands in the 1250–909 cm^{-1} region than highly cellular cancerous and normal tissues rich in nucleic acids. (6) The spectrum for tissues in the 1250–909 cm^{-1} region were subject to many sources of interference, such as bands from carbohydrates and phosphoric compounds. Fats and proteins derived from the blood gave bands which were not very pronounced. Bird and Blout (1952) were among the earlier workers to use infrared methods for tissue studies.

Nervous Tissue

Schwarz and his colleagues (Schwarz, 1952; Schwarz *et al.*, 1951*a,b*, 1952) have reported that the tissues from different areas of the brain gave the same spectra. They used human autopsy specimens and tissues from rats, rabbits, and dogs, and observed no spectral changes when they used a variety of methods of killing the animals. Thomas *et al.* (1954) observed that spinal cord and cerebellum tissue of fowl and rat gave qualitatively identical spectra, and Woernley (1952) and Schwarz *et al.* (1951*b*) also found this to be the case in mouse and human brains, respectively.

Coates *et al.* (1953) have examined the gray and white matter of rat hypothalamus with an infrared microscope and observed additional bands, which may have been the result of using a spectrometer with higher resolution than in previous studies. White matter displayed a band at 1300 cm^{-1} as an inflection point, while gray matter showed one at 1163 cm^{-1}.

Grenell and May (1958) have described a procedure for obtaining the infrared spectra of freeze-dried tissue sections of brain, and found that the spectra of homogenate film and lyophilized tissue film were identical with the spectra of sections. Extraction of the tissue sections at room temperature with lipid solvents (chloroform–methanol, chloroform, and methanol) did not remove all lipids. The extracted residues had more absorption bands than the unextracted tissue sections and the extractants were removing different substances, as indicated by the changes in appearance or disappearance of certain bands with the type of solvent used. Extraction of nerve fiber by chloroform caused decreased intensities of the bands at 2900, 1460, 1375, 1235, and 1080 cm^{-1} in the spectrum of what remained of the fiber. The first three bands are due to CH vibrations and the last from esters containing C—O—C. Lipids contain these groups, so that the decreases in intensity were not unexpected. The band near 1235 cm^{-1} was probably polyamide instead of lipid (Barer *et al.*, 1949), but nucleic acids can also absorb in this region (Blout and Fields, 1948, 1949; Fraser and Fraser, 1951) [see also Table 3 in Grenell and May, (1958)]. Grenell and May (1958) found an increase in the number of bands after extracting brain tissue with various solvents and presented a table of the additional bands and possible correlations with known spectra.

Mills *et al.* (1969) have recently used infrared spectroscopy on lyophilized specimens to distinguish brain tissue of patients with the hereditary atrophic disease known as Huntington's chorea from brain tissue of patients with nonhereditary degenerative disease. Brain tissue samples from patients of the former group differed from those from the latter group in five of the eleven spectral peaks studied from caudate, in ten of the eleven peaks studied from putamen, in two peaks for tissue from superior frontal gyrus, and in one peak from corpus callosum. The infrared method, if applied to material obtained from relatively simple brain biopsy, may permit presymptomatic diagnosis of brain disease having strong hereditary etiology from other atrophic brain disease antedating detectable symptoms.

Schwarz (1960) has discussed infrared absorption analysis of tissue constituents, particularly tissue lipids. He has given a method for examining tissue sections, and has discussed the chromatography of extracted phospholipids, sphingolipids, and fatty acids and neutral lipids.

Adrenal Gland Cortex

Wegmann (1956) has studied the histochemistry of lyophilized human and freshly dehydrated rat adrenal gland cortex sections by using a microscopic attachment to an infrared spectrophotometer and compared the spectra of these glands with that of a cholesterol fraction of a lipid extract of the human adrenal. The bands obtained on the adrenal glands corresponded for the most part to steroid substances, probably steroid hormones, and to unsaturated lipid compounds.

Microsomes and Mitochondria

Wegmann *et al.* (1956) have published infrared spectra of rat liver microsomes and of a mixture of microsomes and mitochondria. The characteristic property of the microsomal spectra was that of phospholipoproteins. The presence of sucrose in the preparations obscured the spectral characteristics of nucleoproteins.

Comparative Spectra of Vertebrate Red Blood Cells

Stewart and Erley (1961) have demonstrated that comparative infrared spectra of the red blood cells of mammals, birds, reptiles, amphibians, and fish possess significant absorption differences; and have shown that a differential technique can be used to detect the absorption differences that exist between the species in the five groups even when such differences are not immediately apparent in the spectrum recorded in a nondifferential manner. All the mammalian red blood cells studied had virtually identical infrared spectra *when scanned with water in the reference beam* (e.g., human adults, infants, beagles, rabbits, cats, a monkey, sheep, guinea pigs, rats, and mice). However, it is interesting to note that when scanned *differentially with human red cells in the reference beam*, small, but significant differences were observed. A striking difference was found at 1075 cm^{-1} in the differential spectrum of human fetal *vs* adult red blood cells.

It is of note that erythrocytes of each vertebrate group studied, composed of complex systems of many compounds, display specific infrared absorption curves. While it is well known that the infrared spectrum of a single organic compound is one of its most characteristic physical properties, it is rather unusual to think in terms of a specific infrared spectrum of the many biochemical compounds in a living erythrocyte, but such was the case.

Studies on Bone and Collagen

Furedi and Walton (1968) have compared transmission and ATR spectra of collagen and normal adult human bone with data in the literature (Beer *et al.*, 1959; Emerson and Fischer, 1962; Underwood *et al.*, 1955; Posner and Duyckaerts, 1954; Caglioti *et al.*, 1954, 1955).

Furedi and Walton pointed out that sampling techniques for infrared transmission spectroscopy involve considerable difficulty because of the extreme hardness of the mineral and the presence of the protein matrix. Also, when applying any kind of sampling techniques to biological material, the possibility of denaturation has to be overcome. The application of multiple ATR spectroscopy (Fahrenfort, 1961, 1962; Harrick, 1963) to biological material (Hermann, 1965; Parker and Ans, 1965, 1967; Parker and D'Agostino, 1967) is particularly attractive since infrared spectra can be obtained from whole samples with a minimum of preparation. Furedi and Walton used a double-beam attachment (Wilks Scientific Co.) with KRS-5 crystals of 2-mm thickness and incident angle of 45° in the sample and reference beams.

Table 19.8 shows absorption maxima of collagen spectra recorded by absorption spectroscopy (KBr pellet of powdered rat skin), and by ATR (rat tail tendon). These data were compared with those of Beer *et al.* (1959) and assignments were given. Generally, the absorption and ATR spectra coincide fairly well. All absorption maxima which are specified by Beer *et al.* (1959) as characteristic protein bands, were observed in both cases, except for an ill-defined maximum between 650 and 700 cm^{-1}, specified by Beer *et al.* as an out-of-plane NH deformation frequency.

Table 19.8. Absorption Maxima of Collagen Spectra (Furedi and Walton, 1968)

KBr pellet	ATR	Beer et al. (1959)	Assignments (Beer et al., 1959)
—	1741m	—	
1662vs	1654⎱ 1632⎰ vs	near 1650vs	CONH amide I CO stretching frequency
	1699wsh		
1537s	1522⎱ 1538⎰s 1560⎰	1535vs	CONH amide II NH deformation and CN stretching frequency
1449m	1464⎱ 1451⎰m	1447m	CH$_2$ deformation, CH$_3$ asymmetric deformation
1409sh	—	—	
1378w	1388w	1385m to w	CH$_3$ symmetric deformation
1334w	1337vw	—	
—	1306vw	1310	CH$_2$ wag
1277sh	1281vw	—	
1233m	1232m	1240m	CONH amide III CN stretching and NH deformation frequency
1206sh	1201w	—	
1160sh	1158w	1170w	
1080w	1079vw	1075w	
1028w	1028w	—	
975sh	975sh	—	
—	—	650– 700bm	NH out-of-plane deformation
—	475–620m Max. 548	—	

Key: b, broad; m, medium; s, strong; sh, shoulder; v, very; w, weak.

Instead, the ATR spectrum shows a broad band between 475 and 620 cm^{-1}. Also, a medium absorption band at 1741 cm^{-1} was observed in the ATR spectrum only. No explanation was offered for these discrepancies.

Ambrose and Elliott (1951) and Beer et al., using polarized infrared spectroscopy, have observed structure in the NH and CO stretching and deformation frequencies. Beer et al. reported that for collagen, both the 1650 and 1535 cm^{-1} bands showed structure, while nylon gave a doublet at 1650 cm^{-1} and the 1535 cm^{-1} band appeared to be a triplet. Ambrose and Elliott (1951) observed a weak shoulder at 1695 cm^{-1} in some α helical proteins.

The characteristic bands of the transmission and ATR spectra of the same sample of normal adult human bone are listed in Table 19.9 and compared with data from the literature. The absorption maxima of hydroxyapatite and octacalcium phosphate are listed. Assignments of the bands are also given.

Table 19.9. Absorption Maxima of Adult Human Bone (Furedi and Walton, 1968)

		Literature data			
KBr	ATR	Bone	HA[a]	OCP[a]	Assignment
—	1743	—	—	—	Also found in ATR spectrum of rat tail tendon
1659s	1649s	1653[b]	—	—	CONH amide I CO stretching mode
1545s	(1559) 1547s	1515[b]	—	—	CONH amide II NH deformation and CN stretching mode
1470 ⎱ sh 1480 ⎰	—	nr	—	—	CH$_2$ deformation CH$_2$ asymmetric deformation (compare with Table 19.8)
1445 ⎱ s 1417 ⎰	1449 ⎱ s 1414 ⎰	1450 ⎱[b] 1410 ⎰	—	—	Carbonate CO stretching mode v_3
1344sh	1342sh	nr	—	—	Due to protein matrix
1276sh	—	nr	—	—	Due to protein matrix
1238w	1233w	nr	—	—	CONH amide III CN stretching and NH deformation mode
—	1204w	—	—	—	Due to protein matrix
1099sh ⎱ vs 1030 ⎰ (1040–45 in other preparations)	1094sh ⎱ vs 1011 ⎰	Broad band between 1250 (1100) and 910[b,c,d,e]	1092 1040	1105 1075 1055 1035 1025	PO stretching mode v_3 (PO$_4$)
961w	960w	~962 na[b]	962w	962w	PO$_4$; symmetric PO stretching mode v_1 [a]
879 873	878 871	878d[b] 873	—	—	Carbonate, out-of-plane, CO mode v_2 [b]
—	—	—	631m 601m	630w,sh 599m	OH librational in HA[a]
a) 604s 564s b) 606s 575sh 562s	598s 557s	nr	575m,sh 561m	575w,sh 559m	O—P—O bending modes v_4 [a]
—	—	nr	—	525w,sh	HOPO$_3$ bending mode [a]
—	464w	nr	464w 447w		

Key: d, doublet; HA, hydroxyapatite; m, medium; na, not assigned; nr, not reported; OCP, octacalcium phosphate; s, strong; sh, shoulder; v, very; w, weak.
[a] Fowler et al. (1966).
[b] Emerson and Fischer (1962).
[c] Underwood et al. (1955).
[d] Posner and Duyckaerts (1954).
[e] Caglioti et al. (1954, 1955).

May (1966) has presented CsBr pellet spectra of human bone and cerebrospinal fluid and various rat tissues in the region 700–400 cm^{-1}. The rat tissues were homogenized and lyophilized before being pressed into the pellet. Comparisons with data by Miller *et al.* (1960) indicated that the bands at 700, 600, and 560 cm^{-1} could be ascribed to the inorganic components of bone, phosphate, and calcium carbonate (Table 19.10).

Table 19.10. Infrared Absorption of Various Tissues and Assignments[a] (May, 1966)

Tissue	Band, cm^{-1}	Assignment
Bone	700w	CO_3^{2-}
	600s	PO_4^{3-}
	560s	
Spinal fluid	700w	protein
	600s	PO_4^{3-}
	465m	PO_4^{3-}
Cerebral tissues[b]	700m	protein, DNA
	535w	DNA
Noncerebral tissue[c]	700m	protein, DNA
	676sh	
	535w	DNA
Ossein	700m	protein, DNA

[a] Key: m, medium; sh, shoulder; w, weak.
[b] Cortex, medulla, and white matter.
[c] Blood, spinal cord, and spleen.

Absorption spectra of many inorganic ions have been determined in the region 4000 to 650 cm^{-1} by Miller and Wilkins (1952).

Bones, Teeth, and Minerals

The mineral components of enamel, dentine, and bone, as well as the processes of mineralization and demineralization in normal and abnormal states, have long been a central concern of dental research. Current activity in these areas is widespread and exploits a wide variety of technical approaches.

Enamel and Dentine

Comparison of Enamel and Dentine to Dahllite. Infrared spectrophotometry and X-ray diffraction studies of natural apatites before and after thermal treatment, led Herman and Dallemagne (1963) to conclude that among the apatites they studied, francolite and quercyite are comparable and the results were explained by considering that CO_3^{2-} groups are present within the apatitic lattice by substitution for PO_4^{3-} groups. Dahllite and Curaçao carbonate–apatite contain calcite, but no evidence was found for the presence of calcite in francolite and quercyite. Enamel, dentine, and bone were found to be comparable to dahllite (Fig. 19.11). Spectra of

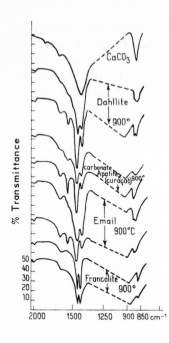

Fig. 19.11. Infrared spectra of calcite, dahllite, Curaçao carbonate–apatite, enamel (email), and francolite before and after heating at 900°C. (Herman and Dallemagne, 1963.)

dentine and bone have the same spectral characteristics but bands are less intense and not as well resolved. The extra-apatitic carbonate present in calcified tissues showed absorption bands at the same frequencies as calcite, but the absence of the band at $710 \, cm^{-1}$ was not explained. Calcified tissues had spectra comparable to mixtures of hydrated tricalcium phosphate and calcite, mixed in a colloidal mill. For these synthetic products, mixing both salts was enough to cause a decomposition of calcite, the carbonate ions of which are partly fixed at the surface of the phosphate microcrystals.

Fluoride and Enamel. The presence or absence of fluoride in the apatite crystallites of enamel is said to be the most important of the environmental factors so far identified as influencing the caries-forming process (Morris and Greulich, 1968). In recent years research has focused on the still obscure mechanism by which the fluoride ion renders the crystalline structure of enamel less susceptible to attack. It has recently been established that the lattice perfection (crystallinity) of enamel improves with increase in fluoride content, and that it is this "more perfect" structure that renders the enamel less soluble in acid. Also, the increased perfection achieved in enamel is not solely a function of lattice a-axis direction, as is the case in bone, but may be related more importantly to the c-axis.

Examination of synthetic fluoride-containing calcium phosphates by X-ray diffraction and infrared spectrophotometry has shown that hydrogen bonding takes place between hydroxyl protons and fluoride anions during the transition. These observations suggest that the hydroxylfluorapatite may be slightly more stable chemically than the hydroxylapatite, as well as more resistant to acid (Morris and Greulich, 1968).

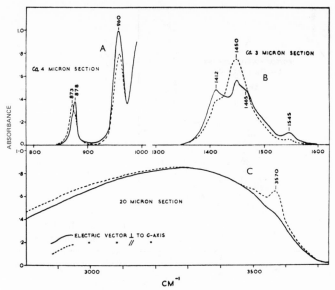

Fig. 19.12. Polarized infrared absorption spectra of longitudinal sections of human enamel. (Elliott, 1965.)

Elliott (1964) has stated that the frequency of the O—H band at 3570 cm^{-1} in hydroxyapatite is lowered if fluorine partially substitutes for the hydroxyl ions. The shift depends on the amount of substitution, and is rather small: a 50% substitution causes a lowering of only about 20 cm^{-1}. However, he believed that the determination of the frequency of the O—H band to measure the amount of replacement of hydroxyl by fluoride ions in hydroxyapatite could be as sensitive as X-ray diffraction techniques and is not subject to the same limitation, namely, that the apatite must be well crystallized.

Carbonate Ions and Enamel. Elliott (1965) has used polarized infrared absorption spectra to study longitudinal sections of human tooth enamel in order to determine whether carbonate ions can substitute for hydroxyl ions in the enamel. His conclusion was that carbonate ions substitute to a very limited extent for hydroxyl ions. The evidence consisted of certain bands in the infrared spectra of enamel (Fig. 19.12) which coincide with those of the synthetic apatite in which this substitution is known to have taken place (Fig. 19.13). In a hydroxyapatite that had been reacted with carbon dioxide at 1000°C, carbonate ion had absorption bands at 878, 1463, and 1528 cm^{-1}, and the hydroxyapatite 3570 cm^{-1} band (OH$^-$) had disappeared (Fig. 19.13). Elliott examined enamel which had been heated at 1000°C in carbon dioxide and measured the dichroism of the out-of-plane deformations at the 879 cm^{-1} mode (Fig. 19.14). From the dichroic ratio he was able to calculate that the plane of the carbonate ion is nearly parallel to the c-axis of the apatite. Elliott *et al.* (1948) have given the dichroic ratio applicable to this case as

$$R = 2 \cot^2 \theta$$

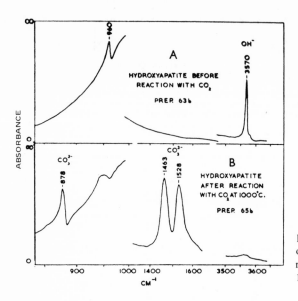

Fig. 19.13. Infrared absorption spectrum of hydroxyapatite before and after the reaction with dry carbon dioxide at 1000°C, bromoform mull. (Elliott, 1965.)

where θ is the angle between the perpendicular from the plane of the carbonate ion and the c-axis. The dichroic ratio is obtained by measuring the magnitude of the absorption at 879 cm^{-1} in the two directions of the polarizer.

The presence of carbonate in dental enamel and dentine is an important factor which contributes to their susceptibility to carious attack (Zapanta-LeGeros *et al.*, 1964). Carbonate increases the solubility of the apatite. The presence of carbonate

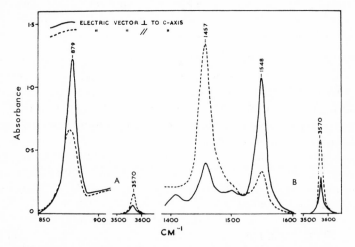

Fig. 19.14. Polarized infrared absorption spectrum of the carbonate ion that has replaced hydroxyl ions in the apatite lattice. (A) A 100-μ longitudinal section of enamel heated at 1100°C in air for 2 hr. (B) A 50-μ longitudinal section of enamel heated at 900°C in CO$_2$ for 30 min. (Elliott, 1965.)

in the apatite causes apatite to form with a shorter a-axis and slightly longer c-axis than the carbonate-free variety. Results of X-ray diffraction and chemical analyses suggest that the carbonate ion substitutes for the phosphate ion. Zapanta-LeGeros *et al.* (1964) have obtained further evidence from infrared studies. One of the absorption bands, due to the angular change of the oxygens in the phosphate ion, is observed at 635 cm^{-1} in mineral and synthetic apatite. This band is only very weakly present and even absent in the spectra of synthetic carbonate apatite, staffelite (carbonate-F-apatite), and biological apatite. They believe that this absence is due to the presence of carbonate ion. Also, another phosphate band at 1100 cm^{-1} (asymmetric stretching of P—O in the phosphate ion) present in the carbonate-free apatite is a much weaker band in staffelite and in biological apatites and is absent in high carbonate-containing synthetic apatites. The infrared data support the concept that carbonate ions substitute for phosphate ions in the apatite structure, and, because of their proximity to the phosphate groups, they cause these anomalies in the phosphate absorption bands. The intensities of the OH bands at 3200–3600 cm^{-1} were the same in the high-carbonate apatites as in the carbonate-free apatites. Carbonate does not substitute for the OH groups, according to these data.

Zapanta-Le Geros *et al.* (1970) have recently reviewed the subject of infrared spectra of carbonate-containing synthetic and biological apatites. Klein *et al.* (1970) have used the polarized infrared specular-reflectance technique to study single crystals of apatites. They found this technique to be a more powerful method for the analysis of the vibration spectra of crystalline structures than the powder absorption techniques.

Decalcification of Dentine. Fukuda (1966) has studied the spectra of normal human dentine and dentine which had been decalcified by treatment with five different agents: EDTA, nitric acid, sulfosalicylic acid, trichloroacetic acid, and lithium lactate. Quick and complete decalcification of dentine was effected by nitric, trichloroacetic, and sulfosalicylic acids, whereas decalcification by EDTA and lithium lactate was slower and less complete. the PO_4^{3-} bands at 1033 and 1095 cm^{-1} were suggested as bands to be examined for decalcification (Fig. 19.15).

Minerals and Bone

Infrared spectra of hydroxyapatite ($Ca_{10}(PO_4)_6(OH)_2$), octacalcium phosphate ($Ca_8H_2(PO_4)_6 \cdot 5H_2O$), and pyrolyzed octacalcium phosphate have been studied by Fowler *et al.* (1966). They investigated by means of infrared techniques, weight loss of sample, and pyrophosphate analysis the conditions under which pyrophosphate formation according to either of the following two reactions would be sufficiently quantitative so that it could be used to measure the amount of octacalcium phosphate in an apatitic sample:

$$Ca_8H_2(PO_4)_6 \cdot 5H_2O = 0.5Ca_{10}(PO_4)_6(OH)_2 + 1.5Ca_2P_2O_7 + 5.5H_2O$$

$$Ca_8H_2(PO_4)_6 \cdot 5H_2O = 2Ca_3(PO_4)_2 + Ca_2P_2O_7 + 6H_2O$$

Simple dehydration was continuous from 50 to about 400°C. The major products between 325 and 600°C were hydroxyapatite and β-$Ca_2P_2O_7$; between 650 and 900°C, the products were β-$Ca_2P_2O_7$ and β-$Ca_3(PO_4)_2$.

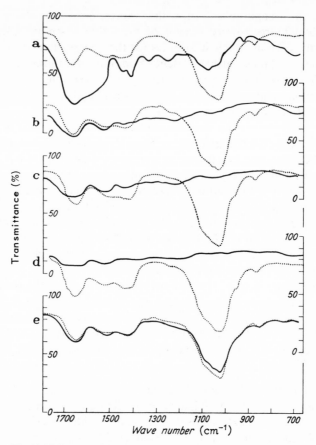

Fig. 19.15. Spectra of decalcified dentine; (———) decalcified dentine,
(·····) normal dentine, (a) with EDTA; (b) with HNO₃; (c) with
CCl₃COOH; (d) with sulfosalicylic acid; and (e) with lithium
lactate. (Fukuda, 1966.)

Posner and Perloff (1957) have introduced the concept of hydrogen bonding
between oxygens of adjacent orthophosphate groups to account for calcium defi-
ciencies in synthetic hydroxyapatites. Infrared evidence supported this concept by
showing a linear relationship between the absorbance of the hydrogen-bonded O—H
stretching frequency at 3400 cm^{-1} and the Ca/P ratio of several synthetic hydroxyapatites (Posner *et al.*, 1960). More recently, Stutman *et al.* (1962) ruled out an objec-
tion to the concept by Winand *et al.* (1961), who had suggested that the band at 3400
cm^{-1} may be caused by adsorbed water.

In the different calcium phosphates, the hydrogen-bonded O—O distance,
obtained from infrared data, decreases from phase to phase as a function of acidity.
This is corroborated by X-ray diffraction data, which indicate that the phosphate
groups in a more acid calcium phosphate occupy a higher percentage of the structural
volume, resulting in shorter hydrogen-bonded O—O distances. Table 19.11 presents

Table 19.11. The Hydrogen-Bonded O—O Distance as a Function of Acidity in Different Calcium Phosphates (Stutman *et al.*, 1962)

Compound	Ca/P	O—H stretching frequency, cm^{-1}	O—O distance from infrared data, Å	PO$_4$ volume index[a]
Ca(H$_2$PO$_4$)$_2$	0.50	2320	2.55	0.0132
Ca(H$_2$PO$_4$)$_2$·H$_2$O	0.50	2320	2.55	
CaHPO$_4$	1.00	2400	2.58	0.0129
CaHPO$_4$·2H$_2$O	1.00	2400	2.58	
Ca$_4$H(PO$_4$)$_3$·3H$_2$O	1.33	3000	2.70	
Ca$_{10-x}$H$_{2x}$(PO$_4$)$_6$(OH)$_2$	1.67	3400	2.85	0.0113

[a] Index is equal to the number of PO$_4$'s per unit cell per unit cell volume.
These values are the ones published in Stutman *et al.* (1962) corrected by multiplying by 0.01 (private communication from A. S. Posner).

these relationships (Stutman *et al.*, 1962). The values in the last column are obtained by dividing the number of phosphate groups per unit cell by the unit cell volume. This calculation yields an index of the number of phosphate groups per Å^3. These "PO$_4$ volume indices" were placed in the table to show that the more acid calcium phosphates have a greater number of phosphate groups per Å^3 than the more basic calcium phosphates. The formula of hydroxyapatite precipitated in a carbonate-free atmosphere is Ca$_{10-x}$H$_{2x}$(PO$_4$)$_6$(OH)$_2$.

It has been suggested that the mineral portions of bone and tooth tissue are calcium-deficient hydroxyapatites (Posner and Perloff, 1957; Neuman and Neuman, 1953). The possible presence of hydrogen bonds in these materials may provide a much needed parameter for an estimation of the stoichiometry of these biological hydroxyapatites under varying conditions (Posner *et al.*, 1960).

Percentage Crystallinity in Minerals and Rat Bone. An infrared method has been used (Termine and Posner, 1966*a*) to determine the percentage of crystallinity in a series of apatitic calcium phosphates. Infrared spectra of the solids formed in the conversion (in a carbonate-free alkaline medium) of amorphous calcium phosphate to crystalline hydroxyapatite show a gradual splitting of the 600 cm^{-1} phosphate ion antisymmetric bending frequency from a broad singlet for amorphous calcium phosphate to a well-defined doublet for the fully crystalline solid (Stutman *et al.*, 1965) (Fig. 19.16).

The method of measuring the splitting of the phosphate singlet at ~600 cm^{-1} is shown in Fig. 19.17. A baseline was drawn, delimiting the area A_2 of the band. The two minimum points of transmission on the band were connected to form the splitting area, A_1. Both areas were determined by planimeter. The amount of splitting was found from the ratio A_1/A_2, and is called the splitting function (SF). A completely amorphous calcium phosphate gives only a single band, and its SF is zero, since A_1 is equal to zero. The precision of the measurement of the SF was $\pm 3\%$.

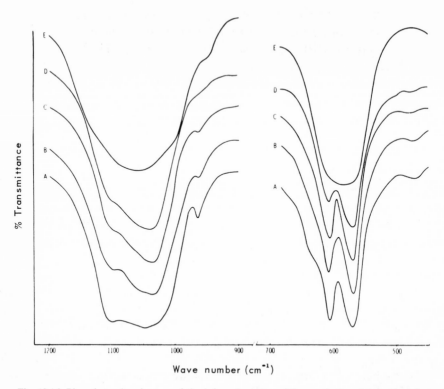

Fig. 19.16. Phosphate absorbances of the infrared spectra of the synthetic calcium phosphate series. The % crystallinity of each sample, as measured by X-ray diffraction, is as follows: A, 100%; B, 81%; C, 81%; D, 30% and E, 0% (amorphous). The Ca/P molar ratios were: A, 1.646; B, 1.627; C, 1.566; D, 1.544; and E, 1.536. (Stutman et al., 1965.)

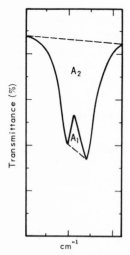

Fig. 19.17. Method of obtaining splitting function. (Termine and Posner, 1966.)

Figure 19.18 shows a plot of SF values *vs* X-ray diffraction-assayed percentage of crystallinity (% C) for two series of mixed (i.e., amorphous and crystalline) phosphates. The series denoted as 1,2,3 was based on amorphous-to-crystalline ratios (by weight, of standard amorphous calcium phosphate and a pure crystalline hydroxyapatite) ranging from 0 to 100% crystalline; the series denoted VI, VII, VIII, XIII was obtained by sampling the preparation of hydroxyapatite at various times during the conversion of the precursor amorphous calcium phosphate to the final crystalline phase (this series was 0–100% crystalline as determined by X-ray diffraction). The plot (after statistical analysis) shows that the infrared splitting function measurement can serve as a quantitative measure of the percentage of crystallinity for the calcium phosphate system. The practical lower limit of detection of the percentage of crystallinity is about 15% by this method. A statistical analysis of variance showed no significant differences in either splitting function or %C between standard mixture sample points 1, 2, and 3 and experimentally obtained points VI, VII, VIII, and XIII. The splitting function values were also found to be useful for obtaining the Ca/P ratio of a given calcium phosphate mixture of amorphous (Ca/P = 1.50) and crystalline hydroxyapatite (Ca/P = 1.67).

The infrared analysis is advantageous over X-ray diffraction in that the former method uses only a few milligrams of sample, while the latter method uses at least 25–50 times as much. Also, the time required for the infrared analysis is much shorter than that required for analysis by X-ray diffraction (Termine and Posner, 1966a).

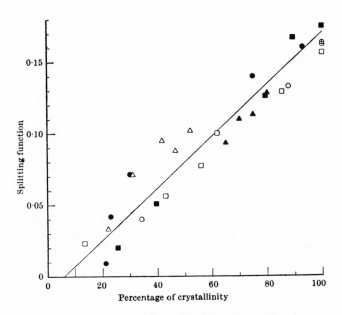

Fig. 19.18. Plot of infrared splitting function *vs* X-ray diffraction percentage of crystallinity of a series of calcium phosphate. ⊕, VI; ○, VII; ●, VIII; □ XIII: △, 1; ■, 2; ▲, 3. (Termine and Posner, 1966.)

Table 19.12 (Termine and Posner, 1966*b*) shows the application of the splitting fraction-method to the analysis of rat bone. Values obtained for percentage crystallinity in 13-, 18-, 29-, and 38-day old femurs were measured by both infrared and X-ray methods. The latter values were within $\pm 2\%$ of the results by the infrared method. It is seen that mature rat bone contains constant levels of both amorphous and crystalline calcium phosphate.

Table 19.12. Analysis of Percentage Crystallinity of Rat Bone by Infrared Spectrophotometric Splitting Fraction (SF) Measurements (Termine and Posner, 1966*b*)

Age of rat femurs, days	Ash, %	Ca/P, molar ratio	SF	Mineral content, %	
				Amorphous	Crystalline
8	18.4	1.41	0.0100	69.5	30.5
13	30.0	1.46	0.0185	56.0	44.0
18	32.3	1.50	0.0240	47.0	53.0
29	44.3	1.55	0.0300	37.5	62.5
32	42.1	1.55	0.0305	37.0	63.0
35	44.8	1.57	0.0315	35.5	64.5
38	43.8	1.56	0.0310	36.0	64.0
41	46.5	1.59	0.0310	36.0	64.0
44	44.7	1.58	0.0310	36.0	64.0
47	46.9	1.59	0.0315	35.5	64.5
50	47.8	1.60	0.0310	36.0	64.0
53	47.1	1.60	0.0310	36.0	64.0
56	50.1	1.61	0.0315	35.5	64.5
264	56.6	1.63	0.0315	35.5	64.5

Wide variations in the proportions of organic to mineral material exist in bone tissue. Microradiography has demonstrated the presence of a range of specific gravities in diaphyseal bone extending from 1.65 to 2.25 g/cm^3. X-ray diffraction and infrared analysis have been applied to bone fractions of progressively increasing specific gravities that had been obtained by density-gradient separation from diaphyseal bone samples in growing rats (Quinaux and Richelle, 1967). These data were related to the sequence of events leading to the mineralization of bone. Table 19.13 gives the frequencies of the absorption bands for the various specific gravity fractions before and after heating at 850°C. The spectra of the unheated materials were poorly resolved. In addition to a broad band centered at 1000 cm^{-1}, a region of phosphate frequencies, two bands are caused by absorbed water (1650 and 3378 cm^{-1}) and four bands are located at 1410, 1450, 1520, and 869 cm^{-1}, attributed to carbonate.

As the specific gravity increases, the intensity of the broad band in the vicinity of 1000 cm^{-1} increases in relation to the intensity of the bands located at 1410, 1450,

Table 19.13. Infrared Absorption Frequencies for the Various Specific Gravity Fractions before and after Heating at 850°C (Quinaux and Richelle, 1967)

Band frequencies, cm^{-1}

Unheated fractions

Specific gravity (mg/mm³)							
1.65	3378s	1650s	1527s	1447m	1408w		875w
1.75	3378s	1650s	1524s	1449m	1408w	1231s	869w
1.85	3378s	1650s	1529s	1445m	1416m	1228m	869m
1.95	3378s	1650s	1534s	1447m	1408m	1235m	869m
2.05	3378s	1645s	1515m	1447w	1442w	1235w	869m
2.15	3375s	1639m	1515w	1440w	1414w	1228vw	869m
2.25	3378s	1639m	1515w	1449w	1412w	1234vw	869m

For all fractions, unresolved bands in the 1050 region.

Fractions heated for 1 hr at 850°C

Specific gravity (mg/mm³)																			
1.65		3401vs	1618w			1400w	1208w	1182w	1168w	1152w	1117s	1075m	1045s	1033s	1011m	981s	976s		941m
1.75	3533vw	3400vs	1623m	1538w	1451m	1402w					1118s	1087s	1044vs	1035vs	1015m	982s		961w	945w
1.85	3533w	3401vs	1618m	1538w	1451w	1400w					1118s	1088s	1038vs		1016vw	982s		961w	945w
1.95	3533w	3401vs	1623m		1449vw	1389m					1119m	1089s	1044vs		1016vw	982m		961m	945vw
2.05	3533m	3390vs	1618m		1449vw	1389vw					1116w	1087s	1044vs		1016vw	981w		961m	945vw
2.15	3533m	3400vs	1623m		1449vw	1389vw					1116vw	1088s	1043vs			982vw		961m	946m
2.25	3533m	3378vs	1623m		1449vw	1389vw					1117vw	1088s	1040vs			982vw		961m	945m

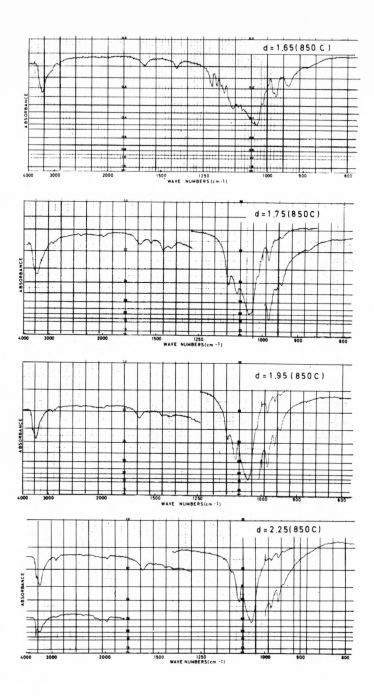

Fig. 19.19. Infrared spectra of the specific gravity fractions 1.65, 1.75, 1.95, and 2.25 after heating for 1 hr at 850°C. (Quinaux and Richelle, 1967.)

and $1520 \, \text{cm}^{-1}$, which decrease. Besides, as reported earlier (Richelle and Quinaux, 1964), in the low-specific-gravity fractions, the band located at $1520 \, \text{cm}^{-1}$ is more intense than the bands at 1450 and $1410 \, \text{cm}^{-1}$. This situation is progressively reversed as the specific gravity increases. The band at $869 \, \text{cm}^{-1}$ shows no apparent difference from one spectrum to another. In the low-specific-gravity fractions, a better defined band occurs at $1230 \, \text{cm}^{-1}$, which is also present, but less intense in the fractions of higher specific gravity.

After the fractions were heated for one hour at 850°C the spectra were better resolved (Fig. 19.19). In the fraction of specific gravity 1.65, bands due to the presence of β-$Ca_2P_2O_7$ are seen at 976, 1152, 1168, 1182, and $1208 \, \text{cm}^{-1}$. Also, bands due to anhydrous tricalcium phosphate (β-TCP) are found at 941, 981, 1011, 1045, 1075, and $1117 \, \text{cm}^{-1}$. In the fraction of specific gravity 1.75, the bands of β-$Ca_2P_2O_7$ are no longer detectable. Two small bands begin to appear at 961 and $3533 \, \text{cm}^{-1}$; they are caused, respectively, by PO_4^{3-} groups in hydroxyapatite and by the stretching vibration of hydroxyl groups, thus indicating the presence of some hydroxyapatite. The bands due to β-TCP still predominate and do not allow the resolution of the 1040 and $1090 \, \text{cm}^{-1}$ superimposed bands. As the specific gravity increases, the bands due to hydroxyapatite become more intense. The bands at 960, 1040, and $1090 \, \text{cm}^{-1}$ become more clearly resolved progressively while the intensity of the bands due to β-TCP weakens more and more. In the fraction of specific gravity 2.25, mainly hydroxyapatite bands are present, with only two bands of β-TCP clearly present. The carbonate bands also vary considerably among the spectra. In the fraction of specific gravity 1.65, carbonate bands appear at 1400 and $868 \, \text{cm}^{-1}$. The fraction of specific gravity 1.75, shows all three bands in the 1400 to $1500 \, \text{cm}^{-1}$ region. The intensity of the bands weakens rapidly as the specific gravity increases, and is almost completely gone in the fraction of specific gravity, 2.25.

Quinaux and Richelle (1967) concluded that the steps leading to the mineralization of bone were the precipitation of an amorphous material, followed by the appearance of an initial and then a final crystalline precipitate. The initial crystalline precipitate is a highly calcium-deficient apatite and the final precipitate is hydroxyapatite.

Relationship of Saliva to Caries Status

In a study of the infrared spectral characteristics of human whole stimulated saliva collected from individuals classified as to dental caries status, a correlation was found between absorption bands at 1389 and $833 \, \text{cm}^{-1}$, and the caries status of 398 airmen of the U.S. Air Force (Harris et al., 1959). Also flow rate and these infrared bands were directly correlated. In order to reinforce such presumptive evidence that the characteristics of the infrared curves might be an indicator of caries-susceptibility status, a series of infrared studies of the saliva of normally caries-resistant rodents and other animals was done and the results were compared with the composite curves of previous human caries-resistant and caries-susceptible groups (Harris et al., 1960). Included in the study were pilocarpine-stimulated salivary specimens of the rat, hamster, rabbit, cat, dog, and monkey. The spectra for all experimental animals within the same species were identical. The spectra of the saliva among the

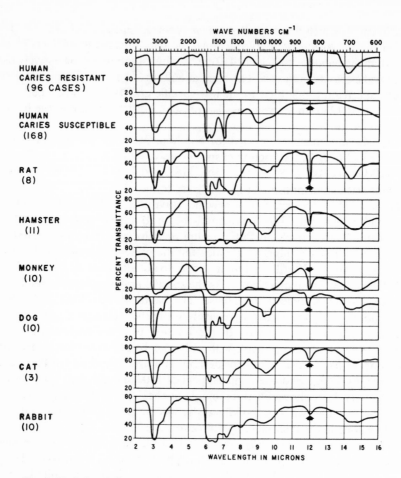

WAVE NUMBERS CM⁻¹

Fig. 19.20. Infrared absorption curves of stimulated saliva of various species compared with that of caries-resistant and caries-susceptible humans. (Harris *et al.*, 1960.)

various species were, generally, also similar to one another although several specific absorption bands were evident (Fig. 19.20). A band is seen at 833 cm⁻¹ for all the species and the human caries-resistant saliva, but not for the human caries-susceptible group. Since the animals selected were normally caries-resistant, the findings indirectly supported the hypothesis that at least the component responsible for the band at 833 cm⁻¹ is related to a human caries-resistant status.

Vitamins, Hormones, and Coenzymes

Rosenkrantz (1957) has published a review of the spectra of vitamins, hormones, and coenzymes in the solid state. Among the spectra of vitamins were those of vitamin A and its acetate, vitamin D_2, α-tocopherol, menadione (the common substitute for vitamin K), thiamine hydrochloride, nicotinamide, niacin, riboflavin, pyridoxine hydrochloride, pyridoxal hydrochloride, pyridoxamine dihydrochloride, calcium pantothenate, folic acid, vitamin B_{12}, biotin, p-aminobenzoic acid, choline chloride, i-inositol, α-lipoic acid, ascorbic acid, and dehydroascorbic acid. Among the hormone spectra were those of epinephrine, norepinephrine hydrochloride, L-thyroxine, 3,5,3′-L-triiodothyronine, estradiol-17β, estrone, progesterone, testosterone, cortisone, cortisol (hydrocortisone), aldosterone, Δ^1-cortisone, Δ^1-cortisol, 9α-fluorocortisol, 2-methyl-9α-fluorocortisol, zinc insulin, glucagon, α-adrenocorticotropin (ACTH), growth hormone, interstitial cell stimulating hormone (ICSH), and lactogenic hormone.

Among the spectra of coenzymes given by Rosenkrantz were glutathione (reduced and oxidized), diphosphopyridine nucleotide (DPN^+ or NAD^+) and its reduced form (Na salt), triphosphopyridine nucleotide and its reduced form (Na salt), flavin mononucleotide, flavin adenine dinucleotide, oxidized cytochrome c (containing some of the reduced form), and thiamine pyrophosphate.

Hvidt and Kägi (1963) have published spectra of NAD^+ and NADH in D_2O.

Parker (1961) has presented spectra of several vitamins in aqueous solution. These include: calcium pantothenate, choline chloride, thiamine hydrochloride, nicotinamide, pyridoxine hydrochloride, and pyridoxal phosphate.

Vitamins D_2 and D_3

Morris *et al.* (1962) have distinguished between vitamins D_2 and D_3 by examination of the infrared spectrum between 1000 and 909 cm^{-1} (10 and 11 μ) (Fig. 19.21). They described a technique for the determination of the proportion of each form of vitamin D present in mixtures of the two by means of a reference curve relating the ratio of absorbance differences $(A_{10.3\mu} - A_{10.5\mu})/(A_{10.4\mu} - A_{10.5\mu})$ to the percentage composition of the mixture. The amount of each form could then be calculated from the total vitamin D content of the sample.

Tocopherols

Csallany and Draper (1963) have elucidated the structure of a metabolite of d-α-tocopherol which had previously been found in rat, mouse, and pig livers after

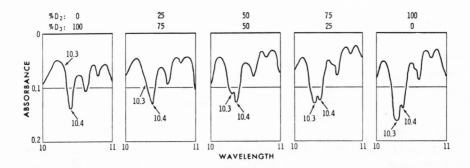

Fig. 19.21. Infrared spectra, between 10.0 and 11.0 μ, of pure vitamin D_2, pure vitamin D_3, and mixtures of vitamins D_2 and D_3. (Morris *et al.*, 1962)

administration of d-α-tocopherol-5-methyl-^{14}C. Infrared, NMR, and ultraviolet spectroscopy and chemical reactions led to the proof of structure of this dimeric form of the vitamin, 2,2′,5,5′,8,8′-hexamethyl-2,2′-(4,8,12-trimethyltridecyl)dichroman-7,7′-ene-6,6′-one, which they abbreviated to di-α-tocopherone (CXXIV, R = $C_{16}H_{33}$).

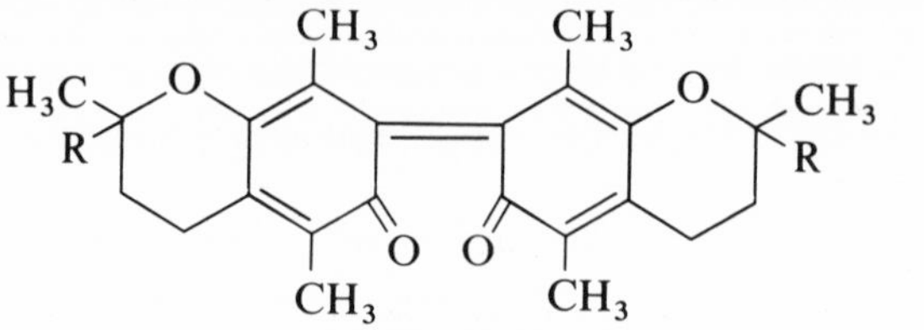

CXXIV

They gave a detailed infrared analysis of the compound, including infrared data for two C=O groups, a C=C conjugated bond, and the position of the C=C bond linking the two halves of the dimer, and rejected on the basis of their data a structure proposed by other workers.

Differences in the infrared spectra of the tocopherols have been observed (Rosenkrantz, 1948; Rosenkrantz and Milhorat, 1950) and have been used to determine the percentage of α-tocopherol in δ-tocopherol (Stern *et al.*, 1947). Morris and Haenni (1962) have applied such spectra (Fig. 19.22 and Table 19.14) to the identification of the form of tocopherol present in different vegetable oils.

Table 19.14. Distinguishing Peaks, in μ, in the Infrared Spectra of the Tocopherols (Morris and Haenni, 1962)

Form of Tocopherol	Type of Peak		
	Singlet	Doublet	Triplet
d-α	7.92	8.55, 8.65	8.97, 9.20, 9.40
	8.25		
d-β	8.10		8.47, 8.55, 8.65[a]
d-γ	8.65	9.07, 9.25	
		(8.05, 8.20)	
d-δ	8.47	(8.05, 8.20)	
	8.75		
	9.55		

[a] The components at 8.47 μ and 8.55 μ are shoulders.

Nair and Luna (1968) have isolated α-tocopherol from rat heart muscle and have characterized it by means of gas–liquid chromatography and infrared and mass spectroscopy. The infrared spectra of trimethylsilyl ethers of known standards and unknown extracts were identical. They were characterized by the lack of an OH

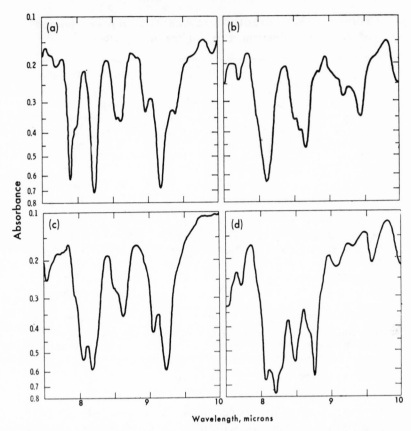

Fig. 19.22. Infrared spectra of the tocopherols. (a) d-α-tocopherol, (b) d-β-tocopherol, (c) d-δ-tocopherol, and (d) d-γ-tocopherol. (Morris and Haenni, 1962.)

absorption band in the 3704 cm^{-1} region and by strong bands at 1250 cm^{-1} for the chroman nucleus and 943, 847, and 752 cm^{-1} for the Si—CH$_3$ and Si—O—C groups.

Characteristic frequencies of the tocopherols appear in Table 19.15.

Vitamins K

Noll (1960) has used pure synthetic reference compounds, vitamins K$_{1(10)}$, K$_{1(15)}$, K$_{1(20)}$, K$_{1(25)}$, K$_{1(30)}$, K$_{2(5)}$, K$_{2(10)}$, K$_{2(15)}$, K$_{2(20)}$, K$_{2(25)}$, K$_{2(30)}$, K$_{2(35)}$, K$_{2(45)}$, and K$_{2(50)}$, and has developed a method for the rapid identification of vitamin K homologues. (The subscript in parentheses indicates the number of carbon atoms of the side chain in the 3-position of the naphthoquinone ring.) Absorption bands near 1100 and 840 cm^{-1}, which are present in K$_2$ spectra but absent in K$_1$ spectra, allow qualitative differentiation to be made between the two homologous series.

The length of the side chain was determined with an accuracy of ± 0.2 isoprenoid unit by measurements in carbon tetrachloride solution of the relative intensities of certain bands. A linear relationship was observed between the number of isoprenoid

Table 19.15. Characteristic Absorption Frequencies[a] of the Tocopherols, cm^{-1} (Pennock, 1965)

Tocopherol	O—H stretch	Tentative assignments (C=O / O—O—CH$_3$,R)		Aromatic absorption[d]
Tocol	3375	1217	1152	1615
5-Methyl-tocol	3440	1245	1155	No band
7-Methyl-tocol	3445	1237[b]	1168[b]	1629
				1587, 1500
8-Methyl-tocol, δ	3345	1209	1140	1605
5,7-Dimethyl-tocol, ζ_2	3443	1222	1151	1629
				1587
5,8-Dimethyl-tocol, β	3356	1222	1145	1595
7,8-Dimethyl-tocol, γ	3390	1220	1143	1613
5,7,8-Trimethyl-tocol, α	3509	1209[c]	1155	1595
		1250		1618
5,8-Dimethyl tocotrienol, ϵ	3378	1225	1159	1592
5,7,8-Trimethyl toco-trienol, ζ_1	3425	1242	1155	None given

[a] Data from Kofler et al. (1962), Green et al. (1959), and Mamalis et al. (1958).
[b] 1237 cm^{-1} band very weak, 1168 cm^{-1} very strong. It looks as though the two bands are condensed into one (for spectrum see Kofler et al., 1962).
[c] Doublet.
[d] All weak except band at 1500 in 7-methyl-tocol.

units in the side chain and the absorbance ratios of the side-chain bands to the naphtho-quinone bands. Noll has given values for these absorbance ratios, which are characteristic for each homologue and independent of the concentration of the substance and the optical path length. The quantitative work involved the use of internally standardized slit widths. The results of this study indicated that, of the available methods for the characterization of K vitamins, quantitative infrared spectrophotometry afforded the greatest precision.

Cobalamins

Hogenkamp et al. (1965) synthesized a series of alkylcobalamins, e.g., ethylcobalamin, carboxyethylcobalamin, carboxymethylcobalamin, methoxycarbonylethylcobalamin, and found pronounced differences in their visible and ultraviolet spectra, which suggested that changes in the inductive character of the alkyl group cause changes in the electronic nature of the corrin ring. In contrast the infrared spectra of these alkylcobalamins, coenzyme B_{12}, cyano- and hydroxocobalamin (all of which have different ultraviolet and visible spectra) are very similar. In the region where the carbon–cobalt absorption is expected, a rather broad band appears at 600 to 400 cm^{-1} in all cobalamins. A weak absorption band is observed in the spectrum of methylcobalamin at 348 cm^{-1}, which can be assigned to the carbon–cobalt stretching vibration (Nakamoto, 1963).

Metabolites of Nicotinamide

After the oral administration of nicotinamide to human subjects, two ultraviolet-absorbing compounds have been regularly detected on chromatograms of plasma extracts (Abelson *et al.*, 1963). The less polar of these compounds is *N'*-methyl-2-pyridone-5-carboxamide. Chemical and infrared spectral data have identified the other more polar metabolite as *N'*-methyl-4-pyridone-3-carboxamide.

Quinones

Pennock (1965) has given a detailed review of the infrared and NMR spectra of quinones and related compounds, covering the prenyl-substituted benzo- and naphthoquinones, related chromenes, chromanes, and some isoprenoid hydrocarbons and alcohols. Table 19.16 presents the main frequencies of the substituted quinone nuclei of some naturally occurring isoprene-substituted quinones. Additional bands are also accounted for in the work of others cited by Pennock. Table 19.17 shows absorption frequencies of quinones in the region $1350-1250\ cm^{-1}$.

Among the quinones that have been discussed in Pennock's review were plastoquinone, β-tocopherolquinone, α-tocopherolquinone, $E_{2(45)}(2,3,5$-trimethyl-6-

Table 19.16. Some Absorption Frequencies, cm^{-1}, of the Various Parts of the Quinone Nuclei of some Naturally Occurring Isoprene-substituted Quinones (Pennock, 1965)

Quinone	C=O stretch of quinone	C=C of quinone ring	Aromatic vibrations	Others
Plastoquinone	1647	1623		
β-Tocopherol-quinone	1649	1617		
α-Tocopherolquinone	1641	—		
$E_{2(45)}$[a]	1639	—		
Rhodoquinone[b]	1640	1600		3580, 3480 (O—H) 1560 (OH in ring)
Ubiquinone (50)	1653	1608		1259, 1153, 1095 cm^{-1}, ether links
Vitamin K_1	1660	1618	1595 ⎫	1299 and 714 cm^{-1} due to aromatic
Vitamin K_2 (MK-7)	1655	1615	1590 ⎭	ring
Chlorobiumquinone[c]	1660	1610		700, 690 cm^{-1} 4-hydrogens in an aromatic ring

[a] No infrared spectrum for $E_{2(50)}$ (Martius, 1962) has been published but that of a synthetic compound, 2,3,5-trimethyl-6-solanesyl 1,4-benzoquinone ($E_{2(45)}$) has been reported (Schudel *et al.*, 1963).
[b] See Glover and Threlfall (1962).
[c] See Frydman and Rapoport (1963).

Table 19.17. Absorption of Quinones in the 1350–1250 cm^{-1} Region (Pennock, 1965)

Quinone	Bands	
Vitamin K$_1$	1330m	1295vs
Vitamin K$_{2(35)}$	1330m	1295vs
Ubiquinone (50)	1282m	1263vs
E$_{2(45)}$	1299	1250
α-Tocopherolquinone	1310s	1280s
β-Tocopherolquinone	1318s	1260m
Plastoquinone	1325m	1266s

Key: m, medium; s, strong; v, very.

Table 19.18. Absorption Frequencies of some Isoprenoid Alcohols^a, cm^{-1} (Pennock, 1965)

Alcohol	O—H stretch	C—H stretch	C=C stretch	C—CH$_2$ CH def.	C—CH$_3$ CH def.	C—O stretch	Skeletal vibrations^c			
Phytol	3401	2976	1675	1466	1381	1006	—	—	—	737
Farnesol	3410	2950	1664	1448	1381	1005	894	838	806~	745
Solanesol	3413	2950	1669	1445	1383	1000	893~	840	805~	746
Spadicol^b	3390	2941	1667	1445	1383	1000	893~	840	800~	752
Castaneol	3413	2976	1669	1447	1379	1002	890	839	—	741
Yeast "dolichol"	3425	3003	1672	1453	1383	1062	892	839	—	741
Dolichol	3401	2976	1672	1451	1383	1060	892	839	—	743
Aspergillus "dolichol"	3310	2960	1669	1450	1378	1058	888	834	—	735
Perhydrosolanesol	3436	3003	—	1464	1381	1058	—	—	—	735

^{a} All the compounds were examined as oily smears. The spectrum of castaneol was recorded by Stevenson (1963) and the aspergillus "dolichol" by P. H. W. Butterworth (cited by Pennock, 1965). The farnesol was natural but may be *trans–trans* or *cis–trans*.
^{b} Synthetic, *trans*-decaprenol (Hemming *et al.*, 1963).
^{c} Inflection in the spectrum is indicated by ~.

solanesyl-1,4-benzoquinone), rhodoquinone, ubiquinone (50), vitamin K$_1$, vitamin K$_2$ (MK-7), and chlorobiumquinone. Another topic included in the review by Pennock was the tocopherols. Table 19.15 presents characteristic frequencies of these compounds. Isoprenoid hydrocarbons and alcohols were also discussed. Table 19.18 gives frequencies of some of the latter type of compounds.

Catechol-Related Polymers

Biological polymers related to catechol have been studied by means of electron paramagnetic resonance and infrared spectroscopy in order to characterize certain chemical properties (Tollin and Steelink, 1966). Among the polymers were squid melanin, melanin formed from dihydroxyphenylalanine by tyrosinase, gambir tannin, catechin, humic acid, lignin from several sources, and various quinone products. All the macromolecules contained stable free-radical moieties and all displayed infrared absorption bands which are consistent with hydroxyquinone structures in the free acid form and in the sodium salt form. For example, DOPA–melanin in the free-acid form has bands at 3448, 1695, and 1639 cm^{-1}, and in the salt form, at 3448 and 1587 cm^{-1}.

Synthetic Blood Substitutes

Zhbankov *et al.* (1963) have studied the spectra of polyglucine films (a glucose polymer of molecular weight 6000 obtained from the hydrolysis and fractionation of native dextrin, a by-product of the metabolism of the microbe *Leuconostoc mesenteroides*, under certain conditions). Polyglucine is a synthetic blood-substitute polymer. A thin layer ($\sim 3.5\,\mu$) of the solution to be analyzed was applied directly onto a thallium bromide–iodide plate and the spectra of the films showed definite regular changes with addition of salts to the solution. The changes were independent of the salt added (the 870, 950, 1240, and 1420 cm^{-1} bands increased in intensity, and the 850 cm^{-1} band grew weaker when either NaCl, KCl, or KBr was added) and had nothing in common with the spectral features of these salts.

Behen *et al.* (1964) have developed an infrared method for the quantitative determination of the blood plasma volume expander, polyvinylpyrrolidone, in serum and urine from experimentally infused dogs.

Botanical Products, Botany, and Plant Physiology

Polyisoprenoids

Stone *et al.* (1967) have presented evidence from mass, NMR, and infrared spectrometry in favor of the presence of *cis-trans*-decaprenol, -undecaprenol, and -dodecaprenol in the mixture of polyprenols they isolated from the leaves of the decorative rubber plant, *Ficus elastica*. The infrared spectra of these compounds, to which the trivial names ficaprenol-10, -11, and -12 were given, are presented in Fig. 19.23. The only significant differences among the spectra are the intensities of the O—H stretching bands at 3575 and 3310 cm^{-1} and the C—O stretching bands at 1000 cm^{-1} relative to the intensities of the other bands in each spectrum. These bands are progressively less intense, relatively, as the size of the molecules increases.

The strong band at 1000 cm^{-1} shows that each compound is a primary allylic alcohol and the bands at 835 cm^{-1} (C—H deformation of a trisubstituted olefin), at 1660 cm^{-1} (C=C stretching) and at 3024 cm^{-1} (C—H stretching of =CH), together with the relative intensities of the bands at 1450 cm^{-1} (C—H deformation of —CH$_2$ and —CH$_3$) and at 1365 cm^{-1} (C—H deformation of —CH$_3$) are consistent with a polyisoprenoid structure (Bellamy, 1958). Also, weak bands appear at 888, 1089, 1130 (with a weaker band at 1149 cm^{-1}), at 1238 (with a shoulder at 1221 cm^{-1}) and at 1307 cm^{-1} (with a shoulder at 1330 cm^{-1}). This pattern, and the absence of a shoulder at 795 cm^{-1} on the side of the intense band at 835 cm^{-1} is in keeping with each polyprenol having more *cis* than *trans* isoprene residues.

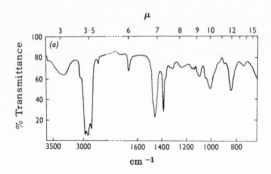

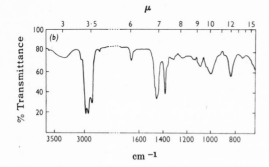

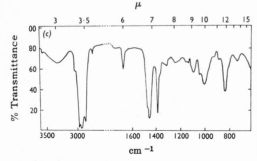

Fig. 19.23. Infrared absorption spectra of fica-prenols as solvent-free films between rock salt prisms. (a) Ficaprenol-10; (b) ficaprenol-11; and (c) ficaprenol-12. The films of the different prenols differed in thickness. (Stone *et al.*, 1967.)

Wellburn *et al.* (1967) have described the isolation and purification of a mixture of *cis-trans*-polyprenols from the leaves of the horse chestnut tree. Results of studies involving mass, NMR, and infrared spectroscopy and various chemical reactions showed that each of the prenols contains three *trans* internal isoprene residues and a *cis* "OH-terminal" isoprene residue. They differ from each other only in the number of *cis* internal isoprene residues. The names castaprenol-11, -12, and -13 were given to these compounds.

Lignins and Lignin Model Compounds

Infrared spectra in the O—H and C=O stretching regions of lignin model compounds and of lignins isolated in various ways from *Eucalyptus regnans* F. Muell.

have been obtained (Michell, 1966) for samples both in the solid state and in solution. In the lignin macromolecules all hydroxyl groups appeared to be involved in hydrogen bonds, and the phenolic groups were sterically hindered. The solvent dependence of the 1665 cm^{-1} band in the lignin spectra supported its assignment to a conjugated aldehyde or ketone group, but the dependence of the 1725 cm^{-1} band suggested that both carboxyl and carbonyl groups may contribute to the band envelope.

Gibberellins

In order to clarify the structure of ring A of gibberellins Mori *et al.* (1961) have prepared thirteen lactones of a cyclohexane series. γ-Lactones showed the characteristic absorption band in the range 1775–1782 cm^{-1} in dioxane, while δ-lactones absorbed in the range 1730–1762 cm^{-1}. Since the absorption band due to the lactone carbonyl of gibberellins occurred in the range 1777–1786 cm^{-1} in dioxane, the lactone ring of gibberellins appeared to have a γ-configuration.

Schneider *et al.* (1965) have discussed the gibberellins, their derivatives and degradation products, and have given UV, ORD, NMR, and IR data.

A Germination Inhibitor

Mitchell and Tolbert (1968) have isolated from the sugar beet fruit a compound which inhibits the germination of lettuce seeds. The structure of this germination inhibitor, *cis*-4-cyclohexene-1,2-dicarboximide, is shown in (CXXV).

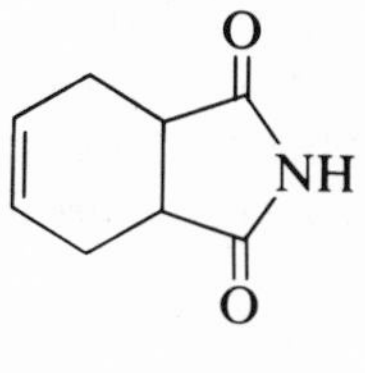

CXXV

Spectra of this compound showed NH next to a strong electronegative atom (3215 cm^{-1}), unsaturated CH stretch (3058 and 2976 cm^{-1}), methylene CH (2924 and 2865 cm^{-1}), C=O of a 5-member cyclic imide ring (1767 and 1704 cm^{-1}) (Nakanishi, 1962), methylene bending (1425 cm^{-1}), and C=C (1637 cm^{-1}).

Conformations of Carvomenthols

Naves (1964) has confirmed with infrared and NMR spectroscopy the stereochemical structures of the following terpenes: carvomenthol, isocarvomenthol, neocarvomenthol, and neoisocarvomenthol. The data are given in Table 19.19; *e* and *a* are equatorial and axial orientations of substituents on the chair-form cyclohexane ring:

Table 19.19. Stereochemical Structures of Carvomenthol and Iso- and Neo-Carvomenthols (Naves, 1964)

Compound	CH_3	OH	Isopropyl
Carvomenthol	e	e	e
Neocarvomenthol	e	a	e
Isocarvomenthol	a	e	e
Neoisocarvomenthol	a	a	e

Distinguishing Nucellar and Zygotic Seedlings

Pieringer and Edwards (1965) have distinguished nucellar and zygotic seedlings within each of nine citrus progenies by infrared analysis of oils distilled from the leaves. One example is presented here of infrared spectral data obtained from seedlings of a Duncan grapefruit crossed with.*P. trifoliata*. The absorption bands (in cm^{-1}) given in the following figures and *not* in parentheses are those for five nucellar seedlings, and the figures in parentheses are those for three zygotic seedlings: 3086 (3096); 3012 (3021); 2740 (2732); 1739 (1745); 1689 (1709, placement questionable); no band (1664); 1650 (no band); 1616 shoulder (no band); no band (1600); 1460 (1477?); 1383 (1389); no band (1370 shoulder); no band (1263); 1233 (no band); 1196 (no band); 1153 (no band); 1117 (1110); no band (1065); 1053 (no band); 1025 (1026); 986 (989); 958 (no band); 914 (no band); 903 shoulder (902.5); 887 (892); 866 (866); 797 (no band); 787 shoulder (no band). These differences and similarities varied among 14 seedlings. In unmarked populations where both types of seedlings were expected, the infrared technique identified more seedlings as zygotic than the usual method of observation based on morphological characteristics.

Measurement of Rates of Photosynthesis and Respiration

Austin and Longden (1967) have described an apparatus for the measurement of rates of photosynthesis of leaves. In principle, the method consisted of interrupting the flow of air over a leaf for 15 sec, during which time air containing $^{14}CO_2$ was passed over it. The amount of ^{14}C taken up by the leaf was then measured and the results were compared with those obtained by means of an infrared gas analyzer. There were discrepancies between the results, the chief cause of which appeared to be preferential fixation of ^{12}C by the leaves and loss of respiratory ^{12}C at low intensities of light.

Richardson (1967) has used an infrared gas analyzer to follow CO_2 exchange (dark respiration and net photosynthesis) through the first 248 hr of germination of cotton seeds.

Using differential infrared gas analysis for measurements of carbon dioxide exchange (grams of $CO_2/cm^2/sec$), Troughton and Cowan (1968) have observed anomalous depressions in carbon dioxide exchange in cotton leaves that were exhibiting oscillations in transpiration under controlled conditions of environment.

The depressions occurred only when leaf temperature exceeded 37.5°C and when the leaf diffusive resistance was minimum.

El-Sharkawy *et al.* (1967) have investigated differences in respiration rates among different plant species (*Sorghum vulgare, Zea mays, Gossypium, Beta vulgaris, Helianthus annuus,* and *Amaranthus edulis*) in CO_2-free air in light and darkness. They used a standard leaf-chamber technique and an infrared carbon dioxide analyzer to measure photosynthesis, transpiration, and respiration. In all the species examined, the rates of respiration were considerably higher in darkness than in light. These workers assumed this effect to be due to reassimilation of the respiratory carbon dioxide.

Infrared gas analysis was said to provide the most rapid and accurate method of measuring carbon dioxide exchange by terrestrial plants in the field or in the laboratory (Bourdeau and Woodwell, 1965). Measurements of carbon dioxide exchange are especially valuable in providing rapid estimates of the productivity of ecosystems. They are also useful in studying the response of plants to environmental and genetic factors, or to various cultural treatments. An infrared method for the continuous measurement of the respiration and assimilation of carbon dioxide in higher water plants has been given by Wattendorff (1963).

Lange (1962) has described a plexiglass "flap-chamber" for registering carbon dioxide exchange by leaves of plants in the open air. The leaf is enclosed only for the short periods of actual measurement, which is done by means of an infrared absorption recorder. Between measurements the chamber is automatically lifted off the base, and the leaf is fully exposed to prevailing conditions.

Brown and Tregunna (1967) have used an infrared carbon dioxide analyzer to measure production and assimilation of carbon dioxide by eleven marine or freshwater algae in acidic media. They found considerable differences among the algae in the concentration of carbon dioxide required to saturate photosynthesis and in the compensation concentration of carbon dioxide. These workers mentioned that the infrared analyzer was being used in laboratories extensively to measure respiration, photosynthesis, and carbon dioxide compensation by intact land plants and leaves. Brown and Tregunna stated that use of the infrared carbon dioxide analyzer to measure the rates of photosynthesis and respiration of algae offers a number of advantages over other techniques. It can be used to measure photosynthesis, respiration, and compensation on the same sample in a short time. The rates can be determined with the use of whole plants rather than disks, and variation in the design of the chamber allows many different sizes and shapes of plants to be used. Also, variations in the solution used allows marine or freshwater plants to be studied. The environmental factors are under close control and variations in light, temperature, gas composition, or pH can be effected while the other factors are held constant. The rates of carbon dioxide assimilation and production can be measured with the neutral to basic pH present during growth of the algae, but with some difficulty due to the buffering effect of bicarbonate.

Infrared gas analyzers have been used to record differential ppm levels of CO_2 and water vapor in the study of photosynthetic mechanisms in investigations of plant growth, and CO_2 has been monitored in the effluent air from bacterial cultures.

Hopkinson (1964) has used an infrared gas analyzer to measure carbon and phosphorus economy of a leaf.

Sparling and Alt (1966) have measured carbon dioxide concentrations in the atmosphere of several Ontario woodlands with an infrared gas analyzer. They found little evidence of seasonal variation in the concentration of carbon dioxide. Measurements over 24-hr periods revealed the existence of high concentrations, frequently exceeding 500 parts per million at night during midsummer. The high concentrations dropped rapidly at sunrise. These workers were not able to confirm the existence of the extreme stratification of carbon dioxide which had been reported by earlier workers.

Prebiology

It is widely believed that the earliest stages in the development of life involved the concurrent evolution of proteins and nucleic acids. Recently it has been proposed that cyanoacetylene had a central role in prebiological synthesis (Sanchez et al., 1966; Ferris et al., 1968). It is one of the major products resulting from the action of an electric discharge on mixtures of methane and nitrogen. Cyanoacetylene has been converted to cytosine on treatment with cyanate or cyanogen and it has been converted to asparagine and aspartic acid with cyanide and ammonia. Ferris (1968) has discussed the conversion of cyanoacetylene ($HC{\equiv}CCN$) to cyanovinyl phosphate and the possible role of this high-energy phosphate compound as a prebiological phosphorylating agent. This compound is an exceptionally stable enol phosphate, $^{2-}O_3POCH{=}CHCN$, whose structure was proven by synthetic chemistry, its ultraviolet spectrum, and infrared absorption at 2235 cm^{-1} (CN) and 1650 cm^{-1} (C=C).

In aqueous solution, condensations of various fundamental starting materials (e.g., H_2O, CH_4, NH_3) to form amino acids have been achieved, and other condensations have taken place in the presence of reagents such as cyanamide and cyanoguanidine to form other biological compounds (Ponnamperuma and Peterson, 1965; Steinman et al., 1964, 1965). Parker (1967) has differentiated by infrared spectroscopy between the dipeptides glycyl-L-leucine and L-leucylglycine in aqueous solution, a problem encountered by Ponnamperuma and Peterson (1965) in trying to differentiate peptide spots on chromatograms. Quite recently Lohrmann and Orgel (1968) have described the effects of each of several individual condensing substances on the formation of uridine 5′-phosphate from uridine and inorganic phosphate in aqueous solution: cyanogen, cyanoformamide, cyanate, cyanamide, thioformate, ethylisocyanide, and a water-soluble carbodiimide.

Paleobiochemistry

Amber

Langenheim (1969) has written a review about amber and its use as an evolutionary framework for interdisciplinary studies of resin-secreting plants. Of the various tools available, infrared spectrophotometry has so far been most extensively used in analyzing amber (Langenheim and Beck, 1965). The 1250–625 cm^{-1} region is more useful than the 4000–1250 cm^{-1} region for detecting differences between resins (Langenheim, 1969). The spectral data in the former region can be used to group fossil resins which not only have similar basic structures, but which sometimes can be related to recent resins. Figure 19.24 shows a comparison of spectra of typical Eocene Baltic amber with those from several Cretaceous ambers. Figure 19.25 shows spectra from representatives of copious resin producers living in southern Mexico today.

Terpenes

Several triterpenes and the tetraterpene perhydro-β-carotene have been identified (Murphy *et al.*, 1967) in the branched-cyclic hydrocarbon fraction of Green River (Eocene) shale. This was the first time that a tetraterpene had been isolated from a geological sample, although sesqui-, di-, and triterpenes had been identified in this and other sediments. Gas–liquid chromatography, mass spectrometry, NMR, and infrared spectroscopy were used for identification.

Odor

Wright (1954) has expanded upon the theory of Dyson (1928; 1937; 1938) that odor is related to a characteristic molecular vibration pattern of molecules in the range 3000–1500 cm^{-1} (so-called "osmic frequencies"). Wright suggested the use of the area below 1000 cm^{-1} for correlation between odors and molecular vibration

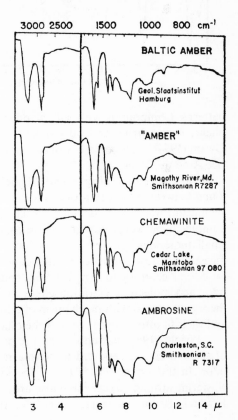

Fig. 19.24. Comparison of spectra of typical Eocene Baltic amber with those from Cretaceous amber from Maryland, Charleston, South Carolina, and Manitoba, Canada. (Langenheim and Beck, 1965.)

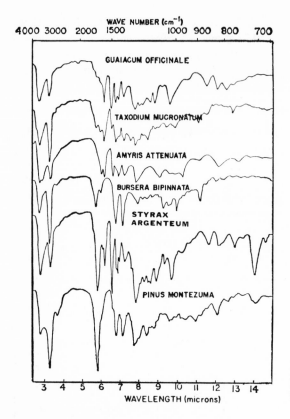

WAVE NUMBER (cm⁻¹)

GUAIACUM OFFICINALE

TAXODIUM MUCRONATUM

AMYRIS ATTENUATA

BURSERA BIPINNATA

STYRAX
ARGENTEUM

PINUS MONTEZUMA

WAVELENGTH (microns)

Fig. 19.25. Comparison of spectra from representatives of copious resin producers living in southern Mexico today. (Langenheim and Beck, 1965.)

patterns. Compounds with similar odors and spectra were nitrobenzene, nitrothiophene, benzonitrile, and butyronitrile. The range of study was 800 to 100 cm⁻¹. Dyson (1954) suggested a few other compounds to be studied to test Wright's theory (see Amoore, 1964).

More recently, Wright and Michels (1964) have stated that the relationship found between olfactory properties and Raman spectra, although statistically highly significant, was not evident by visual inspection of the spectra. Unsaturation can completely outweigh Raman spectra characteristics in determining odor in relatively small molecules, e.g., cyclohexene and cyclopentene. Low molecular energies, as measured by low-frequency Raman bands, did not appear *per se* to be *the* explanation for odor. According to Wright (1962), some relationship does exist, however, between odor and molecular vibration characteristics under 700 cm⁻¹. Another interesting theory concerned with mechanism of odor involves the stereochemistry of molecules and their fit into receptor sites of the olfactory system (Amoore, 1964).

Wright (1964) has compared the positions of absorption peaks of two groups of substances, ten reported to have a musk-like odor and ten with non-musk-like odor. In the region 300–50 cm⁻¹, the musk-like group has absorption frequencies clustering around 90 cm⁻¹ and also around 150 and 180 cm⁻¹, or as an alternative to 150 and 180 cm⁻¹ a single intervening peak between 165 and 170 cm⁻¹. He found indications of other regularities too, but in terms of a vibrational theory, he perceived the musk odor as a result of the simultaneous stimulation of three "primary receptors" which react to frequencies that are not far from 90, 150, and 180 cm^{-1}. He found no member of the non-musk-like group which appeared to include all three of the musk "primary receptors", but each member was capable of stimulating one or two of these "primaries" (along with various other primaries).

Rosano and Scheps (1964) believed that besides the phenomenon of adsorption by the odorant molecules on olfactory receptors (Moncrieff, 1951) another factor must be taken into account, namely, primary interaction between odorant molecules and enzymes on the mucus layer in the nose.

Amoore (1968) has said that the nose responds to 20–30 or more primary odors. However, von Baumgarten (1968) believes that there may be about 250 primary odors, or even a thousand, but admits incomplete information on this point. The presently accepted theory of odor is a stereochemical one developed by Amoore in 1952.

Intermolecular hydrogen bonds may be instrumental in reducing intensity of odor, as Shiftan and Feinsilver (1964) have found in two related terpene alcohols, citronellol (CXXVI) and linalool (CXXVII). These alcohols have the same skeleton,

CXXVI   CXXVII

nearly the same molecular weight and intensity of odor, but they differed in their ability to act as hydrogen-bonding proton donors. A 1% solution of citronellol in a nonpolar solvent, mineral oil, smelled much stronger than when 30% of the nonpolar solvent was replaced by methyl hydroabietate, which was a strong proton acceptor, odorless, and soluble in mineral oil. However, when the same type of experiment was repeated with linalool, a hindered tertiary alcohol, only a very slight weakening of the odor in the presence of the methyl hydroabietate was found. Shiftan and Feinsilver performed infrared spectroscopic experiments to corroborate the relationship of odor to hydrogen bonding, using the free OH stretching band (unassociated alcohol) in the 3600 cm^{-1} region and the OH···ester (OH···hydroabietate) broad band at lower frequencies to measure quantitatively the extent of hydrogen bonding to the methyl abietate. Under all conditions linalool formed intermolecular hydrogen bonds to a lesser extent than citronellol because of the degree to which linalool can hydrogen bond *intra*molecularly by virtue of OH···π bonding and thereby decrease its ability to form a hydrogen bond *inter*molecularly. These authors have also studied self-association of alcohols in relation to their effects upon the odor of certain mixtures, and in general have found greater intensity of odor for less associated alcohols.

Insect Sex Attractants

Methods for the collection, isolation, and identification of insect sex attractants have been discussed in the book by Jacobson (1965). On the basis of NMR, infrared, paper chromatographic, and chemical data the attractant of the American cockroach, *Periplaneta americana*, has been identified (Jacobson *et al.*, 1963) as 2,2-dimethyl-3-isopropylidenecyclopropyl propionate, but this structure has been found to be incorrect by synthesis of an inactive compound with the assigned formula (Day and Whiting, 1964; Jacobson and Beroza, 1965; Wakabayashi, 1965). The structural formula of the sex attractant had not been finally realized (in 1965), but the empirical formula was known to be $C_{11}H_{18}O_2$. Infrared, NMR, and mass spectrometry have been used frequently in these investigations.

Wharton *et al.* (1963) have refuted the paper by Jacobson *et al.* (1963), in which Jacobson *et al.* reported the isolation and identification of the sex attractant of the female American cockroach. The criticism claimed that Jacobson *et al.* showed no proof of correspondence between the attractant isolated and the structure proposed. Doubt was expressed that the structure itself could be validly deduced from data presented: "For example, their assignment of an 800 cm^{-1} (12.5 μ) band to an isopropylidene group is at variance with the reference they cite (Bellamy, 1958), which states, rather, that for a fully substituted ethylene (such as their proposed structure is), 'no band can arise from hydrogen deformations about the double bond.' Moreover, the assignment of the splitting at 1379 cm^{-1} (7.25 μ) to isopropyl in the hydrogenated derivative appears arbitrary, since with respect to the structure proposed, the assignment could as well have been made to the *gem*-dimethyl groups on the ring (Weissberger, 1956)."

The attractant produced by the adult male giant water bug, *Lethocerus indicus*, has been identified as *trans*-2-hexen-1-yl acetate, $CH_3(CH_2)_2CH=CHCH_2OCOCH_3$ (Butenandt and Tam, 1957). *Trans*-2-octen-1-yl acetate, $CH_3(CH_2)_4CH=CHCH_2OCOCH_3$, and *trans*-2-decen-1-yl acetate, $CH_3(CH_2)_6CH=CHCH_2-OCOCH_3$, which were isolated by distillation and gas chromatography from the steam distillates of the bronze orange bug, *Rhoecocoris sulciventris* and *Biprorulus bibax*, respectively, were also believed to be sex attractants (Park and Sutherland, 1962).

The attractant of the silkworm moth, *Bombyx mori*, named "bombykol" was obtained as the 4'-nitroazobenzenecarboxylic acid ester and identified as 10,12-hexadecadien-1-ol, $CH_3(CH_2)_2CH=CHCH=CH(CH_2)_8CH_2OH$ (Butenandt, 1963; Butenandt *et al.*, 1959, 1961*a,b*). Infrared and ultraviolet spectroscopy have been quite helpful in identification procedures. Bombykol has the form *trans*-10, *cis*-12 (Butenandt *et al.*, 1961, 1962; Hecker, 1960; Truscheit and Eiter, 1962).

The structure of the major attractant of the gypsy moth, *Porthetria dispar* has been shown to be (+)10-acetoxy-*cis*-7-hexadecen-1-ol (CXXVIII)

$$CH_3(CH_2)_5CHCH_2CH=CH(CH_2)_5CH_2OH$$
$$|$$
$$OCCH_3$$
$$||$$
$$O \qquad \text{CXXVIII}$$

(Jacobson *et al.*, 1960; 1961). The authors have reported the synthesis of this compound as the first synthesis of a naturally occurring sex attractant.

The attractant found in the mandibular glands of the male bumblebee, *Bombus terrestris*, has been identified as farnesol (Stein, 1963; Steinbrecht, 1964). From the heads of queen honey bees, *Apis mellifera*, Callow *et al.* (1964) have identified many fatty acids including *trans*-9-oxodec-2-enoic acid, which is an attractant to drones (Butler and Fairey, 1964).

Jacobson (1969) has separated a mixture of the *cis* and *trans* forms of "propylure" (10-propyl-*trans*-5,9-tridecadienyl acetate) (CXXIX), the sex pheromone of the female pink bollworm moth (*Pectinophora gossypiella* (Saunders)), into its pure isomers by thin-layer chromatography. The infrared spectrum of the separated *trans*-propylure had a medium band at 965 cm^{-1} (*trans* HC=CH) that was absent in the spectrum of the *cis* isomer. The spectrum of *cis*-propylure had an inflection at 740 cm^{-1}. *cis*-Propylure inhibited or masked the activity of the *trans* isomer, as

$$CH_3CH_2CH_2$$
$$|$$
$$C=CH(CH_2)_2CH=CH(CH_2)_4OCCH_3$$
$$|\qquad\qquad\qquad\qquad\qquad\qquad\parallel$$
$$CH_3CH_2CH_2\qquad\qquad\qquad\qquad O$$

CXXIX

little as 15 % of the *cis* isomer being sufficient to nullify completely the activity of the *trans* isomer. Jacobson mentioned several other cases of pheromone masking by admixed contaminants. In the case of *Trichoplusia ni*, whose sex pheromone is *cis*-7-dodecenyl acetate, the agent responsible for masking is the corresponding *trans* isomer. This is also true for gyplure (12-acetoxy-*cis*-9-octadecen-1-ol), a synthetic sex attractant for the gypsy moth, which is completely inactivated by admixture with 20 % of its *trans* isomer.

Riddiford (1967) has isolated and identified as *trans*-2-hexenal the volatile compound from oak leaves which stimulates the female polyphemus moth (*Antheraea polyphemus*) to release her sex pheromone. The infrared spectrum showed characteristic maxima at 2700, 1680, and 1625 cm^{-1}, indicative of an α,β-unsaturated aldehyde, and a maximum at 970 cm^{-1}, indicative of the *trans* configuration at the double bond.

Insect Defense

The cantharid beetle *Chauliognathus lecontei* defensively secretes from glands in its thorax and abdomen droplets of liquid containing as a major fraction 8-*cis*-decene-4,6-diynoic acid (or 8-*cis*-dihydromatricaria acid) (Meinwald *et al.*, 1968). Ultraviolet and mass spectral data quickly led to the identification of the chromophore $-\overset{|}{C}=\overset{|}{C}-C\equiv C-C\equiv C-$. Infrared and NMR spectroscopy determined the structure as $CH_3-CH=CH-C\equiv C-C\equiv C-CH_2-CH_2-COOH$. Various acetylenic compounds are known from certain fungi and flowering plants (Sörensen, 1963), but Meinwald (1968) believed this to be the first demonstration of an acetylenic acid in an insect.

The defensive glands of *Leichenum caniliculatum variegatum* (Klug) (*Coleoptera*: *Tenebrionidae*) produce a dark odorous secretion. Happ (1967) has identified by thin-layer chromatography (and confirmed by infrared methods) the derivatives of *p*-benzoquinone, ethyl-*p*-benzoquinone, and methyl-*p*-benzoquinone as components of the secretion.

Pesticides

Blinn and Gunther (1963) have reviewed the use of infrared and ultraviolet spectrophotometry in the field of pesticide residue chemistry. Spectrophotometric measurements offer several unique advantages to the analytical chemist: the radiation which is absorbed is characteristic of the material doing the absorbing; the degree of absorption of radiation is directly proportional to the concentration of the material in solution; and spectrophotometric methods are usually essentially nondestructive to the absorbing material and therefore allow recovery for further scrutiny by other types of instrumentation when applicable.

Since the ultimate purpose of spectrophotometry in residue assay is the final interpretive measurement of the compound of interest, the devising of a spectro-photometric residue procedure consists of designing techniques of isolation and ancillary chemical methods which are adequate to eliminate or to compensate for the many interfering materials that occur in foodstuffs and their derived products, and possibly also to enhance the intensity of absorption of the compound of interest. Particular advantages of the infrared region are its abundance of absorption bands which present greater possibilities for specificity and identification, its greater sensitivity to most changes in molecular structure and its response to all organic compounds.

The various problems connected with the use of infrared spectroscopy in pesticide research have been reviewed by Frehse (1963). The paper dealt with qualitative and quantitative analysis, determinations of residues, and special problems such as methods of extraction, cells, solvents, and measuring attachments to be used. Frehse has given many references to the literature concerning the infrared spectroscopic analysis of various food crops for pesticides, e.g., aldrin, alodan, chlorbenside, DDT, dieldrin, endrin, ethion, lindane, malathion, tedion, endosulfan, biphenyl, captan, pentachloronitrobenzene, 2,4-DB, MCPB, and methylisothiocyanate. The infrared band(s) used for the determinations have also been given.

A system that combines gas- and thin-layer chromatography with infrared spectroscopy has been devised (Siewierski and Helrich, 1967) for the separation, quantitation, and tentative identification of DDT and twelve of its possible reaction products (metabolites and degradation products).

The use of microinfrared spectroscopy for the analysis of chlorinated pesticides has been described by Payne and Cox (1966). The trace amounts of pesticides are first isolated by column and thin-layer chromatography.

Metabolism of Herbicides

Low concentrations of a phenolic metabolite accumulated in media containing 2,4-dichlorophenoxyacetate (2,4-D) during the early stages of growth of an *Arthro-*

bacter species. The metabolite of 2,4-D has been identified (Loos *et al.*, 1967) by gas chromatographic, ultraviolet, and infrared analysis, and paper chromatography of a nitro-derivative as 2,4-dichlorophenol, thereby providing evidence that 2,4-D was degraded by the bacterium by way of its corresponding phenol.

The herbicide 3-amino-1,2,4-triazole is effective against developing rather than mature plant tissues. Fredrick and Gentile (1961) have used chemical tests and infrared spectra as evidence for a D-glucose-3-amino-1,2,4-triazole adduct (CXXX) as a product of the metabolism of this herbicide by treated plants.

CXXX

REFERENCES

Abelson, D., Boyle, A., and Seligson, H. *J. Biol. Chem.* **238**, 717 (1963).

Ackermann, T. *Z. Phys. Chem.* **27**, 253 (1961).

Ahlijah, G. E. B. Y. and Mooney, E. F. *Spectrochim. Acta* **22**, 547 (1966).

Ahlijah, G. E. B. Y. and Mooney, E. F. *Spectrochim. Acta* **25A**, 619 (1969).

Al-Kayssi, M. and Magee, R. J. *Talanta* **9**, 667 (1962).

Allerton, S. E., Renner, J., Colt, S., and Saltman, P. *J. Am. Chem. Soc.* **88**, 3147 (1966).

Ambrose, E. J. and Elliott, A. *Proc. Roy. Soc.* (*London*) **A206**, 206 (1951).

Amoore, J. E. *Ann. N. Y. Acad. Sci.* **116**, Art. 2, 457 (1964).

Amoore, J. E., Johnston, J. W., Jr., and Rubin, M. *Scientific American*, Feb. 1964, p. 42.

Amoore, J. E. cited anonymously in *Scientific Research* **3**, 21 (1968).

Anderson, B. M., Cordes, E. H., and Jencks, W. P. *Federation Proc.* **19**, 46 (1960).

Austin, R. B. and Longden, P. C. *Ann. Bot.* (*London*) **31**, 245 (1967).

Barer, R., Coyle, A. R. H., and Thompson, H. W. *Nature* **163**, 198 (1949).

Beer, M., Sutherland, G. B. B. M., Tanner, K. N., and Wood, D. L. *Proc. Roy. Soc.* (*London*) **249A**, 147 (1959).

Behen, J. J., Dwyer, R. F., and Bierl, B. A. *Anal. Biochem.* **9**, 127 (1964).

Bellamy, L. J. *J. Chem. Soc.* **1955**, 4221.

Bellamy, L. J. *The Infrared Spectra of Complex Molecules, 2nd Ed.*, Wiley, New York, 1958.

Bird, G. R. and Blout, E. R. *Laboratory Investigation* **1**, 266 (1952).

Blinn, R. C. and Gunther, F. A. *Residue Rev.* **2**, 99 (1963).

Blout, E. R. and Fields, M. *Science* **107**, 252 (1948).

Blout, E. R. and Fields, M. *J. Biol. Chem.* **178**, 335 (1949).

Bourdeau, P. F. and Woodwell, G. M. *Proceedings Internat. Symp. Arid Zone Res. XXV, Methodology of Plant Ecophysiology*, Montpellier, France, 1962, United Nations Ed., Sci., and Cult. Org., Paris, 1965, p. 283.

Brady, G. W., Kurkjian, C. R., Lyden, E. F. X., Robin, M. B., Saltman, P., Spiro, T., and Terzis, A. *Biochemistry* **7**, 2185 (1968).

Brown, D. L. and Tregunna, E. B. *Can. J. Bot.* **45**, 1135 (1967).

Bunton, C. A. and Hadwick, T. *J. Chem. Soc.* **1958**, 3248.

Butenandt, A. *J. Endocrinol.* **27**, ix (1963) (Proceedings).

Butenandt, A. and Tam, N.-D. *Z. physiol. Chem., Hoppe-Seyler's* **308**, 277 (1957).

Butenandt, A., Beckmann, R., and Hecker, E. *Z. physiol. Chem., Hoppe-Seyler's* **324**, 71 (1961*a*).

Butenandt, A., Beckmann, R., and Stamm, D. *Z. physiol. Chem., Hoppe- Seyler's* **324**, 84 (1961*b*).

Butenandt, A., Beckmann, R., Stamm, D., and Hecker, E. *Z. Naturforsch* **14b**, 283 (1959).

Butenandt, A., Hecker, E., Hopp, M., and Koch, W. *Ann.* **658**, 39 (1962).

Butler, C. G. and Fairey, E. M. *J. Apicult. Research* **3**, 65 (1964).

Butterworth, J., Eley, D. D., and Stone, G. S., *Biochem. J.* **53**, 30 (1953).

Caglioti, V., Ascenzi, A., and Scrocco, M. *Experientia* **10**, 371 (1954).

Caglioti, V., Ascenzi, A., and Scrocco, M. *Arch. Sci. Biol. (Italy)* **39**, 116 (1955).

Callow, R. K., Chapman, J. R., and Paton, P. N. *J. Apicult. Research* **3**, 77 (1964).

Canepa, F. G. and Mooney, E. F. *Nature* **207**, 78 (1965).

Chapman, D., Lloyd, D. R., and Prince, R. H. *J. Chem. Soc.* **1964**, 550.

Chapman, D., Kamat, V. B., deGier, J., and Penkett, S. A. *J. Mol. Biol.* **31**, 101 (1968*a*).

Chapman, D., Kamat, V. B., and Levene, R. J., *Science* **160**, 314 (1968*b*).

Chapman, D., Williams, R. M., and Ladbrooke, B. D. *Chem. Phys. Lipids* **1**, 445 (1967).

Clark, R. J. H. and Williams, C. S. *Spectrochim. Acta* **22**, 1081 (1966).

Coates, V. J., Offner, A., and Siegler, E. H., Jr. *J. Opt. Soc. Am.* **43**, 984 (1953).

Corish, P. J. and Davison, W. H. T. *J. Chem. Soc.* **1955**, 2431.

Cramer, F. and Gärtner, K. *Chem. Ber.* **91**, 704 (1958).

Csallany, A. S. and Draper, H. H. *J. Biol. Chem.* **238**, 2912 (1963).

Davies, M. and Hallam, H. E. *Trans. Faraday Soc.* **47**, 1170 (1951).

Davis, R. L., Rousseau, D. L., and Board, R., a report delivered at the Lehigh University First International Conference on Polywater, quoted in *Chem. Eng. News* **48** (27), 7 (June 29, 1970).

Davis, R. L., Rousseau, D. L., and Board, R. D. *Science* **171**, 167 (1971).

Day, A. C. and Whiting, M. C. *Proc. Chem. Soc.* **1964**, 368.

Deryagin, B. V., Talaev, M. V., and Fedyakin, N. N. *Dokl. Akad. Nauk SSSR* **165**, 597 (1965); *Proc. Acad. Sci. USSR Phys. Chem.* **165**, 807 (1965).

Deryagin, B. V. and Churayev, N. V. *Priroda (Russian)* No. 4, 16 (1968); *Joint Publications Research Service* No. 45, 989 (1968). (This is a summary and review of the work.)

Donohue, J. *Science* **166**, 1000 (1969).

Draegert, D. A., Stone, N. W. B., Curnutte, B., and Williams, D. *J. Opt. Soc. Am.* **56**, 64 (1966).

Dyson, G. M. *Perf. Essent. Oil Record* **19**, 456 (1928).

Dyson, G. M. *Perf. Essent. Oil Record* **28**, 13 (1937).

Dyson, G. M. *Chem. Ind. (London)*, **16**, 647 (1938).

Dyson, G. M. *Nature* **173**, 831 (1954).

Edsall, J. T. *J. Chem. Phys.* **4**, 1 (1936).

Edsall, J. T. *J. Chem. Phys.* **5**, 508 (1937).

Eisenberg, D. and Kauzmann, W. *The Structure and Properties of Water*, Oxford University Press, New York, 1969, p. 228.

Elliott, J. C. *J. Dental Res.* **43**, 959 (1964).

Elliott, J. C. "Tooth Enamel," in *Proceedings of an International Symposium on the Composition, Properties, and Fundamental Structure of Tooth Enamel*, 6 and 7 April, 1964, London Hospital Medical College (1965), pp. 20 and 50.

Elliott, A., Ambrose, E. J., and Temple, R. B. *J. Chem. Phys.* **16**, 877 (1948).

El-Sharkawy, M. A., Loomis, R. S., and Williams, W. A. *Physiol. Plant.* **20**, 171 (1967).

Emerson, W. H. and Fischer, E. E. *Arch. Oral. Biol.* **7**, 671 (1962).

Fahrenfort, J. *Spectrochim. Acta* **17**, 698 (1961).

Fahrenfort, J. and Visser, W. M. *Spectrochim. Acta* **18**, 1103 (1962).

Falk, M. and Ford, T. A. *Can. J. Chem.* **44**, 1699 (1966).

Fedyakin, N. N. *Kolloid Zh.* **24**, 497 (1962); *Colloid J. USSR* **24**, 425 (1962).

Ferris, J. P. *Science* **161**, 53 (1968).

Ferris, J. P., Sanchez, R. A., and Orgel, L. E. *J. Mol. Biol.* **33**, 693 (1968).

Fleischer, S., Rouser, G., Fleischer, B., Casu, A., and Kritchevsky, G. *J. Lipid Res.* **8**, 170 (1967).

Fowler, B. O., Moreno, E. C., and Brown, W. E. *Arch. Oral Biol.* **11**, 477 (1966).

Frank, C. W. and Rogers, L. B. *Inorg. Chem.* **5**, 615 (1966).

Fraser, N. J. and Fraser, R. D. B. *Nature* **167**, 761 (1951).

Fredrick, J. F. and Gentile, A. C. *Arch. Biochem. Biophys.* **92**, 356 (1961).

Freeman, J. P. *J. Am. Chem. Soc.* **80**, 5954 (1958).

Frehse, H. *Pflanzenschutz-Nachr. Bayer* **16**, 182 (1963).

Fronaeus, S. and Larsson, R. *Acta Chem. Scand.* **16**, 1433, 1447 (1962).

Frydman, B. and Rapoport, H. *J. Am. Chem. Soc.* **85**, 823 (1963).

Fukuda, K. *Histochemie* **6**, 127 (1966).

Furedi, H. and Walton, A. G. *Appl. Spectrosc.* **22**, 23 (1968).

Furman, N. H. *Standard Methods of Chemical Analysis*, 6th Ed., Van Nostrand, 1962, p. 747.

Glover, J. and Threlfall, D. R. *Biochem. J.* **85**, 14P (1962).

Gordon, A. S., Graham, J. M., Fernbach, B. R., and Wallach, D. F. H. *Federation Proc.* **28**, 404 (1969).

Goulden, J. D. S. *J. Dairy Res.* **31**, 273 (1964).

Goulden, J. D. S. and Manning, D. J. *Nature* **203**, 403 (1964).

Goulden, J. D. S. and Manning, D. J. *Spectrochim. Acta* **23A**, 2249 (1967).

Goulden, J. D. S. and Scott, J. E. *Nature* **220**, 698 (1968).

Goulden, J. D. S., Shields, J., and Haswell, R. *J. Soc. Dairy Tech.* **17**, 28 (1964).

Green, J., McHale, D., Marcinkiewicz, S., Mamalis, P., and Watt, P. R. *J. Chem. Soc.* **1959**, 3362.

Grenell, R. G. and May, L. *J. Neurochemistry* **2**, 138 (1958).

Haba, F. R. and Wilson, C. L. *Talanta* **11**, 21 (1964).

Happ, G. M. *Ann. Entomol. Soc. Am.* **60**, 279 (1967).

Harrell, S. A. and McDaniel, D. H. *J. Am. Chem. Soc.* **86**, 4497 (1964).

Harrick, N. J. *Ann. N. Y. Acad. Sci.* **101**, Art. 3, 928 (1963).

Harris, N. O., Swanson, A., and Segreto, V. *The Infrared Spectral Characteristics of Human Whole Stimulated Saliva Collected from Individuals Classified as to Dental Caries Experience* (Report 56-90) Brooks Air Force Base, Texas; School of Aviation Medicine, Dental Sciences Div., June 1959.

Harris, N. O., Yeary, R., Swanson, A., and Segreto, V. *J. Dental Res.* **39**, 810 (1960).

Hartman, K. A., Jr. *J. Phys. Chem.* **70**, 270 (1966).

Hartwell, E. J., Richards, R. E., and Thompson, H. W. *J. Chem. Soc.* **1948**, 1436.

Hauptschein, M., Stokes, C. S., and Nodiff, E. A. *J. Am. Chem. Soc.* **74**, 4005 (1952).

Hecker, E. "Chemie und Biochemie des Sexuallstoffes des Seidenspinners (*Bombyx mori*)," *Proc. XIth Intern. Cong. Entomol.* **3B**, 69 (1960), published (1961).

Hemming, F. W., Morton, R. A., and Pennock, J. F. *Proc. Roy. Soc.* **B158**, 291 (1963).

Herman, H. and Dallemagne, M. J. *Bull. Soc. Chim. Biol.* **46**, 373 (1963).

Hermann, T. S. *Anal. Biochem.* **12**, 406 (1965).

Herzberg, G. *Infrared and Raman Spectra of Polyatomic Molecules*, Van Nostrand, Princeton, 1945.

Hewkin, D. J. and Griffith, W. P. *J. Chem. Soc. Sec. A* **1966** 472.

Hoagland, M. B., Stephenson, M. L., Scott, J. F., Hecht, L. I., and Zamecnik, P. C. *J. Biol. Chem.* **231**, 241 (1958).

Hogenkamp, H. P. C., Rush, J. E., and Swenson, C. A. *J. Biol. Chem.* **240**, 3641 (1965).

Holley, R. W. and Goldstein, J. *J. Biol. Chem.* **234**, 1765 (1959).

Hopkinson, J. M. *J. Exp. Bot.* **15**, 125 (1964).

Hulcher, F. H. *Arch. Biochem. Biophys.* **100**, 237 (1963).

Hvidt, A. and Kägi, J. H. R. *Compt. Rend. Trav. Lab. Carlsburg* **33**, 530 (1963).

Jackovitz, J. F. and Walter, J. L. *Spectrochim. Acta.* **22**, 1393 (1966).

Jacobson, M. *Insect Sex Attractants*, Interscience, New York, 1965, Chap. 9.

Jacobson, M. *Science* **163**, 190 (1969).

Jacobson, M. and Beroza, M. *Science* **147**, 748 (1965).

Jacobson, M., Beroza, M., and Jones, W. A. *Science* **132**, 1011 (1960).

Jacobson, M., Beroza, M., and Jones, W. A. *J. Am. Chem. Soc.* **83**, 4819 (1961).

Jacobson, M., Beroza, M., and Yamamoto, R. T. *Science* **139**, 48 (1963).

Jencks, W. P. *Biochim. Biophys. Acta* **24**, 227 (1957).

Jencks, W. P., Moore, C., Perini, F., and Roberts, J. *Arch. Biochem. Biophys.* **88**, 193 (1960).

Jenkinson, T. J., Kamat, V. B., and Chapman, D. *Biochim. Biophys. Acta* **183**, 427 (1969).

Jones, L. H. and Penneman, R. A. *J. Chem. Phys.* **22**, 965 (1954).

Jones, L. H. and Penneman, R. A. *J. Chem. Phys.* **24**, 293 (1956).

Kagarise, R. E. *J. Am. Chem. Soc.* **77**, 1377 (1955).

Klein, E., Le Geros, J. P., Trautz, O. R., and Zapanta-Le Geros, R. in *Developments in Applied Spectroscopy, Vol. 7B* (E. L. Grove and A. J. Perkins, eds.) Plenum Press, New York, 1970, p. 13.

Kofler, M., Sommer, P. F., Bolliger, H. R., Schmidli, B., and Vecchi, M. *Vitamins and Hormones* **20**, 407 (1962).

Korn, E. D. *Abstracts of the 154th National Meeting of the American Chemical Society, Biological Division,* Chicago, Illinois, September, 1967, Paper No. 71.

Kurtin, S. L., Mead, C. A., Mueller, W. A., Kurtin, B. C., and Wolf, E. D. *Science* **167**, 1720 (1970).

Lange, O. L. *Ber. Deutsch. Bot. Ges.* **75**, 41 (1962).

Langenheim, J. H. *Science* **163**, 1157 (1969).

Langenheim, J. H. and Beck, C. W. *Science* **149**, 52 (1965).

Lenard, J. and Singer, S. J. *Proc. Natl. Acad. Sci. U.S.* **56**, 1828 (1966).

Lenormant, H. *Compt. rend.* **234**, 1959 (1952).

Lenormant, H. and Chouteau, J. *Compt. rend.* **234**, 2057 (1952).

Lippincott, E. R., Stromberg, R. R., Grant, W. H., and Cessac, G. L. *Science* **164**, 1482 (1969).

Lohrmann, R. and Orgel, L. E. *Science* **161**, 64 (1968).

Loos, M. A., Roberts, R. N., and Alexander, M. *Can. J. Microbiol.* **13**, 691 (1967).

Maddy, A. H. and Malcolm, B. R. *Science* **150**, 1616 (1965).

Mager, J. and Lipmann, F. *Proc. Natl. Acad. Sci. U.S.* **44**, 305 (1958).

Mamalis, P., McHale, D., Green, J., and Marcinkiewicz, S. *J. Chem. Soc.* **1958**, 1850.

Margoshes, M., Fillwalk, F., Fassel, V. A., and Rundle, R. E. *J. Chem. Phys.* **22**, 381 (1954).

Martius, C. *Vitamins and Hormones* **20**, 457 (1962).

May, L. *Nature* **211**, 412 (1966).

Meinwald, J., Meinwald, Y. C., Chalmers, A. M., and Eisner, T. *Science* **160**, 890 (1968).

Michell, A. J. *Australian J. Chem.* **19**, 2285 (1966).

Milch, R. A. *Biochim. Biophys. Acta* **93**, 45 (1964).

Miller, F. A. and Wilkins, C. H. *Anal. Chem.* **24**, 1253 (1952).

Miller, F. A., Carlson, G. L., Bentley, F. F., and Jones, W. H. *Spectrochim. Acta* **16**, 135 (1960).

Mills, J., Orr, A., and Whittier, J. R. *J. Neurochemistry* **16**, 1033 (1969).

Mitchell, E. D., Jr. and Tolbert, N. E. *Biochemistry* **7**, 1019 (1968).

Miyazawa, T. Masuda, Y., and Fukushima, K. *J. Polymer Sci.* **62**, S62, (1962).

Moncrieff, R. W. *The Chemical Senses, 2nd Ed.*, Leonard Hill, Ltd., London, 1951.

Mori, K., Matsui, M., and Sumiki, Y. *Agr. Biol. Chem. (Tokyo)*, **25**, 205 (1961).

Morris, A. L. and Greulich, R. C. *Science* **160**, 1081 (1968).

Morris, W. W., Jr. and Haenni, E. O. *J. Ass. Offic. Anal. Chem.* **45**, 92 (1962).

Morris, W. W., Jr., Wilkie, J. B., Jones, S. W., and Friedman, L. *Anal. Chem.* **34**, 381 (1962).

Murphy, M. T. J., McCormick, A., and Eglinton, G. *Science* **157**, 1040 (1967).

Nair, P. P. and Luna, Z. *Arch. Biochem. Biophys.* **127**, 413 (1968).

Nakamoto, K. *Infrared Spectra of Inorganic and Coordination Compounds*, Wiley, New York, 1963.

Nakamoto, K. in *Spectroscopy and Structure of Metal Chelate Compounds* (K. Nakamoto and P. J. McCarthy, eds.) Wiley, New York, 1968.

Nakamoto, K., Morimoto, Y., and Martell, A. E. *J. Am. Chem. Soc.* **84**, 2081 (1962); **85**, 309 (1963).

Nakanishi, K. *Infrared Absorption Spectroscopy—Practical*, Holden–Day, San Francisco, 1962, p. 47.

Naves, Y.-R. *Helv. Chim. Acta* **47**, 308 (1964).

Neely, W. A., Turner, M. D., and Smith, J. D., discussed anonymously in the Beckman "*Analyzer*," Jan., 1962.

Neuman, W. F. and Neuman, M. W. *Chem. Revs.* **53**, 1 (1953).

Noll, H. *J. Biol. Chem.* **235**, 2207 (1960).

Oestreich, C. H. and Jones, M. M. *Biochemistry* **6**, 1515 (1967).

Otting, W. *Chem. Ber.* **89**, 1940 (1956).

Page, T. F., Jr., Jakobsen, R. J., and Lippincott, E. R. *Science* **167**, 51 (1970).

Pape, L., Multani, J. S., Stitt, C., and Saltman, P. *Biochemistry* **7**, 606 (1968).

Park, R. J. and Sutherland, M. D. *Austral. J. Chem.* **15**, 172 (1962).

Parker, F. S. *Appl. Spectrosc.* **15**, 96 (1961).

Parker, F. S. in *Progress in Infrared Spectroscopy, Vol. 3*, (H. Szymanski, ed.) Plenum Press, New York, 1967, p. 89.

Parker, F. S. and Ans, R. Proc. 4th Meeting Career Scientists, Health Research Council of City of New York, New York Academy of Medicine, December 15, 1965; (see *Bull. N.Y. Acad. Med.* **42**, 415 (1966).

Parker, F. S. and Ans, R. *Anal. Biochem.* **18**, 414 (1967).

Parker, F. S. and D'Agostino, M. *Bull. N.Y. Acad. Med.* **43**, 418 (1967).

Payne, W. R., Jr. and Cox, W. S. *J. Ass. Offic. Anal. Chem.* **49**, 989 (1966).

Penneman, R. A. and Jones, L. H. *J. Inorg. Nucl. Chem.* **20**, 19 (1961).

Pennock, J. F. in *Biochemistry of Quinones* (R. A. Morton, ed.) Academic Press, New York, 1965, p. 67.

Petsko, G. A. *Science* **167**, 171 (1970).

Pieringer, A. P. and Edwards, G. J. *Proc. Amer. Soc. Hort. Sci.* **86**, 226 (1965).

Ponnamperuma, C. and Peterson, E. *Science* **147**, 1572 (1965).

Posner, A. S. and Duyckaerts, G. *Experientia* **10**, 424 (1954).

Posner, A. S. and Perloff, A. *J. Res. Nat. Bur. Std.* **58**, 279 (1957).

Posner, A. S., Stutman, J. M., and Lippincott, E. R. *Nature* **188**, 486 (1960).

Potts, J. E. and Amis, E. S. *J. Am. Chem. Soc.* **71**, 2112 (1949).

Quinaux, N. and Richelle, L. J. *Israel J. Med. Sci.* **3**, 677 (1967).

Rhodes, W. C. and McElroy, W. D. *J. Biol. Chem.* **233**, 1528 (1958).

Richardson, G. L. *Crop. Sci.* **7**, 6 (1967).

Richelle, L. and Quinaux, N. *Nature* **203**, 84 (1964).

Riddiford, L. M. *Science* **158**, 139 (1967).

Rosano, H. L. and Scheps, S. Q. *Ann. N.Y. Acad. Sci.* **116**, Art. 2, 590 (1964).

Rosenkrantz, H. *J. Biol. Chem.* **173**, 439 (1948).

Rosenkrantz, H. in *Methods of Biochemical Analysis, Vol. 5* (D. Glick, ed.) Interscience, New York· 1957, p. 407.

Rosenkrantz, H. and Milhorat, A. T. *J. Biol. Chem.* **187**, 83 (1950).

Rousseau, D. L. *Science* **171**, 170 (1971).

Rousseau, D. L. and Porto, S. P. S. *Science* **167**, 1715 (1970).

Salmi, E. J. and Leimu, R. *Suomen Kemistilehti* **20B**, 43 (1947).

Sanchez, R. A., Ferris, J. P., and Orgel, L. E. *Science* **154**, 784 (1966).

Sawyer, D. T. *Ann. N.Y. Acad. Sci.* **88**(2), 307 (1960).

Schneider, G., Sembdner, G., and Schreiber, K. *Kulturpflanze* **13**, 267 (1965) (German).

Schudel, P., Mayer, H., Metzger, J., Rüegg, R., and Isler, O. *Helv. Chim. Acta.* **46**, 2517 (1963).

Schwarz, H. P. *Appl. Spectrosc.* **6**, 15 (1952).

Schwarz, H. P. in *Advances in Clinical Chemistry, Vol. 3* (H. Sobotka and C. P. Stewart, eds.) Academic Press, New York, 1960, p. 1.

Schwarz, H. P., Riggs, H. E., and Glick, C. F. *Trans. Am. Neurol. Ass.* **1951a**, 90.

Schwarz, H. P., Riggs, H. E., Glick, C. F., Cameron, W., Beyer, E., Jaffe, B., and Trombetta, L. *Proc. Soc. Exp. Biol. Med.* **76**, 267 (1951b).

Schwarz, H. P., Riggs, H. E., Glick, C. F., McGrath, J., Cameron, W., Beyer, E., Bew, E., Jr., and Childs, R. *Proc. Soc. Exp. Biol. Med.* **80**, 467 (1952).

Searles, S., Tamres, M., and Barrow, G. M. *J. Am. Chem. Soc.* **75**, 71 (1953).

Sherebrin, M. H. *Abstracts of the Third International Biophysics Congress of the I.U.P.A.B.*, Cambridge, Mass., Sept., 1969, p. 100.

Shiftan, E. and Feinsilver, M. *Ann. N.Y. Acad. Sci.* **116**, Art. 2, 692 (1964).

Siewierski, M. and Helrich, K. *J. Ass. Offic. Anal. Chem.* **50**, 627 (1967).

Sörensen, N. A. in *Chemical Plant Taxonomy* (T. Swain, ed.), Academic Press, New York, 1963, p. 219 and references.

Sparling, J. H. and Alt, M. *Can. J. Botany* **44**, 321 (1966).

Spinner, E. *J. Chem. Soc.* **1964**, 4217.

Spiro, T. G., Allerton, S. E., Renner, J., Terzis, A., Bils, R., and Saltman, P. *J. Am. Chem. Soc.* **88**, 2721, 3147 (1966).

Staab, H. A. *Chem. Ber.* **89**, 1927 (1956).

Staab, H. A. *Chem. Ber.* **90**, 1320 (1957).

Stadtman, E. R. in *Mechanism of Enzyme Action* (W. D. McElroy and B. Glass, eds.) Johns Hopkins Press, Baltimore, 1954, p. 581.

Steim, J. M. *Abstracts of the 153rd National Meeting of the American Chemical Society*, 1967, Abstract No. H59.

Stein, G. *Biol. Zentr.* **82**, 345 (1963).

Steinbrecht, R. A. *Z. Zellforsch.* **64**, 227 (1964).

Steinman, G., Lemmon, R. M., and Calvin, M. *Proc. Natl. Acad. Sci. U.S.* **52**, 27 (1964).

Steinman, G., Lemmon, R. M., and Calvin, M. *Science* **147**, 1574 (1965).

Stern, M. H., Robeson, C. D., Weisler, L., and Baxter, J. G. *J. Am. Chem. Soc.* **69**, 869 (1947).

Stevenson, J., Ph.D. Thesis, University of Liverpool (1963), cited by Pennock (1965).

Stewart, R. D. and Erley, D. S. *Exptl. Cell Research* **23**, 460 (1961).

Stone, K. J., Wellburn, A. R., Hemming, F. W., and Pennock, J. F. *Biochem. J.* **102**, 325 (1967).

Stutman, J. M., Posner, A. S., and Lippincott, E. R. *Nature* **193**, 368 (1962).

Stutman, J. M., Termine, J. D., and Posner, A. S. *Trans. N.Y. Acad. Sci.*, Series II, **27**, 669 (1965).

Termine, J. D. and Posner, A. S. *Nature* **211**, 268 (1966a).

Termine, J. D. and Posner, A. S. *Science* **153**, 1523 (1966b).

Thomas, L. C., Austin, L., and Davies, D. R. *Spectrochim. Acta* **6**, 320 (1954).

Tollin, G. and Steelink, C. *Biochim. Biophys. Acta* **112**, 377 (1966).

Troughton, J. H. and Cowan, I. R. *Science* **161**, 281 (1968).

Truscheit, E. and Eiter, K. *Ann.* **658**, 65 (1962).

Turner, M. D., Neely, W. A., and Hardy, J. D. *Federation Proc.* **19**, 80 (1960).

Underwood, A. L., Toribara, T. Y., and Neuman, W. F. *J. Am. Chem. Soc.* **77**, 317 (1955).

Verzhbinskaya, N. A. and Sidorova, A. I. *Biofizika* **11**, 101 (1966).

von Baumgarten, R. cited anonymously in *Scientific Research* **3**, 21 (1968).

Waddington, T. C. *Trans. Faraday Soc.* **54**, 25 (1958).

Wakabayashi, N. unpublished results, cited in Jacobson (1965).

Wallach, D. F. H. and Zahler, P. H. *Proc. Natl. Acad. Sci. U.S.* **56**, 1552 (1966).

Wallach, D. F. H. and Zahler, P. H. *Biochim. Biophys. Acta* **150**, 186 (1968).

Wattendorff, J. *Ber. Deutsch. Bot. Ges.* **76**, 196 (1963).

Wegmann, R. *Annales d'Histochimie* **1**, 30 (1956).

Wegmann, R., Hébert, S., and Biez-Charreton, J. *Annales d'Histochimie* **1**, 287 (1956).

Weissberger, A. in *Chemical Applications of Spectroscopy, Vol. 9* (W. West, ed.) Interscience, New York, 1956, p. 337.

Wellburn, A. R., Stevenson, J., Hemming, F. W., and Morton, R. A. *Biochem. J.* **102**, 313 (1967).

Wharton, D. R. A., Black, E. D., and Merritt, C., Jr. *Science* **142**, 1257 (1963).

Williams, D. *Nature* **210**, 194 (1966).

Winand, L., Dallemagne, M. J., and Duyckaerts, G. *Nature* **190**, 164 (1961).

Woernley, D. L. *Cancer Research* **12**, 516 (1952).

Wright, R. H. *Nature* **173**, 831 (1954).

Wright, R. H. "Theory and Methodology in Olfaction," Doctoral Dissertation, Purdue Univ., Lafayette, Indiana, (1962).

Wright, R. H. *Ann. N.Y. Acad. Sci.* **116**, Art. 2, 552 (1964).

Wright, R. H. and Michels, K. M. *Ann. N.Y. Acad. Sci.* **116**, Art. 2, 535 (1964).

Zapanta-Le Geros, R., Le Geros, J. P., Klein, E., and Trautz, O.R. *J. Dent. Res., Suppl. to No. 5,* **43**, 750 (1964).

Zapanta-Le Geros, R., Le Geros, J. P., Trautz, O. R., and Klein, E. in *Developments in Applied Spectroscopy, Vol. 7B* (E. L. Grove and A. J. Perkins, eds.) Plenum Press, New York, 1970, p. 3.

Zhbankov, R. G., Krivosheev, N. P., and Reutovich, G. V. *Dokl. Akad. Nauk Belorussk. SSR* **6**, 592, 1962; CA **58** 1713*e* (1963).

Chapter 20

INFORMATION SOURCES CONCERNING INFRARED SPECTROSCOPY AND SPECTRA

In biochemical and related types of research it is often imperative that one be able to compare the spectrum of an unknown substance with those in a collection. There are many such collections available in books and other commercial sources, but it is equally important to maintain a collection of spectra recorded in the laboratory, since these are specifically related to the research interests of the workers in that laboratory. For references to individual compounds (many of which are biochemicals) the reader is referred to the indexes of this book.

A Partial Listing of General References to Infrared Spectroscopy

Many other general references are given in the text.

National Research Council of Canada, *Multilingual Dictionary of Important Terms in Molecular Spectroscopy*, Ottawa, 1966. Sections in English, French, German, Japanese, and Russian.

G. Herzberg, *Molecular Spectra and Molecular Structure*, Van Nostrand, New York, *Vol. I*, 1950, *Vol. II*, 1945.

N. L. Alpert, W. E. Keiser, and H. A. Szymanski, *IR—Theory and Practice of Infrared Spectroscopy*, 2nd Ed., Plenum Press, New York, 1970.

A. D. Cross and R. A. Jones, *Introduction to Practical Infrared Spectroscopy*, 3rd Ed., Butterworths, London, 1969.

W. Brügel, *An Introduction to Infrared Spectroscopy*, Wiley, New York, 1962.

W. J. Potts, Jr., *Chemical Infrared Spectroscopy, Vol. I, Techniques*, Wiley, New York, 1963.

C. N. R. Rao, *Chemical Applications of Infrared Spectroscopy*, Academic Press, New York, 1963.

W. West, ed., *Chemical Applications of Spectroscopy*, Chap. I by W. West and Chap. III by A. B. F. Duncan, Interscience, New York, 1956.

K. Nakanishi, *Infrared Absorption Spectroscopy—Practical*, Holden-Day, San Francisco, 1962.

J. C. P. Schwarz, ed., *Physical Methods in Organic Chemistry*, Holden–Day, San Francisco, 1964.

R. P. Bauman and C. Clark (Conference Cochairmen) and others, "Biological Applications of Infrared Spectroscopy," *Ann. N. Y. Acad. Sci.* **69**, Art. 1, 1957.

M. Davies, ed., *Infra-red Spectroscopy and Molecular Structure*, Elsevier, New York, 1963.

S. K. Freeman, ed., *Interpretive Spectroscopy*, Reinhold, New York, 1965.

H. A. Szymanski, ed., *Progress in Infrared Spectroscopy, Vol. 1–3*, Plenum Press, New York, 1962, *et seq.*

Developments in Applied Spectroscopy, Vol. 1–7, various editors, Plenum Press, New York, 1961, *et seq.*

R. G. J. Miller, ed., *Laboratory Methods in Infrared Spectroscopy*, Heyden and Son, Ltd., London, 1965. Articles by specialists who have worked in one field for many years.

Spectra References (Including Theory in Some Cases)

The chapter on *Drugs, Pharmaceuticals, and Pharmacological Applications* (Chap. 16) refers to many collections of infrared and other types of spectra. Among these collections are spectra for a very wide variety of biochemical and other compounds, not only drugs and pharmaceuticals.

The Sadtler Research Laboratories of Philadelphia maintains large collections of spectra of substances of high purity and of various classifications. The spectra are cross-referenced to the ASTM (American Society for Testing and Materials) IBM cards, allowing workers with these IBM cards to sort for Sadtler spectra.

The fiftieth edition of the *Handbook of Chemistry and Physics* (1969–1970) contains a spectra index of organic compounds. The index is a listing of the Sadtler Standard Spectra recorded with an infrared prism instrument. The index also contains the Sadtler infrared grating, ultraviolet, and NMR numbers of many compounds listed in the Table of Physical Constants of Organic Compounds of the Handbook.

The Coblentz Society maintains a collection of high quality and verified spectra which are sold by the Sadtler Research Laboratories. A microfilm edition of these spectra is available from Mrs. Clara D. Smith, Editor, Coblentz Society Spectra, P.O. Box 584, Princeton, New Jersey 08540. Volumes 1 through 7 are now available, and they meet the Coblentz Society specifications for Approved Analytical Spectra (see *Anal. Chem.* **38**, No. 9, p. 27A, Aug., 1966).

The following books are very useful:

L. J. Bellamy, *Infrared Spectra of Complex Molecules*, 2nd Ed., Wiley, New York, 1958.

L. J. Bellamy, *Advances in Infrared Group Frequencies*, Methuen, London, 1968.

F. F. Bentley, L. D. Smithson, and A. L. Rozek, *Infrared Spectra and Characteristic Frequencies*, $\sim 700–300 \ cm^{-1}$, Interscience, New York, 1968.

N. B. Colthup, L. H. Daly, and S. E. Wiberley, *Introduction to Infrared and Raman Spectroscopy*, Academic Press, New York, 1964. Contains 624 spectra with interpretations of absorption bands.

K. Nakamoto, *Infrared Spectra of Inorganic and Coordination Compounds*, Wiley, New York, 1963.

K. E. Lawson, *Infrared Absorption of Inorganic Substances*, Reinhold, New York, 1961.

R. J. P. Lyon, *Minerals in the Infrared—A Critical Bibliography*, Stanford Research Institute, Menlo Park, California, 1962.

R. C. Weast, ed., *Handbook of Chemistry and Physics*, 50th Ed., The Chemical Rubber Co., Cleveland, Ohio.

H. M. Hershenson, *Infrared Absorption Spectra, Index for* 1945–1957, Academic Press, New York, 1959.

H. M. Hershenson, *Infrared Absorption Spectra, Index for* 1958–1962, Academic Press, New York, 1964.

An Index to Published Infra-red Spectra, H.M. Stationery Office, London, England, 1960. References to spectra in the literature up to 1957.

H. A. Szymanski, ed., *Infrared Band Handbook*, Plenum Press, New York, 1963; Several supplementary volumes. Presents spectral data arranged by wave number, in steps of 1 cm^{-1}; 2nd Ed., 1970, H. A. Szymanski and R. E. Erickson, eds.

H. M. Randall, R. G. Fowler, N. Fuson, and J. R. Dangl, *Infrared Determination of Organic Structures*, Van Nostrand, New York, 1949. Contains 354 spectra and band interpretations for a large variety of compounds.

M. St. C. Flett, *Characteristic Frequencies of Chemical Groups in the Infra-red*, Elsevier, New York, 1963.

D. W. Mathieson, *Interpretation of Organic Spectra*, Academic Press, New York, 1965. Problem solving with NMR, infrared, and mass spectroscopy.

H. A. Szymanski, *Interpreted Infrared Spectra, Vol. 1–3*, Plenum Press Data Division, 1964, *et seq.*

T. Cairns, *Spectroscopic Problems in Organic Chemistry*, Heyden and Son, Ltd., London, 1964. Sixty problems using infrared, NMR, and ultraviolet spectra.

A. J. Baker and T. Cairns, *Spectroscopic Techniques in Organic Chemistry*, Heyden and Son, Ltd., London, 1965. A practical laboratory manual. Uses IR, NMR, UV, and mass spectra.

A. J. Baker, T. Cairns, G. Eglinton, and F. J. Preston, *More Spectroscopic Problems in Organic Chemistry*, Heyden and Son, Ltd., London, 1967.

R. M. Silverstein and G. C. Bassler, *Spectrometric Identification of Organic Compounds*, Wiley, 1963, Uses IR, NMR, UV, and mass spectroscopy, 2nd Ed., 1967.

Of course, if one can find an expert who can interpret a spectrum or give a lead to its solution, an immediate advantage is gained. Then, there are the unmechanized procedures such as the Sadtler *Spec Finder**, the *Peek-a-boo* (Termatrex)†, or *DMS*

---

*Sadtler Research Laboratories, Inc., 3316 Spring Garden Street, Philadelphia, Pennsylvania 19104.
†Jonker Business Machines, Inc., Gaithersburg, Maryland.

Cards‡ methods. Mechanical sorters or computers can be used to search through mechanized indexes such as the IBM cards of the ASTM.

Unmechanized Spectra Indexes

The Termatrex "Peek-a-boo" System

The Peek-a-boo system is easy to use in the laboratory, and has a finite number of cards, each corresponding to a specific band interval of the spectrum. A card can have as many as 10,000 holes, and the locations of the holes correspond to serial numbers of the coded spectra. For example, the card for 6.5 μ has holes punched to correspond to the serial numbers of all the compounds displaying an absorption band at 6.5 μ. Each of the spectral characteristics to be coded is accounted for by similar cards.

One chooses the cards corresponding to the characteristics he wants to identify, places the cards in a stack above a light source, and searches for the spots of light that identify those spectra having all the characteristics in common. To produce the Termatrex cards one can obtain information from the ASTM card index.

The Documentation of Molecular Spectroscopy—Spectra (The DMS System)

This is a service which supplies cards containing spectra and other information, and is available from Butterworths Scientific Publications, London, and Verlag Chemie, West Germany. Steel needles are used in the sorting process. The coding system is rather detailed (Thompson, 1955), allowing for classification of structural features so that groups of compounds with structural units common to them can be chosen from the whole field of organic chemistry.

Figure 20.1 shows the coding of a card for 1,6-dimethylpiperid-2-one. Figures 20.2a and 20.2b show both sides of the coded card. The DMS system can be used with IBM cards for mechanical sorting.

Mechanized Use of Spectral Indexes

Baker *et al.* (1953) and Kuentzel (1951) have described machine methods of forming spectral indexes. The ASTM has adopted (1958) the Kuentzel method, which is the basis for the Termatrex and various computer methods.

The Kuentzel System

Figure 20.3 shows the typical IBM punched card used in the Wyandotte–ASTM spectral indexing system. The card gives the "characteristics" of the spectrum and of the compound by indicating the presence of a band, or the presence of a functional molecular characteristic, by the presence of a hole punched at the proper position on the card. For example, a card for a spectrum having a band at 8.2 μ would be punched in column 8, position 2, etc. The ASTM supplies a book of codes and instructions which describes the complete code. Published spectra serve as the source of the data for these cards, and the last columns indicate the origin.

‡The Documentation of Molecular Spectra, Butterworths Scientific Publications, 88 Kingsway, London, W.C.2, England, and Verlag Chemie, GMBH, Weinheim/Bergstrasse, West Germany.

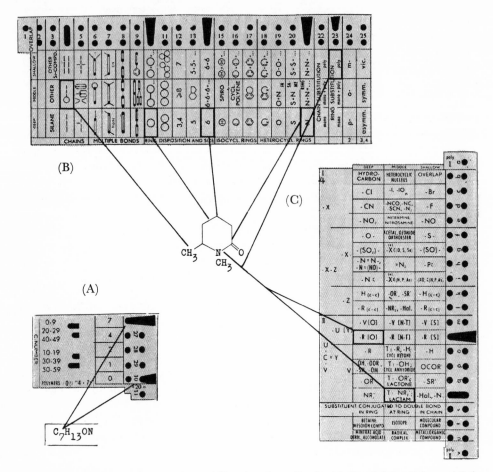

Fig. 20.1. The DMS spectral code. The card illustrates the coding for 1,6-dimethylpiperid-2-one: (A) the number of carbon atoms, (B) the basic skeleton, and (C) the substituents. (Thompson, 1955.)

An IBM card sorter is needed for searching through the deck of cards. Each "characteristic" of the spectrum to be matched requires a complete run-through of the cards. Since this can be a difficult and time-consuming task, many workers presort their cards into piles, one pile containing, say, all ketones, another all hydrocarbons, another nitrogen only.

The process of elimination can be a strong tool here. One looks for all the cards on which a certain characteristic is missing. For example, if a substance displays no band at 7.3 μ, one can remove all cards having a band at 7.3 μ and thus leave fewer cards remaining to be searched through.

Computer Techniques for Using the IBM Card File

Smithson et al. (1966) have produced the earliest program for the IBM 7090 computer. Since then their method has been adapted to other computers (Savitzky,

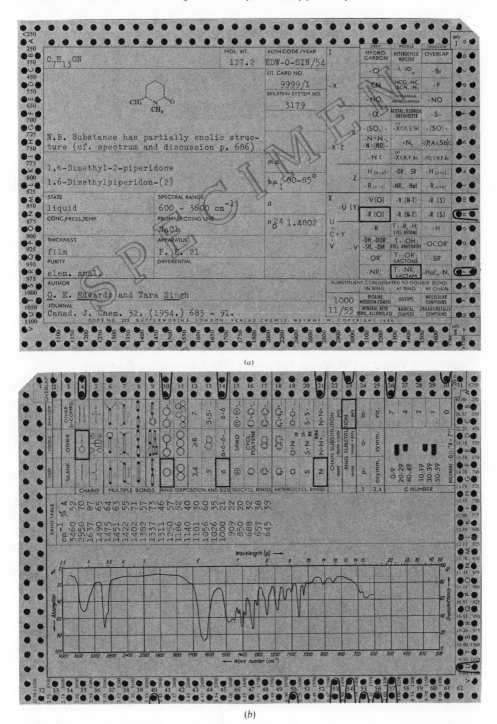

(a)

(b)

Fig. 20.2. A typical DMS coded card, at 60% of original size: (a) front, (b) back. (Thompson, 1955.)

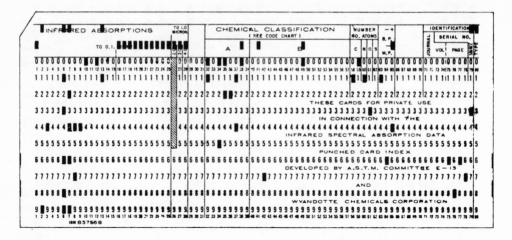

Fig. 20.3. A typical IBM spectral card, at 75 % of original size. (Szymanski, 1964.)

1966). For information on the availability of programs for particular computers it is helpful to write to ASTM.

"SIRCH," The ASTM-Dow Infrared File Searching System

This system is used on the IBM-1130. Band–no-band data are entered via the computer keyboard in either microns or wave numbers. Other physical data are entered by code numbers 1 through 32. Searching a single disk file of about 40,000 spectra takes only 40 sec. Data can be added to selectively, or deleted from the search "masks" after each search. No re-entry of data is necessary to search additional disk files.

Output from the program consists of the ASTM assigned serial numbers which may be used to look up the name of the compound, and the reference to the location of the spectrum. The data coded for any reference spectrum index may also be printed by entering its serial number. The program is a modified form of the one developed by Erley (1968), and inquiries should be made of the ASTM, 1916 Race Street, Philadelphia, Pa. 19103.

Erley (1970) has recently introduced SIRCH-III, an improved system.

FIRST-1

This is a computerized system produced by DNA Systems Inc. of Flint, Michigan. Erley (1971) has compared the performances of SIRCH-I, SIRCH-III, and FIRST-1 using the ASTM infrared data file. He used a test group of 138 unknown spectra selected at random from the literature. Features introduced into the FIRST-1 system increased the accuracy and precision above those of the other systems. Also, a novice could obtain successful results about as well as an expert. The cost per performance (taking all the factors affecting the search into account) was lowest for the FIRST-1 system.

Eastman Infrared Spectral Data Retrieval Service

Anderson and Covert (1967) have described a computer search system for the retrieval of infrared data. The system uses an IBM 7080 computer to search the data on magnetic tape for about 100,000 compounds, 90,000 of which are from the ASTM collection. Eastman Organic Chemicals, Rochester, New York 14650, runs a service based on this system. One can send copies of the infrared curve or curves of unknown compounds and ask for service under the catalogue number Eastman No. 6070. The computer output is a printed report listing in order of best probable match the names of up to 25 compounds most closely resembling the unknown compound. This search service supplies names of compounds only, not infrared spectra.

Computer Programs of the National Research Council of Canada

The following references contain computer programs published by the National Research Council of Canada:

R. N. Jones, T. E. Bach, H. Fuhrer, V. B. Kartha, J. Pitha, K. S. Seshadri, R. Venkataraghavan, and R. P. Young, *Computer Programs for Absorption Spectrophotometry*, National Research Council of Canada, Ottawa, 1968.

J. Pitha and R. N. Jones, *Optimization Methods for Fitting Curves to Infrared Band Envelopes. Computer Programs.*, National Research Council of Canada, Ottawa, 1968.

R. N. Jones and R. P. Young, *Additional Computer Programs for Absorption Spectrophotometry and Band Fitting*, National Research Council of Canada, Ottawa, 1969.

REFERENCES

American Society for Testing and Materials, *Codes and Instructions for Wyandotte–ASTM Punched Cards Indexing Spectral Absorption Data*, Philadelphia, 1958.
Anderson, D. H. and Covert, G. L. *Anal. Chem.* **39**, 1288 (1967).
Baker, A. W., Wright, N., and Opler, A. *Anal. Chem.* **25**, 1457 (1953).
Erley, D. S. *Anal. Chem.* **40**, 894 (1968).
Erley, D. S., Pittsburgh Conf. Anal. Chem. Appl. Spectrosc., Cleveland, March, 1970.
Erley, D. S. *Appl. Spectrosc.* **25**, 200 (1971).
Kuentzel, L. E. *Anal. Chem.* **23**, 1413 (1951).
Savitzky, A. in *Applied Infrared Spectroscopy* (D. N. Kendall, ed.), Reinhold, New York, 1966, p. 509.
Smithson, L. D., Fall, L. B., Pitts, F. D., and Bauer, F. W. "Storage and Retrieval of Wyandotte–ASTM Infrared Spectral Data Using an IBM 7090 Computer," Technical Documentary Report RTD-TDR-63-4265, cited in Savitzky (1966).
Thompson, H. W. *J. Chem. Soc.* **1955**, 4501.
Weast, R. C., ed. *Handbook of Chemistry and Physics*, 50th Ed., The Chemical Rubber Company, Cleveland, 1969–1970.

Appendix

FRACTIONATION METHODS BEFORE THE USE OF INFRARED SPECTROSCOPY

It is not difficult to apply infrared spectroscopy once a good sample is obtained. For the practicing biochemist this is often a most tedious and challenging part of his work—to obtain a compound in pure form from a very complex mixture of substances. A study of any modern biochemistry textbook indicates the enormous variety of compounds found together in living tissues. The problem is to separate them, get rid of unwanted material, and then to fractionate further to obtain a "pure" sample. The sensitivity, however, of infrared spectroscopy *per se* to impurities is not as troublesome as in ultraviolet spectroscopy, and this has advantages if low concentrations of contaminants are not bothersome to the investigation (see *Sampling Methods*, Chap. 3).

Many fractionation techniques have been refined in recent years. This chapter has the purpose of naming types of compounds that can be separated by some of the methods commonly used for the isolation of biochemical substances or for the examination of their purity. Discussions of theory, equipment, and applications will be found in many of the references given in this chapter.

Rosenkrantz (1957) has written one of the earlier articles on the utilization of fractionation procedures with infrared analysis and has listed several types of fractionation techniques: chromatography, countercurrent distribution, preferential solvent extraction, sublimation, fractional crystallization, molecular distillation, dialysis, centrifugation, electrophoresis, diffusion, and freeze-drying. He has also given references to work in which these methods have been used to fractionate a large variety of biological compounds. Elvidge and Sammes (1966) have discussed many of the techniques mentioned above.

Miscellaneous Types of Chromatography

The kinds of chemical substances that can be separated by the various types of chromatography are very numerous. Within the field of chromatography alone many methods are applicable to such separations, for example, adsorption, thin-layer, and partition chromatographies (on paper and on ion-exchange columns), and gel filtration. Morris and Morris (1964) have discussed the various techniques and their applications.

Thin-layer chromatography (TLC) has been discussed by Maier and Mangold (1964) and by Northcote (1966). Methods for recovering thin-layer and gas chromatography fractions for infrared spectroscopic examination have been discussed by Grasselli and Snavely (1967) and Snavely and Grasselli (1964). Rice (1967) has described a transfer technique which allows the sample to be transferred quickly and directly from the thin-layer plate to KBr powder without removing the TLC support material from the plate, thus reducing or eliminating sample loss and contamination before infrared spectroscopy. McCoy and Fiebig (1965) have described a technique for obtaining infrared spectra of microgram amounts of compounds separated by thin-layer chromatography.

Thin-layer chromatography, alone and in combination with other techniques, e.g., gas–liquid chromatography and mass and ultraviolet spectrometry, has been extensively applied to the isolation and identification of the prostaglandins and related factors. Some of the various solvent systems, adsorbents, methods of detection, and compounds studied have been summarized by Ramwell et al. (1968).

Williams et al. (1966a) have identified various phospholipids isolated from human red cells. Thin-layer chromatography and infrared spectroscopy were used to identify and confirm the phospholipids in each fraction eluted from silicic acid columns. The identification and establishment of the purity of the isolated phospholipids by these methods had been described earlier by Kuchmak and Dugan (1963) and Williams et al. (1966b).

In the examination of plant waxes various analytical methods have been used, including ultraviolet and infrared spectroscopy, and thin-layer, column, and gas–liquid chromatographies (Eglinton and Hamilton, 1967). The infrared spectrum has been used to indicate hydroxy compounds, ketones, ethers, and lactones, as well as the more common esters and hydrocarbons. In the use of thin-layer or column chromatography, infrared spectra can be used to make sure that there is no carryover of one class of components into another.

Paper and thin-layer chromatography have been discussed by Smith (1969), column chromatography by Edwards (1969), and horizontal circular chromatography by Kawerau (1960). Among the types of separable compounds listed in Smith (1969) are: amino acids, amines, and related compounds; indoles; imidazoles; guanidines; purines and pyrimidines, sugars, and related compounds; keto-acids; organic acids of the citric acid cycle and related acids; phenolic acids; phenols and tannins; cholesterol, cholesterol esters and their fatty acids; phospholipids and their derivatives; barbiturates; alkaloids; glutarimides and other neutral drugs; steroids; and inorganic ions. Infrared spectroscopy can be used to identify substances directly on paper chromatograms (Bellamy, 1953; Goulden, 1954). Horowitz (1971) has discussed partition chromatography, gel-filtration, adsorption, ion-exchange, and gas–liquid chromatographies as applied to carbohydrates.

Kesner and Muntwyler (1969) have separated compounds of the citric acid cycle and related substances by partition column chromatography. The theory and practice of paper chromatography have been discussed by Macek (1963). The separation of lipids by chromatography has been reviewed in detail by Nichols et al. (1966). The separation of viruses has been discussed by Markham (1966). A detailed review of the separation of peptides by various chromatographic methods has been given by Morris (1966) and the separation of amino acids has been discussed by Eastoe (1966). The separation of polymers from bacterial cell walls by chromatographic techniques has been reviewed by Rogers (1966).

A recent review on biochemical analysis (D'Eustachio, 1968) has given references to work in dialysis, ultrafiltration, centrifugation (including density-gradient separations), electrophoresis, column chromatography, ion-exchange, gel-filtration, gas, paper, and thin-layer chromatographies. Another review (Strickland, 1968) has listed over 2,700 papers on the subject of electrophoresis, including hundreds of papers on biological fluids, human serum (normal ones and those in many types of diseases), mammalian serum, blood proteins, animal tissues, lipoproteins, glycoproteins, histones, plants, microorganisms, enzymes, biochemicals, and forensic, toxicological, and pharmaceutical applications.

Zweig (1968) has reviewed some of the recent methods in column chromatography, gel-filtration, and gel-permeation chromatography, and paper and thin-layer chromatographies. He has listed several papers dealing with forensic and toxicological TLC analyses.

Hundreds of applications have been mentioned in the Zweig (1968) review: acids, alkaloids, amino acids, antibiotics, antioxidants, food and feed additives, bases and amines, bile acids, carbonyls, dyes, enzymes, lipids, hydrocarbons, hormones, indoles, natural products, peptides, proteins, pesticides, plant growth regulators, pharmaceutical products, phenols, pigments (chlorophylls, xanthophylls, porphyrins, melanin, pterins, pteridines, anthocyanins, flavonoids, etc.), polymers, purine and pyrimidine derivatives, quinones, RNA, DNA, organic sulfur compounds, steroids, sugars, toxins, vitamins, inorganic ions, and others.

Zweig et al. (1970) have recently reviewed the subjects of liquid column chromatography (excluding ion-exchange), paper chromatography, and thin-layer chromatography. Among the substances to which these methods have been applied are various acids, alkaloids, antibiotics, amino acids, bile acids, carbohydrates, proteins, lipids, hormones, steroids, pharmaceuticals, pigments, nucleic acids, toxins, and vitamins.

A general quantitative procedure using spectrophotometric analyses following thin-layer chromatography has been given by Lehmann et al. (1967). The final method of analysis may involve infrared spectroscopy, as was illustrated for the determination of oral contraceptives (Beyermann and Roeder, 1967).

Many recent biochemical applications of gas chromatographic techniques have been given in a review by Juvet and Dal Nogare (1968). The treatment of steroids was particularly prominent. Other substances mentioned were drugs, vitamins, amino acids, peptides, carbohydrates, strains of bacteria, and various types of milk.

Preparative gas chromatography has been reviewed by Verzele (1968) and by Sawyer and Hargrove (1968). Discussions of other topics concerned with gas chromatographic applications have appeared in the volume by the latter authors.

The book by Burchfield and Storrs (1962) on biochemical applications of gas chromatography contains a goldmine of information about the separation of substances of interest to investigators in many fields, for example, fatty acids and esters, volatile amines and amino alcohols, aldehydes, ketones, thiols, sulfides, and volatile alcohols; many of these chemicals occur in tissues and biological fluids. Alkyl benzenes such as p-cymene and polycyclic aromatic compounds occur naturally and their separations are of much interest. Naphthalene has been found in certain essential oils, and anthracene, 1,2 -benzanthracene, pyrene, and isomeric benzopyrenes are found in tobacco smoke. Among the essential oils group are many geometric, positional and optical isomers, particularly among the terpenes and sesquiterpenes. Gas chromatography separates many of these components, is a fast technique, and requires small samples. Abietic and pimaric resin acids (from pine rosin) can be fractionated by gas chromatography as their methyl esters.

Lipids are separated into constituent fractions by ancillary methods before gas chromatography, and must often be converted to more volatile derivatives before chromatography. Among this group are the higher aliphatic hydrocarbons, higher fatty alcohols, O-alkylglycerols, and higher fatty aldehydes (see also Nichols et al., 1966). Sterols, higher fatty alcohols, and triglycerides may be chromatographed either unchanged or as derivatives. Pregnanediol and 17-ketosteroids from a urine specimen have been separated (Scott, 1966). Techniques have been described (Sparagana, 1966) for obtaining good infrared spectra from gas chromatographic effluents with as little as one microgram of initial material.

Nonvolatile compounds of tissue have also been fractionated once a change to a derivative with higher vapor pressure has been accomplished. Such compounds are amino acids, certain organic acids, and sugars. Horowitz (1971) has discussed the theory of gas-liquid chromatography and particularly applications to separations of carbohydrates. The per(trimethylsilyl) ether derivatives of these compounds are easy to prepare, and such derivatives have been used for glycosides, deoxy sugars, cyclitols, hexoses, amino sugars and sialic acids. (Dimethylsilyl ethers may be especially useful in separating compounds of higher molecular weight whose volatilities are low in the form of trimethylsilyl ethers). Monosaccharide derivatives have been separated from mixtures of pure sugars, from hydrolysates of crude wood cellulose, and from glycolipids, gastric juice and other body fluids. Horowitz (1971) has also mentioned methyl ether derivatives; acetates; ethylidene, isopropylidene, and benzylidene acetals; and nucleosides and sugar phosphate esters. Derivatives of amino acids that have been used for gas–liquid chromatographic separation are methyl esters and N-acetyl alkyl esters, such as N-acetyl-n-amyl esters.

Fales and Pisano (1964) have discussed the gas chromatography of amines, alkaloids, and amino acids. Pollock and Kawauchi (1968) have resolved derivatives of serine, hydroxyproline, tyrosine, and cysteine, as well as racemic aspartic acid and tryptophan. VandenHeuvel and Horning (1964) have listed derivatives of steroids that can be separated. VandenHeuvel et al. (1960) first described the separation of bile acid methyl esters and Sjövall (1964) has extended the methods to bile acids. Gas–liquid chromatography (GLC) is useful in the analysis of pesticides, herbicides, and pharmaceuticals (Burchfield and Storrs, 1962). Analysis of alkaloids, steroids, and mixtures of anesthetics and expired air are other examples of the application of this very useful technique. Beroza (1970) has discussed the use of gas chromatography for the determination of the chemical structure of organic compounds at the microgram level.

Williams and Sweeley (1964) have given methods for the chromatographic separation of many urinary aromatic acids and have discussed diagnostic applications to (1) secreting tumors, e.g., malignant carcinoid, pheochromocytoma, and neuroblastoma, and (2) inborn errors of metabolism, e.g., tyrosinosis, phenylketonuria, Hartnup disease (involves aminoaciduria), and other inherited diseases. These authors referred to the use of infrared spectroscopy for verification of the identity of fractions of volatile organic anesthetics in blood. Chlorpromazine, pentobarbitone, and amphetamine, are examples of pharmacological substances that have been separated (Scott, 1966).

Chemically unidentified metabolites of viral infections have been detected (Mitruka *et al.*, 1968) in chromatograms of extracts of tissue cultures of dog kidney that had been inoculated with viruses causing canine hepatitis, herpes, and distemper, and a parainfluenza virus similar to simian virus-5.

Combined Use of Gas–Liquid Chromatography and Infrared Spectroscopy

Most infrared spectra of organic compounds contain a multiplicity of absorption bands—enough, frequently to characterize a given molecule. However, when several different molecules are present, the resulting spectrum is so complex it is difficult to assign the bands to the individual molecules and the calculation of the quantity of each present becomes complex. The gas chromatograph, on the other hand, can separate each component, and permits precise quantitative measurement. However, since retention time is a single-valued characteristic, little can be learned directly about the nature of each molecule in the mixture. The combined use of gas chromatography and infrared spectroscopy makes a powerful analytical tool. Practical methods have been devised for condensing a fraction out of the gas stream in such a manner that each molecule is exposed to infrared radiation in a spectrometer. For efficient absorption of infrared radiation use has been made of internal reflection techniques [commonly known as attenuated total reflection (ATR) or frustrated multiple internal reflection (FMIR)] developed around the same time independently by Harrick (1960) and by Fahrenfort (1961). This technique and some of its applications are further discussed in Chap. 3. A gas chromatographic fraction is fed to the surfaces of precooled thallium bromide–iodide (or other) crystalline plates, where the sample then condenses. The infrared energy is attenuated (absorbed) by the sample in contact with the plate, and infrared spectra are obtained that are essentially in excellent agreement with spectra run on pure samples by the usual transmission techniques. The attenuated total reflection method can be applied in such a way that the presence of relatively few molecules produces a strong spectrum. The temperature of the plate on which the sample has condensed can be raised to blow off used sample and the next fraction can then be analyzed.

Entrapment of GLC effluents on cold thin (25 μ) Millipore cellulosic membranes and use of the same type of membrane in the reference beam for compensation, allows good infrared spectra to be recorded (Thomas and Dwyer, 1964). A gas chromatographic fraction can be vacuum deposited onto KBr powder in a die, and pressed into a pellet which can then be examined in a spectrophotometer (Copier and van der Maas, 1967). Grasselli and Snavely (1967) have discussed liquid traps that can be used along with gas chromatographic instruments.

A dual-column gas chromatography system using a fraction collector has been developed (Giuffrida, 1965) to isolate one or more pesticide components in a mixture for infrared analysis. Compounds in the vapor state from the GLC column are condensed directly on KBr powder to obtain infrared spectra of micro-quantities of pesticides.

Rapid-scanning dispersion instruments have been developed to speed the recording of infrared spectra of gas chromatographic fractions (Bartz and Ruhl, 1964; Wilks and Brown, 1964; Wilks, 1968). Propster (1965) has discussed the application of rapid-scan infrared spectroscopy to gas-chromatographic fractions. Low and Freeman (1967) have used a multiple-scan interference spectrometer to measure infrared spectra of such fractions. The instrument scans the 2500–250 cm^{-1} range in one second with a spectral resolution of 18 cm^{-1}. Consecutive scans of either a flow-through cell and/or scans of a fraction trapped within a cell could be summated to yield spectra (Low and Coleman, 1966), although the trapped samples yielded better results than did the flowing streams, because more scans could be made (Low and Freeman, 1967). Block Engineering Company (1969) has available a more highly resolving interference spectrometer, which can resolve down to better than 0.5 cm^{-1} on the spectrum of ammonia and better than 1.7 cm^{-1} on survey runs, e.g., on indene. Krakow (1969) has investigated rapid, on-the-fly infrared analysis of gas-chromatographic effluents. He used a grating spectrometer that continually produced spectra from 3704 cm^{-1} to 1111 cm^{-1} at the rate of 1.6 spectra per second, and found that the instrument had the potential ability to detect unresolved chromatographic peaks. Freeman (1969) has recently discussed the combined use of gas chromatography and infrared and Raman spectrometry.

Ion-Exchange Chromatography

In a review of ion-exchange chromatographic methods Walton (1968) has given an extensive table of types of organic and biochemical compounds that can be fractionated by such methods. He has listed the compound, the type of exchanger, the eluent, the method, and special notes, along with references to the literature. A more recent review by Walton (1970) also contains a table of organic and biochemical applications of ion-exchange chromatography. The fractionation of tRNA's and of oligonucleotides has been discussed by RajBhandary and Stuart (1966).

Countercurrent Chromatography

Ito and Bowman (1970) have described a liquid–liquid partition chromatographic system (without solid support), which involves a long helix of narrow-bore tubing with an inner diameter of less than 0.5 mm. When the coiled tube is filled with one phase of a two-phase system and fed with the other phase, phase interchange takes place in each turn of the coil, leaving a segment of the former phase as the stationary phase. Solutes present in either phase are consequently subjected to a multistep partition process. To demonstrate the capability of the method Ito and Bowman used a two-phase system of chloroform, glacial acetic acid, and 0.1N

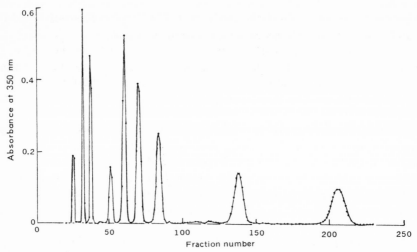

Fig. A.1. Chromatogram for the separation of DNP amino acids. Peaks identified in order of elution and their partition coefficients from left to right are: N-dinitrophenyl-δ-L-ornithine (>100). N-dinitrophenyl-L-aspartic acid (3.8), N-dinitrophenyl-D-L-glutamic acid (1.9), N,N'-dinitrophenyl-L-cystine (0.94), N-dinitrophenyl-β-alanine (0.71), N-dinitrophenyl-L-alanine (0.56), N-dinitrophenyl-L-proline (0.45), N-dinitrophenyl-L-valine (0.26), and N-dinitrophenyl-L-leucine (0.18). (Ito and Bowman, 1970.)

aqueous hydrochloric acid ($2:2:1$) for the separation of dinitrophenyl amino acids. Figure A.1 shows a chromatogram for the separation of the DNP amino acids. The column efficiency for these amino acid derivatives was comparable to that of gas chromatography. These workers claimed that the technique is more versatile than refined gas-chromatographic methods in that nonvolatile materials may be resolved and the phase system used may be changed for programmed separations. Because no solid support was used, the method avoided tailing of the solute, the procedure was highly reproducible and free from contamination, and the separated solutes were readily recovered.

Molecular-Sieve Chromatography (Gel Filtration)

The gels used in molecular-sieve chromatography are made of cross-linked dextrans (Sephadex type), polyacrylamide, and agar and agarose (Bio-gel type).

Desalting and Related Separations

The separation of salts or other compounds of low molecular weight (distribution coefficient, K_d, about 1) from material of high molecular weight ($K_d = 0$) is achieved more rapidly by molecular-sieve chromatography than by dialysis (Andrews, 1966). The procedure requires gels of small pore size, and by proper choice of conditions separation may be accomplished with little dilution of the sample (Flodin, 1961; Morris and Morris, 1964). Apart from desalting (Flodin, 1961), the many separations of this type include the removal of interfering substances from enzymes and other proteins (Bucovaz and Davis, 1961; Ledvina, 1963; Patrick and Thiers, 1963) and the separation of cofactors from enzymes (Kisliuk, 1960). Heparin, a polysaccharide, has been separated from NaCl (Lasker and Stivala, 1966), and nucleotides have been separated from desalted contaminants of low molecular weight by gel filtration (Uziel, 1967). Desalting and related techniques have also been discussed by Smith (1969).

Peptides and Proteins

Molecular sieving has achieved the following: fractionation of the heterogeneous peptide mixtures obtained by hydrolysis of proteins or glycoproteins (Bennich, 1961; Guidotti *et al.*, 1962; Lee and Montgomery, 1962; Nolan and Smith, 1962; Press and Porter, 1962; Carnegie, 1965) and of complex, naturally occurring peptide mixtures (Bagdasarian *et al.*, 1964). Molecular sieving has been used in the purification of peptide and polypeptide hormones (Porath and Schally, 1962; Porath, 1963; Rasmussen *et al.*, 1964) and for extracting peptide hormones from blood before their bioassay (Folley and Knaggs, 1965). Appropriate choice of solvent may overcome troublesome molecular interactions occurring in peptide fractionation (Bagdasarian *et al.*, 1964; Synge, 1964; Carnegie, 1965).

Rather extensive use of molecular-sieve chromatography has been made in the fractionation of protein mixtures, and methods for the purification of enzymes and other proteins often include a molecular-sieving step [enzyme mixtures: Gelotte (1964) and Downey and Andrews (1965); blood proteins: Roskes and Thompson (1963) and Killander (1964)]. Glycopeptides have been isolated from glycoprotein on Sephadex columns (Marks *et al.*, 1962; Lee and Montgomery, 1962). A table of protein fractionations carried out by gel filtration methods appears in Morris and Morris (1964). See also Smith (1969).

Carbohydrates, Nucleotides and Nucleic Acids

Flodin and Aspberg (1961) have reported a marked separation of oligosaccharides on a column of gel of small pore size. Ringertz (1960) has reported the fractionation of acid polysaccharides from mouse tumors on a Sephadex column. The fractions differed in several properties, including the type of amino sugar present. Tanaka (1966) has reported on the fractionation and isolation of acid mucopolysaccharides.

Low-molecular-weight carbohydrate components of normal urine have been separated by Lundblad and Berggård (1962), Flodin *et al.* (1964), and Lee and Ballou (1965). Gels of larger pore size have been used for molecular-sieve separations of mixtures of heterogeneous polysaccharides (Granath and Flodin, 1961; Andrews and Roberts 1962; Heller and Schramm, 1964; Roberts and Gibbons, 1966). Urinary fractions containing hexose, hexosamine, sialic acid, fucose, glucuronic acid, and tyrosine have been obtained by gel filtration (Miettinen, 1962).

Gels of relatively small pore size have been used to separate nucleotides, oligonucleotides, and other constituents of nucleic acid (Ingram and Pierce, 1962; Hohn and Pollmann, 1963; Hayes *et al.*, 1964; Lipsett, 1964), and to investigate the heterogeneity of polynucleotide preparations (Jones *et al.*, 1964) and soluble ribonucleic acid preparations (Richards and Gratzer, 1964; Schleich and Goldstein, 1964; Röschenthaler and Fromageot, 1965). Aminoacyl-tRNA has been purified by molecular-sieve chromatography on Sephadex (Moldave and Sutter, 1967).

Miscellaneous

Plant estrogens and metabolites of these compounds that have been excreted by animals have been fractionated on dextran gel columns (Nilsson, 1962). Gel filtration has been used in an investigation of polyphenols from brewery wort (Woof, 1962). The catecholamines adrenalin and noradrenalin can be separated from plasma proteins on Sephadex (Marshall, 1963). Various constituents of *Staphylococcus aureus* have been separated on Sephadex, namely, teichoic acid, D-alanine, N-acetylglucosamine, and N-acetylglucosaminylribitol (Sanderson *et al.*, 1962).

Electrophoretic Techniques

Many types of compounds have been separated from one another by electrophoresis or have mobilities that indicate that separation is possible by this means. Also, many variations of electrophoretic techniques are available. High-voltage electrophoresis on filter paper complements partition and adsorption chromatography for the separation of minute quantities of closely related substances of low molecular weight. Efron (1968) has discussed the separation of amino acids and other ninhydrin-reacting substances by the former process. She has presented a map of 47 such related substances separated by a combination of electrophoresis for the first separation and chromatography for the second separation run in a direction perpendicular to the electrophoresis. Efron (1968) has also discussed the separation of indoles in urine, sugars, purines, pyrimidines, and related derivatives, acids of the citric acid cycle, phenolic acids, keto-acids, and imidazoles. A study of the separation of sugars has also been made by Consden and Stanier (1952).

Horowitz (1971) has discussed in detail the reagents used for detecting minute amounts of carbohydrate on paper. He has discussed separations by electrophoresis and chromatography as well as elution and quantitative collection of these substances. Wilks and Hirschfeld (1967) have discussed the practice of eluting sample "spots" from paper electrophoretic and paper chromatographic patterns and allowing the eluted material to fall onto the internal-reflector plate of an attenuated-total-reflection (infrared) unit (see Chap. 3). Rübner and Albers (1967) have measured the infrared absorptions of electrophoresed carbohydrates on transparent acetate film strips *versus* transparent strips as blanks. These workers have detected separated substances without the necessity of staining the strips.

Discussions of electrophoresis of the following types of substances are given by Zweig and Whitaker (1967): amines, amino acids, peptides and related compounds, proteins, nucleic acids, nucleoside derivatives, carbohydrates, and organic acids. A chapter especially useful for the preparation of samples includes aldehydes, ketones, alkaloids, antibiotics, choline esters and phospholipid derivatives, coenzymes and vitamins, natural pigments, and steroids. A more recent review on biological applications of electrophoresis is the one by Strickland (1970). Among the many topics included are biological fluids, human serum, hemoglobin, lipoproteins, glycoproteins, cells, microorganisms, enzymes, hormones, pharmaceuticals, and food.

Shaw (1969) has discussed (among many other topics concerning electrophoresis) the technique of isoelectric focusing, which is not electrophoresis in the conventional sense. He has shown the separation of a hemoglobin preparation by this method [see also, Haglund (1967)]. Wrigley (1968) has discussed the analysis of multiple protein samples by isoelectric focusing, and Jonsson and Pettersson (1968) have discussed the application of this method to pigments of the red beet, bilberry, black currant, strawberry, and black tea. Giorgio and Tabachnick (1968) have used several types of fractionation methods in the isolation and characterization of a thyroxine-binding globulin (present in trace quantities) from human plasma. These methods included zone electrophoresis on a cellulose column and preparative electrophoresis in polyacrylamide gel.

Differential Dialysis

Various ways of applying dialysis for the fractionation of substances have been given by Craig (1965). A list of model substances studied in his laboratory for the calibration of membranes is given here (the figure after each substance is its molecular weight): glycine, 75; tryptophan, 204; oxidized glutathione, 612; bacitracin, 1422; subtilin, 3200; ribonuclease, 13,600; chymotrypsinogen, 25,000; pepsinogen, 45,000; and serum albumin, 68,000. Sugars, amino acids, peptides, proteins, and other molecules can be fractionated. Countercurrent dialysis can be applied, for example, to separate sucrose, a disaccharide, from stachyose, a tetrasaccharide. Parallel rotatory evaporators are required to concentrate diffusate at several stages of the separation.

Ultrafiltration and dialysis can be combined. An arrangement suggested by Berggård (1961) made use of both techniques. The filter used provided an excellent way to concentrate dilute solutions of proteins with molecular weights larger than 20,000, or to remove such solutes from considerably smaller ones.

Craig (1968, 1969) has described a rapid countercurrent dialyzer with analytical significance. The dialyzer can quantitatively separate small solutes from mixtures of large solutes such as proteins or nucleic acids.

Density-Gradient Separations in the Ultracentrifuge

The general principles and some examples of density-gradient separation have been discussed by Charlwood (1966). The technique of using sucrose gradients (Pickels, 1943) to stabilize weak sedimenting boundaries has been furthered by Brakke (1951, 1953) and others.

Density-gradient separations of cytoplasmic fractions from liver have been carried out to study the distribution of activity of several enzymes (Thomson, 1959). Ribosomes and polysomes have been fractionated to elucidate the mechanisms of protein synthesis (Watson, 1964). Various viruses have been studied with the density-gradient technique: polio virus (Levintow and Darnell, 1960), Rous sarcoma virus (Crawford, 1960), Shope papilloma virus (Williams et al., 1960), and adenoviruses (Allison et al., 1960). Density-gradient fractionation has been particularly useful for separation of DNA molecules from various species of bacteria (Rolfe and Meselson, 1959) and from animal cells (Kit, 1961). Other types of molecules, e.g., antibodies, lipoproteins, and rheumatoid factor have been isolated by density-gradient methods [see Charlwood (1966) for examples].

Countercurrent Distribution

Craig and his co-worker (Craig, 1944; Craig and Post, 1949) were the originators of the multistage technique of countercurrent distribution. Early reviews have been presented by Hecker (1955) and by Craig and Craig (1956). Morris and Morris (1964) and Fleetwood (1966) have published later reviews.

Some of the substances that have been separated by this method are given in papers referred to by Morris and Morris (1964): amino acids, peptides (particularly those having molecular weights ranging from 500 to 5000), polypeptide antibiotics, proteins (including enzymes), carbohydrates (although for most compounds in this chemical class other fractionation methods are much more frequently applied), purines, pyrimidines, nucleic acid derivatives, tRNA's that are specific for various amino acids, organic acids, steroids, lipids, antibiotics that are not peptides, porphyrins, pterins, vitamin B_{12} and other vitamins, lipoic acid, and alkaloids. The countercurrent-distribution procedure of Holley et al. (1965) is widely used, sometimes with modifications. Korte et al. (1965) have separated three isomers of tetrahydrocannabinol.

Use of the Solubility Diagram for Determining the Presence of Impurities in Soluble Materials

Herriott (1957) has reviewed the use of the solubility diagram as a tool for detecting the presence of impurities in soluble substances, particularly proteins. In these days of advanced types of instrumentation and "sophisticated" methods, it is worth recalling the use of a method that depends on solubility principles and is not difficult to apply. The method is based on application of the phase rule, but one need not apply the phase rule or even understand it to determine whether a preparation contains one or more components. Herriott has discussed the general procedure, the interpretation of results, the quantitative limitations of the method, the conditions and details of technique, and the separation of protein components by methods based on the solubility diagram.

REFERENCES

Allison, A. C., Pereira, H. G., and Farthing, C. P. *Virology* **10**, 316 (1960).
Andrews, P. "Molecular-Sieve Chromatography," *Brit. Med. Bull.* **22**, 109 (1966).
Andrews, P. and Roberts, G. P. *Biochem. J.* **84**, 11P (1962).
Bagdasarian, M., Matheson, N. A., Synge, R. L. M., and Youngson, M. A. *Biochem. J.* **91**, 91 (1964).
Bartz, A. M. and Ruhl, H. D. *Anal. Chem.* **36**, 1892 (1964).
Bellamy, L. J. *J. Appl. Chem.* **3**, 421 (1953).
Bennich, H. *Biochim. Biophys. Acta* **51**, 265 (1961).
Berggård, I. *Ark. Kemi.* **18**, 291 (1961).
Beroza, M. *Accounts of Chemical Research* **3**, 33 (1970).
Beyermann, K. and Roeder, E. *Fresenius' Z. Anal. Chem.* **230**, 347 (1967).
Block Engineering Company, Brochure 1969, Cambridge, Massachusetts. Instrument distributed by Sadtler Research Labs., Philadelphia.
Brakke, M. K. *J. Am. Chem. Soc.* **73**, 1847 (1951).
Brakke, M. K. *Arch. Biochem. Biophys.* **45**, 275 (1953).
Bucovaz, E. T. and Davis, J. W. *J. Biol. Chem.* **236**, 2015 (1961).

Burchfield, H. P. and Storrs, E. E. *Biochemical Applications of Gas Chromatography*, Academic Press, New York, 1962.

Carnegie, P. R. *Nature* **206**, 1128 (1965).

Charlwood, P. A. "Density-Gradient Separations in the Ultracentrifuge," *Brit. Med. Bull.* **22**, 121 (1966).

Consden, R. and Stanier, W. M. *Nature* **169**, 783 (1952).

Copier, H. and van der Maas, J. H. *Spectrochim. Acta* **23A**, 2699 (1967).

Craig, L. C. *J. Biol. Chem.* **155**, 519 (1944).

Craig, L. C. "Differential Dialysis," in *Advances in Analytical Chemistry and Instrumentation, Vol. 4* (C. N. Reilley, ed.), Interscience, New York, 1965, p. 35.

Craig, L. C. *Appl. Spectrosc.* **22**, 231 (1968), Abstract, Pittsburgh Conference on Analytical Chemistry and Applied Spectroscopy, March 3–8, 1968, Cleveland, Ohio.

Craig, L. C. and Chen, H.-C. *Anal. Chem.* **41**, 590 (1969).

Craig, L. C. and Craig, D. in *Technique of Organic Chemistry, 2nd Ed., Vol. 3, Part I* (A. Weissberger, ed.), Interscience, New York, 1956, p. 149.

Craig, L. C. and Post, O. *Anal. Chem.* **21**, 500 (1949).

Crawford, L. V. *Virology* **12**, 143 (1960).

D'Eustachio, A. J. "Biochemical Analysis," in *Anal. Chem.* **40**, 19R (1968).

Downey, W. K. and Andrews, P. *Biochem. J.* **94**, 642 (1965).

Eastoe, J. E. "Separation of Amino Acids," in *Brit. Med. Bull.* **22**, 174 (1966).

Edwards, R. W. H. in *Chromatographic and Electrophoretic Techniques, 3 Ed., Vol. I-Chromatography* (I. Smith, ed.) Interscience, New York, 1969, p. 940.

Efron, M. L. "High Voltage Paper Electrophoresis," in *Chromatographic and Electrophoretic Techniques, 2 Ed., Vol. II-Zone Electrophoresis* (I. Smith, ed.) Interscience, 1968, p. 166.

Eglinton, G. and Hamilton, R. J. *Science* **156**, 1322 (1967).

Elvidge, J. A. and Sammes, P. G. *A Course in Modern Techniques of Organic Chemistry, 2nd Ed.*, Butterworths, London, 1966.

Fahrenfort, J. *Spectrochim. Acta* **17**, 698 (1961).

Fales, H. M. and Pisano, J. J. in *Biomedical Applications of Gas Chromatography* (H. A. Szymanski, ed.), Plenum Press, New York, 1964, p. 39.

Fleetwood, J. G. "Recent Developments in Liquid–Liquid Countercurrent Distribution," in *Brit. Med. Bull.* **22**, 127 (1966).

Flodin, P. *J. Chromatogr.* **5**, 103 (1961).

Flodin, P. and Aspberg, K. in *Biological Structure and Function, Vol. I* (T. W. Goodwin and O. Lindberg, eds.) Academic Press, New York, 1961, p. 345.

Flodin, P., Gregory, J. D., and Rodén, L. *Anal. Biochem.* **8**, 424 (1964).

Folley, S. J. and Knaggs, G. S. *J. Endocrinol.* **33**, 301 (1965).

Freeman, S. K. in *Ancillary Techniques of Gas Chromatography* (L. S. Ettre and W. H. McFadden, eds.), Interscience, New York, 1969, p. 227.

Gelotte, B. *Acta Chem. Scand.* **18**, 1283 (1964).

Giorgio, N. A., Jr. and Tabachnick, M. *J. Biol. Chem.* **243**, 2247 (1968).

Giuffrida, L. *J. Ass. Offic. Agr. Chem.* **48**, 354 (1965).

Goulden, J. D. S. *Nature* **173**, 646 (1954).

Granath, K. A. and Flodin, P. *Makromolek. Chem.* **48**, 160 (1961).

Grasselli, J. G. and Snavely, M. K. in *Progress in Infrared Spectroscopy, Vol. 3* (H. A. Szymanski, ed.), Plenum Press, New York, 1967, p. 55.

Guidotti, G., Hill, R. J. and Konigsberg, W. *J. Biol. Chem.* **237**, 2184 (1962).

Haglund, H. *Science Tools* **14**, 17 (1967), L. K. B. Instruments, Inc.

Harrick, N. J. *Phys. Rev. Letters* **4**, 224 (1960).

Hayes, F. N., Hansbury, E., and Mitchell, V. E. *J. Chromatogr.* **16**, 410 (1964).

Hecker, E. *Verteilungsverfahren im Laboratorium*, Verlag Chemie, Weinheim, 1955; cited by E. Hecker, *Z. Analyt. Chem.* **181**, 284 (1961).

Heller, J. and Schramm, M. *Biochim. Biophys. Acta* **81**, 96 (1964).

Herriott, R. M. "The Solubility Diagram as a Criterion of Protein Homogeneity," in *Methods in Enzymology* **4**, 212 (1957).

Hohn, T. and Pollmann, W. *Z. Naturforsch.* **18b**, 919 (1963).

Holley, R. W., Apgar, J., Everett, G. A., Madison, J. T., Marquisee, M. J., Merrill, S. H., Penswick, J. R., and Zamir, A. *Science* **147**, 1462 (1965).

Horowitz, M. I. "Gas–Liquid Chromatography of Carbohydrates," in *The Carbohydrates, Vol. 1B* (W. Pigman and D. Horton, eds.) Academic Press, New York, in Press.

Horowitz, M. I. "Separation Methods: Chromatography and Electrophoresis," in *The Carbohydrates, Vol. 1B* (W. Pigman and D. Horton, eds.), Academic Press, New York, 1971.

Ingram, V. M. and Pierce, J. G. *Biochemistry* **1**, 580 (1962).

Ito, Y. and Bowman, R. L. *Science* **167**, 281 (1970).

Jones, O. W., Townsend, E. E., Sober, H. A., and Heppel, L. A., *Biochemistry* **3**, 238 (1964).

Jonsson, M. and Pettersson, E. *Science Tools* **15**, 2 (1968), L. K. B. Instruments, Inc.

Juvet, R. S., Jr. and Dal Nogare, S. "Gas Chromatography," in *Anal. Chem.* **40**, 33R (1968).

Kawerau, E. in *Chromatographic and Electrophoretic Techniques, Vol. I—Chromatography* (I. Smith, ed.), Interscience, New York, 1960, p. 546.

Kesner, L. and Muntwyler, E. "Separation of Citric Acid Cycle and Related Compounds by Partition Column Chromatography," in *Methods in Enzymology, Vol. 13* (J. M. Lowenstein, ed.) Academic Press, New York, 1969, p. 415.

Killander, J. *Biochim. Biophys. Acta* **93**, 1 (1964).

Kisliuk, R. L. *Biochim. Biophys. Acta* **40**, 531 (1960).

Kit, S. *J. Mol. Biol.* **3**, 711 (1961).

Korte, F., Haag, M., and Claussen, U. *Angew. Chem.* **77**, 862 (1965).

Krakow, B. *Anal. Chem.* **41**, 815 (1969).

Kuchmak, M. and Dugan, L. R., Jr. *J. Am. Oil Chemists' Soc.* **40**, 734 (1963).

Lasker, S. E. and Stivala, S. S. *Arch. Biochem. Biophys.* **115**, 360 (1966).

Ledvina, M. *J. Chromatogr.* **11**, 71 (1963).

Lee, Y.-C. and Ballou, C. E. *Biochemistry* **4**, 257 (1965).

Lee, Y.-C. and Montgomery, R. *Arch. Biochem. Biophys.* **97**, 9 (1962).

Lehmann, G., Hahn, H. G., and Martinod, P. *Fresenius' Z. Anal. Chem.* **227**, 81 (1967).

Levintow, L. and Darnell, J. E., Jr. *J. Biol. Chem.* **235**, 70 (1960).

Lipsett, M. N. *J. Biol. Chem.* **239**, 1250 (1964).

Low, M. J. D. and Coleman, I. *Spectrochim. Acta* **22**, 369 (1966).

Low, M. J. D. and Freeman, S. K. *Anal. Chem.* **39**, 194 (1967).

Lundblad, A. and Berggard, I. *Biochim. Biophys. Acta* **57**, 129 (1962).

Macek, K. in *Paper Chromatography* (I. M. Hais and K. Macek, eds.) Academic Press, New York, 1963.

Maier, R. and Mangold, H. K. in *Advances in Analytical Chemistry and Instrumentation, Vol. 3* (C. N. Reilley, ed.) Interscience, New York, 1964, p. 369.

Markham, R. "Separation of Viruses," in *Brit. Med. Bull.* **22**, 153 (1966).

Marks, G. S., Marshall, R. D., Neuberger, A., and Papkoff, H. *Biochim. Biophys. Acta* **63**, 340 (1962).

Marshall, C. S. *Biochim. Biophys. Acta* **74**, 158 (1963).

McCoy, R. N. and Fiebig, E. C. *Anal. Chem.* **37**, 593 (1965).

Miettinen, T. A. *Scand. J. Clin. Lab. Invest.* **14**, 380 (1962).

Mitruka, B. M., Alexander, M., and Carmichael, L. E. *Science* **160**, 309 (1968).

Moldave, K. and Sutter, R. P. in *Methods in Enzymology, Vol. XII, Part A* (L. Grossman and K. Moldave, eds.), Academic Press, New York, 1967, p. 598.

Morris, C. J. O. R. "Separation of Peptides," in *Brit. Med. Bull.* **22**, 168 (1966).

Morris, C. J. O. R. and Morris, P. *Separation Methods in Biochemistry*, Interscience, 1964.

Nichols, B. W., Morris, L. J., and James, A. T. "Separation of Lipids by Chromatography," in *Brit. Med. Bull.* **22**, 137 (1966).

Nilsson, A. *Acta Chem. Scand.* **16**, 31 (1962).

Nolan, C. and Smith, E. L. *J. Biol. Chem.* **237**, 446 (1962).

Northcote, D. H. "Carbohydrates and Mucoid Substances," in "The Separation of Biological Materials," *Brit. Med. Bull.* **22**, 180 (1966) (Partition, thin-layer, gas–liquid, absorption, and ion-exchange chromatographies; electrophoresis; gel filtration; and fractional precipitation).

Patrick, R. L. and Thiers, R. E. *Clin. Chem.* **9**, 283 (1963).

Pickels, E. G. *J. Gen. Physiol.* **26**, 341 (1943).

Pollock, G. E. and Kawauchi, A. H. *Anal. Chem.* **40**, 1356 (1968).

Porath, J. *Pure Appl. Chem.* **6**, 233 (1963).

Porath, J. and Schally, A. V. *Endocrinology* **70**, 738 (1962).

Press, E. M. and Porter, R. R. *Biochem. J.* **83**, 172 (1962).

Propster, G. D., Pittsburgh Conference Anal. Chem. Appl. Spectroscopy, March, 1965, Pittsburgh, Pa.

RajBhandary, U. L. and Stuart, A. *Ann. Rev. Biochem.* **35**, Part II, 630 (1966).

Ramwell, P. W., Shaw, J. E., Clarke, G. B., Grostic, M. F., Kaiser, D. G., and Pike, J. E. "Prostaglandins," in *Progress in the Chemistry of Fats and Other Lipids, Vol. 9, Part 2*, Pergamon Press, Ltd., Oxford, 1968, p. 233.

Rasmussen, H., Sze, Y.-L., and Young, R. *J. Biol. Chem.* **239**, 2852 (1964).

Rice, D. D. *Anal. Chem.* **39**, 1906 (1967).

Richards, E. G. and Gratzer, W. B. *Nature* **204**, 878 (1964).

Ringertz, N. R. *Acta Chem. Scand.* **14**, 312 (1960).

Roberts, G. P. and Gibbons, R. A. *Biochem. J.* **98**, 426 (1966).

Rogers, H. J. "Separable Polymers in Bacterial Cell Walls," in *Brit. Med. Bull.* **22**, 185 (1966).

Rolfe, R. and Meselson, M. *Proc. Natl. Acad. Sci. U.S.* **45**, 1039 (1959).

Röschenthaler, R. and Fromageot, P. *J. Mol. Biol.* **11**, 458 (1965).

Rosenkrantz, H. "Utilization of Fractionation Procedures with Infrared Analysis," *Ann. N. Y. Acad. Sci.* **69**, Art. 1, 5 (1957).

Roskes, S. D. and Thompson, T. E. *Clinica Chim. Acta* **8**, 489 (1963).

Rübner, H. and Albers, P. *J. Chromatogr.* **27**, 510 (1967).

Sanderson, A. R., Strominger, J. L., and Nathenson, S. G. *J. Biol. Chem.* **237**, 3603 (1962).

Sawyer, D. T. and Hargrove, G. L. in *Progress in Gas Chromatography* (J. H. Purnell, ed.), Interscience, New York, 1968, p. 325.

Schleich, T. and Goldstein, J. *Proc. Natl. Acad. Sci. U.S.* **52**, 744 (1964).

Scott, R. P. W. "Gas–Liquid Chromatography," in *Brit. Med. Bull.* **22**, 131 (1966).

Shaw, D. J. *Electrophoresis*, Academic Press, New York, 1969.

Sjövall, J. in *Biomedical Applications of Gas Chromatography* (H. A. Szymanski, ed.) Plenum Press, New York, 1964, p. 151.

Smith, I. *Chromatographic and Electrophoretic Techniques, Vol. I, Chromatography, 3 Ed.* (1969) and *Vol. II, Zone Electrophoresis, 2 Ed.* (1968), Interscience, New York.

Snavely, M. K. and Grasselli, J. G. in *Developments in Applied Spectroscopy, Vol. 3* (J. E. Forrette and E. Lanterman, eds.) Plenum Press, New York, 1964, p. 119.

Sparagana, M. *Steroids* **8**, 219 (1966).

Strickland, R. D. "Electrophoresis," in *Anal. Chem.* **40**, 74R (1968).

Strickland, R. D. *Anal. Chem.* **42**, 32R (1970).

Synge, R. L. M. *Metabolism* **13**, 969 (1964).

Tanaka, Y. and Gore, I. *J. Chromatogr.* **23**, 254 (1966).

Thomas, P. J. and Dwyer, J. L. *J. Chromatogr.* **13**, 366 (1964).

Thomson, J. F. *Anal. Chem.* **31**, 836 (1959).

Uziel, M. in *Methods in Enzymology, Vol. XII, Part A* (L. Grossman and K. Moldave, eds.) Academic Press, New York, 1967, p. 407.

VandenHeuvel, W. J. A. and Horning, E. C. in *Biomedical Applications of Gas Chromatography* (H. A. Szymanski, ed.) Plenum Press, New York, 1964, p. 89.

VandenHeuvel, W. J. A., Sweeley, C. C., and Horning, E. C. *Biochem. Biophys. Res. Commun.* **3**, 33 (1960).

Verzele, M. in *Progress in Separation and Purification, Vol. I* (E. S. Perry, ed.) Interscience, New York, 1968, p. 83.

Walton, H. F. *Anal. Chem.* **40**, 51R (1968).

Walton, H. F. *Anal. Chem.* **42**, 86R (1970).

Watson, J. D. *Bull. Soc. Chim. Biol.* **46**, 1399 (1964).

Wilks, P. A., Jr. *Abstracts Seventh Natl. Meet., Soc. Appl. Spectroscopy*, Chicago, Illinois, May, 1968, paper 199.

Wilks, P. A., Jr. and Brown, R. A. *Anal. Chem.* **36**, 1896 (1964).

Wilks, P. A., Jr. and Hirschfeld, T. *Appl. Spectrosc. Rev.* **1**, 99 (1967).

Williams, C. M. and Sweeley, C. C. in *Biomedical Applications of Gas Chromatography* (H. A. Szymanski, ed.) Plenum Press, New York, 1964, p. 225.

Williams, J. H., Kuchmak, M., and Witter, R. F. *Lipids* **1**, 391 (1966*a*).

Williams, J. H., Kuchmak, M., and Witter, R. F. *Lipids* **1**, 89 (1966*b*).

Williams, R. C., Kass, S. J., and Knight, C. A. *Virology* **12**, 48 (1960).

Wrigley, C. *Science Tools* **15**, 17 (1968), L. K. B. Instruments, Inc.

Woof, J. B. *Nature* **195**, 184 (1962).

Zweig, G. "Chromatography," in *Anal. Chem.* **40**, 490R (1968).

Zweig, G. and Whitaker, J. R. *Paper Chromatography and Electrophoresis*, Vol. *I*, Academic Press, New York, 1967.

Zweig, G., Moore, R. B., and Sherma, J. *Anal. Chem.* **42**, 349R (1970).

ADDITIONAL REFERENCES

The following are additional references which may be useful to the reader.

Bier, M., ed. *Electrophoresis: Theory, Methods, and Application*, Vol. *I* (1959), Vol. *II* (1967), Academic Press, New York.

Block, R. J., Durrum, E. L., and Zweig, G. *A Manual of Paper Chromatography and Paper Electrophoresis*, 2nd Ed., Academic Press, New York, 1958.

Bloemendal, H. *Zone Electrophoresis in Blocks and Columns*, Elsevier, New York, 1963.

Bobbitt, J. M. *Thin-Layer Chromatography*, Reinhold, New York, 1963.

Bobbitt, J. M., Schwarting, A. E., and Gritter, R. J. *Introduction to Chromatography*, Reinhold, New York, 1968. (Concentrates on three techniques: thin-layer, column, and gas chromatography.)

Craig, L. C. and King, T. P. "Dialysis," in *Methods of Biochemical Analysis*, Vol. *10* (D. Glick, ed.), Interscience, New York, 1962, p. 175.

Determann, H. *Gel Chromatography: Gel Filtration, Gel Permeation, Molecular Sieves*, Springer Verlag, New York, 1968. A laboratory manual translated from the German by E. Gross.

Ettre, L. S. and Zlatkis, A., eds. *The Practice of Gas Chromatography*, Wiley, New York, 1967. (A survey of practical aspects of gas chromatography.)

Foster, A. B. "Zone Electrophoresis of Carbohydrates," in *Advan. Carbohydrate Chem.* **12**, 81 (1957).

Giddings, J. C. "Concepts and Column Parameters in Gas Chromatography," in *Advances in Analytical Chemistry and Instrumentation Vol. 3* (C. N. Reilley, ed.) Interscience, New York, 1964, p. 315.

Giddings, J. C., ed. *Separation Science*, Marcel Dekker, New York, 1966 *et seq.*

Giddings, J. C. and Keller, R. A., eds. *Advances in Chromatography*, Vol. *I* (1966)–Vol. *8* (1969), Marcel Dekker, New York.

Hamilton, P. B. "Biochemical Analysis," in *Anal. Chem.* **38**, Ann. Rev. (1966); pp. 19R and 20R have references to various separation methods: dialysis, countercurrent distribution, liquid–liquid extractions, and ultrafiltration; p. 21R: column chromatography, electrophoretic methods, and paper and thin-layer chromatography; p. 22R: ion-exchange and gas chromatography.

Heftman, E. "Chromatography," in *Anal. Chem.* **38**, Ann. Rev. (1966). The history, theory, apparatus, and techniques of various types of chromatography are described starting on p. 31R. Applications of chromatography to various compounds are described as follows: p. 34R, inorganic compounds; p. 35R, hydrocarbons, alcohols, and carbohydrates; p. 36R, phenols, fatty acids, and carboxylic acids; p. 37R, amino acids and peptides, proteins, nucleic acid components, and alkaloids; p. 38R, porphyrins, nitrogen compounds, and terpenoids; p. 39R, steroids, other lipids, and vitamins; p. 40R, antibiotics, other pharmaceuticals, pesticides, dyes, miscellaneous compounds, narcotics, local anesthetics, barbiturates, other sedatives, phenothiazines, other tranquilizers, sulfonamides, antihistamines, anticoagulants, emodipodophyllin, ephedrine, and other compounds.

Kirchner, J. G. *Thin-Layer Chromatography*, Vol. *XII* of *Technique of Organic Chemistry* (E. S. Perry and A. Weissberger, eds.) Interscience, New York, 1967.

Methods in Enzymology, contains many sections in various volumes which give detailed discussions of fractionation methods, Academic Press, New York, 1957, *et seq.*

Pataki, G. *Techniques of Thin-Layer Chromatography in Amino Acid and Peptide Chemistry*, Ann Arbor
    Science Publ., Ann Arbor, Michigan, 1968. (Mainly for beginners.)
Porath, J. "Molecular Sieving and Adsorption," *Nature* **218**, 834 (1968).
Samuelson, O. "Partition Chromatography of Sugars, Sugar Alcohols, and Sugar Derivatives," in *Ion
    Exchange Vol. 2*, (J. A. Marinsky, ed.) Marcel Dekker, New York, 1969, p. 167.
Schachman, H. K. *Ultracentrifugation in Biochemistry*, Academic Press, New York, 1959.
Scott, R. M. *Clinical Analysis by Thin-Layer Chromatography Techniques*, Ann Arbor–Humphrey Science
    Publishers, Inc., Box 1425, Ann Arbor, Michigan, 1969.
Sherma, J. and Zweig, G. *Paper Chromatography and Electrophoresis. Vol. II. Paper Chromatography*,
    Academic Press, New York, 1971.
Stahl, E., ed. *Thin-Layer Chromatography*, Academic Press, New York, 1965.
Stahl, E., ed. *Thin-Layer Chromatography, A Laboratory Handbook, 2nd Ed.*, Springer Verlag, New York,
    1969.
Szymanski, H. A., ed. *Biomedical Application of Gas Chromatography*, Plenum Press, New York, Vol. 1,
    1964, Vol. 2, 1968.
Tranchant, J., ed. *Practical Manual of Gas Chromatography*, Elsevier, New York, 1969.
Wolf, F. J., ed. *Separation Methods in Organic Chemistry and Biochemistry*, Academic Press, New York,
    1969.

Appendix 2

A TABLE OF RECIPROCALS

SUBTRACT

	0	1	2	3	4	5	6	7	8	9	1	2	3	4	5	6	7	8	9
1.0	1.0000	.9901	.9804	.9709	.9615	.9524	.9434	.9346	.9259	.9174	9	18	27	36	45	55	64	73	82
1.1	.9091	.9009	.8929	.8950	.8772	.8696	.8621	.8547	.8475	.8403	8	15	23	30	38	45	53	61	68
1.2	.8333	.8264	.8197	.8130	.8065	.8000	.7937	.7874	.7813	.7752	6	13	19	26	32	38	45	51	58
1.3	.7692	.7634	.7576	.7519	.7463	.7407	.7353	.7299	.7246	.7194	5	11	16	22	27	33	38	44	49
1.4	.7143	.7092	.7042	.6993	.6944	.6897	.6849	.6803	.6757	.6711	5	10	14	19	24	29	33	38	43
1.5	.6667	.6623	.6579	.6536	.6494	.6452	.6410	.6369	.6329	.6289	4	8	13	17	21	25	29	33	38
1.6	.6250	.6211	.6173	.6135	.6098	.6061	.6024	.5988	.5952	.5917	4	7	11	15	18	22	26	29	33
1.7	.5882	.5848	.5814	.5780	.5747	.5714	.5682	.5650	.5618	.5587	3	7	10	13	16	20	23	26	30
1.8	.5556	.5525	.5495	.5464	.5435	.5405	.5376	.5348	.5319	.5291	3	6	9	12	15	18	20	23	26
1.9	.5263	.5236	.5208	.5181	.5155	.5128	.5102	.5076	.5051	.5025	3	5	8	11	13	16	18	21	24
2.0	.5000	.4975	.4950	.4926	.4902	.4878	.4854	.4831	.4808	.4785	2	5	7	10	12	14	17	19	21
2.1	.4762	.4739	.4717	.4695	.4673	.4651	.4630	.4608	.4587	.4566	2	4	7	9	11	13	15	17	20
2.2	.4545	.4525	.4505	.4484	.4464	.4444	.4425	.4405	.4386	.4367	2	4	6	8	10	12	14	16	18
2.3	.4348	.4329	.4310	.4292	.4274	.4255	.4237	.4219	.4202	.4184	2	4	5	7	9	11	13	14	16
2.4	.4167	.4149	.4132	.4115	.4098	.4082	.4065	.4049	.4032	.4016	2	3	5	7	8	10	12	13	15
2.5	.4000	.3984	.3968	.3953	.3937	.3922	.3906	.3891	.3876	.3861	2	3	5	6	8	9	11	12	14
2.6	.3846	.3831	.3817	.3802	.3788	.3774	.3759	.3745	.3731	.3717	1	3	4	6	7	8	10	11	13
2.7	.3704	.3690	.3676	.3663	.3650	.3636	.3623	.3610	.3597	.3584	1	3	4	5	7	8	9	11	12
2.8	.3571	.3559	.3546	.3534	.3521	.3509	.3497	.3484	.3472	.3460	1	2	4	5	6	7	9	10	11
2.9	.3448	.3436	.3425	.3413	.3401	.3390	.3378	.3367	.3356	.3344	1	2	3	5	6	7	8	9	10
3.0	.3333	.3322	.3311	.3300	.3289	.3279	.3268	.3257	.3247	.3236	1	2	3	4	5	6	7	9	10
3.1	.3226	.3215	.3205	.3195	.3185	.3175	.3165	.3155	.3145	.3135	1	2	3	4	5	6	7	8	9
3.2	.3125	.3115	.3106	.3096	.3086	.3077	.3067	.3058	.3049	.3040	1	2	3	4	5	6	7	8	9
3.3	.3030	.3021	.3012	.3003	.2994	.2985	.2976	.2967	.2959	.2950	1	2	3	4	4	5	6	7	8
3.4	.2941	.2933	.2924	.2915	.2907	.2899	.2890	.2882	.2874	.2865	1	2	3	3	4	5	6	7	8
3.5	.2857	.2849	.2841	.2833	.2825	.2817	.2809	.2801	.2793	.2786	1	2	2	3	4	5	6	6	7
3.6	.2778	.2770	.2762	.2755	.2747	.2740	.2732	.2725	.2717	.2710	1	2	2	3	4	5	5	6	7
3.7	.2703	.2695	.2688	.2681	.2674	.2667	.2660	.2653	.2646	.2639	1	1	2	3	4	4	5	6	6
3.8	.2632	.2625	.2618	.2611	.2604	.2597	.2591	.2584	.2577	.2571	1	1	2	3	3	4	5	5	6
3.9	.2564	.2558	.2551	.2545	.2538	.2532	.2525	.2519	.2513	.2506	1	1	2	3	3	4	4	5	6
4.0	.2500	.2494	.2488	.2481	.2475	.2469	.2463	.2457	.2451	.2445	1	1	2	2	3	4	4	5	5
4.1	.2439	.2433	.2427	.2421	.2415	.2410	.2404	.2398	.2392	.2387	1	1	2	2	3	3	4	5	5
4.2	.2381	.2375	.2370	.2364	.2358	.2353	.2347	.2342	.2336	.2331	1	1	2	2	3	3	4	4	5
4.3	.2326	.2320	.2315	.2309	.2304	.2299	.2294	.2288	.2283	.2278	1	1	2	2	3	3	4	4	5
4.4	.2273	.2268	.2262	.2257	.2252	.2247	.2242	.2237	.2232	.2227	1	1	2	2	3	3	4	4	5
4.5	.2222	.2217	.2212	.2208	.2203	.2198	.2193	.2188	.2183	.2179	0	1	1	2	2	3	3	4	4
4.6	.2174	.2169	.2165	.2160	.2155	.2151	.2146	.2141	.2137	.2132	0	1	1	2	2	3	3	4	4
4.7	.2128	.2123	.2119	.2114	.2110	.2105	.2101	.2096	.2092	.2088	0	1	1	2	2	3	3	4	4
4.8	.2083	.2079	.2075	.2070	.2066	.2062	.2058	.2053	.2049	.2045	0	1	1	2	2	3	3	3	4
4.9	.2041	.2037	.2033	.2028	.2024	.2020	.2016	.2012	.2008	.2004	0	1	1	2	2	2	3	3	4
5.0	.2000	.1996	.1992	.1988	.1984	.1980	.1976	.1972	.1969	.1965	0	1	1	2	2	2	3	3	4

A Table of Reciprocals

SUBTRACT

	0	1	2	3	4	5	6	7	8	9	1	2	3	4	5	6	7	8	9
5.0	.2000	.1996	.1992	.1988	.1984	.1980	.1976	.1972	.1969	.1965	0	1	1	2	2	2	3	3	4
5.1	.1961	.1957	.1953	.1949	.1946	.1942	.1938	.1934	.1931	.1927	0	1	1	2	2	2	3	3	3
5.2	.1923	.1919	.1916	.1912	.1908	.1905	.1901	.1898	.1894	.1890	0	1	1	1	2	2	3	3	3
5.3	.1887	.1883	.1880	.1876	.1873	.1869	.1866	.1862	.1859	.1855	0	1	1	1	2	2	3	3	3
5.4	.1852	.1848	.1845	.1842	.1838	.1835	.1832	.1828	.1825	.1821	0	1	1	1	2	2	2	3	3
5.5	.1818	.1815	1812	.1808	.1805	1802	.1799	.1795	.1792	.1789	0	1	1	1	2	2	2	3	3
5.6	.1786	.1783	.1779	.1776	.1773	.1770	.1767	.1764	.1761	.1757	0	1	1	1	2	2	2	3	3
5.7	.1754	.1751	.1748	.1745	.1742	.1739	.1736	.1733	.1730	.1727	0	1	1	1	2	2	2	2	3
5.8	.1724	.1721	.1718	.1715	.1712	.1709	.1706	.1704	.1701	.1698	0	1	1	1	1	2	2	2	3
5.9	.1695	.1692	.1689	.1686	.1684	.1681	.1678	.1675	.1672	.1669	0	1	1	1	1	2	2	2	3
6.0	.1667	.1664	.1661	.1658	.1656	.1653	.1650	.1647	.1645	.1642	0	1	1	1	1	2	2	2	3
6.1	.1639	.1637	.1634	.1631	.1629	.1626	.1623	.1621	.1618	.1616	0	1	1	1	1	2	2	2	2
6.2	.1613	.1610	.1608	.1605	.1603	.1600	.1597	.1595	.1592	.1590	0	1	1	1	1	2	2	2	2
6.3	.1587	.1585	.1582	.1580	.1577	.1575	.1572	.1570	.1567	.1565	0	0	1	1	1	1	2	2	2
6.4	.1563	.1560	.1558	.1555	.1553	.1550	.1548	.1546	.1543	.1541	0	0	1	1	1	1	2	2	2
6.5	.1538	.1536	.1534	.1531	.1529	.1527	.1524	.1522	.1520	.1517	0	0	1	1	1	1	2	2	2
6.6	.1515	.1513	.1511	.1508	.1506	.1504	.1502	.1499	.1497	.1495	0	0	1	1	1	1	2	2	2
6.7	.1493	.1490	.1488	.1486	.1484	.1481	.1479	.1477	.1475	.1473	0	0	1	1	1	1	2	2	2
6.8	.1471	.1468	.1466	.1464	.1462	.1460	.1458	.1456	.1453	.1451	0	0	1	1	1	1	2	2	2
6.9	.1449	.1447	.1445	.1443	.1441	.1439	.1437	.1435	.1433	.1431	0	0	1	1	1	1	1	2	2
7.0	.1429	.1427	.1425	.1422	.1420	.1418	.1416	.1414	.1412	.1410	0	0	1	1	1	1	1	2	2
7.1	.1408	.1406	.1404	.1403	.1401	.1399	.1397	.1395	.1393	.1391	0	0	1	1	1	1	1	2	2
7.2	.1389	.1387	.1385	.1383	.1381	.1379	.1377	.1376	.1374	.1372	0	0	1	1	1	1	1	2	2
7.3	.1370	.1368	.1366	.1364	.1362	.1361	.1359	.1357	.1355	.1353	0	0	1	1	1	1	1	2	2
7.4	.1351	.1350	.1348	.1346	.1344	.1342	.1340	.1339	.1337	.1335	0	0	1	1	1	1	1	1	2
7.5	.1333	.1332	.1330	.1328	.1326	.1325	.1323	.1321	.1319	.1318	0	0	1	1	1	1	1	1	2
7.6	.1316	.1314	.1312	.1311	.1309	.1307	.1305	.1304	.1302	.1300	0	0	1	1	1	1	1	1	2
7.7	.1299	.1297	.1295	.1294	.1292	.1290	.1289	.1287	.1285	.1284	0	0	0	1	1	1	1	1	1
7.8	.1282	.1280	.1279	.1277	.1276	.1274	.1272	.1271	.1269	.1267	0	0	0	1	1	1	1	1	1
7.9	.1266	.1264	.1263	.1261	.1259	.1258	.1256	.1255	.1253	.1252	0	0	0	1	1	1	1	1	1
8.0	.1250	.1248	.1247	.1245	.1244	.1242	.1241	.1239	.1238	.1236	0	0	0	1	1	1	1	1	1
8.1	.1235	.1233	.1232	.1230	.1229	.1227	.1225	.1224	.1222	.1221	0	0	0	1	1	1	1	1	1
8.2	.1220	.1218	.1217	.1215	.1214	.1212	.1211	.1209	.1208	.1206	0	0	0	1	1	1	1	1	1
8.3	.1205	.1203	.1202	.1200	.1199	.1198	.1196	.1195	.1193	.1192	0	0	0	1	1	1	1	1	1
8.4	.1190	.1189	.1188	.1186	.1185	.1183	.1182	.1181	.1179	.1178	0	0	0	1	1	1	1	1	1
8.5	.1176	.1175	.1174	.1172	.1171	.1170	.1168	.1167	.1166	.1164	0	0	0	1	1	1	1	1	1
8.6	.1163	.1161	.1160	.1159	.1157	.1156	.1155	.1153	.1152	.1151	0	0	0	1	1	1	1	1	1
8.7	.1149	.1148	.1147	.1145	.1144	.1143	.1142	.1140	.1139	.1138	0	0	0	1	1	1	1	1	1
8.8	.1136	˙1135	.1134	.1133	.1131	.1130	.1129	.1127	.1126	.1125	0	0	0	1	1	1	1	1	1
8.9	.1124	.1122	.1121	.1120	.1119	.1117	.1116	.1115	.1114	.1112	0	0	0	1	1	1	1	1	1
9.0	.1111	.1110	.1109	.1107	.1106	.1105	.1104	.1103	.1101	.1100	0	0	0	1	1	1	1	1	1
9.1	.1099	.1098	.1096	.1095	.1094	.1093	.1092	.1091	.1089	.1088	0	0	0	0	1	1	1	1	1
9.2	.1087	.1086	.1085	.1083	.1082	.1081	.1080	.1079	.1078	.1076	0	0	0	0	1	1	1	1	1
9.3	.1075	.1074	.1073	.1072	.1071	.1070	.1068	.1067	.1066	.1065	0	0	0	0	1	1	1	1	1
9.4	.1064	.1063	.1062	.1060	.1059	.1058	.1057	.1056	.1055	.1054	0	0	0	0	1	1	1	1	1
9.5	.1053	.1052	.1050	.1049	.1048	.1047	.1046	.1045	.1044	.1043	0	0	0	0	1	1	1	1	1
9.6	.1042	.1041	.1040	.1038	.1037	.1036	.1035	.1034	.1033	.1032	0	0	0	0	1	1	1	1	1
9.7	.1031	.1030	.1029	.1028	.1027	.1026	.1025	.1024	.1022	.1021	0	0	0	0	1	1	1	1	1
9.8	.1020	.1019	.1018	.1017	.1016	.1015	.1014	.1013	.1012	.1011	0	0	0	0	1	1	1	1	1
9.9	.1010	.1009	.1008	.1007	.1006	.1005	.1004	.1003	.1002	.1001	0	0	0	0	0	1	1	1	1

GENERAL INDEX

CHEMICAL COMPOUND INDEX

The letter *r* with a page number means that a reference to a spectrum (or to spectral data) is given. The letter *s* with a page number means that a spectrum or spectral data are supplied.

The reader should also consult the *General Index* and compounds listed under *Spectra*. For listings of *Drugs* and *Pharmaceuticals* not given in this index, see page numbers given in General Index.